SWINE SCIENCE

SEVENTH EDITION

Palmer J. Holden

Iowa State University

M. E. Ensminger

PEARSON

Prentice
Hall

Upper Saddle River, New Jersey 07458

Library of Congress Cataloging-in-Publication Data

Holden, Palmer J. (Palmer Joseph), 1943–
 Swine science / Palmer Holden.— 7th ed.
     p. cm.
 Rev. ed. of: Swine science / M.E. Ensminger. 6th ed. 1997.
 Includes index.
 ISBN 0-13-113461-2
 1. Swine. 2. Swine—United States. I. Ensminger, M. Eugene. Swine science. II. Tltle.

 SF395.H74 2005
 636.4—dc22

                                                    2004018854

Executive Editor: Debbie Yarnell
Associate Editor: Maria Rego
Managing Editor: Mary Carnis
Director of Production and Manufacturing: Bruce Johnson
Editorial Assistant: ReAnn Davies
Production Editor: Heather Willison, Carlisle Publishers Services
Production Liaison: Janice Stangel
Manufacturing Manager: Ilene Sanford
Manufacturing Buyer: Cathleen Peterson
Creative Director: Cheryl Asherman
Design Coordinator: Christopher Weigand
Cover Designer: Christopher Weigand
Cover Art: Courtesy of Getty Images
Composition: Carlisle Communications, Ltd.
Marketing Manager: Jimmy Stephens
Marketing Coordinator: Melissa Orsborn

Pearson Prentice Hall™ is a trademark of Pearson Education, Inc.
Pearson® is a registered trademark of Pearson plc
Prentice Hall® is a registered trademark of Pearson Education, Inc.

Pearson Education Ltd.   Pearson Education Australia Pty. Limited
Pearson Education Singapore Pte. Ltd.   Pearson Education North Asia Ltd.
Pearson Education Canada, Ltd.   Pearson Educación de Mexico, S. A. de C.V.
Pearson Education—Japan   Pearson Education Malaysia Pte. Ltd.

10 9 8 7 6 5 4 3 2 1
ISBN: 0-13-113461-2

DEDICATION

I wish to dedicate this book to my many colleagues who have mentored me in my career as an extension swine specialist and professor at Iowa State University for more than 30 years, as well as pork producers and industry personnel who provided me the wealth of experiences necessary to develop a successful and enjoyable career. It is only as a result of this background that I am able to revise and develop this swine science book.

CONTENTS

Chapter 6   Fundamentals of Swine Nutrition 101

Chapter 7   Nutrient Requirements and Allowances, Diet Formulation, and Feeding Programs 135

Chapter 8   Alternative Feeds, Additives, and Feeding Management . . .164

Chapter 9   Grains and Other High-Energy Feedstuffs 200

Chapter 10   Protein and Amino Acid Feedstuffs for Swine 216

PREFACE

The pork industry is changing rapidly. Selecting for desirable traits by mating superior animals is being replaced by searching for genes that affect those phenotypic traits. Marketing has transformed from buying stations procuring live pigs to direct producer shipments to packing plans with prices based on carcass weight and merit. The farrow-to-finish farm is being replaced by specialized units operating only a portion of the production continuum and increased citizen interest in the quality of the environment and in the methods that livestock are produced is evident.

Industry groups, such as the National Pork Board, and state pork organizations are responding with voluntary programs that ensure pork to the consumer is free of antibiotic residues; pork's eating and nutritional qualities are improved; pigs are produced under humane conditions; and swine operations are managed to maintain, and even enhance, the quality of the environment. The cooperation between pork commodity organizations and universities for research and education is at an all-time high as both try to maintain and grow programs with declining budgets.

As an agricultural industry we must be prepared to respond, and even lead, in growing a swine segment vital to the ability of agriculture to survive and grow. This requires knowledge of the issues facing livestock agriculture and working to develop enterprises allowing farm families to garner a profitable and satisfying lifestyle.

WHAT IS AHEAD FOR PORK?

As the pork industry changes, challenges and opportunities will emerge. Following are five of these that will interact to define pork production:

1. A continuing trend toward fewer and larger farms. As expenses increase for all people, the same happens to farm families. The cost to feed, clothe, educate, and plan for retirement demands an increasing level of expenditure each year. To maintain a lifestyle requires an increase in income.

2. Niche markets are growing and providing producers wishing to produce specialty pork products, such as "free range," organic, or "antibiotic-free," larger incomes per specialty pig marketed.

3. Animal welfare issues will continue to direct production methods. I believe that the vast majority of producers know that the welfare of their charges is integral to the success of their operation. However, there are many cultural and ethical differences as to what constitutes the welfare of animals. Hopefully, sound science will be deferred to in directing regulations affecting production methods.

4. Environmental issues will become more prevalent in guiding the conditions under which hogs are produced and the resulting manure and gases. National and state regulations are already dominant forces in livestock production.

5. Feeding pigs specialized diets to produce pork with specific intrinsic characteristics, such as types of fatty acids and other healthful components is being studied.

Producers with the necessary vision will be able to compete and thrive. I hope this text will provide a sound base for educating people new to the industry, and be a valuable resource to current participants.

Palmer J. Holden
Ames, Iowa, 2004

I would like to thank the reviewers for *Swine Science*, 7th ed. including:

Lee E. Anderson
Florida A & M University

W. L. Flower
North Carolina State University

Edward S. Fonda
California State
Polytechnic University-Pomona

Rick Jones
University of Georgia

Lyle G. McNeal
Utah State University

Jodi Sterle
Texas A & M University

Harold Thirey
Wilmington College

ABOUT THE AUTHORS

Palmer J. Holden is an Emeritus Professor of Animal Science at Iowa State University (ISU). He received his undergraduate B.S. from North Dakota State University and his M.S. and Ph.D. from Iowa State University in animal nutrition. Following graduation Dr. Holden served in the U.S. Army in Vietnam and left the service with the rank of captain. He joined the ISU Animal Science Department in 1972 and rose through the professorial ranks until his retirement in 2002.

Dr. Holden established a world-renowned extension program in swine nutrition and livestock agriculture, including presentations and consulting in 19 foreign countries. ISU's *Life Cycle Swine Nutrition* has had an international impact and is a major factor in Iowa State's recognition as the center for swine recommendations. It is in its 17th edition with translations into 4 languages. Dr. Holden served as an ISU representative on the *Pork Industry Handbook* from 1990 until his retirement. The International Satellite Symposium that released the National Research Council's *Nutrient Requirements of Swine* was organized by Dr. Holden and linked to 26 sites in the Western Hemisphere. He has contributed to 33 videotapes, many in the series have aired on Iowa Public Television.

In 2000, Dr. Holden was named a Fellow of the American Society of Animal Science (ASAS) as well as receiving the ASAS Extension Award. The Iowa Pork Producers Association named him an Honorary Master Pork Producer in 1986 and gave him their Feeder Pig Hall of Fame Award in 1997.

M. E. Ensminger (1908–1998) grew up on a Missouri farm and was educated at the University of Missouri (B.S. and M.S.) and the University of Minnesota (Ph.D.). He was on the staffs of the Universities of Massachusetts, Minnesota, and Washington State University (serving as department head). He was president of the Agriservices Foundation.

Dr. Ensminger authored 22 books on animal, dairy, beef, sheep, and swine science and nutrition (some translated into several languages) and numerous articles. His international recognition stemmed from his service to the animal industry, his books, and his many International Stockman's Schools and International Ag-Tech Schools. He received many awards during his distinguished career, including considerable international recognition.

1

Historical Perspectives

1

History and Development of the U.S. Swine Industry

"The police, under the direction of Inspector Downing, clearing the piggeries of Bernard Riley." In an effort to make their city more sanitary, New York City police remove piggeries. The pigs were chased out, the pens burned, and the offal was covered with lime. *(© Bettmann/CORBIS)*

Objectives

After studying this chapter, you should:

1. Understand the origin of domestic swine.
2. Know the zoological derivation of swine.
3. Know when and why pigs were brought to America.
4. Understand westerly migration of hogs in the United States and why it occurred.

Nomadic peoples could not move swine about with them as easily as they could domesticated cattle, sheep, or horses, which were more adapted to herding. Because swine do not migrate great distances under natural conditions and because the early nomadic peoples could not move them about easily, it appears that swine were domesticated in several regions and that each region or country developed a characteristic type of hog.

Moreover, close confinement of swine was invariably accompanied by the foul odors and flies of the pig sty. For this reason, the early keepers of swine were often regarded with contempt. Swine were often fed garbage that could infect them with trichina, which was a human health issue. This may have been the origin of the Hebrew and Moslem dislike of swine, later fortified by religious precept.

THE HOG IN THE ZOOLOGICAL SCHEME

The following outline shows the basic position of the domesticated hog in the zoological scheme:

Kingdom *Animalia*: Animals collectively; the animal kingdom.

Phylum *Chordata*: One of approximately 21 phyla of the animal kingdom, in which there is either a backbone (in the vertebrates) or the rudiment of a backbone (in the cephalochordates).

Class *Mammalia*: Mammals or warm-blooded, hairy animals that produce their young alive and suckle them for a variable period on a secretion from the mammary glands.

Order *Artiodactyla*: Even-toed, ungulate (hoofed) mammals.

Suborder *Suina*: Hippopotamuses, pigs, peccaries are included.

Family *Suidae*: Nonruminant, hoofed animals, consisting of wild and domestic swine but, in modern classifications, excluding the peccaries, which belong to the family *Tayassuidae*.

Genus *Sus*: The typical genus of swine, formerly comprehensive but now restricted to the European Wild Boar and its allies, with the domestic breeds derived from them.

Species *Sus scrofa*: The wild boar (hog) of Europe and Asia from which most domestic swine have been derived.

Subspecies include (1) the Central European Wild Boar, *Sus scrofa*, and (2) the Southeast Asian pigs, *Sus vittatus* and *Sus cristatus*. It is thought that these three subspecies formed the development of (3) the domestic pig, *Sus domesticus*.

Other major species in the *Sus* genus include the wild bearded pig, *Sus barbatus*; and the Javanese Warty pig, *Sus verrucosus*.

Within the pig family, there are other genera, including the warthog, *Phacochoerus aethiopicus*; the babirussa, *Babyrousa babyrussa*; the giant forest hog, *Hylochoerus meinertzhageni*; and the bush pigs, *Potamochoerus porcus*.

ORIGIN AND DOMESTICATION OF SWINE

Wild pigs of the present day represent as many as 6 genera and 31 species. The ancestors of the domestic pig are traceable to the genus *Sus* and several species, which are recognized under three common names: pigs, hogs, and swine—terms that are used interchangeably.

Figure 1.1 Babirussa, Wild Pig, 1941. The Babirussa originated in Indonesia and is one of the five genera of pigs. *(Courtesy, The Field Museum, Chicago, IL)*

The genus *Sus scrofa* contributed to American breeds of swine, primarily through the following subspecies, or races: (1) *Sus scrofa vittatus* of the Malayan region; (2) *Sus scrofa cristatus*, the Indian Wild Boar; and (3) *Sus scrofa scrofa*, the European Wild Boar.

Archeological evidence indicates that swine were first domesticated in the East Indies and southeastern Asia in the Neolithic Period, or New Stone Age (1) beginning about 9000 B.C., in the eastern part of New Guinea (now known as Papua New Guinea), an island in the Pacific Ocean just north of Australia; and (2) about 7000 B.C., in Jericho, which lies in Jordan, north of the Dead Sea. The domestication of the European Wild Boar came independently and later than the East Indian pig (Figure 1.1).

The East Indian pig was taken to China about 5000 B.C. Thence, Chinese pigs were taken to Europe in the 19th century, where they were crossed with the descendants of the European Wild Boar, thereby fusing the European and Asiatic strains of *Sus scrofa* and creating the foundation of present-day breeds.

In their wild state, pigs were gregarious animals, often forming large herds. Their feed consisted mostly of roots, mast (especially acorns and beechnuts), and such forage as they could glean from the fields and forests. Although their diets were primarily vegetarian, they also ate carrion (dead animals), small or injured animals, young birds, eggs, lizards, snakes, frogs, rooted worms, and fish. Because of their roving nature, diseases and parasites were almost unknown. Swine seem to have been especially variable under domestication and especially amenable to human selection. But, when given the opportunity, pigs promptly revert within only a few generations to a wild or feral state in which they acquire the body form and characteristics of their wild progenitors many generations removed. The self-sustaining razorback of the United States is an example of this reversion (Figure 1.2).

Figure 1.2 Texas Razorback boar. *(Courtesy, USDA, Washington, DC)*

Figure 1.3 Southeast Asian pig. *(Courtesy, Pork Productions Systems by Pond, Maner and Harris, 1991)*

Eurasian Wild Boar (*Sus scrofa*)

The Eurasian Wild Boar (European and Asian area) is distributed in Europe, North Africa, and Asia. Although much reduced in numbers in the past few hundred years, it appears unlikely that the famous wild boar will become extinct like the Aurochs (the chief progenitor of domestic cattle). In comparison with the domestic pig, this race of hogs is characterized by its coarser hair (with an almost manelike crest along the back), larger and longer head, larger feet, longer and stronger tusks, narrower body, and greater ability to run and fight. The color of mature animals is nearly black, with a mixture of gray and rusty brown on the body. Very young pigs are striped. The ears are short and erect.

These sturdy ancestors of domestic swine are extremely courageous, are stubborn fighters, and are able to drive off most of their enemies, except humans. If attacked, they will use their tusks with deadly effect, although normally they are as shy as most animals and prefer to avoid people. The wild boar hunt has been regarded as a noble sport throughout history. Ancient custom decreed that the hunt shall be on horseback and with dogs and that the quarry shall be killed with a spear. The Eurasian Wild Boar will cross freely with domestic swine, and the offspring are fertile.

Southeast Asian Pig

At one time, the Southeast Asian pig (Figure 1.3) was considered to be a separate species that included a number of wild stocks of swine native to the East Indies and southeastern Asia. In appearance, these pigs are smaller and more refined than the Eurasian Wild Boar. Also, they have a white streak along the sides of the face and the crest of hair on the back is absent. Today, these pigs are clas-

sified primarily as *Sus scrofa vittatus* and *Sus scrofa cristatus*, eastern subspecies of the wild boar, because all intermediates between western and eastern specimens have been determined.

HOW BACON GOT ITS NAME

The word bacon is said by one authority to have been derived from the old German word *baec*, which means *back*. However, others express the opinion that the word may have been derived from the noted Englishman, Lord Bacon, because the crest of Lord Bacon depicts a pig. In support of the latter story, it can be said that bacon has for many years been a favorite item in the English menu. It has also been suggested that the expression "bringing home the bacon" is of English origin, having grown out of an ancient English ceremony, which annually took place in a village about 40 miles from London. In this ceremony, it was the custom to award a "flitch of bacon" to each couple who, after a year of married life, could swear that they had been happy and had not wished themselves unwed. This 700-year-old tradition was revived in 1949.

INTRODUCTION OF SWINE TO AMERICA

Although many wild animals were widely distributed over the North American continent prior to the coming of Europeans, the wild boar was unknown to American Indians.

Columbus first brought hogs to Cuba on his second voyage in 1493. According to historians, only eight head were landed as foundation stock. However, these hardy animals must have multiplied at a prodigious rate because 13 years later the settlers of this same territory found it necessary to hunt the ferocious wild swine with dogs; the hogs had grown so numerous that they were destroying crops, attacking humans, and even killing cattle.

Although swine were taken to other Spanish settlements following the early explorations of Columbus (Cortez brought pigs to the North American mainland in 1519), pigs first saw American soil when touring the continent with Hernando de Soto. The Spanish explorer arrived in Tampa Bay (now Florida) in 1539. On his several vessels (between 7 and 10), he had 600 or more soldiers, some 200 or 300 horses, and 13 sows to provide "walking food" for the expedition.

This hardy herd of pigs traveled with the army of the Spanish explorer from the Everglades of Florida to the Ozarks of Missouri. In spite of battles with hostile Native Americans, difficult travel, and other hardships, the pigs thrived so well that at the time of de Soto's death in 1542 on the upper Mississippi, 3 years after the landing at Tampa, the hog herd had grown to 700. De Soto's successor then ordered that the swine be auctioned off among the men.

It is reasonable to assume, therefore, that the cross-country tour of de Soto's herd of pigs was the first swine enterprise in America. It was believed that the Indians love of pork and DeSoto's unwillingness to increase his trade of pigs to them resulted in several early conflicts with his expeditionary force. No doubt some of de Soto's herd escaped to the forest, and perhaps still others were traded to the Indians. At any rate, this sturdy stock served as foundation blood for some of the early American breeds.

COLONIAL SWINE PRODUCTION IN THE UNITED STATES

Sir Walter Raleigh brought sows to the Jamestown Colony in 1607. Their semiwild progeny made such rampages that, in order to prevent too much damage from rooting, some New England towns designated a "hog ringer," whose duty it was to ring all swine above a certain height (Hadley, Massachusetts, drew the line at 14 in.). On Manhattan Island, a long solid wall was constructed on the northern edge of the colony to control the roaming herds of hogs. This area is now known as Wall Street.

One of the historical documents of 1633 refers to the innumerable swine. John Pynchon founded the first meat packing plant in Springfield, Massachusetts, in 1641. He packed salted pork in barrels for shipment to the West Indies, most of which was exchanged for sugar and rum. Pynchon's records described the early New England Colony hogs as black or sandy in color, with razorback build; they also indicated that the hogs were said to be speedy runners (as might be inferred from the fact that

they ranged the woods in a half-wild state). With their huge tusks, the boars were believed to be quite capable of taking care of any wolves that might attack them. Pynchon's book records the weight of one lot of 162 hogs as 27,409 lb, an average of only 170 lb per animal. Sixteen of these weighed less than 120 lb, 25 weighed more than 200 lb; and the 2 heaviest tipped the scales at 270 and 282 lb. Others followed Pynchon's lead in packing pork, and in 1790 it was reported that 6 million pounds of pork and lard were exported from the United States.

Unlike the cattle, sheep, and horses, which were largely confined to the town commons for pasturage, the hogs of early New England roamed the surrounding countryside. Many were caught by hound dogs at marketing time. Usually a dog held on to each ear of the hog until the animal could be tied and pitched into a properly enclosed wagon. Few hogs were marked because these animals were so numerous that nobody minded the theft of a pig. The only care ever given the hogs, and then infrequently, was at farrowing time when a sow might be given the privilege of using a shelter or be allowed to crawl under a barn or house.

DEVELOPMENT OF AMERICAN BREEDS OF SWINE

In no other class of animals have so many truly American breeds been created. These facts probably result from (1) the suitability of native maize or Indian corn as a swine feed, (2) the ease with which pork could be cured and stored prior to refrigeration, and (3) the need for fats and high-energy foods for laborers engaged in the heavy development work of a frontier country (Figure 1.4).

Unlike the beef and dairy producers, who sent their native cattle to slaughter and imported whole

Figure 1.4 Before the advent of refrigeration, pigs were butchered outside on a daily basis. *(Courtesy, USDA, Washington, DC)*

herds of blooded cattle from England, the American hog raiser was content to use the mongrel sow descended from colonial ancestry as a base, on which were crossed imported Chinese, Guineas, Neapolitan, Berkshire, Tamworth, Russian, Suffolk Black, Byfield, and Irish Grazier boars. These importations began as early as the second quarter of the 19th century. Out of the various crosses, which varied from area to area, were created the several genuinely American breeds of swine. Breeds that remained as purebreds include the Berkshires arriving in 1823, Tamworths in 1882, Large Yorkshires in 1892, and Hampshires in 1920. Danish Landrace were imported in 1934 but could only be used for crossbreeding.

Structurally, the creation of the modern hog has involved developing an animal that would put muscle on the sides and quarters, instead of growing bone, fat, and a big head (Figure 1.5). Physiologically, breed improvement has resulted in an elongation of the intestine of the hog, thus enabling it to consume more feed for conversion into meat. According to naturalists, the average length of the intestine of a wild boar compared with its body is in the proportion of 9:1; whereas, in the improved American breeds, it is in the proportion of 13.5:1.

RISE OF CINCINNATI AS A PORK-SLAUGHTERING CENTER

With the expansion of farming west of the Allegheny Mountains, corn was marketed chiefly through hogs and cattle. At the time of the first U.S. census in 1840, the leading hog-producing state was Tennessee, followed by Kentucky and Ohio. Tennessee retained the lead in 1850, Indiana took it in 1860, followed by Illinois leading in pork production in 1870. Beginning in 1880 Iowa led the nation in pork production and retains that lead today.

Cincinnati became the earliest and foremost pork-packing center in the United States. By 1850, it was known throughout the land as "Porkopolis." Cincinnati originated and perfected the system that packed 15 bushels of corn into a pig, packed that pig into a barrel, and sent it over the mountains and over the ocean to feed humanity.

Some idea relative to the rapid rise in pork slaughtering in Cincinnati and the price fluctuations of the period may be gained from Table 1.1. The inflated price of 1865–66, followed by the slump in 1967, were the result of the demands caused by the Civil War. History repeated itself during World Wars I and II.

Cincinnati was favored as an early-day, pork-packing center because (1) it was then near the center of U.S. hog production and (2) it was strategically located from the standpoint of distribution. Large quantities of cured pork were shipped in flat boats to southern points via the Ohio and Mississippi Rivers, and both pork and lard were exported to the West Indies, England, and France.

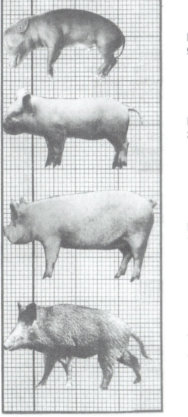

DOMESTIC PIG FETUS
9 WEEKS — ½ LB

DOMESTIC PIG
3 WEEKS — 15 LB

DOMESTIC PIG
15 WEEKS — 100 LB

WILD BOAR
ADULT — 300 LB

ALL SCALED TO THE SAME HEAD SIZE

Figure 1.5 This illustrates how years of selection have changed the conformation of the pig to meet changing market demands. Each animal is reduced to the same head size. As shown, when an improved breed matures, the proportion of loin to head and neck increases greatly; but an unimproved type such as the wild boar matures without much change in body proportions. *(From:* Farm Animals, *by John Hammond, published by Edward Arnold, Ltd., London, 1983)*

TABLE 1.1 Hogs Slaughtered on the Cincinnati Market

Year	No. Slaughtered	Year	Price/100 lb
1833	85,000	1855	$5.75
1838	182,000	1860	6.21
1843	250,000	1862	3.28
1853	360,000	1865	14.62
1863	606,457	1866	11.97
		1867	6.95

Source: USDA.

CHICAGO AS THE PORK-SLAUGHTERING CENTER

By 1860 Indiana led the country in pork production followed by Illinois in 1870. As the center of pork production shifted westward so did the packing industry and Chicago became the foremost packing center (Figure 1.6). Many railroads converged in Chicago, bringing hogs from the productive regions of the East and South. Initially, each of the five major railroads of the time built yards as an enticement to business. Then in 1865, the Illinois legislature incorporated the Union Stockyards and Transit Company—a single facility to accommodate all rail lines. Between 1865 and 1907, the Union Stockyards received about 241 million hogs, or an average of about 6 million pigs per year (Coburn, 1910). For many years, Chicago held the position of the major market center. In 1914, the poet and native son of Illinois, Carl Sandburg, penned these words to describe Chicago:

> "Hog Butcher for the World,
>
> Tool Maker, Stacker of Wheat,
>
> Player with Railroads, and the Nation's Freight Handler;
>
> Stormy, husky, brawling,
>
> City of the Big Shoulders. . . ."

Until after World War II, centralized terminal marketing remained the dominant method of marketing. Gradually, however, marketing decentralized. Packing plants were built nearer the areas of production, and producers began to market hogs direct to the packer bypassing the commission houses. After World War II, the number of hogs received at Chicago declined from about 3.5 million in 1945 to about 1.4 million in 1965 to only about 300,000 in the first part of 1970. In May 1970, the "Hog Butcher for the World" stopped accepting hogs at the Union Stockyards thus ending another era. In 2003 only three public stockyards—South St. Paul (Minnesota), St. Joseph (Missouri), and Sioux Falls (South Dakota)—continue to receive live hogs.

GROWTH OF THE U.S. SWINE INDUSTRY

The growth of hog production has paralleled very closely the production of corn in the north central or Corn Belt states, and these states produce nearly three-fourths of the corn grown in the United States. Although there still is a hog cycle, the swings are not as great as they have been historically. Pig numbers remain fairly constant as more pigs are raised on farms where they are the only source of farm income.

As shown in Figure 1.7, hog inventory numbers change sharply from year to year. A sharp increase occurred during World War II, with an all-time peak of 83,741,000 head on January 1, 1944. In recent years, however, hog numbers have oscillated below 60 million. Generally, Americans consume about 55 to 60 pounds of pork per year. When production drops below that level, the price increases, and when production exceeds that amount, prices fall. Exports are variable depending on the world economy.

The prices of hogs fall into approximately three stages. From 1895 to about 1941 the price of live pigs was about $10 per 100 lb, from 1942 to 1972 prices

Figure 1.6 An 1867 Armour and Company slaughter plant. *(Courtesy, Armour & Co.)*

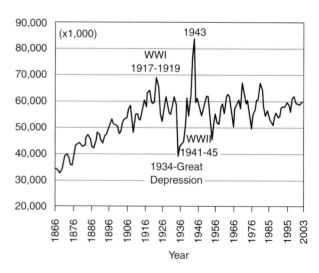

Figure 1.7 U.S. hog inventory numbers, 1866–2003. *(Courtesy, Census of Agriculture and USDA)*

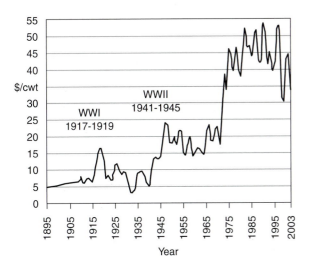

moved up to about $15 to $25, and from 1973 to 2003 prices were from $40 to $50 (Figure 1.8).

More information of the United States and world swine industry is presented in Chapter 2, "World and U.S. Swine and Pork—Past, Present, and Future."

Figure 1.8   U.S. hog prices per 100 lb of live weight, 1895 to 2003. *(Courtesy, NASS and USDA)*

QUESTIONS FOR STUDY AND DISCUSSION

1. How does the hog fit into the zoological scheme?
2. Why did nomadic peoples not move swine to the same extent they moved cattle, horses, and sheep?
3. What prompted early explorers to take swine with them?
4. More American swine breeds were developed than was the case with cattle, sheep, and horses. Explain this.

SELECTED REFERENCES

Encyclopaedia Britannica, Encyclopaedia Britannica, Chicago, IL

History of Livestock Raising in the United States, 1607–1860, J. W. Thompson, Agri. History Series No. 5, U.S. Department of Agriculture, Washington, DC, November 1942

Natural History of the Pig, The, I. M. Mellen, Exposition Press, New York, NY, 1952

Pigs—A Handbook to the Breeds of the World, Valerie Porter, Cornell University Press, Cornell, NY, 1993

Pigs from Cave to Corn Belt, C. W. Towne and E. N. Wentworth, University of Oklahoma Press, Norman, OK, 1950

Pork Facts 2002–2003, Staff, National Pork Producers Council, Des Moines, IA, 2003, http://www.porkboard.org/publications/pubIssues.asp?id=65

Swine in America, F. D. Coburn, Orange Judd Company, New York, NY, 1910

Swine Nutrition, E. R. Miller, D. E. Ullrey, and A. J. Lewis, Butterworth-Heinemann, Stoneham, MA, 1991

2

World and U.S. Swine and Pork—Past, Present, and Future

The changing types of hogs in 1954. *(Courtesy, USDA, Washington, DC)*

Objectives

After studying this chapter, you should:

1. Understand where hogs are raised in the world.
2. Understand why they are raised in those regions.
3. Know why pork production evolved in different parts of the United States.
4. Know the importance of hogs to agriculture.
5. Understand the hog–corn ratio and its relative importance to profitability.
6. Know the factors favorable or unfavorable to the growth of pork production.

Swine are produced most numerously in the temperate zones and in those areas where the population is relatively dense. There is reason to believe that these conditions will continue to prevail. But the future can often be more clearly determined by studying some historical trends.

WORLD SWINE DISTRIBUTION AND PRODUCTION

Most of the pigs in the world are raised in China, which has long had the largest hog population of any nation, followed by the United States. But because of the large human population, production in China is largely domestic, with very negligible quantities of pork entering into world trade. Although China's pigs are primarily scavengers, concentrated production for sales is growing. Also, the value of the manure is one of the main incentives for keeping them.

In South America, hog numbers have advanced at a rapid pace since the late 1950s. Brazil accounts for most of the production in that area.

Population and population density are correlated with pork production (Table 2.1). Of the 10 countries with the most population or population density, all but Canada and Poland are also in the top 10 hog producing countries. World population in 2002 was 6.302 billion people and world pork production was 1.126 billion head.

TABLE 2.1   Relationship of Population, Population Density, and Swine Production, 2002

Country	Population (number)	Density[1]	Production (1,000 head)	Slaughter (1,000 head)
Australia	19,731,984	3	5,458	5,639
Brazil	182,032,604	22	32,455	30,500
Bulgaria	7,537,929	68	2,590	2,401
Canada	32,207,113	4	29,630	22,134
China	1,286,975,468	138	574,367	565,000
Czech Republic	10,249,216	130	5,600	4,700
European Union[2]	820,204,576	121	214,750	202,750
Hungary	10,045,407	109	7,300	5,846
Japan	127,214,499	322	17,100	16,100
Korea, Republic	48,289,037	492	15,906	15,338
Mexico	104,907,991	55	15,250	13,840
Philippines	84,619,974	284	21,500	19,500
Poland	38,622,660	127	25,000	22,300
Romania	22,271,839	97	5,700	5,360
Russian Federation	144,526,278	9	34,200	29,000
Taiwan	22,603,000	701	10,100	9,700
Ukraine	48,055,439	80	8,500	6,900
United States	290,342,554	32	100,759	100,263
WORLD	6,302,309,691	48	1,126,165	1,077,271

[1] Density equals persons per square kilometer. To convert to persons per square mile, multiply by 2.6.
[2] The European Union consists of Austria, Belgium, Denmark, Finland, France, Germany, Greece, Ireland, Italy, Luxembourg, The Netherlands, Portugal, Spain, Sweden, and the United Kingdom.
Source: U.S. Bureau of the Census and USDA Foreign Agricultural Service.

Hog numbers in the European countries were closely related to the development of the dairy industry and the production of barley and potatoes—in much the same manner as the distribution of swine in the United States is closely related to the acreage of corn. Dairy by-products, skim milk, buttermilk, and whey have long been important swine supplements in Denmark, the Netherlands, Ireland, and Sweden. In Germany and Poland, potatoes have always been extensively used in swine feeding. Today's pork production is certainly independent of other species, but of course it is related to feed production. Most North American swine are produced in the United States, where the hog and corn combination go together (Figure 2.1).

Although China leads in pork production, it exports very little compared to the European Union (EU), the United States, and Canada, which are leading exporters (Table 2.2). Japan is the world's major pork importer. The last column compares the net consumption of pork (production plus imports minus exports) to the population. Eastern and western European countries, Chinese (China, Hong Kong, and Taiwan) and North Americans are the major consumers of pork.

From 1923 to the 1950s, except for the increases occurring during World War II, there was a downward trend in the exports of pork and lard from the United States. This was because of a marked increase in production in Canada and in the European countries, particularly in Denmark, Germany, and Ireland, and because of various trade restrictions imposed by the importing countries. In no sense was this decrease due to any lack of capacity to produce on the part of the U.S. farmer. Moreover, the general trend in exports of pork has been upward since the early 1960s, but lard production and exportation has declined as a result of changes in consumer preferences.

World hog numbers fluctuate rather sharply on the basis of available feed supplies. However, production in developed countries is more stable because of larger farm sizes and vertical integration. Also, the annual per capita consumption of pork in different countries of the world varies directly with production and availability, cost, the taste preference of the people, and in some cases with the religious beliefs that bar the use of pork as a food.

U.S. PRODUCTION AND DISTRIBUTION

The contribution of the humble pig to U.S. agriculture is expressed by its undisputed title as the "mortgage lifter." No other animal has been of such importance to the farmer, but the number of hogs and their value fluctuate.

The point and purpose of swine is the production of meat. Figure 2.2 compares pork production with that of beef, poultry and turkey, and lamb and mutton. The 2002 data indicate that pork accounts for about 23% of the poultry and red meat produced in the United States.

Areas of Swine Production

The geographical distribution of swine in the United States coincides with the acreage of corn, the principal swine feed. Normally, one-half of the corn crop is fed to hogs. It is not surprising, therefore, to find that about 63% of the hog production is located in the Corn Belt states of Iowa, Illinois, Indiana, Kansas, Minnesota, Missouri, Nebraska, and Ohio. It should not be concluded, however, that sections other than the Corn Belt are not well adapted to pork production. As a matter of fact, any area that produces small grains is adapted to the production of quality pork.

All 50 states produce some hogs (Figure 2.3), but Iowa has held the undisputed lead in hog numbers since 1880. The rank of the other states has shifted considerably. Growing corn and producing pork have contributed largely toward the profitability of upper Midwest agriculture. It is worth noting that even though the Corn Belt produces the most hogs, North Carolina ranks second in the number of pigs produced, and that is not a grain-producing

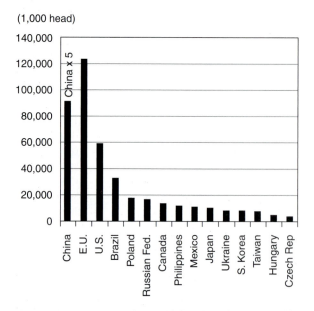

Figure 2.1 The leading countries in pork production inventory, 2001. Note China's hog inventory needs to be multiplied by 5. When Western Europe (EU) is combined, it becomes the second-largest pork producing region, followed by the United States and Brazil. *(Courtesy, USDA Foreign Agricultural Service)*

TABLE 2.2 World Pork Production and Consumption per Capita

Country	Production	Imports	Exports	Total	per Capita	
		1,000 metric tons carcass wt. (2003 preliminary)			(kg) (2004 preliminary)	(lb)
China	44,600	1,150	282	44,467	40.4	88.9
European Union	17,800	70	1,194	16,750	41.2	90.6
United States	9,073	538	779	8,833	29.4	64.7
Japan	1,259	1,133	0	2,372	17.5	38.5
Russian Federation	1,710	600	1	2,309	15.1	33.2
Brazil	2,560	0	603	1,957	10.1	22.2
Poland	1,740	40	120	1,640	37.2	52.4
Mexico	1,100	371	48	1,423	10.0	22.0
South Korea	1,153	155	14	1,255	23.8	52.4
Philippines	1,145	10	0	1,155		
Canada	1,895	91	974	1,026	29.0	63.9
Taiwan	890	54	0	944	42.2	92.8
Ukraine	765	7	12	760		
Czech Republic	583	31	34	586		
Romania	435	50	0	487		
Hong Kong	146	302	0	448	52.2	114.8
Hungary	480	23	90	413	45.8	100.8
Australia	365	67	74	358		
Bulgaria	145	22	0	172		
Singapore	19	40	0	61		

Source: Economic Research Service/USDA International Agricultural Baseline Projections to 2005/AER-750.
http://www.fas.usda.gov/psd/complete_files/default.asp; http://www.ers.usda.gov/publications/aer767/aer767h.pdf

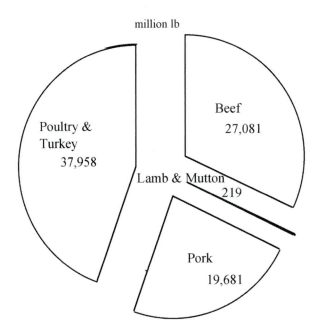

Figure 2.2 U.S. poultry and red meat production, 2002. *(Courtesy, USDA Livestock, Dairy, & Poultry Outlook/LDP-M-103/January 16, 2003)*

area. Many of the pigs farrowed in North Carolina are transported to the Corn Belt for finishing.

Since 1920, there has been a significant growth in pork production in the northwestern Corn Belt and in the northern and western Great Plains area. This has been attributed to the increased corn and barley production in these areas. But the most phenomenal recent increase in swine production has occurred in North Carolina (Table 2.3). North Carolina pork production varies significantly from Corn Belt production in that the majority of production is vertically integrated. Further growth in North Carolina is currently limited by a moratorium on new units unless they can demonstrate new techniques for managing the manure.

Vertically integrated units are also growing in Colorado, Oklahoma, Texas, and Utah. Even though these are grain-deficit areas and most of the feed must be shipped in, the low human population density is a driving factor.

The eastern, New England, and western states of the United States are pork-deficit areas. Despite the greatly expanded human population of the Pacific Coast, it is noteworthy that hog numbers in this area have remained about the same since 1900.

Historically, many of the live hogs slaughtered in West Coast plants were shipped from the Corn Belt by train. They were transported greater distances as live animals than a century ago when the eastern packers were prompted to move their slaughtering plants from the East Coast to Chicago. Today, packing plants are located near the areas of

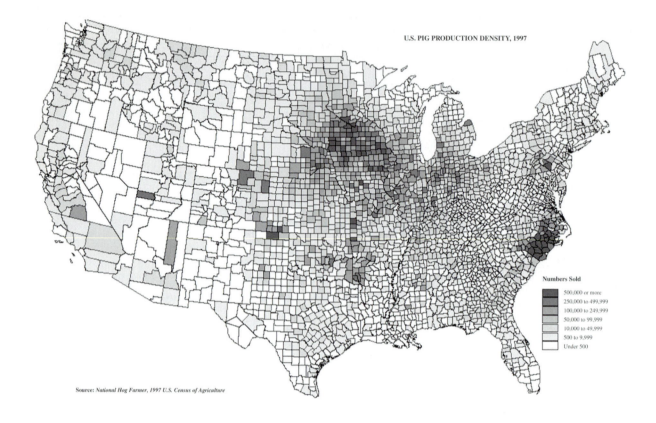

Numbers Sold

500,000 or more
250,000 to 499,999
100,000 to 249,999
50,000 to 99,999
10,000 to 49,999
500 to 9,999
Under 500

Source: *National Hog Farmer, 1997 U.S. Census of Agriculture*

Figure 2.3   U.S. pig production density. Most of the production is located in the Corn Belt and eastern North Carolina. *(Courtesy, U.S. Census of Agriculture, 1997)*

TABLE 2.3   Leading States in Total Hog Inventory and Breeding Stock Inventory, 2003

State	Total (1,000 head)	Breeding (1,000 head)
Iowa	15,800	1,060
North Carolina	9,900	1,000
Minnesota	6,400	600
Illinois	3,950	410
Indiana	3,100	300
Missouri	2,950	340
Nebraska	2,900	370
Oklahoma	2,340	360
Kansas	1,630	160
Ohio	1,520	160
South Dakota	1,260	145
Pennsylvania	1,100	120
Michigan	950	110
Texas	930	105
Colorado	770	110
United States	60,040	5,966

Source: USDA National Agricultural Statistics Service (NASS), December 30, 2003.

production, primarily in the Corn Belt and North Carolina, and hogs are trucked shorter distances. Most are sold directly from producer to packer, with the price determined by carcass merit.

Corn–Hog Relationship

With the opening of the central Mississippi Valley region, it became evident that this was one of the greatest corn producing areas in the world. The fertile soil, relatively long growing season, ample moisture, and warm nights were ideal for corn production. Also, it was soon realized that corn was unsurpassed as a hog feed. Here appeared to be an invincible combination and the early-day hog proved quite competent in the conversion of corn into meat and lard, particularly lard.

After World War II, the demand for tallow and lard plummeted, and U.S. swine producers promptly set about improving existing swine, as well as developing several new and distinctly American breeds. Thus, "King Corn" and the U.S. hog played a major

Figure 2.4 Feeding ear corn to pigs in 1955. *(Courtesy, USDA, Washington, DC)*

role in the development of U.S. agriculture and the prosperity of the farmers. The hog created a channel of disposal, or market, for corn and supplied people of the United States and other countries with highly palatable and nutritious meat at a moderate price (Figure 2.4).

Swine Production in the Southeastern States

Since 1980, there has been a concerted effort to diversify southern agriculture away from the cotton, peanut, and tobacco farming traditional to that area. In part, this movement has been motivated by the recognized need for greater attention to improved soil conservation practices and unprofitable prices sometimes encountered from the sale of cotton, peanuts, and tobacco. In particular North Carolina made a concerted effort to develop pork production. Also, vertical integration has speeded the movement by providing necessary capital, specialization of labor, central management services, and economies of scale (i.e., buying supplies in bulk).

The southern states are by no means uniform in the crops that they grow. A great variety of suitable swine feeds is produced from area to area but not in sufficient quantities to sustain a major feeding region. Many feeder pigs produced in the Southeast are transported to the Corn Belt for finishing and marketing.

Swine Production in the Western and Southwestern States

The western United States is a pork-deficit area, largely because other enterprises have been more remunerative, and, over a long period of time, farmers and ranchers tend to do those things that are

most profitable for them. Wheat, the leading grain crop of the area, is frequently too high in price to use as swine feed because it is considered primarily a human cereal, and often federally subsidized. Expansion of swine production in the West in the 1990s was driven by large swine operations seeking wide-open spaces in Utah and Colorado to escape the increasing legislation and litigation against odor and contamination nuisances in the Corn Belt, and to isolate their operations from other hogs and swine diseases.

Decline in Hog Farms; Increase in Size

The number of U.S. farms raising hogs has declined sharply in recent years (Figure 2.5). During the period from 1968 to 2002, the number of hog farms dropped from 967,580 to 75,400, more than a 90% decrease.

U.S. pork production has not changed significantly as the number of farms decreased; therefore, there are more pigs present on the remaining farms (Figure 2.6). Until 1980 the number of pigs on farms in the December inventory was about 100 or less. By 1990 the number had increased to about 200 head, in 2000 it was 685, and in December of 2002 it was 782.

There has been a notable change in the percentage of farms as categorized by the number of pigs in inventory. In 1988 almost 68% of the farms had less than 100 pigs present and only 3.1% had more than 1,000 hogs (Figure 2.7). By 2002 the number of farms with less than 100 pigs had decreased to 56.7%, and 16.8% had more than 1,000 pigs. In fact, by 1996

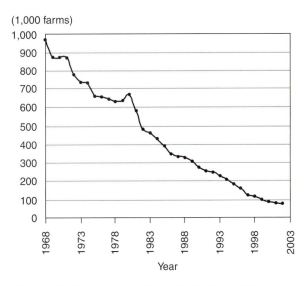

Figure 2.5 The number of farms in the United States producing hogs has declined. *(Courtesy, USDA, National Agricultural Statistics Service)*

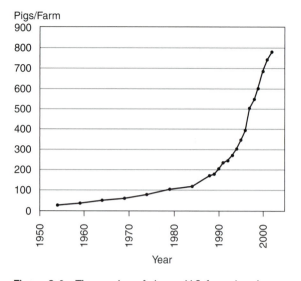

Figure 2.6   The number of pigs on U.S. farms has risen continuously, from 26 in 1954 to 782 in 2002. *(Courtesy, USDA, Washington, DC)*

Percent of Operations by Size

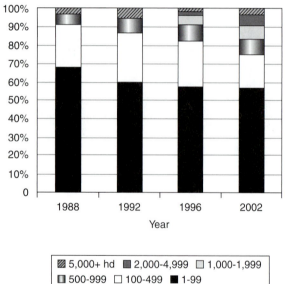

| ▨ 5,000+ hd | ■ 2,000-4,999 | ☐ 1,000-1,999 |
| ▥ 500-999 | ☐ 100-499 | ■ 1-99 |

Figure 2.7   The change in the number of hog operations by size from 1988 to 2002. The category "1,000–1,999" for 1988 and 1992 also includes all inventories greater han 1,000 hogs. *(Courtesy, USDA, Washington, DC)*

two more categories had been added: 2,000 to 4,999 head per farm and 5,000+ head per farm. Clearly, this shows that many small hog farms are exiting pork production that the large farms are increasing in size.

The number of hogs in the United States has been relatively stable in recent years. The number produced by the 56.7% of the hog operations with less than 100 hogs declined from 8.4% in 1988 to 1.0% in 2002. Conversely, the number of hogs on farms with more than 1,000 head increased from 36.1% in 1988 to 87% in 2002, including 53% of the

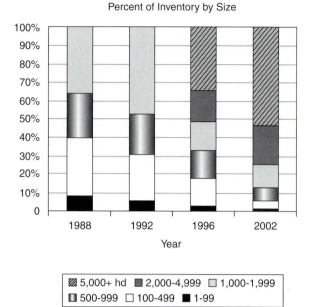

| ▨ 5,000+ hd | ■ 2,000-4,999 | ☐ 1,000-1,999 |
| ▥ 500-999 | ☐ 100-499 | ■ 1-99 |

Figure 2.8   The percentage of the total inventory found on farms of different sizes from 1988 to 2002. In 2002, for example, 53.0% of the pigs were on farms with more than 5,000 head. The category "1,000–1,999" for 1988 and 1992 also includes all inventories greater than 1,000 hogs. *(Courtesy, USDA, Washington, DC)*

inventory found on farms with more than 5,000 head (Figure 2.8).

Cash Receipts

The relative percentage of animal income varies over the years, depending on production levels and consumer demand. The pie graph in Figure 2.9 presents the proportion of animals and their products derived from each class of livestock in 2002. Swine accounted for 10.8% of the cash receipts from animals and animal products in 2002, and that income from swine was exceeded by cattle and calves; dairy; and poultry, turkey, and eggs. As would be expected, the proportions are somewhat changeable from year to year, depending on the relative value of the various farm products and the amount produced.

WORLD PORK CONSUMPTION

In general, pork consumption (and production) is highest in the temperate zones of the world and in those areas where the population is relatively dense. In many countries, such as China, pigs were primarily scavengers; in others, hog numbers are closely related to corn, small grain, or dairy production. As would be expected, the per capita consumption of pork in different countries of the world varies directly with its production and availability (Table 2.2). Food habits and religious re-

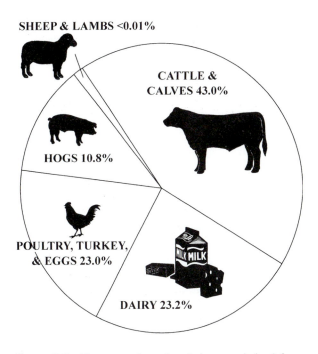

Figure 2.9 The proportion of cash income derived from each class of farm animals and their products in 2002. The 2002 total cash income from farm animals was $89.1 billion. *(Courtesy, USDA, Washington, DC)*

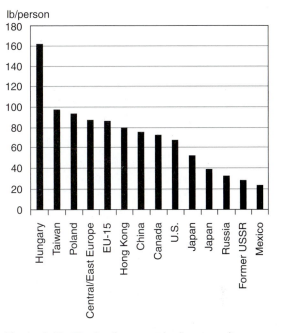

Figure 2.10 The leading countries in per capita consumption of pork, 2002. *(Courtesy, USDA, Economic Research Service)*

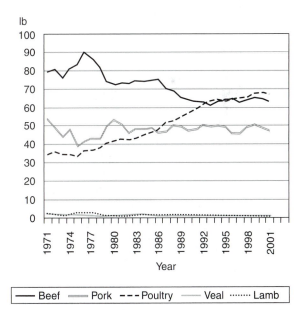

Figure 2.11 Per capita consumption of boneless meat and poultry products in the United States. *(Courtesy, USDA, Economic Research Service)*

strictions also affect the amount of pork consumed. For example, Islam and Judaism both prohibit the consumption of pork.

The leading countries in pork consumption are Hungary, Taiwan, Poland, Central and Eastern Europe, the European Union, Hong Kong, and China. Figure 2.10 gives a summary of per capita meat consumption in the leading meat-eating countries of the world and shows the position of pork. However, the greatest per capita consumers of pork are not necessarily big eaters of all meats. Hungary and Poland are listed separately in addition to Central and Eastern Europe because of their exceptionally high consumption of pork.

U.S. PORK AND OTHER ANIMAL PRODUCT CONSUMPTION

Beef was the preferred meat of Americans until it was surpassed by poultry in the early 1990s. Pork was second until the middle 1980s when it was also surpassed by poultry. Since 1970, pork consumption has fluctuated between 40 and 50 pounds per capita. Veal and lamb consumption have always been low (Figure 2.11).

U.S. Pork Imports and Exports

Hog producers are prone to ask why the United States, with a successful productive swine industry,

imports pork. Conversely, consumers sometimes wonder why we export pork. Occasionally, there is justification for such concerns, on a temporary basis and in certain areas, but as shown in Table 2.4, the United States neither imports nor exports large quantities of pork relative to production. The United States was a net importer of pork until 1995 when exports first exceeded imports, and the United States has been a growing net exporter of pork since that time.

TABLE 2.4   Pork Production and Imports and Exports

Year	Production	Imports	Exports
	(million lb carcass wt)		
1971	16,006	496	88
1981	15,873	542	307
1991	15,999	775	290
1994	17,658	744	549
1995	17,849	664	787
1996	17,085	620	970
2000	18,928	967	1,287
2001	19,160	950	1,563
2002	19,165	1,071	1,612
2003	19,982	1,185	1,717

Source: USDA-ERS, Cumulative U.S. Meat and Livestock Trade; http://usda.mannlib.cornell.edu/data-sets/livestock/ 94006/

The majority of pork imports include high-value cuts from Europe. Exports are divided between high-value products shipped to Japan and lower-value products shipped primarily to Eastern Europe. In the 6 months ending June 30, 2004, Japan accounted for about 47% of U.S. pork exports and 61% of the total value, followed by Mexico receiving 23% of U.S. pork (Figure 2.12).

The United States exports more lard than any other nation. The leading destination countries of lard are Mexico, Taiwan, Belize, and Japan. Mexico buys about one-half U.S. lard exported.[1]

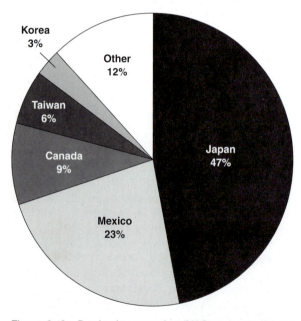

Figure 2.12   Destination countries of U.S. pork exports, percentage of total exports from July 2003 to June 2004. *(Courtesy, USDA, Foreign Agricultural Service)*

[1] USDA Foreign Agricultural Service.

FUNCTIONS OF SWINE

The average person is aware, at least in part, of the basic utility function of swine in contributing food. Few recognize, however, that—because of their added functions—swine are an integral part of a sound, mature, and permanent agriculture. The primary functions of swine are to

1. Produce food.
2. Provide profitable returns.
3. Convert inedible feeds into valuable products.
4. Enhance soil fertility.
5. Companion of feed grain production.
6. Provide unique products, such as heart valves, pharmaceuticals, and skin to protect burns. Additionally pigs are used as nutritional and medical research animals to study human health issues.

More information can be found throughout this book on many of the useful functions of swine.

Produce Food

Pork is the preferred meat of many groups of people, particularly Hispanics and Asians. Growing ethnic markets for pork are recognized as important outlets for U.S. pork.

The United States per capita consumption of selected foods and products is listed in Table 2.5. In the United States, consumption of pork ranks third behind poultry and beef in terms of meat consumption. Americans consume about the same quantity of fats and oils as any of the red meat or poultry foods.

From 1992 to 2001 the consumption of red meats has been relatively constant, and poultry, eggs, and milk have shown some increase. Consumption of fruits and vegetable is more than 680 pounds per person per year, followed by flour and cereal products, and caloric sweeteners (refined sugar and corn sweetener).

Americans, because of increasing disposable income and relatively low cost food, spend less and less of their incomes on food (Table 2.6). Prior to 1960 more than 20% of disposable income was spent on food. Today, this amount is only about 10%. The big savings is in money spent for at-home food, declining from 21.2 to 6.2%. Expenditures for away-from-home meals were only 12% of food costs (3.1 ÷ 24.3), whereas in 2002 they are almost 40%. This indicates a large share of today's meals are eaten in restaurants and so forth.

In the United States, pork makes a significant contribution toward meeting the nutritional needs of people. It supplies about 21% of the protein, 16% of the phosphorus, 16% of the iron, 20% of

TABLE 2.5   Per Capita Consumption of Major Food Commodities[1]

	1992	1997	2001
Red meat[2]	113.4	109.0	111.3
Beef	62.4	62.6	63.1
Veal	0.8	0.8	0.5
Lamb and mutton	1.0	0.8	0.8
Pork	49.1	44.7	46.9
Poultry[2]	60.4	63.1	66.2
Chicken	46.4	49.4	52.4
Turkey	14.0	13.6	13.8
Fish and shellfish[2]	14.6	14.3	14.7
Eggs	30.1	30.2	32.4
All dairy, milk basis	562.6	567.2	587.2
Cheese (excluding cottage)	25.9	27.5	30.0
Cottage cheese	3.1	2.6	2.6
Beverage milks	216.3	201.9	189.8
Fluid cream products	7.9	8.8	10.6
Yogurt (excluding frozen)	4.5	5.8	7.0
Ice cream	16.2	16.1	16.3
Low-fat ice cream	7.0	7.8	7.3
Frozen yogurt	3.1	2.0	1.5
Fats and oils	66.4	63.7	—
Butter and margarine	15.2	12.5	—
Shortening	22.3	20.5	—
Lard and edible tallow	3.5	4.0	—
Salad and cooking oils	27.0	28.0	—
Fruits	282.1	290.3	275.7
Vegetables	394.6	418.0	412.9
Peanuts (shelled)	6.2	5.7	—
Tree nuts (shelled)	2.2	2.1	2.2
Flour and cereal products	184.6	197.3	195.7
Caloric sweeteners[3]	136.1	148.1	147.1
Coffee	10.0	9.1	9.4
Cocoa	4.5	4.0	4.5

— = Not available.

[1] In pounds, retail weight unless otherwise stated. Consumption normally represents total supply minus exports, nonfood use, and ending stocks. Totals may not add due to rounding.
[2] Boneless, trimmed weight.
[3] Dry weight equivalent.

Source: USDA-ERS, Table 39—Per Capita Consumption of Major Food Commodities, Agricultural Outlook: Statistical Indicators, May 2003. http://www.ers.usda.gov/topics/view.asp?T=101402

TABLE 2.6   Share of Disposable Income Spent for Food

Year	Disposable Personal Income (billion $)	At Home (%)	Away from Home (%)	Total[1] (%)
1930	74.6	21.2	3.1	24.3
1940	76.7	17.6	3.1	20.7
1950	210.6	16.9	3.6	20.5
1960	366.2	14.1	3.4	17.5
1970	736.5	10.3	3.6	13.8
1980	2,019.8	8.9	4.2	13.2
1990	4,293.6	7.0	4.1	11.2
2000	7,120.2	6.1	4.0	10.2
2002	7,815.5	6.2	4.0	10.1

[1] Total may not add due to rounding.
Source: USDA-ERS, CPI Food And Expenditures Data, Table 7. http://www.ers.usda.gov/briefing/CPIFoodAndExpenditures/Data/table7.htm

Provide Profitable Returns

Although margins have narrowed in recent years, historically pork production has been an excellent complement to a grain farm operation. Most producers raise pigs because they find them a valuable source of farm income. Primary factors contributing to a profitable swine enterprise include a relatively high labor return in comparison with other types of livestock production and a fairly uniform labor requirement throughout the year. However, not all of the money the consumer pays at the retail counter goes to the farmer who raised the livestock.

In recognition of the importance of food to the U.S. economy and the welfare of its people, it is important to know where the consumer's food dollar goes—what proportion goes to the producer and what proportion goes to the middleman. Furthermore, it is important to understand the prices and profits of the producer and some of the middlemen—packers and retailers. Often misunderstandings arise and consumers seek to blame someone, frequently the producer, for high prices. For example, some consumers may compare what the packer is paying for hogs on foot to what the consumers are paying for a pound of pork over the counter.

In 1970 the farmer received 37% of the retail money spent for all of the food products in a market basket at the grocery store; in 2001 the farmer received only 21%. The pork producers' shares of the retail dollar dropped from 51% in 1970 to about 30% in 2000 and 2001 (Table 2.7). This means that

the thiamin, and 13% of the niacin that is required for good health. Also, pork supplies 100% of the requirement for vitamin B_{12}—a vitamin occurring only in animal food sources and fermentation products. It is noteworthy that the availability of iron in meat is about twice as high as in plants. (For details regarding the nutritive qualities of pork, see Chapter 20.)

TABLE 2.7   **Farmer's Share of the Retail Dollar (percentages)**

	1970	1980	1990	2000	2001[1]
Market basket	37	37	30	19.9	21.0
Beef, choice	64	63	60	48.6	45.8
Poultry	46	54	44	39.3	41.0
Eggs	63	64	56	39.3	35.0
Dairy products	48	52	39	29.5	34.0
Meat products	53	51	46	29.8	31.0
Pork	51	50	45	30.8	30.1
Fresh vegetables	32	27	28	18.8	19.1
Processed fruits and vegetables	19	23	26	16.5	16.1
Fresh fruit	28	26	23	15.7	15.8
Fats and oils	30	29	23	14.8	13.3
Cereal and bakery products	11	14	8	4.9	5.0

[1] 2001 data is preliminary.

Source: Briefing Room. Food Marketing and Price Spreads: USDA Market Basket. ERS-USDA, 2003. http://www.ers.usda.gov/Briefing/FoodPriceSpreads/basket/table2.htm

70% of today's consumer pork dollar goes for preparing, processing, packaging, and selling—and not for the food itself. One reason the farmer's share had dropped is that meat products undergo much more processing today to make them consumer-friendly to prepare and serve.

Overall, the farmer received 21% of the food (market basket) dollar in 2001, down from more than 30% from 1970 to 1990. The producer's share of the retail price of meat, poultry, eggs, and dairy is relatively high because processing is simpler and inexpensive, and transportation costs are low because of the concentrated nature of the products. However, the farmer's share of vegetables, fruit, oils, and bakery and cereal products is low because of the high processing and container costs and the bulky, costly transportation. A good example of the cereal and bakery products is to compare the price of a loaf of bread. A one pound loaf that sells for more than $1.00 probably has less than one pound of wheat, for which the farmer received about 5¢. Obviously, a lot of processing occurred between the farm and the consumer purchase.

Convert Inedible Feeds into Valuable Products

Only a small percentage of U.S. corn is graded for human consumption, the rest must be fed to livestock. Additionally, pigs are able to convert to edible food the numerous by-products of the meat-packing, fishery, grain milling, and vegetable oil processing industries. Some of these residues have been used for animal feeds for so long, and so extensively, that they are commonly classed as feed ingredients, along with such things as the cereal grains, without reference to their by-product origin. Most of these processing residues have little or no value as a source of nutrients for human consumption. It must be noted that as a result of the 2003 identification of a cow in Oregon with bovine spongiform encephalopathy (BSE) the use of ruminant by-products in meat and bone meal for pigs may be affected.

Swine can also use many wastes not suited for human consumption, including garbage, bakery wastes, garden waste, cull or damaged grain, and root crops and fruit. See Chapter 9 for a discussion of garbage processing and feeding.

Enhance Soil Fertility

Livestock manure is an excellent source of nutrients for grain or forage production. In addition to the basic nitrogen, phosphorus, and potassium in manure, it also provides many micronutrients as well as organic material.

Cash crops, whether they are grains or forages, result in the marketing of soil fertility. Although most farmers use commercial fertilizers to maintain soil fertility, it is beneficial to replace part of this lost fertility through the application of manure. If feed efficiency on a farrow to finish farm is 4:1, then 75% of the feed intake is excreted in the feces and urine and not retained by the pig.

Historical support of this situation is found in the fact that, despite the predominantly cereal grain diet of its people, China has the world's largest swine population. Every Chinese peasant recites the following teaching: "The more pigs, the more manure; and the more manure, the more grain." Indeed, animal manure is very precious in China; it is carefully conserved and added to the land. Manure is used as a way in which to increase yields of farmland already under cultivation. With proper conservation, therefore, this fertility value may be returned to the soil (see Chapter 16 for more information).

Companion of Feed Grain Production

Swine give elasticity and stability to grain production by providing a large and flexible outlet for the year-to-year changes in grain supplies. When there is a large production of grain and low prices, more sows can be bred to farrow, and market hogs can be carried to heavier weights. When grain production is low and prices are high, sows can be marketed without too great a sacrifice in price, and market hogs can be slaughtered at lighter weights.

Corn is the primary ingredient of the pig's diet in the majority of feeding areas in the United States, and feed constitutes a major portion of the total production cost. The hog–corn ratio is an example of how grain production determines pork production (Table 2.8). The hog–corn ratio is calculated simply by dividing the price received for live hogs per hundredweight by the current price for corn per bushel at that time.

Therefore, when corn is cheap, relative to hog prices, many producers in traditional nonintegrated operations opt to add value by feeding it to hogs rather than selling it outright. When the corn price is high, relative to hog prices, farmers may choose to sell the corn. Therefore, above a certain ratio, producers will more often expand production of hogs, whereas below that level, they will usually cut back production.

TABLE 2.8   Hog–Corn Ratio[1]

Year	Hog–Corn
1993	20.5
1994	16.4
1995	16.4
1996	15.4
1997	20.1
1998	14.7
1999	17.3
2000	23.3
2001	23.4
2002	15.9
2003	19.1

[1] The hog–corn ratio is the price per 100 lb of live market hogs divided by the price of 1 bushel of corn.
Source: USDA-NASS, Chapter VII, Statistics of Cattle, Hogs, and Sheep. Table 7–34—Hogs and Corn: Hog–corn ratio.
http://www.usda.gov/nass/pubs/agr04/acro04.htm

The hog–corn ratio is still important because corn is the major cost of producing pigs. However, the feed cost of pork production has declined from more than 75% to about 60% as the other costs of production, such as facilities and utilities, have increased.

Additionally, the following points are important when considering grain production as a companion to pork production:

1. Grains, such as corn and barley, have limited value if they are restricted solely to direct human consumption, but, if eventually they can be marketed as animal products, their value becomes immensely greater. A distinction needs to be made, therefore, between food grains for people and feed grains for livestock.

2. There is a savings in market transportation costs because the 10 to 12 bushels of grain consumed by the pig in growing to market weight require only about one-fourth the space on the four legs of the pig as would be needed in marketing the grains. Even with animal transportation rates about double that of grain per hundredweight, the transportation cost of marketing the grain through animals is reduced.

3. In outdoor swine production systems, hogs may supplement crop production through hogging-down crops or gleaning fields after they have been harvested. In addition the fertility value of the manure is conserved. This contribution of pigs is valuable especially where crops have been damaged or lodged. However, hogging-down (grazing unharvested fields) and gleaning (grazing harvested fields) is becoming rare as fences are removed to facilitate large farming machinery.

Provide Unique Products

Swine have additional value because they provide unique products for the consumer. Every day millions of people use swine products for their health, enjoyment, amusement, beautification, and general happiness. These products are further described in Chapter 20 and include heart valves, sutures, paint brushes, glandular extracts, pigskin for leather and as an aid in burn treatment, gelatin, and pepsin.

The list is long, but it could be made longer simply by citing the hog's medical usefulness, including as an experimental animal to study conditions affecting humans, because physiologically hogs and humans are quite similar. These innumerable by-products, many of which would not be available without a swine industry, contribute to the quality of human life.

FACTORS FAVORABLE TO SWINE PRODUCTION

The important position that the hog occupies in U.S. agriculture is based on various factors and economic conditions favorable to swine production.

1. Hogs are adapted to the practice of self-feeding, thereby minimizing labor (Figure 2.13).

2. Compared with many other agricultural enterprises, the initial investment for a beginning producer to get into the business can be small, and the returns come quickly. A gilt may be bred at 6 to 8 months of age and the pigs marketed as feeder pigs in 3 to 8 weeks or as slaughter hogs 5 to 6 months after farrowing.

3. Swine excel in dressing percentage, yielding 72 to 76% of their live-weight when dressed packer style (head, leaf fat, kidneys, and ham facings removed). However, cattle dress about 62 to 63% and sheep and lambs 50 to 57%. Moreover, because of the small proportion of bone, the percentage of edible meat in the carcass of hogs is greater.

4. Hogs may be sold at weights ranging from 230 to 270 lb without penalty in price. Also, sows that have outlived their usefulness in the breeding herd may be marketed without difficulty.

5. Swine are prolific, commonly farrowing from 8 to 14 pigs, and producing 2 litters per year.

6. The pig is adapted to both diversified and intensified agriculture.

7. Hogs are efficient converters of wastes and byproducts into pork.

8. Swine are more efficient than beef cattle and sheep in converting concentrates to food though they are not as efficient as dairy cattle, fish, or poultry. Feedstuffs high in fiber are poorly used by pigs and are best fed to ruminants.

FACTORS UNFAVORABLE TO SWINE PRODUCTION

It is not recommended that hogs be raised under any and all conditions. There are certain limitations that should receive consideration if the venture is to be successful. Some of these reservations follow:

1. Because of the nature of the digestive tract, growing and lactating pigs must be fed primarily concentrates and a minimum of roughages. Where or when grains are scarce and high in price, this may result in high production costs.

2. Because of the nature of their diets and their rapid growth, hogs are extremely sensitive to poorly balanced diets, poor quality or spoiled ingredients, and poor feed management.

Figure 2.13   Feeding pigs by hand in 1942 near Ames, Iowa. *(Photo by Jack Delan, Courtesy, USDA, Washington, DC)*

3. Swine are susceptible to numerous diseases and parasites, as are all livestock. However, intensively raised swine are more apt to be affected than typically extensively raised ruminants.

4. Sows should have skilled attention at farrowing time.

5. Hogs are not adapted to types of agriculture where grazing areas are extensive and vegetation is sparse (e.g., western rangeland). Neither are they best suited to the utilization of permanent pasture areas.

6. Pasture fences are more expensive for hogs than for cattle or sheep.

7. Because of their rooting and close-grazing habits, hogs are hard on pasture.

8. Increasing energy costs will benefit extensive production enterprises.

9. Increasing numbers of urban dwellers moving to the country are not accustomed to the odors and noise associated with swine, or agricultural, production.

THE FUTURE OF THE SWINE INDUSTRY

Some of the factors affecting the future of the American swine industry include the following:

1. Competition for grain.
2. Foreign competition.
3. Increased pork consumption.
4. Competition among geographic areas.
5. High energy costs.
6. Pork's image.
7. Animal welfare.
8. Concentration of U.S. hog production.
9. Packer availability and market access.
10. Manure management.
11. Energy utilization.
12. Animal rights.

Each of these factors is discussed individually in the following sections.

Competition for Grain

Many countries experience grain shortages and require imported grain to feed their growing human populations. By the year 2020, it is estimated that the United States and Canada, and possibly Latin America and Southeast Asia, will be the only areas producing substantially more grain than they consume. Hogs will be in direct competition with humans for food.

Grain shortages favor ruminants—Typically, 95% or more of swine feed is derived from concentrates and less than 5% from roughages, whereas beef cows can survive totally on roughages and feedlot steers consume about 75% concentrates. Thus, grain shortages place hogs in a less favorable position than ruminants (Figure 2.14).

Grain shortages favor cereal grain diets for humans and less grain for animals—cereal grain is the most important single component of the world's food supply, accounting for 30 to 70% of the food produced in all world regions. It is the major, and sometimes almost exclusive, source of food for many of the world's poorest people, supplying 60 to 75% of the total calories many of them consume. With sporadic food shortages and famine in different parts of the world, competition for grain increases. At some time urban sprawl will occupy farmland in some regions.

Disregarding the high nutritive value of meat, milk, and eggs, more hunger can be alleviated with a given quantity of grain by not feeding it to animals. Obviously, the efficiency of producing pork is about 7 or 8 units of feed for 1 unit of edible muscle. Even though the grain nutrients are concentrated in the muscle, the grain would feed more people. This inefficiency is a result of unavoidable nutrient losses in all animal feeding, and no return is received from that portion of the animal's feed which goes for maintenance.

The cost of purchasing animal protein is much higher than that of grain protein and energy. This is precisely why people in developing countries consume large quantities of plant protein and little animal protein. However, typically as income increases, so does the demand for animal protein.

Foreign Competition

Several European and Scandinavian countries are pork exporters, in some cases because of their animal welfare regulations. South American countries developing swine industries will become exporters, an encouraging market being the only needed incentive. Canada is making great progress in swine production in both quality and quantity. The U.S. pork industry with its huge corn and soybean production capability has been the world's low-cost producer. However, Brazil with its growing grain and soybean production is becoming a competitive low-cost pork producer.

Increased Pork Consumption

Increased U.S. consumption of pork has accrued from two sources: (1) increased population and (2) increased per capita consumption of pork brought about through the production of a much leaner and more versatile product than was available in the 1950s, as well as more advertising exposure. Recent issues concern the eating quality of pork, which may diminish if pigs get too lean. One very strong growth area is the introduction of pork products to the fast-food industry.

Figure 2.14 Ranchers have the option of feeding roughage or grazing cattle when grain prices are high. *(Photo by Keith Weller, Courtesy, USDA, Washington, DC)*

Competition among Geographic Areas

The Corn Belt remains the major hog-producing area, though there will be certain shifts in production. Thus, it is noteworthy that three states—North Carolina, Minnesota, and South Dakota—not in the Corn Belt were among the top 10 pork-producing states in 1994 (see Figure 2.3).

Some expansion of swine production in the West occurred in the 1980s and 1990s, driven primarily by very large operations seeking wide open spaces (1) to escape the increasing legislation and litigation against odor and contamination nuisances in more populated areas and (2) to isolate their operations from other hogs and swine diseases.

Energy Costs

The future of the U.S. swine industry will be affected by energy costs because a modern swine farm has a large variety of energy-consuming items—heating, cooling, ventilation, and so on. Efficient energy use is important to the success or failure of a swine operation. The energy crisis of the early 1970s caused producers and the world in general to seek alternatives and more efficient use of energy.

Swine producers must find other forms of energy, such as solar, wind, methane, and geothermal, as well as construction and management techniques that will conserve energy. Insulation and heat exchangers are two energy conserving techniques.

Pork's Image

Despite the excellent nutritive qualities of pork, it has long been portrayed as being an unhealthy food by some of the media and by some self-proclaimed nutritionists. That image has been reversed in recent years, primarily by the "Pork, the Other White Meat" campaign sponsored by the National Pork Board. Pork now compares very favorably with poultry and other meats as a lean, nutritious, and versatile product. Consumers viewing a retail meat counter are presented with a tremendous variety of pork products, varying from lesser quantities of fresh pork and growing choices of (microwave)-oven-ready options (Figure 2.15).

Animal Welfare

To all stockmen, the principles and application of animal behavior and environment depend on understanding and realizing that humans should provide as comfortable an environment as feasible for their animals, for both animal welfare and economic reasons. This requires that attention be paid

Figure 2.15   A Cajun spiced pork roast is the result of "the Other White Meat" program. *(Courtesy, National Pork Board, Des Moines, IA)*

to environmental factors that influence the behavioral welfare of their animals as well as their physical comfort, with emphasis on the two most important influences of all in animal behavior and environment—feed and facilities.

Concentration of U.S. Hog Production

Concentrated hog production involves using specialized facilities staffed with specialized labor, often vertically integrated. Formerly, hog production per unit varied with the acres of associated crop land. Today, the number of hogs per unit varies with the size of facilities.

The trend toward intensive hog production began about 1970 with the rapid movement of hog production into partial or total indoor facilities. The driving forces have been new technologies, new managerial techniques, lower cost of production, and increased quality control. Adverse consequences are being felt by displaced hog producers, displaced agribusinesses, and deteriorating rural communities.

As hog production is concentrated in fewer and larger production units, communities and states debate whether to welcome or try to bar such growth. Although trying to save family farms by anti-incorporation legislation at the state level does not work, a new force is the growing trend of neighbors and communities legislating and litigating against odor and contamination nuisances. Opposition has become very real in the Corn Belt and North Carolina. This is driving some hog operations to the open spaces of the West and Southwest, which may accelerate a more rapid concentration of ownership and vertical integration.

Packer Availability and Market Access

As packers merge, fewer become available in certain production areas, often requiring producers to transport slaughter hogs longer distances to be marketed. Access in many areas is limited by packer–producer contracts specifying times of delivery, and often signed with large producers. Non-contract producers may become residual suppliers, with their pigs only purchased to complete a packer's processing needs for a specified time period. This is particularly true in areas where production is vertically integrated.

Manure Management

The future of the U.S. swine industry will be impacted by the ability and willingness of producers to handle manure in an environmentally responsible manner. If it is not handled properly, swine manure,

gases/odors, dust, and flies and other insects will impact neighbors, surface water, and aquifers.

Energy Utilization

Production of swine in enclosed facilities requires either supplemental heat or ventilation during the production phases. Energy costs are likely to increase so producers must incorporate energy saving features, such as added insulation, bedding, and natural ventilation. Producers in some climates may be able to efficiently convert digest manure to methane gas that can be used to power generators or provide heat. Optimally stocking barns can be a major energy saver.

Animal Rights

Livestock producers depend on the productivity of their animals for the enterprise to be profitable, and these stockmen will modify the conditions of production to benefit both the animals and their own welfare. However various animal rights groups suggest that any type of animal agriculture is improper and promote vegetarianism; they may even destroy animal production facilities. As the public becomes further distanced from where and how their food is produced, false issues raised by animal rights groups become reality to unknowing consumers.

QUESTIONS FOR STUDY AND DISCUSSION

1. Why are there so many hogs in Asia (Table 2.2 and Figure 2.1)?
2. Discuss the factors that account for each of the four leading swine producing countries (Figure 2.1) holding their respective ranks.
3. Discuss several reasons why 60% of the U.S. hog production is located in seven Corn Belt states (Figure 2.3).
4. Select a hog farm, your own or one with which you are familiar. Discuss the relationship of swine production on this farm to local agriculture.
5. Assuming you had no "roots" in a particular location, in what area of the United States would you recommend the establishment of a swine enterprise? Justify your answer.
6. Iowa has a commanding lead as a swine-producing state (Table 2.3). Why, and do you think it will change?
7. North Carolina ranks a solid second as a U.S. swine-producing state (see Figure 2.3). What prompted the pork industry growth in this state?
8. Why has there been such a big decline in the number of U.S. hog farms, accompanied by an increase in size (Figures 2.5 and 2.6)? Is this good or bad?
9. The pie diagram in Figure 2.9 shows that swine rank fourth in U.S. cash receipts from animal commodities. Do you foresee the ranking of swine moving higher than beef, dairy, or poultry? Justify your answer.
10. As a food, what position does pork hold around the world (Table 2.2)? What factors and changes have contributed to this position?
11. What is the importance of pork as an export and as an import, and what determines the amount of pork that the United States exports and imports (Table 2.4)? Why has the United States become a net exporter? What may affect this trend?
12. List and discuss the functions of swine. What do you consider to be the most important function of swine?
13. Looking to the future, do you believe the factors favorable to swine production outweigh the unfavorable factors? Justify your answer.
14. How would you answer the question, "Who should eat grain—people or animals?"
15. Discuss the animal welfare and animal rights movements. Do you support either of them?
16. Discuss the industrialization of U.S. hog production. What are the benefits and costs?

17. How will handling manure in an environmentally responsible manner impact the future of the U.S. swine industry?

18. On the whole, do you believe that the future of U.S. swine production warrants optimism or pessimism? Justify your answer.

SELECTED REFERENCES

Agricultural Statistics, USDA—National Agricultural Statistics Service, http://www.usda.gov/nass/pubs/agstats.htm

Economics and Statistics System, USDA, http://www.usda.mannlib.cornell.edu

Fasonline, USDA Foreign Agricultural Service, http://www.fas.usda.gov

Feeds & Nutrition, 2nd ed., M. E. Ensminger, J. E. Oldfield, and W. W. Heinemann, The Ensminger Publishing Company, Clovis, CA, 1990

Pork Facts 2002/2003, Staff, National Pork Board, Des Moines, IA, 2002

Reports by Commodity—Index of Estimates, USDA—National Agricultural Statistics Service, http://www.usda.gov/ nass/pubs/estindx.htm#A

Swine Production and Nutrition, W. G. Pond and J. H. Maner, AVI Publishing Company, Westport, CT, 1984

3

Breeds of Swine[1]

Piglets, the results of a crossbreeding program. *(Courtesy, Palmer Holden, Iowa State University, Ames, IA)*

Objectives

After studying this chapter, you should:

1. Define types of hogs.
2. Recognize characteristics of various purebreds.
3. Know the significance of the importation of Danish Landrace in the 1930s.
4. Know the significance of the Chinese pigs imported by the USDA to the United States in 1989.

---

[1] The only legal basis for recognizing a breed is contained in the Tariff Act of 1930, which provides for the duty-free admission of purebred breeding stock provided they are registered in the country of origin. But that applies to imported animals to the United States only. Descriptions of breeds in books or by the U.S. Department of Agriculture are not required for a breed to be approved.

A breed is defined as a group of animals or plants presumably related by descent from common ancestors and visibly similar in most characteristics, especially such a group under domestication as differentiated from the wild type, or a group within a species developed by artificial selection and maintained by controlled propagation. Worldwide, there are more than 400 breeds of swine, and an excellent description of many can be found in *Pigs: A Handbook to the Breeds of the World* by Valerie Porter (Cornell University Press). Presently there are active registries for only the following nine breeds in the United States: Berkshire, Chester White, Duroc, Hampshire, Hereford, Landrace, Poland China, Spotted, and Yorkshire. In the past, several other breeds, such as the Tamworth, Ohio Improved Chester (OIC), Conner Prairie, English Large Black, Kentucky Red Berkshire, and Wessex Saddleback, maintained registries, but these do not appear to be active at the present time.

DEVELOPMENT OF AMERICAN SWINE

With the exception of the Berkshire, Landrace, and Yorkshire, the breeds of swine common to the United States are strictly American breeds. This is interesting in view of the fact that only one of our breeds of draft horses and few of our better known breeds of sheep and beef cattle were developed in America. With the exception of the Hereford breed and some of the newer hybrid breeds of swine, the American breeds came into being in the period from 1800 to 1880—an era that was characterized by the production of an abundance of corn for use by hogs and by consumer demand for fat, heavy cuts of pork. The European breeds did not meet these requirements.

Producers have a wide selection choice of keeping meat-type gilts from large litters. Continued selection each year with emphasis on the same characteristics and using boars of the same type can completely alter the conformation of the hogs in a herd in the short period of 4 to 5 years. Despite this fact, progress in producing meat-type hogs was often slow and painful throughout the 1930s and 1940s. As a result, (1) pork gradually lost its place as the preferred meat, with beef taking the lead in the early 1950s, and (2) a number of new "tailor-made" American breeds of swine evolved, most of them carrying some Landrace breeding.

It must be remembered, however, that the American breeds of swine were not developed without recourse to foreign stock. Prior to de Soto's importation, no hogs were found on the American continent (Figure 3.1). The offspring of de Soto's sturdy razorbacks, together with subsequent im-

Figure 3.1 Hernando de Soto's discovery of the Mississippi River. De Soto first brought 13 pigs to America at Tampa Bay (now Florida) in 1539. He took them as "walking food" for his expedition westward. (*© Bettman/CORBIS*)

portations of European and Oriental hogs, served as the foundation stock for the American breeds that followed. Out of these early-day, multiple-colored, and conglomerate types of swine, the swine producers of different areas of the United States, through the tools of selection and controlled matings, gradually molded uniform animals, which later became known as breeds. It is to be noted, however, that these foundation animals carried a variable genetic composition. This made them flexible in the hands of breeders and accounted for the radical subsequent shifts in swine types that have been observed within the pure breeds.

TYPES OF HOGS

Swine types are the result of three contributing factors: (1) the demands of the consumer; (2) the character of the available feeds; and (3) the ability of the breeder to develop pigs that can meet consumer demands, use the available feeds, and be productive and profitable.

Historically, three distinct types of hogs have been recognized: (1) lard type, (2) bacon type, and (3) meat type. At the present time, however, the goal for all U.S. swine breeds is for a meat-type hog producing quality pork, although it is obvious that some breeds have more nearly achieved this than others.

Lard Type

Originally, breeders of hogs stressed immense size and scale and great finishing ability. This general type persisted until the latter part of the 19th century. Beginning about 1890, breeders turned their attention to the development of early maturity, great refinement, and a very thick finish. In order to obtain these desired qualities, animals were developed that were smaller in size, thick, compactly built, and very short of leg. In the Poland China breed, this fashionable fad was carried to the extreme. It finally culminated in the development of the "hot bloods." Hogs of this thick, fat-type, sometimes called "cob rollers," were notoriously lacking in prolificacy. They often farrowed twins and triplets; and, when they were carried to weights in excess of 200 lb, their gains were very expensive. Small, refined animals of this type dominated the show-ring from about 1890 to 1910.

In order to secure increased utility qualities, breeders finally, about 1915, began the shift to the big-type strains. Before long, the craze swept across the nation, and again the pendulum swung too far. Breeders demanded great size, growthiness, length of body, and plenty of bone. The big-type animal was rangy in conformation and slow in maturity. Many champions of the show-ring included as their attri-butes long legs, weak loins, and "cat hams." Inasmuch as this type failed most miserably in meeting the requirements of either the packer or the producer—being too slow to reach maturity and requiring a heavy weight in order to reach market finish—another shift in ideals became necessary (Figure 3.2).

Bacon Type

The countries of Denmark, Canada, and Ireland have long been noted for the production of high-quality bacon. In the past, the surplus pork produced in these countries has found a ready market in England, largely selling as Wiltshire sides. Bacon-type hogs (Figure 3.3) are more common in those areas where the available feeds consist of dairy by-products, peas, barley, wheat, oats, rye, and root crops. Instead of producing a great amount of lard, they build sufficient muscle for desirable bacon.

In emphasizing the importance of character of feeds as a factor influencing the production of bacon-type hogs, it is not to be inferred that there

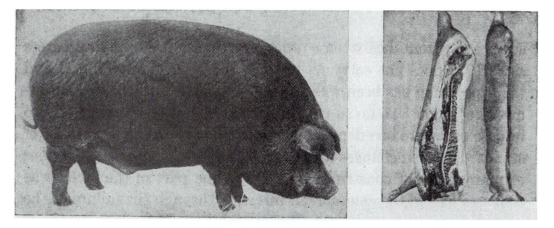

Figure 3.2   The old-fashioned extreme lard type; too lardy for the present market demands. *(Courtesy, U.S. Bureau of Animal Industry, USDA, Washington, DC, 1930)*

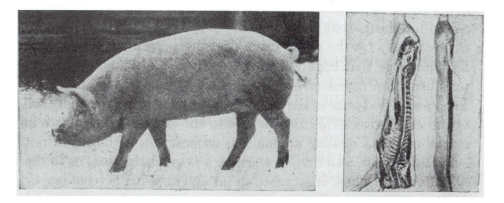

Figure 3.3   The bacon type. Selected by the Industrial and Development Council of Canadian Meat Packers to demonstrate the type and condition required to produce English bacon or the Wiltshire side. *(Courtesy, U.S. Bureau of Animal Industry, USDA, Washington, DC, 1930)*

is no hereditary difference. That is to say, when bacon-type hogs are taken into the Corn Belt and fed largely on corn, they never entirely lose their bacon qualities. See the Tamworth discussion in this chapter for an early bacon-type pig.

Meat Type

Since about 1925, U.S. swine breeders had been moving to develop meat-type hogs—animals that are intermediate between the lard and bacon types (Figure 3.4). The best specimens of the meat type combine muscling, length of body, balance, and the ability to reach market weight and finish without excess fat. In achieving the meat type, the selection and breeding programs of producers were augmented by meat certification programs, livestock shows, and swine-type conferences. It was not until after World War II, when the demand for lard dropped significantly, that producers were forced to quickly develop leaner pigs for a much more discriminating market.

BREEDS OF HOGS IN THE UNITED STATES

Table 3.1 shows the 1994 and 2002 registrations of the common breeds of hogs. The registrations are indicative of the current popularity of each of the major pure breeds. In 1910 the most popular breeds, in order, were the Poland China, Berkshire, Duroc-Jersey, and the Chester White.

Although there are many breed differences and most breed associations are constantly extolling the virtues of their respective breeds, it is perhaps fair to say that there is more difference within than between breeds from the standpoint of efficiency of production and carcass quality. Without doubt, the future and enduring popularity of each breed will depend on how well it fulfills these two primary requisites.

To be eligible for registry in the various breed associations, generally animals possessing a swirl on the upper half of the body, a hernia, retained testes (cryptorchid), absence of anal opening, a rectal or uterine prolapse, or fewer than 12 teats are disqualified.

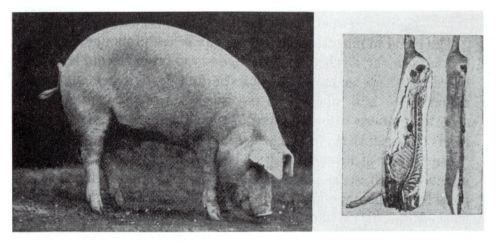

Figure 3.4 The modern meat type; smooth, trim, neat, of the type and condition to furnish the type demanded by the present U.S. market. *(Courtesy, U.S. Bureau of Animal Industry, USDA, Washington, DC, 1930)*

TABLE 3.1 Leading Breeds in Registration in the United States[1]

	1994		2002	
Breed	*Litters*	*Individuals*	*Litters*	*Individuals*
Berkshire	1,792		12,334	96,593
Chester White		27,342	1,634	14,172
Duroc		130,625	11,933	
Hampshire		122,002	7,031	
Landrace	4,714		3,364	
Poland China		14,872	976	7,468
Spotted	26,930		2,109	17,324
Yorkshire		208,662	12,767	

[1] Information provided by respective breed registries.

Berkshire

The Berkshire is one of the oldest of the improved breeds of swine. The striking style and carriage of the Berkshire has made it known as the aristocrat among the breeds of swine.

Origin and Native Home

The native home of the Berkshire is in south central England, principally in the counties of Berkshire and Wiltshire. The old English hog, a descendant of the wild boar, served as foundation stock, and these early animals were improved by introducing Chinese, Siamese, and Neapolitan blood. In 1789, the Berkshire was described as reddish brown in color, with black spots; large drooping ears; short legs; fine bone; and the disposition to fatten at an early age.

Early American Importations

The earliest importation of Berkshire hogs into the United States, of which there is authentic record, was made by John Brentnall of New Jersey in 1823. This importation was followed by those of Bagg and Wait of Orange County, New York, in 1839, and A. B. Allen of Buffalo, New York, in 1841.

Although there have been many constructive Berkshire breeders in the United States, certainly the name of N. H. Gentry of Sedalia, Missouri, is among the immortals. Few breeders either here or abroad achieved the success he did. He was truly a master breeder. The majority of the best Berkshires of today trace to Gentry breeding, particularly to the great and prepotent boar, Longfellow.

Berkshire Characteristics

The distinct peculiarity of the Berkshire breed is the short and sometimes upturned nose. The face is somewhat dished, and the ears are erect but inclined slightly forward. The color is black with six white points—four white feet, some white in the face, and a white switch on the tail. However, splashes of white may be located on any part of the body.

The conformation of the Berkshire may be described as excellent meat type. Fortunately, the big-type craze did not gain great momentum among Berkshire breeders. Thus, there seems to be pronounced uniformity within the breed at the present time.

The modern Berkshire is long bodied, with a long, deep side; moderately wide across the back; smooth throughout; well balanced; and medium in length of leg (Figure 3.5). The meat is exceptionally fine in quality, well-streaked with lean, and has no heavy covering of fat. The breed has established an enviable record in the barrow and carcass contests of the country. Currently, many Berkshires are marketed under the name "Berkshire Gold," commanding a premium at market for those that are 50% or Berkshire breeding. Many of these carcasses are shipped to Japan, which has an excellent demand for quality Berkshire pork.

The Berkshire is not as large as most of the other breeds. However, because of its great length, depth, and balance, it is likely to be underestimated in weight. Overall, Berkshires are hardy, rugged hogs that are able to withstand cold weather.

A Berkshire shall be a black and white animal with erect ears exhibiting Berkshire character. It must have white on all four legs, the face, and tail (unless the tail is docked). It is permissible for one of the white leg points to be missing. It must not have a solid white or a solid black face from the ears forward and must not have a solid black nose (rim of nose). White is allowed on the ears, but no solid white may appear on the ears. An occasional splash of white may appear on the body (Figure 3.6).

Berkshire boar Model Duke at 6 years and 8 months. Sold for $750. Note the six white points. *(Courtesy,* Diseases of Swine, *1914)*

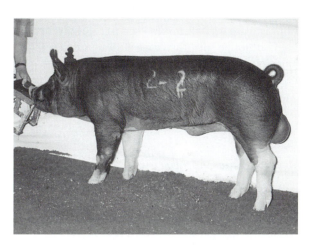

Figure 3.5 Berkshire boar champion at 2002 World Pork Expo. *(Courtesy, Iowa State University College of Agriculture, Ames, IA)*

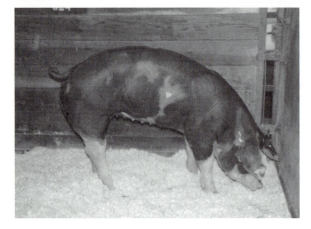

Figure 3.6   Berkshire gilt at the 2003 Iowa State Fair, Des Moines, Iowa. Note the varied color pattern accepted for the breed. *(Courtesy, Palmer Holden, Iowa State University, Ames, IA)*

Grand champion Chester White boar at the 1941 Iowa State Fair. *(Courtesy,* Breeding and Improvement of Farm Animals, *1942)*

Chester White

The Chester White breed was very popular on the farms of the northern part of the United States, and barrows of this breed had an enviable reputation in the barrow classes of livestock shows. Today the numbers of Chester Whites are quite limited.

Origin and Native Home

The Chester White had its origin in the fertile agricultural section of southeastern Pennsylvania, principally in Chester and Delaware counties. Both of these counties border on Lancaster County, one of the most noted agricultural and livestock counties in the United States. The breed seems to have originated early in the 19th century from the amalgamation of several breeds, most of which were white in color. The foundation stock included imported pigs of English Yorkshire, Lincolnshire, and Cheshire breeding. In 1818, Captain James Jeffries of Chester County imported a pair of white pigs from Bedfordshire, England. The boar in this importation was destined to exert a marked refining influence on the foundation stock of the Chester White breed. By 1848, the breed had reached such a degree of uniformity and purity that it was named Chester County White. The word "county" was soon dropped, and the present name became established.

Chester White Characteristics

As the name indicates, the breed is white in color. Although small bluish spots, called freckles, are sometimes found on the skin, such spots are to be discriminated against.

In general, type changes within this breed have followed those of the Poland China and the Duroc,

Figure 3.7   Modern working Chester White boar on an Iowa breeder's farm, 2001. *(Courtesy, Palmer Holden, Iowa State University, Ames, IA)*

although they have been less radical in nature. No doubt, some Yorkshire blood was infused into certain herds during the big-type craze. If so, perhaps it can be added that this blood brought real improvement and benefit to the breed.

Chester White sows are very prolific and are exceptional mothers. The pigs adapt well to a variety of conditions; they mature early; and the finished barrows are very popular on the market.

A Chester White must possess Chester White breed character and must be solid white in color. No color on the skin can be larger than a silver dollar and no colored hair is permitted. The ears must be down and of medium size (Figure 3.7).

Duroc

Currently, the Duroc ranks third in annual litter registrations, being exceeded by Yorkshires and Berkshires.

Figure 3.8 An excellent pair of Duroc-Jerseys. *(Courtesy, F. D. Coburn, 1910)*

Origin and Native Home

The Duroc breed originated in the northeastern United States. Although several different elements composed the foundation stock, it is reasonably certain that the Durocs of New York and the Jersey Reds of New Jersey contributed most to the ancestry.

The Jersey Reds were large, coarse, prolific red hogs that were bred in New Jersey early in the 19th century. The Durocs were smaller in size, more compact, and possessed great refinement. The red hogs of uncertain origin, which were found in New York, were named after the famous stallion, Duroc, a noted horse of that day. Beginning about 1860, these two strains of red hogs were systematically blended together, and thus there was formed the breed that is known at the present time as the Duroc (Figure 3.8).

Duroc Characteristics

The Duroc is red in color, with the shades varying from light to dark. Although a medium cherry red is preferred by the majority of breeders, there is no particular discrimination against lighter or darker shades as long as they are not too extreme. Duroc ears are of medium size and tipped forward (Figure 3.9).

During the big-type era, the Duroc also had its change in type. The show-ring was dominated by tall, rangy individuals—animals possessing narrow loins, light hams, shallow bodies, and slow maturity. Although some lack of uniformity still exists within the breed as a result of this radical shift, today the best representatives are of the most ap-

Figure 3.9 Working Duroc boar. *(Courtesy, Palmer Holden, Iowa State University, Ames, IA)*

proved meat type. The popularity of the breed may be attributed to the valuable combination of size, feeding capacity, growth rate, prolificacy, and hardiness.

Any of the following are disqualifications for registry: white feet or white spots on any part of the body; any white on the end of the nose; black spots larger than 2 inches diameter on the body; swirls on upper half of the body or neck; or ridgeling (one or more undescended testicles) boars or fewer than 6 udder sections on either side.

A Duroc must be red and possess Duroc breed character, including ears that must be down and of medium size, and no white or black hair located on the animal. Pigs must not have more than three black spots on the skin and none of these spots can be larger than 2 inches in diameter. There must not be any shading or indication of a belt (Figure 3.10).

Figure 3.10 Modern Duroc gilt. *(Courtesy, Palmer Holden, Iowa State University, Ames, IA)*

Gloucester Old Spots

Gloucestershire Old Spots (Figure 3.11) is a black and white breed that is predominantly white in color. In recent years, selection has been toward less black and now only a spot or two are usually found. The breed also has a heavy drooped ear. They originated in the Berkeley Valley region of England and have now spread throughout the United Kingdom. The origin of the breed is unknown but is probably from the native stock of the area along with introductions of various breeds. In 1855, Youatt and Martin mentioned there was a native stock in Gloucestershire that was of an unattractive dirty white color.

The Old Spots are among the large-size pigs in England. At one time, they were called the Orchard Pig or Cottage Hog because they were partially raised on windfall apples and waste agricultural products of the area. They are said to be good foragers or grazers. The sows of the breed are known for large litters and high milk production.

A Gloucestershire Old Spots breed society was formed in 1913. And although this breed has never become dominant in its native country or in any other country, it has had an influence on the world's swine production. There is little doubt the breed contributed more than just some influence on the color pattern to the Spot and was also used in the development of the Minnesota 3 breed in the United States.

Hampshire

The Hampshire is one of the youngest breeds of swine, but its rise in popularity has been rapid. It is widely distributed throughout the Corn Belt and the South.

Figure 3.11 Gloucester Old Spots boar at Iowa State University, 1987. *(Courtesy, Palmer Holden, Iowa State University, Ames, IA)*

Figure 3.12 Champion pen of Hampshire barrows at the 1913 International Stock Show. *(Courtesy, Diseases of Swine, 1914)*

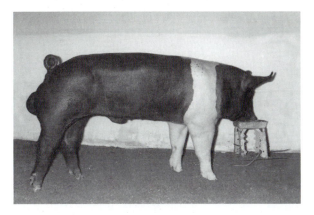

Figure 3.13 Modern Hampshire boar. *(Courtesy, National Swine Registry, West Lafayette, IN)*

Figure 3.14 Hampshire sows on an Iowa breeder's farm. *(Courtesy, Palmer Holden, Iowa State University, Ames, IA)*

Origin and Native Home

Accounts of the origin of the Hampshire breed vary (Figure 3.12). It has similar color patterns to the south England Essex and Wessex Saddleback breeds and may have been brought to the United States between 1825 and 1835 by a Mr. McKay. Major improvements were accomplished in Boone County, Kentucky, just across the Ohio River from Cincinnati. It is said that butchers from Ohio traveled to Kentucky to purchase these hogs at a premium price. The foundation stock consisted of 15 head of belted hogs, generally known as Thin Rinds and Ring Middles. This original herd was purchased in Pennsylvania in 1835 by Major Joel Garnett, who had the animals driven to Pittsburgh and then sent by boat to Kentucky. Years later, in 1893, 6 Boone County (Kentucky) farmers organized the Thin Rind Association, which, in 1904, became the Hampshire Swine Association.

Hampshire Characteristics

The most striking characteristic of the Hampshire is the white belt around the shoulders and body, including the front legs (Figure 3.13). The black color with the white belt constitutes a distinctive trademark; breed enthusiasts refer to it as the million dollar trademark.

Hampshire breeders have always stressed great quality and smoothness. The jowl is trim and light, the head refined, the ears erect, the shoulders smooth and well set, and the back well arched. An effort is now being made to secure bigger-framed, leaner, deeper-bodied, and sounder boars and sows that can carry heavier weights yet remain lean and efficient. Over many years in testing stations the Hampshire pigs have been noted for having the least backfat and largest loin eyes. Animals of this breed are active, and the sows have a reputation of hardiness and raising a high percentage of the pigs

farrowed. Also, they are of medium size and adapt well to outdoor environments.

A Hampshire to be used for breeding must be black in color with a white belt starting on the front leg. The belt must entirely encircle the body (including front legs and striking both front legs and feet). A breeding animal is ineligible if it is more than two-thirds white or has any white on the head. White is allowed on front of the snout surrounding the nostrils but must not break the rim of the nose. When the mouth is closed, white on the lower lip cannot be more than what a quarter can cover. The white on the hind leg must not extend above the hock, there must be no red on any part of the animal or any of its littermates and the animal must show no evidence of an extra dewclaw (Figure 3.14).

Hampshire market animals shall be eligible for registration as off belts regardless of their color marking or the presence of swirls. The evidence of a blaze on the face excludes that animal from off-belt registration.

Hereford

The Hereford is one of the less widely distributed breeds of swine. It has enjoyed a special attraction because the color markings emulate those of Hereford cattle.

Origin and Native Home

The Hereford breed of hogs was founded by R. U. Webber of LaPlata, Missouri. He conceived the idea of producing a white-faced hog with a cherry-red body color that would resemble the markings of Hereford cattle (Figure 3.15). The foundation stock included Chester Whites, OICs (Ohio Improved Chesters), Durocs, and other pigs of unknown origin. By inbreeding and selection, the breed was further developed over a period of 20 years. In 1934,

Figure 3.15 Hereford barrow. *(Courtesy, Palmer Holden, Iowa State University, Ames, IA)*

Figure 3.16 Danish Landrace gilt imported to Iowa State College Agricultural Experiment Station, 1938. *(Courtesy, Iowa State University College of Agriculture, Ames, IA)*

the National Hereford Hog Association was organized under the sponsorship of the Polled Hereford Cattle Registry Association.

Hereford Characteristics

The most distinctive characteristic of the Hereford breed of hogs is their color marking, which is similar to that of Hereford cattle. In order to be eligible for registry, animals must be at least two-thirds red in color, either light or dark, and must have faces that are four-fifths white; the ears can be red or white, or red and white; and white must appear on at least three feet and extend at least an inch above the hoof, all the way around the leg. The ideal colored Hereford has a white head and ears, four white feet, white switch, and white markings on the underline.

In size, the Hereford is smaller than the other breeds of swine. In the past, many specimens of the breed were considered too lardy, heavy shouldered, and rough, but great improvement was made through selection.

Any of the following disqualify an animal from registry or exhibiting: absence of some white in the face and not having a minimum of two-thirds red; a white belt extending over the shoulders, back, or rump; less than two white feet; boar with one testicle; permanent deformities of any kind; or any animal whose dam was under 10 months of age at time of farrowing.

Landrace

As was originally true of Spain with its Merino sheep, Denmark long held a monopoly on the Landrace breed of swine. In 1934, Landrace hogs were shipped to the United States and Canada, but, by government agreement, for several years thereafter they could not be released as purebreds, their use be-

ing restricted to crossbreeding (Figure 3.16). Subsequently, an agreement was reached with Denmark whereby surplus purebred Landrace swine could be released. Thereupon (in 1950), with additions of Norwegian and Swedish Landrace, the American Landrace Association, Inc., was organized, and the breed became known as the American Landrace.

Origin and Native Home

The Landrace breed of hogs is native to Denmark, where it has been bred and fed to produce the highest quality bacon in the world. With the aid of government testing stations, the Landrace breed has long been selected for improved carcass quality and efficiency of pork production. In addition to improved breeding, the feeds common to Denmark— small grains and dairy by-products—are also conducive to the production of high-quality bacon. For many years, the chief outlet for the surplus pork of Denmark has been the London market, where it was sold primarily in the form of Wiltshire sides.

Landrace Characteristics

The Landrace breed is white in color, although black skin spots or freckles are acceptable. The breed is characterized by its very long side, level top, well-defined underline, deep flanks, trim jowl, straight snout, and medium lop ears. It is noted for prolificacy and for efficiency of feed utilization (Figures 3.17, 3.18).

No animal to be used for breeding purpose shall be eligible, (1) that has any hair other than white on any part of the animal's body; (2) that has upright ears; (3) that has less than 6 functional teats on each side of the underline or has any inverted teats; (4) or that shows evidence of an extra dewclaw. Black spots in the skin are very objectionable and

Figure 3.17 Landrace boar, double wide. *(Courtesy, National Swine Registry, West Lafayette, IN)*

Figure 3.19 Mulefoot barrow. *(Courtesy, Palmer Holden, Iowa State University, Ames, IA)*

Figure 3.18 Landrace gilt, 2002. *(Courtesy, National Swine Registry, West Lafayette, IN)*

Figure 3.20 Close view of single toes of a Mulefoot barrow. *(Courtesy, Palmer Holden, Iowa State University, Ames, IA)*

any large spots or numerous black spots located on any part of the hog makes the pig ineligible for registry. However, a small amount of black pigmentation is allowed on the body of the animal.

Mulefoot

The origin of the American Mulefoot breed is not clear, but it has a well-documented history over the past century. F. D. Coburn, in his classic 1916 book *Swine in America,* notes that the Mulefoot hog was found in Arkansas, Missouri, Iowa, Indiana, across the Southwest, and in some parts of Mexico. The National Mulefoot Hog Record Association was organized in Indianapolis, Indiana, in January 1908. Two additional registries were also founded. In 1910 there were 235 breeders registered in 22 states.

Coburn describes Mulefoot hogs as mainly black, with occasional animals having white points;

medium flop ears; and a soft hair coat. The hogs were of fairly gentle disposition, fattened quite easily, and weighed from 400 to 600 lb at 2 years of age. They were considered the highest quality "ham hogs" and were fed to great weights before slaughter. For some years breeders claimed that Mulefoots were immune to hog cholera although that claim has been disproved (Figure 3.19).

The most distinctive feature of the Mulefoot hog is the solid hoof, which resembles that of a mule. Pigs with solid hooves (also called syndactylism) have attracted the interest of many writers over the centuries, including Aristotle and Darwin. Although only a single toe appears on the top, the foot actually has two pads on the bottom (Figure 3.20).

The Mulefoot is classified as "critical" by the American Livestock Breeds Conservancy (fewer than 200 in annual registration). It is now being raised by breeders across the state of Missouri and additional stock has recently been sent to Georgia.

Figure 3.21   A pair of 1908 Kansas-raised Poland Chinas. *(Courtesy,* Swine in America, *F. D. Coburn, 1916)*

Poland China

Origin and Native Home

The Poland China (Figure 3.21) breed of swine originated in southwestern Ohio in the fertile area known as the Miami Valley, particularly in Warren and Butler counties. In the early part of the 19th century, the Miami Valley was the richest corn-producing section of the United States. Moreover, prior to the Civil War, Cincinnati was the pork-packing metropolis of the country ("Porkopolis"). Thus, the conditions were ideal for the development of a new breed of hogs.

The common stock kept by the early settlers of the Miami Valley were described as being of mixed color, breeding, and type. The foundation animals were crossed with the Russian and Byfield hogs that were introduced into the valley early in the 19th century. In 1816, the Shaker Society, a religious sect, introduced the Big China breed. This breeding and improvement gave rise to the so-called Warren County hog, which gained considerable prominence as a result of its huge size and great fattening ability. Later, hogs of Berkshire and Irish Grazer breeding were introduced into the area and were crossed with the Warren County hogs. This improved the quality and refinement of the stock and resulted in earlier maturity. It is generally agreed that no outside blood was brought into the Miami Valley after about 1845.

The name, *Poland China*, was established in 1872 by the National Swine Breeders at a convention in Indianapolis, Indiana. In view of the fact that

Figure 3.22   Poland China gilt at the 2003 Iowa State Fair, Des Moines, Iowa. *(Courtesy, Palmer Holden, Iowa State University, Ames, IA)*

it seems to be definitely established that no breed known as Poland hogs was ever used as foundation stock, it is of interest to know how the name Poland China was selected. A Mr. Asher, a prominent Polish farmer in the valley, was supposedly responsible for the word *Poland*. As China hogs had been introduced by the Shaker Society, the name of *Poland China* was adopted.

Poland China Characteristics

Modern Poland Chinas are black in color with 6 white points—the feet, nose, and tip of tail—but prior to 1872, they were generally mixed black and white and spotted. Absence of 1 or 2 of the 6 white points is of small concern, and a small white spot on the body is not seriously criticized (Figure 3.22).

Figure 3.23 Poland China boar. Artist's view of the ideal type of the breed. *(Courtesy, National Pork Board, Des Moines, IA)*

Figure 3.24 Poland China packers' model. Winner $700 Pork Packers' Prize at St. Louis in 1869. Note the spotted appearance. *(Courtesy,* Diseases of Swine, *1914)*

Until the latter part of the 19th century, Poland China hogs were noted for their immense size and scale. Beginning about 1890, breeders turned their attention to the development of greater refinement, earlier maturity, and smaller size. This fashionable fad finally culminated in the development of the "hot bloods," an extremely small, compact, short-legged type that dominated the showring from 1900 to 1910.

Finally, in order to secure increased utility qualities, breeders began the shift to the big-type strains. The craze swept across the United States, and again the pendulum swung too far. However, since 1925, the efforts of breeders have been toward development of the more conservative meat type. Poland Chinas now yield a high-quality carcass with a high percentage of lean (Figure 3.23).

Producers often pick Polands as the sire breed in a terminal cross because their dark color is dominated by the color of the second breed, usually white. White pigs are preferred in U.S. slaughter houses because of ease of dehairing during processing.

Poland Chinas must possess Poland China breed character with ears down, and black with 6 white points (face, feet, and tail). Although an occasional splash of white is permitted on the body, it may not possess more than one black leg and be registered as a Poland China. (Tail docking is permissible, eliminating that white point.) There must be no evidence of belt formation and it cannot have red or sandy hair or pigment.

Spotted

For many years this breed was called Spotted Poland Chinas. The popularity of the "Spots" is chiefly attributed to the success of breeders in preserving the utility value of the old Spotted Polands while making certain improvements in the breed.

Origin and Native Home

The Spotted was developed in the north central part of the United States, principally in the state of Indiana. As has been indicated, the foundation stock of the Miami Valley Poland Chinas was frequently of a black and white spotted color. In fact, the winner of the Pork Packers' Prize at St. Louis in 1869 was a Poland China (Figure 3.24). A note on the bottom of the photo said, "Note spotted appearance." Moreover, these animals had a reputation for size, ruggedness, bone, and prolificacy.

During the era of the "hot bloods," many breeders forsook the Poland China breed and attempted to revive the utility qualities of the original spotted hogs of the Miami Valley. Thus, the Spotted was established. In addition to utilizing selected strains of the Poland China, breeders introduced, in 1914, the blood of the Gloucester Old Spots of England. Although the Spotted breed had its official beginning with the organization of record associations in 1914, liberal infusion of big-type Poland China blood was continued until 10 years later. Again in the 1970s and 1980s the Spotted Registry was opened to Poland Chinas (Figure 3.25).

Spotted Characteristics

At the present time, there is no great difference between the most approved type of Poland China and the Spotted. The former has a little more size, but in general the conformation is much alike.

Spotted swine must be black and white and possess Spotted breed character. The ears cannot be erect. Any pigs with red tinted or brown spots are ineligible and the head may not be solid black from ears forward. There can be no distinct white belt

Figure 3.25   Modern Spotted boar. *(Courtesy, National Spotted Swine Record, 6320 N. Sheridan Road, Peoria, IL)*

pattern (hair or skin) encircling and extending down and onto each shoulder.

Tamworth

The Tamworth is an English breed of hog that was of distinctly bacon type (Figure 3.28).

Origin and Native Home

The exact origin of this old English breed is not definitely known, but a Tamworth Swine Association booklet says "The Tamworth originated in Ireland and may have descended from 'The Irish Grazer' and is closely linked to the Berkshire breed. About the year 1812 it is said that Sir Robert Peel, being impressed with the characteristics of them, imported some of them and started to breed them on his estate at Tamworth, England. They have been bred quite extensively ever since they were imported into that country."

Tamworth is in Staffordshire, England, and the major improvement in the breed took place in that county and the surrounding counties in central England. The Tamworth is one of the oldest and probably one of the purest of all breeds of hogs. It is also recognized as the most extreme bacon type of any breed. There is record of pure breeding and careful selection dating back more than 100 years.

Early American Importations

The earliest importation of Tamworths into the United States, of which there is authentic record, was made by Thomas Bennett, of Rossville, Illinois, in 1882.

Tamworth Characteristics

The color of the breed is red, varying from light to dark. The conformation may be described as that of extreme bacon type. More than 5% black disqualifies pigs from registration.

Yorkshire

Origin and Native Home

The Yorkshire is a popular English bacon breed that had its origin nearly a century ago in Yorkshire and neighboring counties in northern England. In England the Yorkshire breed is known as the Large White.

Early American Importations

Although Yorkshires were probably first brought to the United States early in the 19th century, representatives of the modern type were first introduced by Wilcox and Liggett of Minnesota in 1893. A great many importations were made into Canada at an early date, and the breed has always enjoyed a position of prominence in that country (Figure 3.26).

Modern Tamworth boar, 2002. *(Courtesy, Mapes Livestock Photography, Milford Center, OH)*

Figure 3.26   Champion Yorkshire boar at the 1904 St. Louis World's Fair. *(Courtesy, Coburn, 1910)*

Yorkshire Characteristics

Yorkshires should be entirely white in color. Although black pigment spots, called "freckles," do not constitute a defect, they are frowned on by breeders. The face is slightly dished, and the ears are erect.

Yorkshire sows are noted as good mothers. They not only farrow and raise large litters but also they are excellent milking females. The pigs are excellent foragers and compare favorably with those of any other breed in economy of gains (Figure 3.27).

Yorkshires must be white in color and possess Yorkshire breed character with erect ears (Figure 3.28). The presence of one or more of the following disqualifies Yorkshire pigs from registry: any hair other than white, black spots in the skin are very objectionable and large or numerous black spots make the pig ineligible for registry, or evidence of an extra dewclaw. However, a small amount of black pigmentation is allowed on the body of the animal.

INBRED SWINE BREEDS

Beginning in the 1930s and continuing through the 1950s, there was interest in developing new breeds of swine that would be prolific, gain rapidly and efficiently, and produce a high-quality carcass. New breeds were developed at Beltsville, Maryland, at several of the state agricultural experiment stations, by private breeders, and in Canada. Many of these were based on crosses with the Danish Landrace imported in 1934 and later. As the numbers of these inbred (new breeds) swine increased, a demand for registration arose. Therefore, in 1946 the Inbred Livestock Registry was organized.

Some of these "tailor-made pigs" included breeds developed by Agricultural Experiment Stations in Minnesota (Minnesota No. 1, No. 2, and No. 3) and Montana (Montana No. 1, or Hamprace), and at the USDA Agricultural Research Center at Beltsville, Maryland. The Minnesota No. 1 (Figure 3.29) was a cross of the Danish Landrace and the Tamworth, and the Montana 1 or Hamprace (Figure 3.30) was the

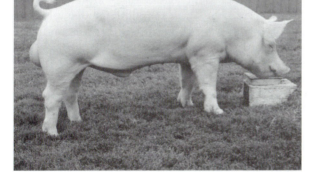

Figure 3.27 Yorkshire boar, Altoulf. *(Courtesy, National Swine Registry, West Lafayette, IN)*

Figure 3.29 Minnesota No. 1 inbred line at the University of Minnesota, 1956. *(Courtesy, College of Agricultural, Food and Environmental Sciences, University of Minnesota)*

Figure 3.28 Yorkshire gilt. *(Courtesy, National Swine Registry, West Lafayette, IN)*

Figure 3.30 Montana No. 1 (Hamprace) sows, 1936. *(Courtesy, Inbred Livestock Registry Association)*

result of a black Hampshire boar crossed with the Danish Landrace. Other breeds were developed by private breeders in Indiana (the San Pierre), Washington (the Palouse), and Canada. Although most of these newer breeds of swine did not survive, it is to their credit that they shook their older counterparts out of their lethargy.

CHINESE PIGS IMPORTED INTO THE UNITED STATES

In 1989 several breeds from China were introduced into the United States under a cooperative effort of the USDA, the University of Illinois, and Iowa State University. A total of 144 pigs of the Fengjing, Meishan, and Minzhu breeds were brought to the United States in this program. These breeds are slow growing and fat but have very good taste. They are considered to be resistant to some diseases and are able to consume large amounts of roughage. Their Chinese diet consists of concentrates, farm by-products, and water plants.

These pigs are so genetically diverse from the conventional "western" pigs that they should enhance research by, for example, defining genes that affect litter size and other traits that are of interest in pork production.

Fengjing

The Fengjing breed is recognized by its wrinkled face and skin (Figure 3.31). Sows grow to be about 27 in. (69 cm) high and 153 lb (69.6 kg) live-weight. They have a backfat thickness of 1.4 in. (3.5 cm) and a dressing percentage of 66.0%.

Fengjings reach puberty at 2.5 to 3.0 months of age. This breed is among the most prolific pig breeds in the world and have high embryo survival rates and large litter sizes. Third and later parities of Fengjing pigs had 17.0 pigs born, with 12.1 being weaned. Their 240-day weight was 174 lb (79 kg), with an average daily gain (ADG) of 0.75 lb (0.35 kg).

Meishan

Meishans are considered Taihu pigs, deriving their name from the Taihu Lake in their region of origin between north and central China, in the Lower Changjiang River Basin and southeast coast. This area has a mild climate.

The Meishan breed is known for its wrinkled face and skin (Figure 3.32). Sows grow to be about 23 in. (58 cm) high and 135 lb (62 kg) live-weight. They have a back fat thickness of 1.0 in. (2.5 cm) and a dressing percentage of 66.8%. They are perhaps one of the most prolific breeds of pig in the world. They reach puberty at 2.5 to 3.0 months of age, achieve high embryo survival rates, and a large litter size of 15 to 16 pigs.

Minzhu

Minzhu pigs come from far northern China where the cold and dry climate makes the Minzhu very tolerant to cold temperatures and harsh feeding conditions. They are also known as the Ming, Min, or Da Min. Minzhu is said to mean "folk pig."

The Minzhu breed can be identified by its very long black hair. This hair has coarse, long bristles and a dense woolen undercoat in the winter. This hair allows sows to farrow in an open shed at 39°F (4°C) with no problems. The body size of this breed is relatively large, with a narrow, level back and loin. Minzhu pigs are able to store 10 lb (4.6 kg) of body

Figure 3.31 Fengjing boar imported to Iowa State University in 1989. *(Courtesy, Palmer Holden, Iowa State University, Ames, IA)*

Figure 3.32 Meishan sow in China. *(Courtesy, Max Rothschild, Iowa State University, Ames, IA)*

fat in the abdomen. Sows grow to be about 34 in. (87.5 cm) high and 194 lb (88.3 kg) live-weight. They have a back fat thickness of 1.2 in. (3.2 cm) and a dressing percentage of 72.2%.

The Minzhu is a prolific breed, though not as prolific as other imported breeds. They reach puberty at 3 to 4 months of age, achieve high embryo survival rates, and a large litter rate of 15 to 16 pigs. Third and later parities of this breed had 15.5 pigs born, with 11.0 being weaned. Their 240-day weight was 233 lb (106 kg) with an ADG of 1.1 lb (0.5 kg).

SPECIALIZED BREEDING STOCK SUPPLIERS

Until about 1950, typical U.S. commercial swine producers of market hogs selected most replacement female breeding stock from their own herds, whereas almost all boars were purchased from purebred breeders. But the entry of specialized breeding stock suppliers has changed many breeding programs. A recent survey showed that specialized breeding stock suppliers were the source of 14.1% of the nation's gilts and of 27.7% of the nation's boars. Also, the survey revealed a trend of large commercial hog producers to buy breeding stock from these new specialized breeding stock suppliers (Figure 3.33).

Specific Pathogen Free

Specific pathogen free (SPF) is not a breed or genetic program. SPF producers raise pigs of many breeds, but their main feature is the development of replacement animals free of certain "specific" diseases. See the discussion in Chapter 15 for information on this herd health program.

Figure 3.33 Boar Power boar. Farmers Hybrid was probably the first commercial company to sell special lines of pigs for use by commercial producers. It was started in Iowa in 1945 and produced 6 lines of breeding swine. The business ended in the late 1990s. *(Courtesy, Boar Power, Des Moines, IA)*

Specialized Breeding Stock

Most of the specialized breeding stock suppliers breed and market one or more of the traditional purebred breeds. In addition, they usually produce and market, often under their own genetic names, one or more hybrids, crossbreds, inbreds, maternal female lines, or terminal sire lines. They have well-developed sales and testing programs, which emphasize performance and carcass quality. Some of them also provide financial assistance to their customers. Several of the specialized breeding stock suppliers are provided in the appendix.

MINIATURE SWINE

During the 1950s, the need developed for a smaller breed of swine for use in biomedical studies. An awareness of the need arose from some of the work of the U.S. Atomic Energy Commission. Swine were selected as the animals of choice for certain studies. But the large size and the amount of manure produced by standard swine created a disposal problem and discouraged their use for long-term studies. In 1949, the Hormel Institute of the University of Minnesota initiated the development of genetically small pigs, the Hormel Miniature; they are now extinct in the United States. These were used in some of the first Atomic Energy Commission studies. Later, other small strains were developed, such as the Pitman-Moore, the Hanford Miniature Swine, and the Gottingen Miniature. These miniatures are much smaller than standard swine. They are good experimental animals for biomedical studies, offering economic and convenience advantages over standard swine. They also require less housing space, eat less, are easier to handle, and produce less manure.

> **Vietnamese pot-bellied pigs**—These native Asian pigs, which are normally less than one-fifth the size of traditional U.S. hogs, were first brought into the United States in 1988 for zoos. But they very quickly became novelty pets, selling for as much as $5,000 or more. Normally, they are 18 to 24 in. tall, and they have pot-bellies, swayed backs, and straight tails that they wag when they are happy. By 1994, it was estimated that the nation's pot-bellied pig population numbered 200,000, and prices plummeted to $50 to $100.

Local ordinances often restrict the keeping of livestock within city or corporate limits. Because pot-bellied pigs are subject to the same state veterinary regulations as U.S. commercial swine, keepers of pot-bellied pigs should contact a veterinarian for assistance in their testing and health care. Pot-bellied pigs should be fed similar feeds to those fed to their bigger counterparts, but in much smaller quantities. Feed one-half to three-fourths lb of feed per 50 lbs of body weight, depending on the condition of the pig.

QUESTIONS FOR STUDY AND DISCUSSION

1. Must a new breed of swine be approved by someone, or can anyone start a new breed? Justify your answer.
2. Trace shifting of swine types throughout the years, including the factors that prompted such shifts.
3. Why have U.S. swine types shifted more rapidly and more radically than cattle and sheep types?
4. Most American-created breeds of swine evolved during two periods, namely, (a) 1800 to 1880 and (b) since 1940. What is the explanation for this?
5. Table 3.1 gives the annual registrations of breeds of swine. These figures show that currently three breeds, Berkshire, Duroc, and Yorkshire, dominate the purebred swine breeds. Why?
6. List the place of origin and the distinguishing characteristics of the major U.S. breeds of swine. Discuss the importance of each.
7. Obtain breed registry association literature and a sample copy of a magazine of your favorite breed of swine. Evaluate the soundness and value of the material that you receive. (See the appendix for addresses.)
8. Justify any preference that you may have for one particular breed of swine.
9. During the 1930s and 1940s, several new American inbred breeds of swine evolved, most of them carrying some Landrace breeding. What was the reasoning behind their development? Was their development and cost justified?
10. How do the specialized breeds of swine differ from the inbred breeds and from the traditional breeds?
11. Why are miniature swine more suitable for biomedical studies than the standard breeds of swine?
12. How important are breed characteristics? Can breed differences in hams and pork chops be detected?

SELECTED REFERENCES

Breeds of Livestock, Oklahoma State University, Department of Animal Science, http://www.ansi.okstate.edu/breeds/swine/

Pigs: A Handbook to the Breeds of the World, Valerie Porter, Cornell University Press, Ithaca, NY, 1993

Pigs from Cave to Corn Belt, C. W. Towne and E. N. Wentworth, University of Oklahoma Press, Norman, OK, 1950

Stockman's Handbook, The, 7th ed., M. E. Ensminger, Interstate Publishers, Inc., Danville, IL, 1992

Swine in America, F. D. Coburn, Orange Judd Publishing, Bennington, VT, 1916

Breed literature pertaining to each breed may be secured by writing to the respective breed registry associations (see the appendix for the name and address of each association).

P A R T

2

Basic Production Considerations

4

Principles of Swine Genetics

Symbol II, the suggested standard for the year 2000 and beyond. (*Courtesy, National Pork Board, Des Moines, IA*)

Objectives

After studying this chapter, you should:

1. Be aware of the work of early geneticists.
2. Know the basic units of inheritance.
3. Understand the genetics of dominant and recessive genes.
4. Understand the difference between qualitative and quantitative traits.
5. Know how sex is determined in swine (and humans).
6. Know the difference between prepotency and nicking.
7. Know the differences between inbreeding and linebreeding and the risks and benefits.
8. Know the relative advantages and disadvantages of different crossbreeding systems.

A background of genetic principles is essential to understanding the basics of factors that influence the genetic components of animals. The combining of genes from a sire and dam becomes more complicated as science progresses to the stage where individual genes are located on a chromosome. Utilizing these genetic variables in defined pure or crossbreeding programs allows the breeder to enhance the occurrence of desirable traits and eliminate or reduce the occurrence of undesirable traits.

SYMBOL II, THE IDEAL MARKET HOG[1]

The National Pork Board (NPB) periodically revises the standards for the ideal market hog. The hog referred to as Symbol II, defined in 1996, is longer, leaner, and more muscular than its predecessor introduced in 1983. The standards for Symbol II are hogs marketed at 260 lb in 156 days for a barrow or 164 days for a gilt.

Both barrows and gilts are to have a live-weight feed efficiency of 2.4 lb feed or less for each pound of gain. Loin eye area at 260 lb is expected to be at least 6.5 square in. for barrows and 7.1 square in. for gilts and have acceptable color, marbling, and ultimate pH levels. The intramuscular fat should be greater than or equal to 2.9% for barrows and 2.5% for gilts.

Barrows are expected to have a fat-free lean index of at least 49.8 and gilts should be 52.2 or better. Symbol II is expected to be free of the stress gene and is the result of a terminal crossbreeding program with a maternal line that can wean 25 pigs per sow per year. Symbol II should be the product of a high-health production system operated by an environmentally assured producer. Many of the standards for Symbol II have already been exceeded by parts of the industry, but a significant portion still needs work to meet these goals.

The economic justification for improved breeding is that quality pigs make more money. Through the years, the swine industry has been responsive to this motivating force. For example, economics caused breeding programs to shift from lard-type to meat-type hogs. In addition to changes in body type and cutout value, it is important that producers initiate breeding and selection programs to improve the performance traits of economic importance, such as litter size, growth rate, and feed efficiency. The bottom line is this: With each improvement gained through breeding programs, the cost of producing pork declines.

Purebred hogs are generally raised to be sold to commercial producers as seedstock for crossbreeding programs. These crossbreeding programs strive to combine the best traits of several breeds to produce superior offspring with the sought-after traits of leanness, meatiness, good feed efficiency, fast growth, and durability. Often today, the commercial producer has the option of purchasing maternal line gilts that will be mated to paternal line boars, and the producer purchases all of the needed replacements.

Any herd improvement means more profit. Also, any permanent herd changes made by individual swine producers inevitably contribute to permanent breed improvement.

Swine breeders should have clear goals in mind and a definite breeding program directed toward obtaining those goals. Despite the remarkable progress of the past, significant opportunities remain. A casual glance at the daily receipts of any hog

---

[1]http://www.porkboard.org/docs/
2002-3%20PORK%20FACTS%20BK.pdf

market is convincing evidence of the task ahead. The challenge is primarily that of improving the great masses of animals in order that more of them may approach desired specimens. Also, in this computer age, there must be greater efficiency of production, and this means more rapid growth, less feed to produce meat, and lifting of the pig crop percentage well above the present U.S. average. With the experience of the pioneers to guide us and with our present knowledge of genetics and physiology of reproduction, our progress should now be much more certain and rapid. Animal breeding certainly has moved from an art to both an art and a science.

EARLY ANIMAL BREEDERS

Until recent times, the general principle that like begets like was the only recognized law of heredity. The application of this principle over a long period of time has been effective in modifying animal types in the direction of selection and is evident from a comparison of present-day types and breeds of swine.

There can be little doubt that men such as Robert Bakewell, the English patriarch who is known as the Father of Animal Breeding, and other 18th-century breeders had made tremendous contributions in pointing the way toward livestock improvement before Gregor Mendel's laws were developed in the 19th century. Bakewell's use of progeny testing and controlled breeding was truly epoch making, and hiring out his prize New Leicester rams to farmers and his improvement of Shire horses and Longhorn cattle were equally outstanding. He and other pioneers had certain ideals in mind, and, according to their standards, they were able to develop some nearly ideal specimens. These men were intensely practical, never overlooking utility value or market requirements. No animal met with their favor unless such favor was earned by meat on the block, milk in the pail, weight and quality of wool, pounds gained for pounds of feed consumed, draft ability, or some other performance of practical value. Their ultimate goal was that of furnishing better animals for the market and lowering the cost of production. It must be so with the master breeders of the present and future.

Others took up the challenge of animal improvement where Bakewell and his contemporaries left off, slowly but surely molding animal types. Armed with a better understanding of genetics, during the past 100 years, remarkable progress has been made in breeding better meat animals—animals that are more efficient and, at the same time, that produce cuts of meat of better eating quality as demanded by the consuming public.

The Wild Boar, the Arkansas razorback, and the lard-type hog have been replaced by modern meat-type swine.

The laws of heredity apply to swine breeding exactly as they do to all classes of farm animals. But the breeding of swine is more flexible because (1) hogs normally breed at an earlier age, thus making for a shorter interval between generations; (2) they have a relatively short gestation period; and (3) they are litter-bearing animals. Because of these factors, together with the available feeds and the types of pork products desired by consumers, the world's swine breeders have created more new breeds and made more rapid shifts in hog types than in any other class of farm animals.

MENDEL'S CONTRIBUTION

Modern genetics was really founded by Gregor Johann Mendel (Figure 4.1), an Austrian monk who conducted breeding experiments with garden peas from 1857 to 1865, during the time of the Civil War in the United States. In his monastery at Brunn (now Brno, in the Czech Republic), Mendel applied a powerful curiosity and a clear mind to reveal some of

Figure 4.1 Gregor Mendel at work in his genetic strain garden. Gregor Mendel was an Austrian scientist who experimented with garden peas to establish the principle of heredity. *(© Bettmann/CORBIS)*

the basic principles of hereditary transmission. He set out to examine and quantify the physical traits in pea plants (he chose pea plants because of their speedy reproductive cycles) in an attempt to predict the traits that would occur in future generations.

Mendel concluded that certain factors were transmitted from parent to offspring, providing a connection from one generation to the next. His interpretation suggested that each individual had not one, but two factors for each trait, and that these factors interacted to produce the final physical characteristics of the individual. In 1866, he published in the proceedings of a local scientific society a report covering 8 years of his studies, but for 34 years his findings went unheralded and ignored. Finally, in 1900, 16 years after Mendel's death, 3 European biologists independently duplicated his findings, and this led to the dusting off of the original paper published by the monk 34 years earlier.

The essence of Mendelism is that inheritance is by particles or units (called genes), that these genes are present in pairs—one member of each pair having come from each parent—and that each gene maintains its identity generation after generation. Thus, Mendel's work with peas laid the basis for two of the general laws of inheritance: (1) the law of segregation and (2) the independent assortment of genes. Later, genetic principles were added, yet all the phenomena of inheritance, based on the reactions of genes, are generally known under the collective term *Mendelism.*

Thus, modern genetics is really unique in that it was founded by an amateur who was not trained as a geneticist and who did his work merely as a hobby. During the years since the rediscovery of Mendel's principles in 1900, many additional genetic principles have been added, but the fundamentals as set forth by Mendel have been proven correct in every detail. It can be said, therefore, that inheritance in both plants and animals follows the biological laws discovered by Mendel.

SOME FUNDAMENTALS OF HEREDITY IN SWINE

The modern breeder knows that the job of transmitting qualities from one generation to the next is performed by the germ cells—a sperm from the male and an ovum or egg from the female. All animals, therefore, are the result of the union of two such tiny cells, one from each of its parents. These two germ cells contain all the anatomical, physiological, and psychological traits that the offspring will inherit.

In the somatic (body) cells of an animal, the chromosomes occur in pairs (diploid numbers), whereas in the formation of the sex cells, the egg and the sperm, a reduction division occurs (a process called meiosis) and only one chromosome and one gene of each pair go into a sex cell (haploid number). This means that only half the number of chromosomes and genes present in the body cells of the animal go into each egg and sperm, but each sperm or egg cell has genes for every characteristic of its species. As is explained later, the particular half that any one germ cell gets is determined by chance. When mating and fertilization occur, the single chromosomes from the germ cell of each parent unite to form new pairs, and the genes are again present in duplicate in the somatic cells of the embryo. When these new cells divide, in a process called mitosis, the chromosomes within the cell are duplicated so the new cells are exactly like the parent somatic cell.

The bodies of all animals are made up of millions of microscopic cells. Each cell contains a nucleus in which there are a number of pairs of bundles called chromosomes. The central inner portion of each chromosome contains a long, double-helical molecule called deoxyribonucleic acid, DNA for short. The DNA molecule is the genetic material—the genetic code that is responsible for the information for protein synthesis. Genes form a portion of each DNA molecule, each in a fixed or special position, called a locus, of each chromosome.

Chromosomes and Genes

Chromosomes carry all the hereditary traits, or genes, of animals, from the body type to the color of the hair. Mammals have between 30,000 and 40,000 different genes, the functional units of inheritance controlling the transmission and expression of one or more traits.

Within each cell nucleus are many small X-shaped objects, called chromosomes, that occur in pairs. Because the chromosomes appear in pairs, the genes are also in pairs. A pig cell carries 18 pairs of chromosomes plus the sex chromosomes, X and Y, or a total of 38 (Figure 4.2). (*Note:* humans have 23 pairs of chromosomes, cattle have 30, and sheep have 27 pairs). A female will have two X chromosomes, and the male will have an X and a Y chromosome. Thousands of genes are on the chromosomes and they determine all the hereditary characteristics of living animals via the coded information in their DNA. Thus, inheritance is transmitted by units (chromosomes) rather than by the blending of two traits.

Deoxyribonucleic Acid (DNA)

The most important genetic material in the nucleus of the cell is deoxyribonucleic acid (DNA), which

Figure 4.2 Chromosome pairs of the pig as they would appear in a photomicrograph made from the chromosome of the metaphase of cell division when they can be stained and observed. There are 18 pairs of autosomes and one pair of sex chromosomes (XY). The chromosomes contain all the genetic information necessary to make a male pig. *(Courtesy of Pearson Education)*

serves as the genetic information source. The genetic code contained on the DNA strand can be translated by cells (1) to make structural proteins, hormones, and enzymes of specific structure that determine the basic morphology and functioning of a cell; (2) to control differentiation, the process by which a group of cells becomes an organ; or (3) to control whether an embryo will become a pig, a

cow, or a human (Figure 4.3). A gene is a very small segment of the DNA strand.

The genetic code consists of a two-stranded DNA molecule. At each nucleotide residue along the DNA molecule, the nucleotides are complementary. Thus, when a cell duplicates, all chromosomes duplicate and divide so that each new cell contains the same number of chromosomes and, therefore, the same heritable information. In most cases the double-stranded DNA molecule folds to form a helical structure, which resembles a spiral staircase. This is why DNA has been referred to as the "double helix."

The double-stranded DNA helix has four nucleotide bases: adenine (A), guanine (G), cytosine (C), and thymine (T). Adenine pairs with thymine and cytosine pairs with guanine (Figure 4.4). The sequence of these four bases along the DNA strand acts as a code in which messages can be transferred from one cell to another during the process of cell division. It is estimated that there are approximately 3 billion nucleotides.

Linkage Map

A gene is a set of information that codes for certain proteins. They are small segments on the chromosomes and are identified as quantitative trait loci (QTLs; locus is singular), which define the location of each gene. Thus, a linkage map gives a specific location, or QTL, to a gene by assigning distances between genes. However, this location may not start at the beginning of a chromosome or run all the way through it. Instead, it is a portion of the chromosome that has located genes and distances between genes, much like that of a road map that indicates towns along an interstate. This map is used when obtaining an overall view of a large portion of DNA on a chromosome and to select for genes that influence certain traits. The lesser the distance (in units called

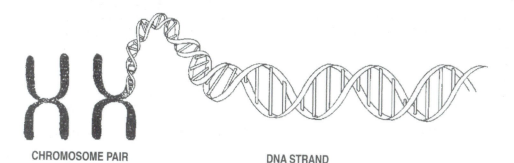

CHROMOSOME PAIR   DNA STRAND

Figure 4.3 Chromosomes consist of deoxyribonucleic acid. (DNA) and a type of protein coiled into a tightly packaged structure. The DNA molecule (uncoiling left to right) has a double-helical structure. *Note: The concept is greatly exaggerated in this diagram.* *(Courtesy of Pearson Education)*

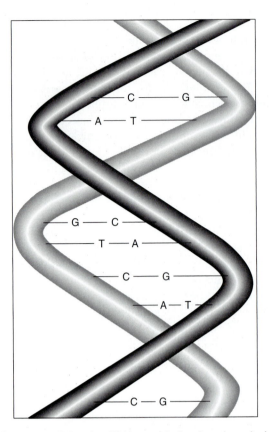

Figure 4.4 Part of a DNA strand showing the adenine (A)–thymine (T) and cytosine (C)–guanine (G) linkages. *(Courtesy, U.S. Pig Genome Mapping Coordination Program, Iowa State University, Ames, IA, 1997–2003)*

centimorgans or cM), the closer the genes are to each other (Figure 4.5).

Genetic markers are detectable differences in a DNA sequence that can be used to mark or "tag" a certain location on the chromosome. Geneticists use these markers in combination with phenotypic (visual) observations to estimate breeding values and to select animals to be parents of the next generation. Because many genes may affect any trait, the identification of genetic markers associated with a certain trait, for example, litter size, can be accomplished.

Once a QTL has been detected for a trait, that area still may consist of thousands of genes. To further locate the specific genes, fine-mapping is employed. This is expensive and time consuming. However, gene locations on chromosomes are often similar between species. Thus, the well-mapped genome of the mouse and human can be used to locate similar genes on the pig chromosomes.

Genetic Diversity

For each QTL or gene in one of the members of a chromosome pair there is a corresponding QTL in the other member of that chromosome pair. These

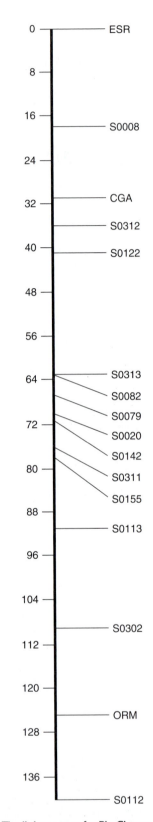

Figure 4.5 The linkage map for Pig Chromosome 1. Various genes are identified along the right side and the distances from the beginning of the chromosome are on the left. The first gene is the estrogen receptor (ESR), which is highly correlated with large litter size. *(Courtesy, U.S. Pig Genome Mapping Coordination Program, Iowa State University, Ames, IA, 1997–2003)*

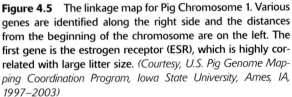

genes located at corresponding loci in chromosome pairs may be identical in the way they affect a trait or they may contrast in the way they affect a trait. If the genes are identical, then the individual is said to be homozygous at that QTL. If they contrast in the way they affect a trait, the individual is said to be heterozygous at that QTL. It follows that homozygous individuals are genetically pure because the genes for a trait passed on in the egg or sperm will always be the same. But heterozygous individuals produce sperm or eggs bearing one of two types of genes.

Homozygous and heterozygous may also refer to a series of hereditary factors. However, few if any animals are entirely homozygous, producing totally uniform offspring as a result of one sperm or egg being just like any other. Rather, animals are heterozygous, and there is often wide variation within the offspring of any given sire or dam. The true breeder recognizes this fact and insists on keeping production records of all offspring rather than only those of selected individuals.

With all possible combinations in 19 pairs of chromosomes in swine and the number of genes they bear, it is not strange, therefore, that no two animals within a given breed are exactly alike (except identical twins from a single egg that split after fertilization). Even between such closely related individuals as full sisters, it is possible that there will be wide differences in size, growth rate, temperament, conformation, and in almost every conceivable trait. Admittedly, many of these differences may be a result of undetected differences in environment, but it is still true that much of the variation is because of hereditary differences.

Genomics is the study of how specific genes are linked to observable traits. It attempts to identify traits that are expensive or difficult to define, for example, litter size or muscle quality. The importation of the Chinese pigs to the United States in 1989 was to study a diverse population when compared to U.S. hogs and try to locate specific genes or QTLs on chromosomes that affect these traits.

Because of the variation in the number of genes affecting a trait, the mating of a sow with a fine production record to a superior boar will not always produce pigs of merit equal to that of their parents. The pigs could be markedly poorer than the parents or, happily, they could in some traits be better than either parent.

Selection and inbreeding or linebreeding are tools with which the swine breeder can obtain boars and sows whose chromosomes and genes contain similar hereditary determiners—animals that are genetically more homozygous.

Mutations

Gene changes are technically known as mutations. A mutation may be defined as a sudden variation that results from changes in a gene or genes that is later passed on through inheritance. Mutations are not only rare, random events, but they are prevailingly harmful. For all practical purposes, therefore, the genes can be thought of as unchanged from one generation to the next.

Once in a great while a mutation occurs in a farm animal, and it produces a visible effect in the animal carrying it. These animals are sometimes called *sports*. Mutations are occasionally of practical value. The occurrence of the polled trait within the horned Hereford and Shorthorn breeds of cattle is an example of a mutation with economic importance. Out of this has arisen polled Hereford and polled Shorthorn cattle.

Gene changes can be accelerated by exposure to ionizing radiation (x-rays, radium), a wide variety of chemicals (mustard gas, LSD), and ultraviolet light rays.

Geneticists have always dreamed of producing specific or directed mutations thereby creating new varieties. Whether this will be achieved in farm animals is uncertain, but it is an enticing possibility, and recombinant DNA techniques suggest its eventuality.

Simple Gene Inheritance (Qualitative Traits)

In the simplest type of inheritance, only one pair of genes is involved. This type of inheritance can be used to demonstrate how genes segregate in the sperm and egg at random. Therefore, the possible gene combinations are governed by the laws of chance (probability) operating in much the same manner as the results obtained from flipping coins. Relatively large numbers are required for certain proportions to be evident. For example, if a penny is flipped often enough, the number of times heads and tails will result is about even. However, with the laws of chance in operation, it is possible that out of any 4 tosses one might get all heads, all tails, or even 3 to 1.

Dominant and Recessive Factors

When genes at corresponding loci on chromosome pairs are unlike, one of the genes often overpowers the expression of the other. This gene is referred to as the dominant gene, and the gene whose expression is prevented is called the recessive gene.

The pair of genes responsible for erect and lop ears in swine may be used as an example of dominant and recessive characteristics and as an example to illustrate how the law of probability operates

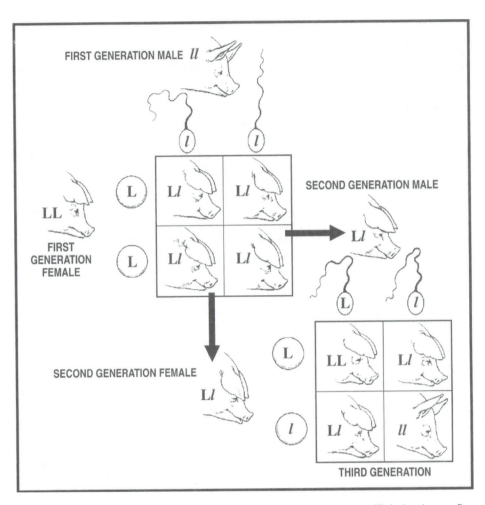

Figure 4.6 Initially, a first-generation male homozygous for erect ears (ll) is bred to a first-generation female homozygous for lop ears (LL). Because lop ears are dominant, all of their offspring will be lop eared (Ll) as they carry the L gene. When these second-generation male and female lop-eared offspring are mated to each other, they produce spermatozoa or eggs for lop ears and erect ears in equal proportion. These third-generation offspring will, on the average, consist of 3 lop ears (dominant gene) to 1 erect ear (two recessive genes). *(Courtesy of Pearson Education)*

in inheritance. This situation is illustrated in Figure 4.6. In this example, the gene for lop ears has dominance, or its full effect, regardless of whether it is present with another just like itself or is paired with a recessive gene. An erect-eared boar possesses only recessive genes for lop ears and can only form sperm bearing recessive genes. A lop-ear sow bearing only the dominant genes for lop ears can only produce eggs bearing the dominant gene.

All the first-cross progeny from the first-generation male are hybrid lop eared, that is, they appear lop eared but possess a dominant gene for lop ears and a recessive gene for erect ears. In scientific terms, their phenotype (how they appear) is lop ear, whereas their genotype is lop erect. If mated, these offspring produce sperm and eggs bearing either a dominant or a recessive gene for lop ear. When large numbers are involved, the results of this mating would be that phenotypically, one-quarter of the off-

spring would have lop ears and three-quarters would have erect ears. Genotypically, one-quarter would be lop–lop (homozygous dominant); one-half would be lop–erect (heterozygous); and one-quarter would be erect–erect (homozygous recessive).

It is clear, therefore, that a dominant trait will cover up a recessive. Hence, a hog's breeding performance cannot be recognized by its phenotype (how it looks), a fact that is of great significance in practical breeding.

As can be readily understood, dominance often makes the task of identifying and discarding all animals carrying an undesirable recessive factor a difficult one. Recessive genes can be passed on from generation to generation, appearing only when two animals both of which carry the recessive factor happen to mate. Even then, only 1 out of 4 offspring produced will, on the average, be homozygous for the recessive factor and show it. If the recessive gene is

Figure 4.7 The Hampshire-marked pig has an umbilical hernia. It caused a weakened abdominal wall but the expression is often affected by environmental effects such as chilling as the pigs pile to keep warm. Most reach market weight with no adverse affects. Neither they nor their siblings should be kept for breeding. *(Courtesy, Palmer Holden, Iowa State University, Ames, IA)*

of economic importance, no animals from the affected litter should be selected as herd replacements and the sire and dam should be culled (Figure 4.7).

Color Inheritance

In swine, white is dominant to colored hair, with the result that when white hogs are crossed with colored hogs, the first crosses are white, although the colored breeds sometimes transmit some of their skin pigmentation. Some examples of color inheritance in swine follow:

1. **Black breed (Poland China) × white breed (Chester White, Yorkshire, or Landrace).** The offspring are usually white with small black spots, although there may be roans in some cases.
2. **Black breed (Poland China) × red breed (Duroc).** The offspring will be black-and-red spotted.
3. **Red breed (Duroc) × white breed.** The offspring are usually white, although there may be roans in some cases.
4. **Belted breed (Hampshire) × red or black breed.** The offspring are generally colored with white belts.
5. **Belted breed (Hampshire) × white breed.** The offspring are usually white with some black spots, or there may be some degree of roan. Ghost patterns often show.

Recessive Defects

Examples of undesirable recessives in animals include scrotal hernia and inverted nipples (blind teats) in pigs. When these conditions appear, one can be very certain that both the sire and dam contributed equally to the condition and that each of them carries the respective recessive gene. This fact should be given strong consideration in the culling program.

Assuming that a hereditary defect or abnormality has occurred in a herd and it is recessive in nature, the breeding program to be followed to prevent or minimize the possibility of its future occurrence will depend somewhat on the type of herd involved—especially on whether it is a commercial or purebred herd. In an ordinary commercial herd, the breeder can usually guard against further reappearance of the undesirable recessive simply by using an outcross (unrelated) sire within the same breed or by crossbreeding with a sire from another breed. With this system, the breeder is fully aware of the recessive being present but has taken action to keep it from showing up.

However, if such an undesirable recessive appears in purebred or seedstock herds, the action should be more drastic. A reputable breeder has an obligation not only to himself/herself but also to customers among both the purebred and commercial herds. Purebred and seedstock animals must be purged of undesirable genes and lethals. This can be done by

1. Eliminating sires and dams that are known to have transmitted the undesirable recessive trait.
2. Culling both the abnormal and normal offspring produced by these sires and dams because approximately half of the normal animals will carry the undesirable trait in the recessive condition.
3. In some instances, mating a prospective herd sire to several females known to carry the undesirable recessive, and observing the offspring, can ensure that the new sire is free from the recessive.

Such culling and test matings in a purebred or seedstock herd is expensive, and it calls for considerable courage. Yet it is the only way in which the purebreds of any farm animal species can be freed from such undesirable genes.

Multiple-Gene Inheritance (Quantitative Traits)

Relatively few traits of economic importance in farm animals are inherited in as simple a manner as the ears described. Important traits, such as litter size, daily gain, feed intake, backfat, and muscle quality, are affected by many genes; thus, they are called multiple-factor traits or multiple-gene traits. Because such traits show all manner of gradation—from high to low performance, for example—they

Figure 4.8 Spraddle or splay legs are one of the most common of leg disorders in pigs. It is probably caused by multiple genetic factors. Piglets are born with weakness in the rear legs and the splayed legs can often be corrected by putting tape around them for the first few days of life. *(Courtesy, Palmer Holden, Iowa State University, Ames, IA)*

are sometimes referred to as quantitative traits. Still the mechanism of inheritance is the same whether a few or multiple genes are involved (Figure 4.8).

In quantitative inheritance, the extremes (either good or bad) tend to swing back to the average. Thus, the offspring of a high-producing boar and a high-producing sow are not apt to be as good as either parent. Likewise, and happily so, the progeny of two very mediocre parents will likely be superior to either parent.

Estimates of the number of pairs of genes affecting each economically important trait vary greatly, but the majority of geneticists agree that for most such traits 10 or more pairs of genes are involved. Growth rate in swine, for example, is affected by (1) the animal's appetite or feed consumption; (2) the efficiency of assimilation—that is, the proportion of the feed eaten that is absorbed into the blood stream; and (3) the use to which the nutrients are put after assimilation—growth or finishing. This example should indicate clearly enough that a trait such as growth rate is controlled by many genes and that it is difficult to determine the mode of gene action of such traits.

HEREDITY AND ENVIRONMENT

A sleek, lean hog, with an ideal body conformation, is undeniably the result of two forces—heredity and environment. If turned to the forest, a littermate to the sleek pig would present an entirely different appearance. By the same token, optimum environment could never make an attractive animal out of a pig with scrub ancestry.

These are extreme examples, and they may be applied to any class of farm animals, but they do emphasize the fact that any particular animal is the product of heredity and environment. Stated differently, heredity may be thought of as the foundation, and environment as the structure. Heredity has already made its contribution at the time of fertilization, but environment works ceaselessly away until death.

Admittedly, after looking over an animal, a breeder cannot with certainty know whether it is genetically a high or a low producer, and there can be no denying the fact that environment plays a tremendous part in determining the extent to which hereditary differences that are present will be expressed in animals. The animal's phenotype (visual appearance) is the combination of its genetic makeup and the environment in which it lives.

Experimental work has shown conclusively that the vigor and size of animals at birth depends on the environment of the embryo from the minute the ovum or egg is fertilized by the sperm. Some evidence indicates that newborn animals are affected by the environment of the egg and sperm long before fertilization has been accomplished. In other words, perhaps because of the storage of substances, the kind and quality of the diet fed to young females may later affect the quality of their progeny. Generally speaking, then, environment may inhibit the full expression of potentialities from a time preceding fertilization until physiological maturity has been attained.

It is generally agreed, therefore, that maximum development of traits of economic importance—growth, body form, milk production, and so on—cannot be achieved unless there are optimum conditions of nutrition, health, environment, and other management factors. However, the next question is whether a breeding program can make maximum progress under conditions of suboptimal nutrition as is often found under some farm conditions. One school of thought is that selection for such factors as body form and growth rate in animals can be most effective only under nutritive conditions promoting the near maximum development of those traits of which the animal is capable.

The other school of thought is that genetic differences affecting usefulness under suboptimal conditions will be expressed under such suboptimal conditions, and that differences observed under forced conditions may not be correlated with real utility under less favorable conditions. Those favoring the latter thinking argue, therefore, that the production and selection of breeding animals for suboptimal nutritive conditions should be under less favorable conditions and that the animals should not be highly fed.

Within the pure breeds of swine—managed under average or better-than-average conditions—it has been found that, in general, only 5 to 40% of the observed variation in a trait is actually brought about by hereditary variations. To be sure, if we contrast animals that differ very greatly in heredity—for example, a top-producing hog and a mediocre hog—90% or more of the apparent differences in type may be a result of heredity. The point is, however, that extreme cases, such as the one just mentioned, are not involved in the advancement within improved breeds of livestock. Here the comparisons are between animals of average or better-than-average quality, and the observed differences are often very minor.

The problem of progressive breeders is that of selecting the very best animals available genetically for parents of the next generation of offspring in their herds. Because only 5 to 40% of the observed variation is a result of differences in inheritance, and because environmental differences can produce misleading variations, mistakes in the selection of breeding animals are inevitable. However, if the purebred or seedstock breeder has clearly in mind a well-defined ideal and adheres rigidly to it in selecting breeding stock, very definite progress can be made.

SEX DETERMINATION

On the average, and when considering a large population, approximately equal numbers of males and females are born in all common species of animals. To be sure, many notable exceptions can be found in individual herds or flocks.

Sex is determined by the chromosomal makeup of the individual. One particular pair of the chromosomes is called the sex chromosomes. In mammals, the female has a pair of similar sex chromosomes, called X chromosomes; the male has a pair of unlike sex chromosomes, called an X chromosome and a Y chromosome. In avian species, this condition is reversed, the female having the unlike pair and the male having the like pair.

The pairs of sex chromosomes separate when the germ cells are formed. Thus, each of the ova or eggs produced by the sow contains the X chromosomes; whereas the sperm of the boar are of two types, one-half containing the X chromosome and the other half the Y chromosome. Because, on the average, the eggs and sperm unite at random, it can be understood that half of the progeny will contain the chromosomal makeup XX (females) and the other half XY (males) (Figure 4.9).

Sex Ratio

Through the ages humans have desired to select the sex of their offspring. Rulers were always anxious to

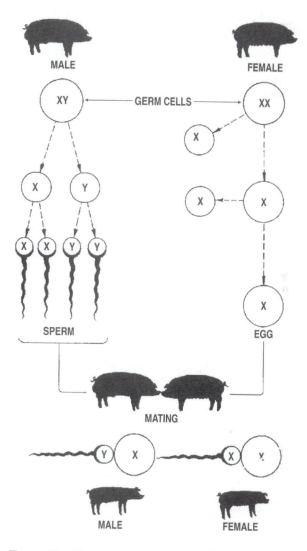

Figure 4.9 Diagrammatic illustration of the mechanism of sex determination in hogs, showing how sex is determined by the chromosomal makeup of the individual. The sow has a pair of like sex chromosomes, called X chromosomes; the boar has a pair of unlike sex chromosomes, called X and Y chromosomes. Thus, if an egg and a sperm of like sex chromosomal makeup unite, the offspring will be a female, and if an egg and a sperm of unlike sex chromosomal makeup unite, the offspring will be a male. *(Courtesy of Pearson Education)*

have sons, and some rulers desired that their servants have fewer sons. For the producers, controlling the sex of the offspring of farm animals is of great interest because of the economic advantage of being able to produce all males or females depending on the need. The normal sex ratio (males to females) is about 50:50 because whether an X- or Y-bearing sperm will fertilize an egg is decided by chance, just as is flipping a coin for heads or tails.

Many approaches have been tried to alter the sex ratio. Some of these approaches are sophisticated and scientific, whereas others are myths that develop from casual observations. Most scientific efforts have focused on eliminating or altering the X or Y sperm in semen before artificial insemination. The three main approaches are (1) sedimentation, in which it is hoped that the heavy X sperm will settle to the bottom and the Y sperm float to the top; (2) electrical charge, in which it is assumed that X and Y sperm have a different net charge, and chemical or electrical methods could bind or attract either negatively or positively charged sperm; and (3) treatment of sperm with a substance deadly to X or Y sperm. Potential methods of sex ratio control include (1) immunological, in which antibodies would prevent Y-bearing sperm from fertilizing an egg; (2) embryo transfer, where only embryos of the desired sex were transferred; and (3) cloning, where the whole genetic makeup including sex could be preselected.

It appears that nature is about to yield to sex control. Although not commercially available yet, predetermination of the sex of 6- to 12-day-old embryos is a reality, and progress is being made in the separation of sperm cells containing X chromosomes from those containing Y chromosomes.

A cogent question is which is preferred, males or females? Boars usually are castrated before feeding and marketing because of off-flavors that occur as they mature, and as a result of castration, they are fatter than gilts but grow faster. Gilts are more efficient and leaner and, of course, have the ability to produce litters of pigs. Very few boars are needed for mating anymore because of the widespread use of artificial insemination. As a result, very few male pigs are desired or needed in the swine industry.

GENETIC DEFECTS

The term *lethal* refers to a genetic factor that causes death of the young, either during prenatal life or shortly after birth. Other defects occur that are not sufficiently severe to cause death but that do impair the usefulness of the affected animals. Some of the lethals and other abnormalities that have been reported in swine are summarized in Table 4.1.

Many such abnormal animals are born on U.S. farms and ranches each year. Unfortunately, some purebred or seedstock suppliers, whose chief business is that of selling breeding stock, may unethically not reveal the appearance of defective animals in their herds because of the justifiable fear that it may hurt their sales. With the commercial producers, however, the appearance of such lethals simply involves so much economic loss that they generally, openly, and without embarrassment admit the presence of the abnormality and seek correction, usually by making changes in their sources of seedstock.

The embryological development—the development of the young from the time that the egg and the sperm unite until the animal is born—is very complicated. Thus, the oddity probably is that so many of the offspring develop normally rather than that a few develop abnormally.

Many such abnormalities (commonly known as monstrosities or freaks) are hereditary, being caused by certain "bad" genes. Moreover, the bulk of such lethals are recessive and may, therefore, remain hidden for many generations. The prevention of such genetic abnormalities requires that the germ plasm be purged of the "bad" genes. This means that, where recessive lethals are involved, the producer must be aware of the fact that both parents carry the gene. For the total removal of the lethals, test matings and rigid selection must be practiced. The best test mating to use for a given sire consists of mating him to some of his daughters or half-sibling matings in which a son and daughter of the boar, from different litters, are mated. Thus, where there is suspicion of scrotal hernia, it is recommended that a boar be bred back to some of his daughters.

Anatomical defects may be caused by either genetics or the environment. Although the frequency of most defects is less than 1%, they need to be evaluated to prevent their perpetuation in the herd. Whereas in some selection criteria there is a floating scale, for example, growth rate, some criteria are all or none.

Some genetic defects are simple recessive which means both parents must carry the gene for it to be expressed. Simple recessive genes are difficult to remove from the herd because they are not always visible in the offspring. Simple dominant genes are expressed in the offspring even if only one parent carries the gene. If both carry the dominant gene, the defect is often more severe and natural selection tends to remove them from the herd.

Some genes are sex linked, that is, usually only found on the X chromosome. They are, therefore, more likely to be expressed in males. Sex-limited traits may be either recessive or dominant and are limited to one sex. An example would be that scrotal hernias only appear in male hogs.

Structural Defects

Structured defects that are a result of genetics should preclude the use of those animals in a selection program. Most of these are clearly visible to the breeder and these animals, and possibly their parents and littermates, should not continue in the breeding herd.

Scrotal hernias and umbilical hernias result from weakened musculature surrounding the inguinal canal or the navel area, respectively. Both situations can be corrected surgically. For scrotal hernias neither the corrected animals nor their parents or littermates

TABLE 4.1 Hereditary Lethal and Nonlethal Abnormalities in Swine

Type of Abnormality	Description of Abnormality	Probable Mode of Inheritance
A. Lethal		
Atresia ani	No anal opening.	Undetermined
Bent legs	Legs bent at right angle and stiff.	Recessive
Brain hernia	Skull fails to close and brain protrudes.	Probably recessive
Catlin mark	Incomplete development of the skull.	Recessive
Cleft palate	Pigs are born alive but unable to nurse; harelip results.	Recessive
Excessive fatness	Pigs become excessively fat at 70 to 150 lb (32 to 68 kg) and die.	Undetermined
Fetal mortality	Born dead or are reabsorbed.	Recessive
Hydrocephalus	Fluid on the brain, head enlarged; often accompanied by short tail.	Recessive
Legless	Pigs born alive but without legs.	Recessive
Muscle contracture	Usually only forelegs affected, but sometimes rear legs are involved; forelegs rigid; animals usually stillborn or live only a short time.	Recessive
Paralysis	Complete paralysis of rear legs; born alive but starve unless given special care.	Recessive
Split ears	Ears split usually associated with cleft palate and deformed rear legs.	Probably recessive
Thickened forelimbs	Thickening of forelegs caused by infiltration of connective tissues that replace the muscle fibers; pigs usually born alive.	Recessive
B. Nonlethal		
Blood warts (melanotic tumors)	Moles or skin tumors that increase in size with age; common in Durocs and Hampshires.	Unknown
Cryptorchidism	One or both testicles retained in body cavity.	Recessive
Gastric ulcers	Erosion of epithelial lining of the stomach, generally in the esophageal region.	Recessive
Hairlessness	Animals born with little or no hair (not to be confused with hairlessness caused by an iodine deficiency).	Recessive
Hemophilia (bleeders)	Blood fails to clot promptly when wounds are inflicted.	Recessive
Hermaphrodites	Animals that possess characteristics of both sexes.	Sex-limited recessive
Humpback	Crooked spine behind shoulder.	Unknown
Inverted nipples (blind teats)	Teats inverted and nonfunctional.	Undetermined
Lymphosarcoma (leukemia, lymphoma)	Malignant tumors of the lymph nodes; stunted growth and death before 15 months.	Autosomal recessive
Motor neuron disease	Distinctive locomotor disorder of nursery pigs, characterized by inability to coordinate muscle movements and slight paralysis.	Autosomal dominant inheritance
Oedema (myxoedema, dropsy, hydrops)	Abnormal accumulation of fluid in tissue and body cavities.	Autosomal recessive disorder
Persistent frenulum	A close attachment of the prepuce to the body by a mucous membrane resulting in inadequate protrusion of the penis and inability to breed.	Unknown
Polydactyl	Extra toes on forefeet.	Undetermined
Porcine stress syndrome	Sudden death of heavily muscled pigs or production of pale, soft, exudative musculature of carcass.	Autosomal recessive
Pseudovitamin D deficiency (rickets)	Indistinguishable from nongenetic lack of vitamin D; most noticeable effect is bowing of the legs.	Autosomal recessive
Rectal prolapse	Protrusion of the terminal part of the rectum and anus.	Mainly environmentally influenced but a genetic liability

(continued)

TABLE 4.1　Hereditary Lethal and Nonlethal Abnormalities in Swine *(continued)*

Type of Abnormality	Description of Abnormality	Probable Mode of Inheritance
B. Nonlethal		
Red eyes	Observed in Hampshires; affected animals also have light brown hair coat.	Probably recessive
Screw tail (kinky tail)	Flexed, crooked, or screw tail caused by fusion of caudal vertebra.	Multigenic recessive
Scrotal hernia	Ruptured; intestines extending into scrotum.	Result of two pairs of recessive factors
Swirls (hair whorls)	Hair forms a cowlick or swirl on the neck or back.	At least 2 pairs of recessive genes are involved
Syndactyl (mule foot)	Only one toe instead of two.	Dominant
Umbilical hernia	Weakness at umbilicus; intestines protrude.	Dominant
Wattles	Skinlike flaps hanging from throat near lower jaw.	Dominant
Wooly	Kinky hair.	Dominant

Source: Courtesy of Pearson Education.

should be used for breeding. Because umbilical hernias may be magnified by environmental conditions, culling of related individuals is not recommended.

Atresia ani is characterized by pigs born without a rectal opening. Cull the parents and do not use littermates for breeding.

Underdeveloped vulva is a condition in which the exterior view indicated a very small vulva. Gilts with this condition should not be saved for breeding because there is a possibility that the reproductive system may be also underdeveloped (Figure 4.10).

Cryptorchidism is the condition where one or both testicles are retained in the body cavity. Cull the parents and do not use littermates for breeding.

Hermaphrodites carry some of the sex organs of both males and females. Cull the parents and do not use littermates for breeding.

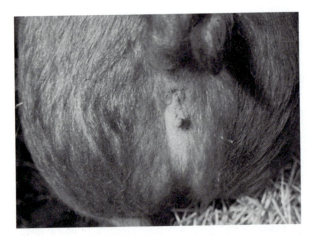

Figure 4.10 Underdeveloped vulva. *(Courtesy, Maynard Hogberg, Iowa State University, Ames, IA)*

Other conditions include nipple abnormalities such as undeveloped or inverted, tremors, and various leg defects. A complete discussion is available in the *Pork Industry Handbook*.

Pork Quality Defects

Porcine Stress Syndrome (PSS)

In the 1960s, Dr. David Topel of Iowa State University coined the term Porcine Stress Syndrome (PSS). PSS is a recessive trait. Pigs that carry the gene are superior in muscling, and, as a result, producers often select these pigs for replacements, thus increasing the incidence of PSS and the production of pale, soft, exudative (PSE) meat. Parents and littermates of affected pigs should be eliminated from the herd. There is an accurate DNA test for the gene. Most commercial boar studs report the results of the test for their sires.

PSS is characterized by the sudden death of heavily muscled pigs when stressed and/or the production of PSE pork muscle observed in the carcasses. PSS is inherited as an autosomal recessive. Birth of a PSS pig incriminates both parents as carriers of the recessive gene (heterozygous) and possibly having two copies of the recessive gene (homozygous). Pigs suffering from PSS demonstrate the following sequence of events when subjected to stresses such as loading and transport to market, overcrowding and fighting, vaccinations, or a sudden change in the weather:

1. Muscle and tail tremors.
2. Dyspnea (shortness of breath).
3. Cyanosis (bluish or purplish discoloration of the skin and mucous membranes as a result of

deficient oxygenation), especially seen in white hogs.

4. Rapid rise in body temperature with signs of heat stress even in cool weather.
5. Total collapse, marked muscle rigidity, and hyperthermia (increase in body temperature).
6. Death in a shocklike state.

At slaughter, PSS pigs produce

1. Pale, soft, watery musculature.
2. Pork showing protein denaturation and lowered quality.

Although less than 1% of the swine in the United States actually die of PSS under normal marketing and management practices, the incidence of low-quality pork in ham and loin muscles runs about 8 to 10%. Pigs homozygous for the PSS gene exhibit about 15% death loss and a greater than 95% incidence of PSE.

PSS animals may appear shorter-bodied and smaller than their normal herd mates. Also, they may be more muscular in appearance than normal animals, with groove-shaped loins, indentations in their rumps, and circular shape to the hams. A separation between the major muscles of the hams is often evident. Dilation of the pupils and tremors of the tail are often observed following exposure to physical stress. However, not all heavily muscled pigs are affected by PSS.

Several reports from Iowa State University and one from Denmark indicate that carriers are 1.5 to 4% higher than normal pigs in lean percentage, and that carriers are equal, if not superior, to normal pigs in growth rate. All of these studies indicate that the muscle quality of carrier animals is generally undesirable.

Common ameliorating factors include removing exposure to stress and cooling pigs that appear to be affected. There is no cure for PSS, except for selection against the trait.

Currently, the DNA probe test, known as the HAL-1843 DNA test, is licensed and is 100% accurate, barring human error. Creatine phosphokinase (CPK) and halothane anesthesia were originally used to evaluate pigs for PSS. The CPK test involves catching a small drop of blood from the pig's ear on a special card and sending the card to a laboratory where it is analyzed for the activity of creatine phosphokinase, a serum enzyme that is abnormally high in PSS swine.

The halothane anesthesia test requires sedating the pig with halothane gas. PSS animals respond to the halothane by showing signs of extreme muscle rigidity within 5 minutes from the start of the treatment (Figure 4.11). This test provides immediate

Figure 4.11   Upon exposure to halothane gas, stress-positive pigs develop a rigid posture and usually die. Note the very heavy muscling typical of PSS-positive pigs. *(Courtesy, Palmer Holden, Iowa State University, Ames, IA)*

results, but the equipment is expensive and must be used under the direction of a trained technician.

The National Genetic Evaluation Program (NGEP) initiated in 1990, and completed in 1995, evaluated the effects of the stress gene. They reported the following:

1. In the homozygous mutant form (*nn*), the stress gene can have some devastating effects. If physical stress does not cause death, it is almost sure to cause PSE pork.
2. The presence of the normal allele (N) will usually prevent stress death and the rigidity response to the anesthetic halothane.

A comparison of the 2,863 normal (NN) pigs and the 391 carriers (N*n*) revealed the following:

1. Growth rate, leg soundness, and backfat thickness were about the same.
2. The 391 stress gene carrier group had a 0.29 sq in. advantage in loin muscle area, a 0.4% higher dressing percentage, and slightly more lean gain. But these pigs produced paler and less tender meat that possessed less intramuscular fat and greater drip loss. *Note:* A 5% shrink per carcass, a conservative estimate for a PSS carcass, results in approximately $10/pig loss. Based on this study, it is clearly evident that the gene should be eliminated from the pig population.
3. Even without the PSS stress gene, 22.5% of the normal loins were disqualified because of lack of quality. So the PSS gene does not account for all of the industry's muscle quality problems. Therefore, a program to monitor pork quality, and to select for the improvement of quality traits is needed.

It must be noted that most participants in the pork industry agree that the stress gene needs to be eliminated. The National Swine Registry has ruled that any Yorkshire, Landrace, Duroc, or Hampshire farrowed after January 1, 2002 (or earlier depending on the breed), proven to be a stress positive or stress carrier by a documented DNA test from a certified lab will have its pedigree cancelled. This responds to packer concerns with downer pigs, producer concerns with producing quality pork, and retailers concerns with marketing quality pork.

Rendement Napole

Rendement napole (RN) is reported to be a major dominant gene influencing meat quality, particularly drip loss and cooking loss during processing of pork. The highest frequency of the gene is reported to be in the Hampshire breed, although it is probable that other breeds also have some incidence as well.

Hereditary or Environmental

In addition to hereditary abnormalities, there are certain abnormalities that may be due to nutritional deficiencies, or to accidents of development—the latter including those that appear to occur sporadically and for which there is no well-defined reason. When only a few defective individuals occur within a particular herd, it is often impossible to determine whether their occurrence is because of (1) defective heredity, (2) defective nutrition, (3) viral infection of the dam, or (4) accidents of development. If the same abnormality occurs in any appreciable number of animals, however, it is probably hereditary or nutritional. In any event, the diagnosis of the condition is not always a simple matter. The following conditions would tend to indicate a hereditary defect:

1. The defect had previously been reported as hereditary in the same species.
2. The defect occurred more frequently within certain families or when there had been inbreeding or linebreeding.
3. The defect occurred in more than one season and when different diets had been fed.

The following conditions might be accepted as indications that the abnormality was the result of a nutritional deficiency:

1. The defect previously had been reliably reported to be the result of a nutritional deficiency.
2. The defect appeared to be restricted to a certain area (mycotoxins for example).
3. The defect occurred when the diet of the mother was known to be deficient.

A toed-out pig. The inside toe is smaller than the outside toe and the toes on the front legs are pointing out at an angle instead of straight ahead. Pigs with this condition do not make good breeding animals because of potential lameness problems. *(Courtesy, Palmer Holden, Iowa State University, Ames, IA)*

4. The defect disappeared when an improved diet was fed.

If there is suspicion that the diet is defective, it should be improved not only to prevent such deformities but also for good and efficient management.

If there is good and sufficient evidence that the abnormal condition is hereditary, the steps to be followed in purging the herd of the undesirable gene are identical to those already outlined for ridding the herd of an undesirable recessive factor. An inbreeding program, of course, is the most effective way in which to expose hereditary lethals in order that purging may follow.

RELATIVE IMPORTANCE OF THE BOAR AND THE SOW

Because a boar can have so many more offspring during a given season or a lifetime than a sow, he is, from a hereditary standpoint, a more important individual than any one sow as far as the whole herd is concerned, although both the boar and the sow are of equal importance as far as concerns any one offspring. Because of their wider use, therefore, boars are usually culled more rigidly than sows, and the breeder can afford to pay more for an outstanding boar than for an equally outstanding sow.

Experienced swine producers have long believed that daughters often resemble their sires more closely than do their sons, whereas sons resemble their dams. Some boars and sows, therefore, enjoy a reputation based almost exclusively on the

The front legs and pasterns on this pig are too straight, causing undue pressure on the joints. Pigs with this defect often develop lameness when raised on hard surfaces such as concrete. *(Courtesy, Palmer Holden, Iowa State University, Ames, IA)*

merit of their sons, whereas others owe their prestige to their daughters. Although this situation is likely to be exaggerated, any such phenomenon that may exist is based on sex-linked inheritance, which may be explained as follows: The genes that determine sex are carried on one of the chromosomes. The other genes that are located on the same chromosome will be linked or associated with sex and will be transmitted to the next generation in combination with sex.

Thus, because of sex linkage, there are more color-blind men than color-blind women. In poultry breeding, the sex-linked factor is used in a practical way for the purpose of distinguishing the pullets from the cockerels early in life, through the process known as *sexing* the chicks. Thus, when a black cock is crossed with barred hens, all the cocks come barred and all the hens come black. It should be emphasized, however, that under most conditions it appears that the influence of the sire and dam on any one offspring is about equal. Most breeders, therefore, will do well to seek excellence in both sexes of breeding animals.

Prepotency

Prepotency refers to the ability of the animal, either male or female, to stamp its own characteristics on its offspring. The offspring of a prepotent boar, for example, resemble both their sire and each other more closely than usual. The only conclusive and final test of prepotency consists of the inspection of the offspring.

From a genetic standpoint, there are two requisites that an animal must possess in order to be prepotent: (1) dominance and (2) homozygosity. Every offspring that receives a dominant gene or genes will

show the effect of that gene or genes in the particular trait or traits that result from those genes. Moreover, a perfectly homozygous animal would transmit the same kind of genes to all of its offspring. Although entirely homozygous animals probably never exist, a system of inbreeding is the only way to produce animals that are as nearly homozygous as possible.

Despite beliefs to the contrary, there is no evidence that prepotency can be predicted by the appearance of a trait in an animal. To be more specific, there is no reason why a vigorous, masculine-appearing boar will be any more prepotent than one less desirable in these respects.

It should also be emphasized that it is impossible to determine just how important prepotency may be in animal breeding, although many sires of the past have enjoyed a reputation for being extremely prepotent. Perhaps these animals were prepotent, but there is also the possibility that their reputation for producing outstanding animals may have rested on the fact that they were mated to some of the best females of the breed.

In summary, it may be said that if a given boar or sow possesses a great number of genes that are completely dominant for desirable type and performance and if the animal is relatively homozygous, the offspring will closely resemble the parent and resemble each other, or be uniform. Fortunate, indeed, is the breeder who possesses such an animal.

Nicking

If the offspring of certain matings are especially outstanding and in general better than their parents, breeders are prone to say that the animals nicked well. For example, a sow may produce outstanding pigs to the service of a certain boar, but when mated to another boar of apparent equal merit as a sire, the offspring may be disappointing. Or sometimes the mating of a rather average boar to an equally average sow will result in the production of a most outstanding individual both from the standpoint of type and performance.

Genetically speaking, so-called successful nicking is a result of the right combination of genes for good traits being contributed by each parent, although each of the parents may be lacking in certain genes necessary for excellence. In other words, the animals nicked well because their respective combinations of good genes complemented each other.

The history of animal breeding includes records of several supposedly favorable nicks. Because of the very nature of successful nicks, however, outstanding animals arising from those matings must be carefully scrutinized from a breeding standpoint because with their heterozygous origin, it is quite unlikely that they will breed true.

Family Names

In animals, depending on the breed, family names are traced through either the males or females. Unfortunately, the value of family names is generally grossly exaggerated. Obviously, if the foundation boar or sow, as the case may be, is very many generations removed, the genetic superiority of this "head of a family" is halved so many times by subsequent matings that there is little reason to think that one family is superior to another. The situation may be further distorted by breeders placing a premium on family names of which there are few members, little realizing that, in at least some cases, there may be unfortunate reasons for the scarcity in numbers, such as poor reproductive performance or livability.

Family names lend themselves readily to speculation. Because of this, the history of livestock breeding has often been blighted by instances of unwise pedigree selection on the basis of not-too-meaningful family names. Fortunately, for swine producers, there has been less worshipping of family names in hogs than in certain other classes of livestock.

Of course, certain linebred families—line bred to a foundation sire or dam so that the family is kept highly related to it—do have genetic significance. Moreover, if the programs involved have been accompanied by rigid culling, many good individuals may have evolved, and the family name may be in good repute.

SYSTEMS OF BREEDING

The many diverse types and breeds among each class of farm animals in existence today originated from only a few wild types within each species. These early domesticated animals possessed the pool of genes, which, through controlled matings and selection, proved flexible in the hands of breeders.

Perhaps at the outset it should be stated that there is no one best system of breeding or secret of success for any and all conditions. Each breeding program is an individual case, requiring careful study. The choice of the system of breeding should be determined primarily by the size and quality of the herd, by the breeder's expertise and knowledge of the principles of animal breeding, by the finances, and and by the ultimate goal desired.

Pure Breeding

A purebred animal may be defined as a member of a breed, the animals of which possess a common ancestry and distinctive characteristics, and either registered or eligible for registration in that breed. The

Figure 4.12   A mature Yorkshire herd boar. *(Courtesy, Palmer Holden, Iowa State University, Ames, IA)*

breed association consists of a group of breeders banded together for the purposes of (1) recording the lineage of their animals, (2) protecting the purity of the breed, and (3) promoting the interest of the breed (Figure 4.12).

The term *purebred* refers to animals whose entire lineage, regardless of the number of generations removed, traces back to the foundation animals accepted by the breed or to animals that have been subsequently approved for infusion.

The terms *pure breeding* and *homozygosity* may bear very different connotations. Yet there is some interrelationship between purebreds and homozygosity. Because most breeds had a relatively small number of foundation animals, the unavoidable inbreeding and linebreeding during the formative state resulted in a certain amount of homozygosity. Moreover, through the normal sequence of events, it is estimated that purebreds become more homozygous by from 0.25 to 0.5% per animal generation.

It should be emphasized that being a purebred animal does not necessarily guarantee superior type or high productivity. That is to say, the word purebred is not, within itself, magic, nor is it sacred. Many a person has found, to much sorrow, that there are such things as purebred never-do-wells. Yet purebred animals are superior to nonpurebreds in transmitting desired traits, whereas crossbreds may actually perform better.

For the producer with experience and adequate capital, the breeding of purebreds may offer unlimited opportunities. It has been well said that honor, fame, and fortune are all within the realm of possible realization of the purebred breeder, but it should also be added that only a few achieve this high calling.

Purebred breeding is a highly specialized type of production. Generally speaking, only the experienced breeder should undertake the production of purebreds with the intention of furnishing foundation or replacement stock to other purebred breeders, or purebred boars for crossbreeding programs. Although we have had many constructive swine breeders and great progress has been made, only a few achieve sufficient success to classify as master breeders.

Inbreeding

Inbreeding is the mating of animals more closely related than the average of the population from which they came, such as sire to daughter, son to dam, and (brother–sister) full-sib matings. It is rarely practiced among present-day swine producers, though it was common in the foundation animals of most of the breeds.

With inbreeding there are a minimum number of different ancestors. In the repeated mating of a brother with his full sister, for example, there are only 2 grandparents instead of 4, only 2 great-grandparents instead of 8, and only 2 different ancestors in each generation farther back—instead of the theoretically possible 16, 32, 64, 128, and so on. The most intensive form of inbreeding is self-fertilization that occurs in some plants, such as wheat and garden peas, and in some lower animals, but domestic animals are not self-fertilized. The reasons for practicing inbreeding are as follows:

1. It increases the degree of homozygosity within animals, making the resulting offspring purer or homozygous in a larger proportion of their gene pairs than in the case of linebred or outcross animals. In so doing, the less desirable recessive genes are brought to light so that they can be more readily culled. Thus, inbreeding, together with rigid culling, affords the surest and quickest method of eliminating undesirable genes and fixing or perpetuating a desirable trait or group of traits.

2. If carried on for a period of time, inbreeding tends to create lines or strains of animals that are uniform in type and in other characteristics.

3. Inbreeding maintains the highest relationship to a desirable ancestor.

4. Because of the greater homozygosity, inbreeding makes for greater prepotency. That is, selected inbred animals are more homozygous for desirable genes (genes that are often dominant), and they, therefore, transmit these genes with greater uniformity.

5. Through the production of inbred lines or families by inbreeding and the subsequent crossing of certain of these lines, inbreeding affords a modern approach to livestock improvement. Moreover, the best of the inbred animals are likely to give superior results in outcrosses.

6. Where a breeder is in the unique position of having a herd so far advanced that to go on the outside for seedstock would merely be a step backward, inbreeding offers the only sound alternative for maintaining existing quality or making further improvement.

The precautions in inbreeding may be summarized as follows:

1. Because inbreeding greatly enhances the chances that recessives will appear during the early generations in obtaining homozygosity, it is almost certain to increase the proportion of undesirable breeding stock produced. This may include such so-called degenerate traits as reduction in size, fertility, and general vigor. Lethals and other genetic abnormalities often appear with increased frequency in inbred animals.

2. Because of the rigid culling necessary in order to avoid the "fixing" of undesirable traits, especially in the first generations of an inbreeding program, it is almost imperative that this system of breeding be confined to a relatively large herd and to instances when the owner has sufficient finances to stand the rigid culling that must accompany such a program.

3. Inbreeding requires skill in making planned matings and rigid selection, thus being most successful when applied by master breeders.

4. Inbreeding should not be adapted for use by the breeder with marginal stock because the very fact that the animals are average means that a minimal number of desirable genes are present. Inbreeding would merely make the animals more homozygous for undesirable genes.

Judging from outward manifestations alone, it might appear that inbreeding is predominantly harmful in its effects—often leading to the production of defective animals lacking in the vitality necessary for successful and profitable production. But this is by no means the whole story. Although inbreeding often leads to the production of animals of low value, the resulting superior animals can confidently be expected to be homozygous for a greater-than-average number of good genes and thus more valuable for breeding purposes. Figuratively speaking, therefore, inbreeding may be referred to as "trial by fire," and the breeder who practices it can expect to obtain many animals that fail to measure up and that have to be culled. However, if inbreeding is handled properly, the breeder can also expect to secure animals of exceptional value.

Although inbreeding has been practiced less during the past century than in the formative period of the different pure breeds of livestock, it has merit when its principles and limitations are fully understood. Perhaps inbreeding had best be confined to use by the skilled master breeder who is in a sufficiently sound financial position to endure rigid and intelligent culling and delayed returns and whose herd is both large and above average in quality. If it were not for inbreeding, breeds would not have been developed.

Linebreeding

Linebreeding is the mating of animals more distantly related than with inbreeding and in which the matings are usually directed toward keeping the offspring closely related to some highly admired ancestor, such as half-brother to half-sister, female to grandsire, and cousins. From a biological standpoint, inbreeding and linebreeding are the same, differing only in intensity. In general, inbreeding has been frowned on by swine producers, but linebreeding (the less intensive form) has been looked on with favor.

In a linebreeding program, the degree of relationship is not closer than half-brother and half-sister or matings more distantly related: cousin matings, grandparent to grand offspring, and so on.

Linebreeding may be practiced in order to conserve and perpetuate the good traits of a certain outstanding boar or sow. Because such descendants are of similar lineage, they have the same general type of germ plasm and, therefore, exhibit a high degree of uniformity in type and performance (Figure 4.13).

In a more limited way, a linebreeding program has the same advantages and disadvantages of an inbreeding program. Stated differently, linebreeding offers fewer possibilities both for good and harm than inbreeding. It is a middle-of-the-road program that the vast majority of average and small breeders can follow safely to their advantage. Through it, reasonable progress can be made without taking any great risk. A greater degree of homozygosity of certain desirable genes can be secured without running too great a risk of intensifying undesirable ones.

Usually a linebreeding program is best accomplished through breeding to an outstanding sire rather than to an outstanding dam because of the greater number of offspring of the former. If a swine breeder is in possession of a great boar—proved great by the production records of a large number of his offspring—a linebreeding program might be initiated in the following way: Select two of the best sons of the noted boar and mate them to their half-sisters (half-sibs), balancing all possible defects in the subsequent matings. The next generation matings might well consist of breeding the daughters of one of the boars to the son of the other, and so on.

If, in such a program, it seems wise to secure some outside blood (genes) to correct a common defect or defects in the herd, this may be done through selecting a few outstanding proven sows from the outside—animals whose offspring are strong where the herd may be deficient—and then mating these sows to one of the linebred boars with the hope of producing a son that may be used in the herd.

The owner of a few sows can often follow a linebreeding program by breeding sows to a boar purchased from a large breeder who follows such a program, thus in effect following the linebreeding program of the larger breeder.

Naturally, a linebreeding program may be achieved in other ways. Regardless of the actual matings used, the main objective in such a system of breeding is to render the animals homozygous in desired type and performance to some great and highly regarded ancestor, while at the same time weeding out homozygous undesirable traits. The success of the program, therefore, depends on having desirable genes with which to start and an intelligent intensification of these good genes.

It should be emphasized that there are some types of herds that should almost never inbreed or linebreed. These include herds of only average quality, those whose performance is no better than the mean of the breed. The owners of grade or commercial herds run the risk of undesirable results, and, even if they are successful as commercial breeders, they cannot sell their stock at increased prices for breeding purposes.

With purebred herds of only average quality, more rapid progress can usually be made by introducing superior outcross sires. Moreover, if the ani-

Figure 4.13 The reserve grand champion Hampshire gilt at a 2002 swine show. *(Courtesy, National Swine Registry, West Lafayette, IN)*

mals are of only average quality, they must have a preponderance of "bad" genes that would only be intensified through an inbreeding or linebreeding program.

Outcrossing

Outcrossing is the mating of animals that are members of the same breed but that show no relationship close up in the pedigree (for at least the first 4 to 6 generations). Most of our purebred animals of all classes of livestock are the result of outcrossing. It is a relatively limited risk system of breeding because it is unlikely that two such unrelated animals will carry the same undesirable genes and pass them on to their offspring.

Perhaps it might well be added that the majority of purebred breeders with average or below-average herds had best follow an outcrossing program because in such herds the problem is that of retaining a heterozygous type of germ plasm with the hope that genes for undesirable traits will be counteracted by genes for desirable traits. With such average or below-average herds, an inbreeding program would merely make the animals homozygous for the less desirable traits, the presence of which already makes for their mediocrity. In general, continued outcrossing offers neither the hope for improvement nor the hazard of retrogression in linebreeding or inbreeding programs.

Judicious and occasional outcrossing may well be an integral part of linebreeding or inbreeding programs. As closely inbred animals become increasingly homozygous with germ plasm for good traits, they may likewise become homozygous for certain undesirable traits even though their general overall type and performance remains well above the breed average. Such defects may best be remedied by introducing an outcross through an animal or animals known to be especially strong in the trait or traits that need strengthening. This having been accomplished, the wise breeder will return to the original inbreeding or linebreeding program, realizing full well the limitations of an outcrossing program.

Grading Up

Grading up is that system of breeding in which a purebred or seedstock sire is mated to a native or grade female. Its purpose is to impart quality and to increase performance in the offspring.

The greatest single step toward improved quality and performance occurs in the first cross. The first generation from such a mating results in offspring carrying 50% of the hereditary material or genes of the purebred or seedstock parent. The next generation gives offspring carrying 75% of the "blood" of the purebred or seedstock parent, and in subsequent generations the proportion of inheritance remaining from the original grade parent is halved with each cross. Later crosses usually increase quality and performance still more, though in less marked degree. After the third or fourth cross, the offspring compare very favorably with purebred or seedstock in conformation, and only exceptionally good sires can bring about further improvement. This is especially so if the boars used in grading up successive generations are derived from the same strain.

CROSSBREEDING

Crossbreeding is the mating of different breeds. It is used by swine producers to (1) increase productivity over purebreds because of the resulting hybrid vigor or heterosis, (2) produce commercial hogs with a desired combination of traits not available in any one breed, and (3) produce foundation stock for developing new breeds.

The motivating forces behind increased crossbreeding in farm animals are (1) more artificial insemination, thereby simplifying the rotation of sires of different breeds, and (2) the necessity for swine producers to become more efficient in order to meet their competition, both from within their industry and from without. Crossbreeding will play an increasing role in the production of market animals in the future because it offers several advantages as discussed in the following sections.

Hybrid Vigor or Heterosis

Heterosis, or hybrid vigor, is the name given to the biological phenomenon that causes crossbreeds to out produce the average of their parents. For numerous traits, the performance of the cross is superior to the average of the parental breeds. This phenomenon has been well known for years and has been used in many breeding programs. The production of hybrid seed corn by developing inbred lines and then crossing them is probably the most visible attempt to take advantage of hybrid vigor. Heterosis is also being used extensively in commercial swine, cattle, sheep, poultry layer, and broiler production. The National Hog Farmer 2002 Producer Profile estimated 86% of the pork producers raised crossbred hogs, which probably account for more than 95% of the hogs marketed.

The genetic explanation for the hybrid's extra vigor is basically the same, whether it be plant or animal. Heterosis is produced because the dominant gene of a parent is usually more favorable than its recessive partner. When the genetic groups differ in the frequency of genes they have and dominance exits, then heterosis will be produced.

Heterosis is measured by the amount the crossbred offspring exceeds the average of the two parent breeds or inbred lines for a particular trait, using the following formula for any one trait:

$$\frac{\text{Crossbred average} - \text{Purebred average}}{\text{Purebred average}}$$

$$\times 100 = \text{Percent heterosis or hybrid vigor}$$

Thus, if the average of the two parent populations for litter weaning weight is 284 lb and the average of their crossbred offspring is 336 lb, application of the formula shows that the amount of heterosis is 52 lb or 18%.

$$\frac{336 - 284}{284} \times 100 = 18\% \text{ heterosis or hybrid vigor}$$

Traits high in heritability, such as carcass length, backfat thickness, and loin eye area, respond consistently to selection but show little response to hybrid vigor. Traits low in heritability, such as litter size, litter weaning weight, and survival rate, usually demonstrate a stronger response to hybrid vigor.

Complementary

Complementary refers to the advantage of one cross over another cross or over a purebred, resulting from the manner in which two or more traits combine or complement each other. It is a matching of breeds so that they compensate each other, the objective being to get the desirable traits of each. Thus, in a crossbreeding program, breeds that complement each other should be selected, thereby maximizing the desirable traits and minimizing the undesirable traits. Because breeds that are selected because they tend to express a maximum of some trait will have some undesirable traits, different breeds must be selected for different purposes.

No one breed or strain has a monopoly on all the desired traits. Therefore, producers must study their operation and the merits of different breeds or seedstock before choosing a breed.

Crossbreeding provides a way in which to introduce new and desired genes quickly—at a faster rate than can be achieved by selection within a breed. Because undesirable qualities are often recessive, crossbreeding offers the best way in which to improve certain traits merely by hiding them with dominants.

Except for a two-breed cross, crossbreeding offers an opportunity to have hybrid vigor expressed in breeding females. This is most important in the swine herd where it results in increased fertility, survivability of piglets, litter weaning size, and pig growth rate—all factors that mean more profit for the producer.

Factors Influencing Crossbreeding

Many other examples of each of the advantages of crossbreeding could be cited. It should be noted, however, that the total magnitude of the advantage of these factors—achieving the 15 to 25% potential immediate increase in yield per female unit through continuous crossbreeding compared to continuous straight breeding—depends on the following:

1. Making wide crosses. The wider the cross, the greater the heterosis (Figure 4.14).
2. Selecting breeds that are complementary. A crossbreeding program should involve breeds that possess the favorable expression of traits desired in the crossbred offspring that will be produced.
3. Using high-performing stock. Once a crossbreeding program is initiated, further genetic improvement primarily depends on the use of superior production-tested boars.
4. Following a sound crossbreeding system. For a continuous high expression of heterosis and maximum output per female, a sound system of crossbreeding must be followed. This should include the use of crossbred females because research clearly indicates that more than one-half the higher profits from a crossbreeding program results from them.
5. Tapping purebreds constantly. Purebreds must be constantly tapped to renew the vigor of crossbreds; otherwise, the vigor is dissipated.

Figure 4.14 Crossbred Hampshire-Duroc barrow and Hampshire-Yorkshire gilt exhibited at a county fair. *(Courtesy, USDA, Washington, DC)*

Disadvantages of Crossbreeding

Crossbreeding does possess certain disadvantages that should be understood:

1. Generally speaking, a crossbred hog lacks the uniformity in color and general attractiveness of purebreds.
2. Desirable boars of the two or three breeds or strains must be located and purchased, although artificial insemination has greatly alleviated this concern.
3. It should not be assumed that the virtues of crossbreeding are sufficiently powerful to alleviate the necessity of selecting outstanding boars.
4. Crossbreeding should not be looked on as a panacea for neglect of sound practices of breeding, feeding, management, and sanitation.

Crossbreeding Systems[2]

Crossbreeding is a very useful means for the pork producer to increase the efficiency and profit of an operation. Full benefits can be gained only by careful combination of available breeds and selection of outstanding replacements from within those breeds. Crossbreeding enables the producer to take advantage of heterosis and it must be systematic and well planned to take full advantage of breed differences.

Desirable traits of different breeds can be used if some breeds can be identified as good maternal breeds and others as good paternal breeds. A system where males from paternal breeds (superior growth and carcass) are mated with females from maternal breeds (superior reproductive and mothering ability) takes advantage of the strengths of both breeds while minimizing some of the weaknesses. The types of crossbreeding systems and percentages of each used in the United States are listed in Table 4.2.

Crossbreeding does not change the genes that are present in a population, but it arranges them in more favorable combinations. It follows that the initial boost from crossbreeding can be maintained by continued crossing. Permanent improvement can result only through selection.

Two basic systems of crossbreeding exist: the rotational cross system and the terminal cross system. Also, the two systems can be used in combination. Rotational cross systems combine two or more breeds, where the breed of boar used is different from the previous generation and replacement crossbred females are retained from each cross. In

TABLE 4.2   Types of Crossbreeding Systems on U.S. Farms

Type of System	Percentage
Farm produced F1 females bred to terminal boars	34.2
Purchased F1 females bred to terminal boars	31.8
Three-breed rotational cross	16.8
Purebred	14.0
Two-breed rotational cross	9.9
Rotational cross using two-breed rotation for females and terminal boars	6.5
Rotational cross using three-breed rotation for females and terminal boars	5.1
Four-breed rotational cross	2.7
Other	5.5
Total	100.0%

Note: Percentages may reflect multiple answers.
Source: National Hog Farmer 2002 Reader Profile 292 responses.

the terminal cross, female replacements are usually purchased or produced by maintaining purebred herds that emphasize reproductive performance.

Rotational Cross

The two-breed rotational cross uses boars of two different breeds in alternate generations, and retains crossbred females for maternal stock (Figure 4.15). This system is fairly simple to follow once the producer chooses two breeds. Breeds used in rotation should be productive because, over time, each breed contributes equally to both the production traits of the market offspring and the reproductive traits of replacement females. Therefore, reproductively sound breeds with adequate growth and carcass traits should be used.

In this system purebred boars are mated to sows with a certain percentage of the same breed as the boars. Therefore, the maximum response from heterosis will not be realized (Table 4.3). In fact, the actual heterosis retained changes a little each generation until the sixth generation, after which the two-breed rotation realizes about two-thirds of the total advantage obtained from crossbreeding.

More heterosis can be realized with the addition of a third breed to the rotation (Table 4.3 and Figure 4.16). A three-breed rotation levels out at 86% of the advantage obtained from crossbreeding after the 7th and 8th generations compared to only 67% with the two-breed crosses. Again, each breed contributes as both a sire and a dam so reproductively sound breeds with adequate growth and carcass traits should be used. Three-breed rotations are recommended over two-breed rotations because a higher percentage of the total

---

[2] Adapted from Oklahoma Extension Facts, No. 3603, by D. S. Buchanan, et al., Oklahoma State University, Stillwater, OK.

advantage obtained from crossbreeding is realized.

A fourth breed could be added to the rotation, in which case about 92% of the total advantage from crossbreeding would be realized. Even though a higher percentage of the total heterosis is realized with a four-breed rotation, it is generally not recommended over a three-breed rotation because of

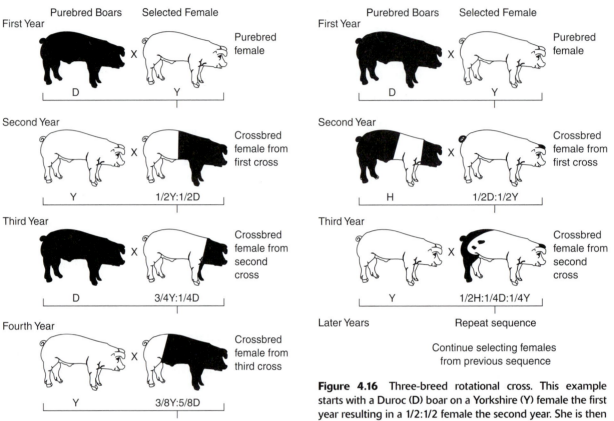

Figure 4.15 Two-breed rotational cross system. This example begins with a purebred Duroc (D) mated to a purebred Yorkshire (Y). Each succeeding generation continues with a sire rotation of Duroc and Yorkshire boars. *(Courtesy of Pearson Education)*

Figure 4.16 Three-breed rotational cross. This example starts with a Duroc (D) boar on a Yorkshire (Y) female the first year resulting in a 1/2:1/2 female the second year. She is then mated to a Hampshire (H) boar resulting in the three-way cross in the third year. At this time a Yorkshire boar is mated to her and the sequence is repeated in later years. *(Courtesy of Pearson Education)*

TABLE 4.3   **Breed Composition and Percentage of Maximum Heterosis from Two-Breed and Three-Breed Rotational Crossing Programs**

| | *Two-Breed Crosses* | | | | | *Three-Breed Crosses* | | | | |
| | *Blood (%)* | | *Expected Heterosis* | | | *Blood (%)* | | | *Expected Heterosis* | |
Generation Number	*A*	*B*	*Offspring*	*Dam*	*A*	*B*	*C*	*Offspring*	*Dam*
1	50	50	100	0	50	50[1]	0	100	0
2	75[1]	25	50	100	25	25	50[1]	100	100
3	38	62[1]	75	50	63[1]	12	25	75	100
4	69[1]	31	62	75	31	56[1]	12	88	75
5	34	66[1]	69	62	16	28	56[1]	88	88
6	67[1]	33	66	69	58[1]	14	28	84	88
7	33	67[1]	67	66	29	57[1]	14	86	84
8	67[1]	33	67	67	14	29	57[1]	86	86

[1]Breed of sire used to produce offspring.
Source: Courtesy of Pearson Education.

the difficulty in finding a fourth breed with a higher average level of productivity. The number of breeding groups that need to be maintained also increases with the number of breeds included in the rotation because females of different parities will be at different stages of the rotation and will require different breeds of boars or mates.

The rotational cross system is more popular and thought by many to be more practical than the terminal cross system because the only outside breeding stock that needs to be purchased once the program is established are boars. Thus, a producer does not have the difficulty of obtaining replacement females at a reasonable cost. Also, there is less risk of introducing disease to the herd because only boars are purchased.

Combinations of desirable traits of breeds cannot be fully used in a rotational system. Ideally, market hogs should be from sows with good mothering abilities and sired by boars that excel in growth and carcass traits. This is difficult to achieve in the framework of a rotational crossing system.

Terminal Cross

Terminal cross systems may involve two, three, or four breeds (Table 4.4). A two-breed cross will use purebred boars on one breed mated to purebred sows of another breed. For example, Hampshire boars and Yorkshire females could be used to produce every pig crop. This system allows the producer to combine a dam breed superior for reproductive performance with a sire breed superior for growth and carcass traits. All pigs produced are crossbred; therefore, this system reaps all the advantages obtained by having crossbred pigs. The sows, however, are purebred and none of the superiority obtained from crossbred sows will be realized.

A three-breed terminal cross will use purebred boars of one breed mated to crossbred sows of two other breeds. For example, Duroc boars and Landrace × Yorkshire cross females could be used to

produce each pig crop. In this system, sows that are a cross between two breeds superior for maternal traits are mated to a third breed of sire that is superior for growth and carcass traits. All pigs produced are crossbreeds. Crossbred sows are used, thus this system maximizes heterosis and is one of the most productive systems.

A four-breed terminal cross has all of the same advantages on the maternal side as the three-breed cross. In addition, there are some advantages in conception rate associated with use of crossbred boars. A disadvantage is that crossbred boars may be more difficult to obtain than purebred boars. However, because some seedstock producers maintain purebred herds of two or more breeds, they could easily produce some crossbred boars if there should be a demand for them.

In general, terminal crosses capitalize on the strengths of each breed and realize maximum gains from heterosis. A major disadvantage can be the difficulty of obtaining replacement females. If they are purchased, there is a health risk associated with the introduction of outside breeding stock. If they are raised within the herd, the commercial producer will need at least one purebred herd to produce replacement stock.

The trend in recent years toward the purchase of replacement females and boars from breeding companies and artificial insemination instead of raising replacements has greatly facilitated the terminal cross system. Producers can purchase crossbred gilts from a genetic supplier several times each year. These gilts should come from the same grandparent herd each time and the receiving producer should have isolation facilities. The breeding supplier may also provide either boars or semen, usually from boars that have bred from grandparent stock emphasizing the desired paternal characteristics. Data in Table 4.2 indicate that 31.8% of U.S. pork producers are purchasing F1 (crossbred) females and breeding them to a terminal boar.

TABLE 4.4   Percentage Heterosis Maintained by Terminal Crosses and Its Effect on One Trait

Sire Breed	Dam Breed	Pig	Dam	Sire	Litter Weight (lb) at 21 Days per Female Exposed
Purebred					
A	A	0	0	0	72.0
Two-breed cross					
A	B	100	0	0	80.1
Three-breed cross					
C	AB	100	100	0	93.8
Four-breed cross					
CD	AB	100	100	100	97.0

Source: Courtesy of Pearson Education.

Combination/Rotational

The advantages of both systems can be used if all replacement females are produced in a rotational cross of prolific, highly productive breeds and the market hogs are then sired by a boar of another breed (Figure 4.17). The producer would maintain a small portion of the herd (10 to 15%) in the two-breed rotational cross. The best females would be kept in the rotation, whereas most of the remaining rotation females would be mated to the terminal cross sire. The rotational breeds should possess strong maternal abilities and the terminal sire should be from a breed with good growth and carcass traits plus rank highly for those traits as an individual. All pigs are from crossbred dams so much of the maternal heterosis can be used and 100% of the individual heterosis used in the terminal cross pigs. Also, only boars or semen need to be purchased, minimizing disease risks.

This system has the disadvantages of being rather complicated and requires large numbers to allow it to operate efficiently. Unless a producer farrows at least 200 litters per year the terminal sire on a rotational female system will be hard to maintain unless semen is purchased for the rotational breeds.

When considering a crossbreeding system, there are many choices one can make. Each has several advantages and disadvantages. Some programs are simple but do not make maximum use of heterosis. Others are more complex, require more time to manage, but should have higher average levels of performance because they take more advan-

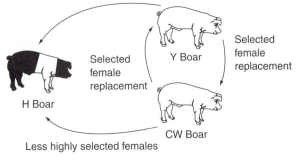

Less highly selected females

Selected female replacement

Y Boar

Selected female replacement

H Boar

CW Boar

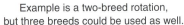

Less highly selected females

Example is a two-breed rotation, but three breeds could be used as well.

Figure 4.17 Terminal sire on rotational female system. This example uses a Hampshire (H) terminal sire (meat-type) on females resulting from a rotational cross from two breeds known for mothering ability, Yorkshire (Y) and Chester White (CW). Landrace would also fit well into the female replacement rotation as the second or third breed. *(Courtesy of Pearson Education)*

tage of the strengths of breeds and retain a higher percentage of heterosis. No one system will be best for every producer.

Producers who spend considerable time managing their swine operations can use more complicated systems that have higher expected levels of performance. In addition, facilities, source of breeding stock, disease control programs, and perhaps other factors are important when deciding which system is best adapted to a particular production unit.

QUESTIONS FOR STUDY AND DISCUSSION

1. Why is swine breeding more flexible than cattle and sheep breeding?
2. What unique circumstances surrounded the founding of genetics by Mendel?
3. Explain the relationships between chromosomes, genes, and DNA.
4. Under what conditions might a theoretically completely homozygous state in hogs be undesirable and unfortunate?
5. Give two examples each of (a) dominance and (b) recessive phenomena in swine.
6. In order to make intelligent selections and breed progress, how important is the environment?
7. Explain how sex is determined. Is it possible to alter the sex ratio?
8. When abnormal pigs are born, what conditions tend to indicate each of the following: (a) a hereditary defect or (b) a nutritional deficiency?

9. The "sire is half the herd"! Is this an understatement or an overstatement?
10. Define (a) prepotency, (b) nicking, (c) outcrossing, and (d) grading up.
11. What system of breeding do you consider to be best adapted to your herd, or to a herd with which you are familiar? Justify your choice.
12. Define and explain heterosis or hybrid vigor.
13. Discuss the advantages and the disadvantages of crossbreeding swine.
14. Which of the following crossbreeding system would you recommend: two-breed rotational cross, three-breed rotational cross, or terminal cross? Justify your choice.
15. Why is it important that a swine producer keep good records?
16. What characteristics would cause you to suspect the presence of the PSS gene?

SELECTED REFERENCES

Animal Breeding Plans, J. L. Lush, Collegiate Press, Inc., Iowa State University, Ames, IA, 1963

Animal Science, M. E. Ensminger, Interstate Publishers, Inc., Danville, IL, 1991

Genetic Evaluation, Bob Uphoff, Chairman of NPPC Genetic Program Committee, National Pork Producers Council, 1995

Pork Industry Handbook, Cooperative Extension Service, Purdue University, West Lafayette, IN, 2003

U.S. Livestock Genome Mapping Projects, http://www.genome.iastate.edu/, Iowa State University, Ames, IA

5

Applied Swine Genetics

One of many considerations when developing a herd is the environment in which the pigs will be raised. A different type of pig may be chosen for a hoop barn as opposed to an environmentally regulated, slatted floor building. *(Courtesy, Palmer Holden, Iowa State University, Ames, IA)*

The objective of applied swine breeding is to mate individuals whose offspring will possess the necessary genes to (1) produce the maximum amount of pork, (2) develop the desired body type, (3) perform at the desired level, and (4) adapt to their environment. These animals should be fed and managed to allow maximum phenotypic expression of their genetic potential because all animals are the product of heredity and environment.

SELECTION BASES

Many factors need to be considered in the selection of foundation sows, boars, and replacement gilts. Several of the important ones are discussed in this section. It must be noted that visual appraisal of animals may have little worth in evaluating their genetic merit unless all pigs were raised in a contemporary group. However, some visual appearances, such as soundness, need to be evaluated visually.

Breed Type or Individuality

Selection based on breed type or individuality implies the selection of those animals that approach an ideal or standard of perfection most closely and the culling out of those that fall short of these standards. Different types of pigs may be desired for different production conditions or markets. For example pigs produced on pasture will be selected on some traits that would not be used in indoor production.

The ideal lean/meaty hog is properly balanced with a blending of all parts. The head and neck are clean-cut and free of fat. The back is moderately and evenly arched, and of adequate width. The loins are wide and strong, and the hams are deep, thick, slightly bulging, and muscled well down to the hocks. The shoulders are well laid in and smooth, and the sides are long, deep, and smooth. The legs are straight and set squarely on the corners of the body, and the pasterns are strong. The walk is free and easy, not stiff. Additionally, the ideal meat-type hog possesses adequate size for age.

Production or Performance Testing

No criterion that can be used in selecting an animal is as accurate or important as its own past performance. Progressive purebred and seedstock producers have production records on their animals. Successful swine producers will make increasing use of such records as a basis for selection.

Production testing systems and the relative merits of each are presented later in this chapter. Also, suggested record forms are included.

Mechanical Measures

Methods of measuring backfat and loin muscle area in swine are very important because of value-based marketing and the demand for lean breeding stock.

The metal backfat probe was developed at Iowa State University about 1952 to assess backfat depth. Drs. Lanoy Hazel and Ed Kline went to a local hardware store and purchased a narrow metal ruler. They made an incision in the pig's skin over the loin and forced the ruler through the fat to the top of the loin muscle. The metal probe has been used for years, and it is still the most economical means of measuring backfat.

Ultrasound technology, for example, measuring body composition and observing developing fetuses in pregnant women, has been used by the medical profession for more than 30 years. Because of improvements made in scanners and transducers used in medicine, the meat animal industry has been able to use ultrasound technology for the past 20 years to measure backfat, loin muscle depth or area, and pregnancy evaluation.

Today, the following methods are available and are used in the swine industry to measure body composition: (1) backfat probe, (2) lean meters, and (3) ultrasonics.

Backfat Probe

The simple backfat probe has been the most influential development in estimating the leanness of hogs. The 3 historical probing sites on the live hog are marked on the hog in Figure 5.1. These 3 locations correspond to the 3 locations where backfat determinations are made on the carcass.

The only equipment needed for probing is a snare to restrain the hog, a sharp knife or scalpel blade, and a narrow 6-in. metal ruler with 0.10 in. gradations (Figure 5.2). The steps and technique in probing are as follows:

1. Wrap the knife or scalpel with several layers of tape about 3/8 in. from the tip to prevent the blade from going too deep.

Figure 5.1 The 3 historical points of a hog on which backfat measurements were made, 2 in. off the midline—the 1st rib, last rib, and last lumbar vertebra. *(Courtesy, Maynard Hogberg, Iowa State University, Ames, IA)*

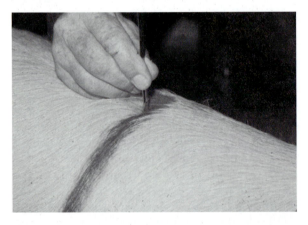

Figure 5.2 A metal ruler is inserted through a cut in the skin, perpendicular to the length of the pig and about 2 in. off of the midline down to the loin muscle. At that point the depth is measured. *(Courtesy, Maynard Hogberg, Iowa State University, Ames, IA)*

2. Weigh the hog.
3. Restrain the hog with a nose snare or use a weighing crate.
4. Push the knife blade through the skin perpendicular (right angle) to the hog's body at 2 in. to one side of the midline.
5. Insert the probe in the cut and slant it so that it points toward the center of the hog's body.
6. Force the probe through the fat down to the loin muscle. When the probe reaches the loin muscle, a firm resistance will be noted.
7. Push the clip on the probe snugly down to the skin line. Remove the ruler and read the measurement.

Historically, backfat measurements were taken at 3 points, off the midline at the 1st rib, last rib, and last lumbar vertebra. Currently, the recommendation is to take 1 measurement at the 10th rib. This reading correlates higher with the amount of muscle found in the carcass. Often the 1st rib reading is difficult to get accurately because of layers of false lean. Slaughter plants use metal rulers to take backfat measurements on the split areas of the midlines on the carcasses and at the last rib. The last rib is much easier to find in a carcass than the 10th rib.

Backfat thickness should be adjusted to a common live-weight basis, for example, 250 lb, to make valid comparisons and selections among animals. Adjustment formulas are presented later in this chapter.

Lean Meters

Many packers are now using various lean meters to evaluate pork carcasses. Most involve a probe inserted through the backfat and muscle, and provide a reading of the fat and loin depths by optical differences. A common probe is the Fat-O-Meter, used on the carcass line in many slaughter plants to assess meatiness.

A highly sophisticated measurement used is known as the total body electrical conductivity (TOBEC). TOBEC equipment consists of a tunnel that is surrounded by a copper coil (Figure 5.3 and Figure 5.4). When a current is applied to the coil, an electromagnetic field of a very low, safe strength is induced. The technology was developed for use on humans. A conveyor passes through the tunnel and when a carcass or box of meat is carried through the tunnel it absorbs energy from the field in proportion to its conductivity. Detectors measure and record this energy absorption as the object passes through the field. Lean tissue has a very high water content and is, therefore, a very good conductor.

Fat and bone are poor conductors. The signal from the TOBEC is, therefore, strongly related to the lean content, and by difference to the fat content of the carcass or box of meat.

It is receiving limited use today because it cannot measure carcasses as fast as the pigs are being slaughtered. It is targeted for use in packing plants where precision is important.

Ultrasound

Ultrasound, or real-time ultrasound, technology uses high-frequency sound waves to map tissue boundaries. Because different tissues, for example, fat and lean, have different properties, the sound waves are reflected back at different speeds. A central processor collects this information and develops an ultrasound image that can be measured by a technician. When the machine is properly calibrated, the fat depth and loin depth or area can be determined.

The primary differences between ultrasound machines are the number of quartz crystals that are used to create the ultrasound image and the sophistication of the information displayed. Basically, two types of ultrasound equipment are in use in the swine industry—the A-mode machine and the B-mode, or real-time, machine.

The A-mode machines have been used since the 1950s, use a single quartz crystal in the ultrasound probe, and give a single point estimate of backfat depth and loin muscle depth. A-mode machines can be accurate in measuring backfat, but they are not as good for loin area (Figure 5.5). Basically they take a single depth measurement through the loin muscle, which can be highly variable depending on the angle of the probe as well as on variations in the shape of loin muscles between pigs and breeds, for example, round versus oval.

B-mode machines were first used in swine evaluation in the 1980s and are both more costly and

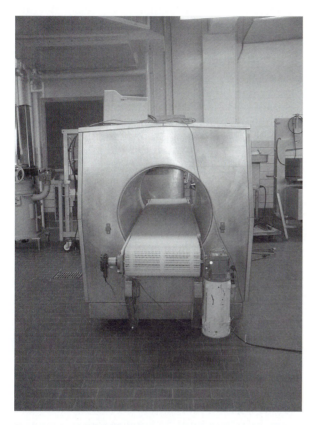

Figure 5.3 A TOBEC analyzer at the University of Nebraska Meat Laboratory. Carcasses or pieces of meat can be put through it to analyze differences in carcass composition. *(Photo by Jennie James, University of Nebraska, Lincoln, NE)*

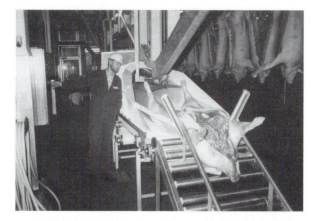

Figure 5.4 A TOBEC analyzer in a slaughter house indicating how the carcass remains attached to the gambrels pulling it through the chamber. *(Courtesy, Vande Berg Scales, Sioux Center, IA)*

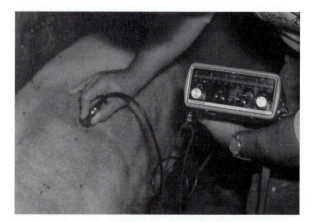

Figure 5.5 An A-mode ultrasound probe measures the depth of the backfat and the depth of the loin muscle. Scanners often multiply the loin depth measured in inches times 2.54 to estimate the loin area. This is at best an estimate assuming an oval-shaped loin. It underestimates kidney-shaped loins and overestimates round loins. *(Courtesy, Palmer Holden, Iowa State University, Ames, IA)*

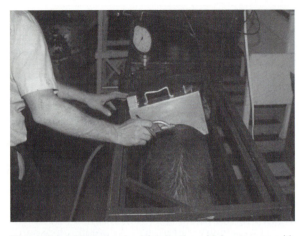

Figure 5.6 Estimating backfat depth and loin eye area with a B-mode ultrasound scanner. B-mode scans provide cross-section views of the pig's loin, allowing tracing of the loin muscle. *(Courtesy, Palmer Holden, Iowa State University, Ames, IA)*

accurate. Rather than measuring with a single crystal, multiple crystals are aligned in probes up to 7 in. long. They provide an ultrasonic image of the backfat and loin area that is displayed on a video monitor within a central processing unit. The image is described as "real time" because you can observe the muscle move as the images are being taken. Real-time machines have been proven highly accurate when used by trained technicians and are the preferred method of ultrasound evaluation (Figure 5.6).

The level of sophistication adds greatly to the cost of the machine and its attachments. The A-mode scanners sell for between $500 and $2,000, whereas the B-mode scanners sell for $10,000 to $25,000.

One of the greatest variables in ultrasonic accuracy is the technician effect. The National Swine Improvement Federation (NSIF) has implemented programs to standardize ultrasonic measurement for backfat and loin muscle area. These programs consist of workshops and training sessions, scanning practicums for participants, and written exams. Participants are evaluated for their ability to predict carcass data, repeatability of their measurements, and bias of live measurements as compared to carcass data.

Current recommendations relative to ultrasound—Because there is considerable cost differential between A-mode and B-mode equipment, producers must determine the level of accuracy that is desired in measuring fat thickness and loin muscle area. A-mode equipment can be effectively used (1) to sort hogs for backfat when marketing on a grade and yield basis and (2) to select replacement gilts for fat thickness. Although the accuracy of A-mode devices is often less than ideal, the relatively low investment cost and ease of operation have made them widely used for seedstock selection.

Specialized seedstock producers should consider B-mode equipment because it allows greater accuracy in both fat thickness and loin muscle area. When data is used across herds, reliability is important; hence, B-mode equipment by trained technicians is recommended. Also, B-mode equipment is recommended for scan data collected at central test stations and for on-farm data that is used by national breed registries.

Mechanical methods provide necessary adjuncts to visual appraisal and scales in the selection of breeding animals and the production of higher-quality pork carcasses. Increasingly, they will be used in the selection programs of purebred, seedstock, and commercial producers, and in value-based marketing.

Pedigree

In the selection of breeding animals, the pedigree is a record of the individual's heredity or inheritance. If the ancestry is good, it lends confidence in projecting how well young animals may breed. It is to be emphasized, however, that mere names and registration numbers are meaningless. A pedigree may be considered as desirable only when the ancestors close up in the lineage, the parents and grandparents, were superior individuals and outstanding producers. Too often, purebred breeders are prone to play up 1 or 2 outstanding animals back in the 3rd or 4th generations. If pedigree selection is to be of any help, one must be familiar with the individual animals listed therein.

Sires should be of known ancestry. This alone is not enough, for they should also be good representatives of the breed or strain selected. Sires of pigs raised for market should be evaluated on a Terminal Sire Index (TSI) that ranks boars for use in a terminal crossbreeding program. TSI weights estimated progeny differences (EPDs) for backfat, days to 250 lb, pounds of lean, and feed/pound of gain relative to their economic values.

Likewise, it is important that the sows be of good ancestry, regardless of whether they are purebreds or crossbreds. Sows are typically evaluated on a Maternal Line Index (MLI), a bioeconomic index for seedstock that are used to produce replacement gilts. MLI weights EPDs for both terminal and maternal traits relative to their economic values in a crossbreeding program, placing twice as much emphasis on reproductive traits as on postweaning traits.

Such ancestry, breeding, and selection indexes give more assurance of the production of high-quality pigs that are uniform and true to type. Also, a registration certificate is greatly enhanced by being a performance pedigree. Currently, the major breed associations attach reproductive and performance information to the pedigree.

Show-Ring Winners

Until about 1970, most swine producers looked favorably on using the show-ring as a basis of selection. Purebred breeders were quick to recognize this appeal and to extol their champions through advertising. Perhaps the principal value of selections based on live exhibitions lies in the fact that they directed attention to those types and strains of hogs meeting the approval of the experienced breeders and judges. In most instances, the selection of breeding hogs on the basis of these standards was effective for type and structure but limited for productivity factors, such as backfat or daily gain.

With the vast amount of production data available, both from off- and on-farm testing, swine shows have lost favor with many producers. Properly conducted junior livestock shows will continue to have values for FFA, 4-H Club, and other junior exhibitors in preparing them for a lifetime of competition. It teaches them fundamentals of selecting, raising, training, and exhibiting animals, and sportsmanship. Junior exhibitors likely do not have the numbers of purebred animals necessary for a rigorous selection program and, therefore, probably are not good sources of seedstock.

PRODUCTION TESTING SWINE

Today, Swine Testing and Genetic Evaluation System (STAGES), estimated breeding value (EBV), estimated progeny difference (EPD), and best linear unbiased prediction (BLUP) overshadow the once-thriving production testing by breed registries. Several breed registries provide these evaluations for their members. Purebred breeders interested in production testing by their swine registry can contact the executive secretary or a field representative of their breed registry.

As a basis for selection, emphasis is placed on production testing that embraces both (1) performance testing (sometimes called individual merit testing) and (2) progeny testing. The distinction between and the relationship of these terms are set forth in the following definitions:

1. Performance testing is the practice of evaluating and selecting animals on the basis of their individual merit or performance.
2. Progeny testing is the practice of selecting animals on the basis of the merit of their progeny.
3. Production testing is a more inclusive term, including performance testing and progeny testing.

Production testing involves the taking of accurate records—from birth on—rather than casual observation. Also, in order to be most effective, the accompanying selection must be based on traits of economic importance and high heritability (Table 5.1), and an objective measure, such as gain, backfat, etc., should be placed on each of the traits to be measured. Finally, those animals that fail to meet the high standards set forth must be removed from the pool of potential replacements or breeding stock sales. Breeding animals can only transmit desirable qualities unfailingly to all their offspring when they themselves have been rendered relatively homozygous or pure for the necessary genes.

This testing process can be made more rigid and certain through securing and intelligently using production records, but a knowledge of what records to keep and how to use them is necessary. The NSIF has developed guidelines for various types of evaluations.

Central Testing Stations

Boar test stations, first established in Denmark in 1907, have been in operation in the United States since 1954. They have provided uniform environments for evaluating the genetic merit of swine for three of the most important economic traits: growth rate, feed efficiency, and backfat thickness. However the only remaining central testing station is in Iowa and functions as a progeny testing station.

On-Farm Testing

Performance tests can be conducted on the farm—regardless of the facilities. The prime requirement is that all animals be handled similarly. Giving a small group of pigs preferential treatment leads to false conclusions and stymies, if not regresses, genetic improvement. A good on-farm testing program promotes rapid genetic improvement because it permits testing of a larger sample than is possible in central testing stations. The on-farm testing program is designed to assist breeders in evaluating their herds in a systematic manner. The program will (1) identify superior individuals, strains, lines, or breeds; (2) assist breeders in the selection of boars and gilts; (3) provide a means of following up on the breeding value of boars purchased from central test stations or seedstock producers; and (4) allow breeders to use common terminology and guidelines in selection of breeding stock.

TABLE 5.1 Economically Important Traits in Swine and Their Heritability[1]

Trait	Approximate Heritability[2]	Comments
Number born alive	10%	A sow consumes about 900 lb of feed from breeding until weaning; more pigs reduces the sow feed cost per pig.
Birth weight	5%	Very light pigs have a reduced chance of surviving.
21-day litter weight	15%	Weaning weight is highly affected by the number born alive and the sow's milking ability.
Litter size weaned	5%	Although influenced by stockmanship, piglet survival is a measure of the sow's mothering ability.
Age at puberty	35%	Early puberty allows animals into production sooner, lowering production costs.
Days to 250 lb	35%	Days to 250 lb is highly correlated with feed efficiency and feed intake.
Average daily gain	30%	High daily gains reduce age at market and fixed costs.
Feed efficiency	30%	The most profitable animals generally are more efficient.
Nipple number	18%	Minimum of 12 functional nipples.
Carcass length	60%	Length is highly heritable and rarely a problem.
Backfat	40%	Backfat is highly negatively correlated to muscling.
Loin area	50%	Loin area is highly positively correlated to muscling.
Percent ham	50%	Ham is a highly valued cut.
Percent lean cuts	50%	A high yield of lean cuts means more edible meat.
Eating quality[3]	20%	Eating quality of pork is indicated by tenderness, color, marbling, firmness, and palatability.

[1]These heritability estimates apply to within-herd and within-breed variations. Variations between breeds are much higher in heritability than the variations within breeds. Adapted from *Pork Industry Handbook PIH 106*, (1998) and M. E. Ensminger's *Swine Science* (1997).

[2]The rest is due to environment. The heritability figures given herein are averages based on large numbers; thus, some variation from these may be expected in individual herds.

[3]From: *Pigs-Misset*, August 1994, pp. 14–15, "Pork Quality: Only 20% Genetic Influence," by de Vries, A. G., et al., DLO Research Institute for Animal Production, Zeist, The Netherlands.

Figure 5.7 An on-farm Duroc boar testing facility. *(Courtesy, Palmer Holden, Iowa State University, Ames, IA)*

Objectives

The objectives of the on-farm testing program can be met completely only if the whole herd is tested. Testing a selected sample of the herd yields limited and biased information. EPDs that describe the genetic merit of each animal as a parent can be estimated most accurately by combining all of the available information on a pig and its relatives (Figure 5.7).

Accuracy is an extremely important part of any testing program. Most producers have the ability to conduct performance tests adequately. However, professional assistance from breed associations, testing organizations, Cooperative State Extension Services, and private commercial concerns is available to aid producers in the mechanics, record processing, and reporting of data connected with a testing program.

Pigs should be evaluated within test groups and divided by farrowing group, month, or season. These test groups should be managed and fed uniformly. All pigs in the test group should be given an equal opportunity. Contemporary groups should consist of at least 20 pigs from at least 5 litters and at least 2 sires. Testing smaller groups will lead to inadequate comparisons and less reliable EPDs.

Each record should be expressed as a ratio of the test group or herd average. The individual's index should be used when ranking possible herd replacements.

Procedures

The NSIF recommends the following on-farm testing procedures:

1. **Identification of all pigs in the herd.** It is recommended that the ear notching system identify the litter in the right ear and the individual pig in the left ear. Ear tags or tattoos may be used to supplement ear notches.
2. **Live evaluation.** Any genetic defects should be recorded. Structural soundness and underline soundness are examples of visual traits that affect production and reproduction. Breeding animals must be structurally correct and mobile to carry out their normal functions, and sows must have functional nipples to raise pigs. The animals should be evaluated at or near the end of testing and the date of evaluation recorded.
3. **Birth record.** The number of pigs farrowed alive and dead should be recorded. Within 3 days of birth all live pigs must be individually identified and the following information recorded: gender, birth date, weight, and identification and breed of parents.
4. **Record litter weight before weaning.** A breeder may wean at any time, but litter weight should be recorded before weaning and as near to 21 days of age as possible for the most accurate assessment of sow-milking ability. NSIF recommends weighing pigs between 14 and 28 days and adjusting to a 21-day basis.

ADJUSTMENT FACTORS

Sow Productivity Index

The Sow Productivity Index (SPI) may be used as a quick on-farm selection tool for selecting replacement females. Sows and gilts need to be evaluated separately because sows have a marked advantage over gilts in both litter size and milking ability. The SPI index is as follows:

Sow Productivity Index (SPI)
= 6.5 × Adjusted number born alive
+ Adjusted 21-day litter weight (lb)

The number born alive should be adjusted to a mature sow equivalent by adding the adjustments in Table 5.2 to the record based on the parity of the female. If possible, litters should be standardized to between 8 and 12 pigs per litter within 24 hours, but not later than 48 hours, after birth. Pigs selected for transfer should be average in size. Males, rather than females, should be transferred if possible, allowing the females to nurse their natural dam. After standardization, the number of pigs reared in the litter should be recorded, including foster pigs. Standardization of the postfarrowing weight record will prevent discrimination against a good milking sow or gilt that has a lesser opportunity because of smaller-than-optimum litter size.

TABLE 5.2 Parity Adjustment Factors for Number Born Alive (add the appropriate value to litter size)

Parity	Born Alive
1	1.2
2	0.9
3	0.2
4–5	0.0
6	0.2
7	0.5
8	0.9
9+	1.1

Source: Swine Improvement Program Guidelines, National Swine Improvement Federation, http://www.nsif.com

TABLE 5.3 Factors for Adjusting Litter Weight to a 21-Day Basis (multiply the weaned litter weight by the appropriate factor)

Age Weighed	Factor	Age Weighed	Factor
10	1.50	20	1.03
11	1.46	21	1.00
12	1.40	22	0.97
13	1.35	23	0.94
14	1.30	24	0.91
15	1.25	25	0.88
16	1.20	26	0.86
17	1.15	27	0.84
18	1.11	28	0.82
19	1.07		

Source: Swine Improvement Program Guidelines, National Swine Improvement Federation, http://www.nsif.com

Pigs should be weaned and weighed, preferably no earlier than 10 days of age. The closer to 21 days of age pigs are weaned, the more accurate the adjustment will be. Postfarrowing litter weights may be adjusted to a 21-day basis by using the multiplicative factors in Table 5.3, permitting sows to be ranked on the SPI. For example, a litter that weighed 100 lb at 16 days of age would have an adjusted 21-day weight of 120 lb (100×1.20).

The following formula also can be used to directly adjust litter weights to a 21-day basis:

Adjusted 21-day litter weight
= weight × $[2.218 - .0811(\text{age}) + .0011(\text{age}^2)]$

Example:

A litter weighed 100 lb when weaned at 16 days of age. What is the adjusted 21-day litter weight?

Adjusted 21-day litter weight
= 100 lb × $[2.218 - .0811(16)$
+ $.0011(256)]$
= 120 lb

Example:

A second parity sow farrowed 10 live pigs and the litter was weaned at 16 days of age weighing 100 lb.

SPI = 6.5 × Adjusted number born alive + Adjusted 21-day litter weight (lb)

SPI = 6.5 × (10 pigs born alive × 0.9) + (100 × 1.2) = 178.5

If the producer desires to compare all of the litters on a basis of weaning 10 pigs, the adjustments in Table 5.4 may be used. The litter weight (already adjusted to a 21-day basis) should be standardized to 10 pigs by adding the appropriate value from Table 5.4. For example, if only 8 pigs were weaned, 21 lb would be added to the adjusted litter weight.

Production Adjustments

Adjustment formulas for growth, backfat, loin muscle area, and predicted percent lean follow.

1. **Growth.** Growth rates must be measured on all intact males or all gilts by 1 of 2 procedures.

   a. Age at a constant weight. If pigs are not weighed on test but only a final weight is taken, weights should be taken at or near 250 lb or some other comparable constant weight. The equation for adjusting days to a constant weight is the same as under the earlier section on central testing stations.

$$\text{Adjusted days to 250 lb} = \text{Actual days} = [(250 - \text{actual wt.}) \times \frac{(\text{actual days} - a)}{\text{actual wt.}}]$$

   where a = 50 for boars and barrows, and 40 for gilts.

TABLE 5.4 Factors for Adjusting 21-Day Litter Weight for Number of Pigs after Transfer (number allowed to nurse)

Pigs after Transfer	Adjustment Factor for 21-Day Litter Weight
1–2	104
3	76
4	61
5	51
6	41
7	30
8	21
9	17
≥10	0

Source: Swine Improvement Program Guidelines, National Swine Improvement Federation, http://www.nsif.com

Example:

A gilt weighed 270 lb at 165 days of age. What was her adjusted days to 250 lb?

$$\text{Adjusted days to 250 lb} = 165 + [(250 - 270) \times \frac{(165 - 40)}{270}] = 155.7 \text{ days}$$

   b. On-test gain. Pigs should be weighed on test at an average pig weight consistent with the management program of the operation; approximately 70 lb are recommended. Ranges in starting weights among individual pigs should be minimized. Off-test pig weight should average at least 160 lb more than starting weight. If pigs being tested have undergone segregated early weaning (SEW), the test may be started at an average starting weight of 40 lb and off-test pig weights should average at least 190 lb more than starting weight. The gain should be reported as average daily gain on test.

2. **Backfat.** Backfat should be measured at the 10th rib at the time pigs are weighed off test at 250 +/− 15 lb. The average of 2 measurements taken 2 in. off the midline on both sides of the pig should be recorded if a metal probe or an A-mode ultrasound machine is used. Both the metal probe and A-mode ultrasounds provide only a depth measurement.

   If a B-mode (real-time) machine is used a cross-section of the backfat and loin muscle can be seen on the screen. When using a B-mode ultrasound, a single-side measurement is sufficient. With the B-mode the backfat depth should be measured at the midpoint of the loin and should include the skin. NSIF recommends pigs be measured by a certified technician. (See Selected References at end of the chapter.) All measurements should be reported as adjusted to a constant basis, using the following formula.

$$\text{Adjusted 10th rib backfat} = \text{Actual backfat} + [(250 - \text{actual wt.}) \times \frac{(\text{actual backfat})}{(\text{actual wt.} - b)}]$$

   where b = −20 for boars, +30 for barrows, and +5 for gilts.

Example:

A barrow had a 10th rib backfat depth of 1.0 in. when probed at 230 lb. Adjust the backfat to 250 lb.

Adjusted 10th rib backfat
$$= 1.00 + [(250 - 230)$$
$$\times \frac{1.00}{(230 - (+30))}] = 1.10 \text{ in.}$$

3. Loin muscle area (LMA). The LMA should be measured on pigs when they are weighed off test, within a 30 lb range of the desired weight endpoint (e.g., 250 +/− 15 lb). Loin muscle area should be measured over the 10th rib at a location 2 in. off the midline. The equation for adjusting LMA to a constant weight basis (250 lb) is

$$\text{Adjusted LMA} = \text{Actual LMA}$$
$$+ [(250 - \text{actual wt.}) \times \frac{\text{actual LMA}}{(\text{actual wt.} + 155)}]$$

Example:
A 270 lb pig had a 10th rib loin muscle area of 6.5 in.^{2} Adjust the LMA to 250 lb.

$$\text{Adjusted LMA} = 6.5 + [(250 - 270)$$
$$\times \frac{6.5}{(270 + 155)}] = 6.2 \text{ in.}^2$$

4. Predicted percent lean (PPL). This trait can be used in place of backfat in a selection index. Use the following equation to calculate PPL if pigs are weighed off test at 250 +/− 15 lb.

$$\text{Adjusted PPL} = [80.95$$
$$- (16.44 \times \text{Adjusted backfat})$$
$$+ (4.693 \times \text{Adjusted LMA})] / \text{Live weight}$$

Example:
A 250 lb live-weight pig had an adjusted backfat of 0.8 in. and an adjusted LMA of 6 in.^{2} What is the adjusted predicted percentage lean?

$$\text{Adjusted PPL} = [80.95 - (16.44 \times 0.8 \text{ in.})$$
$$+ (4.693 \times 6.0 \text{ in.}^2)] / 250 \text{ lb} = 38.4\%$$

5. Feed efficiency. Feed consumption should be measured on an individual basis if possible. If group fed, pigs should be tested in progeny groups. With group feeding, the number of pigs per pen, sex, and relationship among pigs in the pen should be noted.

Feed efficiency should be computed at the average off-test weight of 240 to 250 lb. Stations should report whether a meal or pelleted ration was fed and the method used to weigh the feed. Volumetric measurements are not appropriate because environmental factors affect the accuracy of such measures. Report whether the feed was transported to the station in bag or bulk form. Summaries should include season and average temperature during the test.

SELECTION INDEXES

Environmental influences make comparison of pigs tested at different locations, at different times, or under different management difficult. Using selection indexes that are based on deviations from contemporary group averages will provide valid comparisons of animals within the same test group. These comparisons provide a basis for assessing genetic merit. Keep in mind, however, that indexes are not appropriate for making between-group comparisons of animals. Note, also, that the following indexes are to be used strictly for evaluating performance and carcass traits and not for ranking animals for any maternal traits. Data must be adjusted to a constant endpoint before using in a selection index.

1. Traits included in the indexes are defined as follows:

   G = Average daily gain (ADG) recorded on an individual minus the mean ADG of contemporary group.

   B = Backfat depth measured on an animal minus mean backfat depth of contemporary group.

   F = Feed-to-gain ratio calculated for an individual or pen of animals minus mean F/G for all pens or animals (if fed individually) in the contemporary group.

   M = Predicted percentage lean calculated for an individual minus the mean predicted percentage lean for the contemporary group.

2. Feed conversion measured[1]

   A-mode backfat:
   Index = $100 + 68(G) - 142(B) - 80(F)$

   B-mode backfat[a]:
   Index = $100 + 52(G) - 92(B) - 68(F)$

   PPL:
   Index = $100 + 55(G) + 11(M) - 76(F)$

3. Feed Conversion Not Measured[1]

   A-mode backfat:
   Index = $100 + 100(G) - 194(B)$

   B-mode backfat[a]:
   Index = $100 + 78(G) - 115(B)$

   PPL:
   Index = $100 + 83(G) + 14(M)$

   [a] *Also use this index if backfat is measured with a metal probe.*

---

[1] For details relative to computing these, see Guidelines for Uniform Swine Improvement Programs, National Swine Improvement Federation, 2002–2003.

Carcass Evaluation

Collecting carcass data on 1 or 2 pigs provides little useful selection information. Realistically, data needs to be collected on 20 or more to be significant. Swine producers can obtain carcass evaluation data on their hogs through slaughter plants, exhibitions, and home slaughter. An impartial and experienced individual should collect or advise in the collection of these data.

The procedures to evaluate market hogs include (1) identification by tattoo, (2) inspection, (3) hot carcass weight, and (4) ribbing the carcass.

Quantitative Characteristics

To determine the proportionate amount of lean or muscle, several methods are recommended, with the specific methods often depending on circumstances in the cooperating slaughtering facility or an estimate of the muscle in the live pig using ultrasound.[2]

1. Method 1 includes hot carcass weight, 10th[2] rib fat depth, and 10th rib loin eye area. The standardized fat free lean (SFFL) is estimated from the following formula:

    SFFL, lb = 8.588 + (0.465 × hot carcass wt., lb) − (21.896 × 10th rib fat, in.) + (3.005 × LMA, in.[2])

2. Method 2 combines hot carcass weight and last rib fat and is used when carcasses cannot be ribbed.

    SFFL, lb = 23.568 + (0.503 × hot carcass wt., lb) − (21.348 × last rib fat, in.)

3. Method 3 combines ultrasound and live weight.

    Lean, lb = − 0.534 + (0.291 × live wt., lb) − (16.498 × 10th rib fat, in.) + (5.425 × 10th rib LMA, in.[2]) + (0.833 × sex of pig)

    (sex of pig: barrow = 1, gilt = 2)

Qualitative Carcass Characteristics (Pork Quality)

In addition to measuring efficiency in terms of producing large, healthy litters that gain rapidly on minimum of feed, producers also should be concerned about how much lean, edible pork is produced and how desirable it is for consumption.

A carcass evaluation also includes the estimation of (1) muscle color, (2) muscle firmness, and (3) loin muscle marbling (Figure 5.8). Desirable lean quality in pork is reddish-pink in color, slightly firm in tex-

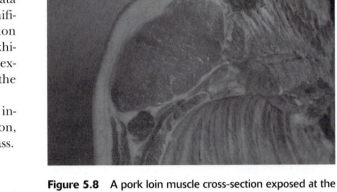

Figure 5.8 A pork loin muscle cross-section exposed at the 10th rib. Backfat, loin area, color, marbling, and firmness can now be estimated. *(Courtesy, Palmer Holden, Iowa State University, Ames, IA)*

ture and practically free of surface exudation (watery). Quality variations from this ideal result in less desirable conditions of pale, soft, and exudative (PSE), and dark, firm, and dry (DFD) lean. It is not unusual to find a wide range of lean quality in pork cuts displayed in a retail meat case, including varying degrees of the PSE and DFD conditions. The "NPPC Official Color and Marbling Standards" describing these conditions is available from the National Pork Board, P.O. Box 10306, Des Moines, IA 50306.

LEAN PORK GAIN[3]

For the most complete and comprehensive evaluation of market hogs, both live hog production and carcass merit should be included. Also, wherever possible, producers should determine the pounds of quality lean pork gain per day on test. The specific steps to determine pounds of quality lean pork gain per day on test follow:

1. Tattoo hog or use ear tag to identify prior to slaughter.
2. Visually evaluate warm carcass and eliminate it from consideration if it is condemned or has excessive trim (greater than 5%) from bruises or loss of body parts as a result of infection and so on.
3. Obtain carcass weight (lb), adjust for minor trim losses (if necessary), and express on a warm, skin-on basis. Eliminate from consideration if carcass weight is less than 150 lb.
4. Measure carcass length (in.) and eliminate from consideration if less than 29.5 in. This is rarely a problem in today's hogs.

---

[2] Based on *Procedures for Estimating Pork Carcass Composition*, National Pork Board, Des Moines, IA, 2002.

[3] Adapted from *NPPC Procedures to Evaluate Market Hogs*, 3rd ed., 1991, pp. 4–5, National Pork Board, Des Moines, IA.

5. Cut carcass side between the 10th and 11th ribs and measure fat depth (in.).
6. Determine loin muscle area (sq in.) and eliminate from consideration if less than 4.5 in.^{2}
7. Assess firmness/wetness score of loin muscle. Eliminate from consideration if score is either PSE or DFD.
8. Assess color score of loin muscle with a range of 1 to 6. Eliminate from consideration if score is either (1) pale pinkish-gray to white or (6) dark purplish-red.
9. Assess marbling score of loin muscle with a range from 1 to 10, approximating the percentage of intramuscular lipid. Eliminate from consideration if score is devoid to practically devoid of lipid.
10. For a carcass meeting minimum qualifications as established above, and for purposes of ranking, determine pounds of fat-free lean pork gain per day on test. Use the following formula:

> Lean gain/day = (Final carcass lean − initial lean)/days
> Final carcass lean, lb = 8.588
> +0.465 × hot carcass wt, lb
> −21.896 × 10th rib fat, in.
> +3.005 × 10th rib LMA, in.^{2}
> Inital lean, lb = (0.418 × live weight, lb) − 3.650

If the test period is from birth, simply divide the final carcass lean by the pig's age in days.

USING HERD RECORDS IN SELECTION

It is important that all animals be evaluated for their performance in terms of economically important traits. This requires a good record-keeping system in which an animal's performance becomes part of a permanent record. Consistently good production is to be desired. It is all too easy for a breeder to remember the good individuals produced by a given sow or boar and to forget those that were mediocre or culls.

A prerequisite for any production data is that each animal be positively identified by means of ear notches. For purebred breeders, who must use a system of animal identification anyway, this does not constitute any additional detail. But the taking of weights, grades, and notes does require additional time and labor.

Information on the productivity of close relatives (the sire and the dam and the brothers and sisters) can supplement that on the animal itself and thus be a distinct aid in selection. The production records of more distant relatives are of little significance because individually, due to the sampling nature of inheritance, they contribute only a few genes to an animal many generations removed.

Finally, it should be recognized that swine are raised primarily for profit, and profit depends on efficiency of production and market price. Fortunately, the factors making for efficiency of production—including litter size and survival, growth rate, and feed efficiency—do not change with type fads. For this reason, emphasis should be placed on proper balance of the production factors. It might be added that type changes, quite likely, would not be so radical as in the past if they were guided by market demands based on carcass values.

Most swine records are now computerized and several quality programs are available either for on-farm maintenance of the data or off-farm record services. In order not to be burdensome, the record forms should be relatively simple. Figure 5.9 is an individual sow record designed for use in recording the lifetime production record of one sow, and Figure 5.10 is a litter record form for use in recording detailed information on one litter.

A good plan for progeny testing boars consists of mating potential boars as soon as they will mate to a limited number of females. The progeny are then tested and evaluated, and only those boars that prove to be best on the basis of their progeny are retained for further breeding purposes. If boar pigs are each mated to 6 or 8 sows, pigs should be born 114 days later, and the progeny can be tested. Thus, with good fortune, it is possible to have progeny data on a boar when he is approximately 12 months of age.

Herd records are, however, of little value unless they are intelligently used in culling operations and in deciding on replacements. Also, most swine producers can and should use production records for purposes of estimating the rate of progress and for determining the relative emphasis to place on each trait.

ECONOMICALLY IMPORTANT TRAITS AND HERITABILITIES

It is generally recognized that swine exhibit considerable variation in economically important traits. The problem is measuring these differences from the standpoint of discovering the most desirable genes and then increasing their concentration and, at the same time, purging the herd of the less desirable traits.

Table 5.1 gives the economically important traits in swine and their estimated heritability. It should be understood that a heritability estimate indicates the percentage of a trait based on heredity, whereas the remaining portion of the variance is based on the environment. Lowly inherited traits are generally less than 20% heritable, moderate traits from 20 to 30% heritable, and highly heritable traits are greater than 30% heritable. For example, a heritability estimate of 10% for number of pigs born alive indicates that about 10% of larger-than-average litters selected is a

SWINE
Individual Sow Record

Breed _____ Name and registration no. _____

Date farrowed _____ Identification _____
(ear notch, tattoo)

Bred by _____
(Name and address)

Sow's pedigree _____ { _____
(Sire)   _____

_____ { _____
(Dam)   _____

Record of litter of which the sow was a member

No. in litter _____ No. of pigs weaned _____

Weaning wt. at _____ days of age
(fill in)

Her own wt. _____ Avg. wt. of litter _____

Litter mate carcass record, if any:

No. carcasses _____ ; avg. backfat _____ ; loin eye _____ ; length _____
(in.)   (sq in.)   (in.)

Number of teats _____

Production Record of Sow

	1	2	3	4	5	6	7	8
Litter no.								
Sire								
No. services								
Farrowing data								
Date								
Temperament of sow (gentle, nervous, cross)								
No. pigs born: Alive								
Dead								
Mummies								
Total								
Avg. birth wt.								
No. functioning teats								
Weaning data: Age								
No. weaned								
Avg. weaning wt.								
Offspring saved for breeding: No. gilts								
No. boars								

Disposal of Sow

Date _____ Reasons _____

Sold to _____
(Name and address)

Price $ _____

Figure 5.9 Individual permanent sow record with lines for name, pedigree, and history of each litter. *(Courtesy of Pearson Education)*

result of genetic influences, but it also indicates 90% is a result of environmental influences—whatever they may be.

Heredity and Environmental Effects

Swine producers are aware that there are differences in litter size, in weaning weight, in body type, and so on. If those animals that excel in the desired traits would, in turn, transmit without loss these same improved qualities to their offspring, progress would be simple and rapid. Unfortunately, this is not the case. Such economically important traits are greatly affected by environment (feeding, growth, health, management, etc.). Thus, only part of the apparent improvement in certain animals is hered-

SWINE
Litter Record

Breed _____ Litter No. _____
(notch, tattoo)

Data on Dam:

    Pedigree _____ { _____
                (name, reg. no., and ear notch)   (Sire)

    Birth date _____   _____
                   (date and year)   (Dam)

    Litter mate carcass data, if any:

        No. carcasses _____ ; avg. backfat _____ ; loin eye _____ ; length _____
                                (in.)   (sq in.)   (in.)

    Sow's _____ litter
              (1st, 2nd, etc.)

Data on Sire:

    Pedigree _____ { _____
                (name, reg. no., and ear notch)   (Sire)

    Birth date _____   _____
                   (date and year)   (Dam)

    Litter mate carcass data, if any:

        No. carcasses _____ ; avg. backfat _____ ; loin eye _____ ; length _____
                                (in.)   (sq in.)   (in.)

Date of Birth _____ Health Services:

No. Pigs Born: Date cholera vaccinated _____

    Alive _____   Date erysipelas vaccinated _____

    Dead _____   Date wormed _____

    Mummies _____   Other, including iron pills or shots (list) _____

    Total _____   _____

No. Pigs Weaned _____   _____

Individual Pig Record

Pig's No.	Sex	No. Teats	Birth Wt.	Off-Color Markings	Defects & Abnormalities	Weaning Wt. ___ days (fill in)	Date Castrated	Date & Cause of Death	Disposal Date & To Whom	Remarks

Figure 5.10 Litter record with lines for name, pedigree, litter health treatments, and individual piglet data. This information is transferred to the permanent sow record after the litter is weaned. *(Courtesy of Pearson Education)*

itary, and can be transmitted on to the next generation (Figure 5.11).

As would be expected, improvements based on environment are not inherited. This means that if most of the improvement in an economically important trait is the result of an improved environment, the heritability of that trait will be low and little progress can be made through selection. However, if the trait is highly heritable, marked progress can be made through selection. Thus, color of hair in swine is a highly heritable trait, for environment appears to have little or no part in determining it. However, a

Figure 5.11 A properly adjusted feeder with less than 25 percent of the feeding area covered with feed. Poorly adjusted feeders can reduce feed intake or overestimate feed intake by allowing feed wastage. *(Courtesy, Palmer Holden, Iowa State University, Ames, IA)*

trait such as weight per pig at weaning is of low heritability because, for the most part, it is affected by environment (by the nursing ability of the sow).

There is need, therefore, to know the approximate amount of percentage of change in each economically important trait that is a result of heredity and the amount that is a result of environment (Table 5.1). These heritability figures are averages based on large numbers, thus some variations from these may be expected in individual herds. Even though the heritability of many of the traits listed is disappointingly small, it is gratifying to know that much of it is cumulative and permanent.

Ratios

To compare animals from different environments, and to determine if an animal is better because of genetics or environment, ratios are often used. Ratios simply involve the expression of an animal's performance relative to the herd or test group average. They are calculated as follows:

$$\frac{\text{Animal's performance} \times 100}{\text{Average performance of all animals in the group}}$$

For example, a boar gaining 2.20 lb per day from a group averaging 2.00 lb per day has a daily gain ratio of $(2.20 \times 100) \div 2.00 = 110$. A ratio of 110 implies that boar is 10% above the test group average for that trait. Similarly, a ratio of 90 would indicate the boar performed 10% below the average of contemporary pigs tested. A ratio of 100 indicates the animal is average.

Ratios remove differences in average performance levels among groups (and generally among traits). Thus, they allow for a more unbiased comparison among individuals that were tested in different groups. This is only true to the extent that average differences among groups are not genetic. Most differences among groups are a result of feeding, weather, housing, management, and so on. Ratios allow individuals to be compared relative to contemporary groups. They are a useful method for comparing individuals that are not contemporaries, such as those in different tests or herds, or for comparing different traits. The NSIF recommends that in production testing programs, the level of performance for various traits should be expressed as a ratio.

Estimating Rate of Progress

For purposes of illustrating the way in which the heritability values in Table 5.1 may be used in practical breeding operations, the following example is given:

In a certain herd of swine, the litters in a given year average 7 pigs, with a range of 4 to 15 pigs. There are available sufficient pigs of the larger litters (averaging 12 pigs) from which to select replacement breeding stock. What amount of this larger litter size (12 vs. 7) is likely to be transmitted to the offspring of these pigs? Step by step, the answer to this question is secured as follows:

1. $12 - 7 = 5$ pigs, the number by which the selected litter size exceeds the average of the herd from which they arose.
2. By referring to Table 5.1, it is found that the number of pigs born alive is 10% heritable. This means that 10% of the extra 5 pig increase can be expected because the superior heredity of the stock saved as breeders, and that the other 90% is because of environment (feed, care, management, etc.).
3. $5 \times 10\% = 0.5$ pig; which means that for litter size the stock saved for the breeding herd is 0.5 pig per litter superior, genetically, to the stock from which it was selected.
4. $7 + 0.5 = 7.5$ pigs per litter; which is the expected performance (litter size) of the next generation.

It is to be emphasized that the 7.5 pigs per litter is merely the expected performance. The actual outcome may be altered by environment (feed, care, management, etc.) and by chance. Also, it should be recognized that where the heritability of a trait is low, less progress can be made. This explains why the degree to which a trait is heritable has a very definite influence on the effectiveness of mass selection.

TABLE 5.5   Estimating Rate of Progress in Swine

Economically Important Items	Herd Average	Individuals Selected for Replacements	Average Selection Advantage[1]	Heritability Percentage	Expected Performance of Next Generation[2]
Number born alive	7	12	5	10	7.5
Litter size weaned	6	10	4	12	6.48
21-day litter weight	180	400	220	15	213
Average daily gain	1.2	1.6	0.4	30	1.32
Feed efficiency	450	375	75	30	427.5

[1] Average selection advantage = Individuals selected for replacements − Herd average
[2] Expected performance = Herd average + (Average selection advantage × Heritability percentage)
Source: Courtesy of Pearson Education.

Using these heritability values, and assuming certain herd records, the progress to be expected from one generation of selection in a given herd of swine might appear somewhat as summarized in Table 5.5. Naturally, the same procedure can be applied to each of the traits listed in Table 5.1.

Swine producers need to know the factors that influence the rate of progress that can be made through selection. They include the following:

1. **The heritability of the trait.** When heritability is high, much of that which is selected will appear in the next generation, and marked improvement will be evident.
2. **The number of traits selected for at the same time.** The greater the number of traits selected, the slower the progress in each. In other words, greater progress can be attained in one trait if it is the only one considered. For example, if selection of equal intensity is practiced for four independent traits, the progress in any one will be only one-half of that which would occur if only one trait were considered; whereas selection for nine traits will reduce the progress in any one to one-third. This emphasizes the importance of limiting the traits in selection to those that have greatest importance as determined by economic value and heritability. At the same time, it is recognized that it is rarely possible to select for one trait only, and that economic value is usually dependent on several traits.
3. **The genotypic and phenotypic correlation between traits.** The effectiveness of selection is lessened by (a) negative correlation between two desirable traits or (b) positive correlation of desirable with undesirable traits.
4. **The amount of heritable variation measured in such specific units as pounds, inches, numbers, and so on.** If the amount of heritable variation—measured in such specific units as pounds, inches, or numbers—is small, the animals selected cannot vary much above the average of the entire herd, and progress will be slow. For example, there is much less spread in the birth weights of pigs than in the 154-day weights (usually there is less than a 2-lb spread in weights at birth, whereas a spread of 30 to 40 lb is common at 154 days of age). Therefore, more marked progress in selection can be made in the older weights that exhibit more range than in birth weights of pigs.
5. **The accuracy of records and adherence to an ideal.** It is an established fact that a breeder who maintains accurate records and selects consistently toward a certain ideal or goal can make more rapid progress than one whose records are inaccurate and whose ideals change.
6. **The number of available animals.** The greater the number of animals available from which to select, the greater the progress that can be made. In other words, for maximum progress, enough animals must be born and raised to permit rigid culling. For this reason, more rapid progress can be made with swine than with animals that often have only one offspring per year, and more rapid progress can be made when a herd is either being maintained at the same numbers or reduced than when it is being increased in size.
7. **The age at which selection is made.** Progress is more rapid if selection is practiced at an early age. This is so because more of the productive life is ahead of the animal, and the opportunity for gain is then greatest.
8. **The length of generation.** Generation interval refers to the period of time required for parents to be succeeded by their offspring, from the standpoint of reproduction. It is equal to the average age of the parents at the time of birth of their first offspring. The minimum generation interval of farm animals is about as follows: horses, 4 years; cattle, 3 years; sheep, 2 years; and swine, 1 year. By way of comparison, the average length of a human generation is more than 30 years. Shorter generation lengths will result in greater

progress per year, provided the same proportion of animals is retained after selection.

Usually it is possible to reduce the length of the generation of sires, but it is not considered practical to reduce materially the length of the generation of females. Thus, if progress is being made, the best young males should be superior to their sires. Then the advantage of this superiority can be gained by changing to new generations as quickly as possible. To this end, it is recommended that the breeder change to younger sires whenever their records equal or excel those of the older sires. In considering this procedure, it should be recognized, however, that it is very difficult to compare records made in different years or at different ages.

9. **The quality of the sires.** Because a much smaller proportion of males than of females is normally saved for replacements, it follows that selection among the males can be more rigorous and that most of the genetic progress in a herd will be made from selection of males. Thus, if 2% of the males and 50% of the females in a given herd become parents, then about 75% of the hereditary gain from selection will result from the selection of males and 25% from the selection of females, provided their generation lengths are equal. If the generation lengths of males are shorter than the generation lengths of females, the proportion of hereditary gain as a result of the selection of males will be even greater.

Relative Trait Emphasis

A replacement animal seldom excels in all of the economically important traits. The producer must decide, therefore, how much importance shall be given to each factor. Thus, the swine producer will have to decide how much emphasis shall be placed on litter size, litter survival, rate of gain, efficiency of feed utilization, and carcass characteristics.

Perhaps the relative emphasis to place on each trait should vary according to the circumstances. Under certain conditions, some traits may even be ignored. Among the factors that determine the emphasis to be placed on each trait are the following:

1. **The economic importance of the trait to the producer.** Table 5.1 lists the economically important traits in swine, and summarizes (see comments column) their importance to the producer. By economic importance is meant their dollars and cents value. Thus, those traits that have the greatest effect on profits should receive the most attention.

2. **The heritability of the trait.** It stands to reason that the more highly heritable traits should receive higher priority than those that are less heritable because then more progress can be made.

3. **Selection differential,** or the amount of variation present for each trait. Obviously, if all animals were exactly alike in a given trait, there can be no selection for that trait. Likewise, if the amount of variation in a given trait is small, the selected animals cannot be very much above the average of the entire herd, and progress will be slow.

4. **The level of performance already attained.** If a herd has reached a satisfactory level of performance for a certain trait, there is not much need for further selection for that trait.

5. **The genetic correlation between traits.** One trait may be so strongly correlated with another that selection for one automatically selects for the other. For example, rate of gain and efficiency of gain are correlated to the extent that selection for rate of gain tends to select for efficiency as well. Conversely, one trait may be negatively correlated with another so that selection for one automatically selects against the other, for example faster growing pigs often have slightly more backfat.

SYSTEMS OF SELECTION

Hand-in-hand with the breeding system and production testing, the swine producer needs to follow a system of selection that will maximize total progress over a period of several years or animal generations.

Among the several selection systems used by swine producers are the following (each of which is detailed): tandem, minimum standards, or selection indexes; STAGES, EBV, EPD, and BLUP. In addition pigs should be genetically free of the porcine stress syndrome (PSS).

Tandem, Independent Culling, or Selection Indexes

1. **Tandem selection.** Tandem selection refers to that system in which there is selection for only one trait at a time until the desired improvement in that particular trait is reached, following which selection is made for another trait, and so on. This system makes it possible to make rapid improvement in the trait for which selection is being practiced, but it has two major disadvantages: (a) usually it is not possible to select for one trait only and (b) generally income depends on several traits. Tandem selection is recommended only in those rare herds where one trait only is primarily in need of im-

provement, for example, where a certain herd of swine needs improving primarily in litter size.

2. **Independent culling or establishing minimum standards for each trait, and selecting simultaneously but independently for each trait.** This system, in which several of the most important traits are selected for simultaneously, is without doubt the most common system of selection. It involves establishing minimum standards for each trait and culling animals that fall below them. For example, it might be decided to select no replacement pigs from litters of fewer than 7 pigs, or weighing less than 120 lb at weaning, or gaining less than 2 lb per day from weaning to 250 lb. The minimum standards may have to vary from year to year if environmental factors change markedly (for example, if pigs average light at weaning time because of a disease). The chief weakness of this system is that an individual may be culled because of being faulty in one trait only, even though he is well nigh ideal otherwise.

3. **Selection indexes.** Selection indexes combine several important traits into one overall value or index. Theoretically, a selection index provides a more desirable way in which to select for several traits than either (a) the tandem method or (b) the method of establishing minimum standards for each trait and selecting simultaneously but independently for each trait. Selection indexes are designed to accomplish the following:

   a. To give emphasis to the different traits in keeping with their relative importance.

   b. To balance the strong points against the weak points of each animal.

   c. To obtain an overall total score for each animal, following which all animals can be ranked from best to poorest.

   d. To assure a constant and objective degree of emphasis on each trait being considered, without any shifting of ideals from year to year.

   e. To provide a convenient way in which to correct for environmental effects, such as feeding differences, and so on.

Several selection indexes are presented earlier in this chapter.

Evaluation Indexes (STAGES, EBV, EPD, and BLUP)

Swine Testing and Genetic Evaluation System (STAGES) is an integrated genetic evaluation system that uses advanced BLUP genetic technology to evaluate seedstock. It was initiated in 1985 as a joint venture between Purdue University, the USDA, the National Association of Swine Records, and the Na-

tional Pork Producers Council. It records performance data provided by swine producers.

Yorkshire, Duroc, Hampshire, and Landrace breeds participate and it represents the largest database available in the swine industry. The 2003 edition of the Trait Leader List features the top boars from the largest data bank in the swine industry. Nearly 37,000 sires have been evaluated, and more than 1.2 million records are represented in these calculations.

An individual's genetic value is affected by an additive and a nonadditive component. The nonadditive component is a result of the interactions of the genes and is not inherited, but its value can be improved through crossbreeding. The additive component is referred to as the breeding value. Animals that excel in this component produce offspring with high-breeding values because their offspring get one-half of their genetic material from each parent.

Estimated breeding value (EBV) is an estimate of an animal's breeding value as a source of genetic material. Stated differently, the EBVs represent the value of an individual as a source of genetic material for the herd. For example, the breeding value of progeny from a mating of a boar with an EBV of -6.0 days and a sow with an EBV of -2.0 days would result in offspring with EBV of -4.0 days.

Estimated progeny difference (EPD) is a prediction of the progeny performance of an animal compared to the progeny of an average animal in the population, based on all information currently available. EPDs are equal to one-half the EBV and are reported in the units of measure of the trait (e.g., pounds, inches, square inches, days). They are adjusted for the differing amounts of information available for each animal. BLUP (best linear unbiased prediction) procedures are used, ranking the animals according to their genetic merit and allowing direct comparison of animals within a breed. Whether $+$ or $-$ values are more desired depends on the trait (negative EPDs are desired for days and backfat; positive EPDs are desired for number born alive and litter weight).

EPDs may be calculated within herd or across herd. Calculation of EPDs on all pigs in a herd using all available information is a requirement. In addition to data on the individual, information should include full-sib, half-sib, parental, and progeny data updated regularly. Computer programs are commercially available for the calculation of within-herd EPDs. Across-herd breeding value estimation should use multiple-trait animal model procedures and genetic parameters derived from the data. An accuracy value that reflects the amount of information used in the genetic evaluation should also be made available. Purebred breeders can participate in

TABLE 5.6   Examples of EPDs for Terminal Sire Index (TSI), April 2003[1]

Sire Name / Owner	Pigs / Herds	BF (acc)[2]	Days (acc)[2]	Lb of Lean	TSI	MLI
Duroc						
SDF0 next level 120–2	202	−0.04	−8.75	0.57	150.3	Only 1 record
Stewarts Duroc Farm	4	(.80)	(.79)			
WFD9 Kobe 241–1	573	−0.02	−8.02	0.55	147.2	111.7
SGI, & others	14	(.96)	(.95)			
Yorkshire						
NAY9 Saturn 503–5	135	0.01	−9.12	−0.49	146.2	121.2
Swine Genetics Int. Ltd.	7	(.77)	(.75)			
Landrace						
BCF0 Omega 46–10	161	0.01	−5.30	−0.46	132.0	123.3
Whiteshire/Hamroc	2	(.78)	(.77)			

[1] BF = expected backfat change in in., negative values are best; Days = days to 250 lb from birth, negative values are best; Lb of Lean = added amount of lean in the carcass in lb, positive value are best.

[2] acc = Accuracy of the EPD. An accuracy of 1.00 is the maximum.

Source: Purdue University, West Lafayette, Indiana, www.ansc.purdue.edu/stages/index.htm

across-herd genetic evaluation programs through the National Association of Swine Records STAGES program.

The accuracy of the EPD is a measure of the precision with which genetic merit is predicted. Accuracy ranges from .01 (low), if no information is available, to .99 (high), if there is a large amount of performance information on the individual and its relatives. It is an expression of the reliability of the EPD. Accuracies indicate the level of confidence that the predicted EPD is near the true genetic potential of that animal. The values for accuracy are more reliable if they exceed 0.5. The accuracy of the EBVs of −6.0 and −2.0 in the preceding example may or may not vary.

The decision as to which boar to use should be based on the EPD of each individual boar. The accuracy value should be used to determine how extensively the animal should be used. Boars with favorable EPDs and high accuracies can be used with confidence; they will contribute to the genetic improvement of the herd. Accuracy is not constant because considerable new information is added for each subsequent analysis (Table 5.6).

Negative EPDs are desired for days to 250 lb and backfat. Positive EPDs are desired for pounds of lean, number born alive, litter weight, Terminal Sire Index, Sow Productivity Index, and Maternal Line Index.

Examples of EPDs from three breeds are shown in Table 5.6 for Terminal Sire Index. These values are from among the top sires in these breeds. First, observe the accuracies in parentheses. The first Duroc boar, SDF0, has performance records on 202 pigs in 4 herds and accuracies for backfat of 0.80

and days to 250 lb of 0.79. The second Duroc boar, WFD9, has records from 573 pigs in 14 herds and accuracies of 0.96 and 0.95.

WFD9 is owned by a boar stud and used in many more herds, resulting in higher accuracies. Even though SDF0 is predicted to reduce the days to 250 lb by 8.75, the second boar's greater accuracy of 0.95 for reducing the days to 250 by 8.02 is slightly more reliable. However, both accuracies are very acceptable.

The top TSI sires for Yorkshire and Landrace are also presented. The Yorkshire sire's days to 250 lb is −9.12, better than the Durocs. However, this does not mean that offspring sired by this Yorkshire will grow faster than those from the Durocs because the comparisons are only within breeds and represent only that their pigs will reach market weight faster than pigs of other Yorkshire parents with average EPDs. Note that the top ranking Yorkshire's TSI is 4 less than the top ranking Duroc.

The Landrace breed, a maternal breed, obviously is not selected based on their TSIs as they are selected on maternal traits, such as litter size and milking ability (litter weaning weight). Maternal Line Indexes (MLIs) are presented as a comparison.

Best linear unbiased prediction (BLUP) is a set of statistical qualities that describe the methodology utilized in calculating STAGES genetic evaluations. BLUP uses an animal's own record (if available) along with all relatives' records, including ancestors, siblings, and progeny. Thus, it takes into account genetic relationships as well as the relative merit of an animal within its contemporary group. The major feature is that the evaluations are unbiased or uninfluenced by environmental factors, such as age differences of animals, management difference, and litter and parity effects. Thus, genetic trends free of these environmental effects can be estimated.

BLUP makes it possible for breeders (1) to make a more accurate assessment of the animals that they are considering for selection and culling and (2) to evaluate the genetic progress that they are making in their breeding programs, which is particularly important when market conditions are changing.

FACTORS TO CONSIDER IN ESTABLISHING THE HERD

At the outset, it should be recognized that the vast majority of the swine producers keep hogs simply because they expect them to be profitable. That hogs have usually lived up to this expectation is attested to by their undisputed decades-long claim to the title of "mortgage lifter." For maximum profit and satisfaction in establishing the herd, the individual swine producer must give consideration to the type, breeding, breed, size of herd, uniformity, health, age, price, and suitability of the farm.

Modern Meat Type

A bulletin from the USDA in 1930 defined three types of pigs. One was the old-fashioned extreme lard type, even then noted as "too lardy for present market demands." Second was the "modern meat-type; smooth, trim, neat, of the type and condition to furnish the type demanded by the present American market." Third was the bacon type, selected by the Canadian Meat Packers to "demonstrate the type and condition required to produce English bacon or the Wiltshire side" (Chapter 3).

After World War II the demand for lard dropped rapidly, becoming an unwanted product, often selling for less per pound than the price of slaughter hogs on foot. As a result, breeders of meat-type breeds stressed leanness, a minimum amount of fat, and the maximum cutout value of primal cuts. The U.S. swine associations began certified litter programs, recognizing breeders of pigs that met the early standards of meat-type hogs.

Today, the vast majority of market hogs are lean and meaty. However, through the years, most purebred breeds of swine have run the gauntlet in types, producing animals of the short and thick, rangy and lean, and medium types. It is evident even today that these breeds possess the necessary store of genes through which such shift in types may be made by breeding and selection. Even so, most pork producers prefer the medium or intermediate type to either the short and thick or rangy types.

Short, thick animals usually are also short in prolificacy and rapidity of gains and tend to be early

maturing, putting on backfat at an earlier weight and age. Rangy hogs can be carried to excessive weights but often fail to ever develop adequate muscling. The packers and consumers object to the short, thick animals because of their excess lardiness and to the rangy ones because of their large cuts. It may be concluded, therefore, that most successful swine producers of the present day favor hogs that are lean and meaty.

Another change that occurs as the pig was selected for improved carcass and less fat was the change in the ratio of the fore-end of the pig to the back-end (Figure 5.12).

Purebreds, Seedstock, or Crossbreds

Generally speaking, only the experienced breeder should undertake the production of purebreds with the intention of eventually furnishing superior foundation or replacement stock to other purebred breeders, seedstock producers, or commercial producers (Figure 5.13). They must be willing to invest the time to keep detailed records, performance test

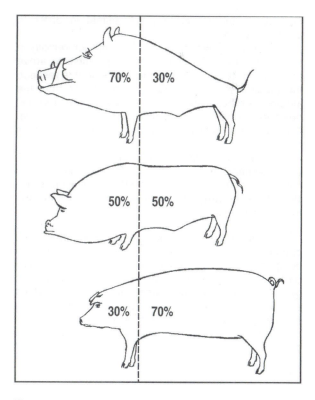

Figure 5.12 This line drawing included the following caption by Sir John Hammond of Cambridge, England, "The object of breeders should be to strive for a light fore-end." According to him, the European Wild Boar has a very heavy fore-end and light hams; so, unless constant selection is carried out for those characters there tends to be a reversion to the primitive type. *(From "The Growth of the Pig," in* Pig Progress, The Netherlands, July, 1957)*

Figure 5.13 A pen of registered Duroc gilts. If you decide to raise purebreds it is essential that you raise a strain that excels in performance and that you have a market for purebreds or possibly F1 females that will support your swine enterprise. *(Courtesy, Palmer Holden, Iowa State University, Ames, IA)*

offspring, use ultrasound techniques to evaluate fat and muscle in live pigs, make a variety of matings to develop desired characteristics and have access to computer evaluations of matings and breeding values. Much of the development of new genetics is being done by large breeding companies using computer-aided selection and sampling genetic material from the purebred populations.

Seedstock production is discussed in Chapter 3 under the heading "Specialized Breeding Stock Suppliers."

Crossbreeding of swine is fully discussed in Chapter 4. At this point, it is sufficient to say that this type of breeding program is more widely used in the production of hogs than with any other class of livestock.

Breed or Strain

No one breed or strain of hogs will excel all others in all points of swine production and for all conditions. It is true, however, that particular breed or strain characteristics may result in a certain breed or strain being better adapted to given conditions, for example, hogs of light color are subject to sunburn if they are raised outdoors in hot, sunny areas. Usually, however, there is a greater difference among individuals within the same breed or strain than between the different breeds or strains; this applies both to type and efficiency of production.

Size of Herd

Hogs multiply more rapidly than any other class of farm animals. They also breed at an early age, can produce litters twice each year, and typically bear litters of 8 to 15 piglets. It does not take long, therefore, to get into the hog business.

The eventual size of the herd is best determined by the following factors: (1) proximity to the neighbors; (2) suitable manure handling facilities; (3) adequate land for manure application; (4) type of production facilities; (5) kind and amount of labor available; (6) the disease and parasite situation, including isolation; (7) the market(s); and (8) comparative profits from hogs and other types of enterprises.

Uniformity

Uniformity of type and ancestry gives assurance of the production of high-quality pigs that are alike and true to type. This applies both to the purebred and the commercial herd. Uniform offspring sell at a premium at any age, whether they are sold as purebreds for foundation stock, as feeder pigs, or as slaughter hogs. With a uniform group of sows, it is also possible to make a more intelligent selection of herd boars.

Health

Producers should match the health status of incoming stock to that of their existing herd. Incoming breeding animals should be in thrifty, vigorous condition and should have been raised by suppliers who have exercised great care in the control of diseases and parasites. All purchases should be made subject to the animals being free from contagious diseases. They should have been developed under conditions similar to which are being used on the purchaser's farm.

Age

In establishing the herd, the beginner may well purchase a few bred gilts that have grown from a local, reputable pork producer, are uniform in type, and of good ancestry and that have been mated to a proven sire. It is difficult to purchase proven sows because most producers keep their productive sows until they market them.

The same is true when purchasing boars. Usually, young tested boars purchased for new sires as mature boars are too heavy to mate gilts. Usually, a wide selection of new boars are available from breeders at prices matched to their evaluated genetic merit. In addition, the younger animals, either gilts or boars, have a longer life of usefulness ahead. However, with boars, the opportunity to use artificial insemination must be very seriously considered.

Price

The beginner should always start in a conservative way. However, this should never be cause for the purchase of poor individuals—animals that are high at any price. Purchasing selected gilts from a commercial pork producer at market weight can be an economical means of getting started. Tell the producer you are interested in gilts that have 50% of their genetics derived from maternal lines or breeds.

Suitability of the Farm

At one time a swine producer may have one neighbor feeding cattle, a second who operates a dairy, a third with a farm flock of sheep, a fourth with light horses for recreation and sport, and a fifth whose chief source of income is from poultry. All may be successful and satisfied with their respective livestock enterprises. This indicates that several types of livestock farming may be equally well adapted to an area or region. Therefore, the selection of the dominant type of livestock enterprise should be analyzed from the standpoint of the individual farm.

Usually a combination of several factors suggests the livestock enterprise or enterprises best adapted to a particular farm and farmer. Some of the things that characterize successful major swine enterprises include the following:

1. Swine knowledge, interest, and skill of the operator.
2. Type of land available. Is it suitable for grain production or for pasture?
3. A plentiful supply of grains or other high-energy feeds in the immediate area.
4. Suitable manure handling facilities, including land available for manure application.
5. Adequate distance from neighbors or public areas.
6. Adequate and convenient, but not elaborate, buildings and equipment.
7. Available labor skilled in caring for swine, especially at farrowing time.
8. A satisfactory market outlet.

CHOOSING THE SEEDSTOCK SUPPLIER

The role of the seedstock industry is to provide the commercial swine industry with healthy, genetically superior breeding animals. If the commercial industry is to make genetic progress, it must come through the seedstock producers. Thus, seedstock producers must concentrate their selection efforts on economically important traits, base their selection on measured performance, and maintain the use of visual appraisal where it is appropriate and effective. Therefore, a planned, effective breeding program is essential to genetic progress. Furthermore, seedstock producers must maintain sufficient production volume to meet the needs of their customers as well as provide the genetic diversity between lines that will allow the commercial producer to maximize heterosis and utilize the superior characteristics of each breed or strain through the systematic use of a crossbreeding program. Thus, it may be concluded that when choosing the seedstock supplier the commercial producer should evaluate both the hogs and the supplier.

Selecting Gilts

The productivity of the sow herd is the foundation of commercial pork production. Also, the sow herd contributes half of the genetic makeup of growing finishing hogs. Together, these factors indicate the importance of careful selection of replacement gilts.

In the following list of characteristics to consider, the first three are "keep or cull." Either the underline, reproductive, and feet and leg soundness are acceptable or the animal should not be selected for the breeding herd. After selecting from the first three categories, select from the best performing gilts. Usually, all of the females still meeting the first four standards will need to be retained as potential replacements. If any excess gilts are available, select the leanest.

1. **Well-developed underline.** Replacement gilts should possess a sufficient number of functional teats to nurse a large litter. Six or more functional and uniformly spaced teats on each side is a minimum standard. Gilts with inverted or scarred nipples should not be saved.
2. **Reproductive soundness.** Most anatomical defects of the reproductive system are internal and hence not visible. However, gilts with small vulvas are likely to possess infantile reproductive tracts and should not be kept.
3. **Feet and legs.** A gilt should have legs that are set wide out on the corners of the body and the legs should be heavy boned with a slight angle to the pasterns. *Pork Industry Handbook* leaflet PIH-101 has an excellent description and diagrams of feet and leg soundness.
4. **Performance.** Select from fastest-growing gilts which are from large litters from good milking mothers (heavy 21-day litter weights). Standardization of weight for age may be accomplished by using the formula for adjusted days to 250 lb in the earlier section comparing animals on the basis of backfat and loin area measurements. Usually all of the females meeting the selection criteria of being the fastest-growing females and from the largest and heaviest 50% of the litters at 21 days will be needed to meet the sow replacement rate.

5. **Backfat.** When selecting from within the herd, replacement gilts should be lean, having 1.0 in. or less of 10th rib backfat, adjusted to a 250 lb basis.

Selecting Boars

Commercial producers will not select boars from their own herd. Most animals farrowed on the farm are crossbreds, and although they are prolific breeders, they are probably related to many of the gilts that will be selected. To prevent inbreeding, purchasing boars from an outside supplier is recommended. Purebred breeders can place tremendous genetic selection pressure when selecting boars to be retained as potential herd sires within their farm or for sale to commercial operations.

Greater selection reach is possible with boars than with gilts because in most herds using natural service 1 boar is purchased for every 15 to 20 sows in the herd. With the use of artificial insemination, either with purchased semen or with that collected from boars on the farm, this number can be expanded 5 or 10 times.

When selecting boars, some traits are apparent from the records of relatives, whereas other traits are apparent, and may be selected for, from the boar's own record. Also, the value of the traits varies depending on whether the boar is being used to farrow replacement gilts or as terminal market hogs. In selecting boars, the following traits or standards should be examined:

1. **Behavior.** Behavioral traits express themselves as docility, temperament, sex characteristics, maturity, and aggressiveness. These are associated with reproductive potential.

2. **Dam productivity.** Dam productivity traits include such things as reproductive ability, litter size, milking ability, and mothering ability. Boars, to be used to produce gilts, should be selected only from those litters of 10 or more pigs farrowed and 8 or more pigs weaned. Therefore, when selecting a boar for these traits, use boars that excel in the MLI.

3. **Performance.** Performance traits include (a) growth rate measured to 250 lb, (b) feed intake, and (c) feed conversion. These traits are above average in economic value. As guidelines, boars should (a) reach 250 lb at 155 days, or less, of age; (b) consume about 275 lb of feed, or less, per 100 lb of weight gain, between the weights of 60 and 250 lb; and (c) gain 2 lb or more per day during this same time. When selecting for these items, one should place more emphasis on the boar's own record and less emphasis on records of relatives.

4. **Backfat.** Carcass merit is best evaluated by taking measurements of backfat thickness, loin muscle area, or the muscle in the animal. These traits are highly heritable. As a guide, the 10th rib backfat of a boar adjusted to 250 lb should be 0.8 in. or less.

5. **Reproductive soundness.** Characteristics associated with soundness include the spacing, number, and presentation of the teats; genetic abnormalities, such as hernia and cryptorchidism; and mating ability. Boars should possess 12 or more evenly spaced teats. Genetic abnormalities and mating ability traits have very high economic importance. For these traits, insist that relatives of these boars be free of these defects and rely on the breeder's integrity. Physical soundness of the feet and legs, and bone size and strength, are also important. Feet and legs should demonstrate medium to large bone, wide stance both front and rear, freedom in movement, good cushion to both front and rear feet, and equal-size toes.

6. **Conformation.** Conformation includes body length, depth, height, and skeletal size; muscle size and shape; boar masculinity characteristics; and testicular development. Conformation traits such as length and height have high heritability values. It is important to select boars on the basis of their own records for these characteristics.

Boars should be selected and purchased by 6 to 7 months of age for use beginning at a minimum of 8 months of age. It is recommended that all replacement boars be purchased at least 45 to 60 days before needed. This allows them to be isolated and checked for health, conditioned to your farm's health status, and test mated or evaluated for reproductive performance. The primary consideration of producers is to select only boars that will increase the present production level of the herd and at the same time reduce weaknesses in the herd.

What's a good boar worth?—Saving $100 or more when buying a boar may be costly in the long run. Here's why: Assume a boar breeds 4 sows a week. With 90% conception and weaning 8 pigs per litter, that boar will sire 187 litters, or 1,497 pigs per year. The assumed cost of labor and facilities is $.17/day, feed cost is $.07/lb, the market premium to reduce backfat by 0.1 in. is $1.70, and the live market price is $.45/lb.

Let's also assume that the producer used the top Duroc boar in Table 5.6 and the offspring are marketed at 250 lb. Boar SDF0 next level has the ability to reduce the days to 250 lb by 8.75 days. If we assume no selection for days to 250 lb on the sow side, his offspring will go to market 4.375 days faster.

1,497 pigs × 4.375 days × \$.17/day =
\$1,113 return over an average Duroc boar

This boar has an EPD for backfat of −0.04. Again, assuming no selection on the sow side, his offspring will have 0.02 in. less backfat.

1,497 pigs × 0.02 × \$17.00/1.0 in. = \$509
return over an average Duroc boar

EPDs for feed efficiency were not reported in STAGES. However, each 1% improvement in feed efficiency over the last 200 lb of gain will save about 6 lb of feed per pig.

1,497 × 6 lb of feed × \$.07/lb = \$629
return over an average Duroc boar

Most people would agree that these are fairly attainable goals. When these three improvements in the boar's progeny are added together, a superior boar can increase profitability by more than \$2,200 during 1 year! Also, this same boar may be used for 2 or 3 years.

This doesn't mean that a producer should buy the most expensive boar available. Rather, performance and carcass data should justify the higher price.

JUDGING AND EVALUATING SWINE

It is to the everlasting credit of purebred swine breeders, however, that they have been very progressive evaluating swine. The type conferences sponsored by the various breed associations have made unique contributions. Through bolstering live-animal work with performance and carcass data, these evaluations have set standards for both the producer and the packer.

The discussion that follows represents a further elucidation of the first point discussed under selection—individuality. In addition to individual merit, the word *judging* implies the comparative appraisal or placing of several animals. Judging swine, as with all livestock judging, is an art achieved through patient study and long practice. Master breeders throughout the years have been competent livestock judges. The essential qualifications that a good judge of swine must possess, and the recommended procedure to follow in the judging assignment, are as follows:

1. Knowledge of the parts of an animal. A good judge has mastered the language that describes and locates the different parts of an animal (see

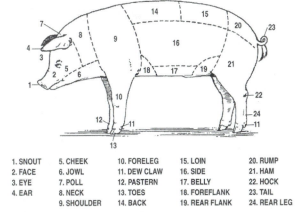

1. SNOUT	5. CHEEK	10. FORELEG	15. LOIN	20. RUMP
2. FACE	6. JOWL	11. DEW CLAW	16. SIDE	21. HAM
3. EYE	7. POLL	12. PASTERN	17. BELLY	22. HOCK
4. EAR	8. NECK	13. TOES	18. FOREFLANK	23. TAIL
	9. SHOULDER	14. BACK	19. REAR FLANK	24. REAR LEG

Figure 5.14 Parts of a hog. The first step in preparation for judging live hogs consists of mastering the language that describes and locates the different parts of the animal. *(Courtesy of Pearson Education)*

Figure 5.14). In addition, it is necessary to know which of these parts are of major economic or soundness importance, that is, what comparative evaluation should be given to the different parts.

2. A clearly defined ideal or standard of perfection. The successful swine judge must know what to look for, that is, the judge must have in mind an ideal or standard of perfection.

3. Keen observation and sound judgment. The good judge possesses the ability to observe both good conformation and defects, and to weigh and evaluate the relative importance of those features.

4. Knowledge of performance standards. The judge must be able to incorporate the relative importance of backfat and loin scans and performance data with the merits of the live animals being observed.

5. Honesty and courage. The good judge of any class of livestock must possess honesty, integrity, and courage, whether it be in making a show-ring placing or conducting a breeding and marketing program. For example, it often requires considerable courage to place a class of animals without regard to (a) placings in previous shows, (b) ownership, and (c) public applause. It may take even greater courage and honesty to discard from the herd a costly animal whose progeny has failed to measure up.

6. Systemic procedure in examining. There is always great danger of beginners making too close an inspection, oftentimes getting "so close to the trees that they fail to see the forest." Good judging procedure consists of the following three separate steps: (a) observing at a distance and securing a panoramic view where several animals are involved, (b) using close inspection, and (c) moving the animal in order to observe action.

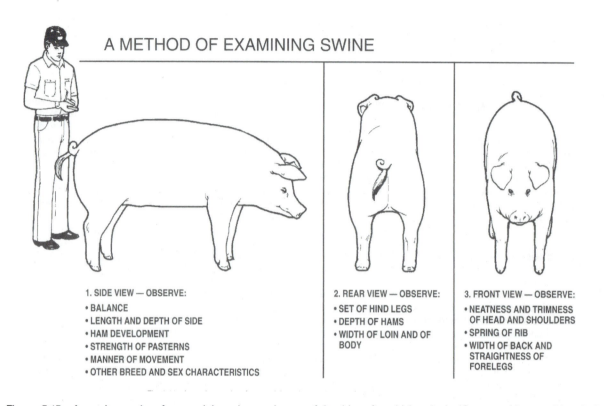

A METHOD OF EXAMINING SWINE

1. SIDE VIEW — OBSERVE:
- BALANCE
- LENGTH AND DEPTH OF SIDE
- HAM DEVELOPMENT
- STRENGTH OF PASTERNS
- MANNER OF MOVEMENT
- OTHER BREED AND SEX CHARACTERISTICS

2. REAR VIEW — OBSERVE:
- SET OF HIND LEGS
- DEPTH OF HAMS
- WIDTH OF LOIN AND OF BODY

3. FRONT VIEW — OBSERVE:
- NEATNESS AND TRIMNESS OF HEAD AND SHOULDERS
- SPRING OF RIB
- WIDTH OF BACK AND STRAIGHTNESS OF FORELEGS

Figure 5.15   A good procedure for examining a hog and some of the things for which to look. *(Courtesy of Pearson Education)*

Because a pig will neither stand still nor remain in the same vicinity for long, it is not possible to arrive at a set procedure for examining swine. In this respect, the judging of hogs is made more difficult than the judging of other classes of livestock. Where feasible, however, the steps for examining as illustrated in Figure 5.15 are very satisfactory, and perhaps as good as any.

7. Tact. In discussing either (a) a show-ring class or (b) animals on a producer's farm or ranch, it is important that the judge be tactful. Owners are likely to resent any remarks that imply that their animals are inferior.

Having acquired this knowledge, the judge must spend long hours in patient study, observing other judges and practicing comparing animals. Even this will not make expert and proficient judges in all instances. Nevertheless, training in judging and selecting animals is effective when it is directed by a competent instructor or an experienced producer.

Ideal Type and Conformation

A major requisite in judging or selection is to have clearly in mind a standard or ideal. Presumably, this ideal should be based on a combination of (1) the

efficient performance of the animal from the standpoint of the producer and (2) the desirable carcass characteristics of the market animals as determined by the consumer. Additionally, when evaluating purebred animals the standard characteristics required to represent that breed must be known and adhered to.

The most approved meat-type breeding animals combine growth, trimness, and quality. The offspring possess the ability to develop adequate muscle during the growing period without producing an excessive amount of lard. The head and neck should be trim and neat; the back smoothly arched and of ample width; the sides long, deep, and smooth; and the hams well developed and deep. The legs should be of medium length, straight, and squarely set; the pasterns should be short and strong; and the bone should be ample and show plenty of quality. With this splendid meat-type animal, there should be style, balance, and symmetry and an abundance of quality and smoothness.

Sows should show femininity and broodiness; and the udder should be well developed, with at least 6 teats on each side. The herd boar should show great masculinity as indicated by strength and character in the head, a somewhat crested neck, well-developed but smooth shoulders, a general ruggedness throughout, and an energetic disposi-

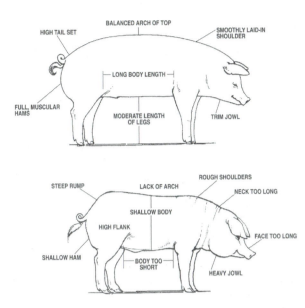

Figure 5.16 Ideal meat-type hog (top) versus common faults (bottom). The successful judge must know what to look for and be able to recognize and appraise both the good points and the common faults. *(Courtesy of Pearson Education)*

tion. The reproductive organs of the boar should be clearly visible and well developed.

The ideal meat-type hog versus some of the common faults are depicted in Figure 5.16. Because no animal is perfect, the proficient swine judge must be able to recognize, weigh, and evaluate both the good points and the common faults. In addition, the judge must be able to arrive at a decision as to the degree to which the given points are good or bad.

QUESTIONS FOR STUDY AND DISCUSSION

1. Figure 5.12 shows that selection has changed the conformation of improved swine so that they have greater proportion of loin to head than the European wild boar. Why have producers made such a change?
2. Why is pork quality important? Describe high-quality pork.
3. What is lean pork gain? How is it determined?
4. Why is it important that a swine producer keep good records?
5. Based on (a) heritability and (b) dollars and cents value, what traits should receive greatest emphasis in a swine production testing program?
6. What is the ratio of a boar that has an average daily gain of 2.3 lb per day where the test group average is 2.1 lb per day?
7. List and discuss some factors influencing the rate of progress that can be made through selection.
8. Is the porcine stress syndrome (PSS) good or bad? Justify your answer.
9. Discuss each of the various bases of selection of swine. How would you rank their value? What do you think are the minimum selection bases?
10. Why are backfat and muscling such important considerations when selecting swine?
11. Cite examples of how purebred and commercial breeders alike have come to regret selections based on show-ring winnings.
12. What do STAGES, EBV, EPD, BLUP, and SPI stand for? Detail how you should use EPDs in selection.
13. Select a certain farm (yours or one with which you are familiar). Assume that there are no hogs on it at the present time. Outline, step by step, (a) how you would go about establishing a herd and (b) the factors that you would consider. Justify your decisions.
14. In establishing a herd, why are the following factors important: type, breeding, breed or strain, size of herd, uniformity, health, age, price, and suitability of the farm.
15. What factors should a commercial swine producer consider when choosing a seedstock supplier?
16. What important traits or standards would you look for in (a) replacement gilts and (b) replacement boars?
17. Why is it important to know the parts of a hog?
18. Why is it difficult to arrive at a set procedure for examining a pig?

SELECTED REFERENCES

Animal Breeding Plans, J. L. Lush, Collegiate Press, Inc., Iowa State University, Ames, IA, 1963

Animal Science, M. E. Ensminger, Interstate Publishers, Inc., Danville, IL, 1991

Genetic Evaluation, Bob Uphoff, Chairman of NPPC Genetic Program Committee, National Pork Producers Council, 1995

National Swine Improvement Federation, http://www.nsif.com, Certified technicians: http://www.nsif.com/ certif.htm

Pork Composition and Quality Assessment Procedures, National Pork Board, Des Moines, IA, 2000

Pork Industry Handbook, Cooperative Extension Service, Purdue University, West Lafayette, IN, 2003

Stockman's Handbook, The, 7th ed., M. E. Ensminger, Interstate Publishers, Inc., Danville, IL, 1992

Swine Improvement Program Guidelines, National Swine Improvement Federation,http://www.nsif.com

Swine Testing and Genetic Evaluation System (STAGES), National Association of Swine Records, http://www.nationalswine. com/ or http://www.ansc.purdue.edu/stages

6

Fundamentals of Swine Nutrition

A stationary feed preparation mill on a family swine farm.
(Courtesy, Palmer Holden, Iowa State University, Ames, IA)

> **Objectives**
>
> After studying this chapter, you should:
>
> 1. Know the relative body composition changes that occur in pigs as they mature.
> 2. Be able to describe the digestive system of the pig.
> 3. Understand the basic maintenance and production requirements for swine.
> 4. Be able to define the major nutrient categories and provide examples of each.
> 5. Understand the various definitions of energy and their applicability to swine nutrition.
> 6. Know the essential amino acids and the effects of "first-limiting" amino acids.
> 7. Know the trace minerals and vitamins that must be routinely added to swine diets.
> 8. Know why plant phosphorus may not be a good source of phosphorus for pigs.
> 9. Know the importance and needs for water in swine feeding.
> 10. Know common feed evaluation techniques.

Efficient and profitable swine production depends on an understanding and application of the fundamentals of swine nutrition for two primary reasons:

1. Feed represents 55 to 75% of the total cost of production. Therefore, pork producers should provide diets that are both satisfactory and as low cost as possible, and that allow the maximum production of quality pork per unit of feed consumed.

2. Today, most of the hogs produced in the United States are raised indoors. As a result, they have minimal selection of feed choices compared to animals that have access to pasture or dirt lots. For the most part, they are able to consume only what the caretaker provides. This consists largely of concentrated feeds with little or no forage. A knowledge of the nutritional needs of swine is especially important because hogs grow much faster in proportion to their body weight than the larger farm animals and reproduce at an earlier age—factors that have been accentuated with modern genetics and optimal environments.

PERSPECTIVE OF NUTRITION

Nutrition is more than simply feeding; it is the science of the interaction of a nutrient with some part of a living organism. It begins with knowledge of the fertility of the soil and the composition of plants, and it includes the ingestion of feed, the liberation of energy, the elimination of wastes, and all the syntheses essential for maintenance, growth, reproduction, and lactation. Several factors affect the nutritive requirements of swine, including the following:

1. Breed or strain, sex, and genetics of pigs.
2. Age and weight of the pigs.

3. Health status of the herd.
4. Environment (temperature, weather, housing, and competition for feed).
5. Variability of nutrient content of the feed.
6. Availability and absorption of dietary nutrients.
7. Energy concentration of the diet.
8. Level of feed additives or growth promotants.
9. Level of feeding, such as limit feeding versus *ad libitum.*
10. Presence of molds, toxins, or inhibitors in the diet.

BODY COMPOSITION OF SWINE

Nutrition encompasses the various chemical and physiological reactions that change feed elements into body elements. It follows that changing body composition is useful in understanding swine responses to nutrition.

In 1843, Lawes and Gilbert, famed English scientists, initiated at the Rothamsted Station the pioneering and laborious task of analyzing the entire bodies of farm animals—studies that extended for more than a half century. Other similar studies involving pigs, cattle, and sheep followed throughout the world.

An estimate of the chemical composition of pigs at different weights are presented in Table 6.1 and indicates the following:

1. **Ash.** The percentage of ash changes very little. However, it will decrease slightly as pigs fattened because fat tissue contains less mineral than lean.
2. **Lipid.** The percentage of lipid or fat increased with growth and fatness, ranging from 2% in the newborn pig to 35% in fat market hogs.
3. **Protein.** The percentage of protein remains rather constant during growth, but decreases as

TABLE 6.1   Chemical Composition of Empty Body Weight of Pigs

Item (%)	Birth	6.8 kg (15 lb)	25 kg (55 lb)	110 kg (242 lb) Lean pigs	110 kg (242 lb) Fat pigs
Ash	3	3	3	3	3
Lipid	2	15	12	15	35
Protein	18	16	16	18	14
Water	77	66	69	64	48

Source: de Lange, C. F. M., S. H. Birkett, and P. C. H. Morel. 2000. "Protein, fat, and bone tissue growth in swine." In A. J. Lewis and L. L. Southern (eds.) Swine Nutrition (pp. 65–81, 2nd ed.). CRC Press, Boca Raton, FL.

pigs fatten. On the average, there are 3 to 4 lb of water per 1 lb of protein in the pig's body.

4. **Water.** On a percentage basis, swine have a marked decrease in water content, ranging from 77% in newborn pigs to 48% in fat market weight animals.

The gain in weight of pigs tells nothing about the composition of the gain. Composition is important from a nutrition standpoint because efficiency of feed utilization (units of feed per unit of body gain) is greatly influenced by the proportion of lean and fat produced. As the percentage of fat increased, the percentage of water decreased.

Body composition as a parameter—Although the general relationships presented in Table 6.1 remain, genetic selection has shifted the tissue growth of modern swine, with less fat and more protein being produced. In addition to genetic background, the following factors influence the body composition of today's pigs: weaning age, genetics, sex, additives, temperature, diet, and compensatory growth. The NRC Nutrient Requirements of Swine (1998) allows for the input of the composition of gain to influence the nutrient requirements. Accomplishing this involves serial slaughter or ultrasound evaluation to estimate lean and fat composition.

Also, the chemical composition of the body varies widely between organs and tissues and is more or less localized according to function. Thus, water is an essential of every part of the body, but the percentage composition varies greatly in different body parts; blood plasma contains 90 to 92% water, muscle 72 to 78%, bone 45%, and the enamel of the teeth only 5%. Proteins are the principal constituents, other than water, of muscles, tendons, and connective tissues. Most of the fat is localized under the skin, near the kidneys, and around the intestines. But it is also present within the muscles (marbling), bones, and elsewhere.

Table 6.1 does not estimate the very small amount of carbohydrates (mostly glucose and glycogen) present in the bodies of animals and found principally in the liver, muscles, and blood. Although these carbohydrates are very important in animal nutrition, they account for less than 1% of the body composition. It is noteworthy, too, that the carbohydrate content is one of the fundamental differences between the composition of plants and animals. In animals, the walls of the body cells are made chiefly of protein, whereas in plants they are composed of cellulose and other carbohydrates. Also, in plants most of the reserve food is stored as starch, another carbohydrate, whereas in animals nearly all the reserve energy is stored as fat.

DIGESTIVE SYSTEM OF THE PIG

Meat animals can be divided into two broad classifications—ruminants and nonruminants. Pigs are nonruminants or monogastrics (Figure 6.1). They have a single stomach, in contrast to ruminants, such as cattle and sheep which have a stomach divided into 4 compartments.

To grow rapidly and efficiently, swine must receive a high-energy, concentrated grain diet that is low in fiber. Ruminants (cattle and sheep), however,

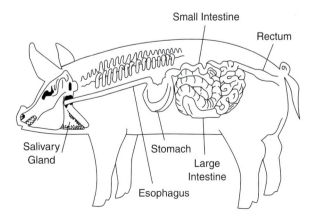

Figure 6.1 The digestive tract of the pig—a nonruminant. *(Courtesy of Pearson Education)*

can digest large quantities of fibrous feeds, such as hay and pasture, because of significant differences in their digestive tracts.

The digestive tract (or gastrointestinal tract) can be considered a continuous hollow tube—open at both ends—with the body built around it. It is a factory assembly line in reverse; instead of building something, it takes things apart. The digestive tract of the pig includes 5 main parts: the mouth, esophagus, stomach, small intestine, and large intestine.

The mouth is the first part of the digestive system and contains 3 organs associated with digestion, the tongue for mixing and moving the food; the teeth for chewing the particles into smaller units; and the salivary glands for providing moisture, mucin to aid in swallowing, buffers to regulate pH, and the enzyme amylase.

The esophagus is a muscular tube that transports the food from the mouth to the stomach by peristaltic waves. The cardiac sphincter is a valve at the junction of the esophagus and stomach that prevents regurgitation of the food once it reaches the stomach.

The stomach is a pear-shaped, muscular organ that stores ingested feed, moves the feed, and secretes gastric juices, particularly hydrochloric acid, pepsin, and rennin. The pH of the stomach is quite acidic, approximately 2. The stomach of a mature pig holds about 2 gal (7.5 l). As the feed leaves the stomach it passes through the pyloric sphincter into the small intestine.

The majority of the digestion and absorption of nutrients occurs in the small intestine. It consists of 3 parts: the duodenum, the jejunum, and the ileum. Together they are about 60 ft (18 m) long and hold about 2.5 gal (9 l). The duodenum receives secretions from the pancreas, liver (bile), and intestinal walls, and most of the digestion occurs here. The jejunum and the ileum are active sites of nutrient absorption.

The large intestine is divided into 3 parts: the cecum, the colon, and the rectum. Together they are about 17 ft (5 m) long and hold 2.5 gal (9 l). The large intestine absorbs much of the water from the ingesta, secretes some mineral elements, stores undigested material and is a source of some bacterial fermentation. There is limited absorption of nutrients from the large intestine.

Although ruminants and monogastrics differ in their physical makeup, the job of the digestive tract is the same in all animals. It breaks down feedstuffs into simple chemical components so that the animal can absorb and rearrange them into its own characteristic body composition.

CLASSIFICATION OF NUTRIENTS

Animals do not use feeds as such. Rather, they use those portions of feeds called *nutrients* that are re-

leased by digestion, then absorbed into the body fluids and tissues.

Nutrients are those substances, usually obtained from feeds, that can be used by the animal when made available in a suitable form to its cells, organs, and tissues. They include carbohydrates, fats, proteins, minerals, vitamins, and water. (More correctly speaking, the term *nutrients* refers to the more than 40 nutrient chemicals, including amino acids, minerals, and vitamins.) Energy is frequently listed with nutrients because it results from the metabolism of carbohydrates, proteins, and fats in the body. Knowledge of the basic functions of nutrients in the animal body, and of the interrelationships among various nutrients and other metabolites within the cells of the animal, is necessary before one can make practical scientific use of the principles of nutrition.

FUNCTIONS OF NUTRIENTS

Of the feed consumed, a portion is digested and absorbed for use by the pig. The remaining undigested portion is excreted and constitutes the major portion of the feces. Nutrients from the digested feed are used for a number of different body processes, which vary with the class, age, and productivity of the pig. All pigs use a portion of their nutrients to carry on essential functions, such as body metabolism, maintaining body temperature, and the replacement and repair of body cells and tissues. These uses of nutrients are referred to as *maintenance*. The portion of digested feed used for growing, pregnancy, or the production of milk is known as production requirements.

Based on the quantity of nutrients needed daily for different purposes, nutrient demands may be classed as high, low, variable, or intermediate. Requirements for milk are considered high-demand uses, whereas hair growth is a low-demand use. The last stage of pregnancy has variable requirements. Growing may be classed as intermediate in nutrient demands. Each of these needs is discussed in more detail.

Maintenance

Pigs, unlike machines, are never idle. They use nutrients to keep their bodies functioning every hour of every day, even when they are not being used for production.

Maintenance requirements may be defined as the combination of nutrients that are needed by the animal to keep its body functioning without any gain or loss in body *weight or any productive activity*. Although these requirements are relatively simple, they are essential for life itself. A pig must have (1) heat to maintain body temperature, (2) sufficient energy to keep vital body processes functional, (3) energy for minimal movement, and (4) the necessary nutrients to repair damaged cells and tissues and to replace those that become nonfunctional. Thus, energy is the primary nutritive need for maintenance. Even though the quantity of other nutrients required for maintenance is relatively small, it is necessary to have a balance of the essential amino acids, minerals, and vitamins.

No matter how quietly a pig may be lying in a pen, it requires a certain amount of energy and other nutrients. The least amount on which it can exist is called its *basal maintenance requirement.* With the exception of horses, most animals require about 9% more energy (calories) when standing than when lying and even more is needed when they walk or run.

There are only a few times in the normal life of a pig when only the maintenance requirement needs to be met. Such a status is closely approached by mature males not in service and by mature, dry, nonpregnant females. Nevertheless, maintenance is the standard benchmark or reference point for evaluating nutritional needs.

Even though maintenance requirements might be considered an expression of the nonproduction needs of a pig, there are many factors that affect the amount of nutrients necessary for this vital function. Among them are (1) exercise, (2) weather, (3) stress, (4) health, (5) body size, (6) temperament, (7) individual variation, (8) level of production, and (9) lactation. The first four are *external factors*—they are subject to control to some degree through management and facilities. The others are *internal factors*—they are part of the animal itself. Both external and internal factors influence requirements according to their intensity. For example, the colder or hotter it gets from the most comfortable (optimum) temperature, the greater will be the maintenance requirements.

Growth

Growth may be defined as the increase in size of bones, muscles, internal organs, and other parts of the body. It is the normal process beginning before birth and continuing after birth until the pig reaches its full, mature size. Growth is influenced primarily by nutrient intake. The nutritive requirements become increasingly acute when young animals are at full production, such as when pigs are fed to reach market weight by 150 to 190 days of age. The growth of pigs in relation to daily feed intake, daily gain, feed efficiency, and weight and age are graphed in Figures 6.2, 6.3, 6.4, and 6.5, respectively. Data from the National Research

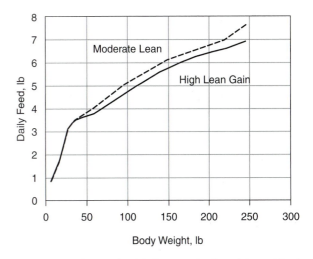

Figure 6.2 The relationship between body weight and feed intake in growing swine. *(Courtesy, Life Cycle Swine Nutrition, 1996)*

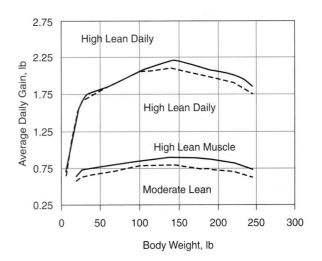

Figure 6.3 The relationship between body weight and average daily total gain and muscle gain in growing swine. *(Courtesy, Life Cycle Swine Nutrition, 1996)*

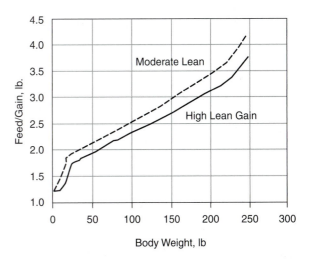

Figure 6.4 The relationship between body weight and feed efficiency in growing swine. *(Courtesy, Life Cycle Swine Nutrition, 1996)*

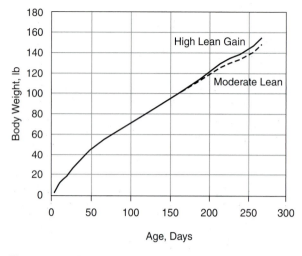

Figure 6.5 The relationship between age and weight in growing swine. *(Courtesy,* Life Cycle Swine Nutrition, *1996)*

Council's (NRC) Nutrient Requirement of Swine estimates the expected performance of pigs from weaning to market (Table 6.2).

Growth is the very foundation of swine production. Young swine will not make the most economical finishing gains unless they have been raised to be thrifty and vigorous. Likewise, breeding females may have their reproductive ability seriously impaired if they have been improperly grown.

Generally speaking, organs vital for the maintenance of life—for example, the brain, which coordinates body activities, and the gut, on which the rest of the postnatal growth depends—are early developing, and the commercially more valuable parts, such as muscle, develop later.

Knowledge of normal growth and development is useful for a variety of purposes. From a nutritional standpoint, growth curves are used primarily as standards against which to gauge the adequacy of nutrient allowances. In fact, such curves are often the entire basis for the allowances set down in dietary and feeding standards. Also, they provide a basis for comparisons of breeding groups and serve as a reference point from which to establish breeding

and management objectives. Economically, growth is important: Young gains are efficient gains. In comparison with older animals, young animals (1) consume more feed according to size; (2) use a smaller proportion of their feed for maintenance; and (3) form relatively more muscle tissue, which has a lower caloric value or requirement than fat. Still, the nutritive needs for growth vary with breed, sex, rate of growth, and health.

Reproduction

Being born and born alive are the first and most important requisites of pork production because if pigs fail to reproduce, the swine enterprise is soon out of business. A perfect pig with the greatest genes in the world is of no value unless these genes result in (1) the successful joining of the sperm and eggs and (2) the birth of live offspring. Still, research shows that embryonic mortality claims 15 to 30% of the swine embryos; of the pigs born alive 10 to 20% die before weaning, and 15% of all sows bred fail to produce litters. Nutrition can be a cause of reproductive failure, both by underfeeding many nutrients but also by overfeeding some nutrients, particularly energy, during gestation. Because swine producers largely determine their own destiny when it comes to feeding, it is important that they know the causes of reproductive failure and how to rectify them.

A review of the literature clearly points to three reproductive difficulties: (1) females failing to show signs of estrus (heat), (2) the low conception rate at first service, and (3) the excessive losses at birth or within the first 2 to 5 days after birth.

With all mammalian species, most of the growth of the fetus occurs during the last third of pregnancy. Additionally, the female must store body reserves during pregnancy because the demands for milk production are usually greater than the sow can obtain by adequate lactation feed intake. Hence, the nutrient intakes are very critical during this period, especially for first parity (litter) females.

It is also known that the diet has a major affect on sperm production and semen quality, primarily by controlling the boar's body condition. Too fat a

TABLE 6.2   Expected Performance of Pigs

Stage	Pig Weight		Feed Intake per Day		Average Daily Gain		Feed/Gain
	(lb)	*(kg)*	*(lb)*	*(kg)*	*(lb)*	*(kg)*	
Prenursery	12–20	5–9	0.7–1.3	0.3–0.6	0.4–0.7	0.2–0.3	1.7–1.8
Nursery	20–60	9–27	1.3–3.5	0.6–1.6	0.7–2.0	0.3–0.9	1.8
Grower	60–130	27–60	3.5–5.3	1.6–2.4	2.0–2.7	0.9–1.2	1.8–2.0
Finisher	130–280	60–127	5.3–7.3	2.4–3.3	2.6–2.7	1.2	2.0–2.9

Source: NRC Nutrient Requirements of Swine (1998).

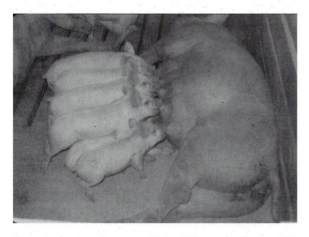

Figure 6.6 Lactation diets along with nursery diets are the two most critical diets in swine nutrition. In both cases adequate intake is the limiting factor. *(Courtesy, Palmer Holden, Iowa State University, Ames, IA)*

condition can lead to temporary or permanent sterility. Greater fertility of herd sires exists under conditions where a well-balanced diet is provided.

Lactation

The lactation requirements of females of all mammalian species for moderate to heavy milk production are much higher than maintenance or pregnancy requirements (Figure 6.6). Fortunately, females can store up body reserves during pregnancy and then draw on them during lactation. But if there has not been proper body storage during pregnancy, something must "give"—and that something will be the mother, for nature ordained that growth of the fetus, and the lactation that follows takes priority over the female's requirements. Hence, when there is a nutrient deficiency, the female's body, particularly if she is young, will be deprived, or even stunted, before the developing fetus or milk production will be materially affected.

NUTRIENTS

Nutrients are used in one of two metabolic processes: (1) for anabolism or (2) for catabolism. Anabolism is the process by which nutrient molecules are used as building blocks for the synthesis of complex molecules. Anabolic reactions are endergonic—that is, they require the input of energy into the system. Catabolism is the oxidation of nutrients, liberating energy (exergonic reaction) that is used to fulfill the body's immediate demands.

Energy

Energy is required for practically all life processes—for the action of the heart, maintenance of blood pressure and muscle tone, transmission of nerve impulses, ion transport across membranes, reabsorption in the kidneys, protein and fat synthesis, and milk production.

A deficiency of energy is manifested by slow or stunted growth; body tissue losses; or lowered production of muscle rather than by specific signs, such as those characterized by many mineral and vitamin deficiencies. For this reason, energy deficiencies often go undetected and uncorrected for extended periods of time.

It is common knowledge that a diet must contain carbohydrates, fats, and proteins. Although each of these has specific functions in maintaining a normal body, all of them can be used to provide energy for maintenance, growth, and reproduction. From the standpoint of supplying normal energy needs the carbohydrates are by far the most important because they are consumed in greater quantities than any other compound, and the fats are second in importance for energy purposes. Carbohydrates are usually more abundant and cheaper, and most of them are very easily digested, absorbed, and transformed into body fat. Also, carbohydrate feedstuffs may be more easily stored than fats in warm weather and for longer periods of time. Excess protein in the diet can also be a source of energy, but proteins are more expensive than carbohydrates and fats.

Carbohydrates

Carbohydrates are organic compounds composed of carbon, hydrogen, and oxygen—formed in plants by the process of photosynthesis. They constitute about 75% of the dry weight of plants and grain and make up a large part of the swine diet. They serve as a source of heat and energy in the pig's body, and surplus carbohydrates are converted into fat and stored in the body.

No appreciable amount of carbohydrate is found in the animal body at any one time. The storage of glycogen (so-called animal starch) in the liver amounts to 3 to 7% of the weight of that organ. It is constantly converted to blood glucose, which is held rather constant at about 0.05 to 0.1% for most animals. This small quantity of glucose in the blood serves as the chief source of fuel with which to maintain the body temperature and furnish the energy needed for all body processes.

Lipids

Lipids (fat and fatlike substances), as do carbohydrates, contain the three elements—carbon, hydrogen, and oxygen. Lipids function much like carbohydrates in that they serve as a source of heat

Figure 6.7 Extruded soybeans are an easy, efficient means of adding fat to swine diets. *(Courtesy, Palmer Holden, Iowa State University, Ames, IA)*

and energy and for the formation of fat. Because of the larger proportion of carbon and hydrogen, however, lipids liberate more heat than carbohydrates when digested, furnishing approximately 2.25 times as much heat or energy per unit on oxidation as do the carbohydrates. A small quantity of lipids is required, therefore, to serve the same function.

Evidence indicates that additions of 3 to 5% fat to growing finishing swine diets will improve feed conversion and often improve average daily gain. Usually, feed efficiency will improve by 2% for each 1% fat added to the diet. Adding fat to diets tends to increase backfat thickness.

Research with sows indicates that the addition of fat to the diet during late gestation or lactation may improve pig survivability, perhaps because of increased milk yield and milk fat content (Figure 6.7).

Feed fats affect body fats—Swine consuming soft fats or oils, such as soybean oil or whole processed soybeans, may produce soft pork and those consuming saturated fats, such as from barley, have relatively firm backfat.

Fatty acids—These are the key components of fats (lipids). Their length and degree of saturation (amount of hydrogen) determine many of the physical aspects—melting point and stability—of fats (lipids).

Linoleic acid is required in swine diets. Arachidonic acid was once considered a dietary essential but research has shown that the pig can convert linoleic acid to arachidonic acid. The linoleic acid requirement is usually sufficient in the cereal grains and protein supplements provided in pig diets. Fatty acid deficiencies are not common but can re-

sult in dermatitis, reduced growth, increased water consumption and retention, impaired reproduction, and increased metabolic rate. They have an important role in the structural integrity of the cell.

Measuring and Expressing Energy Value of Feedstuffs

One nutrient cannot be considered as more important than another because all nutrients must be present in adequate amounts if efficient production is to be maintained. Yet, historically, feedstuffs have been compared or evaluated primarily on their ability to supply energy to animals. This is understandable because (1) energy is required in larger amounts than any other nutrient and (2) energy is the major cost associated with feeding swine.

Our understanding of energy metabolism has increased through the years. With this added knowledge, changes have come in both the methods and terms used to express the energy value of feeds.

The calorie system is commonly used in the United States to measure energy but megajoules are used in many European countries.

Energy Definitions and Conversions

Some pertinent definitions and conversions of energy terms follow:

Calorie (cal)—The amount of energy as heat required to raise the temperature of 1 g of water 1°C (precisely from 14.5° to 15.5°C). It is equivalent to 4.184 joules. In popular writings, especially those concerned with human caloric requirements, the term *calorie* is frequently used erroneously for the kilocalorie (1,000 calories).

Kilocalorie (kcal)—The amount of energy as heat required to raise the temperature of 1 kg (1,000 g) of water 1°C (from 14.5° to 15.5°C). It is equivalent to 1,000 calories.

Megacalorie (mcal)—Equivalent to 1,000 kcal or 1,000,000 calories. The term *megacalorie* is preferred but may be referred to as a *therm*.

Joule (J)—An international unit (4.184 J = 1 cal) for expressing mechanical, chemical, or electrical energy, as well as the concept of heat.

Calorie System of Energy Evaluation

To measure calories, an instrument known as the bomb calorimeter is used, in which the feed (or other substance) to be tested is placed and burned in the presence of oxygen. It represents the gross energy content of the feed.

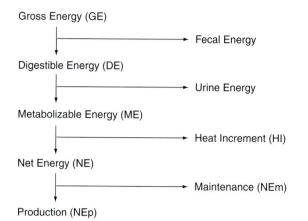

Figure 6.8 Partition of energy in nutrition. The amount of energy lost in each partition depends on the amount of fiber, fat, and protein present in the feedstuffs. *(Courtesy, Palmer Holden, Iowa State University, Ames, IA)*

Through various digestive and metabolic processes, numerous losses occur during the passage of feed through the pig's digestive system. These losses are illustrated in Figure 6.8. Measures that are used to express energy requirements and the energy content of feeds differ primarily in the digestive and metabolic losses that are included in their determination. Thus, the following terms are used to express the energy value of feeds:

Gross energy (GE)—Gross energy represents the total combustible energy in a feedstuff as determined in a bomb calorimeter. It has little value in nutrition as corn and wood have similar gross energy values. It does not differ greatly between feeds, except for those high in fat.

Digestible energy (DE)—Digestible energy is that portion of the GE in a feed that is not excreted in the feces. It probably determines the daily feed intake of pigs.

Metabolizable energy (ME)—Metabolizable energy represents that portion of the GE that is not lost in the feces, urine, and gas. The losses of energy as gas produced in the digestive tract of swine are small; therefore, the ME values are not corrected for this energy loss. Although ME better describes the useful energy in the feed than does GE or DE, it does not take into account the energy lost as the heat increment. The heat increment is the heat released as result of metabolic processes. It has little value except to maintain body temperature in cold environments. ME values may be calculated from several formulae. One of high quality is

$$ME = DE \times (1.012 - (0.0019 \times \% \ CP)),$$
where CP equals crude protien.

Net energy (NE)—Net energy represents the energy fraction in a feed that is left after the fecal, urinary, gas, and heat losses are deducted from the GE, or in other words, the ME minus the heat increment. The net energy, as a fraction of the ME, varies from 27 for wheat midds to 69% for corn and 75% for soybean oil. Although the NE may be the best measure of the energy available for maintenance and production, it is difficult to measure. At present, NE requirements for maintenance and production are not available. Therefore, the energy requirements of swine and the energy values of the feedstuffs for swine are presented in DE and ME values.

Total Digestible Nutrients (TDN)

Total digestible nutrients (TDN) is the sum of the digestible crude protein, crude fiber, nitrogen-free extract (NFE), and ether extract (EE or fat) $\times$ 2.25. It is no longer used for estimating the energy in swine diets, however, a brief description is presented for historical purposes. TDN values are based on the following steps:

1. **Digestibility.** The digestibility of a particular feed for a specific species is determined by a digestion trial.
2. **Computation of digestible nutrients.** Digestible nutrients are computed by multiplying the percentage of the crude protein (CP), crude fiber (CF), nitrogen-free extract (NFE), and ether extract (EE) by its digestion coefficient. The result is expressed as digestible nutrients. For example, if corn contains 8.3% crude protein of which 77% is digestible, the percentage of digestible crude protein is 6.4.
3. **Computation of TDN.** The TDN is computed with the following formula:

$$\% \ TDN = \times 100$$

where DCP = digestible crude protein; DCF = digestible crude fiber; DNFE = digestible nitrogen-free extract; and DEE = digestible ether extract or fat.

TDN is ordinarily expressed as a percentage of the diet or in units of weight (lb or kg), not as a caloric figure.

The disadvantages of the TDN system are as follows:

1. TDN is not an actual total of the digestible nutrients in a feed. It does not include the digestible mineral matter (such as salt, limestone, and defluorinated

phosphate—all of which are digestible), and the digestible fat is multiplied by the factor 2.25 before being included in the TDN figure because its energy value is higher than carbohydrates and protein. As a result of multiplying fat by 2.25, feeds high in fat will sometimes exceed 100 in percentage TDN (a pure fat with a coefficient of digestibility of 100% would have a theoretical TDN value of 225% (100% × 2.25).

2. It is based on chemical determinations that are not related to actual metabolism of the animal.

3. It is expressed as a percentage or in weight (lb or kg), whereas energy is expressed in calories.

4. It takes into consideration only digestive losses; it does not take into account other important losses, such as losses in the urine, gases, and increased heat production (heat increment).

5. It overvalues roughages in relation to concentrates when fed for high rates of production because of the higher heat loss per pound of TDN in high-fiber feeds.

Proteins and Amino Acids

Proteins are complex organic compounds made up chiefly of amino acids that are present in characteristic proportions for each specific protein. This nutrient always contains carbon, hydrogen, oxygen, nitrogen; in addition, it usually contains sulfur and frequently phosphorus. Proteins are essential in all plant and animal life as components of each living cell.

Crude protein refers to all the nitrogenous compounds in a feed. It is determined by analyzing the percentage nitrogen content and multiplying it by 6.25. The nitrogen content of protein averages about 16% (100 ÷ 16 = 6.25).

In plants, the protein is largely concentrated in the actively growing portions, especially the leaves and seeds. Plants have the ability to synthesize their own proteins from such relatively simple soil and air compounds as carbon dioxide, water, nitrates, and sulfates, using energy from the sun. Thus, plants, together with some bacteria, which are able to synthesize these products, are the original sources of all proteins.

Proteins are much more widely distributed in animals than in plants. Thus, the proteins of the animal body are primary constituents of many structural and protective tissues—such as bones, ligaments, feathers, skin, and the soft tissues, including the organs and muscles.

Pigs of all ages require adequate amounts of protein of suitable quality for maintenance, growth, and reproduction. Of course, the protein requirements for growth are the greatest and most critical.

From a nutritional standpoint, the pig's requirement for protein is not for protein per se but

for certain amino acids that the pig cannot synthesize but are necessary for normal growth and development. These amino acids, which must be supplied by the diet, are referred to as *essential* or *indispensable* amino acids. An amino acid is *nonessential* (dispensable) if it can be synthesized in the body. This synthesis requires nitrogen which is a function of the protein in the diet.

In the body, the amino acids function as building blocks for new protein, as well as functioning in some other specific metabolic roles, such as the formation of neurotransmitters, hormones, purines, and urea. The essential and nonessential amino acids are identified as follows:

Essential	**Nonessential**
(Indispensable)	*(Dispensable)*
Arginine	Alanine
Histidine	Asparagine
Isoleucine	Aspartic acid
Leucine	Cysteine
Lysine	Cystine
Methionine	Glutamic acid
Phenylalanine	Glutamine
Threonine	Glycine
Tryptophan	Hydroxyproline
Valine	Proline
	Serine
	Tyrosine

It is essential to specify methionine and lysine levels in the diets of very young pigs and lysine levels when formulating and evaluating all other swine diets. If a diet is inadequate in any essential amino acid, protein synthesis cannot proceed beyond the rate at which that amino acid is available. This is called the first-limiting amino acid. The lysine and other essential amino acid requirements for swine vary as follows:

1. They are higher for young, rapidly growing pigs.
2. They are higher for boars and gilts than for barrows.
3. They are higher for high lean-growth pigs.
4. They are higher in the hot summer months, when the appetite and feed consumption of pigs decreases.

When selecting feeds, producers should be aware that all proteins are not created equal. Cereal grains often contain insufficient quantities of lysine, methionine, cystine, tryptophan, and threonine. Hence, the term quality of protein is often used to describe the amino acid balance of a protein. A protein is said to be of good quality when it contains all the essential amino acids in proper proportions and

Figure 6.9 The Iowa State University Swine Nutrition and Management Research Farm. University research facilities are essential to the development of nutritional requirements and feeding programs. Note also the SlurryStore™ tank for manure storage. *(Courtesy, Iowa State University College of Agriculture, Ames, IA)*

amounts, and to be of poor quality when it is deficient in either content or balance of essential amino acids. From this it is evident that the usefulness of a protein source depends on its amino acid composition because the real need of the pig is for amino acids and not for protein.

Although it is common practice to refer to "percent protein" in a diet, this term has little significance in swine nutrition unless there is information about the feedstuffs used or amino acids present. For swine, quality is just as important as quantity. It is possible for pigs to perform better on a 12% protein diet, well-balanced for amino acids, than on a 16% diet having a poor amino acid balance.

From a practical standpoint, the problem of building a balanced diet for swine centers on correcting the deficiencies of the cereal grains. Although corn, wheat, and barley may contain from 8 to 12% protein, their protein is seriously deficient in the essential amino acid, lysine. Corn is also deficient in tryptophan, as is meat and bone meal. Because protein supplements are more expensive than grain, the tendency is to feed too little of them.

Previously, it was stated that when the pig consumes an excess of energy it is stored in the form of fat. Protein is not stored in the body in appreciable amounts. If an excess of protein is fed, the unused nitrogen portion is discarded as urea in the urine and the carbon fraction is used as a source of energy. From an economic standpoint, it is unprofitable to feed more protein than needed to meet the nutritional requirements of the pig.

The use of crystalline amino acids, particularly lysine and methionine, now makes it possible to provide diets containing less protein than is commonly recommended for swine. Symptoms of protein (amino acid) deficiency are reduced feed intake, reduced growth, poor hair and skin condition, and lowered reproduction.

Management Changes Affect Mineral and Vitamin Needs

Mineral and vitamin additions to the diets of swine have increased since the 1980s along with changes in genetics, housing, feeding, and management (Figure 6.9). Among the more important mineral and vitamin changes are the following:

1. Indoor production has denied swine access to pasture crops and soils, which provided minerals and vitamins.
2. Slotted floors have prevented eating of feces (coprophagy), which are high in B vitamins and vitamin K, synthesized by microorganisms in the large intestine.
3. Reduced use of multiple protein sources in diets, which often compliment each other in providing minerals and vitamins, have lessened proteins as a source of these nutrients.
4. Reduced daily feed intake of sows during gestation calls for increased mineral and vitamin concentration in the diet.
5. The trend to earlier weaning, at 2.5 to 4 weeks of age, calls for quality feedstuffs and all nutrients in baby pig diets.
6. The bioavailability of nutrients in heat-dried grains and feed ingredients varies widely.
7. The presence of inhibitors and molds in feeds may result in reduced absorption, increasing the requirements for certain vitamins.

With modern swine diets, properly supplemented with minerals and vitamins, deficiency symptoms of trace minerals and vitamins are rarely observed. However, a deficiency of salt will be observed very rapidly as noted. Calcium and phosphorus deficiencies become apparent as reduced growth and the potential for weakened and broken bones.

Minerals

Of all common farm livestock, the pig is most likely to suffer from mineral deficiencies. This is because of the following peculiarities of swine husbandry:

1. Hogs are fed principally on cereal grains and their by-products, all of which are relatively low in mineral matter, particularly calcium.
2. The skeleton of the pig supports greater weight in proportion to its size than any other farm livestock.

3. Hogs are fed to grow at a maximum rate for an early market, well before they are mature.
4. Hogs reproduce at a younger age than other classes of livestock.
5. Increased indoor rearing without access to soil or forages tends to balance the mineral deficiencies of the grains.

The functions of minerals are extremely diverse. They range from structural functions in some tissues to a wide variety of regulatory functions in other tissues.

Swine require at least 13 known inorganic elements, including calcium, chlorine, copper, iodine, iron, magnesium, manganese, phosphorus, potassium, selenium, sodium, sulfur, and zinc. Also, cobalt is required in the synthesis of vitamin B_{12}. Pigs may also require other trace elements, such as arsenic, boron, bromine, cadmium, chromium, fluorine, lead, lithium, molybdenum, nickel, silicon, tin, and vanadium, which have been shown to have a physiological role in one or more species. These elements are required at such low levels, however, that their dietary essentiality for the pig has not been proven. Most of them are believed to be present in adequate quantities in natural feed ingredients.

Table 6.3 provides the following pertinent information relative to each mineral: (1) conditions usually prevailing where deficiencies are reported, (2) functions, (3) deficiency symptoms/toxicity, and (4) practical sources.

Major or Macrominerals

Salt (sodium and chlorine), calcium, phosphorus, magnesium, potassium, and sulfur are the major or macrominerals.

Calcium (Ca) and Phosphorus (P)

Calcium and phosphorus are important in skeleton development. Also, they aid in blood clotting, muscle contraction, and energy metabolism. About 99% of the calcium and 80% of the phosphorus in the body are found in the skeleton and teeth.

A deficiency of either calcium or phosphorus in the diet of the pig can result in poor and inefficient gains, rickets or osteomalacia, broken bones, and posterior paralysis. Although severe nutrient deficiencies are rare on swine farms, an example of a phosphorus-deficient pig appears in Figure 6.10. The most common visual symptom of most deficiencies is simply poor growth and feed efficiency.

A large excess of either calcium or phosphorus interferes with the absorption of the other. Thus, it is important to have a suitable ratio between the two minerals. The most favorable calcium to phosphorus ratio is 1.2:1, but any ratio between 1.0:1 and 1.5:1 will not impair absorption of performance. Vitamin D is necessary for the proper utilization of these two minerals.

An excess of calcium interferes with zinc absorption and results in parakeratosis. A combination of a high level of calcium (more than 0.9%) and a marginal zinc level can result in this condition.

Figure 6.10 Phosphorus deficiency. Left: This is a typical phosphorus-deficient pig in advanced stage of deficiency. Leg bones are weak and crooked. Right: This pig received the same diet as the one on the left, except that the diet was adequate in available phosphorus. *(Courtesy, Purdue University, West Lafayette, IN)*

A swine manure lagoon with a small aerator. Much of the nitrogen is lost into the air from a lagoon and the phosphorus and potassium tend to settle in the sludge. Unless it is agitated, the material applied to the cropland is very low in nutrients. *(Courtesy, Palmer Holden, Iowa State University, Ames, IA)*

It is important to supplement swine diets with both calcium and phosphorus. Cereal grains, which make up the bulk of swine diets, are quite low in calcium and are only fair sources of phosphorus. Much of the phosphorus in cereal grains is present in the form of phytate, a form of phosphorus that is poorly digested by the pig. A range of 8 to 60% of phosphorus availability has been reported in cereal grains. Corn for example is 14% available; soybean meal, 23%; and wheat, 50%. Phosphorus of animal origin, such as meat meal and fish meal, is about 90% available and these feeds are also good sources of calcium.

PHYTASE. Much farmland is already high in phosphorus as a result of commercial fertilizer or hog manure applied in excess of crop needs. In Europe, areas of intensive swine production have received much of the blame for excessive phosphorus concentrations of cropland. Several states in the United States are currently restricting manure phosphorus applications or have set a date when manure phosphorus additions to the land will be limited by the amount of phosphorus harvested in the grain or forage.

Supplementing swine diets with phytase enzymes has been effective in improving the availability of phosphorus in corn and soybean meal. This results in less inorganic phosphate in the diet and reduces the phosphorus in the manure, thereby lessening the phosphorus pollution of the land. But, as is the case with all ingredients, producers will need to determine the cost effectiveness of (1) adding phytase versus (2) adding higher levels of inorganic phosphorus to the diet.

Experiments have indicated that with phytase a 30 to 50% reduction of the manure phosphorus is feasible. If much of the phytate phosphorus present in the cereal grains were made available during digestion, it would be sufficient to satisfy the requirements of finishing pigs. Thus, no inorganic phosphorus mineral supplement would need to be added to the diet, resulting in less phosphorus excreted and applied to the land. Phytase is now an accepted, economical addition to swine diets. When used, phytase should be added to diets according to the suppliers' directions.

Salt—Sodium (Na) and Chlorine (Cl)

Salt contains both sodium and chlorine, vital elements found in the fluids and soft tissues of the body. It improves the appetite, promotes growth, helps regulate body pH, and is essential for hydrochloric acid formation in the stomach.

Although swine require less salt than other classes of farm animals, it is generally advantageous to supply them with some, particularly if the protein supplement is not derived from animal or marine sources. A poor and depraved appetite, unthrifty condition, and failure to grow mark a lack of salt. Do not provide pure salt to deprived pigs as they may overconsume it to a toxic level.

Swine can tolerate very high levels of salt if adequate water is available. Some studies have fed as much as 10 or 20% salt to gestating sows. If the pigs' water intake is accidentally restricted, it is important to resupply water under controlled intake conditions.

Magnesium (Mg)

Magnesium is a cofactor in many enzyme systems and a constituent of bone. The magnesium requirement for pigs is about 0.04% of the diet. Corn contains 0.12% magnesium and soybean meal contains 0.30%, easily meeting the requirement without additional supplementation. A toxic level of magnesium is not known, but pigs can tolerate 0.3% magnesium.

Potassium (K)

Potassium is the third most abundant mineral in pig's bodies and the most common mineral in muscle tissue. The potassium requirement for pigs varies from 0.30% of the diet for young pigs to 0.17% for finishing pigs. Corn contains 0.33% potassium and soybean meal contains 2.14%, easily meeting the requirement without additional supplementation. High levels of dietary chloride increase the dietary potassium requirement for the pig. Pigs can tolerate

TABLE 6.3 Essential Minerals for Swine[1]

Mineral	Deficiency Conditions	Functions of Mineral	Deficiency Symptoms/Toxicity	Practical Sources	Comments
Major or Macrominerals					
Salt (NaCl)	Deficiencies may exist when the protein supplement is all or chiefly of plant origin, although herbivorous animals require more salt than swine.	Sodium and chlorine are the principal extracellular cation and anion, respectively, in the body. Chlorine is the chief anion in gastric juice. Improves appetite, promotes growth, helps regulate body pH, and is essential for hydrochloric acid formation in the stomach.	**Deficiency symptoms**—Poor and depraved appetite, unthrifty condition, and failure to grow. **Toxicity**—Nervous, weak, staggering, epileptic seizures, paralysis, and death.	Salt in loose form.	In iodine-deficient areas, stabilized iodized salt should be used. When pigs are salt starved, precaution should be taken to prevent over-consumption of salt.
Calcium (Ca)	When the protein supplements are chiefly of plant origin and little forage is used. When swine are raised in confinement without vitamin D added to the diet. When feed intake is restricted during gestation. When there is a poor Ca:P ratio. Source of dietary protein (or phytic acid content) and the level of magnesium affect retention of calcium.	Bone and teeth formation, nerve function, muscle contraction, blood coagulation, cell permeability. Essential for milk production.	**Deficiency symptoms**—Loss of appetite and poor growth, lack of thrift, lameness and stiffness, weakened bone structure, and impaired reproduction. Severe cases may show reduced serum calcium and tetany. Rickets may develop in young pigs, or osteomalacia in older animals. Paralysis of the hind legs. **Toxicity**—An excess level of calcium tends to reduce the performance of pigs and increase the pig's zinc requirement.	Ground limestone, gypsum, or oyster shell flour. Where both Ca and P are needed, use monocalcium phosphate, dicalcium phosphate, tricalcium phosphate, defluorinated phosphate, or bone meal.	Because cereal grains (the major portion of swine diets) are low in Ca, swine are more apt to suffer from Ca deficiencies than from other minerals except salt. Most favorable Ca:P ratio is between 1:1 and 1.5:1. Sow's milk contains a Ca:P ratio of 1.3:1.
Phosphorus (P)	Diets containing only plant ingredients, late gestation and lactation, high-calcium diets, swine in confinement without the vitamin D added to the diet, poor Ca:P ratio. Source of dietary protein (or phytic acid content) and the level of magnesium affect retention of phosphorus.	Bone and teeth formation, a component of phospholipids which are important in lipid transport and metabolism and cell membrane structure. Energy metabolism. A component of RNA and DNA, the vital cellular constituents required for protein synthesis. A constituent of several enzyme systems.	**Deficiency symptoms**—Loss of appetite, poor growth, lameness and stiffness, weakened bone structure, reduced inorganic blood phosphorus, depraved appetite, breeding difficulties, and rickets in young pigs, or osteomalacia in older animals. Paralysis of the hind legs, which is called posterior paralysis. **Toxicity**—An excess of phosphorus tends to reduce performance but is not toxic.	Where both Ca and P are needed, use monocalcium phosphate, dicalcium phosphate, tricalcium phosphate, defluorinated phosphate, or bone meal. Adding the enzyme phytase to diets will increase the amount of available phosphorus.	About 60–75% of the P in cereal grains and their by-products, and in oilseed meals, is bound as phytate and poorly available to the pig. Most favorable Ca:P ratio is between 1:1 and 1.5:1. Sow's milk contains a Ca:P ratio of 1.3:1. Excess levels of P reduce performance of pigs.

114

Element				
Magnesium (Mg)	Some research suggests that the magnesium in natural ingredients is only 50–60% available to the pig. Practical diets are adequate in magnesium.	Essential for normal skeletal development as a constituent of bone, cofactor in many enzyme systems, primarily in the glycolytic system.	**Deficiency symptoms**—Hyperirritability, muscular twitching, reluctance to stand, weak pasterns, loss of equilibrium, and tetany, followed by death. **Toxicity**—The toxicity level of magnesium is not known.	Magnesium oxide, magnesium sulfate, or magnesium carbonate. Dolomitic limestone.	Milk contains adequate magnesium for suckling pigs.
Potassium (K)	Practical diets are adequate in potassium.	Major cation of intracellular fluid, involved in osmotic pressure and acid–base balance. Electrolyte balance and neuromuscular function. Required in enzyme reaction involving phosphorylation of creatinine. Influences carbohydrate metabolism.	**Deficiency symptoms**—Loss of appetite, slow growth, poor hair and skin condition, decreased feed efficiency, inactivity, lack of coordination, and cardiac impairment. **Toxicity**—The toxic level of potassium is not well established. Pigs can tolerate up to 10 times the requirement with adequate water.	Corn contains 0.33% potassium, and other cereals contain 0.42–0.49% potassium.	Potassium is the third most abundant mineral in the body of the pig, exceeded only by calcium and phosphorus. No estimates available for finishing and breeding swine.
Sulfur (S)		For synthesis of sulfur-containing compounds, such as glutathione, taurocholic acid, and chondroitin sulfate.			The addition of inorganic sulfate to low-protein diets has not been beneficial.

Trace or Microminerals

Element				
Chromium (Cr)	No requirements for chromium have been established.	Possible cofactor with insulin.		Chromium picolinate.	200 ppb may enhance litter size.
Cobalt (Co)	If vitamin B_{12} is limited. Practical diets are adequate in cobalt.	An essential component of vitamin B_{12}.	**Deficiency symptoms**—No deficiency symptoms reported in swine. However, supplemental cobalt prevents lesions associated with zinc deficiency. **Toxicity**—A level of 400 ppm of cobalt is toxic to the young pig and may cause loss of appetite, stiff leggedness, humped back, incoordination, muscle tremors, and anemia.	Cobalt chloride, cobalt sulfate, cobalt oxide, or cobalt carbonate. Also, several good commercial minerals containing cobalt are on the market.	No requirements for cobalt have been established.
Copper (Cu)	Suckling pigs kept off soil.	Essential element in a number of enzyme systems and necessary for synthesizing hemoglobin and preventing nutritional anemia. Hemoglobin serves as a carrier of oxygen throughout the body. When fed at 100 to 250 ppm, copper stimulates growth in pigs.	**Deficiency symptoms**—Slow growth, poor hair and skin condition, lameness and stiffness, weakened bone structure, weak and crooked legs, anemia, and cardiac and vascular disorders. **Toxicity**—Depressed hemoglobin levels and jaundice. 500 ppm is considered toxic.	Copper sulfate, copper carbonate, and copper chloride are about equally effective. The copper in copper sulfide, and copper oxide, is poorly available to the pig.	Beyond the suckling period, natural feedstuffs usually contain enough copper. When fed at a level of 100 to 250 ppm, copper will increase rate and efficiency of gains of pigs to breeding age.

TABLE 6.3 Essential Minerals for Swine (*continued*)

Mineral	Deficiency Conditions	Functions of Mineral	Deficiency Symptoms/Toxicity	Practical Sources	Comments
Trace or Microminerals (continued)					
Iodine (I)	Iodine-deficient areas (northwestern U.S. and the Great Lakes region) when iodized salt is not fed. Where feeds come from iodine-deficient areas.	Needed by the thyroid gland for making thyroxin, an iodine-containing hormone that controls the rate of body metabolism or heat production.	**Deficiency symptoms**—Loss of appetite, slow growth, poor hair and skin condition, impaired breeding or gestation, offspring dead or weak at birth, pigs hairless at birth, and/or enlarged thyroid. **Toxicity**—An 800 ppm iodine level in the diet depresses growth, hemoglobin level, and liver iron in growing pigs. 1,500 to 2,000 ppm was not harmful to sows.	Stabilized iodized salt containing 0.007% iodine. Calcium iodate, potassium iodate, and pentacalcium orthoperiodate.	The majority of the iodine in the bodies of swine is present in the thyroid.
Iron (Fe)	Suckling pigs kept off soil. Newborn pigs require 7 to 16 mg of absorbed iron daily for normal growth.	Iron is a component of hemoglobin in red blood cells. Iron also is found in muscle as myoglobin, in serum as transferrin, in the placenta as interoferrin, in milk as lactoferrin, and in the liver as ferritin and hemosiderin. Iron also plays an important role in the body as a constituent of a number of metabolic enzymes.	**Deficiency symptoms**—Loss of appetite, slow growth, poor hair and skin condition, paleness of mucous membranes, high mortality in young pigs, susceptibility to disease, thumps (characterized by labored breathing), and anemia. The number of grams of hemoglobin per 100 ml of blood is a rapid, reliable indicator of the iron status of the pig. A hemoglobin level of less than 10g/100 ml indicates borderline anemia; a level of 7 g or less/100 ml indicates anemia. **Toxicity**—In 3- to 10-day-old pigs, the toxic oral dose of iron from ferrous sulfate is approximately 600 mg/kg body weight.	A single intramuscular injection of 200 mg of iron, in the form of iron dextran, given the first 3 days of life, or oral administration of iron from iron chelates within the first few hours of life, or daily access to freshly harvested sod. Ferrous sulfate, ferric chloride, ferric citrate, ferric choline citrate, and ferric ammonium citrate are effective in preventing iron deficiency anemia. The iron in ferric oxide is largely unavailable.	Pigs are born with about 50 mg of iron. Iron has a detoxifying effect when added to gossypol-containing diets. Add iron from soluble source to free gossypol at a weight ratio of 1:1. Milk is deficient in iron (sow's milk contains an average of 1 mg of iron/liter). Pigs should be encouraged to eat dry feed as soon as old enough. Natural feed ingredients usually supply enough iron to meet postweaning requirements. The iron requirement of young pigs fed milk or purified liquid diets is 23 to 68 mg/lb (50 to 150 mg/kg) of milk solids. The iron requirement of pigs fed a dry, casein-based diet is about 50% higher/unit of dry matter than for those fed a similar diet in liquid form.

	Functions	Deficiency and Toxicity Symptoms	Source	Comments
Manganese (Mn)	Functions as a component of several enzymes involved in carbohydrate, lipid, and protein metabolism. A component in the organic matrix of bone.	**Deficiency symptoms**—Abnormal skeletal growth, increased fat deposition irregular or absent estrous cycles, resorbed fetuses, small and weak pigs at birth, and reduced milk production. **Toxicity**—The toxic level of manganese is not clearly defined. But 500 to 4,000 ppm result in depressed feed intake, reduced growth, and limb stiffness.	Manganous oxide.	Manganese is usually present in adequate amounts in most swine diets, but it may not be adequate for the optimum reproductive performance of sows.
Selenium (Se)	Functions as a part of glutathione peroxidase, an enzyme that enables the tripeptide glutathione to perform its role as a biological antioxidant in the body. The mutual sparing effect of selenium and vitamin E stems from their shared antiperoxidant roles. But high levels of vitamin E do not completely eliminate the level for selenium. When diets consist almost exclusively of ingredients grown on selenium-deficient soils.	**Deficiency symptoms**—Sudden death, impaired reproduction, reduced milk production, and impaired immune response. **Toxicity**—Loss of appetite, loss of hair, fatty infiltration of the liver, degenerative changes in the liver and kidney, edema, and occasional separation of the hoof and skin at the coronary band. 5 ppm may be toxic.	Sodium selenite or sodium selenate.	Environmental stress may increase the incidence and degree of selenium deficiency. Caution: toxic level of selenium is 5 mg/kg.
Zinc (Zn)	Zinc is a component of many metallo-enzymes and the hormone insulin. So, it plays an important role in protein, carbohydrate, and lipid metabolism. High levels of calcium in relation to zinc levels impair zinc utilization and increase the requirements. The zinc requirement is increased with excessive levels of calcium.	**Deficiency symptoms**—Parakeratosis or swine dermatitis, pigs have a mangy appearance, reduced appetite, unthriftiness, poor growth rate, and diarrhea, and there may be vomiting. It affects swine of all ages. Zinc deficiency results in gilts producing fewer and smaller pigs; in boars with retarded testicular development; and in young pigs with retarded thymic development. **Toxicity**—Growth depression, arthritis, hemorrhage in axillary spaces, gastritis, and enteritis. High dietary calcium reduces the severity of zinc toxicity.	Zinc carbonate or zinc sulfate.	It has been shown that parakeratosis is caused by zinc and calcium forming an unavailable complex. Research indicates 3,000 ppm zinc as zinc oxide improves growth and reduces scours in postweaning pigs.

[1]Requirements per day may be found in Tables 7.4 and 7.9; recommended contents of diets may be found in Tables 7.2 and 7.9; and recommended allowances may be found in Tables 7.14 and 7.22. The distinction between requirements and recommended allowances is as follows: with requirements, no margins of safety are included intentionally, whereas with recommended allowances, margins of safety are provided in order to compensate for variations in feed composition, environment, and possible losses during storage or processing.

Source: Courtesy of Pearson Education.

up to 10 times the dietary requirement of potassium if adequate water is available.

Sulfur (S)

Sulfur is an essential element. However, the sulfur-containing amino acids (cystine and methionine) appear adequate to meet the pig's need for synthesis of sulfur-containing compounds. The addition of inorganic sulfate to low protein swine diets has not been beneficial.

Trace or Microminerals

Minerals that are required in small amounts are known as trace or microminerals. These include copper, iodine, iron, manganese, selenium, and zinc. Another trace mineral, cobalt, is not required in swine diets as long as adequate vitamin B_{12} is present. Others, such as chromium, are usually adequate for normal growth.

Chromium (Cr)

Chromium is involved in the metabolism of several nutrients but its exact function is not known. It may work as a cofactor with insulin. No estimate of the chromium requirement has been made. Added chromium (200 ppb) has been shown in increase litter size following a 6-month loading period. Performance and muscle characteristics of growing pigs have not been consistently enhanced with added chromium.

Cobalt (Co)

Cobalt is a component of vitamin B_{12}. If vitamin B_{12} is added, there is no evidence that pigs have a requirement for cobalt. The intestinal microflora of the pig are capable of synthesizing vitamin B_{12} provided sufficient cobalt is present. But only a minimum level of dietary cobalt is necessary for this process. Intestinal synthesis is of greater importance if preformed vitamin B_{12} is limiting. Cobalt can partially substitute for zinc and will prevent lesions associated with zinc deficiency.

A level of 400 ppm of cobalt is toxic to the young pig. Selenium and vitamin E provide some protection against toxicity from excessive levels of dietary cobalt.

Copper (Cu)

The pig requires copper for the synthesis of hemoglobin and for the synthesis and activation of several oxidative enzymes necessary for normal metabolism. A deficiency of copper leads to poor iron mobilization. A level of 6 ppm in the diet is adequate for baby pigs.

When fed at 100 to 250 ppm, copper stimulates growth in pigs, apparently as a result of the antibacterial action of these high levels of copper. A level of 500 ppm is considered toxic.

Iodine (I)

The dietary iodine requirement is not well established. Moreover, it is increased by goitrogens in certain feedstuffs, including rapeseed, linseed, lentils, peanuts, and soybeans. A level of 0.14 ppm of iodine in a corn-soybean meal diet will prevent goiters in growing pigs, and a level of 0.35 ppm of added iodine will prevent iodine deficiency in sows.

The incorporation of iodized salt (0.007% iodine) at a level of 0.2% of the diet provides sufficient iodine (0.14% ppm) to meet the needs of growing pigs fed grain-soybean meal diets. Calcium iodate, potassium iodide, and ethylenediamine dihydroiodide (EDDI) are considered to be 100% available and stable in salt mixtures.

A dietary level of 800 ppm depresses performance in growing pigs. Levels of 1,500 to 2,500 ppm were not harmful to sows.

Iron (Fe)

Iron is necessary for the formation of hemoglobin in the red blood cells and the prevention of nutritional anemia. Hemoglobin serves as a carrier of oxygen throughout the body.

As the prenatal pig develops, a supply of iron is stored in its body. The amount stored varies greatly between pigs of the same litter. This stored iron is adequate to keep the pig growing at its maximum for no more than 10 to 14 days after birth. Sow's milk is very low in iron and research has not discovered a method of increasing its iron content. Thus, if suckling pigs have no access to soil or feed, serious losses from anemia are likely. Once a pig begins to consume natural feedstuffs, the danger of anemia is practically eliminated because all swine diets are fortified with sufficient amounts of iron to meet the pig's requirement.

The most commonly used iron sources to prevent anemia in newborn pigs are injectable and oral products, with the injectable iron preferred. An intramuscular injection of 200 mg of iron dextran given at 1 to 3 days of age will prevent the anemia problem. Oral iron pastes or fresh sod are also available sources of iron; however, they are not as preferred as an injection because daily intake, biological availability, and absorption are difficult to ascertain.

The need for a second injection depends on how much iron was given in the first injection and how much iron is available to the baby pigs during the lactation period. Once baby pigs begin consuming creep feed, sow feed, or fresh sod, the need for additional injections or oral paste is no longer needed.

The iron injection should be given in the neck muscle. Because the injections may cause permanent staining of the meat, injections in high-value cuts, such as the ham, are discouraged.

Blood hemoglobin is a rapid, reliable indicator of the iron status. Levels of 10 or more g/100 ml (dL) indicate an adequate iron status. Anemic pigs, usually with less than 10 g of hemoglobin/100 ml, are very pale, lose their appetites and become weak and inactive. In more advanced stages of the deficiency, the pig's breathing becomes labored, a condition sometimes called *thumps*. In this condition they are more susceptible to other diseases and parasites. Death may occur in severe cases.

A toxic oral dose in 3- to 10-day-old pigs is about 600 mg/kg of body weight. Injectable iron may be toxic to pigs from vitamin E–deficient sows. The postweaning dietary iron requirement is about 100 ppm for nursing pigs (thus the need for injectable or oral iron because sow milk is very deficient) and decreases to 40 ppm for finishing pigs.

Manganese (Mn)

Manganese functions with many enzymes in soft tissue metabolism and also in bone development. Deficiency symptoms are lameness, weakened bone structure, irregular estrus, offspring born dead or weak, and increased backfat.

The manganese requirement varies from 4 ppm in young pigs down to 2 ppm as the pig ma-

tures. Although dietary manganese is usually present in adequate amounts (corn contains 7 ppm and soybean meal more than 30 ppm) without supplementation, it may not be adequate for the optimum reproductive performance of sows, estimated to be 20 ppm. Reduced performance has been observed in pigs fed 500 to 4,000 ppm manganese.

Selenium (Se)

Selenium functions as a part of glutathione peroxidase, an enzyme that enables the tripeptide glutathione to perform its role as a biological antioxidant in the body, similar to some of the vitamin E functions. This explains why selenium and vitamin E have mutual sparing effects. However, high levels of vitamin E do not completely eliminate the need for selenium.

Currently, the U.S. Food and Drug Administration (FDA) allows the addition of up to 0.3 ppm of selenium in the diets of all pigs. Certain soils in the United States and Canada are low in selenium, particularly around the Great Lakes. Conversely, soils in South Dakota are often high in selenium. Generally, 0.1 ppm added selenium is adequate for all except weanling pigs. In some cases, selenium in the diet at a level of 5 ppm will produce toxic symptoms.

Zinc (Zn)

The requirement for zinc in swine diets is very low, but when high levels of calcium are fed, zinc utilization is impaired and the requirements are increased. A zinc deficiency results in a mangelike skin condition called parakeratosis (Figure 6.11). Other symptoms are

Figure 6.11 Zinc deficiency. Left pig received 17 ppm of zinc and gained only 3 lb in 74 days. Note severe dermatosis ("mangy look"), or parakeratosis. Right pig received the same diet as the pig on the left except that the diet contained 67 ppm of zinc. This pig gained 111 lb in 74 days. *(Courtesy, Purdue University, West Lafayette, IN)*

poor growth, inefficient feed conversion, gilts producing fewer and smaller pigs, boars with retarded testicular development, and young pigs with retarded thymus gland development. Boars have a higher zinc requirement than gilts, and gilts have a higher requirement than barrows.

Toxicity may occur with pigs fed 2,000 to 4,000 ppm zinc. However, recent research has demonstrated a pharmacological effect of feeding 3,000 ppm zinc as zinc oxide for up to 5 weeks resulting in increased postweaning growth and decreased scouring.

Chelated Minerals

A chelated mineral is bound to an organic compound such as an amino acid that enhances absorption. Recent research has shown that chelated minerals are 0 to 15% more available. However, their cost may be 2 to 3 times greater than those of nonchelated minerals.

Feedstuffs as Mineral Sources

The most satisfactory source of minerals for hogs is in the feed consumed. It is important to know that the minerals in the diet are of the right kind and sufficient in amount. Certain general characteristics of feeds in regard to calcium and phosphorus (the two predominating mineral elements of the body) are worth noting:

1. Cereal grains and their by-products and protein supplements of plant origin are low in calcium but fairly high in phosphorus content. However, as mentioned earlier, the phosphorus in plants is not very available to swine unless the phytase enzyme is added.
2. Protein supplements of animal origin (skim milk, buttermilk, tankage, meat scraps, fish meal), legume forage (pasturage and hay), and canola (rapeseed) are all rich in calcium.
3. Most protein-rich supplements are high in phosphorus.
4. The availability of trace minerals in plant sources is variable. For that reason, many of the trace minerals are added to swine diets even if the analyzed levels often appear adequate. Also, because trace minerals are economical additions to swine diets, it is often safer to add them rather than risk deficiencies.

Electrolytes

Electrolytes (minerals) are essential for maintaining water balance in pigs. The major elements involved in electrolyte balance are sodium, chloride, potassium, magnesium, and calcium, with sodium, chloride, and potassium predominating. It is not recommended that electrolytes be included in swine diets at levels exceeding those found in Chapter 7 tables on mineral requirements. This is true even in times of stress, such as those associated with weaning and with feeder pig sales and transfers. Electrolytes may be included in the water in times of stress when feed intake is minimal. Electrolyte balance is particularly important for starting pigs because they are more susceptible to diarrhea, which can cause severe dehydration.

Mineral Sources and Bioavailability

The major sources of the minerals commonly added to swine diets are found in Chapter 22. Information on the bioavailability of minerals from several sources is also provided. Decisions on which source of mineral to use should be based primarily on price per unit of available element.

VITAMINS

Vitamins are complex organic compounds required in minute amounts and are essential for health and normal body functions. Like amino acids, each vitamin has a specific function to perform. Vitamins are classified into two groups—fat soluble and water soluble. The body can store reserves of the fat-soluble vitamins for a considerable period of time. Stores of the water-soluble vitamins are depleted more rapidly.

Some vitamins are present in feed ingredients in adequate amounts. Others are produced in the pig's body in adequate amounts. However, for optimal performance, several vitamins need to be added to swine diets. Because of the prevalence of indoor feeding, swine are more likely to suffer from vitamin deficiencies than livestock raised outdoors.

Table 6.4 provides in summary form the following pertinent information relative to each vitamin listed: (1) conditions usually prevailing where deficiencies are reported, (2) functions, (3) deficiency symptoms/toxicity, and (4) practical sources. Further elucidation of vitamins is contained in the narrative.

Fat-Soluble Vitamins

The primary fat-soluble vitamins of practical importance for swine are vitamins A, D, and E. Vitamin K may be of concern under some circumstances, but is commonly added to swine diets.

Vitamin A

The vitamin A needs of swine can be met by either vitamin A or carotene. Vitamin A as such does not occur in plants. However, green plants and yellow corn contain a pigment called *carotene* that can be converted to vitamin A by the animal body. The combination of vitamin A and carotene present in the diet is referred to as its vitamin A activity.

Carotene is easily destroyed by the ultraviolet rays of the sun and by heat. The carotene content of corn and legume hay usually deteriorates quite rapidly in storage. Therefore, a synthetic concentrate is a more practical and reliable source of vitamin A than natural sources. Most swine feeds are fortified with a stabilized "coated" form of vitamin A ester, which is active over a considerable period of time.

Vitamin A is essential for vision, reproduction, growth, and the maintenance of differentiated epithelia and mucous secretions of swine. Swine are able to store vitamin A in the liver, and to draw from this storage during periods of low intake.

Vitamin A deficiency signs in growing pigs are uncoordinated movement, loss of control of the hind legs, weakness of the back, and night blindness. Sows may fail to come into estrus, may reabsorb their fetuses, or may have young born dead with various deformities and defects. Vitamin A is also needed for normal vision and growth of new cells that line the respiratory, digestive, and reproductive tracts.

Some evidence exists that injected vitamin A may increase litter size by 0.5 to 0.8 pigs per litter. More research is needed to determine proper dose and time of injection.

Vitamin D

Vitamin D is sometimes referred to as the "sunshine" vitamin because the action of sunlight on a compound in the skin will produce it. As long as hogs are exposed to the sun, there is no danger of a deficiency.

Living plants do not contain vitamin D, but plants that mature or are cut and cured in the sun contain some vitamin D as a result of radiation by sunlight. Pigs can utilize equally well either vitamin D_2 (ergocalciferol from plant products) or vitamin D_3 (cholecalciferol from animal products). Irradiated yeast is a good source of vitamin D_2.

Vitamin D is needed for the efficient assimilation of calcium and phosphorus; hence, it is required for the growth of strong bones. A lack of vitamin D will result in stiffness and lameness, broken or deformed bones, enlargement of joints, and general unthriftiness; known as rickets in young pigs and osteomalacia in mature hogs.

It is noteworthy that the vitamin D requirement is less when a proper balance of calcium and phosphorus exists in the diet.

Vitamin E

Vitamin E is a biological antioxidant that protects unsaturated fat against oxidation. Eight naturally occurring compounds called *tocopherols* have vitamin E activity, the most active of which is alpha-tocopherol. Also, alpha-tocopherol is very stable during storage or in mixed feeds. Because cell membranes in the animal body contain unsaturated fat, a vitamin E deficiency may result in oxidative damage to the cell. This is manifested in the pig by liver necrosis, pale muscle, mulberry heart, edema, and sudden death.

For many years, the primary source of vitamin E in feed was the natural form of vitamin E (d-α-tocopherol) found in green plants and seeds. However, oxidation rapidly destroys natural vitamin E. For example, vitamin E losses of 50 to 70% can occur in alfalfa stored at 90°F for 12 weeks, and losses of 5 to 30% can occur during dehydration of alfalfa. Only 20% of the natural vitamin E remains after storage with trace mineral premixes whereas stabilized vitamin E remains stabile for at least 10 weeks. Also, storage of high-moisture grain or its treatment with organic acids greatly reduces the vitamin E content. Therefore, predicting the amount of vitamin E activity in feed ingredients is difficult. One international unit (IU) of vitamin E is the biological equivalent of 0.671 mg of d-α-tocopherol.

The trace element selenium also functions with vitamin E in protecting the body against oxidative damage. The need for vitamin E is more acute when swine feeds are low in selenium. Thus, in areas where feed ingredients are low in selenium and where swine are raised without access to forages, supplemental vitamin E or selenium, or both, are important. Most nutritionists now recommend added vitamin E because of the unpredictability of natural sources in feedstuffs.

Vitamin E toxicity has not been reported in swine. Levels as high as 45 IU/lb (100 IU/kg) of diet have been fed to growing pigs without toxic effects.

Vitamin K

Vitamin K exists in three forms: phylloquinone (K_1), menaquinone (K_2), and menadione (K_3). Menadione is the synthetic form of vitamin K, which has the same cyclic structure as vitamins K_1 and K_2. Vitamin K_1 occurs naturally in green plants. Vitamin K_2 is present in microorganisms and is formed by intestinal bacteria.

Vitamin K is one of the essential factors necessary for proper blood clotting. Generally, sufficient amounts of vitamin K_2 are synthesized by bacteria in the digestive tract to meet the needs of swine, with the vitamin K_2 absorbed directly from the gut or obtained by coprophagy (Figure 6.12). However, this synthesis may be inadequate in situations where high antibiotic levels are used, where clotting inhibitors (dicoumarol) may be present from molds in the feed, or where there is excess calcium.

TABLE 6.4 Essential Vitamins for Swine[1]

Vitamin	Deficiency Conditions	Functions of Vitamin	Deficiency Symptoms/Toxicity	Practical Sources	Comments
Fat-Soluble Vitamins					
A	Absence of green forages or pasture—especially under confined conditions. Where the diet consists chiefly of white corn, milo, barley, wheat, oats, or rye, or by-products of these grains, or yellow corn that has been stored more than one year.	Essential for normal maintenance and functioning of the epithelial tissues, particularly of the eye and the respiratory, digestive, reproductive, nervous, and urinary systems.	**Deficiency symptoms**—Night and day blindness, very irritable, poor appetite and slow growth, lameness, incoordination of movement, loss of control of the hind legs, and back weakness. Low resistance to respiratory infections. Sows may fail to come in heat, may reabsorb their fetuses, and may have young born dead with various deformities and defects. **Toxicity**—A roughened hair coat, scaly skin, hyperirritability and sensitivity to touch, bleeding from the cracks which appear in the skin above the hooves, blood in the urine and feces, loss of control of the legs accompanied by inability to rise, periodic tremors, and death.	Either vitamin A or various provitamins.	Based upon liver storage, the biopotency of 1 mg of carotene in corn fed to weanling pigs is 261 IU of vitamin A. Meals from artificially dehydrated forages are much higher in carotene than sun-cured products. Taken together, liver storage, levels of plasma vitamin A, and pressure of cerebrospinal fluid give reliable estimates of the vitamin A status of the pig.
D	Limited sunlight and/or limited quantities of sun-cured hay in drylot diets.	Aids in assimilation and utilization of calcium and phosphorus, and necessary in the normal bone development of animals—including those of the fetus.	**Deficiency symptoms**—Rickets in young pigs, or osteomalacia in mature hogs. Both conditions result in large joints and weak bones. In severe vitamin D deficiency, pigs may exhibit signs of calcium and magnesium deficiency, including tetany. **Toxicity**—Reduced feed intake, growth rate, and death. Vitamin D_3 is more toxic than D_2 in swine.	Vitamin D_2 (ergocalciferol) and vitamin D_3 (cholecalciferol) are similar in biological activity for swine. The action of ultraviolet light on the ergosterol present in plants forms ergocalciferol; and the photochemical conversion of 7-dehydro-cholesterol in the skin of animals forms cholecalciferol. Irradiated yeast. Exposure to sunlight. Sun-cured hay (10% alfalfa in the total diet will normally supply sufficient vitamin D).	Grains, grain by-products, and high-protein feedstuffs are practically devoid of vitamin D. Unless swine are exposed daily to the ultraviolet rays of the sun, the diet should be fortified with vitamin D. The vitamin D requirement is less when a proper balance of calcium and phosphorus exists in the diet. One IU vitamin D is defined as the biological activity of 0.025 mcg of cholecalciferol.

E	Diets containing excessive amounts of highly unsaturated fatty acids or oxidized fats. Feeds low in selenium, especially where swine are raised indoors without access to forages.	Antioxidant. Muscle structure. Reproduction. High levels of dietary vitamin E may increase the immune response.	**Deficiency symptoms**—Loss of appetite and slow growth. Increased embryonic mortality and muscular incoordination in suckling pigs from sows fed vitamin E–deficient diets during gestation and lactation. A wide variety of pathological conditions. **Toxicity**—Vitamin E toxicity in swine has not been demonstrated.	Alpha-tocopherol. Predicting the amount of vitamin E activity in feed is difficult.	The eight naturally occurring tocopherols differ in biological activities, with α-tocopherol being the most active. One IU of vitamin E is equivalent in biopotency to 1 mg dl-α-tocopherol acetate. Many dietary factors affect the vitamin E requirement, including the selenium level, unsaturated fatty acids, sulfur amino acids, retinol, copper, iron and synthetic antioxidants.
K	Moldy feed. High antibiotic levels, which may make for inadequate intestinal synthesis of vitamin K.	Essential for prothrombin formation and blood clotting.	**Deficiency symptoms**—Bleeding condition in young pigs, which responds to injection or oral administration of vitamin K. Slow growth and hyperirritability. **Toxicity**—Concentrations of 50 mg of menadione pyrimidinol bisulfite (MPB)/lb (110 mg of MPB/kg) of diet were not toxic to weanling pigs.	The following water-soluble forms of menadione are commonly used to supplement swine diets: menadione sodium bisulfite complex (MSB), menadione sodium bisulfate complex (MSBC), and menadione pyrimidinol bisulfite (MPB).	Vitamin K exists in three forms: phylloquinone (K_1), menaquinone (K_2), and menadione (K_3). Under practical conditions, the vitamin K requirement is met by vitamin K in feedstuffs and by intestinal synthesis.

Water-Soluble Vitamins

Biotin	When pigs are fed dried, raw egg white or given sulfa drugs. Marginal deficiency may exist when hogs are fed cereal grain diets, housed in individual stalls or on slotted floors (which lessens coprophagy), and/or have no access to green forage.	Biotin is important metabolically as a cofactor for several enzymes. It may improve sow and litter performance when added to gestation-lactation diets.	**Deficiency symptoms**—Excessive hair loss, skin ulcerations and dermatitis, exudate around the eyes, cracking of the hooves, and cracking and bleeding of the footpads.	Biotin.	The protein avidin in raw egg white makes biotin unavailable to pigs. Heat treatment inactivates avidin and makes egg whites safe for feeding to pigs.
Choline	Baby pigs fed a synthetic milk diet containing not more than 0.8% methionine. Practical diets are adequate in choline.	Involved in nerve impulses, a component of phospholipids, donor of methyl groups, and involved in the mobilization and oxidation of fatty acids in the liver.	**Deficiency symptoms**—Unthriftiness, lack of coordination, fatty infiltration of the liver, poor reproduction, poor lactation, and decreased survival of the young. **Toxicity**—No signs of choline toxicity have been reported in swine.	Choline chlorides or choline dihydrogen.	Choline does not qualify as a true vitamin, because it is required at far greater levels than true vitamins and is not known to participate in any enzyme system. Studies have shown that more live pigs are born and weaned when sows receive supplemental choline throughout gestation.

(continued)

TABLE 6.4   Essential Vitamins for Swine (*continued*)

Vitamin	Deficiency Conditions	Functions of Vitamin	Deficiency Symptoms/Toxicity	Practical Sources	Comments
Water-Soluble Vitamins (continued)					
Folacin (folic acid)	Practical diets plus intestinal synthesis are adequate.	Metabolic reactions involving incorporation of single carbon units into larger molecules. Folacin is involved in the conversion of serine to glycine and homocystine to methionine.	**Deficiency symptoms**—Poor growth, fading hair color, and anemia. Folic acid may increase the number of pigs born alive.	Synthetic folacin.	Folacin includes a group of compounds with folic acid activity.
Niacin (nicotinic acid, nicotinamide)		Niacin is a component of the coenzymes that are essential for the metabolism of carbohydrates, proteins, and lipids.	**Deficiency symptoms**—Loss of appetite and decreased gain, followed by diarrhea, occasional vomiting, dermatitis, and loss of hair.	Nicotinamide. Nicotinic acid.	Niacin occurs in corn, wheat, and milo in bound form; hence, it may be unavailable to the pig. Also, the dietary tryptophan level affects the niacin requirement because of the conversion of tryptophan to niacin.
Pantothenic Acid (B_3)	Long periods of inadequate pantothenic acid intake.	As a component of coenzyme A, pantothenic acid is important in the catabolism and synthesis of 2-carbon units evolved during carbohydrate and fat metabolism.	**Deficiency symptoms**—A goose-stepping gait, loss of appetite, poor growth, diarrhea, loss of hair, reduced fertility, and breeding failure.	Calcium pantothenate (only the D isomer has vitamin activity). Dried milk products, condensed fish solubles, and alfalfa meal.	Widely distributed and occurs in practically all feedstuffs. However, the quantity present may not always be sufficient to meet the needs of the pig.
Riboflavin (B_2)		A component of 2 coenzymes. It is important in the metabolism of proteins, fats, and carbohydrates.	**Deficiency symptoms**—Loss of appetite, poor growth, rough hair coat, diarrhea, cataracts, vomiting, reproductive failure in the sow, pigs dead or weak at birth, and crooked legs and incoordination.	Synthetic riboflavin. Yeast. Milk and milk products. Meat scraps and fish meal.	Riboflavin is apt to be lacking in swine diets that do not contain animal product sources.
Thiamin (B_1)	Thiamin is heat labile. So, excess heat can reduce the thiamin content of the diet ingredient. Practical diets are adequate in thiamin.	As a coenzyme in energy and protein metabolism. Promotes appetite and growth, required for normal carbohydrate metabolism, and aids reproduction.	**Deficiency symptoms**—Loss of appetite and poor growth, diarrhea, dead or weak offspring, slow pulse, low body temperature, and flabby heart.	Thiamin hydrochloride. Thiamin mononitrate. Rice polish. Wheat germ meal. Yeast. Oilseed meals. Distillers' solubles.	

124

Vitamin		Function		Source	Comments
Vitamin B$_6$ (pyridoxine pyridoxal, pyridoxamine)	Practical diets are adequate in vitamin B$_6$.	An important cofactor for many amino acid enzyme systems. Vitamin B$_6$ also plays a crucial role in central nervous system function.	**Deficiency symptoms**—Loss of appetite and poor growth, unsteady gait, anemia, exudate around the eyes, and epilepticlike fits (convulsions).	Pyridoxine hydrochloride. Rich supplemental sources include rice polish, wheat germ, and yeast.	The vitamin B$_6$ content of normal feed is usually sufficient.
Vitamin B$_{12}$ (cobalamine)	Pigs fed ingredients of plant origin and housed on slotted floors.	Vitamin B$_{12}$ contains the trace element cobalt in its molecule. As a coenzyme, B$_{12}$ is involved in the synthesis of methyl groups derived from formate, glycine, or serene, and their transfer to homocystine to reform methionine. It is also important in the methylation of uracil to form thymine, which is converted to thymidine and used for the synthesis of DNA.	**Deficiency symptoms**—Loss of appetite, reduced weight gain, rough skin and hair coat, irritability, hypersensitivity, hind leg incoordination, and anemia.	Synthetic B$_{12}$, which is produced commercially by microbial fermentation. Protein supplements of animal origin. Fermentation products.	Vitamin B$_{12}$ is apt to be lacking in swine diets. Synthesis of vitamin B$_{12}$ by intestinal flora may supplement dietary sources. B$_{12}$ contains the trace element cobalt; hence, the synthesis of B$_{12}$ in the intestines is dependent on the presence of cobalt in the feed. This may be the major, if not the only, function of cobalt as an essential nutrient.
Vitamin C (ascorbic acid, dehydroascorbic acid)	Practical diets plus intestinal synthesis are adequate.	Vitamin C is a water-soluble antioxidant that is involved in the formation and maintenance of collagen, absorption and movement of iron, metabolism of fats and lipids, cholesterol control, sound teeth and bones, strong capillary walls and healthy blood vessels, metabolism of folic acid, and as a general antioxidant.	**Deficiency symptoms**—No specific symptoms noted when there is a deficiency.	Vitamin C (ascorbic acid).	Normally, pigs are able to synthesize vitamin C in amounts sufficient to meet their requirements. However, there is limited evidence that dietary ascorbic acid is beneficial under some conditions.

[1] Requirements per day may be found in Tables 7.4 and 7.9; recommended contents of diets may be found in Tables 7.2 and 7.9; and recommended allowances may be found in Tables 7.14 and 7.22. The distinction between requirements and recommended allowances is as follows: With requirements, no margins of safety are included intentionally, whereas with recommended allowances, margins of safety are provided in order to compensate for variations in feed composition, environment, and possible losses during storage or processing.

Source: Courtesy of Pearson Education.

Figure 6.12 Modern swine pens quickly separate the manure from the pigs, preventing coprophagy (eating feces). The bacterial growth in the feces is an adequate source of several of the vitamins. However, now we must rely on adequately supplementing the diets with essential vitamins. *(Courtesy, Palmer Holden, Iowa State University, Ames, IA)*

The deficiency symptoms of vitamin K are slow growth, hemorrhage, prolonged blood-clotting time, and hyperirritability. It appears that pigs can tolerate at least 1,000 times the recommended level of menadione compounds without toxicity problems.

Water-Soluble Vitamins

Niacin, pantothenic acid, riboflavin, and vitamin B_{12} are the water-soluble vitamins most likely to be deficient in swine diet. However, occasionally the other water-soluble vitamins are deficient. Therefore biotin, choline, folacin, thiamin, vitamin B_6, and vitamin C are also discussed in the sections that follow.

Biotin

Biotin is important metabolically as a cofactor for several enzymes. Biotin is present in adequate amounts in most common feedstuffs, but its bioavailability varies greatly among ingredients. Biotin deficiencies have been associated with foot lesions and toe cracks in sows although biotin additions have not always improved this condition.

In general, biotin supplementation has not improved the performance of baby pigs or of growing-finishing hogs fed a variety of feedstuffs. Results of biotin added to sow diets has not been consistent, sometimes improving reproductive performance and hoof hardness and other times being ineffective.

Choline

Choline is included with the vitamins, but it does not qualify as a true vitamin because it is required at far greater levels than true vitamins and is not known to participate in any enzyme system.

Choline functions as a "methyl donor" in metabolism and can lower the requirement of methionine to the extent that methionine is used for this function. It is also a constituent of some important phospholipids in the body, and it is involved in the mobilization and oxidation of the fatty acids in the liver. Although choline is probably present in adequate amounts in most practical swine diets, studies have shown that adding 400 mg/lb (880 mg/kg) or more during gestation may result in up to one more live pig born per litter. Also, because proteins supply most of the choline in swine diets, a reduction in dietary protein when synthetic lysine is incorporated in the diet will reduce choline intake.

Choline-deficiency symptoms include lowered reproduction, pigs born weak, and lack of coordination. At one time spraddle legs in baby pigs were attributed to choline deficiency. Although the cause of spraddle legs is not fully understood, many factors may contribute to the condition, including genetics, management, slick floors, mycotoxins, or a virus.

Folacin (Folic Acid/Folate)

Folacin includes a group of compounds with folic acid activity. Folic acid participates in many enzymatic reactions that appear to be essential in assuring embryo survival.

A deficiency of folacin causes a disturbance in the metabolism of single-carbon compounds, including the synthesis of methyl groups, serine, purines, and thymine. Folacin is involved in the metabolic conversion of amino acids, of serine to glycine and of homocysteine to methionine. A deficiency in pigs leads to slow weight gain, fading hair color, and anemia. The folacin in feedstuffs commonly fed to swine, along with bacterial synthesis within the intestinal tract, usually adequately meets the requirements for all classes of swine.

Early studies at Kansas State University indicated supplementation of the gestation diet with folic acid increased the number of pigs born alive by approximately one pig per litter. Other studies have not been able to repeat this, and further experiments are needed to determine the value of folic acid supplementation.

Niacin (Nicotinic Acid, Nicotinamide)

Niacin plays an important role in body metabolism as a constituent of two coenzymes, nicotinamide-adenine dinucleotide (NAD) and nicotinamide-adenine dinucleotide phosphate (NADP). These coenzymes in the pig are essential for the metabolism of carbohydrates, proteins, and lipids. A niacin deficiency is characterized by diarrhea, rough skin, and retarded growth.

The niacin in cereal grains is in a bound form, and the niacin of corn may be almost completely unavailable to swine. It should be assumed that all niacin in cereal grains and their by-products is unavailable. The protein source and tryptophan content of the diet also affect the niacin requirement because tryptophan can be converted to niacin.

Pantothenic Acid (B₃)

Pantothenic acid is a constituent of coenzyme A and plays a key role in energy metabolism. An addition of pantothenic acid to swine diets is routinely recommended.

The biological availability of pantothenic acid is high from barley, wheat, and soybean meal, but low from corn and grain sorghum. Dried milk products, condensed fish solubles, and alfalfa meal are good natural sources of pantothenic acid. It is also available in synthetic form as calcium pantothenate.

A lack of this vitamin may result in poor growth, diarrhea, loss of hair, and a high-stepping gait of the hind legs—often called *goose stepping*.

Riboflavin (B₂)

Riboflavin is sometimes referred to as vitamin B₂. It functions as a constituent of two coenzymes, flavin mononucleotide (FMN) and flavin adenine dinucleotide (FAD). Riboflavin is important in the metabolism of proteins, fats, and carbohydrates. An addition of synthetic riboflavin to swine diets is routinely recommended. Milk products and other animal proteins, alfalfa meal, and distillers' solubles are good natural sources of riboflavin.

In growing swine, a deficiency may cause loss of appetite, stiffness, dermatitis, and eye problems. Poor conception and reproduction have been noted in gilts fed riboflavin-deficient rations. Pigs may be born prematurely, dead, or too weak to survive.

Thiamin (B₁)

Thiamin is essential for carbohydrate and protein metabolism. The coenzyme thiamin pyrophosphate is essential for the oxidative decarboxylation of alpha-keto acids.

Pigs must have a dietary source of thiamin because, unlike ruminants, they cannot synthesize sufficient amounts of it in the digestive tract. Normally, the thiamin content of feeds is sufficient to meet the needs of swine because cereal grains, which are major swine feeds, are good sources. However, deficiencies may result from (1) heating feed ingredients excessively in processing because thiamin is heat labile or (2) feeding unprocessed fish or fish scraps of certain types of fish that contain the antithiamin factor known as *thiaminase* (Figure 6.13).

Vitamin B₆ (Pyridoxines)

Vitamin B₆ occurs in feedstuffs as pyridoxine, pyridoxal, pyridoxamine, and pyridoxal phosphate. Pyridoxal phosphate is an important cofactor for many amino acid enzyme systems, and vitamin B₆ plays a key role in central nervous system functions.

Supplementation of grain—soybean meal rations with vitamin B₆—is generally not needed because the concentration and availability of vitamin B₆ in the feed ingredients will meet the pig's requirement.

Figure 6.13 Thiamin deficiency in littermate pigs. Left pig received the equivalent of 2 mg thiamin/100 lb live weight. Right pig received no thiamin. Otherwise, their diets were the same. *(Courtesy, USDA, Washington, DC)*

A deficiency of vitamin B_6 will reduce appetite and growth rate. Advanced deficiency will result in exudate around the eyes, convulsions, poor coordination, coma, and death.

Vitamin B_{12} (Cobalamin)

Vitamin B_{12} was discovered in 1948 and originally was known as the *animal protein factor* because of its association with ingredients of animal origin. Vitamin B_{12} stimulates the appetite, increases rate of growth, improves feed efficiency, and is necessary for normal reproduction.

Pigs require B_{12} but responses to supplementation have been variable, primarily because of the synthesis of vitamin B_{12} by microorganisms in the environment and within the intestinal tract, along with the pig's inclination toward coprophagy (feces eating). The addition of vitamin B_{12} to swine diets, however, is routinely recommended.

Vitamin C (Ascorbic Acid)

Vitamin C is an antioxidant involved in a variety of metabolic processes. It is also essential for hydroxylation of proline and lysine, which are integral constituents of collagen. Collagen is essential for growth of cartilage and bone. Vitamin C enhances the formation of intercellular material, the formation of bone matrix, and the formation of tooth dentin.

A dietary source of vitamin C is essential for primates and guinea pigs, but domestic swine can synthesize this vitamin. Even under the conditions under which supplemental vitamin C may be beneficial, supplementary C has not enhanced performance. Therefore, no recommendation for vitamin C is given for the pig.

Unidentified Growth Factors

Unidentified factors at one time contributed to the performance of swine. Sources of these factors often were distillers' dried solubles, fish solubles, dried whey, and other by-products. As researchers discovered more vitamins and supplemented diets with them, the value of unidentified growth factors gradually disappeared.

Vitamin Sources and Bioavailability

Green leafy plants, grasses, and alfalfa are excellent sources of vitamins for swine. However, with inside rearing very little leafy plant material is available. Additionally, fewer ingredients are being used in diet formulations. So, when formulating swine diets, it is recommended that all vitamin and mineral levels be "added" levels. Further, it is recommended that synthetic vitamins be added.

The major sources of vitamins for swine and their bioavailability are listed in Table 6.5. Although vitamins are present in grains and protein supplements, it is safer to rely on vitamins supplied by the added vitamin sources because those in grains and protein sources may be lost during drying, storage, and processing. *Note:* NRC vitamin requirements are total requirements, including the vitamins present in the feedstuffs (except for niacin). Nutrient allowances for vitamins are levels that should be *added* to the diet, generally discounting the presence of naturally occurring vitamins in the feedstuffs.

Water

Water is so common that it is seldom thought of as a nutrient, but life on earth would not be possible without water. An animal can live longer without feed than without water. However, it is the largest single part of nearly all living things. Water accounts for as much as 80% of the body weight of a pig at birth and declines to approximately 50% in a finished market animal.

Water performs many tasks in the body. It makes up most of the blood that carries nutrients to the cells and carries waste products away from the cells; it is necessary in most of the body's chemical reactions; it is the body's built-in cooling system—it regulates the temperature, and it serves as a lubricant.

The water requirement of hogs is related to feed intake and body weight. Under normal conditions, a pig will consume 2 to 3 units of water for every unit of dry feed. The estimated water consumption of various classes of swine is provided in Table 6.6.

The need for water is increased by diarrhea, high salt intake, high ambient temperature, fever, or lactation. For growing swine, it is recommended that at least 1 watering device be provided for every 15 pigs in the group, with a minimum of 2 devices per group. One drinker should be provided for every 10 pigs in the nursery.

Water can serve as a carrier for dewormers, antibiotics or chemotherapeutics, oral vaccines, or water-soluble nutrients when administered through a properly controlled dispensing system.

FEEDS FOR SWINE

Throughout the world, swine are raised on a great variety of feeds, including numerous by-products. They eat relatively little roughage except when on pasture or when ground dry forages are incorporated in the diet. Only about 4% of the total feed consumed by swine in the United States is derived from roughage.

TABLE 6.5   Vitamin Sources and Bioavailabilities[1]

Vitamin	1 IU Equals	Sources	RB[2] (%)	Comments
Vitamin A	0.3 mcg retinol or 0.344 mcg vitamin A acetate	Vitamin A acetate (all-transretinyl acetate)	Unknown	Use coated form
	0.55 mcg vitamin A palmitate	Vitamin A palmitate	Unknown	Used primarily in food
	0.36 mcg vitamin A propionate	Vitamin A propionate	Unknown	Used primarily in injectables
	1 Retinol equivalent (RE) = 1 mcg all-trans retinol			
Vitamin D		Vitamin D_2 (ergocalciferol)		
	0.025 mcg cholecalciferol	Vitamin D_3 (cholecalciferol)	Unknown	Coated form more stable
Vitamin E	1 mg dl-α-tocopheryl acetate	dl-α-tocopheryl acetate (all rac)	100	
	0.735 mg d-α-tocopheryl acetate	d-α-tocopheryl acetate (RRR)	136	
	0.909 mg d-α-tocopherol	dl-α-tocopherol (all rac)	110	Very unstable
	0.671 mg d-α-tocopherol	d-α-tocopherol (RRR)	244	Very unstable
Vitamin K	1 Ansbacher unit = 20 Dam units = 0.0008 mg menadione	Menadione dimethylphrimidinol bisulfite (MPB)	100	
		Menadione sodium bisulfite complex (MSBC)	100	Legal for poultry only
		Menadione sodium bisulfite (MSB)	100	Coated form more stable
Riboflavin	No IU—use mcg or mg	Crystalline riboflavin	100	
Niacin	No lU—use mcg or mg	Niacinamide	100	
		Nicotinic acid	100	
Pantothenic acid	No lU—use mcg or mg	d-calcium pantothenate	100	
		dl-calcium pantothenate	50	
		dl-calcium pantothenate + calcium chloride complex	50	
Vitamin B_{12}	1 mcg cyanocobalamin or 1 USP unit or 11,000 LLD (*L. lactis* Dorner) units	Cyanocabalamin	Unknown	
Choline	No lU—use mcg or mg	Choline chloride	Unknown	Hygroscopic
Biotin	No lU—use mcg or mg d-biotin	d-biotin		
Folic acid	No lU—use mcg or mg	Folic acid	Unknown	Coated form more stable

[1] Most common sources are in italic.

[2] RB = relative bioavailability

Source: Adapted from Swine Nutrition Guide, *University of Nebraska and South Dakota State University, 2000 and NRC* Nutrient Requirements of Swine, *1998.*

TABLE 6.6   Expected Water Consumption of Swine

Class	Pig Weight		Water Consumption, per Day	
Gestating sows			2–3 gal	(11–20 l)
Lactating sows			4–5 gal	(12–40 l)
Suckling pigs	3–13 lb	(1.5–6 kg)	0.05 gal	(0–0.2 l)
Nursery pigs	13–45 lb	(6–20 kg)	0.5–1 gal	(0.5–1.5 l)
Growing pigs	45–130 lb	(20–63 kg)	1 gal	(4–6 l)
Finishing pigs	130–260 lb	(63–114 kg)	1.5–2.0 gal	(6–9 l)

Source: NRC Nutrient Requirements of Swine, *1998.*

Corn and swine production have always been closely associated. Yet, the agriculture is very diverse, and the diet of the pig is readily adapted to feeds produced locally. Thus, in most sections of the world, swine are fed predominantly on home-grown feeds. Ireland depends largely on potatoes and dairy by-products; the swine industry of Denmark has been built up to augment the dairy industry, with milk and whey supplementing home-grown and imported cereals (mostly barley); and in Germany, the pig is fed on such crops as potatoes, sugar beets, and green forage. In China, pigs on small farms are primarily scavengers, competing very little for grains suitable for human consumption, whereas large-scale farms feed cereal and protein supplement diets.

Thus, a swine producer generally has a wide variety of ingredients from which to choose in formulating a diet. Each ingredient may contain several nutrients in varying amounts. Because ingredients vary in price, and in amount and quality of nutrients contained, judgment must be exercised in the choice made.

The nutrient present in the largest amount determines how an ingredient is classified. Thus, corn is classified as an energy feed because it is high in calories, and soybean meal is classified as a protein supplement because it is high in protein. Of course, both of these feeds provide energy and protein.

Concentrates

Because of their monogastric stomachs, swine consume more concentrates and less roughages than other classes of farm animals except for poultry. This characteristic limits their opportunity to consume large quantities of calcium- and vitamin-rich roughages, necessitating the need for mineral and vitamin supplementation.

Although most concentrate feeds are not suitable as the sole dietary ingredient for hogs, it must be realized that swine can use a larger variety of feeds to a greater advantage than other farm animals. In general, the grain crops—corn, barley, wheat, oats, rye, and sorghum—constitute the major components of the swine diet. However, locally sweet potatoes, cull potatoes, peas, and peanuts are successfully fed. Soybeans are the most common protein source, but in almost every section of the United States by-product feeds, including those of the fishing, meat-packing, milling, and dairy industries, are fed to hogs. Human food wastes, such as refuse or garbage (commercial garbage must be cooked), are also fed.

The protein and vitamin requirements of the pig vary greatly from the ruminant because the rumen microorganisms improve the quality of proteins and create certain vitamins through bacterial synthesis.

Despite all this, it is possible to meet the nutritive needs of the pig with concentrated feeds by keeping in mind the following facts when balancing the diet:

1. Cereal grains and their by-products contain relatively low levels of phosphorus, much of it is unavailable. They are low in calcium and the other minerals.
2. Except for the carotene content of yellow corn and green peas, the grains are very poor sources of the vitamins.
3. Most cereal grains supply proteins of poor quality.
4. Protein sources of animal origin and soybean meal generally supply proteins of high quality, whereas proteins of plant origin other than soybean meal generally supply proteins of low quality, usually limiting in lysine.
5. Because of the inadequacies of most concentrates, it is usually necessary to rely on fortifications with minerals and vitamins.

Energy Feeds

Carbohydrates and fats may be classed together as energy feeds for swine. Ingredients commonly used as sources of energy are corn, barley, sorghum, and wheat.

In the United States corn and swine production have always gone hand-in-hand. The majority of U.S. hogs are fed in the Corn Belt, and about one-fourth of the U.S. corn crop is fed to hogs (Figure 6.14). It is usually the cheapest source of energy, but price fluctuations frequently justify consideration of other energy feeds, such as barley in the upper Midwest and Northwest, wheat in Nebraska and Kansas, and sorghum in the Southwest.

Figure 6.14 Corn is the major grain crop fed to pigs. The Corn Belt often produces more corn than can be promptly stored, fed, or transported, thus, the close relationship between abundant, economical corn in the Corn Belt and pork production. *(Courtesy, Palmer Holden, Iowa State University, Ames, IA)*

Although barley, sorghum, and wheat are higher in protein than corn, it is noteworthy that their protein is generally of the same poor quality as corn. It is noteworthy, too, that all of the cereal grains have similar vitamin and mineral deficiencies.

Further discussion of each of the energy feeds commonly fed to swine is found in Chapter 9, "Grains and Other High-Energy Feedstuffs."

Protein Feeds

Protein is made up of nitrogenous compounds called amino acids, and they vary in the kind and amount of amino acids they contain. During the digestion process, the protein is broken down into the various amino acids and the pig recombines them into the kinds of protein needed for muscle development, repair of worn-out tissue, and so on. Thus, the real need of the pig is for amino acids, not protein.

The pig can synthesize some of the amino acids, with the result that they are not required in the diet. However, 10 of the amino acids are termed essential amino acids because the body either cannot synthesize them at all or cannot synthesize them in adequate quantities to optimize growth. It is important that ingredients rich in the essential amino acids be used in formulating the diet.

Although it is a common practice to refer to "percent protein" in a diet, this term has little meaning in swine feeds unless the ingredients used to formulate the diet are known. A protein feed is considered to be of good quality when it contains all the essential amino acids in the proportions and amounts needed by the pig.

Soybean meal is by far the leading high-protein supplement in the United States. About half of the total high-protein feeds fed to animals (including both oilseed meals and animal proteins) consist of soybean meal. Although soybean meal is marginal in methionine, it is otherwise well balanced in amino acids. It must be supplemented with minerals and vitamins. Usually, it is not fed free choice because of its high palatability and cost relative to grains, which results in pigs eating more than is needed to meet their protein needs.

Although soybean meal is the most widely used protein supplement, other protein feeds are suitable, among them, cottonseed meal, dried skim milk, fish meal, linseed meal, meat meal, meat and bone scraps meal, peanut meal, and canola meal. Further discussion of each of the protein feeds is found in Chapter 10, "Protein and Amino Acid Feedstuffs for Swine."

Pastures

In modern U.S. swine production, pastures have been relegated to a minor role. Nevertheless, good legume pasture provides excellent feed for hogs. With excellent pasture that possesses heavy legume content, sows can be fed 2 lb less grain and 0.5 lb less protein supplement per head per day during gestation.

For growing and finishing hogs, high-quality legume pasture can furnish part of the protein supplement. The protein level in complete ground diets can be reduced by 2% for growing-finishing hogs fed on legume pasture as compared to inside feeding (Figure 6.15).

Pastures and their uses are discussed in Chapter 11, "Forages and Pasture Production," and in Chapter 15, "Swine Production and Management."

Dry Forages

Alfalfa is practically the only dry forage fed to swine in the United States, although other leguminous ingredients are available in other countries. If the price of alfalfa meal provides a cheaper source of protein and energy than corn and other grains, 15 to 50% (or even more) alfalfa meal may be incorporated in gestating sow diets and up to 5% may be satisfactorily used in growing-finishing diets. Further discussion of dry forages is found in Chapter 11.

Silages

Good-quality silage is an excellent feed for brood sows; hence, it may be used to advantage where it is available on dairy and beef farms. Unless the sow herd is very large, however, it will not likely pay to construct a silo especially for hogs. Also, because of silage's tendency to spoil in 1 or 2 days when exposed to air, sufficient quantities must be fed daily to prevent feeding spoiled feed. Corn silage is most

Figure 6.15 A swine pasture in northeast Iowa with A-frame huts for sows. This pasture is used for both gestation and lactation. Note the rolling, erodible ground is better suited to forage than to tilled crops. *(Courtesy, Palmer Holden, Iowa State University, Ames, IA)*

common but alfalfa silage can also be used for gestating sows. Further discussion of silages is found in Chapter 11.

Hogging-Down Crops

Pigs are sometimes allowed to do their own harvesting. Grain crops that have been badly stormed or otherwise damaged, or that cannot be harvested because of weather, may be harvested by hogs. Also, pigs may be used to glean fields postharvest. However, most fences have been removed from farmland and there is the potential for gestating sows to consume too much grain if allowed unlimited access to the damaged fields. Further discussion of hogging-down crops is found in Chapter 9, "Grains and Other High Energy Feedstuffs."

Garbage

From the remote day of domestication forward, swine have been considered scavengers—often fed on table scraps and other wastes. Even today, in most developing countries, pigs, and, to a lesser extent other livestock, consume precious few products that are suitable for human food. All 50 of the United States have laws requiring that municipal garbage be cooked to prevent disease transmission. Further discussion of garbage is found in Chapter 9.

Feed Additives

Certain feed additives have become somewhat standard ingredients of swine diets, especially for pigs from birth to market weight. They are not nutrients as such; hence, they should not be considered as dietary essentials. Feed additives are covered in more detail in Chapter 8, "Alternative Feeds, Additives, and Feeding Management."

FEED EVALUATION

Profit is the ultimate criterion of success in any livestock operation, and the cost of nutrients is an important factor in determining success. The feed composition values presented in this and other books merely represent an averaging of an accumulation of data concerning the nutritive value of feeds. Considerable variation is inherent in the nutrient content of different samples of feeds. Thus, the successful producer must recognize the value of a well-planned feed analysis program.

Feeds may be analyzed by physical, chemical, or biological procedures. Although physical evaluation (appearance, odor, microscopic) may be the least accurate, it provides a quick and easy means of obtaining considerable information about the overall quality of a feed. The chemical procedure is more accurate than a physical evaluation, but it takes time. The biological method (microbiological or feeding animals) necessitates considerable time and expense, and the results are subject to animal variation.

Physical Evaluation

In order to produce or buy superior feeds, swine producers need to know what constitutes feed quality, and how to recognize it. They need to be familiar with those recognizable characteristics of feeds that indicate high palatability and nutrient content. If in doubt, observation of the pigs consuming the feed will tell them because pigs prefer and thrive on high-quality feed. The easily recognizable characteristics of good grains and other concentrates are as follows:

1. Seeds are not split or cracked.
2. Seeds are of low-moisture content—generally containing about 88% dry matter.
3. Seeds have a good color.
4. Concentrates and seeds are free from mold.
5. Concentrates and seeds are free from rodent and insect damage.
6. Concentrates and seeds are free from foreign material, such as screenings or iron filings.
7. Concentrates and seeds are free from rancid odor.

Chemical Analysis

Today, feeds are being analyzed routinely through highly sophisticated chemical procedures. Many agricultural experiment stations, large feed companies, and private laboratories have facilities to analyze feeds for both the prevention and diagnosis of nutritional problems.

A chemical analysis gives a solid foundation on which to start in the evaluation of feeds. Thus, feed composition tables serve as a basis for diet formulation and for ingredient purchasing and merchandising. Commercially prepared feeds are required by state law to be labeled with a list of ingredients and a guaranteed analysis.

Although state laws vary, most of them require that the feed label (tag) show in percentages the minimum crude protein and fat, and the maximum crude fiber and ash. Most feed labels also include minimum and maximum percentages of salt and calcium, and minimums for phosphorus. These figures are the buyer's assurance that the feed contains the minimal amounts of the higher-cost items—protein and fat, and not more than the stipulated amounts of the lower-cost, and less valuable, items—the crude fiber, calcium, and salt.

Proximate Analysis (Weende Procedure)

For more than 100 years, feeds have been analyzed by a method developed by two scientists, Henneberg and Stohmann, at the Weende Experiment Station in Germany. This method is called the proximate analysis, or the Weende system, of feed analysis. Feeds are divided into six components: (1) moisture, (2) ash, (3) crude protein, (4) ether extract, (5) crude fiber, and (6) nitrogen-free extract.

Near Infrared (NIR) Spectroscopy

Near infrared (NIR) is a part of the spectrum of light. Near infrared is between the visible and the infrared. We see in the visible spectrum. The color of an apple as seen in the visible spectrum gives information on a variety of pigments and chemicals in the fruit. However, we cannot "see" things that do not absorb visible light (e.g., a sugar solution). However, it happens that water, sugar, acids, and a range of other organic substances absorb near infrared in proportion to their concentration.

NIR is routinely used to estimate a broad range of components, such as dry matter or, for example, protein or amino acids in grains or forages. Its major benefit is speed, with almost instantaneous results from prepared samples.

Bomb Calorimetry

When compounds are burned completely in the presence of oxygen, the resulting heat is referred to as gross energy or the heat of combustion. The bomb calorimeter is used to determine the gross energy of feed, waste products from feed (for example, feces and urine), and tissues.

The calorie is defined as the amount of heat required to raise the temperature of 1 g of water 1°C (precisely from 14.5° to 15.5°C). With this fact in mind, we can readily see how the bomb calorimeter works. Although the *bomb calorimeter* is useful in obtaining gross energy (GE), in order to determine animal utilization of a feed, animal trials of one type or another must be carried out.

Biological Analysis

Quite often, biological assays are used in the analysis of nutrients in feeds. There are two basic types of biological assays—(1) microbiological assays and (2) the use of nutrient-deficient animals.

Biological assays tend to be laborious and time consuming. Large numbers of samples are needed to produce statistically reliable results, and quite often data obtained from these assays are highly variable. The assay using nutrient-deficient animals is particularly cumbersome because (1) the animals should be of approximately the same age, sex, and weight; and (2) time is required to induce deficient conditions in these animals.

QUESTIONS FOR STUDY AND DISCUSSION

1. Why is knowledge of swine body composition important?
2. Table 6.1 gives the range in body composition of pigs from birth to maturity. Detail the changes in body composition from birth to maturity of water, fat, protein, and ash.
3. Describe the digestive system of the pig. Compare the digestive system of the pig (a nonruminant) and the digestive system of a ruminant.
4. Why is the pig a good model for human nutrition studies?
5. Name the nutrients required by the pig.
6. Discuss the relative nutrient needs for the following body functions:
   a. Maintenance
   b. Growth
   c. Reproduction
   d. Lactation
   e. Finishing
7. Table 6.2 shows a dramatic change in the feed conversion of pigs from prenursery to finisher. How do you explain this?
8. What is the difference between the gross energy of a feed and the digestible energy of a feed? Why is it important?
9. What is an essential amino acid? Which amino acids are essential for swine?
10. What is a limiting amino acid?
11. Explain how mineral and vitamin intakes of swine have been affected by management changes since the 1980s.
12. Distinguish between macrominerals and microminerals.
13. Give (a) the function and (b) the deficiency symptoms resulting from a lack of each of the following minerals in the pig: salt, calcium, phosphorus, iodine, and iron.
14. Discuss the relationships of calcium, phosphorus, and vitamin D in swine nutrition.

15. Why is the availability of phosphorus from plant sources estimated at 30% or less when formulating diets?
16. What are chelated minerals? Should they be recommended?
17. Discuss the sources and bioavailability of minerals.
18. Give (a) the function and (b) the deficiency symptoms resulting from a lack of each of the following vitamins in the pig: A, D, E, choline, pantothenic acid, and riboflavin.
19. What vitamins are most apt to be deficient in grain-soybean meal swine diets?
20. Before the advent of modern feed formulations and vitamins, how did producers ensure that swine received all the necessary vitamins?
21. Discuss the importance of water for the pig.
22. List and describe the ways feeds may be analyzed. Why are feed analyses so important?

SELECTED REFERENCES

Animal Feeding and Nutrition, 9th ed., M. J. Jurgens, Kendall/Hunt Publishing, Co., Dubuque, IA, 2002

Animal Science, 9th ed., M. E. Ensminger, Interstate Publishers Inc., Danville, IL, 1991

Applied Animal Nutrition, Feeds and Feeding, 2nd ed., Peter R. Cheeke, Prentice Hall, Upper Saddle River, NJ, 1999

Bioenergetics and Growth, Samuel Brody, Hafner Publishing Co., New York, NY, 1945

Feeds and Nutrition Digest, M. E. Ensminger, J. E. Oldfield, and W. W. Heinemann, The Ensminger Publishing Company, Clovis, CA, 1990

Fundamentals of Nutrition, L. E. Lloyd, B. E. McDonald, and E. W. Crampton, W. H. Freeman and Co., San Francisco, CA, 1978

Kansas Swine Nutrition Guide, Kansas State University Extension, Manhattan, KS, 2003

Life Cycle Swine Nutrition, 17th ed., Pm-489, P. J. Holden, et al., Iowa State University, Ames, IA, 1996

Livestock Feeds and Feeding, 4th ed., Richard O. Kellems and D. C. Church, Prentice Hall, Upper Saddle River, NJ, 1998

Nontraditional Feed Sources for Use in Swine Production, P. A. Thacker and R. N. Kirkwood, Butterworths, Boston, MA, 1990

Nutrient Requirements of Swine, 10th rev. ed., National Research Council, National Academy of Sciences, Washington, DC, 1998

Pork Industry Handbook, Cooperative Extension Service, Purdue University, West Lafayette, IN, 2003

Stockman's Handbook, The, 7th ed., M. E. Ensminger, Interstate Publishers Inc., Danville, IL, 1992

Swine Nutrition, E. R. Miller, D. E. Ullrey, and A. J. Lewis, Butterworth-Heinemann, Boston, MA, 1991

Swine Nutrition Guide, University of Nebraska, Lincoln, NE and, South Dakota State University, Brookings, SD, 2000

Swine Production and Nutrition, W. G. Pond and J. H. Maner, AVI Publishing Company, Inc., Westport, CT, 1984

CHAPTER

7

Nutrient Requirements and Allowances, Diet Formulation, and Feeding Programs[1]

A modern feed mill on the Iowa State University Swine Nutrition and Management Research Farm. *(Courtesy, Palmer Holden, Iowa State University, Ames, IA)*

Contents	Page
Nutrient Requirements	136
National Research Council (NRC) Requirements	137
Expected Feed Inputs and Product Outputs	137
Recommended Nutrient Allowances	147

[1] Applicable material for this chapter was adapted from four publications: National Research Council's *Nutrient Requirements of Swine* (1998); *Life Cycle Swine Nutrition*, published by Iowa State University, (1996); *Swine Nutrition Guide*, published by the University of Nebraska and South Dakota State University; and *Kansas Swine Nutrition Guide*, published by Kansas State University.

Objectives:

After studying this chapter, you should:

1. Know how to use the National Research Council's nutrient requirements of swine tables.
2. Define the differences between nutrient requirements and recommended nutrient allowances.
3. Know the amount of feed required in each stage of production.
4. Know how to adjust requirements for different environmental or genetic factors.
5. Know the steps necessary to determining a proper diet for growing pigs.
6. Know the steps necessary to determining a proper diet for the breeding herd.

Swine feeding is a complex science. For maximum success the pork producer must follow a scientific feeding program based on the nutritional needs incurred in each stage of the pig's life and how those needs are affected by the environment. These concepts require the producer to define the status of the herd. For example, the lean-gain potential, the existing climatic conditions, and the health status of the herd must be evaluated. Adjustments then can be based on these factors when defining the specific requirements for each stage of the life cycle from the nursing pig through the breeding herd.

Many factors affect swine nutrient requirements, including genetic merit, diet formulation, feeding programs, ingredient availability and quality, feed intake, nutrient interactions, nutrient bioavailability, feed processing, safety margin, nonnutritive additives, environment, health, access to water and economics. These factors interact with each other. Thus, their net output determines the level of swine production and profitability. As a result of research and management, these factors are constantly changing. To maximize performance and profits, the modification of the nutrition and feeding programs requires constant adjustment.

NUTRIENT REQUIREMENTS

Nutrient requirements are found in tables listing the amounts of nutrients required by different species of animals for specific productive functions, such as growth, reproduction, and lactation. Most requirements are expressed in either (1) quantities of nutrients required per day or (2) concentration in the diet. The first type is used where animals are provided a specific amount of a feed during a 24-hour period, and the second is used where animals are provided a diet without limita-

tion on the time in which it is consumed, also known as self-feeding.

Nutrient requirements have been well defined for swine, and, through the use of these standards, the producer can formulate diets designed to meet their specific nutrient requirements. Although nutrient requirements are excellent and needed guides, there are situations where nutrient needs cannot be specified with great accuracy. For example, maximizing carcass leanness may require greater concentrations of certain nutrients than maximizing rate of gain. Lean pigs deposit more protein in their gain, which increases their requirements for protein and individual amino acids. Leanness is associated positively with feed efficiency; thus, requirements for maximal feed efficiency are generally greater than those for maximal weight gain. Maximal blood hemoglobin concentration may necessitate higher levels of iron than those needed for maximal rate or efficiency of gain, and maximal bone ash generally requires higher levels of calcium and phosphorus than is needed for maximal weight gain.

Moreover, nutrient requirements tell nothing about the palatability, physical nature, or possible digestive disturbances of a diet. Nor do they give consideration to individual pig differences, management differences, and the effects of various stresses, such as weather, disease, and parasitism. Thus, there are many variables that alter the nutrient needs and utilization of pigs—variables that are difficult to include quantitatively in developing nutrient requirements, even when feed quality is well known. Ultimately, each swine manager must select a set of standards that will permit the greatest economic return in his or her particular environment. Optimum standards will relate to the genetic potential of the herd and to the price and availability of feedstuffs in the region.

NATIONAL RESEARCH COUNCIL (NRC) REQUIREMENTS

The most up-to-date nutrient requirements are published by the U.S. National Research Council (NRC). Periodically, the National Academy of Sciences appoints a committee of recognized researchers who have worked extensively with a specific species and they review the scientific literature and revise the nutrient requirements for different functions. Thus, the nutritive needs of each livestock species are dealt with separately and in depth.

They are based on published research and should be the basis for estimating nutritional needs. Also, these figures are, for the most part, requirements (rather than allowances); hence, they do not provide for safety margins to compensate for variations in feed composition, environment, and potential nutrient losses during storage or processing. In using Tables 7.1 through 7.10, the following additional points should be recognized:

1. Feedstuffs vary in nutritive value, whether grown in the same locality or in different areas.
2. The environment in which pigs are produced can modify the requirements.
3. Pigs bred with high-genetic ability for lean gain have nutritional needs that are quite different from average- or low-performing pigs.

The NRC nutritional requirements, or standards, are presented in Tables 7.1 to 7.10.

EXPECTED FEED INPUTS AND PRODUCT OUTPUTS

Feed requirements to produce a typical market hog are estimated in Table 7.11 for high lean and moderate lean-gain genotypes. High lean-gain genotype sows require slightly more feed because of higher maintenance requirements, but high lean-genetic nursery and growing-finishing pigs are significantly more efficient and have more lean in the carcass. Information on how the values are obtained is presented after the table. These values include estimated allocations of sow and boar feed to each pig marketed plus the feed consumed by the market pig. The Iowa State University Swine Enterprise Record Summary indicated an average whole herd (farrow to finish) efficiency of 3.7 units of feed/unit of pork produced with a standard deviation of 0.17. Thus, a reasonable goal would be 3.30 units of feed/unit of pork produced, or about 1 standard deviation better than average.

Because grow-finish pigs consume about 80% of the total, this is the portion of the production system that allows for the most savings in efficiency and feed costs. Although starter diets are very expensive they are only 6 or 7% of the total feed. High lean-gain sows have a higher feed requirement because of increased maintenance needs.

Pregestation and Gestation Feed

The estimated number of litters per sow per year is 2.0, including replacement gilts inventoried into the herd at 265 lb (120 kg). The number of pigs marketed per sow per year is 18.1 (9.4 weaned −5% death loss postweaning × 2.0 litters per sow per year). With a 21-day lactation period each sow averages 42 days per year on lactation feed and 323 days on prebreeding and gestation diets (365 − 42). With an average daily feed intake of 4.7 lb (2.1 kg) per day for high-lean sows and 4.5 lb (2.0 kg) per day for moderate-lean sows of pregestation and gestation feed for 323 days, the annual feed usage would be 1,518 (690 kg) and 1,454 lb (661 kg), respectively. This feed divided by 18.1 pigs produced per sow per year results in 84 lb (38 kg) or 80 lb (36 kg) of pregestation and gestation feed charged to each pig.

Lactation Feed

A 21-day lactation period times 2.0 litters per year results in 42 lactating days. If the average daily feed intake is 12 lb (5.5 kg), each sow consumes 504 lb (229 kg) of lactation feed. This feed divided by 18.1 pigs produced results in 28 lb (12.7 kg) of lactation feed charged to each pig.

Boar Feed

The number of boars in the herd is estimated to be 1 for each 20 sows or 362 pigs marketed (20 × 18.1). This number will vary depending on the number of farrowings per year, as well as the method of mating, for example pen mating, hand mating, or artificial insemination. With an average intake of 6 lb (2.7 kg) per day, each boar will consume 2,190 lb (995 kg) of feed per year or 6 lb (2.7 kg) per pig marketed.

Starter Feed

Starter pig feed includes the feed consumed from weaning at 12 to 50 lb (5.5 to 23 kg) body weight. With a feed efficiency of 1.35 for high-lean pigs and 1.65 for moderate lean-growth pigs, they would consume 51 and 63 lb (23 and 28 kg) of starter diet, respectively.

TABLE 7.1   Dietary Amino Acid Requirements of Growing Pigs Allowed Feed Ad Libitum (90% dry matter)[1]

Body Weight (kg)	3–5	5–10	10–20	20–50	50–80	80–120
DE content of diet (kcal/kg)	3,400	3,400	3,400	3,400	3,400	3,400
ME content of diet (kcal/kg)[2]	3,265	3,265	3,265	3,265	3,265	3,265
Estimated DE intake (kcal/day)	855	1,690	3,400	6,305	8,760	10,450
Estimated ME intake (kcal/day)[2]	820	1,620	3,265	6,050	8,410	10,030
Estimated feed intake (g/day)	250	500	1,000	1,855	2,575	3,075
Crude protein (%)[3]	26.0	23.7	20.9	18.0	15.5	13.2

Amino acid requirements[4]
True ileal digestible basis (%)

	3–5	5–10	10–20	20–50	50–80	80–120
Arginine	0.54	0.49	0.42	0.33	0.24	0.16
Histidine	0.43	0.38	0.32	0.26	0.21	0.16
Isoleucine	0.73	0.65	0.55	0.45	0.37	0.29
Leucine	1.35	1.20	1.02	0.83	0.67	0.51
Lysine	1.34	1.19	1.01	0.83	0.66	0.52
Methionine	0.36	0.32	0.27	0.22	0.18	0.14
Methionine + cystine	0.76	0.68	0.58	0.47	0.39	0.31
Phenylalanine	0.80	0.71	0.61	0.49	0.40	0.31
Phenylalanine + tyrosine	1.26	1.12	0.95	0.78	0.63	0.49
Threonine	0.84	0.74	0.63	0.52	0.43	0.34
Tryptophan	0.24	0.22	0.18	0.15	0.12	0.10
Valine	0.91	0.81	0.69	0.56	0.45	0.35

Total basis (%)[5]

	3–5	5–10	10–20	20–50	50–80	80–120
Arginine	0.59	0.54	0.46	0.37	0.27	0.19
Histidine	0.48	0.43	0.36	0.30	0.24	0.19
Isoleucine	0.83	0.73	0.63	0.51	0.42	0.33
Leucine	1.50	1.32	1.12	0.90	0.71	0.54
Lysine	1.50	1.35	1.15	0.95	0.75	0.60
Methionine	0.40	0.35	0.30	0.25	0.20	0.16
Methionine + cystine	0.86	0.76	0.65	0.54	0.44	0.35
Phenylalanine	0.90	0.80	0.68	0.55	0.44	0.34
Phenylalanine + tyrosine	1.41	1.25	1.06	0.87	0.70	0.55
Threonine	0.98	0.86	0.74	0.61	0.51	0.41
Tryptophan	0.27	0.24	0.21	0.17	0.14	0.11
Valine	1.04	0.92	0.79	0.64	0.52	0.40

[1]Mixed gender (1:1 ratio of barrows to gilts) of pigs with high medium lean-growth rate (325 g/day of carcass fat-free lean) from 20 to 120 kg body weight.

[2]Assumes that ME is 96% of DE. In corn–soybean meal diets of these crude protein levels, ME is 94 to 96% of DE.

[3]Crude protein levels apply to corn–soybean meal diets. In 3- to 10-kg pigs fed diets with dried plasma and/or dried milk products, protein levels will be 2 to 3% less than shown.

[4]Total amino acid requirements are based on the following types of diets: 3- to 5-kg pigs, corn–soybean meal diet that includes 5% dried plasma and 25 to 50% dried milk products; 5- to 10-kg pigs, corn–soybean meal diet that includes 5 to 25% dried milk products; 10- to 120-kg pigs, corn–soybean meal diet.

[5]The total lysine percentages for 3- to 20-kg pigs are estimated from empirical data. The other amino acids for 3- to 20-kg pigs are based on the ratios of amino acids to lysine (true digestible basis); however, there are very few empirical data to support these ratios. The requirements for 20- to 120-kg pigs are estimated from the growth model.

Source: Reprinted with permission from Nutrient Requirements of Swine, *10th rev. ed., National Research Council, National Academy Press, 1998, p. 111, Table 10.1.*

TABLE 7.2　Dietary Mineral, Vitamin, and Linoleic Acid Requirements of Growing Pigs Allowed Feed Ad Libitum (90% dry matter)[1]

Body Weight (kg)	3–5	5–10	10–20	20–50	50–80	80–120
		Requirements (% or amount/kg of diet)				
Minerals						
Calcium (%)[2]	0.90	0.80	0.70	0.60	0.50	0.45
Phosphorus, total (%)[2]	0.70	0.65	0.60	0.50	0.45	0.40
Phosphorus, available (%)[2]	0.55	0.40	0.32	0.23	0.19	0.15
Sodium (%)	0.25	0.20	0.15	0.10	0.10	0.10
Chlorine (%)	0.25	0.20	0.15	0.08	0.08	0.08
Magnesium (%)	0.04	0.04	0.04	0.04	0.04	0.04
Potassium (%)	0.30	0.28	0.26	0.23	0.19	0.17
Copper (mg)	6.00	6.00	5.00	4.00	3.50	3.00
Iodine (mg)	0.14	0.14	0.14	0.14	0.14	0.14
Iron (mg)	100	100	80	60	50	40
Manganese (mg)	4.00	4.00	3.00	2.00	2.00	2.00
Selenium (mg)	0.30	0.30	0.25	0.15	0.15	0.15
Zinc (mg)	100	100	80	60	50	50
Vitamins						
Vitamin A (IU)[3]	2,200	2,200	1,750	1,300	1,300	1,300
Vitamin D_3 (IU)[3]	220	220	200	150	150	150
Vitamin E (IU)[3]	16	16	11	11	11	11
Vitamin K (menadione) (mg)	0.50	0.50	0.50	0.50	0.50	0.50
Biotin (mg)	0.08	0.05	0.05	0.05	0.05	0.05
Choline (g)	0.60	0.50	0.40	0.30	0.30	0.30
Folacin (mg)	0.30	0.30	0.30	0.30	0.30	0.30
Niacin, available (mg)[4]	20.00	15.00	12.50	10.00	7.00	7.00
Pantothenic acid (mg)	12.00	10.00	9.00	8.00	7.00	7.00
Riboflavin (mg)	4.00	3.50	3.00	2.50	2.00	2.00
Thiamin (mg)	1.50	1.00	1.00	1.00	1.00	1.00
Vitamin B_6 (mg)	2.00	1.50	1.50	1.00	1.00	1.00
Vitamin B_{12} (mg)	20.00	17.50	15.00	10.00	5.00	5.00
Linoleic acid (%)	0.10	0.10	0.10	0.10	0.10	0.10

[1]Pigs of mixed gender (1:1 ratio of barrows to gilts). The requirements may be slightly higher for pigs having high lean-growth rates (> 325g/day of carcass fat-free lean), but no distinction is made.

[2]The percentages of calcium, phosphorus, and available phosphorus should be increased by 0.05 to 0.1 percentage points for developing boars and replacement gilts from 50- to 120-kg body weight.

[3]Conversions: 1 IU vitamin A = 0.344 ug (g retinyl acetate); 1 IU vitamin D_3 = 0.025 ug cholecalciferol; 1 IU vitamin E = 0.67 mg of d-α-tocopherol or 1 mg of dl-α-tocopheryl acetate.

[4]The niacin in corn, grain sorghum, wheat, and barley is unavailable. Similarly, the niacin in by-products made from these cereal grains is poorly available unless the by-products have undergone a fermentation or wet-milling process.

Source: Reprinted with permission from Nutrient Requirements of Swine, *10th rev. ed., National Research Council, National Academy Press. 1998, p. 115, Table 10.5.*

TABLE 7.3   Daily Amino Acid Requirements of Growing Pigs Allowed Feed Ad Libitum (90% dry matter)[1]

Body Weight (kg)	*3–5*	*5–10*	*10–20*	*20–50*	*50–80*	*80–120*
Average weight in range (kg)	4	7.5	15	35	65	100
DE content of diet (kcal/kg)	3,400	3,400	3,400	3,400	3,400	3,400
ME content of diet (kcal/kg)[2]	3,265	3,265	3,265	3,265	3,265	3,265
Estimated DE intake (kcal/day)	855	1,690	3,400	6,305	8,760	10,450
Estimated ME intake (kcal/day)[2]	820	1,620	3,265	6,050	8,410	10,030
Estimated feed intake (g/day)	250	500	1,000	1,855	2,575	3,075
Crude protein (%)[3]	26.0	23.7	20.9	18.0	15.5	13.2
Amino acid requirements[4]						
True ileal digestible basis (g/day)						
Arginine	1.4	2.4	4.2	6.1	6.2	4.8
Histidine	1.1	1.9	3.2	4.9	5.5	5.1
Isoleucine	1.8	3.2	5.5	8.4	9.4	8.8
Leucine	3.4	6.0	10.3	15.5	7.2	15.8
Lysine	3.4	5.9	10.1	15.3	17.1	15.8
Methionine	0.9	1.6	2.7	4.1	4.6	4.3
Methionine + cystine	1.9	3.4	5.8	8.8	10.0	9.5
Phenylalanine	2.0	3.5	6.1	9.1	10.2	9.4
Phenylalanine + tyrosine	3.2	5.5	9.5	14.4	16.1	15.1
Threonine	2.1	3.7	6.3	9.7	11.0	10.5
Tryptophan	0.6	1.1	1.9	2.8	3.1	2.9
Valine	2.3	4.0	6.9	10.4	11.6	10.8
Total basis (g/day)[5]						
Arginine	1.5	2.7	4.6	6.8	7.1	5.7
Histidine	1.2	2.1	3.7	5.6	6.3	5.9
Isoleucine	2.1	3.7	6.3	9.5	10.7	10.1
Leucine	3.8	6.6	11.2	16.8	18.4	16.6
Lysine	3.8	6.7	11.5	17.5	19.7	18.5
Methionine	1.0	1.8	3.0	4.6	5.1	4.8
Methionine + cystine	2.2	3.8	6.5	9.9	11.3	10.8
Phenylalanine	2.3	4.0	6.8	10.2	11.3	10.4
Phenylalanine + tyrosine	3.5	6.2	10.6	16.1	18.0	16.8
Threonine	2.5	4.3	7.4	11.3	13.0	12.6
Tryptophan	0.7	1.2	2.1	3.2	3.6	3.4
Valine	2.6	4.6	7.9	11.9	13.3	12.4

[1]Mixed gender (1:1 ratio of barrows to gilts) of pigs with high medium lean-growth rate (325 g/day of carcass fat-free lean) from 20- to 120-kg body weight.

[2]Assumes that ME is 96% of DE. In corn–soybean meal diets of these crude protein levels, ME is 94 to 96% of DE.

[3]Crude protein levels apply to corn–soybean meal diets. In 3- to 10-kg pigs fed diets with dried plasma and/or dried milk products, protein levels will be 2 to 3% less than shown.

[4]Total amino acid requirements are based on the following types of diets: 3- to 5-kg pigs, corn–soybean meal diet that includes 5% dried plasma and 25 to 50% dried milk products; 5- to 10-kg pigs, corn–soybean meal diet that includes 5 to 25% dried milk products; 10- to 120-kg pigs, corn–soybean meal diet.

[5]The total lysine estimates for 3- to 20-kg pigs are calculated by multiplying the percentages in NRC Table 10.1 (estimated from empirical data) by the estimated feed intake. The other amino acids for 3- to 20-kg pigs are based on the ratios of amino acids to lysine (true digestible basis); however, there are very few empirical data to support these ratios. The estimates for 20- to 120-kg pigs are from the growth model.

Source: Reprinted with permission from Nutrient Requirements of Swine, *10th rev. ed., National Research Council, National Academy Press. 1998, p. 112, Table 10.2.*

TABLE 7.4 **Daily Mineral, Vitamin, and Linoleic Acid Requirements of Growing Pigs Allowed Feed Ad Libitum (90% dry matter)[1]**

Body Weight (kg)	3–5	5–10	10–20	20–50	50–80	80–120
Average weight in range (kg)	4	7.5	15	35	65	100
DE content of diet (kcal/kg)	3,400	3,400	3,400	3,400	3,400	3,400
ME content of diet (kcal/kg)[2]	3,265	3,265	3,265	3,265	3,265	3,265
Estimated DE intake (kcal/day)	855	1,690	3,400	6,305	8,760	10,450
Estimated ME intake (kcal/day)[2]	820	1,620	3,265	6,050	8,410	10,030
Estimated feed intake (g/day)	250	500	1,000	1,855	2,575	3,075
Requirements (amount/day)						
Minerals						
Calcium (g)[3]	2.25	4.00	7.00	11.13	12.88	13.84
Phosphorus, total (g)[3]	1.75	3.25	6.00	9.28	11.59	12.30
Phosphorus, available (g)[3]	1.38	2.00	3.20	4.27	4.89	4.61
Sodium (g)	0.63	1.00	1.50	1.86	2.58	3.08
Chlorine (g)	0.63	1.00	1.50	1.48	2.06	2.46
Magnesium (g)	0.10	0.20	0.40	0.74	1.03	1.23
Potassium (g)	0.75	1.40	2.60	4.27	4.89	5.23
Copper (mg)	1.50	3.00	5.00	7.42	9.01	9.23
Iodine (mg)	0.04	0.07	0.14	0.26	0.36	0.43
Iron (mg)	25.00	50.00	80.00	111.30	129.75	123.00
Manganese (mg)	1.00	2.00	3.00	3.71	5.15	6.15
Selenium (mg)	0.08	0.15	0.25	0.28	0.39	0.46
Zinc (mg)	25.00	50.00	80.00	111.30	129.75	153.75
Vitamins						
Vitamin A (IU)[4]	550	1,100	1,750	2,412	3,348	3,998
Vitamin D_3 (IU)[4]	55	110	200	278	386	461
Vitamin E (IU)[4]	4	8	11	20	28	34
Vitamin K (menadione) (mg)	0.13	0.25	0.50	0.93	1.29	1.54
Biotin (mg)	0.02	0.03	0.05	0.09	0.13	0.15
Choline (g)	0.15	0.25	0.40	0.56	0.77	0.92
Folacin (mg)	0.08	0.15	0.30	0.56	0.77	0.92
Niacin, available (mg)[5]	5.00	7.50	12.50	18.55	18.03	21.53
Pantothenic acid (mg)	3.00	5.00	9.00	14.84	18.03	21.53
Riboflavin (mg)	1.00	1.75	3.00	4.64	5.15	6.15
Thiamin (mg)	0.38	0.50	1.00	1.86	2.58	3.08
Vitamin B_6 (mg)	0.50	0.75	1.50	1.86	2.58	3.08
Vitamin B_{12} (mg)	5.00	8.75	15.00	18.55	12.88	15.38
Linoleic acid (g)	0.25	0.50	1.00	1.86	2.58	3.08

[1]Pigs of mixed gender (1:1 ratio of barrows to gilts). The daily requirements of certain minerals and vitamins may be slightly higher for pigs having high lean-growth rates (325 g/day of carcass fat-free lean), but no distinction is made.

[2]Assumes that ME is 96% of DE. In corn–soybean meal diets, ME is 94 to 96% of DE, depending on crude protein level of the diet.

[3]The daily amounts of calcium, phosphorus, and available phosphorus are slightly higher in developing boars and gilts from 50- to 120-kg body weight.

[4]Conversions: 1 IU vitamin A = 0.344 mg (g retinyl acetate); 1 IU vitamin D_3 = 0.025 ug cholecalciferol; 1 IU vitamin E = 0.67 mg of d-α-tocopherol or 1 mg of dl-α-tocopheryl acetate.

[5]The niacin in corn, grain sorghum, wheat, and barley is unavailable. Similarly, the niacin in by-products made from these cereal grains is poorly available unless the by-products have undergone a fermentation or wet-milling process.

Source: Reprinted with permission from Nutrient Requirements of Swine, *10th rev. ed., National Research Council, National Academy Press. 1998, p. 116, Table 10.6.*

TABLE 7.5   **Dietary Amino Acid Requirements of Gestating Sows (90% dry matter)**[1]

Body weight at breeding (kg)	125	150	175	200	200	200
Gestation weight gain (kg)[2]	55	45	40	35	30	35
Anticipated pigs in litter	11	12	12	12	12	14
DE content of diet (kcal/kg)	3,400	3,400	3,400	3,400	3,400	3,400
ME content of diet (kcal/kg)[3]	3,265	3,265	3,265	3,265	3,265	3,265
Estimated DE intake (kcal/day)	6,660	6,265	6,405	6,535	6,115	6,275
Estimated ME intake (kcal/day)[3]	6,395	6,015	6,150	6,275	5,870	6,025
Estimated feed intake (kg/day)	1.96	1.84	1.88	1.92	1.80	1.85
Crude protein (%)[4]	12.9	12.8	12.4	12.0	12.1	12.4
Amino acid requirements *True ileal digestible basis (%)*						
Arginine	0.04	0.00	0.00	0.00	0.00	0.00
Histidine	0.16	0.16	0.15	0.14	0.14	0.15
Isoleucine	0.29	0.28	0.27	0.26	0.26	0.27
Leucine	0.48	0.47	0.44	0.41	0.41	0.44
Lysine	0.50	0.49	0.46	0.44	0.44	0.46
Methionine	0.14	0.13	0.13	0.12	0.12	0.13
Methionine + cystine	0.33	0.33	0.32	0.31	0.32	0.33
Phenylalanine	0.29	0.28	0.27	0.25	0.25	0.27
Phenylalanine + tyrosine	0.48	0.48	0.46	0.44	0.44	0.46
Threonine	0.37	0.38	0.37	0.36	0.37	0.38
Tryptophan	0.10	0.10	0.09	0.09	0.09	0.09
Valine	0.34	0.33	0.31	0.30	0.30	0.31
Total basis (%)[4]						
Arginine	0.06	0.03	0.00	0.00	0.00	0.00
Histidine	0.19	0.18	0.17	0.16	0.17	0.17
Isoleucine	0.33	0.32	0.31	0.30	0.30	0.31
Leucine	0.50	0.49	0.46	0.42	0.43	0.45
Lysine	0.58	0.57	0.54	0.52	0.52	0.54
Methionine	0.15	0.15	0.14	0.13	0.13	0.14
Methionine + cystine	0.37	0.38	0.37	0.36	0.36	0.37
Phenylalanine	0.32	0.32	0.30	0.28	0.28	0.30
Phenylalanine + tyrosine	0.54	0.54	0.51	0.49	0.49	0.51
Threonine	0.44	0.45	0.44	0.43	0.44	0.45
Tryptophan	0.11	0.11	0.11	0.10	0.10	0.11
Valine	0.39	0.38	0.36	0.34	0.34	0.36

[1]Daily intakes of DE and feed and the amino acid requirements are estimated by the gestation model.
[2]Weight gain includes maternal tissue and products of conception.
[3]Assumes that ME is 96% of DE.
[4]Crude protein and total amino acid requirements are based on a corn–soybean meal diet.
Source: Reprinted with permission from Nutrient Requirements of Swine, *10th rev. ed., National Research Council, National Academy Press. 1998, p. 117, Table 10.7.*

TABLE 7.6   Daily Amino Acid Requirements of Gestating Sows (90% dry matter)[1]

Body weight at breeding (kg)	125	150	175	200	200	200
Gestation weight gain (kg)[2]	55	45	40	35	30	35
Anticipated pigs in litter	11	12	12	12	12	14
DE content of diet (kcal/kg)	3,400	3,400	3,400	3,400	3,400	3,400
ME content of diet (kcal/kg)[3]	3,265	3,265	3,265	3,265	3,265	3,265
Estimated DE intake (kcal/day)	6,660	6,265	6,405	6,535	6,115	6,275
Estimated ME intake (kcal/day)[3]	6,395	6,015	6,150	6,275	5,870	6,025
Estimated feed intake (kg/day)	1.96	1.84	1.88	1.92	1.80	1.85
Crude protein (%)[3]	12.9	12.8	12.4	12.0	12.1	12.4

Amino acid requirements *True ileal digestible basis (g/day)*						
Arginine	0.8	0.1	0.0	0.0	0.0	0.0
Histidine	3.1	2.9	2.8	2.7	2.5	2.7
Isoleucine	5.6	5.2	5.1	5.0	4.7	5.0
Leucine	9.4	8.7	8.3	7.9	7.4	8.1
Lysine	9.7	9.0	8.7	8.4	7.9	8.5
Methionine	2.7	2.5	2.4	2.3	2.2	2.3
Methionine + cystine	6.4	6.1	6.1	6.0	5.7	6.1
Phenylalanine	5.7	5.2	5.0	4.8	4.6	4.9
Phenylalanine + tyrosine	9.5	8.9	8.6	8.4	7.9	8.5
Threonine	7.3	7.0	6.9	6.9	6.6	7.0
Tryptophan	1.9	1.8	1.7	1.7	1.6	1.7
Valine	6.6	6.1	5.9	5.7	5.4	5.8

Total basis (g/day)[4]						
Arginine	1.3	0.5	0.0	0.0	0.0	0.0
Histidine	3.6	3.4	3.3	3.2	3.0	3.2
Isoleucine	6.4	6.0	5.9	5.7	5.4	5.8
Leucine	9.9	9.0	8.6	8.2	7.7	8.3
Lysine	11.4	10.6	10.3	9.9	9.4	10.0
Methionine	2.9	2.7	2.6	2.6	2.4	2.6
Methionine + cystine	7.3	7.0	6.9	6.8	6.5	6.9
Phenylalanine	6.3	5.8	5.6	5.4	5.0	5.4
Phenylalanine + tyrosine	10.6	9.9	9.6	9.4	8.9	9.5
Threonine	8.6	8.3	8.3	8.2	7.8	8.3
Tryptophan	2.2	2.0	2.0	1.9	1.8	2.0
Valine	7.6	7.0	6.8	6.6	6.2	6.7

[1]Daily intakes of DE and feed and the amino acid requirements are estimated by the gestation model.
[2]Weight gain includes maternal tissue and products of conception.
[3]Assumes that ME is 96% of DE.
[4]Crude protein and total amino acid requirements are based on a corn–soybean meal diet.
Source: Reprinted with permission from Nutrient Requirements of Swine, *10th rev. ed., National Research Council, National Academy Press. 1998, p. 118, Table 10.8.*

TABLE 7.7   Dietary Amino Acid Requirements of Lactating Sows (90% dry matter)[1]

Sow postfarrowing weight (kg)	175	175	175	175	175	175
Anticipated lactation weight change (kg)[1]	0	0	0	−10	−10	−10
Daily weight gain of pigs (g)[1]	150	200	250	150	200	250
DE content of diet (kcal/kg)	3,400	3,400	3,400	3,400	3,400	3,400
ME content of diet (kcal/kg)[3]	3,265	3,265	3,265	3,265	3,265	3,265
Estimated DE intake (kcal/day)	14,645	18,205	21,765	12,120	15,680	19,240
Estimated ME intake (kcal/day)[3]	14,060	17,475	20,895	11,635	15,055	18,470
Estimated feed intake (kg/day)	4.31	5.35	6.40	3.56	4.61	5.66
Crude protein (%)[4]	16.3	17.5	18.4	17.2	18.5	19.2
Amino acid requirements *True ileal digestible basis (%)*						
Arginine	0.36	0.44	0.49	0.35	0.44	0.50
Histidine	0.28	0.32	0.34	0.30	0.34	0.36
Isoleucine	0.40	0.44	0.47	0.44	0.48	0.50
Leucine	0.80	0.90	0.96	0.87	0.97	1.03
Lysine	0.71	0.79	0.85	0.77	0.85	0.90
Methionine	0.19	0.21	0.22	0.20	0.22	0.23
Methionine + cystine	0.35	0.39	0.41	0.39	0.42	0.43
Phenylalanine	0.39	0.43	0.46	0.42	0.46	0.49
Phenylalanine + tyrosine	0.80	0.89	0.95	0.88	0.97	1.02
Threonine	0.45	0.49	0.52	0.50	0.53	0.56
Tryptophan	0.13	0.14	0.15	0.15	0.16	0.17
Valine	0.60	0.67	0.72	0.66	0.73	0.77
Total basis (%)[4]						
Arginine	0.40	0.48	0.54	0.39	0.49	0.55
Histidine	0.32	0.36	0.38	0.34	0.38	0.40
Isoleucine	0.45	0.50	0.53	0.50	0.54	0.57
Leucine	0.86	0.97	1.05	0.95	1.05	1.12
Lysine	0.82	0.91	0.97	0.89	0.97	1.03
Methionine	0.21	0.23	0.24	0.22	0.24	0.26
Methionine + cystine	0.40	0.44	0.46	0.44	0.47	0.49
Phenylalanine	0.43	0.48	0.52	0.47	0.52	0.55
Phenylalanine + tyrosine	0.90	1.00	1.07	0.98	1.08	1.14
Threonine	0.54	0.58	0.61	0.58	0.63	0.65
Tryptophan	0.15	0.16	0.17	0.17	0.18	0.19
Valine	0.68	0.76	0.82	0.76	0.83	0.88

[1]Daily intakes of DE and feed and the amino acid requirements are estimated by the lactation model.
[2]Assumes 10 pigs per litter and a 21-day lactation period.
[3]Assumes that ME is 96% of DE. In corn–soybean meal diets of these crude protein levels, ME is 95 to 96% of DE.
[4]Crude protein and total amino acid requirements are based on a corn–soybean meal diet.
Source: Reprinted with permission from Nutrient Requirements of Swine, *10th rev. ed., National Research Council, National Academy Press. 1998, p. 119, Table 10.9.*

TABLE 7.8   Daily Amino Acid Requirements of Lactating Sows (90% dry matter)[1]

Sow post-farrowing weight (kg)	175	175	175	175	175	175
Anticipated lactational weight change (kg)[2]	0	0	0	−10	−10	−10
Daily weight gain of pigs(g)[2]	150	200	250	150	200	250
DE content of diet (kcal/kg)	3,400	3,400	3,400	3,400	3,400	3,400
ME content of diet (kcal/kg)[3]	3,265	3,265	3,265	3,265	3,265	3,265
Estimated DE intake (kcal/day)	14,645	18,205	21,765	12,120	15,680	19,240
Estimated ME intake (kcal/day)[3]	14,060	17,475	20,895	11,635	15,055	18,470
Estimated feed intake (kg/day)	4.31	5.35	6.40	3.56	4.61	5.66
Crude protein (%)[4]	16.3	17.5	18.4	17.2	18.5	19.2

Amino acid requirements
True ileal digestible basis (g/day)

Arginine	15.6	23.4	31.1	12.5	20.3	28.0
Histidine	12.2	17.0	21.7	10.9	15.6	20.3
Isoleucine	17.2	23.6	30.1	15.6	22.1	28.5
Leucine	34.4	48.0	61.5	31.0	44.5	58.1
Lysine	30.7	42.5	54.3	27.6	39.4	51.2
Methionine	8.0	11.0	14.1	7.2	10.2	13.2
Methionine + cystine	15.3	20.6	26.0	13.9	19.2	24.5
Phenylalanine	16.8	23.3	29.7	14.9	21.4	27.9
Phenylalanine + tyrosine	34.6	47.9	61.1	31.4	44.6	57.8
Threonine	19.5	26.4	33.3	17.7	24.6	31.5
Tryptophan	5.5	7.6	9.7	5.2	7.3	9.4
Valine	25.8	35.8	45.8	23.6	33.6	43.6

Total basis (g/day)[4]

Arginine	17.4	25.8	34.3	14.0	22.4	30.8
Histidine	13.8	19.1	24.4	12.2	17.5	22.8
Isoleucine	19.5	26.8	34.1	17.7	25.0	32.3
Leucine	37.2	52.1	67.0	33.7	48.6	63.5
Lysine	35.3	48.6	61.9	31.6	44.9	58.2
Methionine	8.8	12.2	15.6	7.9	11.3	14.6
Methionine + cystine	17.3	23.4	29.4	15.7	21.7	27.8
Phenylalanine	18.7	25.9	33.2	16.6	23.9	31.1
Phenylalanine + tyrosine	38.7	53.4	68.2	35.1	49.8	64.6
Threonine	23.0	31.1	39.1	20.8	28.8	36.9
Tryptophan	6.3	8.6	11.0	5.9	8.2	10.6
Valine	29.5	40.9	52.3	26.9	38.4	49.8

[1]Daily intakes of DE and feed and the amino acid requirements are estimated by the lactation model.
[2]Assumes 10 pigs per litter and a 21-day lactation period.
[3]Assumes that ME is 96% of DE. In corn–soybean meal diets of these crude protein levels, ME is 95 to 96% of DE.
[4]Crude protein and total amino acid requirements are based on a corn–soybean meal diet.
Source: Reprinted with permission from Nutrient Requirements of Swine, *10th rev. ed., National Research Council, National Academy Press. 1998, p. 120, Table 10.10.*

TABLE 7.9   **Dietary and Daily Mineral, Vitamin, and Linoleic Acid Requirements of Gestating and Lactating Sows (90% dry matter)[1]**

	Gestation	Lactation	Gestation	Lactation
DE content of diet (kcal/kg)	3,400	3,400	3,400	3,400
ME content of diet (kcal/kg)[2]	3,265	3,265	3,265	3,265
DE intake (kcal/day)	6,290	17,850	6,290	17,850
ME intake (kcal/day)[2]	6,040	17,135	6,040	17,135
Feed intake (kg/day)	1.85	5.25	1.85	5.25
	(% or amount/kg of diet)		(amount/day)	
Minerals				
Calcium (%)	0.75	0.75	13.9	39.4
Phosphorus, total (%)	0.60	0.60	11.1	31.5
Phosphorus, available (%)	0.35	0.35	6.5	18.4
Sodium (%)	0.15	0.20	2.8	10.5
Chlorine (%)	0.12	0.16	2.2	8.4
Magnesium (%)	0.04	0.04	0.7	2.1
Potassium (%)	0.20	0.20	3.7	10.5
Copper (mg)	5.00	5.00	9.3	26.3
Iodine (mg)	0.14	0.14	0.3	0.7
Iron (mg)	80	80	148	420
Manganese (mg)	20	20	37	105
Selenium (mg)	0.15	0.15	0.3	0.8
Zinc (mg)	50	50	93	263
Vitamins				
Vitamin A (IU)[3]	4,000	2,000	7,400	10,500
Vitamin D_3 (IU)[3]	200	200	370	1,050
Vitamin E (IU)[3]	44	44	81	231
Vitamin K (menadione) (mg)	0.50	0.50	0.9	2.6
Biotin (mg)	0.20	0.20	0.4	1.1
Choline (g)	1.25	1.00	2.3	5.3
Folacin (mg)	1.30	1.30	2.4	6.8
Niacin, available (mg)[4]	10	10	19	53
Pantothenic acid (mg)	12	12	22	63
Riboflavin (mg)	3.75	3.75	6.9	19.7
Thiamin (mg)	1.00	1.00	1.9	5.3
Vitamin B_6 (mg)	1.00	1.00	1.9	5.3
Vitamin B_{12} (mg)	15	15	28	79
Linoleic acid (%)	0.10	0.10	1.9	5.3

[1]The requirements or daily amounts are based on the daily consumption of 1.85 gestation and 5.25 kg of lactation feed, respectively. If lower amounts of feed are consumed, the dietary percentage may need to be increased.

[2]Assumes that ME is 96% of DE.

[3]Conversions: 1 IU vitamin A = 0.344 ug retinyl acetate; 1 IU vitamin D_3 = 0.025 ug cholecalciferol; 1 IU vitamin E = 0.67 mg of d-α-tocopherol or 1 mg of dl-α-tocopherol acetate.

[4]The niacin in corn, grain sorghum, wheat, and barley is unavailable. Similarly, the niacin in by-products made from these cereal grains is poorly available unless the by-products have undergone a fermentation or wet-milling process.

Source: Reprinted with permission from Nutrient Requirements of Swine, *10th rev. ed., National Research Council, National Academy Press. 1998, p. 121–22, Tables 10.11 and 10.12.*

TABLE 7.10   Dietary and Daily Amino Acid, Mineral, Vitamin, and Fatty Acid Requirements of Sexually Active Boars (90% dry matter)[1]

				% or amt/kg	amt/day
DE content of diet (kcal/kg)	3,400	3,400			
ME content of diet (kcal/kg)	3,265	3,265	**Minerals**		
DE intake (kcal/day)	6,800	6,800	Calcium	0.75 %	15.0 g
ME intake (kcal/day)	6,530	6,530	Phosphorus, total	0.60 %	12.0 g
Feed intake (kg/day)	2.00	2.00	Phosphorus, available	0.35 %	7.0 g
Crude protein (%)[2]	13.0	13.0	Sodium	0.15 %	3.0 g
			Chlorine	0.12 %	2.4 g
	% or amt/kg	**amt/day**	Magnesium	0.04 %	0.8 g
			Potassium	0.20 %	4.0 g
Amino acids (Total basis)			Copper	5 mg	10 mg
Arginine	—	—	Iodine	0.14 mg	0.28 mg
Histidine	0.19 %	3.8 g	Iron	80 mg	160 mg
Isoleucine	0.35 %	7.0 g	Manganese	20 mg	40 mg
Leucine	0.51 %	10.2 g	Selenium	0.15 mg	0.3 mg
Lysine	0.60 %	12.0 g	Zinc	50 mg	100 mg
Methionine	0.16 %	3.2 g	**Vitamins**		
Methionine + cystine	0.42 %	8.4 g	Vitamin A[3]	4,000 IU	8,000 IU
Phenylalanine	0.33 %	6.6 g	Vitamin D$_3$[3]	200 IU	400 IU
Phenylalanine + tyrosine	0.57 %	11.4 g	Vitamin E[3]	44 IU	88 IU
Threonine	0.50 %	10.0 g	Vitamin K (menadione)	0.50 mg	1.0 mg
Tryptophan	0.12 %	2.4 g	Biotin	0.20 mg	0.4 mg
Valine	0.40 %	8.0 g	Choline	1.25 g	2.5 g
			Folacin	1.30 mg	2.6 mg
			Niacin, available[4]	10 mg	20 mg
			Pantothenic acid	12 mg	24 mg
			Riboflavin	3.75 mg	7.5 mg
			Thiamin	1.0 mg	2.0 mg
			Vitamin B$_6$	1.0 mg	2.0 mg
			Vitamin B$_{12}$	15 mg	30 mg
			Linoleic acid	0.1 %	2.0 g

[1]The requirements are based on the daily consumption of 2.0 kg of feed. Feed intake may need to be adjusted, depending on the weight of the boar and the amount of weight gain desired.

[2]Assumes a corn–soybean meal diet. The lysine requirement was set as 0.60% (12.0 g/day). Other amino acids were calculated using ratios (total basis) similar to those for gestating sows.

[3]Conversions: 1 IU vitamin A 4 0.344 mg retinyl acetate; 1 IU vitamin D$_3$ 4 0.025 mg cholecalciferol; 1 IU vitamin E 4 0.67 mg of d-α-tocopherol or 1 mg of dl-α tocopheryl acetate.

[4]The niacin in corn, grain sorghum, wheat, and barley is unavailable. Similarly, the niacin in by-products made from these cereal grains is poorly available unless the by-products have undergone a fermentation or wet-milling process.

Source: Reprinted with permission from Nutrient Requirements of Swine, *10th rev. ed., National Research Council, National Academy Press. 1998, p. 123, Table 10.13.*

Grow-Finish Feed

Lean-growth capacity is an important factor in estimating the feed efficiency of pigs from 50 to 265 lb (23 to 120 kg) body weight. Assuming a feed efficiency of 2.9 for high lean-growth pigs results in a feed requirement of 623 lb (283 kg) during this stage. Moderate lean-growth pigs have a feed efficiency of 3.3 and require 710 lb (322 kg) of grower-finisher diet.

Product Outputs

The feed inputs result in 5 lb (2.3 kg) sow gain per pig marketed (90 lb (41 kg) gain per year ÷ 18.1 pigs) plus the pig market weight for a total gain of 270 lb (122.8 kg) of pork per pig produced. High-lean pigs yield 74% carcass weight containing 50% fat-free lean or about 98 lb (44.5 kg) of lean. Moderate-lean pigs yield 74.2% carcass weight containing 46% fat-free lean or about 90.4 lb (41.1 kg) of muscle.

RECOMMENDED NUTRIENT ALLOWANCES

Because swine producers are interested in optimizing performance, the input of nutrients must be ample to allow this without feeding great excesses. Nutrient requirements generally represent the minimum quantity of the nutrients that should be incorporated in a diet to maximize performance under

TABLE 7.11   Expected Feed Inputs and Product Outputs per 265 lb (120 kg) Pig Produced under Standard Production Conditions[1]

	Lean Growth Capacity			
Item	*High*	*High*	*Moderate*	*Moderate*
Feed inputs, kg	lb	kg	lb	kg
Pregestation, gestation	84	38.1	80	36.3
Lactation	28	12.7	28	12.7
Boar feed	6	2.7	6	2.7
Starter, 12–50 lb (5.5–23 kg)	51	23.2	63	28.6
Grow-finish, 50–265 lb (23–120 kg)	623	283.0	710	322.8
Total	792	359.7	887	403.1
Product outputs				
Sow gain/pig produced	5	2.3	5	2.3
Pig	265	120.5	265	120.5
Total	270	122.8	270	122.8
Carcass lean (fat-free)	98.0	44.5	90.4	41.1
Efficiency of production:				
Feed/unit pig weight produced		2.93		3.28
Feed/unit muscle produced		9.08		9.81

[1] These values are estimated allocations of sow and boar feed to each pig marketed plus the feed consumed by the market pig. The 1994 ISU Swine Enterprise Record Summary indicates an average whole herd efficiency of 3.69 units of feed/unit of pork produced with a standard deviation of 0.17. Thus, a reasonable goal would be 3.30 units of feed/unit of pork produced, or about 1 standard deviation better than average.

Source: Adapted from ISU Life Cycle Swine Nutrition, *1996.*

standard conditions, whereas nutrient allowances take into consideration a safety margin.

Basic nutrient requirements are based on weight and genetic merit. However, adjustments need to be made for conditions that vary from a defined set of standard production conditions. Table 7.12 defines the standard production conditions on which the nutrient allowances in Tables 7.14 (nutrient allowances for growing pigs) and 7.22 (nutrient allowances for the breeding herd) are based. Growing pigs exposed to conditions that vary from these standards should have the nutrient needs redefined using Table 7.17a and 7.17b. Adjustments for sows and boars are in Tables 7.20, 7.21, and 7.23.

DEFINING THE NUTRIENT ALLOWANCES OF GROWING PIGS

Biological Basis for Establishing Nutrient Requirements

Animal growth and reproduction are the result of numerous biological processes regulated by genetic and production conditions. The genotype of the animal determines the maximum rate at which these processes occur. Conditions, such as thermal and social climates, dietary regimens, and health status,

determine the proportion of the genetic potential actually expressed (Figure 7.1). These biological processes require inputs of energy (carbohydrates and lipids), amino acids, minerals, and vitamins. Thus, the rates at which these processes occur are affected by the nutrient intake of the animals.

The major biological processes occurring in the pig are grouped into maintenance and growth. Maintenance includes repair or replacement of body tissues and fluids, energy for voluntary (e.g., walking) and involuntary (e.g., heart contractions) activities, generation of body heat for warmth, and regulation of body defense (e.g., immune) systems. Growth includes accretion of body tissues (e.g., muscle, bone), organs (e.g., mammary gland), fluids (e.g., milk), and fluid components (e.g., red blood cells).

The primary determinants of nutrient needs for maintenance are body weight (tissue mass) and to a lesser degree physical activity, body temperature regulation, and body defense systems. The primary determinants of nutrient needs for growth are tissue (e.g., lean, fetal-placental) accretion rates and milk yield, or pig weight gain preweaning.

Nutrients consumed in excess of the amount needed to support these biological processes are stored in the body or degraded and excreted. Con-

TABLE 7.12 Standard Production Conditions

Herd Factors	Standard Conditions
Pig gender	1:1 ratio of barrows and gilts
Antigen exposure	Modest (Table 7.15)
Thermal climate	Thermoneutral (Figure 7.4)
Social climate	
Pen and feeder space/pig	Adequate or above (Figure 7.5)
Pig weight variation	Less than 20% of mean weight
Dietary ingredients	
Grain source	Corn
Protein source	
2.7–14 kg pigs	Soy and animal proteins
Over 14 kg pigs	Soybean meal, dehulled
Vitamins and trace minerals	Biologically available and stable
Antimicrobial agent	Subtherapeutic levels
Feeding regimen	
Feed intake	Ad libitum
Feed form	Meal
Feed particle size	650 to 750 microns
Feed toxins	None
Growth modifiers	None

Breeding Herd Factors	Standard Conditions
Parity	Gilt (1st parity)
Initial breeding weight	130 kg
Feed particle size	750 to 900 microns
Sow weight gain	
Parities 1–3	37–41 kg
Parities 4–5	34 kg
Parities 6–8	20 kg
Lean-growth capacity	Moderate (Table 7.13)
Sow body condition	Optimum
Lactation weight loss	None postfarrowing to weaning

Source: Adapted from ISU Life Cycle Swine Nutrition, 1996.

Figure 7.1 These bred gilts in open front barns and fed outside have significantly different nutrient requirements than sows housed in gestating stalls, particularly in their energy requirements. *(Courtesy, Palmer Holden, Iowa State University, Ames, IA)*

of the pigs; and (5) the presence of substances that may be harmful to product quality.

Economic Basis for Establishing Nutrient Requirements

Economic returns in the pork industry are the value of the pork products produced minus the costs expended to produce them. The value of pork products is determined by consumer perception of the quality of the food products produced and the acceptability of production methods. Production costs per unit of quality pork produced normally is lowest in pigs provided the amounts and types of nutrients needed to allow each biological process to proceed at the maximum rate possible in pigs maintained under specific production conditions. Deficient or excess intakes of nutrients relative to the animals' biological demands will lower biological and economic efficiency of pork production. Dietary additions of nutrients resulting in only small enhancements in performance may not return the cost of the added nutrients. Currently, dietary regimens that result in the lowest feed cost per pound of muscle of acceptable quality produced yet that maintain the well-being of the animal and the environment are recommended.

Standard Production Conditions

Nutrient requirements should be based on a defined group of pigs raised under a defined set of conditions (Tables 7.12 and 7.14). Growing pigs, sows, and boars exposed to production conditions that deviate from these standard conditions must have their nutrient needs redefined (Tables 7.17a or 7.17b, 7.20, and 7.22).

sequently, nutrients consumed in excess of those needed to maximize tissue growth may contribute to the nutrient content (e.g., vitamin, mineral, and lipid) of edible pork products and to the physical properties (e.g., color, water-holding capacity) of these products. However, a high proportion of excess nutrients are excreted in the urine and feces.

Before formulating a swine diet, it is necessary to know (1) the nutrient requirements of the particular pigs to be fed, including weight and genotype; (2) the availability, nutrient content, and cost of feedstuffs; (3) the acceptability and physical condition of feedstuffs; (4) the average daily feed intake

TABLE 7.13   Pig Lean-Growth Capacities from 40 to 265 lb (18 to 120 kg)

	Fat-Free Lean	
Lean-Growth Capacity	*lb/day*	*kg/day*
High	0.65–0.77	0.29–0.35
Moderate	0.52–0.60	0.24–0.27
Low	0.34–0.47	0.15–0.21

Source: Adapted from ISU Life Cycle Swine Nutrition, *1996.*

Pig Lean-Growth Capacities

The first step in defining nutrient requirements for pigs depends on their genetic capacity for lean growth. The lean-growth capacities on which the nutrient allowances are based are defined in Table 7.13. The lean-growth capacities, high, moderate, and low, arbitrarily separate pigs into three genetic sets to define allowances. These genetic capacities are not based only on leanness and loin eye area because the leanest market hogs are often the slower-growing pigs, and the goal is not to define diets for slow-growing pigs. The more valuable genetic measurement is the ability of the pigs to rapidly grow lean muscle. The values in Table 7.13 are based on pigs in a feeding period from post-nursery to market.

Adjusting the Nutrient Allowances of Growing Pigs

Following is an outline for devising a set of nutrient allowances appropriate for the pigs to be fed. Nutrient requirements for pigs vary depending on their genetic capacities for growth and the production conditions to which they are subjected.

Determine Genetic Capacity

Determine the genetic capacity of the pigs for lean growth measured in lb or kg of lean growth per day. This may be done with any of the following methods:

1. Use published values for the genetic line involved;
2. Estimate the lean-growth capacity of the pigs combining growth data from the swine enterprise and slaughter records using procedures described in the following section, "Estimate Genetic Capacity for Lean Growth"; or
3. Assume the pigs are of moderate lean-growth capacity.
4. These estimations should be made on pigs reared under production conditions approxi-

mating those defined in Table 7.12. Use the estimated lean-growth values from Table 7.13 to categorize the pigs into high, moderate, or low lean-growth capacities.

Estimate Genetic Capacity for Lean Growth

Several procedures exist for determining the rate of lean growth. These provide estimates only and should be periodically reevaluated. The information should be derived from groups of pigs of a common genetic capacity evaluated under standard production conditions similar to those described in Table 7.13. Lean-growth estimates derived under other than the standard production conditions must be adjusted or even discarded. Evaluate the pigs' lean growth from body weights of about 40 to 265 lb (18 to 120 kg) for the best estimate of lean- growth capacity.

Example Estimations of Fat-Free Lean Growth

a. **Example data**

		Loin muscle	
Initial live wt.	43 lb	area (LMA)	5.5 in.[2]
43 lb to market	110 days	Backfat, 10th rib	1.0 in.[1]
Market weight	265 lb	Fat Free Lean	43.8%[2]
Hot carcass wt.			
(HCW)	196 lb	Yield	74%

[1]To convert packer last rib midline fat to 10th rib off-midline fat, subtract 0.15 from last rib midline measurement. If LMA is not known, use 5.5 in.[2]
[2]The fat-free lean percentage is usually provided on slaughter reports.

b. **Estimating pounds of carcass fat-free lean (FFL) and percentage FFL (if not provided on slaughter report)**

$$FFL \text{ (lb)} = (0.465 \times HCW) + (3.005 \times LMA) - (21.896 \times Backfat, 10^{th} \text{ rib})$$

$$FFL \text{ (lb)} = (0.465 \times 196) + (3.005 \times 5.5) - (21.896 \times 1.0) = 85.8 \text{ lb}$$

$$FFL \text{ (\%)} = (FFL \text{ (lb)} \div \text{Hot carcass wt.}) \times 100$$

$$FFL \text{ (\%)} = (85.77 \div 196) \times 100 = 43.8\%$$

c. **Estimating fat-free lean (FFL) in the carcass and in the feeder pig**

$$\text{Carcass FFL (lb)} = FFL \text{ (\%)} \times HCW \text{ (lb)}$$

$$\text{Carcass FFL (lb)} = 43.8\% \times 196 \text{ lb} = 85.8 \text{ lb FFL}$$

$$\text{Feeder pig FFL (lb)} = 43.8 \times \text{Feeder pig weight (lb)} - 3.65$$

$$\text{Feeder pig FFL (lb)} = (41.8\% \times 43 \text{ lb}) - 3.65 = 14.3 \text{ lb}$$

Figure 7.2 Overcrowded pens reduce feed intake and increase aggression. *(Courtesy, Palmer Holden, Iowa State University, Ames, IA)*

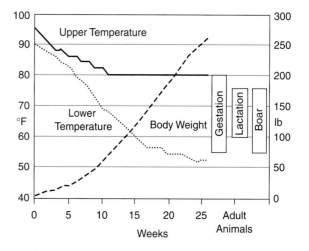

Figure 7.3 Optimum temperature range. The three bars on the right indicate the range for gestation, lactation, and for boars. *(Courtesy, Iowa State University, Ames, IA)*

d. FFL growth (lb)/days on test = FFL gain/day (lb)

FFL growth from feeder pig wt. to market wt.
= Carcass FFL (lb) − Feeder pig FFL (lb)

FFL growth = 85.5 lb − 14.3 lb = 71.2 lb

71.2 lb/110 days = 0.65 lb FFL/day
= High-lean gain (Table 7.13)

e. Quick lean-growth estimator [1]

Avg. daily gain, lb	10th Rib Fat Thickness, in.				
	0.7	0.8	0.9	1.0	1.1
	Daily Fat-Free Lean Growth, lb				
1.60	0.57	0.56	0.54	0.53	0.51
1.80	0.64	0.63	0.61	0.59	0.58
2.00	0.68	0.66	0.65	0.63	0.61
2.20	0.78	0.77	0.75	0.73	0.71

[1] Assumes gain from approximately 40 lb to market weight and 74% dressing percentage.

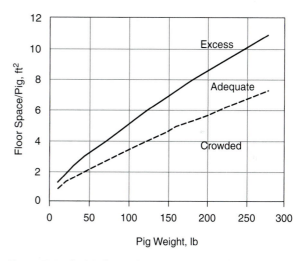

Figure 7.4 Social climate for growing pigs. Restricting space reduces performance. Excess space is inadequate use of the facility. *(Adapted from ISU Life Cycle Swine Nutrition, 1996)*

Identify the Stage of Development

Identify the stage of development and associated nutrient requirements for groups of pigs based on their lean-growth capacity and body weight using Table 7.14, nutrient allowances of growing pigs under standard production conditions.

ADJUSTING FOR NONSTANDARD PRODUCTION CONDITIONS

After defining the nutrient allowances for pigs of certain genetic merit raised under standard conditions it is recommended that adjustments be made for nonstandard conditions. For example temperature, space, and exposure to diseases and the devel-

opment of antigens all will affect the pig's ability to grow and use feed effectively (Figure 7.2).

Climatic, Social, and Antigens

Determine climatic, social, and antigen adjustments for conditions under which the pigs are raised that differ from the standard production conditions. The thermoneutral and social climates for growing pigs are illustrated in Figures 7.3 and 7.4. It is assumed pigs are penned in groups of 10 to 30 on slotted floors, in insulated buildings with drafts less than 50 ft/sec and a relative humidity between 20 and 80%. Pigs housed on wet floors, in poorly insulated buildings, and in drafts require the temperature to be increased 3 to 5°F, 3 to 5°F, and 5 to 7°F, respectively, to maintain thermoneutrality.

TABLE 7.14 Nutrient Allowances of Growing Pigs Reared under Standard Production Conditions

Lean-Growth Capacity

Pig Weights at Which Diets Should Be Fed, lb

		1	2	3	4	5	6	7	8	9	10	11	12	13	14
High		6–12	12–18	18–27	27–39	39–55	55–75	75–100	100–130	130–166	166–208	208–257	257–313		
Moderate		6–8	8–13	13–18	18–26	26–37	37–51	51–69	69–91	91–118	118–150	150–188	188–233	233–283	
Low			6–9	9–14	14–19	19–26	26–36	36–49	49–65	65–85	85–109	109–138	138–162	162–201	201–246
Stage of Development															
Amino acids															
Lysine	%	1.90	1.71	1.54	1.39	1.25	1.13	1.02	0.92	0.83	0.75	0.67	0.60	0.54	0.49
Threonine	%	1.29	1.16	1.05	0.94	0.84	0.74	0.67	0.61	0.55	0.51	0.46	0.42	0.38	0.35
Tryptophan	%	0.38	0.34	0.31	0.28	0.25	0.22	0.20	0.18	0.17	0.15	0.13	0.12	0.11	0.10
Methionine + Cystine	%	1.06	0.96	0.86	0.78	0.70	0.63	0.57	0.52	0.47	0.44	0.39	0.35	0.32	0.29
Crude protein[1]	%	27.70	27.10	26.50	24.00	22.10	20.30	18.80	17.50	16.10	15.10	14.00	13.00	12.20	11.40
Metabolizable energy[1]	kcal/lb	1566	1469	1466	1480	1484	1486	1489	1493	1496	1498	1499	1502	1503	1504
Minerals															
Calcium	%	1.11	1.04	0.95	0.88	0.80	0.75	0.70	0.65	0.61	0.59	0.56	0.52	0.50	0.49
Phosphorus, total	%	0.93	0.86	0.78	0.72	0.66	0.60	0.56	0.52	0.49	0.47	0.45	0.42	0.40	0.39
Phosphorus, available	%	0.70	0.62	0.54	0.48	0.41	0.36	0.31	0.27	0.24	0.21	0.19	0.16	0.14	0.12
Sodium (Na)[2]	%	0.21	0.20	0.19	0.18	0.17	0.16	0.16	0.15	0.15	0.14	0.14	0.13	0.13	0.12
Chlorine (Cl)[2]	%	0.31	0.30	0.28	0.27	0.26	0.24	0.24	0.22	0.22	0.21	0.21	0.20	0.20	0.19
Trace minerals (added)															
Iron	ppm	174	160	147	135	124	114	105	97	89	82	75	69	64	58
Zinc	ppm	174	160	147	135	124	114	105	97	89	82	75	69	64	58
Copper	ppm	10.8	10.0	9.2	8.5	7.8	7.2	6.6	6.1	5.6	5.1	4.7	4.4	4.0	3.7
Manganese	ppm	5.4	5.0	4.6	4.3	3.9	3.6	3.3	3.1	2.8	2.6	2.4	2.2	2.0	1.9
Iodine	ppm	0.22	0.21	0.2	0.19	0.17	0.16	0.15	0.14	0.14	0.13	0.12	0.11	0.11	0.1
Selenium	ppm	0.30	0.30	0.30	0.30	0.28	0.25	0.23	0.21	0.2	0.18	0.17	0.15	0.14	0.13
Vitamins (added/lb of diet)															
Vitamin A	IU	2300	2200	2100	1900	1800	1700	1600	1500	1400	1300	1250	1200	1100	1000
Vitamin D3	IU	230	220	210	190	180	170	160	150	140	130	125	120	110	100
Vitamin E	IU	16	15	14	13	12	12	11	10	10	9	9	8	8	7
Vitamin K (menadione)	mg	0.69	0.65	0.61	0.57	0.54	0.51	0.48	0.45	0.42	0.40	0.37	0.35	0.33	0.31
Niacin	mg	31	28	25	23	20	18	16	14	12	10	9	8	7	6
Pantothenic acid	mg	20	18	16	15	13	12	11	10	9	8	7	6	5	5
Riboflavin	mg	7.1	6.4	5.8	5.3	4.8	4.2	3.7	3.2	2.9	2.5	2.2	1.9	1.7	1.5
Vitamin B12	mcg	35	32	29	26	23	20	18	16	14	12	10	9	8	7
Biotin[3]	mg	0.013	0.012	0.011	0.010	0.009	0.007	0.005	0.002						
Folic acid[3]	mg	0.061	0.055	0.050	0.045	0.036	0.03	0.024	0.012						
Choline[3]	mg	90	80	70	62	55	48	32	16						

[1] The crude protein and metabolizable energy values are the results of formulating the example diets in Table 7.18. They are not minimum levels.

[2] Sodium and chlorine supplementation is provided by salt (NaCl).

[3] Biotin, folic acid, and choline additions are not required for pigs with a moderate or low lean-growth capacity and major or maximal antigen exposure.

Source: Adapted from ISU Life Cycle Swine Nutrition, 1996.

TABLE 7.15   Level of Antigen Exposure

	Antigen Exposure				
	Minimal	*Modest*	*Moderate*	*Major*	*Maximal*
Factor Score =	*1*	*2*	*3*	*4*	*5*
Facility all in/all out[1]	Always	Usually	Some	Rarely	Never
Multiple groups/room[2]	Never	Rarely	Some	Usually	Always
Facility sanitation[3]	Always	Usually	Some	Rarely	Never
Facility biosecurity[4]	High	Moderate	Low	Minimal	None
No. antigens present[5]					
Serum titers	0	1	2	3	>3
Visual symptoms	0	1	2	3	>3
Poor performance[6]	0–2%	2–3%	3–5%	5–7%	7–10%
Mortality[7]	0–1%	1–2%	2–3%	3–5%	4–10%

[1] Facility completely emptied of pigs between pig groups.

[2] Two or more groups from different buildings, farms, or sources placed in the same room.

[3] Facility high pressure sprayed and disinfected between groups; surfaces resealed as needed.

[4] Facility biosecurity: High = separate sites (minimum of 0.5 miles apart) and labor, shower in; Moderate = separate sites (less than 0.2 miles apart), same labor, shower in; Low = same site, separate labor, shower in; Minimal = same site and labor, clean coveralls and boots; None = same site and labor force, no precautions.

[5] Number of antigens present based on 1) serum antibody titers indicating the number of disease antigens that the pigs have been exposed to (e.g., mycoplasmal pneumonia, APP, PRV, PRRS, TGE) and 2) visual symptoms of illness (e.g., coughing, diarrhea, emaciation).

[6] Percentage of pigs in a group that grow 20 to 30 percent below average.

[7] The percentage death loss of pigs in a group.

Source: Adapted from ISU Life Cycle Swine Nutrition, 1996. An antigen is any substance that stimulates the production of antibodies.

TABLE 7.16   Default Dietary Regimens

Lean-Growth Capacity	Pig Weights at Which Diets Should Be Fed, lb					Pig Weights at Which Diets Should Be Fed, kg				
High	12–39	39–100	100–208	208–313	-	5–18	18–45	45–94	94–142	-
Moderate	8–26	26–69	69–150	150–283	-	4–12	12–31	31–68	68–129	-
Low	6–19	19–49	49–109	109–201	201–246	3–9	9–22	22–50	50–91	91–112
Stage of Development	3	6	9	12	14	3	6	9	12	14

Source: Adapted from ISU Life Cycle Swine Nutrition, 1996.

Figure 7.5 Pregnant gilts on pasture will get some of their nutritional needs from the forage. *(Courtesy, John McGlone, Texas Tech, Lubbock, TX)*

Social climate is strongly related to available space per pig, including pen space, feeder space, and waterer space (Figure 7.5). Crowded pens reduce feed intake and increase fighting and tail bit-ing. Excess space is not a performance issue but does reduce the economical use of the available space. When subjected to cold climates, insufficient pigs per room limits the amount of heat they provide to maintain the room temperature.

Antigen Exposure

Exposure to antigens, or substances that stimulate the production of antibodies, has a marked effect on nutrient requirements. Classify groups of pigs at each stage of growth (e.g., nursery and grow-finish) for their level of antigen exposure according to Table 7.15. The footnotes describe how to estimate the value for each factor. Assign a 1, 2, 3, 4, or 5 score, respectively, if the group has minimal, modest, moderate, major, or maximal exposure for each factor listed. Total the numbers for the factors evaluated, divide by the number of factors evaluated, and round off to the nearest whole number. This number represents the

TABLE 7.17a High Lean-Growth Pig Adjustments of Nutrient Allowances (Table 7.14) to Specific Production Conditions

						Pig Weights at Which Diets Should Be Fed						
lb (kg)	6–12 (3–5)	12–18 (5–8)	18–27 (8–12)	27–39 (12–18)	39–55 (18–25)	55–75 (25–34)	75–100 (34–45)	100–130 (45–59)	130–166 (59–75)	166–207 (75–94)	207–257 (94–117)	257–313 (117–142)
	1	*2*	*3*	*4*	*5*	*6*	*7*	*8*	*9*	*10*	*11*	*12*
Stage of Development												
Pig gender												
Boars	0	0	0	0	0	−0.15	−0.35	−0.70	−1.15	−1.60	−1.40	−1.10
Gilts	0	0	0	0	0	−0.10	−0.30	−0.60	−1.00	−1.40	−1.00	−0.70
Barrows	0	0	0	0	0	+0.10	+0.30	+0.60	+1.00	+1.40	+1.00	+0.70
Thermal climate (Figure 7.4)												
Cold −10°F below TN	+0.60	+0.60	+0.50	+0.40	+0.40	+0.70	+0.70	+0.70	+0.80	+0.90	+1.40	+2.00
− 5°F below TN	+0.30	+0.30	+0.25	+0.20	+0.20	+0.35	+0.35	+0.35	+0.40	+0.45	+0.70	+1.00
Hot + 5°F above TN	0	−0.05	−0.05	−0.10	−0.10	−0.15	−0.30	−0.50	−0.60	−0.70	−0.80	−0.90
+10°F above TN	0	+0.05	+0.10	+0.15	+0.20	+0.30	+0.45	+0.60	+0.70	+0.80	+0.90	+1.00
Social climate (Figure 7.5)												
Crowded by 20%	+0.15	+0.12	+0.10	+0.08	+0.06	+0.03	0	−0.05	−0.15	−0.30	−0.50	−0.70
Dietary form												
Particle > 900 microns	+0.30	+0.25	+0.20	+0.15	+0.12	+0.10	+0.08	+0.05	+0.02	−0.01	−0.04	−0.08
Pelleted [1]	−0.15	−0.15	−0.10	−0.10	−0.10	−0.10	−0.10	−0.12	−0.15	−0.25	−0.35	−0.45
Antigen exposure (Table 7.15)												
Minimal	−0.30	−0.40	−0.50	−0.50	−0.55	−0.55	−0.55	−0.55	−0.60	−0.60	−0.60	−0.60
Modest	0	0	0	0	0	0	0	0	0	0	0	0
Moderate	+0.30	+0.40	+0.50	+0.50	+0.55	+0.55	+0.55	+0.55	+0.60	+0.60	+0.60	+0.60
Major	+0.60	+0.80	+1.00	+1.00	+1.10	+1.10	+1.10	+1.10	+1.20	+1.20	+1.20	+1.20
Maximal	+0.90	+1.20	+1.50	+1.50	+1.65	+1.65	+1.65	+1.65	+1.80	+1.80	+1.80	+1.80
Dietary ingredients other than corn and soybean meal	- Maintain same digestible nutrient/ME ratio as in Table 7.14. - If digestibility of nutrients (especially amino acids and phosphorous) in alternative ingredients differ substantially from that of corn and soybean meal, raise and lower dietary nutrient/ME ratios accordingly.											
Growth regulators	- Adjust nutrient concentrations based on the impact of growth regulators on feed intake and lean tissue growth and nutrient digestibility.											

[1] Increase dietary vitamin concentrations by 20% when pelleting the diet.

Source: Adapted from ISU Life Cycle Swine Nutrition, 1996.

TABLE 7.17b Moderate Lean-Growth Pig Adjustments of Nutrient Allowances (Table 7.14) to Specific Production Conditions

						Pig Weights at Which Diets Should Be Fed							
lb (kg)	6–8 (3–4)	8–13 (4–6)	13–18 (6–8)	18–26 (8–12)	26–37 (12–17)	37–51 (17–23)	51–69 (23–31)	69–91 (31–41)	91–118 (41–54)	118–150 (54–68)	150–188 (68–85)	188–233 (85–106)	233–283 (106–129)
	1	2	3	4	5	6	7	8	9	10	11	12	13
Stage of Development													
Pig gender													
Boars	0	0	0	0	0	0	−0.15	−0.30	−0.60	−1.10	−1.60	−1.30	−0.80
Gilts	0	0	0	0	0	0	−0.10	−0.25	−0.50	−0.90	−1.30	−1.00	−0.50
Barrows	0	0	0	0	0	+0.10	+0.10	+0.25	+0.60	+0.90	+1.30	+0.90	+0.50
Thermal climate (Figure 7.4)													
Cold −10°F below TN	+0.60	+0.60	+0.50	+0.40	+0.40	+0.70	+0.70	+0.70	+0.80	+0.90	+1.40	+2.00	+2.60
−5°F below TN	+0.30	+0.30	+0.25	+0.20	+0.20	+0.35	+0.35	+0.35	+0.40	+0.45	+0.70	+1.00	+1.40
Hot +5°F above TN	0	−0.05	−0.05	−0.10	−0.10	−0.15	−0.30	−0.50	−0.60	−0.70	−0.80	−1.00	−1.15
+10°F above TN	0	+0.05	+0.10	+0.15	+0.20	+0.30	+0.45	+0.80	+0.90	+1.00	+1.10	+1.20	+1.35
Social climate (Figure 7.5)													
Crowded by 20%	+0.12	+0.10	+0.08	+0.05	+0.03	−0.05	−0.15	−0.25	−0.40	−0.60	−0.80	−1.00	−1.20
Dietary form													
Particle > 900 microns	+0.30	+0.25	+0.20	+0.15	+0.10	+0.07	+0.04	+0.01	0	0	0	0	0
Pelleted[1]	−0.15	−0.15	−0.10	−0.10	−0.10	−0.10	−0.10	−0.12	−0.15	−0.25	−0.35	−0.50	−0.65
Antigen exposure (Table 7.15)													
Minimal	−0.30	−0.35	−0.35	−0.40	−0.40	−0.40	−0.40	−0.40	−0.45	−0.45	−0.50	−0.50	−0.55
Modest	0	0	0	0	0	0	0	0	0	0	0	0	0
Moderate	+0.30	+0.35	+0.35	+0.40	+0.40	+0.40	+0.40	+0.40	+0.45	+0.45	+0.50	+0.50	+.55
Major	+0.60	+0.70	+0.70	+0.80	+0.80	+0.85	+0.85	+0.85	+0.90	+0.90	+1.00	+1.05	+1.10
Maximal	+0.90	+1.05	+1.05	+1.25	+1.25	+1.30	+1.30	+1.30	+1.45	+1.45	+1.50	+1.60	+1.65
Dietary ingredients other than corn and soybean meal	- Maintain same digestible nutrient/ME ratio as in Table 7.14.												
	- If the digestibility of nutrients (especially amino acids and phosphorous) in alternative ingredients differ substantially from that of corn and soybean meal, raise and lower dietary nutrient/ME ratios accordingly.												
Growth regulators	- Adjust nutrient concentrations based on the impact of growth regulators on feed intake and lean tissue growth and nutrient digestibility.												

[1] Increase dietary vitamin concentrations by 20% when pelleting the diet.
Source: Adapted from ISU Life Cycle Swine Nutrition, 1996.

pigs' levels of antigen exposure to be used in adjusting their nutrient needs in Tables 7.17a and 7.17b.

A simplified antigen exposure estimate may be used. For example, 1 = medicated early weaning (MEW), 2 = segregated early weaning (SEW), 3 = all in/all out (AIAO), 4 = continuous flow, and 5 = chronic disease.

Dietary Adjustments

If you prefer not to make adjustments for environmental factors, use Table 7.16. However, it is recommended for best feeding efficiency to make the appropriate adjustments for specific production situations. Use Table 7.17a to make adjustments for high lean-growth pigs or Table 7.17b for adjustments for moderate or low lean-growth pigs.

Determine the adjustments necessary to select the proper stage of development for nutrient allowances in Table 7.14 reflecting the appropriate gender, climatic, social, dietary form, and antigen exposure factors. Following are steps to follow when adjusting nutrient allowances to specific production conditions in Tables 7.17a and 7.17b:

1. Identify the stage of development (pig weight) to be addressed.
2. Identify the production conditions for pigs in that stage of development that deviate from the standard production conditions defined in Table 7.12.
3. For the production conditions that deviate from standard conditions, total the appropriate numbers for the stage of development being addressed. Round the total to the nearest whole number. Add or subtract that number from the stage of development being addressed. Use the adjusted number to determine the dietary nutrient concentrations in Table 7.14 that most closely match the nutrient needs of pigs reared in these specific production condition.

Example calculation

Pigs with high lean-growth capacities in Stage 6 of development (55 to 75 lb) are maintained in a cold thermal climate (5°F below TN) and experience a moderate level of antigen exposure. Adjustments for these specific conditions total + 0.90 (+ 0.35 + 0.55). This number, rounded to the nearest whole is + 1.0. This number (1) added to stage 6 equals 7 and represents the stage of development in Table 7.14 that defines the nutrient needs of pigs reared under these unique production conditions.

4. Select diets that meet the nutrient needs of the pigs. This may be done by purchasing commercial diets or preparing diets from available ingredients

using the diets in Tables 7.18 or 7.19), corn-soybean meal and premix, or corn and protein supplements. Other good sources of diet formulation information may be found in the *Pork Industry Handbook* and university Extension Services that reflect local feedstuffs.

DEFINING NUTRIENT REQUIREMENTS OF THE BREEDING HERD

The first step in defining nutrient requirements of the breeding herd is to identify the gender (sow or boar) and stage of production (breeding, gestation, lactation) being addressed. Second, identify specific production conditions in which the breeding herd is being reared that differ from the standard production conditions (Table 7.12). Use Tables 7.20 to 7.23 as guidelines to define appropriate nutrient concentrations and diets for breeding/gestating sows, lactating sows, and boars, respectively. Nutrient requirements G2 (gestation) and L5 (lactation) in Table 7.23 may be used as default feeding programs.

Replacement Gilts

If gilts are to be selected on genetic merit they should be full-fed until they reach the heaviest stage in Table 7.14. Record daily gain or days to market and backfat and try to select gilts that are better than average for both traits. For example, high lean-

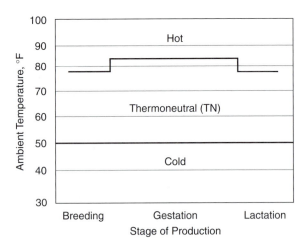

Figure 7.6 The zone of thermoneutrality for the breeding herd. Sows subjected to heat stress during breeding, implantation, and lactation may experience reproductive problems. Excessively high temperatures during breeding and implantation may reduce conception and implantation, causing the sows to recycle. Excessive temperatures during lactation drastically reduce feed intake. Cool temperatures generally are not a problem for sows. When the temperature drops below 50°F, added feed is required to maintain their energy balance. *(Adapted from ISU Life Cycle Swine Nutrition, 1996)*

TABLE 7.18 Diets for Growing Pigs Using Feedstuffs Defined under Standard Production Conditions

							Stage of Development							
Ingredient, lb	1	2[1]	3[1]	4	5	6	7	8	9	10	11	12	13	14
Corn	347.50	339.60	357.60	546.40	600.60	648.40	691.00	728.70	765.80	791.85	823.05	849.90	871.00	891.90
Soybean meal, dehulled	140.00	435.00	460.00	415.00	365.00	320.00	280.00	245.00	210.00	185.00	155.00	130.00	110.00	90.00
Dried whey	175.00	150.00	150.00											
Skim milk, dried	125.00													
Plasma protein, spray dried	75.00	40.00												
Fish meal, menhaden	100.00													
Soybean oil or other fat	25.00													
Limestone	5.40	9.10	7.20	8.50	8.25	8.50	8.90	8.75	8.70	9.00	9.00	9.00	9.00	9.50
Dicalcium phosphate	1.20	20.55	19.75	21.00	17.60	15.25	12.75	11.00	9.50	8.50	7.50	6.00	5.00	4.00
Salt[2]				4.50	4.30	4.10	4.10	3.70	3.70	3.50	3.50	3.30	3.30	3.10
Trace minerals and vitamins[3]	5.90	5.75	5.55	4.60	4.25	3.75	3.25	2.85	2.30	2.15	1.95	1.80	0.95	0.85
Feed additives[4]	—	—	—	—	—	—	—	—	—	—	—	—	—	—
Total	1000.00	1000.00	1000.00	1000.00	1000.00	1000.00	1000.00	1000.00	1000.00	1000.00	1000.00	1000.00	1000.00	1000.00

[1]Piglets nursing the sow can be creep-fed diets from stages 2 or 3.
[2]Added salt (source of sodium and chlorine) is not required in the diets containing dried whey and/or spray dried plasma protein.
[3]Example trace mineral, vitamin, and choline premixes can be found in Chapter 8, Tables 8.7 and 8.8.
[4]Feed additives may be added at various levels as approved by the Food and Drug Administration in Chapter 8, Table 8.2.
Source: Adapted from ISU Life Cycle Swine Nutrition, 1996.

TABLE 7.19 Diets for Growing Pigs Using Alternative Feedstuffs

Lean-Growth Capacity

Ingredient	Stage of Development													
	Antigen Status													
	1	2¹	2	3¹	3	6	6	6	9	9	9	12	12	12
	Minimal	*Minimal*	*Modest*	*Minimal*	*Modest*									
Corn	241.35	226.50	381.00	433.00	497.00	712.05	532.55	647.10	809.40	739.00	761.65	913.90	865.60	863.15
Soybean meal, dehulled	165.00	392.50	300.00	430.00	460.00	217.50		270.00	127.50		162.50	40.00		65.00
Soybeans, full-fat, cooked²							435.00			235.00			112.50	
Meat and bone meal						55.00			56.00			35.00		
Wheat middlings				100.00				50.00			50.00			50.00
Dried whey	300.00	250.00	150.00											
Skim milk, dried	125.00	75.00	50.00											
Plasma protein, spray dried	75.00		60.00											
Fish meal, menhaden	50.00													
L-lysine HCl	0.75	0.50	0.50	0.50	0.50	1.75	1.50	1.50	1.00	1.75	1.50	2.00	2.00	2.00
D, L-methionine		1.00	0.50	0.50	0.50									
Soybean oil	25.00	25.00	25.00											
Limestone	7.00	6.50	10.25	7.50	8.25		9.00	9.20		9.15	9.25	4.00	9.20	9.50
Dicalcium phosphate	5.00	17.50	17.25	21.50	24.00	6.00	14.25	14.50		9.00	9.00		5.60	5.25
Salt³				2.00	4.75	4.00	4.00	4.00	3.75	3.75	3.75	3.30	3.30	3.30
Trace minerals and vitamins⁴	5.90	5.50	5.50	5.00	5.00	3.70	3.70	3.70	2.35	2.35	2.35	1.80	1.80	1.80
Feed additives⁵	-	-	-	-	-	-	-	-	-	-	-	-	-	-
Total	1000.00	1000.00	1000.00	1000.00	1000.00	1000.00	1000.00	1000.00	1000.00	1000.00	1000.00	1000.00	1000.00	1000.00

[1]Piglets nursing the sow can be creep-fed these diets.

[2]Nutrient levels increased in relation to the added energy from full-fat soybeans.

[3]Added salt (source of sodium and chlorine) is not required in the diets containing dried whey or spray-dried plasma protein.

[4]Example trace mineral, vitamin, and choline premixes can be found in Chapter 8, Tables 8.7 and 8.8.

[5]Feed additives may be added at various levels as approved by the Food and Drug Administration in Chapter 8, Table 8.2.

Source: Adapted from ISU Life Cycle Swine Nutrition, 1996.

TABLE 7.20   Gestating Sow Adjustments of Feed Intake and Lysine Requirement to Match Specific Production Situations[1]

		Feed/Day		Lysine
Standard Production Conditions[2]		4.0 lb	1.8 kg	0.51%
Adjustments:				
Parity	2	+0.4	+0.2	0
	3	+0.4	+0.2	−0.07
	4	+0.5	+0.2	−0.09
	5	+0.6	+0.3	−0.10
	6–8	+0.4	+0.2	−0.20
High lean-growth		+0.2	+0.1	0
10°F below thermoneutrality		+0.8	+0.4	−0.07
Overcondition		−0.5	−0.2	+0.03
Weight loss, postfarrow to weaning				
15 lb	7 kg	+0.3	+0.1	0
30	14	+0.6	+0.3	+0.01
45	20	+0.8	+0.4	+0.02
60	27	+1.1	+0.5	+0.03
75	34	+1.4	+0.6	+0.04

[1]See Table 7.12.
Source: Adapted from ISU Life Cycle Swine Nutrition, *1996.*

TABLE 7.21   Selecting Lactating Sow Diets

	Daily Feed Intake [1]						
Pigs Nursed per Litter	10 lb (4.8 kg)	11 lb (5.0 kg)	12 lb (5.5 kg)	13 lb (5.9 kg)	14 lb (6.4 kg)	15 lb (6.8 kg)	16 lb (7.3 kg)
				Diet Number [2, 3]			
8	**L7**	**L6**	**L4**	**L3**	L2	L1	
9	**L8**	**L7**	**L6**	**L4**	L3	L2	L1
10	**L9**	**L7**	**L6**	**L5**	**L4**	L3	L2
11	**L10**	**L8**	**L7**	**L6**	**L5**	**L4**	L3
12	**L11**	**L9**	**L8**	**L7**	**L6**	**L5**	L4

[1]Average daily feed intake over a 21-day lactation.
[2]**Bold diet numbers** indicate that the associated feed intake is inadequate to supply sufficient energy for milk production. In these situations body energy stores will be used.
[3]Example: Sows have an average intake of 11 lb of feed per day over a 21-day lactation and a weaning average of 10 pigs/litter. Diet L7 most closely matches these nutrient requirements.
Source: Adapted from ISU Life Cycle Swine Nutrition, *1996.*

growth gilts should be full-fed until approximately 257 lb (117 kg) body weight. After removal from full-feed, daily feed and nutrient intakes should be based on the adjustments for gestating sows in Table 7.20. Gilts may be flushed 14 to 21 days before breeding by increasing feed intake 50 to 100%. Reduce daily intake to 4 or 5 lb (1.8 or 2.3 kg) postmating.

Breeding and Gestating Sows

Nutrient recommendations for sows are based on the assumption that nonlactating and pregnant sows are maintained under the standard production conditions in Table 7.12 and a thermal neutral climate illustrated in Figure 7.6. Standard conditions include sows and boars penned individually on slotted floors, in insulated buildings, and in the presence of minimal drafts. For sows housed on wet floors, in poorly insulated buildings, or in the presence of drafts, the temperature must be increased 5, 5, and 8 degrees, respectively, to maintain thermoneutrality. For animals penned in groups on solid concrete floors or housed in dry straw, the temperature may be reduced by 4 and 10 degrees, respectively and still maintain thermoneutrality.

The standard sow is a first parity female, of moderate lean-growth capacity and optimum body condition with an expected loss of less than 15 lb

TABLE 7.22 Nutrient Allowances for the Breeding Herd

| | | Stage of Production | | | | | | | | | | | | | |
| | | Breeding/Gestation | | | Lactation | | | | | | | | | | |
Diet Number		G1	G2	G3	L1	L2	L3	L4	L5	L6	L7	L8	L9	L10	L11
Amino acids															
Lysine	%	0.44	0.51	0.56	0.72	0.76	0.81	0.85	0.88	0.94	1.00	1.06	1.11	1.15	1.20
Threonine	%	0.32	0.37	0.41	0.45	0.47	0.50	0.53	0.55	0.58	0.62	0.66	0.69	0.71	0.74
Tryptophan	%	0.10	0.11	0.12	0.14	0.15	0.16	0.17	0.18	0.19	0.20	0.21	0.22	0.23	0.24
Methionine + Cystine	%	0.36	0.41	0.45	0.35	0.37	0.40	0.42	0.43	0.46	0.49	0.52	0.54	0.56	0.59
Crude protein[1]	%	10.60	11.60	12.20	14.70	15.20	15.80	16.40	16.80	17.70	18.50	19.30	20.10	20.60	21.20
Metabolizable energy[1]	kcal/lb	1483	1478	1481	1496	1492	1490	1488	1485	1482	1479	1477	1474	1471	1469
Minerals															
Calcium	%	0.80	0.85	0.90	0.65	0.67	0.70	0.73	0.76	0.80	0.84	0.88	0.91	0.94	0.97
Phosphorus, total	%	0.63	0.68	0.73	0.59	0.61	0.64	0.66	0.69	0.73	0.76	0.80	0.83	0.85	0.88
Phosphorus, available	%	0.36	0.42	0.50	0.31	0.33	0.35	0.37	0.40	0.43	0.46	0.49	0.52	0.54	0.56
Sodium (Na)[2]	%	0.14	0.16	0.18	0.13	0.14	0.15	0.15	0.16	0.17	0.18	0.19	0.20	0.21	0.22
Chlorine (Cl)[2]	%	0.21	0.24	0.26	0.20	0.21	0.22	0.23	0.24	0.26	0.27	0.29	0.30	0.31	0.33
Trace minerals (added)															
Iron	ppm	86	100	110	82	86	92	96	100	107	114	120	126	131	136
Zinc	ppm	73	85	94	70	73	78	82	85	91	97	102	107	111	116
Copper	ppm	5.2	6.0	6.6	4.9	5.2	5.5	5.8	6.0	6.4	6.8	7.2	7.6	7.9	8.2
Manganese	ppm	10	12	13	10	10	11	12	12	13	14	14	15	16	16
Iodine	ppm	0.15	0.17	0.19	0.14	0.15	0.16	0.16	0.17	0.18	0.19	0.20	0.21	0.22	0.23
Selenium	ppm	0.15	0.18	0.20	0.15	0.15	0.17	0.17	0.18	0.19	0.21	0.22	0.23	0.24	0.24
Vitamins (added/lb of diet)															
Vitamin A	IU	1376	1600	1760	1312	1376	1472	1536	1600	1712	1824	1920	2016	2096	2176
Vitamin D$_3$	IU	138	160	176	131	138	147	154	160	171	182	192	202	210	218
Vitamin E	IU	15	18	20	15	15	17	17	18	19	21	22	23	24	24
Vitamin K (menadione)	mg	0.34	0.40	0.44	0.33	0.34	0.37	0.38	0.40	0.43	0.46	0.48	0.50	0.52	0.54
Niacin	mg	6.9	8.0	8.8	6.6	6.9	7.4	7.7	8.0	8.6	9.1	9.6	10.1	10.5	10.9
Pantothenic acid	mg	8.2	9.5	10.5	7.8	8.2	8.7	9.1	9.5	10.2	10.8	11.4	12.0	12.4	12.9
Riboflavin	mg	2.6	3.0	3.3	2.5	2.6	2.8	2.9	3.0	3.2	3.4	3.6	3.8	3.9	4.1
Vitamin B$_{12}$	mcg	10.3	12.0	13.2	9.8	10.3	11.0	11.5	12.0	12.8	13.7	14.4	15.1	15.7	16.3
Biotin	mg	0.12	0.14	0.15	0.11	0.12	0.13	0.13	0.14	0.15	0.16	0.17	0.18	0.18	0.19
Folic acid	mg	0.28	0.34	0.38	0.11	0.12	0.13	0.13	0.14	0.15	0.16	0.17	0.18	0.18	0.19
Choline	mg	301	350	385	287	301	322	336	350	375	399	420	441	459	476

[1]The crude protein and metabolizable energy values are the results of formulating the example diets in Table 7.24. They are not minimum levels.
[2]Sodium and chlorine supplementation is provided by salt (NaCl).
Source: Adapted from ISU Life Cycle Swine Nutrition, 1996.

(7 kg) from postfarrowing to weaning. Table 7.20 indicates adjustments in feed intake and lysine levels for gestating sows maintained in production conditions that deviate from the standard production conditions.

For example, first parity sows (gilts) raised under standard production condition would be fed 4.0 lb (1.8 kg) of feed/day containing 0.51% lysine. However, third parity sows require 4.4 lb (4.0 + 0.4 lb) or 2.0 kg of feed/day containing 0.44% lysine (0.51% – 0.07%). Use the estimated feed intake and lysine needs to identify the appropriate nutrient requirements in Table 7.22 for breeding and gestating sows.

Lactating Sows

Appropriate diets for lactating sows can be estimated using Table 7.21. Because only one diet will likely be fed in a lactation room, the producer should determine the average number of pigs nursed per sow and the average lactation feed intake by season. Then select a diet that most closely matches that litter size and feed intake (Figure 7.7). Ad lib feeding is recommended during lactation and sows should be brought up to maximum feed intake as quickly as possible postfarrowing. Feed fresh diet twice per day and remove stale feed daily. Note most of the diet numbers in Table 7.21 are in bold type, indicating that those sows will be expected to lose weight during lactation. Lactation requirements are in Table 7.22.

Nutrient allowances for the breeding herd are found in Table 7.22. Three gestation diets are presented, G1, G2, and G3. They are based on feed intakes and lysine adjustments from Table 7.20. Generally, gestating sows consuming 4 lb (1.8 kg) per day should receive diet G2 containing 0.51% lysine. Lactation diets, L1 to L11, are selected based on average lactation feed intakes and number of

Figure 7.7 A sow and her nursing litter. *(Courtesy, USDA, Washington, DC)*

pigs nursed from Table 7.21. Lactating sows should be fed as much feed as they will consume on a daily basis. Not all sows will consume an adequate amount of feed to meet their nutrient requirements, therefore consider feeding a higher lysine diet to provide a margin of safety.

Example diets for the breeding herd that match the recommendations from Table 7.22 are presented in Table 7.24. These are based on corn and soybean meal but other feeds may be substituted depending on locally available feeds and ingredient cost considerations.

Boars

Nutrient allowances for growing boars may be found in Table 7.14 after making appropriate adjustments for gender from Table 7.17a. Recommended feed intakes and diet selections for boars of different weights and frequency of matings are in Table 7.23.

TABLE 7.23   Boar Feeding Regimens

		Boar Weight		
Matings/ Week	270–350 lb (123–159 kg)	350–450 (159–204)	450–550 (204–250)	550–650 (250–295)
		Feed/Day, lb (kg)[1]		
1–2	5.3 (2.4)	5.6 (2.5)	5.8 (2.6)	6.2 (2.8)
2–4	5.5 (2.5)	5.8 (2.6)	6.0 (2.7)	6.4 (2.9)
Diet[2,3]	L3	L2	L1	G3

[1]Requirements are based on the standard production conditions in Table 7.12. Boars housed in a colder thermal climate should have the daily feed allowance increased by 0.17 lb (0.08 kg) for each 5°F decrease in temperature below thermoneutrality (Figure 7.3).
[2]If diets differ from corn–soybean meal, adjustments in feed and nutrient allowances may be needed.
[3]Diets represent lactation (L3, L2, and L1) and gestation (G3) requirements in Table 7.22 that most closely match the nutrient requirements of boars.
Source: Adapted from ISU Life Cycle Swine Nutrition, 1996.

TABLE 7.24 Diets for Sows and Boars under Standard Production Conditions

								Stage of Production						
Ingredient	G1	G2	G3	L1	L2	L3	L4	L5	L6	L7	L8	L9	L10	L11
Corn	891.15	862.50	843.70	799.30	781.50	765.10	748.85	737.00	709.95	687.90	665.95	943.95	627.45	611.15
Soybean meal, dehulled	75.00	100.00	115.00	175.00	190.00	205.00	220.00	230.00	255.00	275.00	295.00	315.00	330.00	345.00
Limestone	10.50	10.00	9.00	6.25	7.70	7.70	7.75	7.75	7.80	7.85	7.90	8.00	8.00	8.00
Dicalcium phosphate	16.80	20.00	24.00	13.50	14.50	15.50	16.50	18.00	19.50	21.00	22.50	24.00	25.00	26.00
Salt	3.50	4.00	4.50	3.25	3.50	3.75	3.75	4.00	4.25	4.50	4.75	5.00	5.25	5.50
Sow trace mineral premix[1]	0.90	1.00	1.10	0.85	0.85	0.90	0.95	1.00	1.10	1.15	1.20	1.25	1.30	1.35
Sow fat-soluble vitamin premix[1]	0.30	0.35	0.35	0.30	0.30	0.30	0.35	0.35	0.35	0.40	0.40	0.40	0.45	0.45
Sow B vitamin premix[1]	0.35	0.40	0.45	0.35	0.35	0.40	0.40	0.40	0.45	0.50	0.50	0.50	0.55	0.55
Sow folic acid premix[1]	0.20	0.25	0.25											
Choline premix[1]	1.30	1.50	1.65	1.20	1.30	1.35	1.45	1.50	1.60	1.70	1.80	1.90	2.00	2.00
Feed additives[2]														
Total	1000.00	1000.00	1000.00	1000.00	1000.00	1000.00	1000.00	1000.00	1000.00	1000.00	1000.00	1000.00	1000.00	1000.00

[1]Example trace mineral, vitamin, and choline premixes for growing pigs can be found in Chapter 8, Tables 8.7 and 8.8.
[2]Feed additives may be added at various levels as approved by the Food and Drug Administration in Chapter 8, Table 8.2.
Source: Adapted from ISU Life Cycle Swine Nutrition, 1996.

QUESTIONS FOR STUDY AND DISCUSSION

1. List the primary factors that affect swine nutrient requirements, diet formulation, and feeding programs.
2. Why are nutritional requirements, such as those noted in Tables 7.1 to 7.10 not final?
3. Why should there be swine diets for different stages of production?
4. Explain the difference between nutrient requirements and nutrient allowances.
5. Discuss the lysine requirements of
   a. Lean, rapidly growing pigs
   b. Winter versus summer
   c. Split-sex groups
6. What are growing-finishing pigs? Why should decisions to change or modify their diets be based on economics?
7. At what stage and age should replacement gilts be selected? Why? How should they be fed thereafter?
8. Would you recommend flushing gilts? If so, how would you do it?
9. Limited feeding is necessary for gestating gilts and sows. List and detail four systems of limited feeding.
10. Would you recommend ad libitum (free choice) or limited feeding of sows immediately after farrowing? Why?
11. Estimate the total feed required to produce a 265-lb market pig, then break it down into the amount required for each of the following stages of production: (a) sow gestation diet (including pregestation and breeding), (b) boar diet, (c) lactation diet, (d) starter diet (creep to 40 lb), and (e) grower-finisher diet (40 to 265 lb).
12. Since boars gain faster and are more efficient in feed conversion than barrows or gilts, why aren't more boars fed?

SELECTED REFERENCES

Animal Feeding and Nutrition, 9th ed., M. J. Jurgens, Kendall/Hunt Publishing Co., Dubuque, IA, 2002

Composition and Quality Assessment Procedures, National Park Board, Des Moines, IA, 2000

Feeds and Nutrition, 2nd ed., M. E. Ensminger, J. E. Oldfield, and W. W. Heinemann, The Ensminger Publishing Company, Clovis, CA, 1990

Kansas Swine Nutrition Guide, Kansas State University Extension, Manhattan, KS, 2003

Life Cycle Swine Nutrition, 17th ed., Pm-489, P. J. Holden, et al., Iowa State University, Ames, IA, 1996

Nutrient Requirements of Swine, 10th rev. ed., National Research Council, National Academy of Sciences, Washington, DC, 1998

Pork Industry Handbook, Cooperative Extension Service, Purdue University, West Lafayette, IN, 2003

Swine Nutrition, E. R. Miller, D. E. Ullrey, and A. J. Lewis, Butterworth-Heinemann, Boston, MA, 1991

Swine Nutrition Guide, University of Nebraska, Lincoln, NE, and, South Dakota State University, Brookings, SD, 2000

Swine Production and Nutrition, W. G. Pond and J. H. Maner, AVI Publishing Co., Inc., Westport, CT, 1984

8

Alternative Feeds, Additives, and Feeding Management

A well-managed swine feeding system requires adequate bulk feed storage tanks providing several diets to a building. *(Courtesy, Palmer Holden, Iowa State University, Ames, IA)*

Objectives

After studying this chapter, you should:

1. Know how to select alternative feeds that may economically substitute into swine diets.
2. Know and understand the types of feed additives and where to find information on approved additives.
3. Know how to adjust for different moisture content feeds and when it is important.
4. Be able to balance a diet using the Pearson square.
5. Understand the factors involved in selecting methods of diet preparation, such as the use of base mixes, premixes, and complete protein supplements.
6. Be able to design vitamin and trace mineral premixes.
7. Understand how to read a feed tag.
8. Know the advantages and disadvantages of different feed processing and feeding systems.
9. Know various methods for limiting the feed intake of sows and their advantages.
10. Understand the hog–corn ratio.

ALTERNATIVE FEEDS

Successful swine producers recognize that feeds of similar nutritive properties can and should be interchanged in the diet as price relationships warrant, thus making it possible at all times to obtain a balanced diet at the lowest cost. The feed substitution levels for swine in Table 8.1 provides a summary of the comparative values of common feeds. The maximum recommended inclusion levels are those that can be generally fed while maintaining satisfactory performance. The relative feeding values compare the nutritive value of energy sources to corn, and protein sources to dehulled soybean meal. These values are based on metabolizable energy, lysine, and available phosphorus. When using these feed substitution levels, the following facts should be recognized:

1. Different ages and groups of animals within classes should be fed differently.
2. Individual feeds differ widely in feeding value. Barley and oats, for example, vary widely in feeding value according to the hull content and bulk density (weight per volume).
3. Certain feeds, especially those of medium protein content, such as alfalfa, peanuts, and peas (dried), can be used interchangeably as (a) grains and by-product feeds or (b) protein supplements (Figure 8.1).
4. The feeding value of certain feeds is materially affected by preparation; potatoes and beans should always be cooked for hogs. The values reported are based on proper feed preparation in each case.

Figure 8.1 A canola field in North Dakota ready for combining. Canola is a competitive protein source for pigs when combined with synthetic lysine. *(Courtesy, Palmer Holden, Iowa State University, Ames, IA)*

For these reasons, the comparative values of feeds shown in the feed substitution table are not absolute. They are reasonable approximations based on average quality feeds. Laboratory analyses of feeds will increase the accuracy of feed substitutions.

FEED ADDITIVES

The use of feed additives in swine diets has been extensive in the United States since their discovery in the early 1950s. Most swine producers use additives because of their demonstrated ability to increase growth rate, improve feed utilization, and reduce mortality and morbidity from clinical and subclinical

TABLE 8.1   Feed Substitution Levels for Swine

Feedstuff (as fed basis)	Maximum Recommended Inclusion Level (%)				Relative Feeding Value vs.[1]		Remarks
	Gestation	Lactation	1–3	4–14	Corn	Soybean Meal, Dehulled	
Alfalfa hay, early bloom	0–50	5	0	5	75–85		Low energy, good source of carotene and B vitamins, unpalatable to baby pigs.
Alfalfa meal, dehydrated	50	10	0	5	80–100	30	Low energy, high fiber, carotene.
Bakery by-product	40	40	20	40	95–110		Variable salt content.
Barley[2]	90	85	25	95	100–105		Energy source, moderate fiber, low lysine.
Beans (cull)	5	10	0	7		70	Cook thoroughly; supplement with methionine.
Beet pulp, dried	10	10	0	0	90–105		Bulky, high fiber, laxative.
Blood meal, spray dried	5	5	5	5		205–220	High lysine, low isoleucine, and tryptophan. Low palatability.
Buttermilk, dry	0	0	20	0		85–90	Good amino acid balance.
Buttermilk, liquid (9.7% DM)	Ad lib			Ad lib	27–37	10–12	Usually fed free-choice.
Buttermilk, cond. (29.1% DM)					80–110	31–34	Good amino acid balance.
Canola meal, solvent	5	5	0	5		72–74	High fiber.
Cassava, dried meal	80	70	0	45	85		Low in methionine. Available as a swine feed in the tropics. Hydrocyanic acid risk. Sun drying may inactivate.
Copra meal (coconut meal), (21% CP)	25	5	0	20	50		Expect minor decrease in performance.
Corn and cob meal	80	5	0		75–80		Bulky, low energy. May constitute up to 80% of gestation diets.
Corn, distillers dried grain w/soluble	10	5	5	5		45–50	B vitamin source, low lysine.
Corn, gluten feed, 23% CP	90	10	5	10	110–130	40–50	Low lysine and tryptophan availability.
Corn, gluten meal, 60% CP	5	5	0	5		55–65	Low lysine and tryptophan availability.
Corn, high lysine	95	90	50	95	100–105		Superior to corn in lysine.
Corn, silage (25–30% DM)	90					20–30	Bulky, low energy, gestation only.
Corn, yellow[2]	90	85	35–45	60–90	100		Energy source, low lysine and tryptophan.
Corn, yellow, high oil	90	85	35–45	85	105		Energy source, low lysine and tryptophan.
Cottonseed meal, solvent	5	5	0	5		64–68	Gossypol toxicity, low lysine.
Emmer	80	5	0	5	80–90		Emmer may be used about like barley.
Fat, animal, stabilized	8	8	8	8	140–160		High energy; reduce dust, more than 5% may bridge feed.
Feathers meal	3	0	0	3		85–92	Low lysine, methionine, tryptophan.
Fish meal (60%)	10	10	10	5		165–170	Excellent balance of amino acids, and good source of calcium and phosphorus.
Fish meal, menhaden	10	10	10	5		165–170	Excellent amino acids, calcium, phosphorus source. Potential rancidity.
Fish solubles, condensed (51% DM)	5	5	5	3		65	Approximately 5% salt.

(continued)

Ingredient					Comments
Hominy feed	60	60	0	60	100–105	Energy source, possible rancidity, soft pork at high levels.
Lactose	0	0	20	0	85–95	Energy source, palatable.
Linseed meal (35% CP)	25	5	0	25	60–70	Supplement with lysine; slightly laxative.
Malt sprouts	10	5			100	
Meat and bone meal (50% CP)	10	10	5	5	108–115	Low tryptophan, high calcium and phosphorus.
Meat meal (54% CP)	10	10	5	5	125–135	
Millet (proso)	50	50			85–95	Low lysine.
Molasses, beet, cane, citrus (74% DM)	5	5	5	5	34–38	Palatable, used in pelleting, laxative above 15%.
Oats, groats (dehulled oats)	30	30	30	30	115–125	Energy source, palatable, expensive.
Oats	90	15	0	20	80–85	Value as energy source depends on density, high fiber.
Oats, naked	90	70	40	65–95	105–110	Risk of rancidity.
Peanut meal, solvent (49% CP)	5	5	0	5	65–75	Becomes rancid when stored too long. Low in lysine; very palatable.
Peanuts, kernals	20	12	5	5	60–70	Low lysine, cook for nursery pigs, aflatoxin potential, soft pork.
Peas, seeds	0	0	8	0	70–80	Low methionine.
Plasma protein, spray dried	5	5	0	5	205–215	Immunoglobulins, high salt content.
Potatoes, cooked (20–22% DM)	50	25	0	30–50	20–22	Level as % of DM in diet. Value assumes fed at 3 parts wet potato to 1 part grain.
Potatoes, dried pulp (88% DM)	50	0	0	0–15	92–98	Need to be cooked.
Potatoes, sweet (27–30% DM)	50	0	0	0	20	Need to be cooked.
Potatoes, sweet (90% DM)	40	10	0	40	70	
Poultry by-product meal	5	5	0	5	136–139	Potential of high ash content.
Rice (rough rice)	90	20	0	40	80–85	Low energy, low lysine, equal gain, reduced F/G.
Rice bran	33	5	0	5	100	High fat, rancidity risk, soft pork.
Rice polishings	33	10	0	10	100–120	Rancidity risk, soft pork.
Rye	20–30	0	0	25	90	Should be limited because it is unpalatable. Grind for swine. Watch for ergot.
Safflower meal, with hulls	5	0	0	0	45–55	High fiber, low lysine, tryptophan.
Safflower meal, dehulled	3	0	0	5	80–90	Low lysine, tryptophan.
Shrimp meal, sun dried	50	5–10	0	5–20	85–95	Contains 0–7% salt.
Skim milk, dried (94% DM)	0	0	20	0	100–105	Excellent amino acid balance, palatable, expensive.
Skim milk, liquid (9.5% DM)	Ad lib	Ad lib	0	Ad lib	12–13	Usually fed free-choice.
Sorghum, grain[1]	90	85	45	95	92	Energy source, low lysine, and tryptophan.
Soybean meal, dehulled, solvent[1]	10	25	45	40	100	Good amino acid balance with corn.
Soybean meal, solvent[1]	12	30	48	45	96	Good amino acid balance with corn.
Soybean oil	8	8	8	8	190–215	High energy; more than 5% may bridge feed.
Soybeans, full-fat, cooked	15	32	60	50	65–85	High energy; soft pork.
Spelt	90	60	0	90	90–100	Low energy, low lysine. Value varies according to the amount of hulls.

TABLE 8.1 Feed Substitution Levels for Swine (continued)

Feedstuff (as fed basis)	Maximum Recommended Inclusion Level (%)				Relative Feeding Value vs.[1]		Remarks
	Gestation	Lactation	1-3	4-14	Corn	Soybean Meal, Dehulled	
Sugar (sucrose)	0	0	5	0	95–105		Added for palatability.
Sunflower meal (36–45% CP)	0	10	0	10	90–95		Low lysine.
Sunflower meal (28% CP)	10	0	0	10		45–55	High fiber, low lysine.
Sunflower seed	50	25	0	25	100		High energy, high fiber, low lysine, soft pork.
Triticale	90	85	45	90	90–95		Higher levels may not be palatable. Ergot risk.
Wheat bran	30	10	0	5	95–105		High fiber, laxative.
Wheat midds	30	20	5	5	95–115		Low energy, partial grain substitute.
Wheat, red dog or white shorts	90	20	0	50	115–120		Inconsistent product.
Wheat, hard[2]	90	85	45	95	110–115		Energy source, low fiber; use course grind.
Whey, dried	5	5	30	5	130–160		Excellent amino acid balance, palatable, high salt content.
Whey, dried, delactosed	5	5	30	5	140–200		Same as whey, dried.
Whey, liquid (7.1%DM)	Ad lib			Ad lib	11–15		Usually fed free-choice.
Yeast, dried brewers	3	3	3	3		112–115	Source of B vitamins.

[1] Relative value considers lysine content, net energy, and available phosphorus. The cost of an ingredient can be evaluated by comparing its cost with the cost of corn or dehulled soybean meal times the particular coefficient for the relative value of the ingredient compared to corn or soybean meal. These values assume feeding levels within the suggested limits. High-fiber feeds will decrease in value as their level in the feed is increased.

[2] Corn, barley, sorghum, or wheat should be the basic energy source with other substitutions made within the ranges suggested. Soybean meal is assumed to be the basic protein source with other substitutions made within the ranges suggested. Roots and tubers are of lower value than the grain and by-product feeds due to their higher moisture content.

Source: Values adapted with information from the following sources: Pork Industry Handbook, 2003; Pond and Maner, Swine Production and Nutrition, 1984; ISU Life Cycle Swine Nutrition, 1996; NRC Nutrient Requirements of Swine, 1998; Thacker & Kirkwood, Nontraditional Feed Sources for Use in Swine Production, 1990.

infections. Most, but not all, additives used for swine fall into the following classifications:

1. Antibiotics
2. Chemotherapeutics
3. Anthelmintics or dewormers
4. Copper compounds
5. Zinc oxide
6. Probiotics
7. Other additives

An antibiotic is a compound synthesized by living organisms, such as bacteria or molds, which inhibits the growth of other bacteria or molds.

Chemotherapeutics are compounds similar to antibiotics but are produced chemically rather than microbiologically.

Anthelmintics (dewormers) are compounds added to swine diets to help control internal or external parasites.

Copper compounds, such as copper sulfate, have growth-stimulating properties similar to antibiotics when fed at levels well above the nutritional requirement for copper. Levels of 250 ppm copper enhance performance similar to antibiotics. Copper also is effective as therapeutic treatment for certain intestinal disorders that do not respond satisfactorily to antibiotics or chemotherapeutics.

Zinc oxide fed at 3,000 ppm of zinc appears to enhance performance and reduce diarrheas in nursery pigs.

Probiotics, which means "in favor of life," have an opposite effect to antibiotics on the microorganisms of the digestive tract. Probiotics are normally bacteria such as *Lactobacilli sp.* added to the diet. It is proposed that probiotics populate the gut with desirable microorganisms rather than kill or inhibit undesirable organisms.

There are a wide variety of other additives, such as Ractopamine·HCl® (Paylean™), which enhances performance and muscling in finishing pigs; chromium, which may enhance litter size; and various botanicals, some of which may have antibacterial properties.

A partial list of approved additives for use in swine diets is included in Table 8.2. Information for each additive includes the (1) the chemical name, (2) trade name, (3) approved use level (g/ton), (4) withdrawal period before slaughter, and (5) FDA approved claims. Regardless of the feed additive used, the following safety precautions should be observed:

1. Use only additives approved for swine. Regulations change, so check them often.
2. Follow label directions carefully.
3. Use minimal effective amounts.

4. Add or apply precise quantities.
5. Certain feed additives must be withdrawn from the feed before slaughter to ensure residue-free meat. Consult the feed label for withdrawal time for the specific feed additive that is being fed.

In addition to the additives listed in Table 8.2, numerous other additives are used to increase acceptability of the diet, preserve the quality of the diet, or improve digestion and utilization of the feed. Among such additives are antioxidants, mold inhibitors, flavors, sweeteners, pellet binders, clays, enzymes, organic acids, yucca extract, and electrolytes.

Balancing Diets

Animals in confinement have access only to the feed and water provided by the caretaker. Therefore, it is important to provide balanced diets meeting the requirements for all the essential nutrients. Suggested diets in this chapter generally will meet the animal's needs, but diets need to vary with conditions and should be reformulated to meet the genetic or environmental conditions of a specific farm.

Pork producers should know how to balance diets; select and buy feeds with informed appraisal; check on how well manufacturers, dealers, or consultants meet the producer's needs; and evaluate the results.

Diet formulation consists of combining feeds that will be consumed in the amounts needed to supply the daily nutrient requirements of the animal. This may be accomplished by the methods presented later in this chapter, but first the following pointers are necessary:

1. In computing diets, more than simple arithmetic should be considered because no set of figures can substitute for experience and swine intuition. Formulating diets is both an art and a science—the art comes from animal know-how, experience, and keen observation; the science is largely founded on mathematics, chemistry, physiology, and bacteriology. Both are essential for success.
2. Before attempting to balance a diet, the following major factors should be considered:

Availability and cost of the different feed ingredients. Preferably, ingredient costs should be based on delivery after processing because those costs are quite variable.

Moisture content. Most swine diets area bsed on the "as-fed basis" or 88 to 90% dry matter. When considering costs and balancing diets, feed should be placed on a comparable moisture basis; usually, either "as-fed" or "moisture-free." This is especially important in the case of high-moisture feeds.

TABLE 8.2 Swine Feed Additives with FDA Approved Claims[1]

Chemical Name	Trade Name	Additive Use Level	Withdrawal Period before Slaughter	A	B	C	D	E	F	G	H	I	J	K	L	M	N	O	Q
								FDA Approved Claim[2]											
Apramycin	Apralan	150 g/ton	28 days		B														
Arsanilic Acid	Pro-Gen	45–90 g/ton	5 days	A															
Arsanilic acid & Bacitracin MD or zinc		45–90 g/ton & 10–50 g/ton	5 days	A							H								
Arsanilic acid & Chlortetracycline		45–90 g/ton & 10–50 g/ton	5 days	A							H								
Arsanilic acid & Chlor- or Oxytetracycline		45–90 & 10mb/lb BW daily	5 days	A				E			H								
Arsanilic acid & Penicillin		45–90 g/ton & 100 g/ton	5 days	A				E			H								
Bacitracin MD	BMD	10–30 g/ton	none	A															
Bacitracin MD	BMD	250 g/ton (sole ration)	none	A				E			H		J						Q
Bacitracin MD & Chlortetracycline		10–30 g/ton & 400g/t (max 14 d)	none	A															
Bacitracin Zinc	Albac, Baciferm	10–50 g/ton	none	A															
Bambermycin	Flavomycin	2–4 g/ton	none	A															
Carbadox	Mecadox	10–25 g/ton	42 days[3]	A					F		H								
Carbadox	Mecadox	50 g/ton	42 days[3]	A															
Chlortetracycline	Aureomycin, Aureomix,	10–50 g/ton	none	A													N		
Chlortetracycline	Aureozol, PfiChlor, CLTC	50–100 g/ton	none					E					J			M			Q
Chlortetracycline	CTC	400 g/ton for 14 days	none																Q
Chlortetracycline	CTC	10 mg/lb BW for 14 d	varies					E					J						
Chlortetracycline & Sulfamethazine & Penicillin	Aureo SP 250, PfiChlor 250	100 & 100 & 50 g/ton	15 days	A				E						K			N		
Chlortetracycline & Sulfathiazole & Penicillin	CSP 250	100 & 100 & 50 g/ton	7 days	A			D	E						K			N		
Lincomycin[4]	Lincomix	20 g/ton	none	A															
Lincomycin[4]	Lincomix	40 g/ton	none	A							H								Q
Lincomycin[4]	Lincomix	100 g/ton (sole ration for 21 d)	none									I							Q
Lincomycin[4]	Lincomix	200 g/ton (sole ration for 21 d)	none															O	
Neomycin	Neomix	10 mg/lb BW, max 14 days	3 days		B														
Oxytetracycline	Terramycin, OXTC, OTC	10–50 g/ton	none	A															
Oxytetracycline	Terramycin, OXTC, OTC	10 mg/lb BW, 7–14 days	none	A				E					J			M			
Oxytetracycline & Neomycin base[5]	Neo-Terramycin, NEO/OXTC r/OXY	50–150 g/ton & 35–140 g/ton	5–10 days				D	E		G				K					
Penicillin (from Procaine or G Procaine)	Penicillin P-50, Penicillin G Procaine	10–50 g/ton	none	A															
Rabon	Rabon	0.05 g/100 lb BW/day	none			C													
Ractopamine[6]	Paylean	4.5–18.0 g/ton, >16% CP diet	none	A															
Ractopamine[6] & Tylosin (see Tylosin)		4.5–18.0 g/ton & 100 g/ton	none	A															Q
Roxarsone	3-Nitro-10, −20, −50	22.7–34.0 g/ton, complete feed	5 days	A															
Roxarsone	3-Nitro-10, −20, −50	181.5 g/t, (maximum 6 days)	5 days	A								I							
Roxarsone & Bacitracin MD		22.7–34.1 g/ton & 10–30 g/ton	5 days	A															
Roxarsone & Bacitracin MD		22.7–34.1 g/ton & 250 g/ton	5 days	A							H								

Additive	Product	Use Level	Withdrawal Time	Claims
Roxarsone & Bacitracin MD		181.5 g/ton & 10–30 g/t (6 days)	5 days	a, I
Roxarsone & Bacitracin zinc		22.7–34 g/ton & 10–50 g/ton	5 days	a, I
Roxarsone & Bacitracin zinc		181.5 g/t &10–50 g/t	5 days	a, I
Roxarsone & Chlortetracycline		22.7–34.1 g/t & 400 g/t(14 d)	5 days	a, E, J
Roxarsone & Chlortetracycline		181.5 g/t & 10–50 g/t (6 days)	5 days	a, I, J
Roxarsone & Chlortetracycline		181.5 g/ton & 400 g/t (6 days)	5 days	a, E, I, J
Roxarsone & Penicillin		22.7–34.1 g/ton & 100 g/ton	5 days	a, E
Tiamulin	DENAGARD 10	10 g/ton	none	a
Tiamulin	DENAGARD 10	35 g/ton (feed minimum 10 d)	2 days	H
Tiamulin	DENAGARD 10	200 g/ton (2 wk., then 35 g/ton)	7 days	I, Q
Tiamulin & Chlortetracycline	DENAGARD 10	35 g/ton & 400 g/ton for 14 d	2 days	E, H, I
Tilmicosin	Pulmotil 90	181–363 g/ton	7 days	E, H, J, P
Tylosin	TYLAN 40, TYLAN 100	100g/ton for 3 wk, then 40 g/t	none	G
Tylosin	TYLAN 40, TYLAN 100	100 g/ton	none	K
Tylosin	TYLAN 40, TYLAN 100	100 g/ton for 21 days, sole ration	none	
Tylosin	TYLAN 40, TYLAN 100	10–20 or 20–40 or 20–100 g/t	none	a
Tylosin/Sulfamethazine	TYLAN 40 Sulfa-G	100 g/ton	15 days	a, G, J, K, L
Virginiamycin	V-Max 10, 20, 50, Stafac 10, 20, 500	5–10 g/ton	none	a, H
Virginiamycin	V-Max 10, 20, 50, Stafac 10, 20, 500	25 g/ton	none	H, I, Q
Virginiamycin	V-Max 10, 20, 50, Stafac 10, 20, 500	100 g/ton for 2 wk, then 50 g/t	none	H, I

[1]This is a partial list of feed additives and combinations approved by the FDA. This list does not contain the complete use levels approved for each additive. Users are advised to read the product label and adhere to use recommendations of the additive manufacturer.

[2] The claims, use levels, and limitations are those for which FDA clearance was obtained. Iowa State University disclaims all responsibility for any results that may occur from use of this table. It is essential that the rules and regulations governing the use of feed additives be followed. The following claims are indicated for the appropriate additive:

a. Increase rate of gain and improve feed efficiency.
b. Prevention of postweaning colibacillosis.
c. Control of fecal flies in manure of treated swine.
d. Prevention of bacterial enteritis (scours).
e. Treatment of bacterial enteritis (scours).
f. Control of bacterial enteritis (scours).
g. Prevention of swine dysentery (*Serpulina hyodysenteriae*).
h. Control of swine dysentery (bloody scours).
i. Treatment of swine dysentery (bloody scours).
j. Control of swine bacterial pneumonia caused by (*Pasteurella multocida* and/or *C. pyogenes*).
k. Maintenance of weight gains in presence of atrophic rhinitis.
l. Lower incidence and severity of *Bordetella bronchiseptica* rhinitis.
m. Aid in prevention and treatment of leptospirosis.
n. Reduce cervical abscesses.
o. Reduction of severity of Mycoplansma pneumonia caused by *M. hyopneumonia*.
p. Control of swine respiratory disease (*Actinobacillus pleuropneumoniae* and *P. multocida*).
q. Prevention/control of porcine proliferative enteropathies (ileitis) associated with *Lawsonia intracellularis*.

[3] Do not feed Carbadox to pregnant swine or swine intended for breeding purposes. Do not mix in feeds containing bentonite.

[4] Swine fed Lincomycin may, within the first 2 days, develop diarrhea, and/or swelling of the anus. These conditions have been self-correcting within 5 to 8 days without discontinuing the Lincomycin treatment. Do not allow rabbits, hamsters, guinea pigs, horses, or ruminants access to feeds containing Lincomycin. Not to be fed to swine that weigh more than 250 lb.

[5] Neomycin use levels are expressed as Neomycin base per ton (70% neomycin sulfate level). For example, 140 g Neomycin base is equal to 200 g of neomycin sulfate. Withdraw from feed 10 days before slaughter at 140 g/ton and 5 days when below 140 g/ton.

[6]Ractopamine at 4.5 – 18.0 g/ton increased rate of gain and increased carcass leanness. Pigs fed Ractopamine HCl are at an increased risk for exhibiting the downer pig syndrome (also referred to as slows, subjects, or suspects). Pig handling methods to reduce the incidence of downer pigs should be thoroughly evaluated prior to initiating use of Ractopamine HCl. Not for use in breeding swine.

Source: Information in this table is from the June 2004 Feed Additive Compendium *published by the Miller Publishing Co., 12400 Whitewater Drive, Suite 160, Minnetonka, MN 55343.*

Feed composition. Feed composition tables should be considered as guides because of the wide variations in the composition of feeds. Wherever possible, especially with large purchases, it is best to take a representative sample of major feed ingredients and have done a chemical analysis of the nutrients of concern, such as, lysine, fat, and moisture, and often calcium and phosphorus. Ingredients such as oil meals and prepared supplements, which must meet specific standards, need not be analyzed often, except as quality control measures. A protein analysis has minimal value for a pig feed because the pig requires amino acids, not protein. Although there are equations estimating relationships between protein and lysine, they are subject to variation.

Despite the recognized value of a chemical analysis, it does not provide information on the availability of nutrients, variation from sample to sample, obtaining a representative sample, or anything about the associated effect of feedstuffs. A chemical analysis tells nothing about taste, palatability, texture, and undesirable physiological effects, such as laxative effects.

Nevertheless, a chemical analysis provides a solid foundation in evaluating feeds, and bearing in mind that it is the composition of the total feed (the complete diet) that counts, the person formulating the diet can intelligently determine the kinds and quantities of protein to buy, and the amounts of calcium and phosphorus to add. Trace minerals and vitamins are not typically analyzed, because of the cost and the fact that essential trace minerals and vita-

mins that may be limiting are routinely added to the diet, regardless of the amounts in the energy and protein sources.

Feed quality. Numerous factors determine the quality of feed, including

a. Stage of harvesting forages—Early cut forages are of higher quality than those that are mature.

b. Freedom from contamination—Contamination from foreign substances, such as dirt, sticks, and rocks, can reduce feed quality, as can aflatoxins, pesticide residues, and a variety of chemicals (Figure 8.2).

c. Uniformity—Does the feed come from one particular area or does it represent a conglomerate of several sources?

d. Length of storage—When feed is stored for extended periods, some quality is lost because of time and exposure to the elements. This is particularly true with forages.

Degree of processing. Often, the value of feed can be either increased or decreased by processing. For example, heating some types of grains makes them more readily digestible to swine and increases their feeding value but overheating may reduce the availability of some of the amino acids. Grinding always enhances nutrient availability.

Soil analysis. If the origin of a given ingredient is known, a knowledge of the soils of the area can be helpful, for example, (a) the phosphorus content of soils, (b) soils high in molybdenum and selenium, (c) iodine-deficient areas that are important in animal nutrition, (d) high nitrogen fertilization that can increase the protein content, and (e) other similar soil–plant–animal relationships that exist.

3. In addition to providing the proper quantity of feed and to meeting the amino acid and energy requirements, a balanced and satisfactory diet should be

a. Palatable and digestible.

b. Economical. This generally calls for the maximum use of locally available feeds.

c. Adequate in amino acid content, but not higher than is actually needed. Generally speaking, medium- and high-protein feeds are higher in price than energy feeds. Lower-protein content diets may be used successfully, provided they are of good quality and fortified, properly limiting amino acids and needed vitamins and minerals.

d. Well fortified with the needed minerals, but mineral imbalances should be avoided.

Figure 8.2 Feed samples being stored on a swine farm as checks for quality and feed additive residues. Samples can be discarded about 2 months after the pigs are sent to slaughter if there is no contact regarding illegal residues. *(Courtesy, Palmer Holden, Iowa State University, Ames, IA)*

e. Well fortified with the needed vitamins.

f. Fortified with antibiotics, or other antimicrobial agents, as justified.

g. Able to maintain or enhance, rather than impair, the quality of pork produced.

4. In addition to considering changes in availability of feeds and feed prices, diet formulation should be altered at stages to correspond to changes in weight and productivity of animals.

Steps in Diet Formulation

The ideal diet is one that optimizes production at the lowest cost. A costly diet may produce superior gains, but the cost per unit of production may make the diet economically infeasible. Likewise, the cheapest diet is not always the best because it may be marginal in many nutrients and not allow for maximum production.

Therefore, the cost per unit of production is the ultimate determinant of what constitutes the best diet. Awareness of this fact separates successful producers from marginal or unsuccessful ones. The following four steps should be taken to formulate an economical diet:

1. **Find and list the nutrient requirements or allowances for the specific animal to be fed.** Nutrient requirements generally represent the minimum quantity of the nutrients that should be incorporated, whereas allowances take into consideration a margin of safety. Factors to be considered include the following:

   a. Weight or age.

   b. Sex.

   c. Feed intake.

   d. Genetic merit.

   e. Type of production. Is the animal being fed for maintenance, growth, reproduction, or lactation?

   f. Intensity of production. Is the animal gaining 0, 1, 2, or 3 lb per day? Is the gain high in fat or lean? Is the lactating animal at the peak of milk production?

2. **Determine what feeds are available and their respective nutrient compositions.** In diets for swine, amino acids, energy, calcium, phosphorus, sodium and chlorine (salt), vitamins A, D, E, and B-complex, and several trace minerals are generally considered. Because of these many considerations, it is easy to see why most producers use computers for diet formulation.

3. **Determine the cost of the feed ingredients under consideration.** Not only should the cost of the feed be considered but also the cost of transportation, mixing, and storage. Some feeds require an-

tioxidants to prevent spoilage. Others lose nutritive value when stored for extended periods.

4. **Consider the limitations of the various feed ingredients and formulate the most economical diet.** Remember that the ultimate goal is to formulate a diet that minimizes the cost per unit of production.

Adjusting Moisture Content

The majority of feed composition tables for swine diets are given on an "as-fed" basis, including the National Research Council tables, but the dry matter content of the feed is also included. Nutrient requirement tables for cattle feeds are usually on a dry basis because silages, forages, and grains are often quite high and variable in moisture.

The significance of water content of feeds becomes obvious in the examples given in Table 8.3. When comparing the lysine content of whey and potatoes to corn, milk has a much higher lysine content on a dry matter basis and potatoes are similar to corn. The same principle applies to other nutrients, also.

It is often desirable to convert as-fed diets to a dry matter basis. This may be done by using the following formulas:

Formula 1

When the diet is listed on an as-fed basis, and the producer wishes to compare the content of the various ingredients with the requirements on a dry matter basis, the equation is

$$\text{nutrient in dry ingredient (total)} = \frac{\% \text{ nutrient in as-fed diet (total)}}{\% \text{ dry matter in diet (total)}} \times 100$$

For example, an ingredient, such as high-moisture corn containing 75% dry matter containing 6.4% protein on an as-fed basis, becomes an 8.5% protein diet on a dry matter basis.

Formula 2

If the dry matter content of the ingredient, the percentage of the ingredient in the as-fed diet, and the percentage dry matter wanted in the diet are known, it is possible to calculate the amount of that ingredient in the diet on a dry matter basis.

$$\text{Amount of ingredient in dry diet} = \frac{\% \text{ of ingredient in as-fed diet}}{\% \text{ dry matter wanted in diet}} \times \% \text{ ingredient dry matter}$$

For example, if an 88% dry matter diet is wanted, and if an ingredient containing 75%

TABLE 8.3   Comparing Protein Content between As-Fed and Dry Matter Basis

Feed	Water	Dry Matter	Lysine As-fed	Lysine Dry Matter
Corn, grain	11.0%	89.0%	0.26%	0.29%
Whey, liquid	93.4%	6.6%	0.07%	1.06%
Potatoes, cull, cooked	78.0%	22.0%	0.06%	0.27%

dry matter is incorporated at a level of 80% of the wet diet, the ingredient constitutes 68% of the dry matter weight in the diet.

Formula 3

If the producer wants to change the amounts of the ingredients from an as-fed basis to a dry matter basis, the following equation should be used.

Parts on an as-fed basis
= % ingredient in as-fed diet
× % dry matter of the ingredient

This calculation should be done for each ingredient and the products added. Each product should then be divided by the sum of the products.

- To convert the components of a dry diet to that of an as-fed diet having a given percentage of dry matter, use the following equation:

Parts of ingredient in as-fed diet
= % ingredient in dry diet
$\times \dfrac{\text{\% dry matter desired in diet}}{\text{\% dry matter in ingredient}}$

The total number of parts should be summed and water added to make 100 parts.

METHODS OF FORMULATING DIETS

In the sections that follow, 4 different methods of diet formulation are presented: (1) the computer method, (2) the Pearson square method, (3) the simultaneous equation method, and (4) the trial-and-error method. Despite the sometimes confusing mechanics of each the end result of all 4 methods is the same—to produce a diet that provides the desired allowance of nutrients in correct proportions.

Ideally, the method will help design a diet so as to achieve the greatest net returns—for it is net profit, rather than cost, that counts. Because feed represents by far the greatest cost item in swine production, the importance of balanced diets is evident. An exercise in diet formulation follows for purposes of illustrating the application of each of these five methods.

Computer

Practically all swine operations use computers for diet formulations. Although computers can alleviate many human errors in calculations, the data produced on a computer are no better than those that go into it. The producer and the nutritionist who prepare the inputs and evaluate and apply the results become more important than ever. They must provide common sense on (1) feed palatability; (2) limitations that must be imposed on certain feeds; (3) by-product feeds that may be locally available; (4) feed processing and storage facilities; (5) the health, environment, and stress of the animals; and (6) those responsible for actual feed preparation and feeding.

The computer demands much of the nutritionist in terms of exact information relative to nutrient composition, availability, requirements or allowances, and cost. A quality program should evaluate at least 4 or 5 amino acids; digestible, metabolizable, and/or net energy content; major minerals; 8 to 11 vitamins; and 6 trace minerals. With manual calculation methods, it is impractical to consider more than 3 or 4 nutrients at a time. With computers, all nutrients can be considered simultaneously.

Computer software and hardware—Personal computers have adequate capacity to run all diet formulation software available. Several universities have developed low-cost spreadsheets that can formulate diets using simultaneous equations, the Pearson square for the first limiting amino acid, and trial and error for the other essential nutrients. Most consider costs of ingredients, and some project performance and cost of gain.

Numerous commercial companies market computer software for diet formulation. The software varies from the very simple and straightforward to very complex packages intended for large feed manufacturers. University personnel and nutrition consultants are good sources of information on software.

Information provided by a computer-based feed formulation program—Tables 8.4 and 8.5

TABLE 8.4 Ingredient Output from a Feed Formulation Program

Ingredient	Amount	Minimum %	Maximum %
Corn, yellow	499.50		
Barley	300.00		30.00
Soybean meal, dehulled	174.00		
Lysine, synthetic	1.20		0.50
Dicalcium phosphate	9.30		1.00
Limestone (calcium carbonate)	9.40		1.00
Salt (NaCl)	4.00	0.25	0.50
G-F Trace Mineral Vitamin Premix	2.60	0.026	0.026
Total	1000.00		

outline the type of information provided by a computer feed formulation program.

The ingredients selected and the amount of each required in the diet to meet nutrient requirements are itemized in Table 8.4. Some programs allow the user to set minimum and maximum levels to control ingredient usage. For example barley is limited to 30% in this diet. Limits may also be set on other ingredients. The feeder may want to maximize alfalfa meal to 5% or provide a minimum of a certain ingredient regardless of costs. The outputs from least-cost computer formulations also show the competitive price of each ingredient and are very useful in determining when an ingredient should be considered.

A summary for several amino acids, minerals, trace minerals, and vitamins is provided in Table 8.5.

TABLE 8.5 Nutrient Output from a Feed Formulation Program

Diet Analysis		Diet	Recommended	Ratio
Amino Acids				
Lysine	%	0.87	0.87	1.00
Threonine	%	0.58	0.58	1.00
Tryptophan	%	0.19	0.17	1.09
Methionine + Cystine	%	0.54	0.49	1.08
Crude protein	%	15.8		
Metabolizable energy	kcal/lb	1449		
Energy to lysine relationship		1661	1711	0.97
Minerals				
Calcium	%	0.63	0.63	1.00
Phosphorus, total	%	0.53		
Phosphorus, available	%	0.25	0.25	1.00
Sodium (Na)	%	0.16	0.15	1.05
Chlorine (Cl)	%	0.24	0.22	1.10
Trace Minerals (added)				
Iron	ppm	98	93	1.06
Zinc	ppm	98	93	1.06
Copper	ppm	6.2	5.8	1.05
Manganese	ppm	3	3	1.05
Iodine	ppm	0.14	0.14	1.00
Selenium	ppm	0.21	0.20	1.03
Vitamins (added/lb of diet)				
Vitamin A	IU	1610	1449	1.11
Vitamin D_3	IU	161	145	1.11
Vitamin E	IU	11	10	1.12
Vitamin K (menadione)	mg	0.49	0.43	1.13
Niacin	mg	16	13	1.20
Pantothenic acid	mg	10	9	1.05
Riboflavin	mg	3.5	3.0	1.15
Vitamin B_{12}	mcg	18	15	1.17

This example also includes crude protein and total phosphorus as reference values. However, the real requirement is for amino acids and available phosphorus. The dietary level of each nutrient is given in the first column, the recommended level next, and finally the ratio of dietary to recommended.

Nutrients that are at their requirement are forcing the cost of the diet up. For example, lysine and available phosphorus are at their lower limits which means that if any of these could be lowered, the cost of the diet would be reduced. If a reduction in these ingredients causes animal performance to suffer, reducing the dietary cost would result in lower performance.

Other computer formulation programs allow the user to specify minimums and/or maximums for each nutrient as well as various ratios. For example a maximum calcium:phosphorus ratio may be set or a desired energy to lysine relationship.

Pearson Square

The square method is simple, direct, and easy. Also, it permits quick substitution of feed ingredients in keeping with market fluctuations, without disturbing the lysine content. In balancing diets by the Pearson square, it is recognized that one specific nutrient alone receives major consideration. Correctly speaking, therefore, it is a method of balancing for one nutrient, with no consideration given to the vitamin, mineral, and other nutritive requirements. The following example shows how to use the square method in formulating a swine diet:

Example 1   *A swine producer has 40-lb pigs to which he/she desires to feed a 0.75% protein lysine. The corn on hand contains 0.26% lysine. The producer can buy a 35% protein (2.25% lysine) supplement, which is fortified with minerals and vitamins. What percentage of the diet should consist of each corn and the 36% protein supplement?*

Step by step, the procedure in balancing this diet is as follows (Figure 8.3):

1. Draw a square, and place the number 0.75 (desired lysine level) in the center.

2. At the upper left-hand corner of the square, write *corn* and its lysine content (0.26); at the lower left-hand corner, write *protein supplement* and its lysine content (2.25).

3. Subtract diagonally across the square, the smaller number from the larger number, and record the difference at the corners on the right-hand side ($0.75 - 0.26 = 0.49$; $2.25 - 0.75 = 1.50$). The number at the upper right-hand corner gives the parts of corn by weight, and the number at the lower right-hand corner gives the parts of supplement by weight to make a diet with 0.75% lysine.

4. To determine what percentage of the diet would be corn, divide the parts of corn by the total parts: $1.50 \div 1.99 = 75.4\%$ corn. The remainder, 24.6%, would be supplement.

The square method may also be used to balance diets using grain, soybean meal, and a mineral–vitamin premix. In this case the amounts of grain and soybean meal must be adjusted because they will only account for a portion of the diet.

Example 2   *A swine producer wants a corn–soybean meal diet, with 3% premix added to provide the minerals and vitamins, to furnish 0.75% lysine to grower pigs. What percentage of the diet should consist of corn and soybean meal?*

The following steps are involved:

1. Because the lysine must come from the corn and soybean meal, 97% of the diet ($100 - 3$), the value in the center of the square is 0.77 ($0.75 \div 97\%$).

2. The corn contains 0.26% lysine, and the soybean meal contains 3.04% lysine.

3. Set up square as shown in Figure 8.4.

4. Calculate the parts of corn and soybean meal as in Example 1, resulting in 81.7% corn

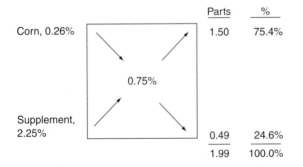

Figure 8.3   The Pearson square method for balancing grain and a protein supplement. *(Courtesy, Palmer Holden, Iowa State University, Ames, IA)*

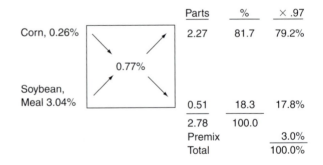

Figure 8.4   The Pearson square method for corn, soybean meal, and premix. *(Courtesy, Palmer Holden, Iowa State University, Ames, IA)*

and 18.3% soybean meal. However, because they only compose 97% of the diet, they must be multiplied by 97%, resulting in 79.2% corn, 17.8% soybean meal, and 3% premix. Multiplying the lysine in corn times the percentage of corn (0.26 × 79.2%) plus the lysine in soybean meal (3.04 × 17.8%) results in a diet containing 0.75% lsyine.

Simultaneous Equations

In addition to the square method, it is possible to formulate diets involving two sources and one nutrient quickly through the solving of simultaneous equations:

Example 3   *A producer has on hand corn containing 9% protein, and can buy a 40% protein supplement, which is fortified with minerals and vitamins. A diet for 30-lb pigs, containing 18% protein, is desired.* Step by step, the procedure in balancing this diet is as follows:

1. Let X = amount of corn to be used in 100 lb of mixed feed, and Y = amount of 40% protein supplement to be used in 100 lb of mixed feed. We know that the corn contains 9% protein and the protein supplement 40% and that the diet should be 18% protein. Therefore, the equation we must solve is as follows:

$0.09X + 0.40Y = 18$ (lb of protein in 100 lb of feed)

2. In order to solve for two unknowns (X and Y), create a "dummy equation." This can be done in the following manner:

$X + Y = 100$ lb of feed

3. Multiply the dummy equation by 0.09 in order that the X term will cancel out with the original equation. Therefore

$X + Y = 100$ becomes $0.09X + 0.09Y = 9$

4. Subtracting the new dummy equation from the original equation, solve for Y as shown below:

Original equation: $0.09X + 0.40Y = 18$
Dummy equation: $-0.09X - 0.09Y = -9.0$
$\overline{0.00X + 0.31Y = 9.0}$

$Y = \dfrac{9}{0.31} = 29.03$ lb of 40% protein supplement per 100 lb of feed or 29.03%

5. We can now substitute the newly acquired value for Y in the original equation and solve for X, as follows:

$X = 100 - 29.03 = 70.97$ lb of corn per 100 lb of feed or 70.97%

Trial-and-Error Method

In the trial-and-error method, consideration is given to meeting whatever allowances are decided on for each of the nutrients that one cares to list and consider.

Example 1   *A producer decides to use a diet containing 0.60% lysine, 1,500 kcal of metabolizable energy per lb, 0.52% calcium, and 0.16% available phosphorus.*

1. Considering available feeds and common feeding practices, the first step is arbitrarily to set down a diet and see how well it measures up to the desired allowances. The approximate composition of the available feeds may be arrived at using the feed composition tables (Chapter 22). Where commercial supplements are used, the guarantee on the feed tag may be used.

Try the following diet per 1,000 units as shown in Table 8.6a. Multiply the amount of each ingredient times its nutrient analysis and sum the values for each of the four nutrients.

2. Compare the desired allowances to that supplied by the proposed diet. Calculations show that this diet supplies 0.46% lysine, 1,515 kcal of metabolizable energy per pound, 0.29% calcium, and 0.13% available phosphorus. Therefore, it falls somewhat short and the producer decides to make changes, including the addition of limestone, giving the new diet shown in Table 8.6b.

Through calculations based on the composition of these feeds, the producer finds that this diet will now supply 0.60% lysine, 1,504 kcal of metabolizble energy per pound, 0.55% calcium, and 0.16% available phosphorus. So, the producer decides it approximates the desired allowances and may be considered satisfactory.

Formulation/Mixing Worksheet

When formulating diets, it is advisable to record the diet on a worksheet similar to that in Figure 8.5. A similar sheet can be worked up for micronutrient composition of premixes for trace minerals, vitamins, and feed additives. The worksheet serves four purposes:

1. It provides a means of reviewing and double checking the calculations used to formulate the diet. If there is a gross error, it should become obvious when listed on the worksheet.

TABLE 8.6a   Trial-and-Error Method, Estimated Diet

Ingredient	Amount	Lysine, %	ME/lb	Calcium, %	Av Phos, %
Corn	920	0.26	1535	0.03	0.04
Soybean meal, dehulled	50	3.04	1540	0.26	0.16
Meat and bone meal	25	2.89	1025	9.90	3.50
Limestone				38.00	
Vitamin and mineral premix	5				
Diet	1,000	0.46	1,515	0.29	0.13
Desired		0.60	1,500	0.52	0.16
Difference		**−0.14**	15	**−0.23**	**−0.03**

TABLE 8.6b   Trial-and-Error Method, Adjusted Diet

Ingredient	Amount	Lysine, %	ME/lb	Calcium, %	Av Phos, %
Corn	864	0.26	1,535	0.03	0.04
Soybean meal, dehulled	95	3.04	1,540	0.26	0.16
Meat and bone meal	31	2.89	1,025	9.90	3.50
Limestone	5			38.00	
Vitamin and mineral premix	5				
Diet	1,000	0.60	1,504	0.55	0.16
Desired		0.60	1,500	0.52	0.16
Difference		0.00	4	0.03	0.00

2. It can be used to organize mixing procedures. It is vital that the person mixing feed refer to a worksheet on which can be recorded what has been mixed.
3. The worksheet can be filed for future reference. If questions should arise when the feed is being used, the worksheet provides an orderly record of the content of the feed, the date of preparation, and any mixing notes.
4. Any withdrawal periods for feed additives should be noted.

Each diet should be assigned a number for future reference, and the date of formulation or mixing should be recorded. In addition to listing the feed ingredients and their respective amounts, the nutrient requirements to be fulfilled by the diet should be listed on the worksheet immediately below the totals of various components of the feed. By subtracting the totals contained in the feed from the nutrient requirements, the person formulating the diet can then determine if there are significant excesses or deficiencies (Figure 8.5).

Pointers in Formulating Diets

In formulating diets and in feeding swine, the following points are noteworthy:

1. Feeds of similar nutritive properties can be interchanged in the diet as price relationships warrant.
2. If wheat, barley, oats, and/or grain sorghum are used instead of corn as the grain in a diet, the protein supplement may be slightly reduced because these grains have a higher lysine content than corn.
3. Grains from different regions vary in nutrient content so it is important to have them analyzed or to use local nutritive values.
4. When proteins of animal origin are used, less limestone and phosphorus supplementation is necessary.
5. Because of the variability of vitamins in feedstuffs, several should always be supplemented.
6. Except for gestating sows and boars of breeding age, hogs are generally self-fed. It is preferred that all ingredients be mixed together and placed in the same self-feeder as opposed to free-choice feeding.
7. Full-fed finishing hogs will consume diet at 4 to 5% of their live-weight until they weigh 100 lb (45kg). They will eat 3 to 4% daily per 100 lb (45 kg) live-weight from this stage until marketing.

HOME MIXED VERSUS COMMERCIALLY PREPARED DIETS

Pork producers have the choice of mixing feeds on the farm or purchasing various complete feeds. Producers who feed home-grown grains are more likely

NUTRIENT WORKSHEET

Diet Number: **Date:**

Ingredient	✓	Amount lb	Lysine lb	Thr lb	Trypt lb	Met + Cys lb	Crude Protein lb	Met Energy kcal	Ca lb	Phos lb	Avail Phos lb
Total											
Percent (lb/Total)											
Requirement											
Balance (% − Requirement)											

✓ when ingredient is added to mix.

To calculate lb of each nutrient, multiply the lb of the ingredient by the nutrient percentage composition.

Percent = the lb of each nutrient divided by the Total lb.

Figure 8.5   A formulation worksheet for mixing a diet with analyses for some of the major nutrients. (*Courtesy, Palmer Holden, Iowa State University, Ames, IA*)

to mix on the farm, purchasing commercial protein supplements or oilseed meals and minerals and vitamins. Many types of mixers are available for on-farm mixing, including both portable grinder-mixers and various stationary mills. The swine producer has the following options from which to choose for home mixing feeds:

1. Purchase a commercially prepared protein supplement (fortified with vitamins and minerals), which may be blended with local or home-grown grain.
2. Purchase a commercially prepared vitamin or trace mineral premix, which may be mixed with an oil meal, a calcium–phosphorus source, limestone, or salt and then blended with local or home-grown grain.
   a. A suggested trace mineral premix is given in Table 8.7. *Note:* This mineral premix is designed to be fed to all ages of growing swine by adjusting its inclusion rate.
   b. Suggested vitamin premixes are given in Table 8.8. They are divided into the fat-soluble vitamins, B vitamins, biotin and folic acid, and choline. Normally, the fat-soluble and B vitamin premixes can be combined although the pigs' requirements for each group change as they mature. Biotin, folic acid, and choline should be separate premixes because they are not required to be added to the diets of all pigs and certainly not to those of pigs weighing more than 50 to 100 lb (23 to 45 kg). *Note:* These vitamin premixes are designed to be fed to all ages of growing swine by adjusting the inclusion rates.
   c. These premixes are examples only and will need to be adjusted if different nutrient allowances are used. Also different premixes would be designed for the breeding herd.
3. Use a base mix.
   a. Base mixes contain all needed ingredients (calcium, phosphorus, salt, trace minerals and vitamins) except grain and protein and usually account for 2.5 to 5.0% of the diet. An example base mix is shown in Table 8.9.
   b. Some base mixes may contain amino acids and feed additives. These will be listed on the feed tag along with mixing instructions as shown in the Table 8.9 footnotes.
   c. This base mix is an example only and will need to be adjusted if different nutrient allowances are followed. A different base mix would be designed for the breeding herd.

TABLE 8.7 Example Trace Mineral Premix for Growing Pigs[1]

Trace Mineral	% in Premix	Units/lb	Units/kg
Iron	7.00 g	31.8 g	70.0 g
Zinc	0.44 g	31.8 g	70.0 g
Copper	0.22 g	2.0 g	4.4 g
Manganese	0.010 mg	1.0 g	2.2 g
Iodine [2]	0.012 mg	0.045 mg	0.10 mg
Selenium	7.00 g	0.054 mg	0.12 mg

[1]Several sources of trace minerals may be used. See Table 22.6 for concentrations and bioavailability.
[2]Iodine may be eliminated if iodized salt is used.
Source: Adapted from ISU Life Cycle Swine Nutrition, 1996.

TABLE 8.8 Example Vitamin Premixes for Growing Pigs[1]

	Amount/lb of Premix	Amount/kg of Premix
Fat-Soluble Vitamin Premix		
A	2,300,000 IU	5,060,000 IU
D	230,000 IU	506,000 IU
E	16,000 IU	35,200 IU
K (menadione)	700 mg	1,540 mg
B Vitamin Premix		
Niacin	31,000 mg	68,200 mg
Pantothenic acid	20,000 mg	44,000 mg
Riboflavin	7,000 mg	15,400 mg
Vitamin B_{12}	35,000 mcg	77,000 mcg
Biotin and Folic Acid Premix		
Biotin	13 mg	28.6 mg
Folic acid	60 mg	132 mg
Choline Premix		
Choline (60%)	236,000 mg	519,200 mg

[1]Several sources of vitamins may be used (Table 6.5).
Source: Adapted from ISU Life Cycle Swine Nutrition, 1996.

DESIGNING A BASE MIX OR PREMIX

Designing a base mix or premix is a simple matter of mathematics. You need to define several things:

1. Units of base mix to be added to the complete diet.
2. Units of the nutrient wanted.
3. Quantity of nutrient source needed in the base mix.

Example 1 *Determine the level of a nutrient listed as a percentage, for example salt (Na and Cl).*

Assume 25 units of base mix will be added per 1,000 units of complete diet (2.5%).

TABLE 8.9   Complete Growing Pig Base Mix for Corn–Soybean Meal-Based Diets[1]

Ingredient, lb	1	2
Limestone	340.00	160.00
Dicalcium phosphate	406.00	
Defluorinated phosphate		485.00
Salt	150.00	150.00
Trace mineral premix[2]	50.00	50.00
Fat-soluble vitamin premix[2]	24.00	24.00
3 vitamin premix[2]	16.00	16.00
Biotin and folic acid premix[2]	(8.00)	(8.00)
Choline premix[2]	(6.00)	(6.00)
Carrier, wheat midds		101.00
Feed additive[3]		
Total	1000.00	1000.00

Calculated Analysis, Minimum

Calcium	21.57%	Vitamin A	55,200 IU/lb
Phosphorus, total	7.59%	Vitamin D$_3$	5,520 IU/lb
Phosphorus, available	7.59%	Vitamin E	384 IU/lb
Sodium (Na)	6.00%	Vitamin K	16.8 mg/lb
Chlorine (Cl)	9.00%	Niacin	496 mg/lb
Iron	0.350%	Pantothenic acid	320 mg/lb
Zinc	0.350%	Riboflavin	112 mg/lb
Copper	0.022%	Vitamin B$_{12}$	560 mcg/lb
Manganese	0.011%	Optional:	
Iodine	0.0005%	Biotin	0.10 mg/lb
Selenium	0.0006%	Folic acid	0.46 mg/lb
		Choline	1416 mg/lb

[1]Due to the instability of vitamins in the presence of trace minerals, this premix should be used within 30 days of preparation. Complete premixes are effective over a limited range of diets. Diets for stages lower than 6 will have excess calcium and deficient available phosphorus. Stages greater than 12 will be deficient in calcium and excessive in available phosphorus.
Example mixing directions:

Stage	Growing Pig Premix	Corn, lb	Soybean Meal, Dehulled
6	35	645	320
8	27	725	245
9	25	765	210
10	22	793	185
12	21	849	130

[2]Trace mineral and vitamin premixes are in Tables 8.7 and 8.8. Biotin, folic acid, and choline additions are optional. If they are not added, replace the pounds of each with carrier. See Chapter 7, Table 7.14 for requirements.
[3]Feed additives may be added at various levels as approved by the Food and Drug Administration in Table 8.2.
Source: Adapted from ISU Life Cycle Swine Nutrition, 1996.

The final diet requires 0.4% added salt and the source of salt is 100% salt.

$$\frac{\% \text{ nutrient in base mix}}{= \frac{\% \text{ nutrient in final diet}}{\text{Base mix (\% of diet)}}}$$

$$\% \text{ salt in base mix} = \frac{0.4\%}{2.50\%/100}$$

$$= 16.0\% \text{ salt}$$

If the nutrient source is not 100% pure, an adjustment must be made. For example, if the base mix required 8% phosphorus and the source is dicalcium phosphate (18.5% phosphorus), the base mix will require (8/0.185) 43.2% dicalcium phosphate.

Example 2   *Determine the level of a nutrient required in parts per million (ppm), for example iron.*

Assume 25 units of base mix will be added per 1,000 units of complete diet (2.5%). The final diet requires 100 ppm (0.01%) added iron and the source of the iron is ferrous sulfate (20.1% Fe).

To convert ppm to a percentage, divide the percentage by 10,000.

$$\% \text{ nutrient in the base mix}$$
$$= \frac{\% \text{ nutrient in final diet}}{\text{Base mix (\% of diet)}}$$
$$= \frac{\text{ppm}/10,000}{\text{Base mix (\% of diet)}}$$
$$\% \text{ iron in base mix}$$
$$= \frac{100 \text{ ppm}/10,000}{2.50\%/100} = 0.4\% \text{ iron}$$

The source of iron (ferrous sulfate) contains 20.1% iron. Therefore, the base mix would need to contain (0.4%/.201) 1.99% ferrous sulfate.

Example 3 *Determine the level of a nutrient listed as units/weight.*

Assume 25 units of base mix will be added per 1,000 units of complete diet (2.5%).
The final diet requires 12 mg niacin/unit of diet, for a total mix of 1,000 units and the source of niacin is 99.5% niacin.

Total mg of niacin needed
= 12 mg × 1,000 units of feed
= 12,000 mg niacin

The 12,000 mg of niacin will be in 25 units (lb or kg) of base mix. Therefore, the base mix requires (12,000 mg/25) 480 mg niacin per unit (lb or kg) of base mix. Because the source of niacin in 99.5% pure niacin, 482 mg (480/.995) of the niacin source needs to be added per unit. If there is a total of 100 units or percent, (482 × 100), 48,200 mg needs to be added to the base mix.

Summary of Base Mix

Ingredient	Amount or %
Salt	16.0%
Dicalcium phosphate	43.2%
Ferrous sulfate	0.4%
Niacin source	48.2 g (48,200 mg)
Other ingredients	?
Total	100.0%

QUALITY CONTROL

Quality control is an essential component of the manufacture of swine feeds. It is particularly important in the processing and mixing of baby pig diets. A sound quality control program assures that the feed contains the desired concentration of nutri-

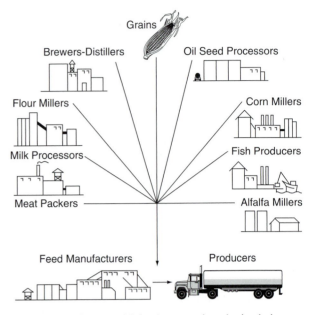

Figure 8.6 Commercial feed companies obtain their raw material for feeds from many sources. More than 100 different ingredients are processed into various complete feeds or concentrates that are fed with home-grown grains. *(Courtesy of Pearson Education)*

ents. Because feed usually represents 60 to 70% of the total cost of producing swine, it makes good business sense to ensure pigs receive feed that has been properly formulated using quality ingredients, and manufactured properly.

SELECTING COMMERCIAL FEEDS

Commercial feeds are just what the term implies—instead of being home mixed, these feeds are mixed by commercial feed manufacturers. The commercial feed manufacturer has the distinct advantages of (1) purchasing feedstuffs in quantity lots, allowing for possible price advantages; (2) economical and controlled mixing; (3) hiring scientifically trained personnel for assistance in determining the diets; and (4) quality control (Figure 8.6). Because of these advantages, commercial feeds are of interest to swine producers.

Large swine operations that do not raise grain are very likely to have a commercial mill prepare and deliver the needed diets. Midsize producers that are limited in labor availability may also wish to purchase this service rather than hiring additional labor. Also, producers that have pigs on scattered production sites are more apt to purchase commercial feeds because they do not have feed trucks to haul diets to several locations.

Numerous types of commercial feeds, ranging from additives to complete diets, are on the market, and they are designed for specific species, ages, or needs. Among them are complete diets, concentrates, pellets, protein supplements (with or without added vitamins and minerals), vitamin or mineral supplements, additives, milk replacers, starters, growers, finishers, diets for different levels of genetics, diets for gestation and lactation, and medicated feeds.

In summary, there are two alternative sources of most feeds and diets—home mixed or commercial—and the able manager will choose wisely between them or even combine them.

State commercial feed laws—All states have laws regulating the sale of commercial feeds. These benefit both producers and reputable feed manufacturers. States laws require that every commercial feed sold be labeled, and that the chemical composition be guaranteed.

Samples of commercial feed are taken periodically and analyzed in the state's laboratory to determine if manufacturers are meeting their guarantees. Additionally, skilled microscopists examine the samples to ascertain that the ingredients present are the same as those guaranteed. Flagrant violations may be prosecuted.

There is a difference in commercial feeds! The most important factors to consider or look for in buying a commercial feed are as follows:

1. **Reputation of the manufacturer.** This should be determined by (a) conferring with other producers who have used the particular products and (b) checking on whether the commercial mill under consideration has consistently met its guarantees.

The cost of the diet may indicate the use of higher-quality ingredients, supporting research, consulting services of well-trained and experienced staff, publications, or other benefits. Therefore, increased costs may be justified by the increased performance of the animals to which the diet is fed.

2. **Specific needs.** Feed needs vary according to (a) the class, age, and productivity of the animals and (b) whether the animals are fed primarily for maintenance, growth, finishing (or show-ring fitting), reproduction, or lactation. The operator will buy different formula feeds for different needs.

Feeding swine has become a sophisticated and complicated process. Feed manufacturers have extensive resources with which to formulate and test diets for different needs. As a result, most manufacturers have a large selection of feeds—one of which should be applicable to the needs of the producer. It is essential that the producer make clear to the feed salesperson the needs of the animals to be fed.

3. **Labeling.** State laws require that mixed feeds carry labels guaranteeing the ingredients and the chemical makeup of the feed (Figure 8.7). The feed tag should contain the following information:

Net weight of the feed—lb or kg.

Brand name and product name—The brand name refers to any word, name, symbol, or logo that identifies the feed of a distributor and distinguishes it from the feeds of other manufacturers. The product name identifies the specific use for the feed.

Guaranteed analysis—Most feed labels give minimum and/or maximum guarantees of certain nutrients within the feed. Laws vary from state to state as to what analyses must be guaranteed, but most, if not all, of the following analyses are generally listed: (a) dry matter, (b) crude protein, (c) crude fat, (d) crude fiber, (e) ash, (f) nitrogen-free extract, (g) calcium and phosphorus, (h) any feed additives, and (h) other nutrients. Many labels now list some added amino acids. Generally, feeds with more protein and fat and less fiber indicate quality. Most labels list minimum values of crude protein, crude fat, and phosphorus and maximum values of crude fiber. Both maximum and minimum guarantees are listed for calcium and salt on most feed tags.

A high-fiber content often indicates a low-feeding value. Feeds for swine generally contain negligible amounts of fiber, although gestation diets are good places for utilizing these lower-energy feedstuffs.

Listing of ingredients—The feed tag lists the ingredients of the feed in descending order of quantity used. Although the exact quantities of the ingredients are not generally given, the buyer can obtain a rough idea as to the composition of the feed. Often proteins are grouped as "Plant protein sources" or "Animal protein sources" rather than listing individual ingredients.

Directions for use—If the feed is to be used for a specific purpose, directions may be given on the label. They should specify for what kind of animal and for what particular purpose the feed was formulated.

Name and mailing address of the manufacturer—The manufacturer is responsible for the quality of the feed. Any failure to meet guarantees or any contamination problems incurred with the feed makes the manufacturer liable for damages incurred from the feed. For this reason, the manufacturer must be identified clearly on the product.

Warnings—When medications have been added to commercial feed, the feed label must clearly indicate that it is medicated. States require that the name of the medication with a listing of the quantity of active ingredients added be stated on the feed tag. Also, the purpose of medication, any restrictions of use, and withdrawal period for the medicated feed must be stated.

EZ Grow Pig Supplement

EZ Grow Pig Supplement

50 lb. Net

EZ - CTC

Medicated

An aid to increase rate of gain and feed efficiency, prevent or treat bacterial enteritis, maintain weight gain in the presence of atrophic rhinitis when fed according to mixing directions on back of tag.

ACTIVE DRUG INGREDIENTS

Chlortetracycline	2 gms/lb

GUARANTEED ANALYSIS

Crude Protein, min	40.0%
Crude Fat, min	1.0%
Crude Fiber, max	6.0%
Calcium (Ca), min	2.5%
Calcium (Ca), max	3.5%
Phosphorus (P), min	1.7%
Salt (NaCl), min	1.6%
Salt (NaCl), max	2.6%
Zinc (Zn), min	0.054% (540 ppm)
Vitamin A, min	12,000 USP Units/lb
Lysine, min	2.35%

INGREDIENTS

Soybean Meal, Meat Meal, L-Lysine, etc.
EZ Grow, Inc
Anywhere, USA

Mixing Directions for Base Diets

Pig Weights, lb

	60–100	100-150	150-Mkt.
Gr. Corn or Milo	450	1635	1750
EZ Grow Pig Supp	450	350	225
(Un-medicated)			
EZ Mineral	–	15	25
Total (lb)	2000	2000	2000
Protein (%)	16	14	12

To provide Chlortetracycline in the above diets substitute the following pounds of EZ-CTC Grow Pig Supplement for the un-medicated EZ Grow Pig Supplement to provide the following levels of Chlortetracycline.

	Grams/Ton Complete Feed
EZ-CTC	Chlortetracycline
25 lb/ton	50
50 lb/ton	100
75 lb/ton	150

Feeding Directions: Feed above diets as sole feed to swine.

Figure 8.7   A typical commercial feed tag. *(Courtesy, Palmer Holden, Iowa State University, Ames, IA)*

4. **Quality control.** A good commercial feed manufacturer will follow a sound quality control program, including mixing procedures and sequencing of additions of feeds to the mixer, clean-out procedures between batches of feed, cleanliness, records, sample storage, and so on. Producers should look for and evaluate this program.

5. **Flexible formulas.** Feeds with flexible formulas are usually the best buy. This is because the price of feed ingredients in different feeds varies considerably from time to time. Thus, a feed manufacturer, having access to least-cost computer formulation programs, will shift formulas as prices change, in order to minimize diet cost to the producer. This is as it should be because (a) there is no one best diet, and (b) if substitutions are made wisely, the price of the feed can be kept down and the feeder will continue to get equally good results.

FEED PROCESSING

Most swine feeds are processed before distribution and consumption. Common methods include grinding or rolling, pelleting, and heat treating. The selection of one or more processes will depend on the ingredients employed, the age of the pig being fed, and the cost–benefit relationship. Poorly processed feeds result in particle size variation, inadequate ingredient blending, particle sorting by the pig, impaired feed utilization, poor performance, and, in extreme instances, serious health problems.

Grinding or Rolling

Grinding grain is the most common process used. It is simple, relatively inexpensive, and has a definite cost–benefit relationship. Grinding reduces particle size and thereby exposes a larger area to the digestive enzymes in the alimentary tract.

Either a hammer mill or a roller mill is satisfactory for processing. Hammer mills can change from grinding one grain to another by changing screens, but they require more energy than roller mills and produce a higher percentage of fines and dust. Hammers need to be replaced or reversed as they become worn and worn screens need to be replaced. Roller mills have a lower production rate than hammer mills. Pigs fed either hammered or rolled grains will perform equally if the equipment is in proper condition.

Traditionally, the grinding of feedstuffs produces particle sizes that may be described as fine, medium, or coarse. This terminology leaves much to be desired for definitive particle size measurement because the terms mean different things to different people. Often particle size is expressed by stating the size opening in the hammer mill screen, such as "through a 3/16 inch opening." This method is one step better; however, many factors, such as peripheral speed, hammer–screen clearance, and moisture content of the grain affect the particle size.

The preferred method of describing particle size is by sieve screen analysis. This procedure uses a set of preselected screens with specific-size openings. The ground sample is sieved through these screens and an average particle size is calculated and expressed in units of microns (one millionth of a meter). The particle sizes can be approximately defined into the following classifications:

Classification	Particle size		Hammer mill screen size	
	microns	inches	mm	inches
Fine	<700	<0.0276	3.2 – 4.75	1/8 – 3/16
Medium	800–1,000	0.0315 – 0.0394	6.35 – 9.52	1/4 – 3/8
Coarse	>1300	>0.0512 in	12.7	1/2

Grinding ingredients to a uniform particle size prevents sorting by the pig and allows more uniform mixing. Generally, feed efficiency improves as feeds are ground more finely. Benefits are greatest in high-fiber diets. However, grinding grain too fine can reduce palatability, cause bridging in self-feeders, increase dusty conditions in confinement facilities, and increase the incidence of gastric ulcers. Although there is not full agreement relative to particle size, most swine producers favor an average particle size of 650 to 900 microns for all grains except wheat for growing-finishing pigs and the breeding herd. Diets for nursery pigs may be more finely ground. If gastric ulcers are a problem, the breeding herd should be fed diets more coarsely ground to 750 to 900 microns. Finely ground wheat creates palatability problems.

Pelleting

Pelleting is accomplished by agglomerating ground feed by compacting and forcing it through die openings with a combination of heat, moisture, and pressure. Pellets can be made in varying lengths, diameters, and degrees of hardness. Pigs prefer a pelleted diet over a meal diet, and nursing and newly weaned pigs prefer a small size pellet.

Pelleting swine diets improve growth rate and feed efficiency. Other advantages include reduced dustiness, reduced storage space, less feed wastage, and reduced sorting of the diet by the pig. Pelleting temperatures in excess of 180 to 190°F destroy some feed-borne pathogens, such as, salmonella. Disadvantages include increased processing cost, increased incidence of gastric ulcers, partial vitamin destruction, and difficulty in producing quality pellets in diets containing more than 6% fat.

The use of a complete pelleted diet for growing-finishing hogs increases daily gain by 2 to 5% and improves feed efficiency by 5 to 10%. This is because of improved nutrient digestibility, probably as a result of partial gelatinization of starch (rupture of starch molecule) and reduced feed wastage. The higher improvements come with pelleting less dense, higher-fiber diets, such as those containing barley or oats.

When a complete diet is purchased, buying a pelleted feed may be more economical than buying a meal. But the advantage of pelleting will usually not be sufficient to offset the cost of hauling grain to the mill and having a pelleted diet made. Also, pellet machines are costly; hence, the purchase of such equipment cannot be justified with the volume of hogs handled by most swine producers.

Heat Processing

Heat processing certain feedstuffs improves pig performance by increasing the bioavailability of nutrients, destroying enzyme inhibitors, or eliminating toxins. Processing methods include heat; heat and moisture; or heat, moisture, and pressure. Excessive heat, however, causes carbohydrates, such as glucose, to react with free amino groups to form a bond (Maillard or "browning" reaction) that is not hydrolyzed by digestive enzymes. Because digestive enzymes cannot release these bonded amino acids, particularly lysine, they become unavailable to the pig. Consequently, the time and temperature of the heating process can increase or decrease the digestibility and availability of nutrients.

Some plant protein sources must be heat treated to be effectively used by swine. Soybeans are an excellent example. The major objectives in the processing of soybeans for use by swine are to destroy the growth inhibitors (antitrypsin) and to inactivate the toxic hemagglutinin. Most soybeans are processed with a solvent to remove the oil and the resulting soybean meal is heated to improve its quality for livestock feed.

Heating time, temperature, and moisture level must be regulated closely to attain maximum product quality. Full-fat soybeans may be either cooked or extruded and both will produce satisfactory results (Figure 8.8). Be sure the equipment is operating properly and minimum and maximum cooking times or extrusion rates are followed. Occasionally, have a sample tested for suitability of processing.

Figure 8.8 An on-farm extruder for processing whole soybeans into a protein product suitable for swine feeding. *(Courtesy, Palmer Holden, Iowa State University, Ames, IA)*

If soybean meal or whole soybeans are undercooked, the antigrowth factors may not be destroyed or deactivated. Likewise, if it has been overcooked, the antigrowth factors are destroyed, but the availability of essential amino acids, particularly lysine, is reduced substantially. A urease, of pH rise test, is used to determine proper heat treatment of soybeans. Ideally, the pH of the sample should increase (rise) between 0.05 and 0.20 pH units. A pH rise of less than 0.05 units indicates the soybeans have been overheated, and a rise of more than 0.20 indicates undercooking.

Considering the cost of heat treating corn and other grains and its small effect on pig performance, there has been little interest in this processing method.

DIET PREPARATION METHODS

Basically, there are four systems of preparing diets for a swine operation:

Complete feed—Complete feeds, which may be mixed by a commercial feed supplier, are complete and ready to feed. Although they are convenient, complete feeds are usually the most expensive because overhead of the mill, delivery, and service are part of the cost. Feed flexibility is only limited by the ability of the mill. Usually a wide range of feeds and ingredient choices are competitively available.

Grain and supplement—Producer-raised grain is mixed with a supplement (most commonly a 35 to 40% protein supplement). This system is usually more expensive than the premix or base mix system and restricts flexibility in feed formulation.

Base mix—A base mix provides minerals, vitamins, and additives. It is mixed with grain and

high-protein sources available locally. Generally, the base mix system is cost effective for farm mixing and fits well into many portable feed systems. (See Table 8.9 for a suggested base mix.)

Premix—The premix consists of minerals and/or vitamins and additives. It is mixed with the main ingredients of the diet. Normally, major minerals and mineral and vitamin premixes are added at 0.1 to 2.0% of the otherwise complete feed. (See Tables 8.7 and 8.8.) Producers with nutritional experience find that this program provides the most flexibility in meeting the requirements of a wide variety of feeding situations.

FEEDING SYSTEMS

The choice of the feeding system(s) and the choice of the diet(s) must go hand-in-hand. For example, if the grain and the protein supplements are to be self-fed in separate feeders or compartments, it is important that they be of equal palatability; otherwise, pigs will consume too much of one and too little of the other. A listing and discussion of each of the common feeding systems follows.

Complete Diets

Complete diets are those in which all feedstuffs are combined into one feed mixture (Figure 8.9). They may be prepared by mixing a balanced supplement with grain; soybean meal and a commercial vitamin–mineral base mix with grain; or by using vitamin and trace mineral premixes on the farm mixed with soybean meal and grain. Complete self-fed diets for growing-finishing swine are recommended because, in comparison with free-choice feeding, they (1) require less labor and management than free-choice feeding, (2) lend themselves to automation,

Figure 8.9 A complete diet being fed through a tube feeder. The water source is found in a trough beside the feed. *(Courtesy, Palmer Holden, Iowa State University, Ames, IA)*

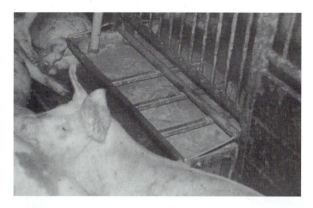

Figure 8.10 A liquid feeding system in Denmark. Liquid feeders work best when a liquid by-product, such as whey, is available for mixing with the dry diet. A risk is adding excessive liquid thereby restricting dry matter intake. *(Courtesy, Palmer Holden, Iowa State University, Ames, IA)*

(3) provide control of nutrient intake, and (4) result in faster gains.

Where producers do their own mixing on-farm, they favor simplified diets. Fortunately, a simple diet of corn and soybean meal, fortified with minerals and vitamins, will generally provide results similar to a complex diet consisting of many energy and protein ingredients. With large volume buying and computerized formulations, large swine operations and commercial feed companies can use more complex diets advantageously, especially from the standpoints of enhancing palatability and balancing amino acids, minerals, and vitamins.

Free-Choice

A **free-choice** feeding program is one in which the pig is free to eat two or more feeds at will. Grain and protein supplements are fortified with minerals and vitamins and may be fed separately from different feeders or compartments within a feeder. Pigs are allowed to eat as much of each component as they desire. Free-choice feeding may be the best feeding system for the small producer who does not have mixing equipment or when high-moisture grain is used. High-moisture grains must be fed fresh daily and high-moisture corn may be fed without grinding. Providing the protein supplement, free-choice facilitates the high-moisture feeding program.

The free-choice system requires more supervision because the palatability of the grain or the protein supplement will vary and the pigs will overeat or undereat the supplement or grain. If poorly supervised, poor feed efficiency and more expensive weight gains can occur. If consumption is ideal, the difference in economy of gain between feeding a free-choice or a complete ground mixed diet will be minimal.

Liquid

Liquid feeding usually involves mixing predetermined amounts of feed and water prior to, or at the time of, feeding (Figure 8.10). When properly used, this method can practically eliminate feed dust in the feeding area and minimize wastage. Ratios of feed and water can be varied to produce a free-flowing liquid or a thick paste, usually accomplished by blending 2 to 3 parts of water with 1 part of dry meal feed. In some cases, feed is automatically dropped into the water in the feed trough. Research has shown no difference in rate of gain or feed efficiency in pigs full-fed on liquid or dry feeds. Neither does liquid feeding have any effect on dressing percentage, carcass measurements, or carcass quality.

Liquid feeding has some benefits for young weanling pigs because they tend to eat wet feed more readily. Maintaining feed freshness is a problem. Starter diets high in milk products tend to develop off-flavors very quickly. However, if the feed delivery system is managed properly and if suitable antioxidants are included in the diet, these concerns can be largely overcome.

FEEDING PROGRAMS AND DIETS

Various commercial feeding systems are available to provide liquid-feed mixtures for all phases of swine production.

Because the nursery pig undergoes dramatic changes in digestive development, dietary changes should be made more often. Table 8.10 lists expected feed requirements of the appropriate starter diets based on various ages at weaning. Adhering to these feed use guidelines will help minimize overfeeding expensive starter diets. By phase feeding, the producer can match the pig's nutrient requirements and digestive capabilities with the most economical diet possible, yet get optimum performance. Although the first nursery diets are very expensive, the small amount of feed consumed and the excellent feed efficiency of young pigs justifies their cost.

As is obvious from the previous discussion, the nutritive requirements of swine vary according to age, weight, and stage of production—gestation

TABLE 8.10   Suggested Phase Feeding Programs for Early Weaned and Conventional Nursery Pigs and Suggested Feed Requirement

Weaning Age	Early Weaned (14–21 day)	Conventional (21+ days)	Feed/Pig
1–11 lb		Diet 1	6 lb
11–17 lb	Diet 2	Diet 3	11 lb
17–30 lb	Diet 4	Diet 5	22 lb
30–50 lb	Diet 6	Diet 6	37 lb

or lactation. Furthermore, diets must be reformulated from time to time in keeping with new developments and changing prices. In the following sections recommendations for feeding the various categories of swine are given.

Suggested diets available in Chapter 7 are intended to serve as useful guides. Wise producers will adapt suggested diets to meet their needs.

Feeding Baby Pigs

Newborn pigs must receive colostrum during the first 24 hours postfarrowing. Colostrum contains the antibodies necessary for protecting the piglet from disease until it can develop its own immune responses. Equalizing litters within 24 hours after birth and transferring pigs so that litters contain pigs of similar weight can improve pig survival.

With the availability of quality prestarter and starter diets and improved feeding and management practices, early weaning at 7 to 14 days of age is feasible. However, weaning between 2 and 3 weeks of age (early weaning) is generally considered more practical. Weaning older pigs still may be advantageous depending on the nursery facilities available and preferred scheduling of the breeding program.

Advantages, disadvantages, and recommendations of early weaning are given in Chapter 15, Swine Production and Management.

Creep Feeding

The practice of self-feeding young pigs in a separate area away from their dams is known as creep feeding. Research has shown that very little creep feed will be consumed before 3 weeks of age. Often, more creep feed is wasted than consumed before 3 weeks of age. It is suggested that a handful of creep diet be fed on a daily basis until pigs are eating well in order to keep the diet fresh (Figure 8.11).

Because of the dramatic changes in the digestive systems of early weaned pigs, a specialized nutritional program should be used, compared to pigs weaned at older ages. Diets and feeding programs for segregated early weaning (SEW) and other nursery pigs are presented in Chapter 7, Tables 7.18 and 7.19.

Feeding Growing-Finishing Pigs

Pigs ranging in weight from about 50 lb (23 kg) to slaughter weights are known as growing-finishing pigs. In a farrow-to-finish operation, grower-finisher diets represent approximately 75 to 80% of the feed usage (Table 7.11). The growing pig, up to about 120 lb (54 kg) deposits lean tissue at a fast rate. Pigs from approximately this weight to market weight are known as finishing pigs. They begin to put on more fat as they mature and consume a large share of the total feed.

Figure 8.11   A small amount of creep feed is placed on the heating pad to stimulate the piglets' interest in starting to consume dry feed. It should be placed once or twice daily to keep it fresh. *(Courtesy, Palmer Holden, Iowa State University, Ames, IA)*

Decisions to change or modify growing-finishing diets must be based on economics as well as performance. It follows that summer versus winter diets or split-sex feeding can be economically justified. Adjustments for lean gain merit, split-sex feeding, environment, and so on are estimated in Chapter 7, Tables 7.17a and 7.17b and can be used to define a stage of development for selecting appropriate diets. Suggested diets for weaning to market using basically corn and soybean meal–based diets are presented in Table 7.18. Other ingredients may be substituted as appropriate based on feedstuff availability and price (Table 7.19).

Split-Sex Feeding

Split-sex feeding refers to sorting boars, gilts, and barrows and feeding each specifically formulated diets. Full-fed boars consume 10 to 15% less feed daily than barrows or gilts and are 10 to 15% more efficient in feed conversion. Also, boars gain faster than barrows or gilts. However, boars are not customarily fed as market hogs in most countries because of the risk of "boar taint" or off-flavor in the meat. Boars in a testing program should be fed higher levels of nutrition to best measure their genetic potential.

Barrows gain approximately 0.1 lb (0.05 kg) faster per day than gilts, which reduces their age at slaughter by 10 days. Feed per unit of gain is slightly favors gilts over barrows. Gilts, compared to barrows, yield carcasses having about 0.1 in. (0.25 cm) less backfat, 0.5 in.^{2} (3.2 cm^2) larger loin eye area, and 1.8% more lean cuts. Dressing percentage usually favors barrows, consistent with their greater backfat.

Because gilts consume about 0.5 lb (0.23 kg) less feed per day than barrows, they need more highly fortified diets to meet their requirements, and because they are leaner and more efficient than barrows, their amino acid requirements are increased. Split-sex feeding is very common in large operations, but anytime a producer weans or purchases enough pigs to fill 2 or more pens and adequate feed storage bins are available for two diets, split-sex feeding should be considered.

Although the nutrient requirements of barrows and gilts vary little before 55 or 75 lb (25 to 34 kg), the most convenient time to segregate the sexes is when the pigs are weaned or when feeder pigs arrive at the feeding facility. Deciding to divide groups of pigs by sex or by weight is a decision the producer needs to make, depending on the weight variation and number of pens availabale because both factors are important in determining requirements.

The effect of various factors affecting nutrient requirements, including rapidly growing pigs, split-sex groups, environmental conditions, and antigen exposure (disease exposure), are presented in Chapter 7. Lean, rapidly growing pigs have a higher amino acid requirement than slow-growing pigs. Additionally, the winter and summer requirements for lysine and energy differ. The dietary nutrients for growing pigs under standard conditions and lean-growth potentials are presented in Table 7.14 and adjustments for temperature and for split-sex feeding are presented in Tables 7.17a and 7.17b.

Limit Feeding

Limit feeding implies feeding animals less than they would voluntarily consume. The primary purpose is to restrict energy intake sufficiently to maintain lean tissue growth but not to allow excess body fat deposition. Usually, it is started when pigs weigh about 100 to 130 lb (45 to 60 kg) and feed is limited to about 90 to 95% of what pigs of comparable age or weight consume when self-fed. Limit feeding market hogs results in slower gains, increased labor, and more mechanization.

This system exhibits greater benefits in pigs of low lean-growth capacity than in pigs with genetic capacity for high lean-growth. The slower growth rates, increased labor or mechanization required, and increased supervision needed probably more than offset the small beneficial effect produced on the carcass fat-to-lean ratio. Thus, unless sufficient premium is paid for the modestly leaner carcasses, limit feeding may not be justified.

Usually limit feeding is accomplished by overhead drop feeders that are programmed to drop feed on the floor from 3 to about 8 times per day (Figure 8.12). Dropping the feed in the sleeping area encourages cleanliness because pigs are less inclined to defecate where they eat. Feed wastage is reduced to a minimum when the animals do not have

Figure 8.12 A drop feeder in a grow-finish feeding barn. A limited amount of feed is dropped on the solid portion of the floor several times per day. The desired restriction is about 90% of full-feed. *(Courtesy, Palmer Holden, Iowa State University, Ames, IA)*

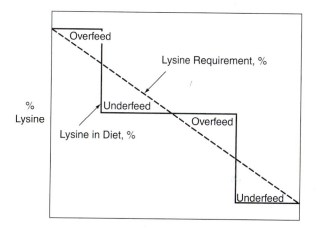

Figure 8.13 Phase feeding program indicating the amount of overfeeding and underfeeding of lysine with three dietary lysine adjustments. The more often the dietary lysine is adjusted, the more precisely the lysine requirements of the pigs can be accomplished. *(Courtesy, Palmer Holden, Iowa State University, Ames, IA)*

more feed available than they will consume at one eating. Even though it is automated, restricted feeding requires close attention because the voluntary feed intake of pigs is affected by weather.

Phase Feeding

Phase feeding involves providing several diets for varying periods of time in order to meet the pig's nutrient requirements.

When one diet is fed for a long period of time, it usually fails to meet the pig's nutrient requirements when the diet is first fed and for overfeeds for older pigs. By phase feeding, the producer can minimize this over- and underfeeding and provide a more economical feeding program. Figure 8.13 illustrates a feeding program with three levels of lysine over a large pig weight range. The result is two large periods

Figure 8.14   Three feed lines allow the producer to select different diets to fill each self-feeder for different stages of production. *(Courtesy, Palmer Holden, Iowa State University, Ames, IA)*

Figure 8.15   Individual feeding stalls in a deep-bedded gestation barn. Sows are locked in the stall until all have finished eating their allotted feed. *(Courtesy, Palmer Holden, Iowa State University, Ames, IA)*

of overfeeding and two of underfeeding. As more diets or phases are fed the lysine level stays closer to the requirement, resulting in smaller amounts of overfeeding and underfeeding. Some phase feeding is practical in all operations (Figure 8.14).

Feeding the Breeding Herd

Producers will want to select replacements that will transmit genetic potential for above-average growth rate and less backfat over a time period relative to the growth period for their offspring. Therefore replacement gilts should be on full-feed until they reach the heaviest stage of development as noted in Table 7.14. For example, high lean-growth gilts may be full-fed until approximately 257 lb (117 kg) body weight and then should be fed about 5 lb (2.3 kg) per day after being removed from the finishing pens.

Feeding Boars

The feed allowance of young boars should vary according to the condition of the animals and their breeding activity. Performance testing of boars is more important than for gilts. Therefore, they should be full-fed over the typical market growing period. After the testing period, limited feeding will be necessary.

The feed requirements of herd boars are similar to those of females of equal weight. They should always be kept in thrifty, vigorous condition and virile. In no case should boars be overfat, nor should they be in a thin, run-down condition. Mature boars should receive 5 to 7 lb of feed daily depending on weight and services or semen collections per week (Table 7.23). A more liberal quantity must be provided in the winter-time and when the sire is in heavy service.

The NRC dietary and daily amino acid, mineral, vitamin, and linoleic acid requirements of sexually active boars are given in Table 7.10. Because the boar population on a farm is quite small, it usually

is not practical to prepare a special diet for them. Usually the diet for lactating sows will easily meet the nutrient requirements of boars. Gestation diets are not adequate.

Limit Feeding

Reduce daily intake to 4 or 5 lb (1.8 to 2.3 kg) postmating and base feed and nutrient intakes on the adjustments in Table 7.20. This table is based on 4 lb of feed per day containing 0.51% lysine and adjustments may be made according to the defined factors. Overfeeding during gestation can cause embryonic death and thus decrease litter size.

Several methods may be used to limit feed intake. They include individual feeding, group feeding, bulky or fibrous feeds, and interval feeding.

Individual feeding provides a measured quantity of feed to each sow. This requires sows to be fed in individual pens, crates, or tie stalls, or by electronic feeders (Figure 8.15). This system offers the best control of intake.

Group feeding provides a measured quantity of feed to several sows. It is hoped that each sow will eat her proper amount of feed. However, this system often results in the aggressive sows overconsuming and the timid sows receiving inadequate feed. The feed for groups needs to be spread over a large area.

Feeding bulky, fibrous feed, such as silage, haylage, or alfalfa. These feeds should constitute at least one-third of the diet. Actually, this is a way to lower the energy content of the diet. Although bulky feeds will hold the weight gain down, they often increase the cost of the feeding program. This system requires more supervision to assure that the level of fibrous feed is sufficient to restrict energy intake and that feed costs are not excessive. Most bulky feeds are more expensive than corn, plus more feed must be fed to meet the sow's daily calorie requirement.

With interval feeding, gilts and sows are allowed access to self-feeders for 2 to 4 hours every second or third day or a set amount of feed is put into each pen. Under this system, gilts will usually eat about 12 lb of feed at a time (or an average of 4 lb per day) and older sows will consume about 15 lb (or an average of 5 lb per day). A major problem is that larger dominant sows will consume more than their share of the daily allotment. The amount of feed consumed in interval feeding may be controlled either (1) by varying the interval, from every other day to twice a week; (2) by varying the length of time that the gilts and sows are left on the self-feeders; or (3) by supplying a fixed amount of feed spread over a large area.

Reproductive performance is similar with either interval feeding or daily hand-feeding a limited amount. Turning sows to self-feeders at intervals requires less labor than daily hand-feeding, but under some conditions it results in greater stress on fences and equipment.

Flushing

Flushing is the practice of feeding sows and gilts more liberally so that they gain weight from 1.0 to 1.5 lb daily from 1 to 2 weeks before mating until they are bred. Usually, this involves a 50 to 100% increased feed allocation compared to the 4 to 5 lb they normally consume. Experiments and experiences indicate that flushing is effective with gilts, but not with second and third litter sows (see the Chapter 12, section "Flushing Sows and Gilts"). If females are flushed, it is important to reduce their feed intake to 4 to 6 lb immediately after breeding. Continuing a flushing diet to pregnant females may decrease embryo survival. Also overfeeding during gestation decreases lactation feed intake, to the detriment of the sow.

Feeding Gestating Sows

The nutrients fed to the pregnant gilt or sow must first take care of the usual maintenance needs. If the gilt is not fully mature, nutrients are required for both maternal growth and growth of the fetuses. Quality and quantity of proteins, minerals, and vitamins are particularly critical in the diet of young, pregnant gilts because their requirements are much greater and more exacting than those of mature sows.

Approximately two-thirds of the fetal growth occurs during the last month of gestation. However, data on the merit of increasing intake the last 30 days of pregnancy is inconsistent. Usually, ensuring 4 to 6 lb per day for the entire gestation is equal to other programs.

During gestation, it is also necessary that body reserves lost during the previous lactation be restored for use during the next lactation. With a large litter and a sow that is a heavy milker, the demands for milk production generally exceed the amount of feed the sow will voluntarily consume during lactation. Although desired gains will vary according to the initial condition, first litter pregnant gilts should be fed to gain about 90 to 100 lb (40 to 45 kg) and mature sows are generally fed to gain about 70 lb (32 kg) during pregnancy. This means that from the day of mating until farrowing time gilts should be fed to gain about 0.9 lb (0.4 kg) per day and mature sows about 0.7 lb (0.3 kg) per day. This requires a daily feed allowance of approximately 4 to 5 lb per head, with variation according to environmental conditions.

The condition of pregnant sows should be regulated so that they are neither too fat nor too thin at farrowing time. Overly fat sows may have difficulty in farrowing and give birth to weak or dead pigs. Sows too thin at farrowing tend to become suckled down during lactation. Thus, one way or another, limited feeding is a must for gestating gilts and sows.

The success of limit-fed gilts and sows depends on regulating the intake of each female. Individual feeding is preferred, but if group feeding is practiced, care must be taken to see that each female gets her share.

A visual appraisal for estimating the adequacy of gestation feed intake and its effect on body condition are illustrated in Chapter 15, Figures 15.3A and 15.3B.

Daily nutrient allowances for during gestation are given in Chapter 7, Table 7.22. Suggested diets are given in Table 7.24, diets G1, G2, or G3. *Note:* Laxative feeds should not be fed to sows unless there is a constipation problem.

Feeding Lactating Sows

Most lactating sows perform best when they are allowed to consume all the feed they can beginning the day they farrow. However, some people find it easier to detect lactational problems in sows if they are limit fed during the first 3 days after farrowing. If limited feeding is practiced after farrowing, provide at least 3 lb of feed the day after farrowing and increase the offering 3 lb/day thereafter. By day 4 postfarrowing, the sow should be given ad libitum (free-choice) access to feed. Sows should be fed no more than they will eat daily because leftover feed can become rancid and reduce feed consumption.

The nutritive requirements of a lactating sow are more rigorous than those during gestation. They are very similar to those of a milk cow, except they are more exacting relative to quality proteins and the B vitamins because of the absence of rumen synthesis in the pig.

At peak production, a good sow will produce 25 lb of milk per day. Sow's milk is richer than cow's milk in all nutrients, especially in fat. Thus, sows suckling litters need a liberal allowance of concentrates rich in essential amino acids, minerals, and vitamins.

It is essential that suckling pigs receive a generous supply of milk because at no other stage in life will they make such efficient and economical gains. The gains made by pigs from birth to weaning are largely determined by the milk production of the sows, and this in turn depends on the diet fed and the sow's inherent ability to produce milk. The lactating sow should be provided with a liberal feed allowance. Generous feeding during lactation, with a small weight loss, is more economical than a restricted allowance of feed because the nutrients in milk must come either from the feed or from the sow's body. Lactating sows are commonly self-fed because even when hand-fed they are practically on full-feed.

Nutrient allowances of females during lactation are given in Table 7.22. Suggested lactation diets are given in Table 7.24, diets L1 to L11, depending on the information from Table 7.21 estimating dietary requirements based on the number of pigs nursed and the average daily feed intake over the entire lactation.

Feeding Show-and-Sale Swine

Any of the diets described in the preceding sections for the respective classes and ages of swine are suitable for use in fitting show animals of similar classification. The recent trend has been toward self-feeding both young breeding animals and market barrows and gilts that are being fitted for show. Production-oriented swine shows include rate of gain or rate of muscle gain as part of the criteria for placing on an equal emphasis as body conformation. Thus production-oriented pigs are left on full-feed right up to show time.

Show pigs evaluated only on conformation and leanness may be hand-fed during the last month or 2 of the fitting period. However, most experienced exhibitors believe that they can get superior bloom and condition by either (1) hand-feeding or (2) using a combination of hand-feeding and self-feeding (hand-feeding twice daily and allowing free access to a self-feeder). This feed restriction will decrease their backfat and make them trimmer appearing.

Additional details regarding fitting and showing swine are given in Chapter 18, Fitting and Showing Swine.

OTHER FEED AND MANAGEMENT ASPECTS

In addition to the subject matter covered in Chapter 7 and this chapter, there are other feed and management aspects of great importance in swine production. Some of these are detailed in the sections that follow.

Nutritional Anemia

Anemia is a blood condition in which there is a deficiency of hemoglobin that transports oxygen to various parts of the body. It is caused by a deficiency of iron and copper in the diet, and it is most likely to occur in nursing pigs that have not received supplemental iron or do not have access to soil. The four basic reasons for anemia in nursing pigs raised in confinement are as follows:

1. Low body storage of iron in the newborn pig.

The pig has a comparatively low tissue level of iron at birth but a high level of iron at maturity, whereas other species only maintain the newborn iron level or see a decrease as they mature. Thus, the iron requirement of the baby pig is greater than for other species (Table 8.11).

Anemic pigs show listlessness, rough hair coat, wrinkled skins, drooping ears and tails, pale membranes around the mouth and eyes, and labored breathing.

2. Low iron content of sow's colostrum and milk. Only approximately 10% of the pig's iron requirements are met by colostrum, and this drops to about 5% about the third day.

3. Elimination of contact with soil iron. Pigs farrowed and raised in confinement have no contact with the soil.

4. Tremendous growth rate of the nursing pig. During the first weeks of life, the pig makes a much greater increase of birth weight than other species. As a result, anemia develops more rapidly in pigs than in other species.

The following nutritional anemia preventive measures may be used:

1. Give each baby pig an intramuscular injection (in the neck) of 200 mg of iron dextran within the first 3 days of life. If less than 200 mg of iron is given in the first injection, a second iron shot may be needed if pigs are not consuming dry feed by 3 weeks of age.

2. Give the pigs iron tablets or paste at 2 to 3 days of age. Repeat the treatment every 7 to 10 days until the pigs are eating creep diet adequately. If pills

TABLE 8.11 Relative Blood–Iron Concentrations of Various Species

Species	Iron Concentration (ppm)	
	Newborn	Adult
Pig	29	90
Human	94	74
Rabbit	135	60
Rat	59	60

Source: Courtesy of Pearson Education.

are given, it is important to see that the pigs swallow them and not spit them out. This method is not as desirable as the iron injection.

3. Provide fresh sod for the piglets to consume.

Scouring

One of the major problems facing the swine producer is scouring (very loose fecal stools) in baby pigs, and it is estimated to cause about 20% of pig losses between farrowing and weaning. Scouring may be caused by feeding practices or ingredients, management practices, the environment, or disease. More information regarding scouring is given in Chapter 14, Swine Health.

Soft Pork

Fats consumed in the diet move to intramuscular fat and backfat without undergoing much change. Thus, when finishing hogs are fed on high-fat content feeds in which the fatty acids are highly unsaturated, or the fat is liquid at room temperatures, soft pork results. This condition prevails when hogs eat whole fat soybeans, peanuts, added unsaturated fats, or garbage. The fat of the cereal grains is liquid at ordinary temperatures, but their fat content is relatively low. When low-fat feeds are fed, most of the carcass fat is formed from the more abundant carbohydrates in these feeds.

Soft pork is undesirable from the standpoint of both the processor and the consumer. It remains flabby and oily even under refrigeration. Soft pork has greater shrinkage in processing; the cuts are soft and unattractive in the meat case; it is difficult to slice the bacon; and the cooking losses are higher. For these reasons, hogs that are liberally fed on feeds known to produce soft pork are usually discounted at the market.

The firmness of pork carcasses may be judged by (1) grasping the flank below the ham, (2) lifting one end of the cut while permitting the other end to rest on the table (a firm pork cut will not bend readily), or (3) applying a slight pressure of the thumb (not gouging) on a cut surface. Experimentally, either the iodine number or the refractive index (measures of the degree of unsaturation) are used in determining the degree of softness.

It is recommended that feeds that normally produce soft pork be fed in quantity only to pigs under 150 lb in weight and to the breeding herd. For growing-finishing pigs more than 150 lb body weight, soybeans and peanuts should not constitute more than 10% of the diet if a soft pork problem is to be averted.

Experimental evidence has shown, however, that when a diet producing hard fat (saturated) is given following a period of feeds rich in unsaturated fats, the body fat gradually becomes harder. This practice is called *hardening off*. Thus, many hogs that are, for practical reasons, finished primarily on such feeds as soybeans, peanuts, or garbage, are hardened off with a diet of barley, corn, or some other suitable grain.

HOG–CORN RATIO

Corn is the primary ingredient of the pig's diet in most of the United States, and feed constitutes 60 to 65% of the total cost of hog production. Therefore, when corn is cheap relative to hog prices, the hog–corn ratio is high and many farmers opt to add value to the corn by feeding it rather than selling it outright. When the corn price is high, relative to hog prices, farmers may choose to sell the corn. Therefore, above a certain ratio, farmers will more often expand production of hogs or feed them to a heavier weight, whereas below that level, they will usually cut back production. In this way, the ratio serves as a general indicator of future trends.

The hog–corn ratio is determined simply by dividing the price received for live hogs per hundredweight by the current price for corn per bushel at that time.

$$\text{Hog–corn ratio} = \frac{\text{Market price of hogs, \$/100 lb}}{\text{Price of corn, \$/bu (56 lb)}}$$

For example if the live price of hogs is \$45/100 lb and corn is \$2.25/bu, the hog–corn ratio is 20. (U.S. hog prices are usually quoted on a carcass basis. To convert a carcass price to a live price multiply the carcass price by 74%.)

$$\text{Hog–corn ratio} = \frac{\$45/100 \text{ lb}}{\$2.25/\text{bu}} = 20$$

The hog–corn ratio is less important as a guide for hog operations that are not associated with a grain farm. A pork producer who has a large investment in facilities for only one use—producing hogs—is not going to use the price of corn as a major factor in pork production decisions.

POISONS AND TOXINS

Swine are susceptible to a number of poisons, any one of which may be disastrous in a herd. Among them are the following: moldy feed, including three species of mycotoxins—aflatoxins, ergot poisoning, and estrogenic syndrome; pitch poisoning; lead poisoning; mercury poisoning; pesticides; plant poisoning, involving several plants that are toxic to swine; and blue-green algae. The source, symptoms and signs, distribution and losses, prevention, and treatment of some potential poisons are covered in Table 8.12.

TABLE 8.12   Poisons and Toxins

Poison	Source	Symptoms and Signs	Distribution and Losses Caused By	Prevention	Treatment	Remarks
Arsenic (As)	Arsenic used to control insects and weeds, and to defoliate crops. Overdosing of phylarsonic compounds as feed additives to swine for growth and disease control.	The onset is sudden; characterized by groaning, restlessness, rapid breathing; muscular incoordination, blindness, and photosensitization. Death in 3–4 hours or, if less material is consumed, in a few weeks. Necropsy reveals severe hemorrhagic inflammation of stomach and intestines, with perhaps areas of erosion on mucous membranes.	Arsenic has long been a leading cause of chemical poisoning.	Keep animals away from arsenic.	Handled by the veterinarian. If caught in time, first remove the material from the animal. Sodium thiosulfate may be used, and supportive treatment may be indicated. British Anti-Lewisite (Dimercaprol) is a specific antidote for some forms of arsenic poisoning.	Accumulation of arsenic in soils may sharply decrease crop growth and yields, but it is not a hazard to animals or humans that eat plants grown in these fields, provided they do not eat the foliage of sprayed plants.
Copper sulfate (CuSO4•5H₂O)	Overdosing with copper sulfate (bluestone), or the use of too concentrated solution, in treating for parasites.	Reduced growth, lower hemoglobin, and death. Abdominal pains, vomiting, and diarrhea. Necropsy shows stomach and intestines inflamed and stomach lining coated with copper sulfate.		Never give copper sulfate in a concentration greater than 2% or in a dose of more than 4 fl oz of a 1% solution.		Dietary Cu toxicosis is closely related to molybdenum concentration. Cu:Mo ratio of 10:1 may be toxic.
Ergot (a parasitic fungus)	It replaces the seed in the heads of grasses and cereal grains. It appears as a purplish-black, hard banana-shaped dense mass from 1/4–3/4 in. (5.6–16.9 mm) long. Most common in rye, wild rye, brome grass, and dallisgrass.	Acute ergot poisoning, caused by large quantities eaten at one time, may produce paralysis of the limbs and tongue, disturbance of the gastrointestinal tract, and abortion. It is a cumulative poison; hence, poisoning may develop from lesser quantities eaten over a long period of time. Chronic poisoning produces gangrene of the extremities, with subsequent sloughing off of hooves, ears, and tail. Delirium, spasms, and paralysis may occur before death.	Ergot is found throughout the world.	Never feed heavily ergot-infested hay or grain.	If noticed in time, stricken animals may recover if put on good feed. Tannin used as a drench is an antidote, and sedations, such as chloral hydrate, may be given to nervous animals.	Six different alkaloids are involved in ergot poisoning.

$CuSO_4 \cdot 5H_2O$

Fluorine (F)	Ingesting excessive quantities of fluorine through either the feed or water.	The water in parts of Alaska, California, South Carolina, and Texas has been reported to contain excess fluorine. Occasionally, throughout the United States high-fluorine phosphates are used in mineral mixtures.	Abnormal teeth (especially mottled enamel) and bones, stiffness of joints, loss of appetite, emaciation, reduction in milk flow, diarrhea, and salt hunger.	Avoid the use of feeds, water, or mineral supplements containing excessive fluorine.	Any damage may be permanent, but animals that have not developed severe symptoms may be helped to some extent, if the source of excess fluorine is eliminated.
					Fluorine is a cumulative poison.
Lead (Pb)	Lead is discharged into the air from auto exhaust fumes and other sources. Lead pollution of feed and food crops as a result of lead being deposited on the leaves and other edible portions of the plant by direct fallout. Inhaling airborne lead. Lead may get into feed or food and water from contact with lead pipes, utensils, or discharged storage batteries.	At one time, it was a component of sprays used to control insects and ant diseases and of paints. Law now bans leaded paint.	Symptoms develop rapidly in young animals, but slowly in mature animals. Loss of appetite and evidence of gastroenteritis. Feces may become very dark gray and be tinged with blood. Salivation, champing of the jaws, frenzy, blindness, convulsions, coma, and death. Mature animals usually have diarrhea, show incoordination, especially in the hind limbs, and prostration.	Avoid sources of lead.	If damage to tissue has been extensive, treatment is of little value. Protein (milk, eggs, blood serum) may reduce gastrointestinal absorption. Calcium disodium EDTA (ethylene diamine tetra acetic acid).
					It is a cumulative poison. When incorporated in the soil, nearly all the lead is converted into forms that are not available to plants. Any lead taken up by the plant roots tends to stay in the roots, rather than move up to the top of the plant. Lead poisoning can be diagnosed positively by analyzing the blood tissue for lead content.
Mercury (Hg)	Mercury is discharged into air and water from industrial operations and is used in herbicide and fungicide treatments. Consumption of seed grains treated with fungicides that contain mercury. Mercury poisoning has occurred where mercury from industrial plants has been discharged into water and then accumulated in fish and shellfish.	When mercury-treated grain is fed to animals.	Gastrointestinal, renal, and nervous disturbances; but impossible, on basis of symptoms, to differentiate mercury from other poisons. Case history of animals consuming mercury-treated grains should be considered strong circumstantial evidence.	Do not feed livestock seed grains treated with a mercury-containing fungicide. Surplus of treated grain should be burned and the ash buried deep in the ground.	Treatment is not too satisfactory. Protein (milk, eggs, blood serum) may reduce gastrointestinal absorption. Sodium thiosulfate and BAL (dimercaprol) may be used.
					Ultimate diagnosis depends on demonstrating the presence of mercury in the tissues, especially in the kidneys and liver. Food and Drug Administration prohibits use of mercury-treated grain for feed or food. Mercury is a cumulative poison. Grain crops produced from mercury-treated seed and crops produced on soils treated with mercury herbicides have not been found to contain harmful concentrations of the element.

(continued)

TABLE 8.12   Poisons and Toxins (continued)

Poison	Source	Symptoms and Signs	Distribution and Losses Caused By	Prevention	Treatment	Remarks
Mycotoxins (toxin-producing molds), e.g., *Aspergillus flavus*, *Penicillium cyclopium*, *P. islandicum*, and *P. palitans*	Aflatoxin (most studied of the group) associated with peanuts, Brazil nuts, silage, corn and most other cereals, hay, and grasses. The mold can produce toxic compounds on virtually any food (even synthetic) that will support growth. Another very troublesome mycotoxin in certain areas is vomitoxin, which is produced by *fusarium* olds. Much of the vomitoxin problem is associated with head scab in small grains.	Molds affect animals in a variety of ways, from decreased production to sudden death. Usually, the first sign is loss of appetite and weight. A few animals will abort, and an occasional animal will die. With high intakes of mycotoxin, or with the several types of molds, any one or a combination of the following symptoms may develop; liver damage, hyperkeratosis, a typical interstitial pneumonia, bloody slimy scours, arched back, dry gangrene at end or tail or top of hoof, hemorrhagic hepatitis, renal damage, lameness, and/or swollen legs. Vomitoxin suppresses the immune system.	Widely distributed throughout the world. In addition milk is contaminated by the mycotoxins or their metabolic products.	The prime cause of aflatoxin is moisture; hence, proper harvesting, drying, and storage are important factors in lessening contamination and toxin production. Propionic and acetic acids, and sodium propionate, will inhibit mold growth; hence, their use in preserving high-moisture grains is encouraged.	Remove the source of the mold. Animals suffering from molds frequently respond to vitamin B injections. Iron therapy may be helpful because hemorrhaging is a frequent problem.	Certain molds produce toxins, or mycotoxins. Aflatoxin has been clearly shown to be a carcinogen (tumor producing). Ultraviolet irradiation and anhydrous ammonia under pressure will reduce the aflatoxin toxicity and, if continued long enough, will deactivate it entirely.
Nitrate-nitrite poisoning	Consuming feeds with a high-nitrate content due to nitrate fertilization, drought, and so on. Eating nitrate or nitrite fertilizer, or drinking pond water containing it.	Accelerated respiration and pulse rate; diarrhea; frequent urination; loss of appetite; general weakness, trembling, and a staggering gait; frothing from the mouth; abortion; blue color of mucous membrane, muzzle, and udder due to lack of oxygen; dark brown blood; and death in 90–150 minutes after eating lethal doses of nitrite.	Excessive nitrate content of feeds is an increasingly important cause of poisoning in farm animals primarily because of high-nitrogen fertilization. Nitrates and nitrites are water soluble and may be leached from the soil into groundwater.	Check potential sources, such as water and feed. When in doubt, have the feed analyzed. Properly store nitrogen fertilizer.	A 4% solution of methylene blue (in a 5% glucose or a 1.8% sodium sulfate solution) administered by a veterinarian intravenously at the rate of 10 mg/kg live-weight.	Nitrate does not appear to cause the actual toxicity, and may only cause gastrointestinal irritation in pigs. Nitrite is far more toxic. Nitrite converts hemoglobin to methemoglobin, which cannot transport oxygen.

	Cause/Source	Signs	Occurrence	Prevention	Treatment	Remarks
Pitch (clay pigeon poisoning; coal tar poisoning)	Expended clay pigeons; roofing material, certain types of tarpaper, and plumbers' pitch.	An acute, highly fatal disease, characterized, clinically, by depression, and pathologically by striking liver lesions. Anemia and jaundice.	Wherever there is pitch.	Do not allow hogs access to pitch-containing or coal tar-containing products.	No known treatment.	Pastures containing clay pigeons are dangerous for years; deaths having been reported 35 years after area was used for trap shooting.
Plants	Pigweed (*Amaranthus retroflexus*), cocklebur (*Xanthium strumarium*), and nightshade (*Solanum nigrum*).	**Pigweed**: Weakness, trembling, and incoordination; paralysis of rear legs; coma and death within 48 hours of onset of signs. **Cocklebur**: Within 8.24 hours after ingestion, depression, nausea, weakness, spasms, labored breathing, and death. **Nightshade**: Anorexia, constipation, depression, incoordination, trembling spasms, coma, and death.	Primarily, pigs on pastures.	Keep weeds down. Be sure pasture contains adequate, good forage.	Remove animals from the source. Oral administration of mineral oil and possibly injection of physostigmine for cocklebur poisoning.	Producers should recognize these plants at all stages of their growth.
Protein poisoning	Heavy feeding of high-protein feeds.	None, providing animal does not have a kidney ailment.	None.			Despite occasional diagnosis of protein poisoning, there is no proof that heavy feeding of high-protein feeds is harmful. Generally, feeds of high-protein content are more costly so there is no reason to feed too much of them.

(continued)

TABLE 8.12   Poisons and Toxins (*continued*)

Poison	Source	Symptoms and Signs	Distribution and Losses Caused By	Prevention	Treatment	Remarks
Salt poisoning (NaCl—sodium chloride)	Brine from cured meats; wet salt. When excess salt is fed following salt starvation. When salt is improperly used to govern self-feeding of concentrate.	Sudden onset 1–2 hours after ingesting salt; extreme nervousness; muscle twitching and fine tremors; much weaving, wobbling, staggering, and circling; blindness; weakness; normal temperature, rapid but weak pulse, and very rapid and shallow breathing; diarrhea; convulsions; death from within a few hours up to 48 hours.	Salt poisoning is relatively rare except in pigs.	If animals have not had salt for a long time, they should first be hand-fed salt, gradually increasing daily allowance until they leave a little in the mineral box; then self-feed.	Provide large quantities of fresh water to affected animals. Those unable to drink should be given water via stomach tube.	Salt poisoning in pigs does not occur unless there is water deprivation. Even normal salt concentration may be toxic if water intake is low.
Selenium poisoning	Consumption of plants grown on soils containing selenium.	A general loss of hair in swine. In severe cases, the hooves slough off; lameness occurs, food consumption decreases, and death may occur by starvation.	Especially certain in South Dakota, Montana, Wyoming, Nebraska, Kansas, and perhaps other states in the Great Plains and Rocky Mountains. Also, in Canada.	Abandon areas where soils contain selenium, because crops produced on such soils constitute a menace to both farm animals and humans.	Although arsenic has been shown to counteract the effects of selenium toxicity, there appears to be no practical method of treating other than removal of animals from affected areas.	Chronic cases occur when animals consume feeds containing 8.5 ppm of selenium over an extended period; acute cases occur on 25–125 ppm. The toxic level of selenium is in the range of 5–10 ppm of feed.
Zearalenone (estrogenic syndrome)	An estrogenic mycotoxin produced by *Fusarium graminearum* in corn.	Swelling and edema of the vagina and possibly mammary development in prepubertal gilts. Preputial enlargement in boars. Abortions or reduction in live pigs born.	Wherever corn is fed.	Prevent favorable conditions when storing corn.	Withdrawal of offending feed.	Associated with vulvovaginitis and rectal prolapses. Wet corn with high temperatures followed by low temperatures favor zearalenone production. Suspected diets or corn should be analyzed for the presence of zearalenone.

Source: Courtesy of Pearson Education.

QUESTIONS FOR STUDY AND DISCUSSION

1. Of what value are feed substitution tables?
2. What factors should be considered in arriving at the comparative values of feeds listed in a feed substitution table?
3. Why are antibiotics used as feed additives? How do antibiotics and probiotics differ in their respective functions?
4. List some factors that must be considered before attempting to balance a diet by any method.
5. What precautions should be taken when using diets formulated by the computer?
6. Select a specific class of swine and formulate a balanced diet using those feeds that are available at the lowest cost.
7. Will the "least-cost diet" always make for the greatest net returns?
8. Under what circumstances would you
   a. Farm mix a hog feed?
   b. Buy a commercial hog feed?
9. What factors should you consider or look for in buying a commercial hog feed?
10. List and discuss the forms of processing feeds for swine.
11. List and discuss the four basic systems of preparing diets for swine.
12. Why is quality control essential in manufacturing swine feeds?
13. What is phase feeding? When and how should it be used?
14. Define and discuss the following:
    a. Creep feeding.
    b. Phase feeding.
    c. Split-sex feeding.
15. At what stage and age should replacement gilts be sorted out from finishing pigs? How should they be fed thereafter?
16. Would you recommend flushing
    a. Second and third litter sows?
    b. Gilts?
    c. Why, and if so, how would you do it?
17. Limited feeding is necessary for gestating gilts and sows. List and detail four systems of limit feeding.
18. How would you feed sows immediately after farrowing and during the following first week?
19. Discuss the causes, symptoms, and prevention of nutritional anemia in pigs.
20. Boars gain faster and are more efficient in feed conversion than barrows or gilts, so why do we not feed more boars?
21. Of what significance is the hog–corn ratio?
22. Name two mycotoxins and indicate some of the symptoms and signs and problems caused by each.

SELECTED REFERENCES

Animal Feeding and Nutrition, 9th ed., M. J. Jurgens, Kendall/Hunt Publishing Co., Dubuque, IA, 2002

Applied Animal Nutrition, Feeds and Feeding, 2nd ed., Peter R. Cheeke, Prentice Hall, Upper Saddle River, NJ, 1999

Feeds and Nutrition, 2nd ed., M. E. Ensminger, J. E. Oldfield, and W. W. Heinemann, The Ensminger Publishing Company, Clovis, CA, 1990

Kansas Swine Nutrition Guide, Kansas State University Extension, Manhattan, KS, 2003

Life Cycle Swine Nutrition, 17th ed. Pm-489, P. J. Holden, et al., Iowa State University, Ames, IA, 1996

Nontraditional Feed Sources for Use in Swine Production, P. A. Thacker and R. N. Kirkwood, Butterworths, Boston, MA, 1990

Nutrient Requirements of Swine, 10th rev. ed., National Research Council, National Academy of Sciences, Washington, DC, 1998

Pork Industry Handbook, Leaflets PIH 4, 23, 31, 71, 73, 86, 108, 112, 126, and 129, Cooperative Extension Service, Purdue University, West Lafayette, IN, 2003

Swine Nutrition, E. R. Miller, D. E. Ullrey, and A. J. Lewis, Butterworth-Heinemann, Boston, MA, 1991

Swine Nutrition Guide, University of Nebraska, Lincoln, NE, and, South Dakota State University, Brookings, SD, 2000

Swine Production and Nutrition, W. G. Pond and J. H. Maner, AVI Publishing Co., Inc., Westport, CT, 1984

9

Grains and Other High-Energy Feedstuffs[1]

A totally slotted double curtain finisher with tube feeders.
(Courtesy, Palmer Holden, Iowa State University, Ames, IA)

---

[1] Applicable material was adapted from the following publications: *Life Cycle Swine Nutrition* (1996), published by Iowa State University; *Kansas Swine Nutrition Guide* (2003), published by Kansas State University; and *Swine Nutrition Guide* (2000), published by the University of Nebraska and South Dakota State University.

Objectives

After studying this chapter, you should:

1. Be familiar with the grains commonly used in pork production, their advantages and any feeding concerns.
2. Be cognizant of the advantages and disadvantages of high-lysine corn.
3. Know the corn milling and grain distilling by-products and the differences in feeding by-products.
4. Know the wheat milling by-products and their advantages and disadvantages in swine feeding.
5. Know the value and influence of fats and oils on pork production.
6. Know why molasses may be used in swine feeding.

Pigs need energy for maintenance, growth, reproduction, and lactation. The bulk of the pig's energy is met by carbohydrates and fats. Fats and oils are dense sources of energy, containing about 2.25 times more calories than carbohydrates. Although cereal grains and fats or oils will meet the pig's energy needs, they must be supplemented with amino acids (protein), minerals, and vitamins.

The energy content of feedstuffs and the energy requirements of pigs are generally expressed as digestible energy (DE), metabolizable energy (ME), and/or net energy (NE). The primary energy sources for swine are the cereal grains corn, sorghum, wheat, barley, and their by-products. Fat is often used to increase the energy density of swine diets. Most common cereal grains and fats are quite palatable and digestible.

The relative feed substitution values of common swine feed energy sources is shown in Chapter 8, Table 8.1. Corn is the most extensively used swine feed in the United States. It is assumed to have a feeding value of 100, and grain sorghum (milo) is assumed to have a feeding value of 95 to 97% that of corn. Thus, sorghum can replace corn in the diet when the price of sorghum is less than 95% of the price of the same weight of corn. For example, if the price of sorghum is less than $0.038/lb, it is a better buy than corn that costs $0.040/lb. *Note:* The feeding value of sorghum is slightly less than that of corn because it has less ME and available lysine.

The relative feeding values of the energy sources given in Table 8.1 apply when ingredients are included in diets in quantities no greater than those shown. When ingredients are included in diets at higher levels than indicated, their feeding value may decrease slightly or considerably depending on the feed. Average daily gain and reproductive performance will not normally be reduced by replacing corn with any of the energy sources in Table 8.1 as long as bulkiness does not reduce intake. A range in feeding value is presented because of the variation in ingredient quality and individual producer goals. *Note:* In addition to the relative feeding value of energy sources, the producer should consider such factors as storage costs, availability, palatability, and carcass quality.

Backfat thickness may decrease slightly when oats, barley, or other lower-energy ingredients replace all of the corn in the diet as the added fiber slightly depresses energy intake. The feeds in Table 8.1 may be augmented with the composition of some of the more commonly used grains and other high-energy feeds in Chapter 22.

GRAINS

Grains are seeds from cereal plants and are members of the grass family, *Gramineae*. They contain large quantities of carbohydrates. Corn, sorghums, barley, and oats are the primary grains fed to swine. Rice and wheat are nutritionally excellent for pigs but are consumed primarily by humans. Many of their milling by-products are fed to pigs. Rye, millet, emmer, spelt, and triticale are fed to hogs in limited amounts. The feed concentrates fed to livestock from 1985 to 2000 in the United States are listed in Table 9.1.

Grains provide the swine producer with an excellent source of highly digestible energy, but it is important that the characteristics of each available grain be thoroughly understood, thus making it possible to formulate economical diets that take full advantage of their nutritive properties and compensate for their deficiencies. The grain and hay production in the United States in 2002 is listed in Table 9.2. A brief discussion of each of the more important grains used by swine follows.

TABLE 9.1 Feed Concentrates Fed to Livestock and Poultry, 1985–2000 (in million metric tons)

Year	Corn	Sorghum	Oats and Barley	Total	Wheat	Rye	By-product Feeds[1]	Total Concentrates
			Feed Grains					
1985	104.7	16.9	13.5	135.2	10.9	0.3	36.4	182.7
1986	118.8	13.6	11.5	144.0	11.3	0.4	37.8	193.5
1987	122.1	14.1	10.5	146.7	5.7	0.3	39.3	191.9
1988	100.3	11.9	7.5	119.6	3.6	0.3	36.0	159.5
1989	111.9	13.2	8.6	133.7	7.5	0.2	37.7	179.1
1990	117.3	10.2	8.8	136.4	12.1	0.2	39.8	188.5
1991	122.1	9.3	8.7	140.2	6.3	0.2	41.2	187.8
1992	133.7	11.6	7.7	153.0	3.9	0.2	42.4	199.5
1993	119.1	11.2	9.5	139.7	9.6	0.2	43.6	193.2
1994	139.0	9.6	7.9	156.5	7.5	0.2	45.3	209.4
1995	119.5	7.5	6.9	133.8	6.2	0.2	44.8	184.9
1996	134.4	13.1	6.3	153.7	7.7	0.1	47.3	208.9
1997	139.5	9.3	6.5	155.4	8.8	0.1	48.4	212.6
1998	139.2	6.6	6.2	152.1	6.5	0.1	49.2	207.8
1999	144.2	7.3	6.0	157.5	9.1	0.1	49.2	215.7
2000	148.6	5.6	5.4	159.6	6.0	0.1	53.2	218.9

[1]Oilseed meals, animal protein feeds, mill by-products, and mineral supplements.
Source: National Agricultural Statistics Service, USDA, Washington, DC Agricultural Statistics 1995 and 2002, Table 71.

TABLE 9.2 Grains and Hay Production Summary, United States, 2002

	Unit	Production (in 1,000s)
Barley	bu	226,873
Corn for grain	bu	9,007,659
Corn for silage	ton	104,979
Hay, All	ton	150,962
Alfalfa	ton	73,824
All Other	ton	77,138
Oats	bu	119,132
Proso millet	bu	2,755
Rice	cwt	210,960
Rye	bu	6,985
Sorghum for grain	bu	369,758
Sorghum for silage	ton	3,360
Wheat, All	bu	1,616,441
Winter	bu	1,142,802
Durum	bu	79,450
Other Spring	bu	394,189

Source: National Agricultural Statistics Service, USDA, Washington, DC.

Corn (Maize)

Corn has long been the most important U.S. cereal grain and swine feed (Figure 9.1). Since 1996 the annual yield, ranging from 9 billion to 10 billion bushels (about 230,000 to 250,000 metric tons), is equal to that of all other cereals combined. About one-fourth to one-half of this production is fed to hogs. For this reason, the feeding value of corn is usually accepted as the standard with which other

Figure 9.1 Hogging down a corn field in the 1930s. *(Courtesy, USDA, Washington, DC)*

cereals are compared. World annual production is about 600,000 metric tons.

Corn is an excellent energy feed for all classes of swine. It is an ideal finishing feed because it is high in digestible carbohydrate (starch), low in fiber, and very palatable.

In spite of its virtues, corn alone will not keep pigs alive. It contains 7 to 9% protein, but the protein is deficient in practically all of the essential amino acids required by the weanling pig, especially lysine and tryptophan. It is also so deficient in calcium and other minerals, and it is so inadequate in vitamin content that pigs will die if they are limited to a diet containing only corn. So, corn must be supplemented with a protein that makes up its amino acid deficiencies. Equally important are the needed minerals and vitamins. When properly supplemented, corn is an excellent energy feed for all classes of swine. *Note:* Researchers have developed equations to estimate the lysine content of corn based on crude protein concentration. However, the poor relationship between lysine and protein concentration makes this method unreliable. It is recommended to use a book value for lysine unless the grain is actually tested for lysine. Corn that is very high or low in protein could be considered a good candidate for a lysine test.

White corn is low in carotene (the precursor of vitamin A). However, when pigs have access to a good pasture, when the diet is supplemented adequately with alfalfa meal, or when vitamin A is added, white and yellow corn appear to be equal in feeding value. It is noteworthy that the carotene content of yellow corn decreases with storage; about 25% of its original vitamin A value may be lost after 1 year of storage and 50% after 2 years.

Corn should be ground and combined in definite proportions with other feeds. Whole kernel corn may be fed if it is high-moisture corn or if fed to sows on outside feeding floors where ground grain may be lost.

Contrary to the opinion prevailing in some quarters, experimental studies have not revealed any difference between hybrid and open-pollinated yellow dent corns in feeding value for growing-finishing hogs.

From a market standpoint, the swine producer should be familiar with the federal grades of corn listed in Table 9.3. A final grade is U.S. sample grade, which is corn that (1) does not meet the requirements for grades from U.S. No. 1 to 5, (2) contains stones, (3) is musty or sour, (4) is heating or has heated, (5) has any commercially objectionable foreign odor, or (6) is otherwise of distinctly low quality.

Although moisture content is the chief criterion in determining federal grades of corn, the percentage of unsound kernels and amount of foreign

Figure 9.2 Corn harvest in Columbia, Missouri. *(Photo by Bruce Fritz, Courtesy, USDA, Washington, DC)*

materials are also factors. From a swine feeding standpoint, the value of the several grades of corn is about equal, provided they are compared on a common dry matter basis. The same situation applies to soft corn (corn frosted before maturity). Of course, handling and storage problems are increased with higher moisture content. Corn with more moisture than in grade No. 2 corn will not keep well in storage (Figure 9.2).

Yellow corn is usually the cheapest source of energy throughout much of the United States. But price fluctuations frequently justify consideration of other energy sources. When rating corn, the primary swine feed, as 100%, Table 8.1 (Chapter 8) indicates how some other grain and high-energy feeds, properly fed, compare with it on an equal unit basis.

• **High-moisture corn**—High-moisture corn may be substituted for dry corn on a dry matter basis with little effect on overall performance, in either feed conversion or rate of gain. Most high-moisture corns, particularly those with more than about 18% moisture, will began to spoil within about 24 hours after exposure to air, such as when grinding or putting in a feeder. Therefore, high-moisture grains

TABLE 9.3 Grade Requirements for Corn

			Maximum Limits of Damaged Kernels		
Grade	Minimum Test Weight per Bushel (lb)	Moisture (%)	Broken Corn and Foreign Material (%)	Total (%)	Heat-Damaged Kernels (%)
U.S. No. 1	56.0	14.0	2.0	3.0	0.1
U.S. No. 2	54.0	15.5	3.0	5.0	0.2
U.S. No. 3	52.0	17.5	4.0	7.0	0.5
U.S. No. 4	49.0	20.0	5.0	10.0	1.0
U.S. No. 5	46.0	23.0	7.0	15.0	3.0

Source: Official U.S Standards for Grain. USDA. Agriculture Marketing Service, Grain Division.

must be processed and fed fresh daily in warm climates, possibly every other day in cold climates.

- **High-lysine corn (opaque-2)**—Corn is now bred that is much higher than normal corn in lysine and tryptophan; hence, it has a better balance of the amino acids for swine. Also, the high-lysine corn (called Opaque-2) is higher in total protein than normal corn.

High-lysine corn is available commercially but suffers from slightly lower yields and is subject to cross-pollination by adjacent normal corn fields. Also the high-lysine kernel is softer and more subject to damage during harvest and handling. Corn breeders are working to enhance Opaque-2, for it would make for improved nutrition of both animals and people wherever corn (maize) is grown. Based on experiments and experiences, the following recommendations are made relative to high-lysine corn:

1. Check the moisture carefully when mechanically drying high-lysine corn because it dries more rapidly than normal corn.
2. Grind high-lysine corn more coarsely than regular corn. The kernels are softer and powder more easily. A grinder screen of 0.5 in. works best. Some producers prefer a roller mill when processing high-lysine corn.
3. Analyze the corn for lysine content.
4. Balance the diet based on lysine content and requirements. This is always recommended.

If swine producers consider growing high-lysine corn, they must evaluate such economic factors as the lower yield of high-lysine corn versus the price of normal corn plus supplemental lysine.

- **Waxy corn**—Waxy corn is the name given to a variety of corn that is sometimes grown for industrial use because of its special type of starch. There appears to be no advantage or disadvantage associated with incorporating waxy corn in the diets of swine. Thus, the decision to produce and feed waxy corn should be based on agronomic and economic factors, and not on the effects of waxy corn on the performance of swine.

Barley

Barley is the world's most ancient and most widely grown cereal. It is grown extensively in sections too cool for corn to thrive, such as in the northern part of the United States. In Canada and northern Europe, it is the most common grain for swine. It is an excellent feed, and because of its saturated fat, it produces firm pork of high quality.

Compared with corn, barley contains somewhat more protein (lysine) and fiber (because of the hulls) and somewhat less carbohydrates and fats (see Chapter 22). Like oats, the feeding value of barley is quite variable because of the wide spread in density (bushel weight or weight per unit of volume). Although barley is lower in energy than corn, when properly supplemented, ground barley is worth about 100 to 105% as much as corn for hogs, primarily because of its higher-lysine content. Fine-grinding (600 to 700 microns) barley improves its feeding value. Barley is less palatable and higher in protein content than corn. Therefore, it is not well adapted to self-feeding free choice because pigs usually eat more of the accompanying protein supplements than is required to balance the diet. For this reason, it is best to feed the barley and supplement together as a mixed diet.

Emmer and Spelt

Emmer and spelt resemble barley somewhat in appearance and feeding value for hogs because the hulls usually are not removed from the kernels in threshing. U.S. production of both grains peaked in the early 1900s and declined steadily thereafter. Limited amounts of emmer are produced in North Dakota and Montana and spelt in the midwestern United States, mainly in Ohio. When ground and limited to not more than one-third of the diet, they are similar to oats in feeding value for hogs, depending on the amount of hulls present.

Millet (Hog Millet, Proso Millet, or Broomcorn Millet)

Millet has been raised since prehistoric times in the Old World as an important bread grain crop. In the United States, hog millet, often called broomcorn millet, is sometimes grown for livestock feeding purposes in the northern part of the Great Plains, where the growing season is too short for the grain sorghums. The yields usually range from 500 to 1600 lb per acre (600 to 1800 kg/ha). Ground millet is worth 85 to 95% as much as corn for hogs.

Oats

Oats contain more lysine than corn or sorghum, but they are high in fiber. Because of the high-fiber content (1) oats are best utilized in sow gestation diets and (2) they should be limited to 20% of the diet of growing-finishing pigs. In larger quantities, oats retard feed intake and thus rate of gain. They have 80 to 85% of the feeding value of corn. For mature gestating sows, oats may be used as the only grain source along with appropriate protein, mineral, and vitamin supplementation.

Grinding increases by 25 to 30% the feeding value of oats for swine. Oatmeal and rolled oats, such as are used for human consumption, are excellent feeds in fitting young pigs for the show, but they are usually expensive.

Dehulled oats are particularly valuable in diets for very young pigs and have a feeding value of 115 to 125% that of corn. However, it takes 155 to 165 units of whole oats to produce 100 units of dehulled oats.

Rice

Rice is one of the most important cereal grains of the world and provides most of the food for half the human population of the earth. It forms a major part of the diets of Asian populations. In the United States, rice is grown extensively for human food in Louisiana, Arkansas, Texas, and California (Figure 9.3). When low priced or off-grade, it is frequently fed to swine. Also, a considerable part of the byproducts of rice processing is profitably used as swine feed.

The whole rice fed to swine is known as ground rough rice (or ground paddy rice). It is finely ground, threshed rice (or rough rice) from which the hull has not been removed. From the standpoint of chemical composition and feeding value for swine, ground rough rice compares favorably with oats. For growing-finishing pigs, it is worth 80 to 85% as much as corn when it is finely ground and limited to 40% of the diet. Ground rough rice produces firm pork of good quality.

Rye

Rye thrives on poor, sandy soils. In such areas, it is sometimes marketed through hogs, though it generally sells at a premium for bread making or brewing.

Rye is similar to wheat in chemical composition (see Chapter 22), but it is less palatable to pigs. For the latter reason, it should be fed in combination with more palatable feeds and limited to no more than 25% of gestating and growing-finishing swine diets. When properly supplemented, ground rye has a feeding value approximately 90% that of corn. Because of the small, hard kernel, rye should be ground for hog feed.

Rye is often mildly infested with ergot (a common fungus disease in rye, Figure 9.4). Rye infested with ergot may be fed to growing-finishing hogs, providing it does not constitute more than 10% of the total diet. Because ergot may cause abortions, ergot-infested rye should never be fed to pregnant sows. Likewise, ergot should not be permitted in the diet of lactating sows because it stops milk production or for suckling pigs.

Sorghum (Kaffir, Milo)

The grain sorghums, which include a number of varieties of earless plants, bearing heads of seeds, are assuming an increasingly important role in American agriculture. New and higher-yielding varieties have been developed and become popular. As a result, more and more grain sorghums are being fed to hogs.

Kaffirs and milos are among the more important sorghums grown for grain. Other less widely produced types include sorgo, hegari, black and red kaffir, feterita, and dari. These grains are generally

Figure 9.3 Lemont was the first high-yielding semidwarf rice variety that matured early and had high-milling yields. *(Photo by David Nance, Courtesy, USDA, Washington, DC)*

Figure 9.4 A head of rye infected with ergot, a common fungus disease. It also can appear on barley and other small grains. *(Courtesy of Pearson Education)*

grown in regions where climatic and soil conditions are unfavorable for corn, primarily in the central and southern plains states.

Like white corn, the grain sorghums are low in carotene (vitamin A value). Also, they are deficient in other vitamins, and in proteins and minerals (see Chapter 22). Because the energy content of corn is slightly higher than that of sorghum, the feed efficiency of pigs fed corn diets will be slightly better than that of pigs fed sorghum, but average daily gains will be the same. The grain sorghums are worth about 92% as much as corn for growing-finishing pigs.

One disadvantage of sorghum is that it is more variable in nutrient content than corn. In addition, because a sorghum kernel is smaller and harder than a corn kernel, finer grinding (1/8- or 5/32-in. screen) or rolling is recommended for best utilization. Sorghum-fed hogs yield firm carcasses of good quality.

Triticale

Triticale is a hybrid cereal derived from a cross between wheat *(Triticum)* and rye *(Secale)*, followed by doubling the chromosomes in the hybrid. The objective of the cross is to combine the grain quality, productivity, and disease resistance of wheat with the vigor and hardiness of rye. The first such crosses were made in 1875 when the Scottish botanist Stephen Wilson dusted pollen from a rye plant onto the stigma of a wheat plant. Only a few seeds developed and germinated, and the hybrids were found to be sterile. In 1937, the French botanist Pierre Givaudon produced some fertile wheat–rye hybrids with the ability to reproduce. Intensive experimental work to improve triticale was undertaken at the University of Manitoba beginning in 1954.

In comparison with wheat, triticale (1) has a larger grain, but there are fewer of them in each head (spike); (2) has a higher protein content, with a slightly better balanced amino acid composition and more lysine; and (3) is more winter hardy. So far, the main use for triticale has been as a feed grain, pasture, green chop, and silage crop for animals. Its relative feeding value is 90 to 95% that of corn.

Wheat

Through the ages, wheat has been made into precious bread for humankind. It was used for this purpose by the aboriginal Swiss lake dwellers; by the ancient Egyptians, Assyrians, Greeks, and Romans; and by the Hebrews, who, according to the Book of Revelations, exchanged "a measure of wheat for a penny." Even today, most U.S. wheat is used for baking.

Figure 9.5   Wheat harvest in the state of Washington in the Palouse Valley. *(Courtesy, USDA, Washington, DC)*

The total annual U.S. wheat production is second only to corn (Figure 9.5). However, because it is produced mainly for the manufacture of flour and other human foods, it is generally too high in price to feed to hogs. When the price is favorable or when the grain has been damaged by insects, frost, fire, or disease, it may be more profitable to market it through hogs than for human consumption.

Compared with corn, wheat is higher in protein (and lysine) and carbohydrates, lower in fat, and slightly higher in total digestible nutrients. Wheat, like white corn, is deficient in carotene (Chapter 22). On an equal weight basis it is 10 to 15% more valuable than corn in producing gains on hogs, thus having the unique distinction of being the only cereal grain that excels corn as a swine feed (Table 8.1).

Because the kernels are small and hard, wheat should be ground. Because wheat tends to flour when processed, it should be coarsely ground (3/16-in. screen) or rolled. If ground too finely, feed intake may be decreased and performance lowered. Although it can completely replace corn in a swine diet, it usually is fed on approximately equal parts with corn. Wheat produces high-quality pork.

BY-PRODUCT FEEDS FROM GRAINS

Most grains are often milled in some manner for the preparation of foods for human consumption. In these milling processes a number of by-products are produced that, although they are often considered to be of better value to humans, can and are used extensively as swine feeds.

Corn Milling By-Products

In the modern manufacture of cornstarch, sugar, syrup, and oil for human food, the following common by-product feeds are obtained: (1) germ meal (dry), (2) germ meal (wet), (3) gluten feed, (4) gluten meal, and (5) hominy feed (grits).

Germ Meal (Dry)

Germ meal (dry) is a by-product of the dry milling of corn for corn meal, corn grits, hominy feed, and other corn products. For best results in swine feeding, it is recommended that corn germ meal (dry milled) be used in the same manner as recommended for corn germ meal (wet milled).

Germ Meal (Wet)

Germ meal (wet) is a by-product of the wet milling of corn for cornstarch, corn syrup, and other corn products. The corn germ obtained in these processing operations are dried, crushed, and the oil extracted. Then the remaining residue or cake is ground, producing corn germ meal.

Corn germ meal contains about 22% protein and 10% fiber. When used in combination with higher-quality proteins, corn germ meal is a satisfactory feed for swine. As a protein feed for pigs, it should not constitute more than half of the supplement, with the other half consisting of a protein source, which supplies the amino acids lacking in germ meal.

When price relationships are favorable, corn germ meal can be used as a partial grain substitute in the diet, but it should not constitute more than 20% of the total feed. When so limited, corn germ meal is equal in value to corn in feeding value.

Gluten Feed

Corn gluten feed is the corn by-product remaining after the extraction of most of the starch and germ in the wet milling manufacture of cornstarch or syrup, with or without fermented corn extractives or corn germ meal. This feed contains approximately 21% protein and 8.7% fiber. Because of its bulkiness, low palatability, and poor quality proteins, it is recommended that corn gluten feed be restricted to 10% of the diet of sows and growing-finishing pigs, or incorporated in the diets of ruminants which will make better use of the crude protein content.

Gluten Meal

Corn gluten meal is similar to corn gluten feed without the bran. It averages about 43% protein and 3.0% fiber.

Because the proteins are of low quality, corn gluten meal should never be fed as the sole protein supplement to swine. When price relationships are favorable, it may be satisfactorily incorporated in confinement diets as part (25 to 50%) of the protein supplement in combination with supplements that supply the amino acids that gluten meal lacks. For pasture feeding, it may constitute 50% of the protein supplement—provided the other half is of animal or marine origin. When used in this manner, corn gluten meal is of about equal value to any of the common protein-rich plant supplements. Corn gluten meal is used more in poultry feeding because the high-carotene content adds yellow pigmentation to the poultry skin and egg yolks.

Hominy Feed (Grits)

Hominy feed, a by-product of corn milling, is a mixture of corn bran, corn germ, and a part of the starchy portion of either white or yellow corn kernels or mixtures thereof, as produced in the manufacture of pearl hominy, hominy grits, or cornmeal. Hominy feed must contain not less than 4% crude fat. If prefixed with the words *white* or *yellow*, the product must correspond thereto.

Hominy feed is a satisfactory substitute for corn in swine feeding, but it is somewhat higher in fiber and less palatable than the grain from which it is derived. Yellow hominy feed supplies carotene (vitamin A), whereas white hominy feed lacks this factor. From the standpoint of gains and feed efficiency, properly supplemented hominy feed is worth 100 to 105% as much as corn, but it will produce soft pork if it constitutes more than half the grain diet because of its high fat content. In order to minimize the chance of hominy feed becoming rancid, it should be used fresh and stored in a cool, well-ventilated building.

Rice Milling By-Products

The common by-products resulting from processing rice are (1) brewers' rice, (2) ground brown rice, (3) polishings, and (4) rice bran. In swine feeding, these products are used as grain replacements or energy feeds. Their compositions are given in Chapter 22.

Brewers' Rice (Chipped or Broken Rice)

Chipped rice consists of the small broken kernels of rice resulting from the milling operations. It is used chiefly in the brewing industry. Chipped rice is similar to corn in composition and feeding value. For best results in swine feeding, it should be finely ground, mixed with more palatable feeds, and included in a properly balanced diet. The carcasses of hogs fed on chipped rice are hard and firm.

Ground Brown Rice

Ground brown rice is the product obtained in grinding the rice kernels after the hull has been removed. It is nearly equal to corn as a swine feed and produces firm pork.

Polishings

Rice polishings are the finely powdered by-product obtained in polishing the rice kernels after the hulls and bran have been removed. They average about 13% protein, 13.7% fat, and 3.2% fiber and are high in thiamin. Because of the high fat content, rice polishings become rancid in storage and can produce soft pork. It is suggested that they comprise no more than 10% of a finishing diet. When incorporated in a properly balanced diet, they are equal or superior in feeding value to corn for all classes of swine, including young pigs and gestating-lactating sows.

Rice Bran

Rice bran is the pericarp, or bran layer, obtained in milling rice for human food, together with such quantity of hull fragments as is unavoidable in the regular milling operations. It must contain not more than 13% crude fiber. When the calcium carbonate exceeds 3% (Ca exceeds 1.2%), the percentage must be declared in the brand name.

Rancidity is a major issue with rice bran and care must be taken to ensure that it is fed fresh. When limited to 5% of the swine diet, rice bran is equal in value to corn. Higher proportions have a lower feeding value and produce soft pork.

Wheat Milling By-Products

In the average wheat milling operations, about 75% of the weight of the wheat kernel is converted into flour, with the remaining 25% going into by-product feeds. Among the latter are the following hog feeds: (1) middlings, (2) mill run, (3) red dog, (4) shorts, and (5) wheat bran.

Middlings

Wheat middlings are a by-product from spring wheat. They consist mostly of fine particles of bran and germ, with very little of the red dog flour. According to the definition of the Association of American Feed Control Officials (AAFCO), wheat flour middlings must not contain more than 9.5% fiber.

In general, this feed is characterized by fair amounts of protein (about 16%) that are somewhat lacking in quality. It is high in phosphorus, low in calcium, and deficient in both carotene and vitamin D. Because of these deficiencies, middlings are best used in combination with protein supplements of animal origin, and then limited to about 5% for growing-finishing diets and up to 30% for gestating sows. When fed to hogs in this manner, they are worth 95 to 115% as much as corn (Figure 9.6).

Figure 9.6 A group of finishing pigs on a multiholed self-feeder and totally slotted concrete floors. *(Courtesy, Palmer Holden, Iowa State University, Ames, IA)*

Mill Run

Wheat mill run consists of the coarse wheat bran, fine particles of wheat bran, wheat shorts, wheat germ, wheat flour, and the offal from the "tail of the mill." According to the AAFCO, this product must contain not more than 9.5% crude fiber. Because of its bulkiness, it is recommended that wheat mill run be fed to the older hogs.

The terms used to designate the common wheat by-products resulting from flour milling differ according to (1) area of the country and (2) whether they are obtained from winter or spring wheat. The by-product feeds resulting from the milling of spring wheat are commonly known as bran, standard middlings, flour middlings, and wheat red dog. The by-products resulting from the milling of winter wheat are known as bran, brown shorts, gray shorts, and white middlings. Wheat flour by-products are distinguished by their fiber content.

Although the wheat milling products are higher in protein content than the cereal grains (and this fact is taken advantage of in formulating diets), in actual swine feeding practice they function more as grain replacements than as protein supplements.

Red Dog

Wheat red dog (red dog flour, or wheat red dog flour), which is a by-product from milling spring wheat, consists chiefly of the aleurone layer together with small quantities of flour and fine bran particles. It must contain not more than 4% crude fiber. Red dog may replace the grain in gestating sow diets and up to 50% in the diets of growing-finishing pigs because of its lower fiber and high digestibility. When

Figure 9.7 A swathed North Dakota wheat field awaiting combining. *(Courtesy, Palmer Holden, Iowa State University, Ames, IA)*

limited to the recommended levels, it is 15 to 20% higher in feeding value than corn.

Shorts

Wheat shorts consist of fine particles of wheat bran, wheat germ, wheat flour, and the offal from the "tail of the mill." The AAFCO states that wheat shorts shall contain not more than 7.0% crude fiber. Feeding value is similar to red dog.

Wheat Bran

Wheat bran is the coarse outer covering of the wheat kernel. It contains a fair amount of protein (about 16%), a good amount of phosphorus, and is laxative in action. Because of its bulk, wheat bran is not used extensively in diets for growing-finishing pigs. It does have a very definite place in the diets of pregnant and lactating sows. Although constipation is not common in lactating sows, adding 10% to the lactation diet should prevent the problem and not significantly reduce feed intake or lactation. Bran may satisfactorily make up 30% of a gestating sow diet and 5% of a finishing diet (Figure 9.7).

Other Milling By-Products

The milling of other grains, such as barley, sorghum, rye, and oats, consists of generally similar procedures and comparable by-products. The usefulness of these by-products for swine feeding must be considered basically in terms of their contributions as sources of energy. Generally, the by-product feeds are higher in protein and fiber than the seeds from which they are derived but, because of the poor amino acid balance of the protein, they have their greatest usefulness as energy substitutes when economics warrant it.

BREWING AND DISTILLING INDUSTRY BY-PRODUCTS

Considerable quantities of grains are used in the brewing of beers and ales and in the distilling of liquors. After processing, the remaining by-products can be fed wet in ruminant feeding programs, but a small portion, if dried, can be used in swine diets. In addition to those feeds, solubles and yeast products from these industries are used in livestock feeds, though to a much lesser extent.

Brewers' By-Products

Barley is the primary grain used for the brewing of beers and ales. The initial step of brewing involves the malting of the barley. The malt sprouts and malt hulls are then separated from the malted barley, forming two by-products that can be used as feed.

The clean malted barley is mixed with other grains (generally corn or rice) and a flavoring agent (hops) to form a mash. This mash is then cooked in water to enhance further enzymatic activity, and, following cooking, it is separated into liquid and solid fractions. The liquid—called *wort*—undergoes yeast fermentation to form the alcoholic beverage of either beer or ale.

Brewers' condensed solubles—Brewers' condensed solubles are obtained by condensing the liquids resulting as by-products of the production of beer or wort. It must contain not less than 20% total solids, and on a dry matter basis it must contain at least 70% carbohydrates. The guaranteed analysis shall include maximum moisture.

Brewers' dried grains—Brewers' dried grain is the dried extracted residue of barley malt, alone or in mixture with other cereal grain or grain products resulting from the manufacture of wort or beer. It may contain pulverized dried spent hops in an amount not to exceed 3% evenly distributed.

Brewers' dried yeast—Brewers' dried yeast is the dried nonfermentative, nonextracted yeast of the botanical classification, *Saccharomyces*, resulting from tile brewing of beer and ale. It must contain 35% crude protein and must be labeled according to its crude protein content. Brewers' dried yeast is an excellent source of highly digestible protein of good quality, with more than 1% lysine, and is a source of B vitamins.

Brewers' wet grains—Brewers' wet grain is the extracted residue resulting from the manufacture of wort from barley malt alone or in mixture with other cereal grains or grain products.

The guaranteed analysis shall include the maximum moisture.

Dried spent hops—Dried spent hops are obtained by drying the material filtered from hopped wort.

Malt cleanings—Malt cleanings are by-products produced in the cleaning of malted barley or from the recleaning of malt that does not meet the minimum standards for crude protein in malt sprouts. It must be designated and sold according to its crude protein content.

Malt hulls—Barley grain is covered by a hull. During the cleaning of malted barley, the hulls are removed and subsequently used as feed. They contain about 22% crude fiber, which greatly limits their use in swine diets, except for gestating sows.

Malt sprouts (malt culms)—Malt sprouts are obtained by removal of the sprouts of malted barley. These sprouts may include some of the hulls and other parts of the malt, but they must contain not less than 24% crude protein. Because only about one-half of the crude protein is true protein, malt sprouts can be used most effectively in ruminant diets.

Distillers' By-Products

A large number of distilled spirits are produced throughout the world, each characterized by (1) the area of origin, (2) type of material used, (3) preparation of those materials, (4) proportions of materials, (5) fermentation conditions, (6) distillation processes, (7) maturation processes, and (8) mixture techniques.

Distillers' grains can be fed fresh, dried, or ensiled. The dried product is by far the easiest to handle and store. This product is less palatable to livestock than brewers' grains, but it contains more crude protein and less crude fiber. Distillers' products can be made from several grains, including barley, corn, rye, sorghum, and wheat, as well as molasses.

Distillers' Dried Solubles (DDS)

Distillers' dried solubles (DDS) are obtained after the removal of ethyl alcohol by distillation from the yeast fermentation of a grain mixture by condensing the thin stillage fraction and drying it by methods employed in the grain distilling industry. The predominating grain must be declared as the first work in the name.

Condensed Distillers' Solubles (CDS)

The by-product feed officially known as condensed distillers' solubles (CDS) is a product obtained following the removal of ethyl alcohol by distillation from the yeast fermentation of a grain or a grain mixture by condensing the thin stillage fraction to a semisolid. Because a variety of grains are used in this process, the predominating grain is declared as the first word in the name—for example, corn condensed distillers' solubles.

Distillers' Dried Grains (DDG)

Distillers' dried grains (DDG) are obtained after the removal of ethyl alcohol by distillation from the yeast fermentation of a grain or a grain mixture by separating the resultant coarse grain fraction of the whole stillage and drying it by methods employed in the grain distilling industry. The predominating grain must be declared as the first work in the name.

Distillers' Dried Grains with Solubles (DDGS)

Distillers' dried grains with solubles (DDGS) are obtained after the removal of ethyl alcohol by distillation from the yeast fermentation of a grain or a grain mixture by condensing and drying at least three-fourths of the solids of the resultant whole stillage by methods employed in the grain distilling industry. The predominating grain must be declared as the first work in the name.

Recent research has indicated marked differences in quality between "old process" ethanol production and the "new generation" ethanol plants being constructed in the Corn Belt (Figure 9.8). New process distillers' products tend to be higher in protein, energy, several amino acids, and available phosphorus.

Feedstuff descriptions do not differentiate between the old and new products and even within the "new" much variation exists depending on the quality of the process. The energy value appears to be

Figure 9.8 A dark color DDGS indicates a lower-quality, less digestible product. The lighter DDGS on the right is indicative of high-quality, highly digestible DDGS. *(Courtesy, Gerald Shurson, University of Minnesota, St. Paul, MN)*

similar to corn. Color may be a differentiating quality as the "new generation" product is more golden and consistent in color. It also has a sweet, fermented odor as opposed to a burnt or smoky color.

Old process DDGS has a feeding value of approximately 45 to 50% that of dehulled soybean meal and should be limited to 10% of the diet for gestating sows and 5% for other stages of production. However, interest in feeding to U.S. swine is increasing with the development of many ethanol plants in the upper Midwest reported by the University of Minnesota (*http://www.ddgs.umn.edu/articles-swine/studies. pdf*). DDGS from these plants can constitute up to 50% of the diet for gestation, 20% for lactation, and 20% for growing-finishing.

> **Distillers' wet grains**—Distillers' wet grains are obtained after the removal of ethyl alcohol by distillation from the yeast fermentation of a grain mixture. The guaranteed analysis shall include the maximum moisture.
>
> **Grain distillers' dried yeast**—Grain distillers' dried yeast is the dried nonfermentative yeast of the botanical classification, *Saccharomyces*, resulting from the fermentation of grains and yeasts that are separated from the mash either before or after distillation. It must contain at least 40% crude protein. This by-product is extremely rich in all of the B complex vitamins except vitamin B_{12}.

DRIED BAKERY PRODUCT

Bakery wastes consisting of bread, cookies, cake, crackers, flours, and doughs can be dried and sold as livestock feed under the name dried bakery product. Because it is high in digestible fat and carbohydrates, it is often used to replace grain when economically feasible. However, like grains, dried bakery product is low in vitamin A, protein, and minerals. It contains rather large amounts of salt, and if the salt content exceeds 3.5%, the maximum amount of salt must be so labeled in the name of the product. It can replace all of the grain in swine diets without affecting palatability.

Typically bakery wastes are worth 110 to 120% as much as corn for hogs, depending on the base grain flours used in processing and the amount of fat added. It generally can be substituted on a weight basis for corn with no credit for added lysine unless the product is analyzed.

OTHER HIGH-ENERGY FEEDS

Although feed grains and their milling by-products comprise the vast majority of the energy feeds, nu-

merous other feeds are routinely used to supply energy to livestock, including swine. Seeds from plants other than *Gramineae* can be used effectively (for example, beans). Fats provide an extremely concentrated source of energy. Molasses is a liquid energy feed that is highly palatable and digestible. When the price and availability are advantageous, roots, tubers, and certain other by-product feeds are fed to swine.

Beans (Cull Beans)

Throughout the world several varieties of beans, including navy, Lima, kidney, pinto, and tepary beans, are raised for human food. Occasionally, low prices result in some of these beans being marketed through livestock channels. Also, cull beans—consisting of the discolored, shrunken, and cracked seeds, together with some foreign materials sorted out from the first-quality dry beans—are frequently cooked and fed to swine.

Chemically, beans of all varieties closely resemble peas, but their feeding value is much lower. For swine, beans should be cooked thoroughly, limited to 5 to 10% of the diet, and supplemented with a good quality protein source. When prepared and fed to swine in this manner, beans are 70% as valuable as dehulled soybean meal.

Buckwheat

Although not a cereal, buckwheat has the same general nutritive characteristics as the cereal grains. It is used chiefly for flour manufacture for making pancakes, a breakfast delicacy. However, when low in price or off-grade, buckwheat is often used for stock feeding. Because it contains about 10% fiber and is somewhat unpalatable, buckwheat should not constitute more than one-third of the swine diet. It should be ground for hogs. Some pigs, especially white ones, suffer from photosensitization (sensitivity to light) when fed buckwheat. Also, limited amounts of buckwheat flour by-product without hulls and buckwheat middlings are fed to pigs.

Citrus Fruits

Citrus fruits—drop and off-grade oranges, tangerines, grapefruit, and lemons—are sometimes used as hog feed in those areas where they are produced. When they are fresh, they are fed free choice along with a balanced grain and a protein supplemented diet.

- **Citrus meal and dried citrus pulp**—Citrus meal consists of the finer particles of dried citrus, whereas citrus pulp is the bulkier portions. Both are by-products of citrus-canning factories, which make citrus fruit juices, canned fruit, and other products. They consist of the dried and ground peel, residue of the inside portions, and occasional cull fruits of the citrus family, with or without extraction of part of the oil of the peel.

These products somewhat resemble dried beet pulp in chemical composition, but they are less palatable. They contain about 6% protein, 3 to 5% fat, and 10 to 12% fiber. Citrus meal and citrus pulp are best suited as cattle feeds, but they may be satisfactorily used for growing-finishing pigs if limited to 5% of the diet. Higher levels result in smaller rate and efficiency of gains, digestive disturbances, and lower dressing percentage and carcass grade.

Fats and Oils

Fats and oils, such as white grease, beef tallow, corn oil, and soybean oil, are excellent high-energy feeds; they contain 2.25 times as much metabolizable energy as most of the cereal grains. Additionally, a small amount of fat in the diet is desirable because fats are the carriers of the fat-soluble vitamins, and there is evidence that pigs require a dietary source of linoleic acid. Research indicates

1. That feeding sows at the rate of 5 lb per day a diet with 5% added fat for 10 days before farrowing has the potential to improve pig survivability if preweaning survival is typically below 80% because of increases in milk yield and milk fat content.
2. That the addition of 3 to 5% fat to growing-finishing diets will improve feed efficiency and often daily gain. A rule of thumb is that feed efficiency is usually improved 2% for each 1% increment of added fat in growing-finishing diets.
3. That adding fats to ad libitum diets generally tends to increase backfat thickness and intramuscular fat.
4. That fats and oils in diets assist in dust control in confinement operations.

Fats are solid at room temperature and oils are liquid. Both are difficult to handle and mix. Fats need to be heated to a liquid before mixing. Generally, adding 1% fat to a diet will reduce dustiness of the feed, resulting in a cleaner mixing room and pig rooms. Adding more than 5 or 6% fat to a diet will increase the risk of the feed bridging or sticking in the bulk bins and feeders.

New commercial products that contain *dried fat* may reduce part of the commercial problems of adding liquid fat on the farm, but the economic feasibility of using these products must be evaluated.

From these facts it may be concluded that adding fat to swine diets should be evaluated in terms of economic considerations. For example, if adding fat will increase diet cost by 5%, there must be more than 5% improvement in feed efficiency before it is practical.

Garbage

The feeding of garbage in the United States is regulated by both federal and state laws. In fact, many states ban the feeding of garbage to pigs. All garbage must be cooked.

Federal rules stipulate no person shall feed or permit the feeding of garbage to swine unless the garbage is treated to kill disease organisms. Inedible material must be rendered. It includes the meat of any animal (including fish and poultry) and other refuse that has been associated with animal material, resulting from the handling, preparation, cooking, or consumption of food. It must be ground and heated to a minimum temperature of 230°F (110°C) to make products such as animal, poultry, or fish protein meal, grease or tallow. Edible waste to be used for animal consumption must be heated throughout at 212°F (100°C) for 30 minutes under the supervision of a licensee.

Material excluded from cooking is garbage consisting of any of the following: rendered products, bakery waste, candy waste, eggs, domestic dairy products (including milk), fish from the Atlantic Ocean within 200 miles of the continental United States or Canada, or fish from inland waters of the United States or Canada that do not flow into the Pacific Ocean (Animal and Plant Health Inspection Service, USDA, §166.4, 2001).

Garbage may be used either as feed for sow and pig enterprises or for finishing feeder pigs that are obtained from other sources. Usually, the venture seems most successful when a combination of grain and garbage feeding is practiced.

It is also observed that the most successful garbage feeders use concrete feeding floors, practice rigid sanitation, and take every precaution to prevent diseases and parasites. Unless considerable grain is fed to market hogs, especially after weights are more than 100 lb, soft pork and paunchiness will result in garbage-fed hogs.

Nutrient analyses of garbage are highly variable because of the dry matter content and the variety of materials that may be present in the product, depending on the various sources as well as the material treating the garbage.

Molasses

Three kinds of molasses are fed to swine, cane molasses, beet molasses, and citrus molasses. The first two are by-products of sugar manufacture; the cane molasses derived from sugar cane and the beet mo-

Figure 9.9　A sugarcane field in Costa Rica. *(Courtesy, Palmer Holden, Iowa State University, Ames, IA)*

lasses from sugar beets. As the name implies, citrus molasses is a by-product of the citrus industry. The different types of molasses are available in both liquid and dehydrated forms.

Molasses is used in the following ways: (1) as an appetizer; (2) to reduce the dustiness of a diet; (3) as a binder for pelleting; (4) in the case of cane molasses, to provide trace minerals; and (5) to provide a carrier for vitamins in liquid supplements. But there are better, more concentrated sources of energy for pigs than molasses.

Cane Molasses (Blackstrap)

A large quantity of this carbohydrate feed is produced in the southern states, and an additional tonnage is imported (Figure 9.9). Molasses must contain not less than 43% total sugars expressed as invert. If its moisture content exceeds 27%, its density determined by double dilution must not be less than 79.5 Brix (measure of specific gravity). Molasses is very low in phosphorus (see Chapter 22), thus indicating the need for feeding this mineral supplement when molasses constitutes much of the diet.

Molasses may cause scouring in pigs unless they are started on it gradually and then fed in limited quantities. Molasses has its highest feeding value when it is used at a level of 3 to 10% of the diet and has a feeding value of 34 to 38% that of corn on an as-fed basis (74% DM (dry matter) molasses). Also, heavier pigs can use molasses more effectively than lighter pigs. Loose feces caused by the molasses are not a problem, providing the animals do well.

Beet Molasses

Beet molasses is a by-product of beet sugar factories. It must contain not less than 48% total sugars expressed as invert and its density must not be less than 79.5 Brix. Beet molasses has a laxative effect,

thus necessitating that pigs be started on it gradually and then fed limited amounts. It may constitute up to 10% of the diet for finishing pigs, but 5% is a recommended level. When properly used, beet molasses is equal to cane molasses as a swine feed, which means that it is 34 to 38% as valuable as corn.

Citrus Molasses

Citrus molasses must contain not less than 45% total sugars expressed as invert and its density must not be less than 17.0 Brix. Citrus molasses is unpalatable to hogs because of its bitter taste. Because cattle do not object to the taste, most citrus molasses is fed to them. When mixed with more palatable feeds, citrus molasses can be fed to swine at about the same levels as cane molasses, and the two products have about the same feeding value.

Root and Tuber Crops

Several root and tuber crops are fed to hogs, including sweet potatoes, white or Irish potatoes, chufas, cassavas, and Jerusalem artichokes. Roots are palatable, succulent, and laxative, but they are low in protein, calcium, and vitamin D, and, except for carrots and sweet potatoes, they have little or no carotene (vitamin A value).

Sweet Potatoes

Cull or unmarketable sweet potatoes are a palatable swine feed. Sometimes sweet potatoes are fed to hogs, usually by allowing the pigs to glean the fields after digging. They are high in carbohydrates in proportion to their protein and mineral content. Best results are usually obtained when pigs grazing sweet potatoes are given one-third to one-half the usual grain allowance plus a protein supplement, a suitable mineral supplement, and a vitamin supplement. Sweet potatoes are too bulky for young pigs. It requires 400 to 500 units of potatoes to equal 100 units of cereal grain when fed to hogs. Cooking improves the feeding value of sweet potatoes. They produce firm pork of good quality, although market animals fed on sweet potatoes are paunchy and have a lower dressing percentage. Dehydrated sweet potatoes have approximately 70% the feeding value of corn when used at levels of about 40% in a well-balanced diet.

Potatoes (Irish Potatoes)

Potatoes are sometimes used as hog feed when they have little value on the market, especially if grain prices are high. Because potatoes are not palatable or as readily digested in the raw state, they must be steamed or boiled (preferably in salt water, with any cooking water discarded) to increase palatability. When properly cooked, and when replacing 30 to 50% of the diet, they have 20 to 22% the feeding value of corn. Potatoes should not be fed to sows during the latter part of gestation or immediately after farrowing.

When limited to 5% of the diet of growing-finishing pigs, dehydrated Irish potatoes (known as potato meal or potato flakes) are worth 92 to 98% of corn. Of course, the cost of dehydrating is too great to consider the process practical, but surplus dehydrated potatoes are sometimes available at a nominal price.

Chufas

Chufas, a southern tuber crop, were formerly planted from April to June for hogging-down. The chufas sedge, frequently a weed in damp fields on southern farms, will remain viable over winter in the ground. Chufas should be supplemented properly with proteins and minerals. They will produce 300 to 600 lb of gain per acre after making allowance for the supplemental feeds consumed. Chufas produce soft pork.

Cassava

Cassava is a fleshy root crop that grows along the U.S. Gulf Coast and many tropical countries, yielding 5 to 6 tons per acre. The roots, containing 25 to 30% starch, are used in the production of the tapioca of commerce and as a swine feed. They usually are sun-dried before feeding.

When they do not constitute more than one-third of the dry matter in the diet, cassavas are a satisfactory feed for hogs. Dried cassava meal is worth about 85% as much as corn for growing-finishing hogs when limited to 5% of the diet. However, much higher levels can be fed with proper protein supplementation.

Jerusalem Artichokes

Jerusalem artichokes, a hardy perennial vegetable, produce tubers resembling the potato in composi-

tion, except that the chief carbohydrate is inulin (a fructan) instead of starch. They yield 6 to 15 tons of tubers per acre. The tubers live over the winter in the ground, and, even when hogged-down or dug in the fall, usually enough tubers will remain to make the next crop. When hogging-down artichokes, pigs should be fed grain and supplement because they will make little gain on the tubers alone.

Other Root Crops

Such root crops as beets, mangels, carrots, and turnips are never produced in the United States specifically for hog feed. However, when they are not salable for human food, they are sometimes fed to hogs. Because of their high water content (usually 80 to 90%), about 9 lb of roots are required to equal the feeding value of 1 lb of grain. Best results are secured when the roots are cut in small pieces, fed raw, and limited to a replacement of not more than one-fourth of the grain diet.

Velvet Beans

Velvet beans are grown chiefly for forage, but occasionally the beans are harvested for livestock feed. When hulled and ground, they become velvet bean meal. When the beans and pod are ground together, they become ground velvet bean and pod. Regardless of the type of preparation, velvet beans should not constitute more than one-fourth of the diet for any class of swine; otherwise, toxicity (due to dihydroxyphenylalanine, which is closely related to epinephrine)—with accompanying severe diarrhea and vomiting—will result. Cooking decreases the toxicity and increases palatability and digestibility, but it does not generally make the velvet beans entirely satisfactory. If used, it is recommended that velvet beans be fed only to heavier pigs and not to brood sows, and that a high-quality protein supplement be fed.

In the Gulf Coast region, velvet beans are sometimes grown with corn or with corn and peanuts for hogging down. Velvet bean pasture is not recommended for swine.

QUESTIONS FOR STUDY AND DISCUSSION

1. For your area, what type of high-energy feeds would you expect to be readily available?
2. What grains are grown primarily as food for humans? What grains are produced chiefly as feed for livestock? Why do you think this is so?
3. What are the advantages of feeding high-lysine corn? List the disadvantages of feeding or producing this kind of corn.

4. The following energy feeds are available at a low cost. What recommendations would you make for the use of each of these in swine diets?
   a. Oats
   b. Rice polishings
   c. Wheat bran
   d. Dried bakery products
   e. Garbage

5. List the milling by-products of corn and wheat that may be used in swine diets.
6. Why are so few by-products from the brewing and distilling industries used in swine diets?
7. High-energy feeds possess certain nutritive deficiencies. Using the information from this chapter and Chapter 22, list some of the deficiencies for the major high-energy feeds: corn, oats, and barley.
8. What are the advantages of adding fat to a swine diet? What classes of swine, and at what levels, would you recommend feeding fats and oils?
9. Why has garbage feeding declined in the United States?
10. Why would molasses be added to the diet? List the sources of molasses.
11. List some feeds that may cause soft pork.

SELECTED REFERENCES

Animal Feeding and Nutrition, 9th ed., M. J. Jurgens, Kendall/Hunt Publishing Co., Dubuque, IA, 2002

Applied Animal Nutrition, Feeds and Feeding, 2nd ed., Peter R. Cheeke, Prentice Hall, Upper Saddle River, NJ, 1999

Association of American Feed Control Officials Official Publication, Association of American Feed Control Officials, Inc., Oxford, IN, 2003

Feeds and *Nutrition Digest*, M. E. Ensminger, J. E. Oldfield, and W. W. Heinemann, The Ensminger Publishing Company, Clovis, CA, 1990

Kansas Swine Nutrition Guide, Kansas State University Extension, Manhattan, KS, 2003

Life Cycle Swine Nutrition, 17th ed., Pm-489, P. J. Holden, et al., Iowa State University, Ames, IA, 1996

Livestock Feeds and Feeding, 4th ed., R. O. Kellems and D. C. Church, Prentice Hall, Upper Saddle River, NJ, 1998

Nontraditional Feed Sources for Use in Swine Production, P. A. Thacker and R. N. Kirkwood, Butterworth Publishers, Stoneham, MA, 1990

Nutrient Requirements of Swine, 10th rev. ed., National Research Council, National Academy of Sciences, Washington, DC, 1998

Overview of Swine Nutrition Research, J. Shurson, University of Minnesota, St. Paul, MN, http://www.ddgs.umn.edu/ articles-swine/studies.pdf

Pork Industry Handbook, Cooperative Extension Service, Purdue University, West Lafayette, IN, 2003

Swine Nutrition Guide, University of Nebraska, Lincoln, NE, and, South Dakota State University, Brookings, SD, 2000

Swine Production and Nutrition, W. G. Pond and J. H. Maner, AVI Publishing Co., Inc., Westport, CT, 1984

10

Protein and Amino Acid Feedstuffs for Swine

Soybeans ready for harvest. *(Photo by Scott Bauer, Courtesy, USDA, Washington, DC)*

Objectives

After studying this chapter, you should:

1. Differentiate the importance of protein and amino acid nutrition.
2. Understand the strengths and shortcomings of plant versus animal proteins.
3. Be able to evaluate the uses of milk proteins.
4. Define the importance of meat-packing by-products.
5. Know what a pulse protein is.
6. Understand the issues on single-cell protein.
7. Understand the benefits and limits of protein supplements.

Prior to 1890, no one was concerned about adding proteins to livestock diets, and vitamins were unknown. The flour mills in Minneapolis dumped wheat bran into the Mississippi River because nobody wanted to buy it. Most of the linseed meal was shipped to Europe. Soybeans were little known outside the Orient, and tankage had not been processed.

Before 1890, swine were fed whatever homegrown grains and farm wastes were available. Corn was usually harvested and stored as ear corn, and fed to hogs and cattle as corn on the cob (Figure 10.1). Other small grains, such as barley and oats, also were fed depending on geographic location. Pigs were marketed at 10 to 16 months of age. Sows farrowed once a year, usually in the spring. Seasonal hog marketings were very uneven, and the various spring market hog shows were begun in the 1940s to stimulate fall farrowing and spring marketings.

PROTEINS

Beginning about 1900, scientists discovered that the kind or quality of protein in livestock feeds was of tremendous importance, thus ushering in the

Figure 10.1 Mechanical corn picker on Grundy County, Iowa, farm. *(Courtesy, USDA, Washington, DC, 1939)*

golden era in nutrition. Soon the race was on for protein-rich feeds. Many of the by-products that once polluted the streams of the nation were in unprecedented demand.

Protein in plants is largely concentrated in the actively growing portions, especially in the leaves and seeds. Plants possess the ability to synthesize their own proteins from such relatively simple soil and air compounds as carbon dioxide, water, nitrates, and sulfates. Thus, plants, together with some bacteria that are able to synthesize these products, are the original sources of all proteins.

Proteins in animals are much more widely distributed than in plants. The proteins of the animal body are primary constituents of many structural and protective tissues—such as the bones, ligaments, hair, hoofs, skin, and soft tissues that include the organs and muscles. The total protein content of the bodies of growing pigs ranges from 14 to 18% (Chapter 6, Table 6.1). By way of further contrast, it is noteworthy that, except for the bacterial action in the rumen, animals lack the ability of plants to synthesize proteins from inorganic materials. Hence, they must depend on plants or other animals as a source of dietary protein.

Proteins are found in most of the feeds commonly fed to animals. The amount of protein, its digestibility, and the balance of essential amino acids are important factors to consider when balancing diets. Recognizing the need for a variety of protein sources, Dr. John Evvard's original mixtures at Iowa State College consisted of 50% tankage, 25% corn oil cake, and 25% alfalfa meal. Linseed oil meal was substituted for the corn oil cake by Russell and Morrison (Wisconsin Farms Bulletin 352, Wisconsin Agriculture Experiment Station, Madison, 1923) and became known through the swine world as "Trinity Mix."

In general, animal proteins are superior in amino acid ratios to plant proteins for swine and other monogastric animals. For example, zein (a corn protein) is a low-quality or unbalanced protein, deficient in the essential amino acids lysine and tryptophan. However, animal proteins, such as milk and eggs, are excellent sources of lysine and tryptophan. Meat by-products are often limiting in tryptophan.

Because protein feeds are among the more costly components of swine diets, it is important to provide enough protein for the animal to perform its assigned function but to avoid feeding more than is necessary.

Historically, it was common practice to use several sources of protein in swine diets so that their respective amino acid profiles would complement each other. Today, the use of the computer in formulating diets and the increased availability of competitively priced synthetic amino acids, such as lysine, methionine, threonine, and tryptophan, allow nutritionists to formulate diets with a minimum number of protein feeds. A diet formulation program can rapidly determine what specific amino acids must be added and at what levels. Thus, the trend is toward fewer protein feed sources, properly supplemented with specific amino acids.

Ingredients that contain more than 20% of their total weight in crude (total) protein are generally classified as protein feeds. Protein supplements may be further categorized according to source of origin as (1) oilseed cake and meal proteins, (2) animal and marine proteins, and (3) mill products. The metric tons of each of these three sources fed to U.S. livestock during the 16-year period from 1985 to 2000 are shown in Table 10.1. During this time the percentage of oilseeds in animal feed increased from 64 to 74%, whereas animal

proteins decreased from 11 to 6%, and mill by-products decreased from 25 to 20%. This suggests that oilseed proteins are either more available or more competitively priced for use in animal feeds.

AMINO ACIDS

Amino acids are the structural units of proteins. Pigs do not have a specific requirement for crude protein, rather, pigs of all ages and stages of the life cycle require amino acids to enable them to grow and reproduce. During digestion, proteins are broken down into amino acids. Then, the amino acids are absorbed into the bloodstream and used to build new proteins, such as muscles.

Diets that are balanced with respect to amino acids contain a desirable level and ratio of the 10 essential amino acids required by pigs for maintenance, growth, reproduction, and lactation. The 10 essential amino acids that must be provided in swine diets are arginine, histidine, isoleucine, leucine, lysine, methionine, phenylalanine, threonine, tryptophan, and valine.

The proteins of corn and other cereal grains are deficient in certain essential amino acids, especially lysine, tryptophan, and threonine. Protein supplements are used to correct amino acid deficiencies in grains. For example, the correct combination of grain and soybean meal provides an excellent balance of amino acids except in starter diets.

Although grain and soybean meal are an excellent combination, other protein sources should be considered. Many acceptable alternatives exist, but

TABLE 10.1   Commercial Feeds: Disappearance for Feed, United States, 1985–2000 (1,000 metric tons)

| | Oilseed Cake and Meal | | | | | | Animal Protein | | | |
Year	Soybean	Cottonseed	Linseed	Peanut	Sunflower	Total	Tankage and Meat Meal	Fish Meal	Dried Milk[1]	Total
1985	17,355	1,382	100	159	314	19,309	2,545	465	375	3,385
1990	20,787	2,421	109	103	306	23,726	2,297	250	416	2,964
1995	24,191	2,692	117	165	435	27,599	2,305	264	382	2,951
2000[2]	28,806	2,585	175	100	451	32,118	2,170	199	298	2,667

| | Mill Products | | | | | |
Year	Wheat Mill Feeds	Gluten Feed and Meal	Rice Mill Feeds	Alfalfa Meal	Total	Total Commercial
1985	5,289	1,057	504	778	7,628	30,322
1990	6,000	165	556	334	7,055	33,745
1995	6,543	801	547	232	8,123	38,672
2000[2]	6,638	1,294	569	NA	8,501	43,286

[1]Includes dried skim milk and whey for feed, but does not include milk products fed on the farm.
[2]Preliminary data.
Source: USDA-NASS Agricultural Statistics 1995 and 2002, Table 69.

items such as amino acid balance and availability and synthetic amino acid sources should be considered.

• **Alternative amino acid sources**—The various protein or amino acid sources that can be used to replace soybean meal are listed in Table 10.2. Note that protein sources are classified into two major categories, plant proteins and animal proteins.

Soybean meal is the only plant protein that compares favorably with animal protein in terms of quality of amino acid content and ratio, and that can be used as the only protein source in most swine diets. Soybean meal may serve as the only protein source, with the exception of starter diets, which should contain some animal or milk products.

In order to determine the relative lysine value of alternative protein sources, it is important to compare the lysine level in the alternative to soybean meal. The relative lysine values of some alternative sources are listed in Table 10.2 and can be used to determine the comparative economic value of the protein source as a partial or complete replacement of dehulled soybean meal. The lysine values were calculated by dividing the lysine content of the feed ingredient by that of dehulled soybean meal (3.02% lysine) and multiplying by 100 to put them on a percentage basis.

Example:

Assuming that 47.5% dehulled soybean meal can be purchased at $250 per U.S. ton, what would a ton of 44% soybean meal be worth? Because the lysine content of 47.5% soybean meal is 3.02% and 44% soybean meal has 2.83% lysine, the lower-protein soybean meal has 94% the feeding value of dehulled soybean meal ($2.83 \div 3.02 \times 100 = 94\%$). Therefore, if 94% is multiplied by the cost of dehulled soybean meal ($94\% \times \$250$), the relative lysine value of the lower-protein soybean meal is $235. If it can be purchased for less than that, it is a better value than dehulled meal priced at $250.

The relative lysine value is not the same as a relative feeding value. A relative feeding value should include other nutrients that have significant value in a feedstuff. Typically a relative feeding value should also include energy and available phosphorus contents because these along with lysine are the three most costly items in a swine diet.

• **"Hidden" costs of alternative amino acid sources**—When selecting alternative amino acid sources, in addition to ingredient cost, the swine producer should consider storage and transportation costs, antinutritional factors, product variability,

TABLE 10.2   Comparison of Alternative Lysine Sources to Soybean Meal

Source	Protein (%)	Lysine (%)	Lysine, % of protein	Lysine vs. Soybean Meal, dehulled (%)
Plant Proteins				
Soy protein isolate	85.8	5.26	6.1	174
Soy protein concentrate	64.0	4.20	6.6	139
Soybean meal, without hulls	47.5	3.02	6.4	**100**
Yeast, brewers' dried	46.4	3.47	7.5	115
Soybean meal	43.8	2.83	6.5	94
Corn gluten meal	43.3	0.83	1.9	27
Peanut meal	43.2	1.48	3.4	49
Cottonseed meal	42.4	1.65	3.9	55
Sunflower meal	42.2	1.20	2.8	40
Canola meal	35.6	2.58	7.2	85
Corn gluten feed	21.5	0.63	2.9	21
Alfalfa meal	17.0	0.74	4.4	25
Wheat middlings	15.9	0.57	3.6	19
Wheat bran	15.7	0.64	4.1	21
Animal Proteins				
Blood meal, spray dried	92.0	8.51	9.3	282
Blood plasma, spray dried	78.0	6.84	8.8	226
Fish meal, mechanically extracted	64.5	5.87	9.1	194
Fish solubles, dried	64.2	2.84	4.4	94
Meat and bone meal	51.5	2.51	4.9	83
Skim milk, dried	34.6	2.86	8.3	95
Whey, dried	12.1	0.90	7.4	30

crude fiber content, spoilage, and under- or overprocessing. These factors are especially problematic with by-product protein sources. Because by-product feed ingredients tend to vary in composition, proper information regarding chemical composition is necessary to ensure optimum pig performance.

• **Maximum level at which some feed ingredients can replace soybean meal**—When substituting other protein sources for soybean meal, it is important to know the maximum level at which the new feed ingredient can be used without seriously affecting performance. The suggested levels of commonly used feedstuffs that can be used in gestation, lactation, starter, and growing-finishing, and diets to replace all or a part of the soybean meal are presented in Chapter 8, Table 8.1. For example, sunflower meal can be used at up to 10% of the diet of gestating sows and growing-finishing pigs. Reformulating the diet then with soybean meal and synthetic lysine will indicate the potential savings in soybean meal.

• **Soybean meal as the sole source of supplemental protein in the diet**—Soybean meal can serve as the sole supplemental protein in the diet for pigs heavier than 25 lb (11 kg). Younger and lighter pigs have a reduced ability to use the complex proteins found in soybean meal. Additionally, starting pigs may develop minor allergic reactions to certain proteins in soybean meal. For this reason, in the diet of starting pigs it is desirable to use less allergenic, highly digestible amino acid sources in diets for starter pigs, such as dried whey, dried skim milk, spray-dried plasma proteins and blood meal, menhaden fish meal, or soybean protein concentrate.

• **Ideal protein or amino acid balance**—The ideal protein or ideal amino acid balance is a protein that provides a perfect pattern of essential and nonessential amino acids in the diet without any excesses or deficiencies. This pattern is supposed to reflect the exact amino acid requirements of the pig. Thus, an ideal protein would provide exactly 100% of the recommended level of each amino acid. A typical corn–soybean meal diet will provide some essential amino acids at more than 2 times their requirement. But there is no evidence to indicate that the performance of pigs fed diets containing an ideal balance of amino acids is better or worse than that of pigs fed practical corn–soybean meal diets.

However, if excess amino acids are reduced, nitrogen excreted through the urine and feces will be reduced, thereby lessening the nitrogen in the manure. Unless there is an economic or environmental reason to reduce the nitrogen in the manure, producers should choose sources of amino acids that will produce lowest cost gains.

• **Limiting amino acid**—If a diet is inadequate in one essential amino acid, protein synthesis cannot proceed beyond the rate at which that amino acid is available. This is called the limiting amino acid. Standard swine diets are formulated to meet the pig's requirements for lysine—the most limiting amino acid. But when formulating diets to meet the lysine requirements, excesses of many other amino acids usually exist.

• **Amino acid availability**—Excessive heat in drying or processing feed ingredients will reduce the availability of amino acids, especially lysine. If soybean meal or dried whey look darker than usual or have a burnt smell, the protein quality likely has been reduced. Also, the presence of an antinutritional factor may affect availability of amino acids. For example, soybeans contain several antinutritional factors that are destroyed with heat. If a high percentage of by-product feed ingredients or feed ingredients that have been overprocessed are being used, consideration should be given to balancing the diet on an available amino acid basis.

• **Formulating diets on an available lysine basis**—Formulating diets on an available lysine basis are more accurate than formulating on a protein basis. When formulated on a protein basis, the diet could be deficient in lysine, resulting in reduced pig performance. Lysine is often the most limiting amino acid in a grain–soybean meal diet. Generally, if the lysine recommendation is met, the other amino acids will be adequate. An exception to this rule occurs when plasma proteins, blood meal, distillers' dried grains with solubles, and other by-products are used in the diet.

• **Crystalline amino acids in swine diets**—Whether it is economical to use crystalline amino acids in swine diets depends on the price of amino acids and the prices of grain and supplemental protein sources. The use of L-lysine · HCl as a source of

Stationary on-farm feed mill with corn and soybean meal tank on the left and small bins for vitamins and minerals. *(Courtesy, Palmer Holden, Iowa State University, Ames, IA)*

crystalline lysine is often economically sound. Crystalline methionine is commercially available and inexpensive. Crystalline tryptophan and threonine can be purchased in feed-grade forms but currently are rather expensive. Crystalline lysine and tryptophan together in the same source are now commercially available. It is anticipated that other sources combining these crystalline amino acids, as well as others, will be developed in the future.

Generally, 3 units of L-lysine · HCl (containing 78% pure lysine) and 97 units of corn contribute the same amount of digestible lysine as 100 units of 44% crude protein soybean meal. The level of crystalline amino acids supplemented depends on the feed used in the formulation and on the second limiting amino acid. In most swine diets, lysine is the first limiting amino acid and either tryptophan or threonine is the second limiting amino acid. When using synthetic lysine, the diet should be formulated to meet the second limiting amino acid and then synthetic lysine should be added to meet the lysine requirement. However, starter pig diets containing large amounts of plasma proteins and blood meal will benefit from supplemental crystalline methionine. In order to assume proper distribution of crystalline amino acids in a complete feed, amino acids must first be combined with a carrier to achieve a minimum volume before they are added to the mixer.

CAUTION:

Research indicates that limit-fed pigs (for example, sows fed once per day) use crystalline amino acids less efficiently than pigs fed free choice (eating several times per day). The synthetic amino acids are absorbed quickly from the small intestine whereas the amino acid–containing proteins must be first digested to release the individual amino acids for absorption. Ultimately, the blood levels of the synthetic amino acids peak before the natural amino acids are digested and absorbed. Under limit-fed circumstances replacement of intact protein (such as soybean meal) with synthetic amino acids may lead to a deficiency of other amino acids even though the formulation appears to be adequate. An amino acid deficiency reduces litter gain and sow lactation feed intake.

PLANT PROTEINS

Even though they are not especially high in protein by comparison with other feedstuffs, the vegetative portions of many plants supply an extremely large portion of the protein in the total diet of livestock, simply because these portions of feeds are consumed in large quantities. Needed protein not provided in these feeds is commonly obtained from one or more of the oilseed by-products—soybean meal, canola meal, cottonseed meal, linseed meal, peanut meal, safflower meal, sunflower seed meal, or coconut (copra) meal. The protein content and feeding value of these products vary according to the seed from which they are produced, the geographical area in which they are grown, the amount of hull or seed coat included, and the method of oil extraction used. Sometimes, the unprocessed seed is used to provide both a source of protein and a concentrated source of energy. The oil-bearing seeds are especially high in energy because of the oil that they contain.

Additional plant proteins are obtained as by-products from grain milling, brewing and distilling, and starch production. Most of these industries use the starch in grains and seeds, then dispose of the residue, which contains a large portion of the protein of the original plant seed.

OILSEED MEALS

Several rich oil-bearing seeds are produced for vegetable oils for human food (margarine, shortening, and salad oil), and for paints and other industrial purposes. In the processing of these seeds, protein-rich products of value in swine feedings are obtained. Among such high-protein feeds are soybean meal, canola meal (rapeseed meal), coconut (copra) meal, cottonseed meal, linseed meal (flax), peanut meal, safflower meal, sesame meal, and sunflower seed meal. Figure 10.2 presents the relative worldwide production of the various oilseeds.

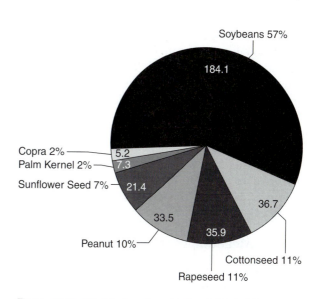

Figure 10.2 World oilseed production, 2001 (million metric tons). *(Courtesy, USDA World Statistics)*

Soybean Meal

Soybeans constitute 57% of the world's oilseed production (Figure 10.2). The United States produces about one-half of the world's soybeans, and soybean meal provides about two-thirds of the protein supplements used in the United States. Table 10.1 illustrates the increasing use of soybean meal from 1985 to 2000 as well as changes in the use of other protein by-products.

Soybean meal is the ground residue (soybean oil cake or soybean oil chips) remaining after the removal of most of the oil from soybeans. The oil is extracted by any one of three processes: (1) the expeller process, (2) the hydraulic process, or (3) the solvent process. Soybean meal produced by each of the extraction processes is of approximately the same feeding quality, but some nutritional differences occur depending on the amount of oil removed. Regardless of the method of extraction employed, reputable manufacturers now use the proper heat treatment so that the protein quality remains high.

Soybean meal normally contains 41 to 50% protein, according to the amount of hull and oil removed. Dehulled soybean meal must contain not more than 3.5% crude fiber and the other meals must contain not more than 7.0% crude fiber. It may contain calcium carbonate or other anticaking agent not to exceed 0.5%.

Although soybean meal is marginal in methionine, it is otherwise very well balanced in amino acids; and the protein of soybean meal is of higher quality than the other protein-rich supplements of plant origin. Like seeds of most plants, it is low in calcium, phosphorus, carotene, and vitamin D.

Soybean meal is extremely palatable to pigs and if self-fed, free choice, as the supplement to grain, they will often eat much more of it than is required. This is uneconomical because of the higher cost of the supplement in comparison with cereal grains and other high-energy feeds. This wasteful practice should be alleviated by mixing the ground grain and soybean meal together and self-feeding as a complete diet.

Soybean meal is excellent as the only protein supplement to corn for swine of all ages above 25 lb (11 kg) weight providing the proper proportions are used to yield an appropriate balance of the amino acids. It must be supplemented with minerals and vitamins.

Canola Meal (Rapeseed Meal)

World canola (rapeseed) production currently ranks third among oilseeds. Canola is grown mainly in Canada, but it is increasing in the United States (Figure 10.3).

Figure 10.3 Canola field in North Dakota ready for combining. *(Courtesy, Palmer Holden, Iowa State University, Ames, IA)*

Canola was created from selected rapeseed varieties by Canadian plant scientists in the 1970s. The old rapeseed was high in glucosinolate compounds, which, when fed to animals at high levels, caused palatability problems and lowered performance because of goitrogenic action. Canola changed this. The new canola is low in glucosinolates in the meal, and low in erucic acid (a long-chain fatty acid) in the oil. The feed ingredient definition of the AAFCO specifies that the oil component of the seed contain less than 2% erucic acid; that the solid component of the seed contain less than 30 micromoles of glucosinolates per gram of air-dry, oil-free solid; and that canola meal must not exceed a maximum of 12% crude fiber.

Canola meal averages between 35 to 43% crude protein and its amino acids compare favorably with soybean meal with the exception of lysine. The diets of sows and growing-finishing pigs may contain up to 5% canola meal and maintain equal performance.

> **CAUTION:**
>
> The reduced levels of erucic acid and glucosinolate in rapeseed (canola) is fairly common. However, care should be exercised when selecting varieties.

Coconut Meal (Copra Meal)

Coconut meal is the ground residue that remains from the extraction of oil from the dried meats of coconuts. The oil is extracted either mechanically or with solvents. Coconut meal averages about 21% crude protein. Because the proteins are of low quality, it is generally recommended that coconut meal not constitute more than 20 to 25% of the diet of

Figure 10.4 Cotton plants. *(Photo by David Nance, Courtesy, ARS-USDA, Washington, DC)*

gestating sows and growing-finishing pigs and up to 5% for lactating sows because the higher-fiber content can reduce consumption.

Cottonseed Meal

Among the oilseed meals, cottonseed meal ranks second in tonnage to soybean meal (Figure 10.4). The processing steps in making cottonseed meal are as follows: (1) cleaning the seeds; (2) dehulling the seeds; (3) crushing the kernels; (4) extracting the oil by (a) mechanical (or screw pressed), or (b) solvent; and (5) grinding the remaining residue or cake, thus forming cottonseed meal.

The protein content of cottonseed meal must contain at least 36% crude protein. For pigs, cottonseed meal is low in lysine and tryptophan and deficient in vitamin D, carotene (vitamin A value), and calcium. Also, it contains a toxic substance known as gossypol, varying in amounts with the seed and the processing. It is rich in phosphorus.

Low gossypol cottonseed meal must contain not more than 0.04% gossypol. It is recommended that the cottonseed meal content of practical swine diets (especially diets for growing swine) be limited to 5% of the diet. At this level, it is unlikely that the total diet will contain more than 0.01% free gossypol. Pig performance begins to be reduced at gossypol concentrations of 0.04% of the diet. More and more glandless cottonseed, free of gossypol, is being produced. It alleviates any restrictions as to levels of meal and the need to add iron.

Linseed Meal (Flax)

Linseed meal is a by-product of flax, a fiber plant that antedates recorded history. In the United States, most of the flax is produced as a cash crop for oil from the seed and the resulting by-product, linseed meal. Most of the U.S. flax is produced in North Dakota, South Dakota, Minnesota, and Texas.

The oil may be extracted from the seed by either the mechanical (old process) or the solvent (new process) method. Producers prefer the commonly used mechanical process because the remaining meal is more palatable. Linseed meal is the finely ground residue (known as cake, chips, or flakes) remaining after the oil has been extracted. It averages about 35% protein content. The AAFCO stipulates that linseed meal must contain no more than 10% crude fiber.

For swine, the proteins of linseed meal do not effectively balance the deficiencies of the cereal grains; linseed meal is low in the amino acids, lysine and tryptophan. Also, linseed meal is lacking in carotene and vitamin D and is only fair in calcium and the B vitamins. Because of its deficiencies, linseed meal should not be fed to swine as the sole protein supplement and should be supplemented with lysine.

Linseed meal is a laxative and in limited quantities can be a valuable addition to the diet of brood sows (about 1 lb daily for mature sows) at farrowing time or for animals that are being fed for show-ring competition as it imparts a desirable "bloom" to the hair of show animals.

Peanuts

Peanuts are an important cash and feed crop of the South and are the leading oil crop produced in the United States, with 3.3 billion pounds produced in 2002. They are grown for human consumption as peanuts, peanut butter, or peanut oil.

Historically, about one-third was harvested by hogging-down (turning pigs into the field to root out the nuts). Generally, Spanish peanuts were grown for hogging-down early in the fall, whereas runner peanuts were grown for winter use. Most practical swine producers used the brood sows and the young pigs for gleaning the fields, saving new fields for finishing hogs. Peanuts were frequently planted with corn when the crop was to be hogged-down. Even where the peanuts were harvested for sale through human channels, it was common practice to allow hogs to glean the fields.

Today, the practice of harvesting peanuts by hogging-down is limited except when the quality of the crop is poor or prices are down, or when gleaning the fields. Pigs should have free access to salt and a calcium supplement when hogging-down peanuts.

With the hulls on, peanuts contain about 25% protein and 36% fat. They are deficient in lysine, carotene, vitamin D, and calcium, and only fair in phosphorus.

Unfortunately, the feeding of large quantities of peanuts will produce soft pork, unless they are not fed to finishing pigs. Under most conditions, such a limitation is impractical. Furthermore, there is some evidence that peanut feeding increases backfat. Some specialty hams may be produced as "peanut fed."

Peanut Meal

Peanut meal, a by-product of the peanut industry, is ground peanut cake, the product that remains after the extraction of part of the oil of peanuts by mechanical means or solvents. It is a palatable, high-quality vegetable protein supplement used extensively in livestock and poultry feeds. Peanut meal ranges from 41 to 50% protein and from 1 to 8% fat. It is low in methionine, lysine, and tryptophan; and low in calcium, carotene, and vitamin D. The AAFCO stipulates that peanut meal must contain no more than 7% crude fiber.

Peanut meal tends to become rancid when held too long, especially in warm, moist climates, and it should not be stored longer than 6 weeks in the summer or 2 to 3 months in the winter.

If a calcium supplement is provided, such as ground limestone or oyster shell flour, peanut meal is satisfactory as the only protein supplement for pigs or brood sows on good pasture. For growing-finishing pigs and gestating-lactating sows in drylot, peanut meal, like other proteins of vegetable origin, should not constitute more than 5 to 10% of the feed, the other protein being composed of a high-quality source such as soybean meal or an animal or marine protein.

Safflower Meal

About 40% of the safflower seed is composed of hull. Once the oil is removed from the seeds, the resulting product contains about 60% hulls and 20 to 24% protein and more than 30% crude fiber. Various means have been tried to reduce this high-hull content. Most meals contain seeds with the hulls removed yielding a product of about 42% protein and 8.5% crude fiber. Safflower meal is low in lysine, methionine, and tryptophan. Its use in swine diets is limited.

Sesame Meal

Little sesame is grown in the United States despite the fact that it is one of the oldest cultivated oilseeds. The oil meal is produced from the entire seed. Solvent extraction yields higher protein (45%) and lower fat levels (1%) than either the screw-press or hydraulic methods, which produce meals containing about 42% protein and 5 to 11%

Figure 10.5 Sunflowers in Fargo, North Dakota. *(Photo by Bruce Fritz, Courtesy, ARS-USDA, Washington, DC)*

oil. Sesame meal is a poor source of lysine but an excellent source of methionine.

Sunflower Seed Meal

Sunflower seed production in the United States ranks third among the oil crops (Figure 10.5). In 2002, 2,200,000 acres of sunflowers were harvested in the United States, and 1,250,000 tons of sunflower seed were produced.

Sunflower oil meal varies considerably depending on the extraction process and whether the seeds are dehulled. Meal from solvent extraction of dehulled seeds contains about 42% protein, as opposed to 28% for whole seeds. Solvent-extracted sunflower seed meal, with hulls, ranges from 25 to 30% protein. It should not be used as the only protein supplement for swine because it is low in lysine.

Because of its high crude fiber content, sunflower meal with hulls should be limited in swine diets. It may replace up to 10% of the diet of gestating sows and growing-finishing pigs. Synthetic lysine will help balance the diet. Dehulled meal may be used up to 10% in lactating diets.

ALFALFA

The advantages of using alfalfa for a protein resource are numerous; among them, it is a perennial crop that can be harvested several times in one growing season and, being a legume, it has the ability to convert nitrogen into protein. Hence, alfalfa yields more protein per unit of cropland than any other crop.

It would appear that the most efficient means of preserving the nutrients in alfalfa would be to process the alfalfa immediately after cutting to reduce field losses. Leaves could then be separated from the stems, thereby yielding a high-protein leaf meal. The stems could then be fed as green chop, ensiled or dehydrated to form a medium protein dry roughage. Unfortunately, this is not economically feasible on a large-scale basis. Until new developments lower the cost of processing, dehydrated alfalfa meal will remain as a protein supplement of secondary importance.

- **Alfalfa meal or pellets**—Alfalfa meal is the aerial portion of the alfalfa plant, either sun cured or dehydrated. The following guarantees are recommended by the AAFCO for the various grades of alfalfa meal or ground alfalfa hay:

Crude Protein (%)	Crude Fiber Not More Than (%)
15	30
17	27
18	25
20	22
22	20

Alfalfa meal contains proteins of the right quality to balance amino acid deficiencies of the cereal grains; it is a rich source of minerals, especially calcium; and it is an excellent source of vitamins. If leafy and green and not more than a year old, alfalfa is high in carotene (provitamin A). Sun-cured hay is also a good source of vitamin D, the antirachitic vitamin. Alfalfa is an excellent source of the B vitamins.

Despite alfalfa's virtues, its high crude fiber (20 to 30%) and low energy density make it unsuited in the diets of nursing pigs and lactating sows and should be restricted to about 5% of the diet of growing-finishing pigs.

CORN PROCESSING BY-PRODUCTS

(See also Chapter 9, Grains and Other High-Energy Feedstuffs.)

Corn Gluten Meal

Corn gluten meal is similar to corn gluten feed without the bran. It averages about 43% protein and 4.5% fiber.

Because the proteins are of low quality, corn gluten meal should never be fed as the sole protein supplement to swine. Corn gluten meal is used more in poultry broiler diets where it makes for a more yellow skin pigmentation. Although usually too expensive for swine diets, when price relationships are favorable, it may be satisfactorily incorporated in confinement diets at up to 5%. For pasture feeding, it may constitute 50% of the protein supplement—provided the other half is of animal or marine origin. Corn gluten meal is worth about 55 to 65% of the value of dehulled soybean meal.

PULSE PROTEINS

Pulses are the seeds of leguminous plants and are used primarily for human consumption, but they can be effectively fed to swine when the price is competitive. Although there are more than 13,000 species within the family *Leguminosae*, only about 20 are used for food or feed. These various pulses include beans (common beans, dry beans, snap beans, kidney beans, navy beans, mung beans, and lima beans), chickpeas, cowpeas, field beans (horse or broad beans), field peas, and pigeon peas. Their crude protein content ranges from 19 to 31%. Soybeans and peanuts are pulses, but they are used almost entirely as oilseed meals in livestock diets.

All of the pulses contain components that possess antinutritional properties. Fortunately, processing procedures, such as cooking, germination, and fermentation, can reduce the risks of feeding pulses to livestock. Among the chemical factors that can create problems in feeding pulses are protease inhibitors, goitrogens, cyanogens, antivitamins, metal-binding factors, lathyrogens, and phytohemagglutinins. When considering pulses for feeding swine, producers should be certain of the necessary processing procedures.

Soybeans

Soybeans, the precious beans from the Orient, have advanced from a minor to a major U.S. grain crop since World War II. For the most part, soybeans are processed, with the oil used for human food and the by-products used for livestock feeds. However, a widespread interest in their use as a home-grown feed for swine has been created.

The protein of raw soybeans is poorly used by pigs because of the presence of a factor that inhibits protein digestion. However, cooking or roasting makes soybeans a satisfactory feed for hogs.

Soybeans, Raw

Raw soybeans are poorly used by young, growing pigs because of the presence of antitrypsin, a growth inhibitor that affects the protein digestion of young pigs. Raw soybeans, especially weather-damaged or low test-weight beans, are often attractive alternatives to add to gestating swine diets. As the pig becomes older, its susceptibility to trypsin inhibitors decreases. When fed to nursery pigs, death would not be uncommon. They should not be fed to growing-finishing pigs because of a significant decrease in performance and the likelihood of soft carcasses.

Raw soybeans may be successfully used in gestation diets because the amount consumed by a limit-fed sow is insufficient for the trypsin inhibitor to be of concern. Because of the large feed intake by lactating sows and the risk of the soybeans being consumed by nursing pigs, they should not be used in lactation diets.

Soybeans, Full-Fat (Heat-Processed Soybeans)

Whole cooked soybeans can be used to replace soybean meal or other forms of protein supplement in swine diets. In order to process whole soybeans, an investment in cooking or roasting equipment is required. Such an investment commits the operator to long-term use even though lower prices of soybean meal, relative to the price of whole soybeans, may not justify the practice.

It is noteworthy that hogs fed cooked whole soybeans have softer carcass fat than those fed a soybean meal diet. This condition may influence the price packers are willing to pay for those hogs.

Cooking (extruding or roasting) at the proper temperature (250°F for 2.5 to 3.5 minutes in a roaster) destroys the antitrypsin factor and makes soybeans a satisfactory feed for all pigs (Figure 10.6). A more precise test is to measure the increase in pH resulting from the addition of urea to a processed soybean slurry. The AAFCO lists the maximum pH rise for heat-processed soybeans to not exceed 0.01 pH units. Suggested pH ranges are shown. Soybeans can be overheated (pH rise of less than 0.5 units), which reduces the availability of the amino acid lysine by tying it up with sugars.

Figure 10.6 A commercial extruder for processing uncooked whole soybeans into an excellent protein supplement. *(Courtesy, Palmer Holden, Iowa State University, Ames, IA)*

Rise in pH units

< 0.05	Overheated
0.05 to 0.20	Properly processed
> 0.20	Underheated

If a pork producer has decided to add fat or oil to certain diets, on-farm extrusion or cooking and feeding full-fat soybeans may be more economical than selling the beans and buying back soybean meal and oil. Because whole or full-fat soybeans have less protein and lysine than soybean meal (35 to 40% protein and 2.1 to 2.4% lysine), it is necessary to add 20 to 25% more whole soybeans than soybean meal to have a similar lysine level in the diet. This will supply approximately 3% added fat to the diet, which will improve feed efficiency by 4 to 6%.

Whole cooked beans have little effect on rate of gain, but they usually improve feed efficiency by 5 to 10% because of the higher fat content of the whole soybeans, which increases the energy in the diet. However, this improvement in feed efficiency may be offset by the lower protein content of whole soybeans; whole cooked beans average about 37% crude protein, whereas soybean meal usually runs 44 to 49%. Also, because of the higher energy of the beans, the protein content of the diet should be formulated 1 to 2% higher than in a soybean meal diet.

Soy Protein Concentrate

Soy protein concentrate is produced by removing most of oil and water-soluble sugars, nonprotein constituents, from high-quality sound, clean, dehulled soybean seeds. Soy protein concentrate must contain not less than 65% protein on a moisture-free basis. It contains approximately 4.2% lysine. Research indicates that soy protein concentrate can effectively replace dried skim milk in starter pig diets.

Soy Protein Isolate

Soy protein isolate is the highest soy protein source (it must contain not less than 90% protein on a moisture-free basis). Defatted soy flakes are insolubilized by reducing the pH to 4.5 (isoelectric point). At this point, the isoelectric proteins are separated from the insoluble materials. Then, the removal of insoluble fibrous material by either decantation or centrifugation completes the protein isolation procedure. The final product can be spray dried to give an isoelectric protein, or neutralized to pH 7.0 and dried to give the common soy protein isolate. Soy protein isolate is an effective replacement for dried skim milk in starter pig diets.

ANIMAL PRODUCTS

Protein supplements of animal origin are derived from (1) meat-packing and rendering operations, (2) poultry and poultry processing, (3) milk and milk processing, and (4) fish and fish processing. Before the discovery of vitamin B_{12}, it was generally considered necessary to include one or more of these protein supplements in the diets of hogs and chickens. With the discovery and increased availability of synthetic vitamin B_{12}, high-protein feeds of animal origin have become less essential, though they still are included in many diets for most monogastric animals.

The use of animal proteins of mammalian origin is restricted to nonruminant feeds unless specifically exempted by U.S. federal rules (2 CFR 589-2000). Feeds containing prohibited substances must be labeled "Do not feed to cattle or other ruminants." The primary concern is to reduce the spread of bovine spongiform encephalopathy (BSE or mad cow disease).

Meat-Packing By-Products

Although the meat of animals is the primary object of slaughtering, modern plants process numerous and valuable by-products, including protein-rich livestock feeds. The Association of American Feed Control Officials' (AAFCO, 2003) description of selected meat-packing by-products follows.

Animal Meat Products

Animal meat products is the clean flesh derived from slaughtered mammals. It includes, but is not limited to, lungs, spleen, kidneys, livers, blood, bone, partially defatted low-temperature fatty tissue, and stomachs and intestines freed of their contents. It does not include hair, horns, teeth, and hoofs.

The label shall include guarantees for minimum crude protein, minimum crude fat, maximum crude fiber, minimum phosphorus, and minimum and maximum calcium. It shall not contain more than 12% pepsin indigestible residue and not more than 9% of the crude protein in the product shall be pepsin indigestible. The calcium level shall not exceed the actual level of phosphorus by more than 2.2 times.

• **Meat meal**—Meat meal is the rendered product from mammal tissues, as defined under animal meat products. It does not contain bone.

• **Meat and bone meal**—When bone is added, the word *bone* must be inserted in the name of both meat meal and meat meal tankage, and they are designated as meat and bone meal and meat and bone

meal tankage, respectively. It shall contain a minimum of 4.0% phosphorus.

• **Meat meal tankage**—Meat meal tankage is similar to meat meal except it may contain added blood or blood meal. It contains about 60% crude protein, whereas meat meal is about 54%.

• **Meat and bone meal tankage**—Meat and bone meal tankage is similar to meat and bone meal except it may contain added blood or blood meal; however, it shall not contain any added extraneous materials not provided for in the preceding definition. It shall contain a minimum of 4.0% phosphorus.

Hydrolyzed Hair

Hydrolyzed hair is prepared from clean, undecomposed hair, by heat and pressure to produce a product suitable for animal feeding. Not less than 80% of its crude protein must be digestible by the pepsin digestibility method.

Glandular Meal and Extracted Glandular Meal

Glandular meal and extracted glandular meal is obtained by drying liver and other glandular tissues from slaughtered mammals. When a significant portion of the water-soluble material has been removed, it may be called extracted glandular meal.

Meat Protein Isolate

Meat protein isolate is produced by separating meat protein from fresh, clean, unadulterated bones by heat processing followed by low-temperature drying to preserve function and nutrition. This product is characterized by a fresh meaty aroma, a 90% minimum protein level, 1% maximum fat, and 2% maximum ash.

Blood Products

Historically, blood has been cooked and dried as blood meal for use in livestock feeds using the traditional vat cooking process. Although such processing results in a high-protein ingredient, it has found limited use in pig diets because of lack of palatability and low availability of lysine. Spray drying and newer flash drying procedures have significantly improved both the palatability and lysine availability of blood meal.

Spray-dried porcine and bovine plasma and spray-dried blood meal, by-products from pig and cattle slaughter, have revolutionized nutritional programs for early weaned pigs.

• **Animal plasma, spray dried**—Spray-dried animal plasma is made up of the albumin-, globulin-, and fibrinogen-type protein. It contains 78% protein and 6.8% lysine. The blood is collected in refrigerated tanks and prevented from coagulating by adding sodium citrate. The plasma fraction is separated from the blood cells by centrifugation and stored at 25°F until the product is spray dried. The product may bear the species name, such as porcine or bovine.

• **Blood meal, flash dried**—Flash-dried blood meal is processed similarly to plasma, except it contains both the plasma and red blood cell fractions. The minimum biological activity of the lysine shall be 80%.

Both spray-dried blood products are effective protein sources in starter pig diets. Synthetic methionine should be added to starter diets containing either spray-dried plasma or spray-dried blood meal.

POULTRY PRODUCTS

Poultry by-product feedstuffs are derived from all segments of the poultry industry—from hatching all the way through processing for market. They come from the broiler and turkey segments of the industry as well as from egg production. Centralization of these industries into large units with enough volume of wastes to make it feasible to process the potential feeds has opened new markets for what was previously a disposal problem. Certain precautions must be included to make the products most useful and safe, but considerable amounts of poultry products are currently being used in the diets of various animals, both ruminants and monogastrics (Figure 10.7).

Figure 10.7 Proper amino acid balance is essential for nursery pigs. *(Courtesy, Palmer Holden, Iowa State University, Ames, IA)*

Poultry

Poultry is the clean combination flesh and skin with or without accompanying bone, derived from the parts or whole carcasses of poultry or a combination thereof, exclusive of feathers, heads, feet, and entrails. It shall be suitable for use in animal food. If it bears a name descriptive of its kind, it must correspond thereto. If the bone has been removed, the process may be so designated by use of the appropriate feed term.

Poultry Meal

Poultry meal is the dry rendered product from a combination of clean flesh and skin with or without accompanying bone, derived from the parts or whole carcasses of poultry or a combination thereof, exclusive of feathers, heads, feet, and entrails. It shall be suitable for use in animal food. If it bears a name descriptive of its kind, it must correspond thereto. If the bone has been removed, the process may be so designated by use of the appropriate feed term.

Hydrolyzed Whole Poultry

Hydrolyzed whole poultry is the product resulting from the hydrolyzation of whole carcasses of culled or dead, undecomposed poultry, including feathers, heads, feet, entrails, undeveloped eggs, blood, and any other specific portions of the carcass. The product must be consistent with the actual proportions of whole poultry and must be free of added parts, including, but not limited to, entrails, blood, or feathers. The poultry may be fermented as a part of the manufacturing process. The product shall be processed in such a fashion as to make it suitable for animal food, including heating (boiling at 212°F, or 100°C, at sea level for 30 minutes; dry extrusion at a minimum temperature of 284°F, or 140°C, for 30 seconds with a pressure differential of approximately 40 atmospheres as the product exits the extruder, or their equivalents) and agitating (except in steam cooking equipment). The product may, if acid or alkaline treated, be subsequently neutralized. If the product bears a name descriptive of its kind, the name must correspond thereto.

Hydrolyzed Poultry By-Products Aggregate

Hydrolyzed poultry by-products aggregate is the product resulting from hydrolyzation, heat treatment, or a combination thereof of all by-products of slaughtered poultry, clean and undecomposed, including such parts as heads, feet, undeveloped eggs, intestines, feathers, and blood. The parts may be fermented as a part of the manufacturing process. The product shall be processed in such a fashion as to make it suitable for animal food, including heating (boiling at 212°F, or 100°C, at sea level for 30 minutes, or its equivalent, and agitated except in steam cooking equipment). It may, if acid or alkaline treated, be subsequently neutralized. If the product bears a name descriptive of its kind, the name must correspond thereto.

Poultry By-Product Meal

Poultry by-product meal consists of the ground, rendered, clean parts of the carcass of slaughtered poultry, such as necks, feet, undeveloped eggs, and intestines, exclusive of feathers, except in such amounts as might occur unavoidably in good processing practices. The label shall include guarantees for minimum crude protein, minimum crude fat, maximum crude fiber, minimum phosphorus, and minimum and maximum calcium. The calcium level shall not exceed the actual level of phosphorus by more than 2.2 times.

Because of the heads and feet, poultry by-products are lower in nutritional value than the flesh of animals, including poultry. Thus, the biological value of the proteins is lower than the other animal proteins. They may be successfully used in pig diets, however, provided they are not the sole source of proteins.

Poultry By-Products, Fresh

Fresh poultry by-products must consist of nonrendered clean parts of carcasses of slaughtered poultry, such as heads, feet, and viscera, free from fecal content and foreign matter except in such trace amounts as might occur unavoidably in good factory practice. Because of the heads and feet, poultry by-products are lower in nutritional value than the flesh of animals, including poultry.

Hydrolyzed Poultry Feathers

Hydrolyzed poultry feathers is the product resulting from the treatment under pressure of clean, undecomposed feathers from slaughtered poultry, free of additives, and/or accelerators. Not less than 75% of its crude protein content must be digestible by the pepsin digestibility method.

Although hydrolyzed feathers meal is high in protein, it is rather low in nutritional value, especially in the amino acids histidine, lysine, methionine, and tryptophan. The amino acids that are present are comparably available relative to grains or plant proteins.

Because of the deficiencies of several amino acids, care must be exercised when incorporating feathers meal into swine feed. The addition of fish meal or meat meal tends to complement feathers meal and facilitates its use. In practice, feathers meal rarely exceeds 3% of swine diets.

Poultry Hatchery By-Product

Poultry hatchery by-product is a mixture of egg shells, infertile and unhatched eggs, and culled chicks, which have been cooked, dried, and ground, with or without removal of part of the fat. Hatchery by-products are the most valuable of the feed by-products from poultry. However, these products deteriorate quite rapidly if not cooled promptly.

Egg Product

Egg product is obtained from egg graders, egg breakers, and/or hatchery operations. Egg products are dehydrated, handled as liquid, or frozen. These shall be labeled as per USDA regulations governing eggs and egg products (7CFR, Part 59). This product shall be free of shells or other nonegg materials except in such amounts that might occur unavoidably in good processing practices and contain a maximum ash content of 6% on a dry matter basis.

MILK PRODUCTS (CATTLE)

Skimmed milk, buttermilk, and whey have long been used on or in close proximity to the farms and plants where they are produced or processed. However, in their liquid form it is impossible economically to ship them long distances or to store them, and it is difficult to maintain sanitary feeding conditions.

About 1910, processes were developed for drying buttermilk. Soon thereafter, special plants were built for dehydrating buttermilk, and the process was extended to skimmed milk and whey. Beginning about 1915, dried milk by-products were incorporated in commercial poultry feeds. Subsequently, they have been used for swine feeding.

The superior nutritive values of milk by-products are because of their high-quality proteins, vitamins, a good mineral balance, and the beneficial effect of the milk sugar, lactose. In addition, these products are palatable and highly digestible. They are an ideal feed for young pigs and for balancing deficiencies of the cereal grains. The chief limitations of their wider use are price, together with the perishability and bulkiness of the liquid products.

Although whole milk is an excellent feed, worth about twice as much as skimmed milk, it is usually too expensive to feed to swine. For this reason, only the milk by-product feeds are discussed. In general, liquid milk products contain 7 to 13% dry matter; semisolid milk products, 30% dry matter; and dried milk products, more than 90% dry matter.

Skimmed Milk

Fresh skimmed milk contains about 7% dry matter and 0.07% lysine. Because of the removal of the fat, skimmed milk supplies little vitamin A value. However, in comparison with whole milk, it is higher in protein, milk sugar, and minerals. Like all milk products, skimmed milk is low in vitamin D and iron. Skimmed milk should be fed consistently sweet or sour, because abrupt changes are apt to produce digestive disturbances. Where a choice is possible, fresh skimmed milk is recommended.

Skimmed milk is the best single protein supplement for swine and when dried is especially valuable for young pigs prior to and immediately after weaning. The addition of either pasture or a choice legume hay will supplement free-choice skimmed milk with the needed vitamins A and D.

The amount of skimmed milk to feed will vary according to (1) the available supply, (2) the relative price of feeds, (3) the kind of grain diet fed, and (4) whether pasture is available.

Under nonpasture conditions and with corn constituting all or most of the grain diet, assume a daily allowance of 1 to 1.5 gal (3.8 to 5.7 l) of skimmed milk per pig. This will allow the replacement of most of the supplemental protein. With the same grain diet and good pasture, the skimmed milk allowance may be cut in half. With barley or wheat (feeds of higher-protein content) replacing the corn, about two-thirds of these amounts of skimmed milk will be adequate. Because the amino acid requirements decrease with the age of the animal whereas total daily feed consumption increases, a constant daily intake of skimmed milk on this basis will approximate the changing needs throughout the life of the pig.

The feeding value of skimmed milk varies with the ages of the pigs, the type of diet, the price of other feeds, and the amount of milk fed. When the supply of milk is abundant and cheap, a larger proportion may be fed profitably—especially if grain or protein are high in price. Roughly, it can be estimated that 6 weight units of skimmed milk will replace 1 unit of complete feed.

Condensed Skimmed Milk

Condensed skimmed milk is the residue obtained by evaporating defatted milk. It contains 27% minimum total solids.

Dried Milk, Feed Grade (Dried Whole Milk)

Dried whole milk is the residue left following the drying of milk or milk of intermediate fat levels other than defatted milk. It contains at least 26% milk fat. The label must contain a guarantee for minimum crude protein and for minimum crude fat.

Skimmed Milk and Buttermilk

Prior to World War II, dried skimmed milk was the most widely used dried milk product included in feeds. Since then much of it has been marketed as a human food.

Liquid buttermilk and skimmed milk have approximately the same composition and feeding value for swine, provided the buttermilk has not been diluted by the addition of churn washings. Dried buttermilk and dried skimmed milk average 32 to 35% protein. Though both are excellent swine feeds, they are generally too high priced to be economical for this purpose except in pig prestarter diets.

Accordingly, the discussion of skimmed milk is also applicable to buttermilk, and it may be considered that 6 weight units of liquid buttermilk will replace 1 unit of complete feed.

Condensed Buttermilk

Condensed buttermilk is the residue obtained by evaporating buttermilk. It contains 27% minimum total solids, 0.055% minimum milk fat for each percent of total solids, and 0.14% maximum ash for each percent of total solids.

Like the other milk by-products, condensed buttermilk is (1) very palatable to pigs, (2) more valuable for young pigs than for older animals, and (3) more valuable for drylot feeding than for pigs on pasture. Perhaps the greatest value of condensed buttermilk is as an appetizer.

Dried Buttermilk, Feed Grade

Feed-grade dried buttermilk is the residue obtained by drying buttermilk. It contains 8.0% maximum moisture, 13% maximum ash, and 5% minimum milk fat.

Dried Skimmed Milk, Feed Grade

Feed-grade dried skimmed milk is the residue obtained by drying defatted milk. It contains 8% maximum moisture.

Dried Cultured Skimmed Milk and Condensed Cultured Skimmed Milk

These two milk products are obtained following the culturing of skimmed milk with lactic acid bacteria. The dried product contains 8% maximum moisture, and the condensed product contains 27% minimum total solids.

Cheese Rind

Cheese rind is obtained by cooking cheese trimming devoid of fat other than milk fat.

Milk Protein Products

The four milk protein products currently available are as follows:

1. Casein is the solid residue that remains after the acid or rennet coagulation of defatted milk. It contains 80% minimum crude protein.
2. Dried lactalbumin is produced by drying the coagulated protein residue separated from whey. It contains 80% minimum crude protein on a moisture-free basis.
3. Dried milk protein is obtained by drying the coagulated protein residue resulting from the controlled coprecipitation of casein, lactalbumin, and minor milk proteins from defatted milk.
4. Dried hydrolyzed casein is the residue obtained by drying the water-soluble product resulting from the enzymatic digestion of casein. It contains at least 74% crude protein.

Although the quality and quantity of protein from milk protein products are excellent, these sources of protein are generally too expensive to be used routinely as swine feeds except for baby pigs.

Whey

Whey is a by-product of cheese making. Practically all of the casein and most of the fat go into the cheese, leaving the whey, which is high in lactose (milk sugar) but low in protein (0.6 to 0.9%) and fat (0.1 to 0.3%). However, its protein is of high quality. In general, whey has a feeding value equal to about one-half that of skimmed milk or buttermilk. Pigs should be allowed to gradually become accustomed to whey, after which the whey may be fed it free choice (Figure 10.8). Because of its low-protein content, growing pigs should receive a protein-rich supplement in addition to grain and whey. In order to prevent the spread of disease, whey should be pasteurized at the factory and fed under sanitary conditions.

Figure 10.8 Liquid whey feeding system provides quality nutrients as well as liquid. *(Courtesy, Palmer Holden, Iowa State University, Ames, IA)*

Numerous whey products are commercially available. Those recognized by the AAFCO follow:

1. Dried (dry) whey is derived from removing water from whey. It contains not less than 11% protein or less than 61% lactose. (It is rich in riboflavin and pantothenic acid. Some pork producers prefer to feed "human-grade" dried whey to young pigs because of its consistent quality. Acceptable quality dried whey should be light and creamy in color. Darkened product, caused by the Maillard Reaction between lysine and the sugars in the whey, likely has reduced lysine availability.)

2. Condensed whey is the product obtained by partially removing water from whey. The minimum percentage total whey solids must be prominently declared on the label.

3. Dried (dry) whey solubles is the product obtained by drying the whey residue after removal of whey protein, with or without partial removal of lactose. The minimum percentage of crude protein and lactose, and the maximum percentage of ash must be prominently declared on the label.

4. Condensed whey solubles is the product obtained by concentrating the whey residue after the removal of whey protein, with or without partial removal of lactose. Minimum percentage of solids, crude protein, and lactose, and maximum percentage of ash must be prominently declared on the label.

5. Dried hydrolyzed whey is the residue obtained by drying lactase enzyme hydrolyzed whey. It contains at least 30% minimum total glucose and galactose.

6. Condensed hydrolyzed whey is the residue obtained by evaporating lactase enzyme hydrolyzed whey. It contains at least 50% total solids and 0.3% minimum total glucose and galactose for each percentage of total solids.

7. Condensed whey product is obtained by evaporating whey from which a portion of lactose, protein, and/or minerals has been removed. The minimum percentage of solids, crude protein, and lactose, and the maximum percentage of ash must be prominently shown on the label.

8. Dried (dry) whey product is obtained by drying whey from which a portion of lactose, protein and/or minerals has been removed. When a portion of the lactose (milk sugar) that normally occurs in whey is removed, the resulting dried residue is called dried whey product. The minimum percentage of crude protein and lactose, and the maximum percentage of ash must be prominently declared on the label. (Dried whey product is a rich source of water-soluble vitamins.)

9. Condensed cultured whey is the product obtained by partially removing water from whey that has been cultured. The minimum percentage of solids must be prominently declared on the label.

Other Dairy Products

Other dairy products regionally available include (1) dairy food by-products, (2) condensed modified whey solubles, (3) dried (dry) whey protein concentrate, (4) dried cultured whey product, (5) dried cultured whey, (6) dried chocolate milk, (7) dried cheese, and (8) dried cheese product. See AAFCO (2003) for product descriptions.

MARINE PRODUCTS

Early fishery processors dumped marine wastes into the sea. Later, some were dried at high temperatures—usually in an open flame dryer—and used as fertilizers. In about 1910, it was discovered that waste materials from fish canning plants and inedible fish were a desirable protein source for livestock feeding (Figure 10.9). Again, experiment stations led the way in determining the feeding value of these new products that are now incorporated in many swine and poultry diets.

Fish Meal

Fish meal is the clean, dried, ground tissue of undecomposed whole fish or fish cuttings, either or both, with or without the extraction of part of the oil. If it contains more than 3% salt, the amount of salt must be a part of the brand name. In no case shall the salt content exceed 7%.

Fish meal is a by-product of the fisheries industry and the feeding value varies according to the following:

1. The method of drying. Drying may be done by vacuum, steam, or flame dried. The older flame drying method exposes the product to a higher tem-

Figure 10.9 Hold of a fishing boat full of anchoveta (*Engraulis ringens*). *(Courtesy, Jose Cort Fisheries Collection, National Oceanic and Atmospheric Administration (NOAA), Washington, DC)*

perature that makes the proteins less digestible and destroys some of the vitamins.

2. The type of raw material used. Raw material may be made from the offal produced in fish packing or canning factories, or from the whole fish with or without extraction of part of the oil. Fish meal made from offal containing a large proportion of heads is less desirable because of the lower quality and digestibility of the proteins. Although few feeding comparisons have been made between the different kinds of fish meals, it is apparent that all of them are satisfactory when properly processed raw materials of good quality and moderate fat content are used. Fish meals containing high fat levels are considered to be of low quality because the high fat content may impart a fishy taste to eggs, meat, and milk, and such a meal is apt to become rancid in storage. The sources commonly used in fish meals include the following:

1. Menhaden fish meal is the most common kind used in the eastern United States (Figure 10.10). It is made from menhaden herring (a very fat fish not suited for human food) caught primarily for its body oil. The meal is the dried residue after most of the oil has been extracted.

Figure 10.10 A menhaden fishing vessel with a lookout in the crow's nest looking for indications of schools of menhaden. *(Courtesy, Fisheries Collection, National Oceanic and Atmospheric Administration (NOAA), Washington, DC)*

2. Sardine meal or pilchard meal is made from sardine canning waste and from the whole fish, principally on the West Coast of the United States.

3. Herring meal is a high-grade product produced in the Pacific Northwest and in Alaska.

4. Salmon meal is a by-product of the salmon canning industry in the Pacific Northwest and in Alaska.

5. White fish meal. This is a by-product from fisheries making cod and haddock products for human food. Its proteins are of very high quality.

Fish meal should be purchased from a reputable company on the basis of protein content. It varies in protein content from 60 to 70%, depending on the variety of fish. When of comparable quality, fish meal is even superior to plant and animal by-products as a protein supplement for swine. The protein of a good-quality fish meal is 92 to 95% digestible. If it is poorly processed or improperly stored, the protein digestibility decreases dramatically. Because fish meals are cooked, there is danger that certain amino acids, notably lysine, cystine, tryptophan, and histidine, will be denatured, but these losses are minimized with proper processing techniques.

Fish meal is an excellent source of minerals. Calcium and phosphorus are especially abundant in the amounts of 3 to 6% and 1.5 to 3.6%, respectively. Many of the trace minerals, especially iodine, required by swine can be supplied in part by fish meal.

Most of the fat-soluble vitamins are lost during the extraction of oil, but a fair amount of the B vitamins remain. However, fish meal is one of the richest sources of vitamin B_{12}.

The difference between fish meal and tankage or meat meal is not so marked for pigs on pasture as for pigs in drylot. When it is fed as the only protein supplement to corn for pigs on pasture, fish meal is worth approximately 10% more than tankage or meat meal. Because of the generally higher cost of fish meal, it is commonly recommended that it be mixed with one or more protein-rich feeds of plant origin in providing supplements for pigs on pasture.

In addition to the various fish meals available, many by-products of processing fish, crabs, and shrimp are available on the market. Some of these are described in this section.

Fish meal residue is the clean, dried, undecomposed residue from the manufacture of glue from nonoily fish. The provisions relative to salt content are the same as for fish meal.

Fish liver and glandular meal is obtained by drying the complete viscera of the fish. At least 50% of dry weight of the product must consist of fish liver and must contain at least 18 mg of riboflavin per lb (40 mg/kg).

Crab meal is the undecomposed ground dried waste of the crab and contains the shell, viscera, and part or all of the flesh. It must contain not less than 25% crude protein. If it contains more than 3% salt, the amount must constitute a part of the brand name.

Mineral content is exceedingly high, about 40%. A rule of thumb for feeding crab meal is that 1.6 parts of crab meal can replace 1 part of fish meal.

Shrimp meal is the undecomposed ground dried waste of shrimp and contains parts and/or whole shrimp. If it contains more than 3% salt, the amount must constitute a part of the brand name. In no case must the salt content of this product exceed 7%.

It is either steam dried or sun dried, with the former method being preferable. On the average, it contains 32% protein and 18% minerals.

Condensed fish solubles is obtained by evaporating excess moisture from the stickwater, aqueous liquids, resulting from the wet rendering of fish into fish meal, with or without removal of part of the oil. A minimum percentage of solids, a minimum percentage of crude protein, and a minimum percentage of crude fat must be guaranteed.

Condensed fish solubles, containing approximately 50% total solids and 30% crude protein, is a rich source of the B vitamins. Solubles are particularly rich in pantothenic acid, niacin, and vitamin B_{12}. Unfortunately, fish solubles cannot be stored successfully in dried form but must be left as a sticky, semisolid form.

Therefore, their use is largely limited to commercially mixed diets or supplements for swine in drylot.

Dried fish solubles is obtained by dehydrating the stickwater. It must contain not less than 60% crude protein.

Fish protein concentrate—Feed-grade is prepared from clean, undecomposed whole fish or fish cuttings using the solvent extraction process developed for the production of edible whole fish protein concentrate. It must contain not less than 70% protein and not more than 10% moisture. If the degree fineness is stated, it must conform thereto. Solvent residues are not to exceed those established in food additive regulations.

Fish by-products must consist of nonrendered, clean, undecomposed portions of fish (such as, but not limited to, heads, fins, tails, ends, skin, bone, and viscera) that result from the fish processing industry.

Dried fish protein digest is the dried enzymatic digest of clean, undecomposed whole fish or fish cuttings using the enzyme hydrolysis process. The product must be free of bones, scales, and undigested solids with or without the extraction of part of the oil. It must contain not less than 80% protein and not more than 10% moisture. If the degree fineness is stated, it must conform thereto.

Condensed fish protein digest is similar to dried fish protein digest except for the dry matter content and it must contain not less than 30% protein.

Dried fish digest residue is the clean, dried, undecomposed residue (bones, scales, undigested solids) of the enzymatic digest resulting from the enzyme hydrolysis process of producing fish protein digest. It must be designated according to its protein, calcium, and phosphorus content.

SINGLE-CELL PROTEIN (SCP)

Single-cell protein (SCP) refers to protein obtained from single-cell organisms, such as yeast, bacteria, and algae, that have been grown on specially prepared media. Of these, only yeasts have successfully become a source of dietary protein for livestock. Production of this type of protein can be attained through the fermentation of petroleum derivatives or organic waste.

Yeast has long been used in the baking, brewing, and distilling industries; in making cheese and other fermented foods; and in storing and preserving foods. Dried brewers' yeast, a residue from the brewing industry, and torula yeast, from the fermentation of wood residue and other cellulose sources, are currently marketed.

A wide variety of materials can be used as substrates for the growth of these organisms. Industrial by-products from the chemical, wood and paper, and food industries have shown considerable promise as sources of nutrients for single-cell organisms, among them, (1) crude and refined petroleum products, (2) methane, (3) alcohols, (4) sulfite waste liquor, (5) starch, (6) molasses, (7) cellulose, and (8) animal wastes.

Yeasts are nonphotosynthetic organisms and have been used as sources of vitamins and, traditionally, unidentified factors. Brewers' dried yeast, torula dried yeast, grain distillers' dried yeast, and molasses distillers' dried yeast are used chiefly as sources of B vitamins and protein.

Yeasts contain about 45% protein, but a disadvantage to their use as protein sources is their deficiency in the sulfur amino acids. Fortunately, synthetic methionine or methionine hydroxy analog (MHA) are economical commercial products that can make up for this deficiency.

Types of Single-Cell Protein

- **Brewers dried yeast** is the dried, nonfermentative, nonextracted yeast of the botanical classification *Saccharomyces,* resulting from the brewing of beer and ale. It must contain not less than 35% crude protein. It must be labeled according to its crude protein content.
- **Irradiated dry yeast** is the dried nonfermentative yeast that has been exposed to ultraviolet rays in order to produce antirachitic potency. When it is used as an ingredient for four-footed animals, the name maybe followed by a parenthetical phrase (Source of Vitamin D_2).
- **Torula dried yeast** is the dried, nonfermentative yeast of the botanical classification (torulopsis) *Candida utilis* (formerly *Torulopsis utilis*) that has been separated from the medium in which it was propagated. It must contain not less than 40% crude protein.

Problems Associated with Single-Cell Protein

The potential of single-cell protein as a high-protein source for both humans and livestock is enormous, but many obstacles must be overcome before it becomes widely used. Although the feeding of SCP to livestock may not produce physiological problems, it will be necessary to demonstrate to the consuming public that problems will not arise from the consumption of animal products produced from the feeding of SCP. These problems include palatability, digestibility, nucleic acid content, toxicities, protein quality, and economics. There is a limited amount of single-cell protein on the market in the form of brewers' yeast and torula yeast, but these products are generally too expensive to use as a major protein source for pigs.

- **Palatability**—Microbial cells must be processed to some extent if they are to be palatable. Otherwise, animals will not eat them. Yeasts are bitter tasting, and algae and bacteria have characteristically unpleasant tastes that can depress intake.
- **Digestibility**—Ways of making single-cell protein more digestible must be developed if it is to be competitive with traditional protein feeds. Digestibility among single-cell product sources tends to be extremely variable. When eaten alone, the digestibility of algae is low, whereas mixing algae with other feeds improves digestibility. If the organisms are not killed prior to being used as a feed, digestibility is dramatically reduced. Certain forms of processing can improve the digestibility of some single-cell protein products. For example, if algae are cooked prior to use, digestibility can be doubled in many cases. However, processing beyond the killing of yeasts does little to improve digestibility.
- **Nucleic acid content**—Much of the nitrogen found in single-cell organisms is in the form of nucleic acids. When the purines of the nucleic acids are metabolized, uric acid is formed. In humans, uric acid is relatively insoluble with the result that uric acid deposits accumulate and lead to kidney stones or gout. Some of the current research is aimed at reducing the nucleic acid concentration.
- **Toxicities**—Toxicities arising from the use of single-cell protein can result from two sources: (1) toxins produced by the microorganisms themselves and (2) contaminated microorganisms. The second type of risk is probably the most likely to occur. Much of the single-cell protein will be derived from processes whereby by-products of industry are used as substrates. Thus, the microorganisms will in some ways reflect the chemical composition of the by-product. For example, if chemical residues, such as pesticides, or large amounts of trace minerals are present in the by-product substrate, the microorganism will, in all likelihood, absorb these chemicals. When livestock consume these contaminated microorganisms, toxicities may result. Hence, before widespread use can be made of single-cell proteins, there is the problem of convincing people and agencies of their wholesomeness.

- **Protein quality**—Research to date has indicated that SCP is deficient in the sulfur-containing amino acids, and possibly in lysine and isoleucine. Although the amino acid profile of SCP is more balanced than that of the cereal grains, it is clearly inferior to the traditional protein supplements. However, this inferiority can often be nullified by adding commercially available methionine, or by combining with other protein sources. Also, the advances in genetic engineering offer hope of developing new microorganisms capable of producing ideal amino acid profiles.
- **Economic problems**—As long as the more traditional sources of protein—such as the oilseed meals, meat, fish, eggs, and milk—are readily available at reasonable prices, single-cell protein will remain a relatively obscure alternative.

This is an era of concern for the maintenance of the quality of our environment. This means that greater emphasis will be placed on the transformation of industrial by-products into commodities that can be used by both livestock and humans. Single-cell protein is one way that this challenge can be met.

MIXED PROTEIN SUPPLEMENTS

With the knowledge of the protein sources available, protein supplements can be mixed that will fill the needs of various ages of swine and at favorable prices. Protein supplements containing added minerals and vitamins can be purchased commercially or mixed on the farm. In turn, these may be combined with different cereal grains to obtain a diet of the desired level of protein. Tables 10.3 and 10.4 give suggested formulas for 35 and 40 protein supplements, whereas Table 10.5 shows how these protein supplements (40 and 35%, respectively) may be combined with different cereal grains to obtain diets with varying levels of protein. (For specific uses of these diets, see Chapter 7.)

The ratios of protein, calcium, and phosphorus requirements change as the pig matures. Unfortunately, the ratios of these nutrients are fixed in a complete protein supplement, limiting their use to a narrow growth range. Footnote 1 in Tables 10.3 and 10.4 observes that these supplements are not balanced for young pigs (stages 1 to 5) because the calcium level in the diet becomes excessive and there are imbalances of several other nutrients as well.

This imbalance leads to the use of mineral–vitamin premixes to blend the grain and protein sources, as well as diets in which particularly the calcium and phosphorus are provided by limestone and dicalcium phosphate or other phosphorus source.

TABLE 10.3   35% Protein Supplements for Growing Pigs

| Ingredient | Protein | Protein Supplement, 2.25% Lysine[1] | | | |
		1	2	3	4
Soybean meal, solvent	44.0%	737.40	578.40	457.90	582.90
Soybean meal, dehulled	47.5%				
Corn distillers dried grain, solubles	27.0%			175.00	
Meat and bone meal	50.9%		150.00	145.00	157.50
Wheat middlings	16.5%	166.00	225.00	170.00	165.00
Fish meal, menhaden	61.2%				50.00
L-lysine·HCl				3.50	
Limestone		44.00	21.50	23.50	19.50
Dicalcium phosphate		27.50			
Salt		15.00	15.00	15.00	15.00
Trace mineral premix		4.50	4.50	4.50	4.50
Vitamin premix		5.60	5.60	5.60	5.60
Feed additives[2]		—	—	—	—
Total		1000.00	1000.00	1000.00	1000.00
Calculated analysis:					
Lysine, %		2.25	2.26	2.26	2.26
Threonine, %		1.35	1.35	1.27	1.34
Tryptophan, %		0.50	0.45	0.40	0.45
Methionine + cystine, %		0.94	0.95	0.91	0.94
Crude protein, %		35.18	36.80	35.41	36.39
Met. energy, kcal/lb		1309	1310	1304	1242
Calcium, %		2.50	2.50	2.51	2.50
Phosphorus, total, %		1.14	1.26	1.23	1.25
Phosphorus, available, %		0.75	0.75	0.75	0.75

[1]These supplements cannot be used for stages 1 to 5 because levels of calcium become excessive and other nutrients become imbalanced.
[2]Feed additives may be added at various levels as approved by the U.S. Food and Drug Administration.
Source: Adapted from ISU Life Cycle Swine Nutrition, 1996.

TABLE 10.4   40% Protein Supplements for Growing Pigs

| Ingredient | Protein | Protein Supplement, 2.55% Lysine[1] | | | |
		5	6	7	8
Soybean meal, solvent	44.0%				
Soybean meal, dehulled	47.5%	824.40	638.15	504.90	633.40
Corn distillers dried grain, solubles	27.0%			100.00	
Meat and bone meal	50.9%		182.50	197.50	200.00
Wheat middlings	16.5%	62.50	127.50	147.50	70.00
Fish meal, menhaden	61.2%				50.00
L-lysine·HCl				3.50	
Limestone		46.50	19.00	18.00	16.00
Dicalcium phosphate		38.00	4.25		2.00
Salt		17.50	17.50	17.50	17.50
Trace mineral premix		5.00	5.00	5.00	5.00
Vitamin premix		6.10	6.10	6.10	6.10
Feed additives[2]		—	—	—	—
Total, lb		1000.00	1000.00	1000.00	1000.00
Calculated analysis:					
Lysine, %		2.55	2.55	2.55	2.55
Threonine, %		1.60	1.58	1.45	1.56
Tryptophan, %		0.58	0.52	0.45	0.51
Methionine + cystine, %		1.19	1.16	1.08	1.15
Crude protein, %		40.19	41.71	39.52	41.42
Met. energy, kcal/lb		1356	1345	1326	1277
Calcium, %		2.80	2.80	2.80	2.80
Phosphorus, total, %		1.29	1.44	1.42	1.42
Phosphorus, available, %		0.87	0.87	0.87	0.87

[1]These supplements cannot be used for stages 1 to 5 because levels of calcium become excessive and other nutrients become imbalanced.
[2]Feed additives may be added at various levels as approved by the U.S. Food and Drug Administration.
Source: Adapted from ISU Life Cycle Swine Nutrition, 1996.

TABLE 10.5   Example Diets Using Supplements in Tables 10.3 and 10.4 [1]

Stage of Development [2] *Pig Wt for Moderate Lean*	6 *37–51 lb*		8 *69–91 lb*		10 *118–150 lb*		12 *188–233 lb*	
Corn	550.0	616.5	665.0	710.0	750.0	780.0	818.5	845.0
35% protein supplement	450.0		335.0		250.0		180.0	
40% protein supplement		383.5		290.0		220.0		153.0
Limestone							1.5	2.0
Total	1000.0	1000.0	1000.0	1000.0	1000.0	1000.0	1000.0	1000.0
Calculated analysis:								
Lysine, %	1.15	1.13	0.92	0.92	0.75	0.76	0.61	0.60
Threonine, %	0.77	0.78	0.66	0.68	0.59	0.60	0.52	0.53
Tryptophan, %	0.23	0.23	0.19	0.19	0.17	0.17	0.14	0.14
Methionine + cystine, %	0.63	0.66	0.57	0.60	0.53	0.55	0.49	0.50
Crude protein, %	20.24	20.09	17.11	17.14	14.80	14.94	12.88	12.79
Met. energy, kcal/lb	1403	1436	1437	1460	1462	1478	1480	1493
Calcium, %	1.14	1.09	0.86	0.83	0.65	0.64	0.53	0.53
Phosphorus, total, %	0.67	0.67	0.57	0.57	0.50	0.50	0.43	0.43
Phosphorus, available, %	0.36	0.36	0.28	0.28	0.22	0.22	0.17	0.17

[1]These supplements cannot be used for stages 1 to 5 because levels of calcium become excessive and other nutrients become imbalanced.
[2]See Chapter 7, Table 7.13 to determine stages of development.
Source: Adapted from ISU Life Cycle Swine Nutrition, 1996.

QUESTIONS FOR STUDY AND DISCUSSION

1. What parts of plants contain the highest concentration of protein?
2. What is "protein quality"?
3. Discuss the relative importance of plant and animal protein sources. Which sources have become increasingly more important in recent years? Why do you think this occurred?
4. Explain why pigs do not have a specific requirement for crude protein.
5. Name five alternate amino acid (protein) sources from (a) plants and (b) animals.
6. Discuss each of the following:
   a. Soybean meal as the sole source of supplemental protein in swine diets.
   b. An ideal protein.
   c. Limiting amino acid.
   d. Amino acid availability.
   e. Formulating swine diets on an available lysine basis.
   f. The use of crystalline amino acids in swine diets.
7. List the commonly used oilseed meals. Why are they called oilseed meals?
8. Discuss the feeding value of oilseeds in comparison to soybean meal.
9. Define *pulses*. List three types of pulses and indicate what precautions should be taken when pulses are to be used as either feed or food.
10. What is alfalfa meal, and what is its value in swine diets?
11. What are the following ingredients and how should they be used: full-fat soybeans and raw soybeans?
12. What are spray-dried porcine plasma, and spray-dried blood meal? How are these products being used?
13. What are the limitations of the use of dairy products?
14. What is whey and what nutritional value does it have?
15. List some common fish meals, and describe the feeding value of fish meals.
16. What are condensed fish solubles?
17. What is a single-cell protein? Is any currently used as animal feed?
18. Briefly discuss the advantages and disadvantages inherent with single-cell protein.
19. Describe 35 and 40% protein supplements and their use and limitations in formulating diets.

SELECTED REFERENCES

Animal Feeding and Nutrition, 9th ed., M. J. Jurgens, Kendall/Hunt Publishing Co., Dubuque, IA, 2002

Applied Animal Nutrition, Feeds and Feeding, 2nd ed., Peter R. Cheeke, Prentice Hall, Upper Saddle River, NJ, 1999

Association of American Feed Control Officials Official Publication, Association of American Feed Control Officials, Inc., Oxford, IN, 2003

Feeds and Nutrition, 2nd ed., M. E. Ensminger, J. E. Oldfield, and W. Heinemann, The Ensminger Publishing Company, Clovis, CA, 1990

Nontraditional Feed Sources for Use in Swine Production, P. A. Thacker and R. N. Kirkwood, Butterworth Publishers, Stoneham, MA, 1990

Pork Industry Handbook, Cooperative Extension Service, Purdue University, West Lafayette, IN, 2003

Swine Production and Nutrition, W. G. Pond and J. H. Maner, AVI Publishing Co., Inc., Westport, CT, 1984

11

Forages and Pasture Production

A-frames on alfalfa pasture. *(Courtesy, Palmer Holden, Iowa State University, Ames, IA)*

Objectives

After studying this chapter, you should:

1. Know where a pasture system fits into pork production.
2. Know the relative advantages of legumes, brassicas, and grasses.
3. Understand why the use of forages should be limited to older animals.
4. Be able to design a radial paddock system.
5. Know the dietary issues involved when using forages in swine diets.

A pig's digestive system does not lend itself to consuming great quantities of pasture or roughage like sheep and cattle. But a 400-lb sow can handle relatively large amounts. Good forage can also provide quality protein and certain vitamins and can reduce total feed requirements. Use of good pasture containing alfalfa, ladino clover, and grass can lower sow feed costs; help maintain high-level reproductive capacity of boars; and, in many cases, increase litter size born as compared to indoor production.

Pigs and pastures are popular in England and Wales, with 29% of the total sow herd in England and Wales kept outdoors in 1998 and 17.5% of the farrowings occurring outdoors.[1] Much of western Europe, with animal welfare laws stipulating access to bedding or outdoors, is increasing the numbers in outside production also. Although pasture production has the potential to be welfare and environmentally friendly, much of that potential lies with the stockman.

Outdoor production with access to forage is a minor part of pork production in the United States. Producers involved usually have hilly land that is best left as pasture to prevent erosion. Many outdoor producers are involved in niche marketing programs for their pork production as a means to recoup more income from a smaller production base.

Research reports on feed savings for pigs on pasture vary considerably, depending on the type of pasture, the age of hogs, and management systems. Data indicate this will amount to 3 to 10 percent of the grain and as much as one-third of the protein needed for growing and finishing hogs. Pastures are particularly beneficial for the breeding herd because they facilitate exercise in addition to providing nutrients.

Pastures can enhance a good swine sanitation and disease control program. Sunshine is an excellent disinfectant as are conditions that dry the soil. Providing more space reduces the incidence of pig-to-pig transmission of disease. After swine have grazed a pasture for 1 season, rotate the pasture to cattle or harvest hay from it for 2 years before using it again for hogs. This break from hog production will reduce the microbial contamination between groups of pigs. Many, but not all, swine diseases are not transmissible to cattle and vice versa.

DESIRABLE CHARACTERISTICS OF A GOOD PASTURE

Although it is recognized that no single forage excels all others in all the desired qualities, the fol-

lowing characteristics may serve as criteria in the choice of pasture crops for swine:

1. Adapted to local soil and climatic conditions.
2. Palatable and succulent.
3. Ability to endure tramping and grazing.
4. Easy to grow, and grown at a nominal cost.
5. Provide tender and succulent growth for a short period or consistent growth over a long period.
6. Highly nutritious; rich in proteins, vitamins, and minerals, and relatively low in fiber compared to other forages.
7. High carrying capacity.
8. Fit satisfactorily into the crop rotation.
9. Uncontaminated with diseases or parasites.

Throughout most of the United States, one or more adapted legumes possess most of these qualities. Fortunately, they can be selected without fear of bloat because swine—unlike cattle and sheep—are not susceptible to this ailment.

No one forage embodies all the desirable characteristics of a good swine pasture. None will grow the year round, or during extremely cold or dry weather. Further, each has a period of peak growth that must be conserved for periods of low growth. Consequently, the progressive swine producer will find it desirable (1) to grow more than one species; (2) to plan for each month of the year; and (3) to secure year-round grazing, or as early spring and late fall grazing as is possible in the particular area. In general, a combination of permanent and temporary pastures will best achieve these ends.

Selected forages should be succulent and capable of high production, palatable, high in protein and vitamins, and produce over a reasonably long growth period. In general, temporary pastures are preferable to permanent pastures for swine, especially from the standpoint of disease and parasite control.

TYPES AND COMPOSITION OF PASTURES

Forage Analysis

Table 11.1 lists the average nutrient composition for various forages. Forage analysis should form the basis for diet formulation whenever practical. Crude protein, calcium, and phosphorus can be assayed at most analytical laboratories at a relatively low cost. In addition, most laboratories offer neutral detergent fiber (NDF) and acid detergent fiber (ADF) analyses.

NDF contains the fibrous components of the cell wall, primarily cellulose, hemicellulose, and lignin. None of the lignin is digestible and only 30

---

[1]Farm Animal Welfare Committee (*http://www.fawc.org.uk/reports/pigs/fawcptoc.htm*)

TABLE 11.1 Nutrient Composition of Some Forage Crops (dry matter basis)[1]

Forage Crop	IFN (International Feed Number)	Dry Matter (%)	Metabolizable Energy (kcal/kg)	Crude Protein (%)	Ether Extract (%)	Crude Fiber (%)	NDF (%)	ADF (%)	Calcium (%)	Phosphorus (%)
Alfalfa, fresh		23.4	2,224	18.9	3.15	26.5	47.1	36.8	1.29	0.26
Alfalfa, fresh, full bloom	2-00-188	23.8	1,181	19.3	2.6	30.4	38.6	35.9	1.19	0.26
Alfalfa hay, sun cured, early bloom	1-00-059	90.5	2,170	19.9	2.9	28.5	39.3	31.9	1.63	0.21
Alfalfa hay, sun cured, full bloom	1-00-068	90.9	1,990	17.0	3.4	30.1	48.8	38.7	1.19	0.24
Alfalfa meal, dehydrated	1-00-022	90.4	2,130	17.3	2.4	29.0	55.4	37.5	1.38	0.25
Alfalfa silage	3-00-216	44.1	2,280	19.5	3.7	25.4	47.5	37.5	1.32	0.31
Barley silage		37.1	2,170	11.9	2.9	—	56.8	33.9	0.52	0.29
Bluegrass, Kentucky, fresh, early vegetative	2-00-777	30.8	2,600	17.4	3.5	25.2	55	29	0.50	0.44
Brome, smooth, early vegetative	2-00-956	26.1	2,168	21.3	4.0	23.0	47.9	31.0	0.55	0.45
Brome, smooth, hay, sun-cured, midbloom	1-05-633	87.6	2,030	14.4	2.2	31.9	57.7	36.8	0.29	0.28
Clover, ladino, early vegetative	2-01-380	19.3	2,460	25.8	4.6	13.9	35	33	1.27	0.35
Clover, ladino, hay, sun cured	1-01-378	89.1	2,170	22.4	2.7	20.8	36.0	32.0	1.45	0.33
Clover, red, fresh, early bloom	2-01-428	19.6	2,490	20.8	5.0	23.2	40.0	31.0	2.26	0.38
Clover, red, hay, sun cured	1-01-415	88.4	1,990	15.0	2.8	30.7	46.9	36.0	1.38	0.24
Corn, dent yellow, silage, well eared	3-28-250	34.6	2,600	8.65	3.09	19.5	46.0	26.6	0.25	0.22
Oats, hay, sun cured	1-03-280	90.7	1,910	9.5	2.4	32.0	63.0	38.4	0.32	0.25
Oats, silage	3-03-296	36.4	2,130	12.7	3.12	31.8	58.1	38.6	0.04	0.05
Orchard grass, fresh, midbloom	2-03-443	27.4	2,060	10.1	3.5	33.5	57.6	35.6	0.23	0.17
Rape (canola), fresh, early bloom		11	1,232	23.5	3.8	—	—	—	—	—
Rye grass, fresh	2-04-073	22.6	3,040	17.9	4.1	20.9	61.0	38.0	0.65	0.41
Rye grass, perennial, hay, sun cured		86	986	8.6	2.2	—	—	—	0.65	0.32
Sorghum, silage	3-04-323	30	2,170	9.39	2.64	26.9	60.9	38.8	0.49	0.22
Sorghum, Sudan grass, fresh, midbloom		23	1,036	8.8	1.8		25		0.43	0.36
Sweet clover, yellow, hay, sun cured		87	886	15.7	2.0		—	34	1.27	0.25
Timothy, fresh, late vegetative	2-04-903	26.7	2,390	12.2	3.8	32.1	55.7	29.0	0.40	0.26
Timothy hay, sun cured, early bloom	1-04-882	89.1	2,130	10.8	2.8	33.6	61.4	35.2	0.51	0.29
Timothy hay, sun cured, full bloom	1-04-884	89.4	2,030	8.1	2.9	35.2	64.2	37.5	0.43	0.20
Trefoil, birdsfoot, fresh	2-20-786	19.3	2,390	20.6	4.0	21.2	46.7	—	1.74	0.26
Trefoil, birdsfoot, hay, sun cured	1-05-044	90.6	2,130	15.9	2.1	32.3	47.5	36.0	1.70	0.23
Wheat, fresh, early vegetative	2-05-176	22.2	2,640	27.4	4.4	17.4	46.2	28.4	0.15	0.13
Wheat, hay, sun cured	1-05-172	88.7	2,100	8.7	2.2	29.0	68.0	41.0	0.15	0.20
Wheat, silage	3-05-184	34.2	2,060	12.5	6.09	26.8	60.7	39.2	0.44	0.29

[1]Metabolizable energy values shown are for beef because comparable values are unavailable for swine. Similarly, amino acid levels are unknown for most forages.
Source: From Pork Industry Handbook, PIH-126, 1998, Purdue University Cooperative Extension Service, West Lafayette, IN; and Nutrient Requirements of Beef Cattle, National Academy Press, Washington, DC, 1996.

TABLE 11.2   Effect of Maturity on the Quality of Sun-Cured Alfalfa Hay[1]

Stage	Dry Matter (%)	Crude Protein (%)	Fiber (%)	NDF (%)[2]	ADF (%)[2]
Early bloom	90.5	19.9	28.5	39.3	31.9
Midbloom	91.0	18.7	28.0	47.1	36.7
Full bloom	90.9	17.0	30.1	48.8	38.7

[1] Values of components are on a dry matter basis.
[2] NDF = neutral detergent fiber, ADF = acid detergent fiber
Source: Nutrient Requirements of Beef Cattle, *National Academy Press, Washington, DC, 1996.*

to 40% of the cellulose and hemicellulose is digestible. As levels of fiber increase, metabolizable energy and crude protein concentrations in the forage will likely be lower. ADF is a subportion of NDF and contains the cellulose and lignin. Low levels of both are desirable for pigs.

Unfortunately, little data is available on the amino acid content of forages. Because forages are generally fed to ruminants, which do not require amino acid information for feed formulation, the data in Table 11.1 allows very minimal inputs for diet formulation. Great care must be exercised when estimating the contribution of most forages to swine diets.

The quality of the forages decreases rapidly as they mature. Table 11.2 illustrates the declining percentage of crude protein and the increased concentration of fiber, NDF, and ADF from alfalfa hay harvested at increasing stages of maturity.

Legumes

Legumes are plants that have the ability to work symbiotically with bacteria to fix nitrogen from the air. The most commonly used legumes for swine are alfalfa, alsike clover, birdsfoot trefoil, crimson clover, ladino clover, lespedeza, red clover, soybean forage, sweet clover, and white clover (Figure 11.1).

As a group legumes have a higher protein, calcium, and carotene content than grasses. They can furnish an adequate supply of most vitamins with the exception of vitamins D and B_{12}. The following types of pastures should be considered.

• **Alfalfa**—Most of the forage research in swine diets has been with alfalfa. It appears to be the most practical forage crop for the pig because it can be used for pasture, silage, and as an ingredient in the diet. Because of their maturity, sows better utilize alfalfa (and other forages) than do growing-finishing hogs. Potential benefits to accrue from feeding alfalfa during gestation include improved survival of the baby pigs during the nursing period and a reduced culling rate in the sow herd.

Some research has shown that alfalfa hay or silage can compose up to 97% of gestation diets

Figure 11.1 Gestating sows on a Texas radial pasture. *(Courtesy, John McGlone, Texas Tech University, Lubbock, TX)*

without impairing reproductive performance, often with only calcium, phosphorus, salt, and some vitamin and trace mineral additions required. Because alfalfa is low in energy compared to grains, more feed must be fed to provide 5,000 to 6,000 kcal of metabolizable energy required per day for a pregnant sow. For commercial operations, however, no more than 50% alfalfa should be used in gestation diets. Dehydrated alfalfa is an expensive addition to swine diets.

Growing pigs have performance comparable to grain–soybean meal diets if the alfalfa provided does not exceed 5% of the diet. At higher levels there will be a depression in feed intake, feed efficiency, and growth rate compared to a grain–soybean meal diet.

• **Alsike clover**—Alsike clover provides a leafy crop with fine stems and grows well in soils that are too acidic or too wet for red clover. It is often used in pasture mixtures.
• **Birdsfoot trefoil**—Birdsfoot trefoil is palatable and similar in nutrient content to alfalfa. Unlike alfalfa, it grows well on poorly drained soils. Although it is not as productive as alfalfa on good soils, birdsfoot trefoil yields have exceeded yields of alfalfa on the wetter soils. Most varieties perform better in cooler climates. It will normally outlive red clover by several years.

- **Crimson clover**—Crimson clover provides a good spring forage and sometimes winter forage in warm climates.
- **Ladino clover**—Under optimum conditions, ladino clover will not produce as much forage per acre as alfalfa. But the protein content of ladino clover is superior to that of alfalfa and the fiber content is lower. Ladino clover works best as an all-summer pasture crop in the northeastern and north central United States or areas of similar climate.
- **Lespedeza**—Also called *Japanese clover*, lespedeza is less palatable to pigs than all clovers except sweet clover. It cannot be grazed until midsummer, but it does grow reasonably well without lime and fertilizer. It will adapt to soils that cannot be used for red clover.
- **Red clover**—Red clover is a short-lived, relatively easy to establish perennial legume that will grow on soils too acidic or too wet for alfalfa. Red clover does not yield as much forage early in the spring as alfalfa and it is not as drought resistant as alfalfa. It is useful for pasture or silage. It will provide good forage through most of the grazing season if it is not overgrazed or allowed to become too mature. Several studies have shown that pigs on red clover forage gain as rapidly as those on alfalfa.
- **Soybean forage**—Soybeans as a green forage are less valuable than alfalfa and red and ladino clovers. The soybeans should be grown in rows to reduce damage from trampling. Unlike most forages, soybeans cannot regenerate new growth from the crowns. Soybeans are less sensitive to nutrient levels in the soil than are alfalfa and clovers. In hot climates, soybeans may outyield other legumes during the same period of time.

Pigs grazing mature soybeans should also have access to grain fortified with vitamins and minerals. However, inhibitors in the raw soybeans will prevent the pig from using the dietary protein efficiently. In addition, the oil contained in the beans tends to make the carcass soft.

- **Sweet clover**—Because pigs find sweet clover unpalatable, it may be more suitable for soil improvement. Sweet clover may be planted on soils not adapted for alfalfa or other clovers. If biennial sweet clover is sown in the spring, the first season's growth is more succulent and palatable than that harvested during the second summer.
- **White clover**—White clover is a practical perennial legume to use with permanent pasture, especially those containing bluegrass. White clover makes a high-quality pasture and it does well in years of frequent rain. Note that ladino clover is a large-type white clover. Dutch or common white clover is a small type.

Brassicas

Brassicas belong to the mustard family and include many vegetables such as broccoli, cabbage, and turnips. However only canola (rape) is of significant importance to swine feeding.

- **Canola (rape)**—Rape is a high-yielding, fast-growing annual forage that belongs to the *Brassica* family. Related species include kale and swede. Canola provides an excellent pasture for swine. When overgrazing is avoided, it provides abundant, palatable forage for a long growing season. Canola can lead to photosensitization (sun burning) when grazed wet. Pigs with white skin are most sensitive.

Grasses

Grasses, compared to legumes and canola, are of much lower value as a swine feed, when grass is young and lush, the protein content is acceptable but the moisture content is very high. As grass matures its nutritive value diminishes rapidly for pigs.

- **Bluegrass**—Bluegrass may serve as a permanent pasture for swine. The pasture can be grazed early, but it contains less protein than do legumes and is usually dormant during the warmest part of the summer.
- **Smooth brome grass**—Brome grass is a palatable crop that withstands heavy grazing. Its early spring growth enables it to be pastured for longer periods than many legumes. Studies show that pigs on brome grass pastures require more grain and supplement than pigs grazing alfalfa. Brome grass can be successfully mixed with legumes.
- **Orchard grass**—Orchard grass, a perennial, is a hardy species that can tolerate trampling. It quickly loses its palatability if not grazed down or clipped to prevent the grass from becoming tall and mature.
- **Sudan grass**—Sudan grass, an annual, is palatable to pigs, and, when seeded thickly, provides ample forage during the hottest part of the summer when other species are dormant. The early growth of Sudan grass contains a cyanogen, which may be converted to prussic acid (extremely toxic to pigs and ruminants) under certain conditions, such as wilting, trampling, chewing, frost, and drought. Because of the near-neutral pH in the rumen, ruminants are more sensitive to cyanogens than are nonruminants. Poisoning can be avoided if the grass is grazed only after it reaches a height of at least 18 to 24 in. Because Sudan grass is low in protein, it is better adapted for sows and older market hogs.
- **Timothy**—Timothy withstands heavy use, but it should be included only as a minor part of a pasture

mixture because it is less desirable than most other pasture crops.

- **Winter rye**—Winter rye seeded during late summer will provide a useful forage crop for winter or early spring grazing. Optimal planting time should provide just enough growth so that seed stems are starting to shoot when the plant enters winter dormancy. When pigs are allowed to graze rye during the winter and spring months, stock the pasture with no more than 8 growing pigs, or 3 to 5 sows, per acre.
- **Winter wheat and barley**—Winter wheat and barley are two cereal grains that are at least as palatable and nutritious as rye, but they do not provide as much fall production as rye and they cannot be grazed as heavily. Note that fresh wheat forage contains significantly more crude protein than is contained in barley forage.

MANAGEMENT OF SWINE AND PASTURES

Good management of both the pastures and the pastured animals go hand in hand; they are inseparable. Both require attention in order to get the highest returns. In brief, good producers, good pastures, and good hogs go together.

Research reports on feed savings on pasture vary considerably, depending on type of pasture, age of hogs, and management systems. Data indicate savings for growing and finishing hogs that amounts to 3 to 10 percent of the grain and as much as one-third of the protein. Pastures are exceptionally beneficial for the breeding herd because they facilitate exercise in addition to providing nutrients. Practical operators attend to the following swine pasture management factors:

1. **Maintaining pasture quality.**
   a. **Provide young, succulent pastures.** As pasture crops mature, they become less palatable and much less nutritious because of higher-fiber content and lower amounts of proteins and vitamins. Clipping at intervals during the growing season is the best method of keeping swine pastures young and succulent. Clipping is also effective in both weed and parasite control (the latter because of exposing the feces to drying). As a rule of thumb, grazing should be regulated to allow the forage to maintain a height of 3 to 6 ins.
   b. **Prevent overgrazing.** Avoid continuous overgrazing. It would be beneficial to have sufficient pastures to allow one to two crops of hay to be harvested each season in addition to the pasture.

TABLE 11.3   Recommended Pasture Stocking Rates[1]

	Per Acre	Per Hectare
Gestating sows	8–12	20–30
Sows with litters	6–8	15–20
Pigs from weaning to 100 lb	15–30	37–75
Pigs from 100 lb to market	10–20	25–50

[1] These recommendations assume the use of good quality legume pasture under conditions of adequate moisture. Stocking rates depend on soil fertility, quality of pasture, and time of year.
Source: University of Missouri, 1993, Ag. Publ. G2360.

Figure 11.2 Hog ringing pliers and hog rings. Usually at least 2 rings are placed in the cartilage of the nose. *(Courtesy, Palmer Holden, Iowa State University, Ames, IA)*

   c. **Stocking rates** are critical to a successful pasture system if it is to be a significant source of nutrition for the pigs grazing it. Suggested stocking rates (Table 11.3) are highly dependent on soil fertility, quality of pasture, time of year, and the pigs' access to feed.
   d. **Ringing.** Hogs rooting the ground can easily destroy the vegetation. When rooting becomes destructive, the animals should be rung. Any one of several types of rings may be used with success if properly inserted in the snout.

   By instinct most hogs do some rooting and it is likely to be especially damaging to pastures. When rooting starts, the herd should be "ringed"; this applies to all hogs past weaning age. Older animals can be restrained by a rope or snare placed around the snout, whereas young pigs can be held.

   Many types of rings can be and are used, but the triangle or the fishhook types are most common (Figure 11.2). Rings (usually

Figure 11.3 A grow-finish pig with 2 rings in its nose. Pigs on pasture usually require rings in their noses to prevent destruction of the forage. *(Courtesy,* Pork Industry Handbook, *Purdue University, West Lafayette, IN)*

1 to 3 rings) are placed in the snout, just back of the cartilage but away from the bone; although some producers prefer to use a ring that is placed through the septum (the partition of the nose, Figure 11.3).

    e. Ponding. Pastures should be protected during periods of heavy rainfall or when irrigating by removing swine until the soil has dried sufficiently to prevent formation of ponds of water. Pigs will rapidly destroy vegetation in ponded areas.

2. Fencing.

    a. Permanent pastures. Many producers install permanent perimeter fences but use temporary divider fences along with low-voltage electricity. Permanent fences for small areas and high concentrations of hogs should be of woven wire 32 to 36 in. high with 6 in. mesh. With larger areas (10 acres or more) and less concentration of hogs, a 26-in. woven wire is satisfactory. All permanent hog fences should include one strand of barbed wire stretched tightly between the woven wire and the ground.

    b. Electric fences. The introduction of electric fences has allowed more flexibility in pasture designs. A more complete description of fences for swine can be found in Chapter 15.

3. Grazing. Early spring and late fall grazing are desirable for swine but detrimental for pastures. The forages need time in the spring to develop adequately for the pigs to prevent overgrazing. In the fall, the plants need protection to build up root reserves for the winter to prevent winter kill. In recognition of this situation, a compromise is usually necessary—pasturing early and late, but not enough to do too great harm to the pastures. Where plenty of pasture areas are available, pasture management in this regard is usually achieved through the rotation of pastures.

    a. Rotating pastures. Modern swine producers realize the importance of rotating pastures in order to control parasites and reduce the disease hazard. Pigs should never be allowed to follow pigs on a particular pasture without plowing the field in the interim. In pastures with heavy parasite infestations, it is recommended that swine be kept off permanent pastures for 2 or 3 years. Also, in rotating pastures, avoid cross-drainage from one field to another.

    b. Year-round grazing. In the temperate and tropical climates, year-round grazing is a reality for many swine farms. By careful planning, other areas can approach this desired goal. Figure 11.4 illustrates the growth period of each of the common swine pasture plants of the Corn Belt and North Central states. As noted, by selecting the proper combination of crops, swine pastures for each month of the year are assured. From this standpoint, the more temperate climates have a very real advantage. Similar graphs for different regions can be developed.

    Some Corn Belt swine producers obtain a 12-month hog pasture by using 2 crops only, ladino clover and rye. In the North, where year-round grazing cannot be secured, it is merely recommended that swine producers attain as long a grazing season as is possible, especially through arranging for early spring and late fall pastures.

4. Environmental needs.

    a. Clean, fresh water. Pasture swine need access to clean, fresh water at all times. Under a system of pasture rotation, it is often difficult and expensive to pipe water to the area. But usually the added expense will be justified. Where possible, automatic waterers should be provided. Hand watering on pasture is inconvenient, labor consuming, and otherwise unsatisfactory. Plastic water lines can be laid aboveground to provide flexibility in pasture systems as long as temperatures remain above freezing.

    b. Shelter. Swine on pastures should be provided with proper shelter, including both huts and shades. For flexibility in rotating pastures, portable equipment is preferable. Sunburn can be a problem with white-skinned hogs.

APPROXIMATE GRAZING PERIOD OF COMMON PASTURE CROPS FOR SWINE IN THE CORN BELT AND NORTH CENTRAL STATES

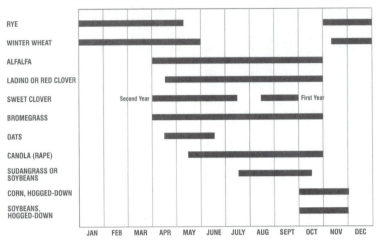

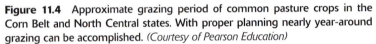

Figure 11.4 Approximate grazing period of common pasture crops in the Corn Belt and North Central states. With proper planning nearly year-around grazing can be accomplished. *(Courtesy of Pearson Education)*

Figure 11.5 Hog wallow for cooling pigs on pasture. *(Courtesy, John McGlone, Texas Tech University, Lubbock, TX)*

c. Pasture cooling. Pasture cooling can be accomplished several ways. Adequate shade over sand that is wet down several times a day on hot days can be used. A fogging nozzle over sand or concrete is more desirable because it eliminates the destruction of forage and is less time consuming. However, the best method of providing summer cooling on pasture is a permanent structure that provides shade, fogging, and a concrete wallow (Figure 11.5). Heat-stressed sows may farrow up to 2 less live pigs and wean about 1 less pig.

5. Limit-feeding sows. Gestating sows must be limit-fed. This can accomplished by hand feeding a quantity of diet each day or by interval feeding. Interval-fed sows are allowed access to a self-feeder once every 2 or 3 days. Adjustments in daily feed intake are made by altering either the time on the feeder (2 to 12 hours) or days between access to the feeder (2 or 3 days). Interval feeding is not recommended for bred gilts because they cannot consume sufficient feed every other or third day. As a result, they will gain less weight and farrow smaller pigs.

Managing sows' conditions on pasture is difficult because the quantity and quality of forage consumed cannot be predicted easily. Monthly body condition scoring followed by adjusting the amount of feed provided will accomplish both optimization of forage consumption and body condition. See the discussion on condition scoring in Chapter 15.

6. Quantity of protein. In arriving at the quantity of protein supplement to feed swine on pasture, consideration should be given to the following:

   a. Pigs full-fed on a good legume pasture require about 50% as much protein supplement as when full-fed in drylot. Pigs limit-fed on good pasture will eat more forage and require only 30 to 40% as much protein supplement as drylot-fed pigs.

   b. Less protein supplement is required for older hogs than for young pigs because mature animals consume more pasture and have lower protein needs—finally reaching the stage where pastures may supply ample proteins to balance their needs. As a result a mineral–vitamin–salt source may need to be provided.

   c. Less protein supplement is required with pastures that are higher in protein content and more palatable.

   d. Less protein supplement will require more pasture acreage because of higher forage consumption per animal and will result in more rooting, as the pigs attempt to meet their requirements from the soil.

7. Minerals. The cereal grains and their by-products as well as most protein-rich supplements are high in total phosphorus, whereas legume and

Figure 11.6 This is not a pasture, it is a dirt lot with much runoff potential. *(Courtesy, Palmer Holden, Iowa State University, Ames, IA)*

canola pastures are a fair source of calcium. Other needed minerals, including iodized salt in iodine-deficient areas, should be incorporated in the diet. Free-choice feeding a mineral–vitamin–salt mix is a safe method to ensure those needs are met. This mix should not contain any energy or protein sources because it will be too palatable and pigs will overconsume an expensive premix.

8. Combination grazing with other classes of animals. Different species of animals should not be commingled on the same pastures within the same grazing season because of the potential spread of diseases.

9. Scattering droppings and handling manure. Proper scattering of the droppings on pasture spreads the fertility value of the manure. It also reduces the pig's disease and parasite exposure by exposing the droppings to the sun and desiccation. The droppings may be scattered easily by harrowing the area at intervals, especially in the early spring and late fall. For best parasite and disease control, swine manure should never be hauled from swine barns or lots and scattered on pastures used by this particular class of livestock.

10. Pollution. Little pollution potential exists from pasture systems with low animal densities or numbers, or where pastures are rotated. However, in high-density pasture systems involving a large number of hogs, a pollution potential does exist. Frequently, there is little vegetation in these pasture lots, and rainfall runoff carries some solids with it. Where the lots are steeply sloping or where a watercourse runs adjacent to or through the lots, the problem can be serious.

Where a pollution problem exists from pasture lots, the site should be abandoned as a hog pasture, or all runoff should be caught in a retention reservoir (Figure 11.6). After each rain, liquid must be removed from the reservoir by pumping or evaporation to ensure sufficient volume to capture runoff from the next rain.

ADVANTAGES OF PASTURE SYSTEMS

Many producers believe that pasture production has certain real advantages over indoor production. Pasture production requires neither gestation stalls nor farrowing crates. Whether this is an advantage or disadvantage depends on the producer. In general, the following advantages may be cited in favor of pasture production over drylot or indoor production of swine:

1. Lower feed costs. Pastures provide savings in feed costs in both grain and protein supplements. With properly balanced diets used in each case, the feed saving effected through the use of swine pastures is about as follows:

   a. Quality pastures will reduce (1) the grain required in producing pork by 15 to 20% and (2) the protein supplement required in producing pork by 20 to 50%.

   b. With mature brood sows, good pastures may lower feed costs by 50%. With pastures that possess a heavy legume content, sows can be fed 2 lb less grain and 0.5 lb less protein supplement per day during gestation. In fact, sows may get the major portion of their feed from good pastures up to 6 to 8 weeks before farrowing. The condition of sows is the best guide as to the amount of concentrate feeding necessary.

   c. When growing-finishing hogs are on high-quality legume pasture, the lysine level of the diet can be reduced by about 0.15% (reduce protein about 2%) as compared to drylot or indoor production. When finishing pigs are on a limited feeding program, only one-half to three-fourths as many animals per acre can be carried as when pigs are full-fed because of their larger consumption of forage.

2. Reduces risk of nutritional deficiencies. High-quality pastures reduce the risk of nutritional deficiencies because of (a) their high-quality proteins, (b) the abundance of carotene and water-soluble vitamins in the forage plus the vitamin D value of the sunlight to which animals on pasture are exposed, and (c) their minerals (especially calcium). Even though these nutrients are present, their quantity and the amount consumed by the pigs is not known. In addition, the soil itself contains many trace minerals that can be used by the pigs.

Even though pasture-raised pigs will have adequate carotene (vitamin A), many of the B vitamins, and vitamin D, purchasing a specifically formulated supplement without these vitamins may cost more than purchasing a generic supplement. Therefore, it is likely more economical to purchase a supplement that meets the requirements for all vitamins and trace minerals. Access to salt is essential on pastures because none of the forages consumed provide adequate sodium and chlorine.

3. Better isolation and disease control. Hogs on pasture come in contact with each other less than hogs in drylot or indoor production, with the result that fewer communicable disease problems are spread than where hogs are in close confinement. For example, it is easier to control the spread of diseases such as TGE (transmissable gastroenteritis) or other types of baby pig scours when pigs are farrowed in individual houses rather than in a central farrowing house.

4. Lower capital investment per production unit. Less expensive buildings (including movable houses) and equipment can be used in a pasture system than in indoor production, with the result of less capital investment on a per hog basis.

5. Greater flexibility. Pasture operations are more flexible than indoor production programs— an important consideration where renters are involved or where other uncertainties exist about a long-range program.

6. Lower levels of skill and management. Pasture rearing does not require as high level skill and management as are necessary to make indoor production work. Also, more reliable labor is required to operate a fully automated, highly mechanized, complex than a pasture operation.

7. Good use of land not suitable for cropping. A pasture system may be the best use of rolling land and well-maintained pastures reduce erosion.

8. Decreased manure management problems. When animals are on pasture, 60 to 75% of the plant nutrients are returned directly to the soil, and pigs will spread their own manure. Manure from indoor production is also returned to the soil but requires labor and there are some losses from lagoon or manure pit storage.

9. Exercise and nutrients provided for breeding sows, improving reproduction. More satisfactory litters are farrowed on pasture and sows provide a more abundant milk flow. Also, boars on pasture are more vigorous and surer breeders. Finishing pigs may develop muscles differently because of the added exercise.

10. Decreased cannibalism. Pigs on pasture have more space and a more enriched environment to provide diversions, making tail or ear biting a negligible problem.

DISADVANTAGES OF PASTURE SYSTEMS

Disadvantages sometimes attributed to pasture systems include the following:

1. More labor required for handling, feeding, and watering. Raising hogs on pasture does not lend itself to automation and labor-saving devices to the extent that indoor production does, primarily in feeding and watering.

2. More labor at farrowing. Sows need to be encouraged to selected individual huts. Postfarrowing litter treatments, such as ear notching, iron shots, and castration, become more difficult.

3. Greater problems with internal parasites. Concrete floors can be washed and disinfected, whereas the eggs of many internal parasites can survive indefinitely in the soil. An aggressive parasite management program is essential, including pasture rotation and animal treatments.

4. Stockmen encounter discomfort and inconvenience during inclement weather. Additional bedding must be provided under wet or cold conditions.

5. Lower rate of gain. The added exercise may stimulate appetite, but growing-finishing hogs usually have lower rates of gain because of the bulkiness and lower-energy density of the forage. Additionally, during extremely hot or cold conditions lasting several days, feed intake likely will be reduced.

6. Possible decrease of cropland. Farm operators can make more money growing corn, soybeans, and other crops on highly tillable land than they can from pasturing hogs.

7. Slightly longer time for hogs to reach similar market weights. Reduced feed intake and more physical activity contributes to hogs taking longer to reach market weight on pasture.

8. Less facilitation of manure handling. Although less manure has to be handled in a pasture system than indoor production, it is more difficult to automate and handle manure where hog shelters are scattered over a large area. Much of the nitrogen content of the manure is volatilized and lost.

9. Prevention of enlarging hog production without enlarging the farm. With high-priced land, this fact must be evaluated when it is desired to increase the size of the hog operation.

10. Lack of environmental control in extreme weather. In many areas the weather is too adverse for pasture facilities to be used during the winter months. Furthermore, there may be more death loss of young pigs during cool or cold weather.

SILAGE AND HAYLAGE

Good-quality silage or haylage is an excellent feed for brood sows; however, unless the sow herd is very large, it will probably not pay to construct a silo especially for hogs. Silage or haylage works best when it is harvested for a beef or dairy operation that can use large quantities of it per day. Usually about 4 in. of surface silage must be removed from a silo or bunker daily to keep it from spoiling. The most common silages are either corn silage or alfalfa haylage. Silage should be fed fresh each day in amounts the sows will clean up in 8 or 10 hours.

A full-feed of corn silage that has a lot of corn kernels in it may meet the gestating sow's energy needs, but she will need 1.0 to 1.5 lb/day of a supplemental protein containing salt, minerals, and vitamins. Sows will usually eat 10 to 15 lb, and bred gilts will usually eat 8 to 12 lb of silage per day. Some wastage can be expected.

A full-feed of alfalfa haylage will meet the protein and amino acid needs of gestating sows, but additional grain may be needed to meet their energy needs as well as supplemental phosphorus, salt, trace minerals, and vitamins. When feeding legume haylage, offer all the haylage sows will clean up, usually 6 to 8 lb per day. Because of the bulkiness, neither is very acceptable for gilts without a large quantity of added grain or supplement. When feeding silage to hogs, the following precautions should be observed:

1. Do not feed silage to growing pigs. It has too much bulk to support good gains.
2. Never feed corn silage without protein.
3. Do not feed any silage to lactating sows because the bulk will restrict the intake of sufficient calories to produce milk.
4. Avoid feeding moldy silage. It can cause sows to abort.

DRY FORAGES

Alfalfa is practically the only dry forage fed to swine in the United States. If its price is right—if it is a cheaper source of lysine and net energy than corn and other grains—up to 50% (or even higher levels) alfalfa meal may, to advantage, be incorporated in gestating sow diets and up to 5% may be used in growing-finishing diets.

The favorable results obtained from feeding a corn–soybean meal diet, properly fortified with minerals and vitamins, causes many nutritionists and pork producers to question the wisdom of using alfalfa meal in the diet. There is no doubt about alfalfa meal being an excellent source of quality protein, carotene (vitamin A), B vitamins (riboflavin, pantothenic acid, and niacin), vitamin D if sun cured, calcium, and other nutrients. But these nutrients often can be purchased more cheaply from other sources. Its high fiber (20 to 30%), low-energy density, and low palatability, make it unsuited in diets for nursery pigs or lactating sows.

GESTATION AND OUTDOOR FARROWING

There are two main types of paddock layouts. The outdoor system of the 1990s, successfully pioneered in Britain and being used now in the United States, is very different from earlier rectangular systems used in the United States. The pasture plots of the modern system being promulgated in the 1990s are of radial design, as opposed to the more conventional rectangular pastures. Normally, paddocks will be allocated for the various parts of the herd, including gestation, farrowing, weaned sows and breeding, and growing-finishing, or some combination of these stages.

In the conventional system paddocks are normally rectangular in blocks of four, separated by roadways. They can be laid out to take into account the natural shape and size of fields.

In the radial system (Figure 11.7), wedge-shaped paddocks are laid out within a large circle and gated into a small handling and access area at

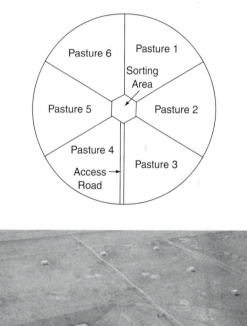

Figure 11.7 The paddocks radiate out from a center hub used as a sorting area. Note the front center paddock contained 7 gestating sows per acre. The pasture to the left had 14 sows per acre. Pastures with 14 sows per acre have lost considerable ground cover compared with pastures with 7 sows per acre. Areas in the paddock that lose ground cover include the feeding area, the hut area, the wallow area, and paths that pigs walk. The wallow area loses all of the ground cover, but has less nutrient build up than in the feeding area. Sows apparently urinate and defecate in and near the wallow but not too frequently compared with other pen regions (such as the feeding area). *(Courtesy, John McGlone, Texas Tech University, Lubbock, TX)*

the hub or center. There is normally a wide track around the perimeter fence and access to the center which allow servicing of the paddocks. Most operations involving pig movement and handling can be carried out single-handed in this system.

The modern outdoor farrowing system can be established and maintained with a minimum of capital—perhaps less than half the per sow cost of a modern confinement facility. Typically, an outdoor sow unit consists of the following:

1. **Hub area.** The hub area provides (a) access to each pen and (b) boar pens.
2. **Gestation pens.** Each gestation pen is equipped with a metal hut or A-frame (wooden), shade, self-feeder, and water.
3. **Farrowing pens.** Each farrowing pen is equipped with a steel or wood hut for each sow and litter, shade, self-feeder, creep-feeder, and water.
4. **Growing-finishing area.** Growing and finishing hogs can also be fed in the paddock system, either as part of the sow radial or in a separate radial that has paddocks to accommodate different groups of weaned pigs. Provide shade, self-feeders, and waterers.

Geographic location is important if it is desired to farrow every month of the year. Climates with severe winters are not conducive to pasture production. Pigs may be weaned at less than 21 days of age but often pasture-farrowed litters are weaned later, depending on the type of nursery they will go to after weaning. A climate that ranges from 25 to 85°F is ideal for outdoor production (Figure 11.8).

• **Feed savings on pasture**—Research reports on feed savings on pasture vary considerably, depending on type of pasture, class and age of hogs, and management system. On the average, however, bred sows and gilts on good legume pasture require about half as much grain and little or no supplemental protein compared to those in drylots, and good pasture for growing-finishing pigs will effect a savings of 3 to 10% of the grain and of as much as 33% of the protein supplement when compared to drylot or indoor production. Thus, the decision on whether to raise swine on pasture or in drylot should be based primarily on (1) net returns and (2) whether the land can be put to a more profitable alternative use.

DIETS FOR GESTATING SOWS ON PASTURE

Suggested diets for gestating sows on high-forage or silage diets are presented in Table 11.4. The calculated compositions of the diets are only for the supplemental concentrate diets, not the entire diet including the forage. Note, the concentrate supplements for legume-based diets, including legume and legume–grass pastures, legume silage, and rape or canola, do not contain any soybean meal because the protein content of these forages should meet the protein or amino acid requirements.

Additionally, the legume forages completely meet the calcium requirement of the sows so no calcium supplementation is needed. That is the reason monosodium phosphate is used as the phosphorus source in place of dicalcium phosphate. Supplemental limestone or dicalcium phosphate likely will create a calcium–phosphorus imbalance.

No diets are suggested for growing pigs because the bulkiness of pasture and forage will depress the total weight of feed intake. If it is desired to finish pigs on legume pastures, provide a self-feeder with a grain–soybean meal–based diet, and reduce the lysine content about 0.10 percentage units. Pigs finished on a nonlegume pasture should be provided the same diet they would receive on drylot or indoors.

FARROW-TO-MARKET ON PASTURE

Completing the hog life cycle on pasture can be accomplished by including nursery and growing-finishing as long as adequate shelter, feeders, and water are available. Often these stages are best managed without the use of pastures because of the large land area required for a large number of pigs compared to the sow unit.

Farrow-to-market on pasture provides an opportunity to phase into hog production gradually, with much less capital investment, allowing the producer to learn and develop the necessary skills before expanding into a production system requiring

Figure 11.8 A row of pasture individual farrowing huts. *(Courtesy, John McGlone, Texas Tech University, Lubbock, TX)*

TABLE 11.4   Suggested Diets to Supplement Pasture or Silage for Gestating Sows[1]

| | Type of Pasture | | | | Type of Silage | |
Ingredients	Legume	Grass	Legume—Grass Mix	Rape (Canola)	Legume (45% DM)	Corn (33% DM)
Corn	975.5	884.5	972.0	949.0	975.5	864.0
Soybean meal, dehulled	—	76.0	—	—	—	94.5
Ground limestone	—	6.5	—	6.5	—	10.5
Dicalcium phosphate	—	21.0	—	30.5	—	19.0
Monosodium phosphate	12.5	—	16.0	—	12.5	—
Salt	6.0	6.0	6.0	6.0	6.0	6.0
Vitamin premix[2]	3.0	3.0	3.0	3.0	3.0	3.0
Trace mineral premix[3]	3.0	3.0	3.0	3.0	3.0	3.0
Totals	1,000.0	1,000.0	1,000.0	1,000.0	1,000.0	1,000.0
Calculated Composition						
Metabolizable energy, (kcal/lb)	1,516	1,492	1,511	1,475	1,576	1,489
Crude protein, %	8.1	11.0	8.1	7.9	8.1	11.7
Lysine (estimated), %	0.25	0.46	0.25	0.25	0.25	0.51
Calcium, %	0.04	0.81	0.04	1.09	0.04	0.91
Phosphorus, %	0.55	0.69	0.62	0.83	0.55	0.66

[1] Assumptions: Sows will consume 3.5 lb pasture or silage dry matter and will be fed 2.5 lb of the diet per day. Pasture and feed together will provide a minimum of 0.65 lb protein, 11 g lysine, 14 g calcium, and 11 g phosphorus per day.
[2] Should provide the following amounts per ton of complete feed: 6,000,000 IU vitamin A, 300,000 IU vitamin D, 64,000 IU vitamin E, 1.0 g vitamin K (menadione), 6 g riboflavin, 15 g niacin, 18 g pantothenic acid, 22 mg vitamin B_{12}, 2,000 g choline, 2 g folacin.
[3] Should provide the following amounts per ton of complete feed: 8 g copper, 120 g iron, 75 g zinc, 30 g manganese, 200 mg iodine, 0.44 g selenium. (Concentrations higher than normal to compensate for reduced dry feed intake.)
Source: Adapted from Pork Industry Handbook, *PIH-126 (1998). Purdue University Cooperative Extension Service, West Lafayette, IN.*

more capital and intensive management. Young or inexperienced swine producers often choose a pasture production system because that can be more "forgiving" than intensive production.

The primary disadvantage with pasture systems is that some producers use the flexibility of the system to move "in and out" of swine production, usually at the wrong time. They are prone to increase production when market prices and profits are high, and to decrease or stop production when prices and profits are low. Usually, they end up cutting production just before the market begins to rebound.

QUESTIONS FOR STUDY AND DISCUSSION

1. List and discuss the advantages and disadvantages of producing hogs on pasture.
2. How do you account for the current trend of indoor rearing of swine by large, highly specialized swine producers?
3. Compute the monetary return that might reasonably be expected from an acre of pasture grazed by hogs versus the return from corn raised for grain.
4. Describe a quality hog pasture.
5. Outline the management—feeding, watering, fencing, shelter, manure handling, grazing, and pollution control—of hogs on pasture.
6. What are the primary differences between good management of hogs on pasture versus good management of hogs in drylot or indoors?
7. What pasture(s) would you recommend for use for swine in your area?
8. Suppose you have access to good quality silage and you wish to feed it to your hogs. What recommendations and precautions would you make?
9. How may alfalfa be used advantageously in feeding swine?
10. Describe and evaluate gestating sows on pasture and outdoor farrowing.
11. Describe and evaluate farrow-to-market on pasture.

SELECTED REFERENCES

Feeds and Nutrition, 2nd ed., J. E. Oldfield, W. W. Heinemann, and M. E. Ensminger, Prentice Hall, Upper Saddle River, NJ, 1989

Forages for Swine, Agricultural publication G2360, H. N. Wheaton and J. C. Rea, University of Missouri-Columbia, 1993, http://muextension.missouri.edu/xplor/agguides/ansci/g02360.htm

Nontraditional Feed Sources for Use in Swine Production, P. A. Thacker and R. N. Kirkwood, Butterworth Publishers, Stoneham, MA, 1990

Plant Growth and Development as the Basis of Forage Management, E. B. Rayburn, Extension Specialist, University of West Virginia, 1993, http://www.caf.wvu.edu/~forage/growth.htm

Report on the Welfare of Pigs Kept Outdoors, Farm Animal Welfare Council, London, UK, 2003, http://www.fawc.org.uk/reports/pigs/fawcptoc.htm

Stockman's Handbook, The, 7th ed., M. E. Ensminger, Interstate Publishers, Inc., Danville, IL, 1992

12

Reproduction in Swine

Detection of estrus by observing sows' mounting behavior.
(Courtesy, Palmer Holden, Iowa State University, Ames, IA)

Objectives

After studying this chapter, you should:

1. Identify the male reproductive organs and define their functions.
2. Identify the female reproductive organs and define their functions.
3. Know the breeding methods used in swine reproduction.
4. Know the reproductive characteristics and their relevance.
5. Know the factors affecting litter size and pigs per sow per year.
6. Understand the factors affecting boar performance.
7. Know what are adequate and appropriate facilities for farrowing and lactation.
8. Know the factors involved in deciding to implement artificial insemination.

Swine have many reproductive problems and reducing them calls for a full understanding of reproductive physiology and the application of scientific practices therein. In fact, it may be said that reproduction is the first and most important requisite of swine breeding because if animals fail to reproduce the breeder is soon out of business.

Many outstanding individuals are disappointments because of reproductive disorders; they are either sterile or reproduce poorly. From 5 to 30% of the fertilized eggs do not develop normally, resulting in embryonic mortality or death, and from 10 to 20% of the live pigs farrowed die within the first 5 days of life. The subject of physiology of reproduction is, therefore, of great importance.

REPRODUCTIVE ORGANS OF THE BOAR

The boar's functions in reproduction are (1) to produce male reproductive cells called the *sperm* or *spermatozoa* and (2) to introduce sperm into the female reproductive tract at the proper time. In order that these functions may be fulfilled, swine breeders should have a clear understanding of the anatomy of the reproductive system of the boar and of the functions of each of its parts. Figure 12.1 is a schematic drawing of the reproductive organs of the boar. A description of each part follows.

1. Testes (testicles). The primary function of the testes (singular testis) is to produce sperm. They are enclosed in the scrotum, a diverticulum of the abdomen.

The chief function of the scrotum is thermoregulatory; to maintain the testes at temperatures several degrees lower than that of the body proper. This temperature regulation is accomplished by the pampiniform plexus, which causes the scrotum to contract in cold environments to keep the testes warmer or to dilate in hot temperatures to allow the testes to cool relative to the boar's internal body temperature.

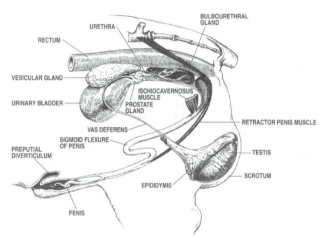

Figure 12.1 The reproductive organs of the boar, showing location in the body. The testes, its ducts, the vesicular gland, and the bulbourethral gland are paired, but for simplicity, only those on the left side are reproduced. *(Courtesy of Pearson Education)*

Cryptorchids are males of which one or both testes have not descended to the scrotum. The undescended testicle(s) is usually sterile because of the higher temperature in the abdomen.

The testes are connected through the inguinal canal with the pelvic cavity, where accessory organs and glands are located. A weakness of the inguinal canal, which is heritable in swine, sometimes allows part of the viscera to pass out into the scrotum—a condition called *scrotal hernia.* The following structures within the testes (see Figure 12.2) contribute to their function:

a. Seminiferous tubules. The seminiferous tubules are the germinal portion of the testis, in which are located the spermatogonia (sperm-producing cells). If laid end to end, it has been estimated that the seminiferous tubules of 1 boar testicle would be nearly 2 miles long. Each gram of testis is capable of producing about 27 million sperm each day.

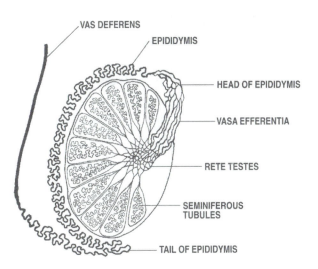

Figure 12.2 Drawing of the structures within the testes (sagittal section). *(Courtesy of Pearson Education)*

Around and between the seminiferous tubules are the *interstitial (Leydig) cells*, which produce the male sex hormone, *testosterone.* Testosterone is necessary for the mature development and continued function of the reproductive organs, for the development of the male secondary sexual characteristics, and for sexual drive.

  b. **Rete testis.** The rete testis is formed from the union of several seminiferous tubules.

  c. **Vasa efferentia (efferent ducts).** The efferent ducts carry the sperm cells from the rete testis to the head of the epididymis. Also, it is thought that their secretions are vital to the nutrition and maturing of the sperm cells.

2. **Epididymis.** The efferent ducts of each testis unite into one duct, thus forming the epididymis. This greatly coiled tube consists of three parts.

  a. **The head.** The head includes several tubules that are grouped into lobules.

  b. **The body.** The part of the epididymis that passes down along the sides of the testis.

  c. **The tail.** The tail is located at the bottom of the testis.

The epididymis has four functions; namely, (1) as a passageway for sperm from the seminiferous tubules, (2) for the storage of sperm, (3) for the maturation of the sperm, and (4) for concentration of sperm.

3. **Vas deferens (ductus deferens).** The vas deferens is a slender tube, lined with ciliated cells, leading from the tail of the epididymis to the pelvic part of the urethra. Its primary function is to move sperm into the urethra at the time of ejaculation.

The vas deferens—together with the longitudinal strands of smooth muscle, blood vessels, and nerves, all encased in a fibrous sheath—make up the spermatic cord (one from each testicle) which passes up through an opening in the abdominal wall, the inguinal canal, into the pelvic cavity.

The cutting or closing off of the ductus deferens, known as *vasectomy,* is the most usual operation performed to produce sterility, where sterility without castration is desired. This is most often done when the breeder wants to have a nonfertile boar (gomer) to assist in estrous detection but not have the ability to impregnate the female.

4. **Vesicular glands.** The vesicular glands are a pair of glands that flank the vas deferens near its point of termination. They are the largest of the accessory glands of reproduction in the male and are located in the pelvic cavity.

The vesicular glands secrete a fluid that provides a medium of transport, energy substrates, and buffers for the spermatozoa. They are not a storage site for sperm.

5. **Prostate gland.** The prostate gland is located at the neck of the bladder, surrounding or nearly surrounding the urethra and ventral (below) to the rectum. The prostate gland contributes fluid and salts (inorganic ions) to semen. It provides bulk and a suitable medium for the transport of sperm.

6. **Bulbourethral gland (Cowper's gland).** The bulbourethral glands, which often reach a diameter of 1.5 in. in the boar, are located on either side of the urethra in the pelvic region. They connect with the urethra by means of a small duct. These glands produce the prominent gel-like component of boar semen, which forms a plug in the vagina of the female.

7. **Urethra.** The urethra is a long tube that extends from the bladder to the end of the penis. The vas deferens and vesicular glands open to the urethra close to its point of origin. The urethra serves for the passage of both urine and semen.

8. **Penis.** The penis is the boar's organ of copulation. It is composed essentially of fibrous tissue. At the time of erection, cavernous spaces in the penis become engorged with blood.

In total, the reproductive organs of the boar are designed to produce and convey semen to the female at the time of mating. The semen consists of two parts: (1) the sperm, which are produced by the testes, and (2) the liquid portion, or seminal plasma, which is secreted by the seminiferous tubules, the epididymis, the vas deferens, the vesicular glands, the prostate, and the bulbourethral glands. Actually, the sperm make up only a small portion of the ejaculate. On the average, at the time of each service, a boar ejaculates 150 to 250 ml, with a range from less than 50 to more than 500 ml.

REPRODUCTIVE ORGANS OF THE SOW

The sow's functions in reproduction are (1) to pro-duce the female reproductive cells (*eggs* or *ova*), (2) to develop the new individual (the *embryo*) in the uterus, (3) to expel the fully developed young at the time of *birth* or parturition, and (4) to produce milk for the nourishment of the young. Actually, the part played by the sow in the generative process is much more complicated than that of the boar. It is imperative, therefore, that swine breeders have a full understanding of the anatomy of the reproductive organs of the sow and the functions of each part. Figures 12.3 and 12.4 show the reproductive organs of the sow; descriptions of some parts follow.

1. **Ovaries.** The two irregular-shaped ovaries of the sow are suspended in the abdominal cavity near the backbone and just in front of the pelvis. The ovaries have three functions: (a) to produce the fe-male reproductive cells (*eggs* or *ova*); (b) to secrete the female sex hormone, *estrogen*; and (c) to form the *corpora lutea*, which secrete the hormone prog-esterone. The ovaries may alternate somewhat ir-regularly in the performance of these functions.

The ovaries differ from the testes in that eggs are produced in very limited numbers and at inter-vals, during or shortly after heat. Each miniature egg is contained in a *follicle*, which surrounds the egg with numerous small cells. Large numbers of

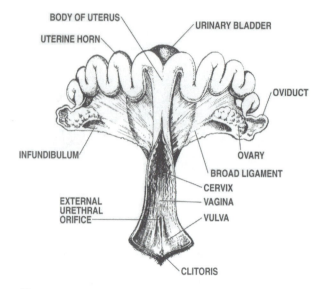

Figure 12.4 The reproductive organs of the sow, as viewed from above. The vagina and cervix are cut open. *(Courtesy of Pearson Education)*

follicles are scattered throughout the ovary. Gener-ally, the follicles remain in an unchanged state un-til the advent of puberty, at which time some of them begin to enlarge through an increase in the follicular liquid within. Toward the end of heat (es-trus), the follicles (which at maturity measure about 0.33 in. in diameter) rupture and discharge the egg. This process is known as *ovulation*.

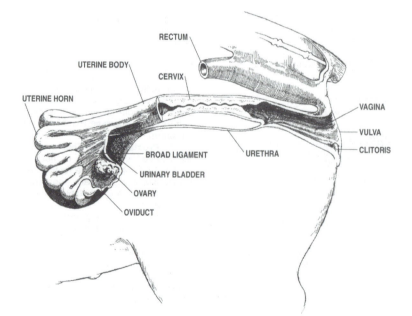

Figure 12.3 The reproductive organs of the sow, showing their location in the body. Note the location of the urethra, the duct to the bladder. When using arti-ficial insemination it is essential that the tube be directed upward in the vagina so the semen is not ejected into the bladder. *(Courtesy of Pearson Education)*

As soon as the eggs are released, the corpora lutea form at the site from which the eggs were released. The corpora lutea secrete a hormone called *progesterone,* which (a) acts on the uterus so that it implants and nourishes the embryo, (b) prevents other eggs from maturing and keeps the animal from coming in estrus (heat) during pregnancy, (c) maintains the female in a pregnant condition, and (d) assists estrogen and other hormones in the development of the mammary glands. If the eggs are not fertilized, however, the corpora lutea atrophy and allow new follicles to ripen and a new heat period to begin.

Occasionally, cystic ovaries develop, inducing temporary sterility. Actually, there are 3 types of disturbed conditions that are commonly called cystic ovaries: (a) cystic follicles, (b) cystic corpus luteum, and (c) persistent corpus luteum. All 3 conditions prevent normal ovulation and constitute major causes of sterility in sows.

The egg-containing follicles also secrete into the blood the female sex hormone estrogen. Estrogen is necessary for the development of the female reproductive system, for the mating behavior or heat of the female, for the development of the mammary

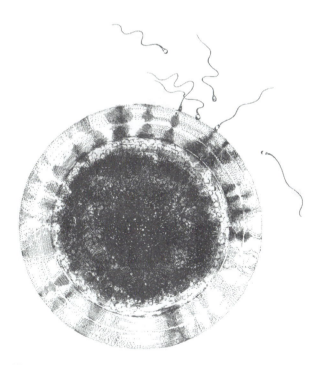

Figure 12.5 The ultimate objective of swine breeding is the union of the sperm from the male with the ovum, or egg, from the female. These two objects transmit to the embryo all of the inheritance from each parent. Each sperm contains 19 half pairs (haploid numbers) of chromosomes as does each ovum; thus the embryo or offspring will have 38 pairs (diploid number) of chromosomes. The sperm and egg are drawn to scale. *(Courtesy of Pearson Education)*

glands, and for the development of the secondary sex characteristics, or femininity, in the sow.

From the standpoint of the practical hog breeder, the ripening of the first Graafian follicle in a gilt generally coincides with puberty, and this marks the beginning of reproduction.

2. Oviducts (fallopian tubes). Oviducts are the small, cilia-lined tubes or ducts that lead from the ovaries to the horns of the uterus. They are about 10 in. long in the sow. The end of each oviduct nearest the ovary flares out like a funnel called *infundibulum.* The infundibulum are not attached to the ovaries but lie so close to them that they seldom fail to catch the released eggs.

At ovulation, the eggs pass into the infundibulum where, within a few minutes, the ciliary movement within the tube, assisted by the muscular movements of the tube itself, carries them down into the oviduct (Figure 12.5). If mating has taken place, the union of the sperm and eggs usually takes place in the upper third of the oviduct. Thence, the fertilized eggs move into the uterine horn. All this movement from the ovary to the uterine horn takes place in 3 to 4 days.

3. Uterus. The uterus is the muscular sac, connecting the fallopian tubes and the vagina, in which the fertilized eggs attach themselves and develop until expelled from the body of the sow at the time of parturition. The uterus consists of the two horns, the body, and the neck or cervix. In the sow, the horns are about 4 to 5 ft long, the body about 2 in. long, and the cervix about 6 in. long.

In swine, the fetal membranes that surround the developing embryo are in contact with the entire lining of the uterus, and there are no *buttons* or *cotyledons* as in the cow and ewe.

4. Cervix. Although it is technically part of the uterus, the cervix is often discussed as a distinct organ. It is a thick-walled, inelastic structure about 6 in. long. The canal of the cervix linking the body of the uterus and the vagina is funnel shaped, with ridges in the canal having a corkscrew configuration which conforms to that of the end of the boar's penis. The main function of the cervix is to prevent microbial contamination of the uterus. In swine, semen is deposited directly into the cervix during natural mating. When birth occurs, the cervix must dilate to allow passage of the piglets.

5. Vagina. The vagina is the canal that admits the penis of the boar at the time of service. At the time of birth, it expands and serves as the final passageway for the fetus.

6. Vulva (or urogenital sinus). The vulva is the external opening of both the urinary and genital tracts. It is about 3 in. in length. During sexual excitement, the vulva of gilt becomes reddened and swollen in response to the hormone estrogen, thus producing one of the signs of heat.

MATING

Mating is a prolonged process in swine, varying widely from 3 to 20 minutes in which waves of high and low sperm concentration exist in the flow of the ejaculate. For this reason, it is important that copulation take place without disturbance. A fractionated ejaculation consists of three phases:

1. The first, or presperm phase, which lasts 1 to 5 minutes, consists of a watery fluid in which there are tapiocalike pellets but no sperm, and comprises 5 to 20% of the ejaculate. Its primary purpose is to flush and irrigate both the urethra and vagina.
2. The second, or sperm-rich phase, which lasts 2 to 5 minutes, consists of a whitish, uniform fluid that contains the sperm, and comprises 30 to 50% of the ejaculate.
3. The last, or postsperm, phase, which lasts 3 to 8 minutes, contains very few sperm and helps form a gelatinous plug in the female tract, and comprises 40 to 60% of the total volume.

The boar may ejaculate as described here or may ejaculate in waves going through the phases more than once in each mating.

BREEDING METHODS

Three breeding methods are used in swine: (1) pen-mating, (2) hand-mating, and (3) artificial insemination (AI). Regardless of the method of breeding, the all-important factor in achieving a high conception rate and good litter size is to get sperm into the female's reproductive tract at the time when pregnancy rate and litter size will be maximized.

Pen-mating is simply turning a boar or group of boars into a pen of females to be bred. Pen-mating is usually practiced by smaller commercial swine producers. They may (1) split the sow herd so as to have 1 young boar per group of 10 to 12 females or (2) alternate boars in the sow herd—that is, use 1 boar or set of boars 1 day and another boar or set of boars the next day. Pen-mating may result in a lower farrowing rate because the time of insemination is not controlled, resulting in being too early or too late for optimum fertility. Additionally, boars may mate with 1 sow several times and ignore other females in estrus. Pen-mating requires less labor and lower-cost facilities than hand-mating. Few or no records are kept of dates that sows are actually bred.

A group of deep-bedded large white sows in Great Britain. *(Courtesy, Palmer Holden, Iowa State University, Ames, IA)*

Hand-mating is observing the matings between the boar and the female. Hand-mating is practiced more in swine than with pastured animals, such as cattle or sheep. It lends itself to record keeping because (1) exact breeding dates are known, (2) there is assurance that each female is bred, and (3) more is known about the breeding performance of the boar. Females may be mated with different boars in multiple inseminations.

Artificial insemination is a more specific version of hand-mating in which the breeder actually inseminates the female with an artificial penis using semen collected on the farm or purchased from a boar stud. Hand-mating or AI are essential for purebred swine breeders who must be able to identify the sire of each litter.

Pen-mating disadvantages include sows not being bred, unknown breeding and projected farrowing dates, inability to identify sows that fail to come into heat and boars that fail to breed. Often a boar, or boars, in a pen-mating system may only mate with one female when several are in estrus.

Hand-mating or AI are commonly used by industrialized swine operations, with AI increasing. The National Hog Farmer 2002 Producer Profile found that 54.5% of producers use AI and that 75% of the litters are sired by AI boars, suggesting AI is used more often by larger producers.

NORMAL REPRODUCTIVE TRAITS OF FEMALE SWINE

The pig lends itself well to experimental study in confined conditions. It is reasonable to expect, therefore, that we should have a considerable store of knowledge relative to the normal breeding habits

TABLE 12.1 Average Female Reproductive Values

Reproductive Trait	Average Time
Age at puberty[1]	4.5–6 months
Weight at puberty	150–230 lb (81–104 kg)
Duration of estrus	2–3 days (1–2 for gilts)
Length of estrous cycle	18–24 days (20–21 average)
Time of ovulation	12 hours before end of estrus (sows 35–40 hrs after onset of estrus) (gilts sometimes at 24 hrs)
Weaning to estrus[2]	3–7 days (5 average)
Time of ovulation	35–40 hrs after onset of estrus
Length of gestation	114 days

[1]Age at puberty is greatly affected by breed, indoor versus outdoor production, nutrition, and boar exposure.
[2]Estrus delayed beyond 7 days is usually a result of excessive lactation weight loss although weaning at 2 weeks or less may play a role.

of swine. Table 12.1 shows the average reproductive traits of female swine.

Age at Puberty

The age of puberty in swine varies from 4 to 8 months. This wide range is a result of differences in breeds and strains, boar exposure, and environment, especially indoor or outdoor exposure and nutrition. In general, boars do not reach puberty quite as early as gilts. It is recommended that gilts be allowed to pass 2 estrous periods and then mate them on their third estrus.

Estrous Cycle

During the estrous cycle, follicles form and ovulate, corpora lutea form, regress, and more follicles form (Figure 12.6). Unless mating and conception occurs, the cycle is normally repeated every 18 to 24 days, typically averaging 21 days. Progesterone is secreted by the corpora lutea, whereas the follicles secrete estrogen. In the event of pregnancy, the corpora lutea do not regress but remain functional throughout gestation. Ultimately, control of the estrous cycle resides with levels of follicle stimulating hormone (FSH) and luteinizing hormone (LH) from the pituitary in the brain as well as control from the hypothalamus gland.

Age to Breed Gilts

Reasonably early breeding has the advantages of establishing regular and reliable breeding habits and reducing the cost of the pigs at birth. However, ovulation rates increase significantly during each of the first two or three cycles, with the potential of far-

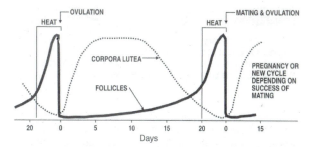

Figure 12.6 The estrous cycle of the sow; follicles form, ovulate, corpora lutea form, regress, and more follicles form. *(Courtesy of Pearson Education)*

rowing 1 or 2 more pigs per gilt litter. Therefore, gilts are usually bred at 7 to 8 months of age and farrow at 11 to 12 months of age.

Mixing pens of indoor-reared gilts and then regrouping them in fence-line boar contact is likely to start heat periods earlier. This may also help synchronize the first heat and the second heat.

Estrus (Heat Period)

The heat period or estrus is the time during which the gilt or sow will accept the boar. It lasts from 1 to 5 days, with an average of 2 days. Older sows generally remain in heat longer than gilts.

Ovulation usually occurs about 12 hours before the end of the heat period. However, that is a difficult time to measure. Therefore, insemination is usually planned for some time after the beginning of estrus and this time varies between gilts and sows. Gilts are in estrus for a much shorter time than sows. Live sperm must be in the female reproductive tract a few hours before ovulation occurs; otherwise, litter size will be reduced.

Optimal breeding is based on the number of times per day that a producer checks the females for standing estrus. With once-a-day detection, females should be bred each day they will accept the boar. With twice-a-day detection, females should be bred at 12 and 24 hours after they are first detected in heat. Gilts will sometimes have heat periods less than 2 days long and may have to be bred as soon as they are detected in heat and then each succeeding 12 hours they will stand for the boar. Heat detection should always be accomplished in the presence of a boar because his presence maximizes the chances of detecting females in heat and enhances uterine contractions which aid in sperm transport (Figure 12.7).

When the heat period lasts longer than 3 days, continued breeding is likely a waste of boar power because conception is doubtful. If not bred, the heat period normally recurs 18 to 24 days later. The effect of insemination timing on conception rates is indicated in Figure 12.8.

Figure 12.7 Providing access to a boar will stimulate the standing reflex in sows. Note the Hampshire boar present in the alleyway next to the group of sows to be bred in a Swedish facility. *(Courtesy, Palmer Holden, Iowa State University, Ames, IA)*

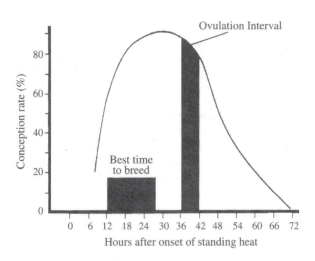

Figure 12.8 Effect of time of insemination on conception rate in swine. Note the best time to breed is 12 to 24 hours after the onset of standing heat. The sow normally ovulates about 8 to 12 hours before the end of standing heat. *(Courtesy, Pork Industry Handbook, Purdue University, West Lafayette, IN)*

The external signs of heat in the sow are restless activity, swelling and pink-red coloring of the vulva, frequent mounting of other sows, frequent urination, and occasional loud grunting. All signs are not always present.

Ovulation Rate

The ovulation rate—the number of eggs (ova) released—is associated with genetic background, age at breeding, weight at breeding, and nutrition. Crossbred females ovulate more eggs than purebreds. In gilts, the ovulation rate may increase 1 or

2 eggs with each successive heat period. Following the first litter, the ovulation rate increases until about the fifth or sixth litter when the rate plateaus. Flushing (increasing the dietary energy intake) the gilts increases their ovulation rates.

Fertilization

It is believed that the eggs are usually ovulated (freed) about 35 to 40 hours after the onset of estrus (standing heat). When the boar mates the female, the sperm are deposited in the cervix and uterus at the time of service and from there ascend the female reproductive tract. Under favorable conditions, they meet the eggs, and one of them fertilizes each egg in the upper part of the oviduct near the ovary.

A series of delicate time relationships must be met, however, or the eggs will not be fertilized. The sperm cells live only 24 to 48 hours in the reproductive tract of the female. Moreover, the eggs are viable for an even shorter period of time, between 8 and 10 hours after ovulation. For conception, therefore, breeding must take place at the right time. When aged sperm or eggs are involved in fertilization, developmental abnormalities may result; hence timing is critical. Multiple inseminations are usually attempted to ensure that viable sperm are available at ovulation.

Gestation Period

The average gestation period of sows is 114 days, though extremes of 98 to 124 days have been reported. An easy way to remember the gestation period is that it is 3 months, 3 weeks, and 3 days. Table 12.2 provides approximate due dates for pigs when the breeding date is known.

Breeding after Farrowing

Although sows do not ovulate while nursing pigs, some may come in heat during the first few days after farrowing, but obviously they cannot conceive if they do not ovulate. With 5 or 6 week lactations some data suggests that removing the piglets from the sow for 24 hours around days 21 to 25 postfarrowing will stimulate a fertile estrus.

The trend of large commercial operators is to wean at 3 weeks or less of age, with weaning to estrus of the sow beginning 3 to 7 days later. So, if pigs are weaned at 21 days of age, sows may be bred from 24 to 28 days after farrowing.

FERTILITY AND PROLIFICACY IN SWINE

Under domestication and conditions of good care, a high degree of fertility is desired. The cost of carrying a litter of 10 or 12 pigs to weaning time is very lit-

TABLE 12.2 Gestation Table Based on 114 Days

Date Bred	Date Due	Date Bred	Date Due
Jan. 1	April 25	July 5	Oct. 27
Jan. 6	April 30	July 10	Nov. 1
Jan. 11	May 5	July 15	Nov. 6
Jan. 16	May 10	July 20	Nov. 11
Jan. 21	May 15	July 25	Nov. 16
Jan. 26	May 20	July 30	Nov. 21
Jan. 31	May 25	Aug. 4	Nov. 26
Feb. 5	May 30	Aug. 9	Nov. 31
Feb. 10	June 4	Aug. 14	Dec. 6
Feb. 15	June 9	Aug. 19	Dec. 11
Feb. 20	June 14	Aug. 24	Dec. 16
Feb. 25	June 19	Aug. 29	Dec. 21
Mar. 2	June 24	Sept. 3	Dec. 26
Mar. 7	June 29	Sept. 8	Dec. 31
Mar. 12	July 4	Sept. 13	Jan. 5
Mar. 17	July 9	Sept. 18	Jan. 10
Mar. 22	July 14	Sept. 23	Jan. 15
Mar. 27	July 19	Sept. 28	Jan. 20
April 1	July 24	Oct. 3	Jan. 25
April 6	July 29	Oct. 8	Jan. 30
April 11	Aug. 3	Oct. 13	Feb. 4
April 16	Aug. 8	Oct. 18	Feb. 9
April 21	Aug. 13	Oct. 23	Feb. 14
April 26	Aug. 18	Oct. 28	Feb. 19
May 1	Aug. 23	Nov. 2	Feb. 24
May 6	Aug. 28	Nov. 7	Mar. 1
May 11	Sept. 2	Nov. 12	Mar. 6
May 16	Sept. 7	Nov. 17	Mar. 11
May 21	Sept. 12	Nov. 22	Mar. 16
May 26	Sept. 17	Nov. 27	Mar. 21
May 31	Sept. 22	Dec. 2	Mar. 26
June 5	Sept. 27	Dec. 7	Mar. 31
June 10	Oct. 2	Dec. 12	April 5
June 15	Oct. 7	Dec. 17	April 10
June 20	Oct. 12	Dec. 22	April 15
June 25	Oct. 17	Dec. 27	April 20
June 30	Oct. 22		

tle greater, requiring only a little more sow feed, compared to the cost of weaning a litter of only 5 or 6. In other words, the maintenance costs on both the sow and the boar remain fairly constant. It must be remembered, however, that in the wild state high fertility may not have been characteristic of swine. Survival and natural selection were probably in the direction of smaller litters, but nature's plan has been reversed through planned matings and selection.

Low fertility in swine must be attributed to both environmental and hereditary factors. Because the heritability of litter size is quite low, environmental factors have the main effect. Maximum prolificacy depends on having a large number of eggs shed at the time of estrus, on adequate viable sperm present for fertilization at the proper time, and on a minimum of embryonic and fetal mortality.

It is well known that some breeds and strains of swine are much more prolific than others and using them in a crossing program can enhance litter size. Litters of 12 or more are considered normal rather than exceptional among Chinese swine. Also, through selection, more prolific strains of swine can be developed. However, the heritability of litter size at birth and at weaning is only 10 and 5%, respectively (Chapter 5, Table 5.1). Yet, because such gains are cumulative, they are highly worthwhile.

The number of pigs produced increases with the age of the sow and plateaus at the fifth or sixth parity. More can be accomplished through proper management and environment to increase litter size than can be done through selection.

Even though many eggs may be shed and fertilized, the size of the litter is affected materially by embryonic and fetal mortality, which ranges from 5 to 30%. Exact causes of this are uncertain, but embryonic and fetal mortality have been attributed to (1) hereditary factors, perhaps recessive lethals; (2) intrauterine migration of embryos or overcrowding in the uterus resulting from a large number of embryos and as a consequence limited uterine surface area available for the nourishment of the individual embryos; (3) nutrition of the dam prior to and during gestation; (4) aged sperm or eggs at fertilization; (5) diseases or parasites; (6) accidents, injuries, or stress resulting from fighting among sows; or (7) hormone imbalances. Because some sows carry most embryos to term without loss, additional studies are determining the cause and prevention of embryonic and fetal mortality.

Certainly, the boar cannot affect the number of eggs shed, and under usual circumstances fertilization is very much an all-or-none phenomenon. Still, some evidence suggests that the boar can have a marked effect on litter size, primarily because embryos and fetuses sired by certain boars are less likely to survive to term. This further supports the contention that embryonic and fetal mortality are caused by genetic and fertilization errors (aged sperm or eggs). In this regard, conception rate and litter size can be increased by using more than one boar on each female. Continued research is needed to determine the extent of the influence of the boar.

It is recommended that, in advance of the regular use, new boars should be test mated to a few gilts. During these test matings, the producer should observe the boar for aggressiveness and desire to mate, and if necessary, give the boar assistance for the first service or two. Also, the producer should check the boar's ability to enter the gilt, which may be hindered by a limp, infantile, or tied penis. Most importantly, serviced gilts should become pregnant. A semen evaluation conducted by a veterinarian or

TABLE 12.3   Analysis of Reasons for Culling Sows

Reasons for Culling	Percentage of Total
Age	41.9
Lameness	16.0
Performance	12.0
Reproductive failure	21.3
Other	8.8
Total	100.0

Source: Swine 2000. National Animal Health Monitoring System. USDA-APHIS, Part I, p. 8.

qualified technician will complement the test matings. Although there is no absolute test for fertility, test matings and semen evaluation can often detect a sterile boar or one of questionable fertility.

The National Animal Health Monitoring System in 2000 (Table 12.3) revealed that age is the primary reason for culling sows in the United States followed by reproductive failure, likely a combination of a failure to breed or failure to maintain pregnancy.

Estrous Synchronization

It is often desirable to synchronize estrus in a group of females so they can be bred and farrowed in a short time period. Synchronization in sows is a relatively simple matter. When litters are weaned from a group of sows at the same time, a high proportion of the sows will come into heat within 4 to 7 days postweaning. Adequate boar power is very essential in order to take advantage of synchronization of postweaning heat.

Often moving gilts from a finishing barn, transporting them to a new site, or exposing them to a mature boar will stimulate the estrous cycle. With gilts naturally synchronizing, estrus is difficult and with variable results. For this reason many producers maintain a large pool of gilts to breed to ensure that a sufficient number will be in estrus when needed to fill out a breeding group of sows.

A commercial product, developed by Intervet, Inc., was approved in 2003 for estrus synchronization in mature, cycling gilts. The product, called Matrix, contains the active ingredient, altrenogest, a synthetic progestagen that mimics progesterone to block follicular development and suppress estrus. When withdrawn, follicular development commences, leading to estrus and ovulation. Treatment is recommended only in gilts that have completed at least 1 estrous cycle.

The product is spray applied on the feed with a device that delivers a fixed 15 mg/dose. Cost per gilt treated for the 14-day period is about $20.50. Individual feeding is recommended to ensure gilts receive the proper dose.

Research indicates that when gilts were fed 15 mg of altrenogest daily for 14 consecutive days, up to 85% of them displayed estrus within 4 to 9 days after withdrawal, providing a predictable stream of gilts for breeding when needed.

Flushing Sows and Gilts

The practice of increasing the energy intake of sows and gilts by 50 to 100% for 2 to 3 weeks before breeding is known as *flushing*. The beneficial effect attributed to flushing is that more eggs are shed with the potential for larger litters. Experiments and experiences do not show any benefits from flushing second and third litter sows, probably because of the short period between weaning and mating. However, flushing does seem to be effective for gilts, especially limit-fed gilts, perhaps because of the higher energy needs of these younger animals.

Two warnings must be heeded when considering flushing. First, animals already on full-feed or a fat animal cannot be successfully "flushed." It is most effective when following a program of restricted-fed gilts. Second, it is important to reduce the daily feed to 4 to 6 lb immediately after breeding. Data indicates continuing a full-feed postmating will negate the benefits of the flush. With individually mated females this should not be a problem. However, when pen-mating gilts, it would be best to flush the gilts until the boars are allowed to breed. At that time begin restricted feeding of the gilts.

Managing the body condition is very important. Sows that are in extremely thin condition following weaning often experience delayed estrus. If this is common, the results usually are delayed estrus and smaller resulting litters. These sows will benefit from either skipping the first delayed estrus and mating 21 days later (usually a normal estrus) or feeding a double ration until the occurrence of the delayed estrus, often 10 or 12 days after weaning.

PREGNANCY DIAGNOSIS

Open (nonpregnant) females are expensive to maintain! With ultrasonic detectors, pregnancy diagnosis is a reality. Producers can determine with a 90 to 95% accuracy the number of females that have settled. These detectors are most accurate when they are used between 30 and 60 days after mating.

The principle of ultrasonic pregnancy detectors is an ultrasonic echo from fluid in the uterus. Uterine fluid increases rapidly following conception and reaches detectable levels 25 to 30 days after breeding. It remains detectable for 80 to 90 days after breeding, after which the mass of pigs in the uterus exceeds the fluid content.

There are several advantages of early pregnancy detection in sows and gilts: (1) It makes it possible to cull or rebreed nonpregnant females; (2) it allows closer grouping of a number of sows for a farrowing period; (3) it gives early warning of breeding troubles, such as infertile boars and cystic ovaries of sows; (4) it enables the producer to make more effective use of breeding facilities and to adequately plan for farrowing, nursing, and finishing requirements; and (5) it makes it possible to guarantee pregnancy on females that are for sale.

MEASURING SOW PRODUCTIVITY

Because of the prolificacy and fecundity of swine, the output of sows is one of the most important economic traits used to evaluate efficiency of production. Sow productivity is usually measured by the following economically important traits:

1. Number of pigs weaned per sow per year.
2. Number of litters per year.
3. Number of pigs born alive.
4. Litter size at weaning.
5. Nonproductive sow days.

Because these traits have low heritability, emphasis on good management and nutrition is needed to minimize the effect of environmental differences on sow productivity. If environmental factors are standardized and optimized, the measured differences between litters can be regarded as genetic and can be more effectively used in selection programs.

The most important measure of sow productivity is the number of pigs weaned per sow per year. The two components affecting it are (1) the number of litters per year and (2) the litter size at weaning. Basically, the number of litters per sow per year depends on the interval between farrowings. Because the gestation period is a constant component, the lactation length (or weaning age) and the weaning-to-mating interval become the main factors that can influence the farrowing interval. Litter size at birth depends on ovulation rate, fertilization rate, embryo losses, and fetal losses.

Litter size at weaning is affected by (1) the number of pigs born alive and (2) the preweaning mortality. Preweaning losses can range from 10 to 20%, caused primarily by overlaying and starvation (Table 12.4). Often the two are difficult to separate because many of the overlain pigs were weak from starvation and unable to get out from under the sow.

To optimize sow productivity and to ensure a high number of pigs at weaning, producers need to take good care of the components affecting it. These include the following:

1. Select sows with proven ability to be prolific.
2. Use boars with maximum proven fertility.
3. Provide optimum housing and management throughout the reproductive cycle.
4. Provide a proper feeding program for sows consisting of (a) low level of feeding throughout the gestation period and (b) usually full-feeding during lactation.
5. Optimize management at farrowing and during early lactation.
6. Keep records of each sow and litter, and use this data in making selections for high-weaning performance.
7. Maintain a sound herd health program.

Reducing lactation length is the most effective way in which to increase sow productivity. With an adequate management program, weaning pigs at 2.5 to 4 weeks of age can be the regular practice. Reducing the nonproductive days is important and to achieve this, the following practices are important:

1. Follow the progress of individual sows from weaning onward.
2. Evaluate the performance of each sow, and determine the best time for culling.
3. Use boars of high fertility, and service each sow at least twice during the heat period.
4. Use proper procedures to check heat and to check returns to service.
5. Use an accurate pregnancy diagnostic technique.
6. Check breeding sows and boars regularly for reproductive diseases.

CARE AND MANAGEMENT OF BOARS

Proper care and management of boars allows long and productive service. This generally involves the proper consideration of the following: (1) the purchase and induction of boars, (2) test mating and

TABLE 12.4   Reasons for Nursing Piglet Mortality

Item	Percent
Scours	8.2
Laid-on	55.5
Starvation	17.0
Respiratory problem	1.6
Other known problem	8.6
Unknown problem	9.1
Total	100.0

Source: Swine 2000. National Animal Health Monitoring System. USDA-APHIS, Part III, p. 7.

semen evaluation, (3) housing and feeding, (4) ranting problem, (5) age and service of the boar, (6) maximizing fertility, and (7) keeping and using adequate records.

Purchase of Boars

Boars should be purchased at least 45 to 60 days before they are needed for breeding. If any purchase adjustments are necessary, the Code of Fair Practices adopted by the National Association of Swine Records contains guidelines (see Appendix). Also, buyers should purchase boars from sellers who can provide health records, and all new boars should be quarantined for at least 30 days and health checked again prior to use. They should be acclimatized to their new herd a few days after arrival by providing contact with cull females to gain exposure to the farm's normal microflora.

Test-Mating and Semen Evaluation

Records indicate that about 1 in 10 young untried boars has a fertility problem that renders them either sterile or subfertile. The simple practice of test-mating to identify a problem boar before the breeding starts can avert lost time and interrupted pig flow. Boars should be test-mated following their isolation period, and at 7 months of age or older.

Housing and Feeding Boars

When using the hand-mating system (individual mating system), boars should be penned separately, commonly in crates that are approximately 28 in. wide by 7 ft long, or in pens about 6 ft by 6 ft. Individual housing of boars eliminates fighting, riding, and competition for feed. Groups of boars to be used in a pen-mating system should be penned together when they are removed from a sow group.

During the initial isolation period, boars should be fed a diet similar to that fed them by the seller. This reduces stress associated with relocation. Gradually, the diet should be changed to match the diet used by the buyer. Young boars that are still growing should be fed at a level that allows for moderate weight gain. Depending on the diet, boar age, boar condition, and housing or climate conditions, the boar should be fed a balanced 16% crude protein diet (0.80% lysine)—5 to 5.5 lb for younger boars, 5.5 to 6.5 lb for mature boars (see Tables 7.10 and 7.23). Boars that are used or collected more often, or that are penned in outside facilities during cold weather, have higher-than-normal feed requirements and their feed levels should be adjusted accordingly. Overfeeding boars can lead to reproductive problems and decrease the length of service in the herd.

When limit feeding any diet to reduce energy intake, be certain that adequate levels of protein (amino acids), vitamins, and minerals are present to meet the boar's requirements for these nutrients. Formulation of special boar diets may be justified where several boars are to be maintained. However, in most herds, a balanced lactation diet is satisfactory.

Ranting Problem

Some boars chop their jaws and slobber. Such action is called ranting. Young boars that take to excessive ranting may go off feed, become "shieldy" (hard, leather shoulders), and fail to develop properly. Although this condition will not affect their breeding ability, it is undesirable from the standpoint of appearance. Isolation from other boars or from the sow herd is usually an effective means of quieting such boars. Should the boar remain off feed, placing a barrow or a bred sow in the pen with him will help to get him back on feed.

Age and Service of Boars

The number of services allowed will vary with the age, development, temperament, health, breeding condition, distribution of services, system of mating (hand-breeding, pen-breeding, or use of artificial insemination), and farrowing schedule. No standard number of services can be recommended for any and all conditions. Yet the practices followed by good swine producers are not far different. Such practices are summarized in Table 12.5.

For best results, the boar should be at least 8 months old and well grown before being put into service. Even then, he should be limited to 1 service per day and a maximum of 5 per week. Turning a young untried boar into a group of just-weaned sows coming into heat may be disastrous. If the estrous cycle of a group of sows is synchronized, more boar power is needed.

The number of boars necessary should be thought of in terms of the services required rather than sows per boar because sows or gilts will breed more than once during a heat period, though hand-mating controls the number of services by a boar. In

TABLE 12.5 Recommended Maximum Number of Services for a Boar

	Individual Matings		Pen-Mating[1]
Age	Daily	Weekly	Females per Boar
Young (8 to 12 months)	1	5	2–4
Mature (12 months +)	2	7	3–5

[1]Assumes that all sows are weaned the same day.
Source: Pork Industry Handbook, PIH-1, 1993.

pen-breeding systems, more boar power will be needed if sows are weaned and bred back by groups.

When fed and cared for by an experienced producer, a strong, vigorous boar from 1 to 4 years of age (the period of most active service) may serve 2 sows per day during the breeding season provided a system of hand-mating is practiced. Excessive service will result in a decreased concentration of sperm as well as immature sperm. With pen-mating, fewer sows can be bred.

A boar should remain a vigorous and reliable breeder up to 6 or 8 years of age or older, provided he has been managed properly throughout his lifetime. A major factor in culling a boar is that after 2 years he will have daughters in the herd to which he could be mistakenly mated.

Maximizing Fertility

Producers should maximize boar fertility by providing adequate boar power, rotating boars or individual-mating, providing an adequate breeding area, keeping boars cool during the summer months, and using other sound management practices. With the ready availability of fresh semen from many boar studs, lack of boar fertility should not occur.

Keeping and Using Adequate Records

In order to use boar power efficiently and identify breeding problems early, records are necessary. Producers should keep records of the frequency of boar services, and if artificial insemination is used, the date and volume of each ejaculate should be recorded. Computer programs are available for use in keeping, organizing, and analyzing records.

CARE AND MANAGEMENT OF PREGNANT SOWS

Without attempting to duplicate the discussion on feeding the gestating sow in Chapters 6 and 7, it may be well to emphasize that there are 2 cardinal principles that the producer should keep in mind when feeding sows during the pregnancy period: (1) to provide a diet that ensures the complete nourishment of the sow and her developing fetal litter and (2) to choose the feeds and adopt a method of feeding that is economical and adaptable to local conditions. Contrary to a common practice, a laxative diet is rarely necessary at farrowing and only if constipation is a problem, and constipation is seldom a problem.

The shelter for bred sows need not be elaborate or expensive. The chief requirements are that it be tight overhead, that it provide protection from inclement weather, and that it be well drained and dry. It should be of sufficient size to allow the animals to move about and lie down in comfort. Historically, except during the most inclement weather, sows were encouraged to run outdoors for exercise, fresh air, and sunshine. But times have changed! The majority of hog producers maintain their bred sows indoors throughout gestation, either in pens or stalls.

High land values, environmental problems, and efficiency in handling and managing large numbers of sows have caused producers to turn to an indoor system of housing. The type of system, however, depends on the manager's skill, ability, and finances. Chapter 15 discusses production systems, including confinement.

Advantages of indoor sow housing include (1) better control of mud, dust, and manure; (2) reduced labor for feeding, breeding, and moving to farrowing house; (3) improved control of internal and external parasites; (4) smaller land requirements; (5) better supervision of the herd at breeding time; (6) use of existing buildings; (7) improved operator comfort and convenience; and (8) opportunities for better all-around management.

Some disadvantages of indoor sow housing include (1) a higher initial investment; (2) possible delayed sexual maturity and increased breeding age, lower conception rates in gilts, and lower rebreeding efficiency in sows; (3) requirement of better management and daily attention to details, such as monitoring temperature and ventilation; and (4) increase in feet and leg problems.

CARE OF SOWS AT FARROWING TIME

The careful and observant stockman realizes the importance of having everything in readiness at farrowing time. If pregnant sows have been properly fed and managed to give birth to a crop of strong, vigorous pigs, the next problem is that of saving the pigs at farrowing time.

It is estimated that from 10 to 20% of the pigs farrowed never reach weaning age, and an additional loss of 3 to 10% occurs after weaning.

Signs of Approaching Parturition

The immediate indications that the sow is about to farrow are extreme nervousness and uneasiness, an enlarged vulva, a possible mucous discharge, and presence of milk in the teats.

Preparation for Farrowing

About 2 weeks prior to farrowing, the sow should be dewormed and treated for external parasites within a few days of moving to the farrowing facility. A variety of products are available and it is essential that the product directions be followed. When farrowing in crates or pens, sows should be moved to these no later than the 110th day of gestation.

If constipation is a problem, substitute 20% wheat bran or 10% dehydrated alfalfa meal or beet pulp for grain in the diet on moving the sow into the farrowing unit. Some producers avoid the constipation problem by adding 15 to 20 lb of magnesium sulfate (Epsom salts) or 15 to 20 lb of potassium chloride per ton of farrowing-lactation diet. If only a few sows need to be treated, top dress the ration with 2 oz per day with magnesium sulfate. This may have a disagreeable flavor, so it should be stirred into the feed.

Sanitary Measures

Performance of pigs can be improved by developing an "all-in, all-out" (AIAO) management system. This system entails scheduling the breeding of the sows so that a clean farrowing facility can be filled with sows within a 2- or 3-day period. This allows clean-up time before another group of sows is brought into the same facility.

Before being moved into the farrowing quarters, sows should be washed with warm water and soap and rinsed with a mild disinfectant to minimize contamination of the clean farrowing quarters.

Cleaning the farrowing facility can be accomplished by scraping, high-pressure cleaners, steam cleaners, or a stiff scrub brush. A complete job is necessary; otherwise, the remaining organic material in the pens greatly reduces the effectiveness of the disinfectant. Many good commercial disinfectants are available, including the quaternary ammonium compounds, iodoform compounds, and lye. Chapter 14 provides information on disinfectants.

Farrowing Facilities

Hogs are sensitive to extremes of heat and cold and require more protection than other classes of farm animals. This is especially true at the time of parturition. It is recommended that the temperature in the sow area be in the range of 55° to 75°F, and the temperature in the baby pig area have supplemental heat to maintain the temperature at 90° to 95°F for the first few days. After this time, the temperature for piglets can be gradually decreased to 70° to 80°F by weaning. Along with this temperature, there should be adequate ventilation but minimal drafts at all times.

Farrowing Crates or Pens

Most producers use farrowing crates because they reduce the number of piglets crushed by the sow. An additional advantage is that the operator is protected when handling the piglets. Most farrowing crates are 5 ft wide by 7 ft long. The width includes

an 18-in. piglet area on both sides of the 24-in. sow stall. Commercial crates adjust to accommodate very large or very small females. Most crates may have slotted floors although some are solid with bedding and the sows are turned out twice daily for feed and exercise (Figure 12.9).

When open-pen farrowing is practiced, a guard rail around the farrowing pen is an effective means of preventing sows from crushing their pigs. The importance of this simple protective measure may be emphasized best by pointing out that more than half of the young pig losses are accounted for by those pigs that are overlaid by their mothers. The rail should be raised 8 to 10 in. from the floor and should be 8 to 12 in. from the walls. It may be constructed of 2-by-4s, 2-by-6s, or strong poles or steel pipe. Often the stalls are designed to restrict the sow for the first 3 days and then 1 side of the sow panel removed or swung back to provide a creep area for the piglets (Figure 12.10).

Figure 12.9 Typical farrowing crate with woven wire floors restricting the sows' movement for the prevention of overlaying the piglets. *(Courtesy, Palmer Holden, Iowa State University, Ames, IA)*

Figure 12.10 Comfort stall allowing the sow access to more free area in the pen. After the piglets are big enough to protect themselves from being overlain by the sow, the restricting panel is opened or removed. *(Courtesy, Palmer Holden, Iowa State University, Ames, IA)*

Bedding

In open-pen farrowing, quarters should be lightly bedded with clean, fresh material. Any good absorbent that is not too long and coarse is satisfactory. Wheat, barley, rye, or oat straw; short or chopped hay; ground corncobs; peanut or cottonseed hulls; shredded corn fodder; shredded newsprint; or wood shavings are most commonly used. Additional information regarding bedding is given in Chapter 16.

Attendant

Attending sows at farrowing decreases the incidence of pigs that appear stillborn, pigs that likely died during the birth process or became entrapped in the afterbirth. The attendant also ensures that each piglet finds the sow's udder, begins nursing colostrum, and then is placed in a heated area of the stall. Moreover, care given during this time improves survival the first few days after farrowing.

On the average, the interval between the birth of pigs is approximately 15 to 20 minutes unless a problem develops. Normal presentation is either head first or tail first. An attendant can (1) free piglets from the membranes, (2) help piglets reach a teat, (3) possibly revive some piglets that are not breathing, and (4) treat the navel cord with tincture of iodine. If farrowing is proceeding normally but slowly, oxytocin may be used to speed the rate of delivery. Oxytocin should not be used if there is any suggestion that a pig is blocking the birth canal.

Continued strong labor for an extended period without the birth of piglets indicates the need for manual assistance by the attendant. A well-lubricated sterile glove worn over the hand and arm should be inserted in the vulva and up the vagina as far as is needed to find the piglet blocking the birth canal. Then the piglet should be grasped and gently but firmly pulled. Because manual assistance increases the chances of complications, such as infecting the sow, it is advisable to use an antibacterial solution as a lubricant. Mechanical implements to help grasp the piglet are available.

Some producers choose to induce farrowing in sows in order to control when parturition will occur. Induced farrowing avoids parturition at inconvenient times, such as late hours or over the weekend. It also (1) results in more efficient use of farrowing crates and barns, (2) facilitates crossfostering of pigs from very large litters to sows that delivered smaller litters, (3) makes it possible to wean pigs together that are more uniform in age and size, and (4) results in sows coming in heat (estrus) about the same time after weaning.

A hormonal injection sometimes used to induce farrowing is prostaglandin $PGF_{2\alpha}$. Lutalyse (Upjohn Co.) is the most commonly used product. All label instructions should be followed when using Lutalyse. It is essential that breeding and farrowing dates be known; sows should not be treated earlier than 110 days of gestation. Usually, sows will farrow within 48 hours after treatment.

As soon as the afterbirth is expelled, it should be removed from the pen and disposed of by composting, burying, or burning. This prevents the sow from eating the afterbirth and prevents the development of bacteria and foul odors. Dead pigs should be removed for the same reason.

If bedding is used, it is also well to work over the bedding; remove wet, stained, or soiled bedding and provide clean, fresh material.

Chilled and Weak Pigs

Pigs arriving in a cold environment are easily chilled. If they are born in an unheated building during cold weather, it may be advisable to take the pigs from the mother as they are born and to place them in a half-barrel or basket lined with straw or rags. A heat lamp or a jug of warm water (properly wrapped to prevent burns) may be placed in the barrel or basket; or the pigs may be taken to a warm room until they are dry and active.

One of the most effective methods of reviving a chilled pig is to immerse the body, except the head, in water as warm as one's elbow can bear. The pig should be kept in this for a few minutes, then removed and rubbed vigorously with cloths.

Orphan Pigs

Pigs may be orphaned either through death of their mother or sickness and fever in the sow, which causes lactation to cease. In either event, the most satisfactory arrangement for the orphans is to provide a foster mother. When it is impossible to transfer the pigs to another sow, they may be raised on cow's milk or milk replacer. The problem will be simplified if the pigs have received a small amount of colostrum (the first milk) from their birth mother. Colostrum can be hand-milked if necessary.

If cow's milk is used, do not add cream or sugar; however, skim milk powder, at the rate of a tablespoonful to a pint of fluid milk may be added, if available. Sow milk replacer should be mixed according to the directions found on the container. For the first 2 or 3 days the orphans should be fed regularly every 2 hours, and the milk should be at 100°F. Thereafter, the intervals may be spaced farther apart. All utensils (pan feeding or a bottle and nipple may be used) should be clean and sterilized.

Orphan pigs should be started on a prestarter or starter diet when they are 1 week old or as soon as they will begin consuming it. Also, an iron injection or dietary source of iron must be provided.

Crossfostering

Sows with exceptionally large or small litters will benefit by having the number of nursing pigs equalized soon after birth. However, they should be allowed to suckle colostrum from their birth mother prior to being moved to another sow. Generally, piglets should be transferred within the first 2 days postfarrowing and it is suggested that female piglets remain with their birth mother to facilitate record keeping for replacement purposes. Some farrowing managers equalize litters when they are temporarily removed from the sow for processing.

Runts

Small pigs or "runts" present a problem for producers. Larger litters have more runts, and these runts are often some of the last pigs born; thus, forcing them to compete for a teat with the larger earlier-born pigs. Because runts often perish, many producers euthanize them rather than trying to save them. Some research indicates that supplemental feeding of underweight (under 2.5 lb) newborn pigs (runts) can reduce their mortality. Supplemental feedings may consist of a commercial milk replacer or a mixture of 1 qt milk, 1 pt "half and half," (or 1 cup cream) and 1 raw egg, which is administered once or twice daily in 15 to 20 ml portions with a soft plastic stomach tube attached to a syringe (Moody et al., 1966).

Artificial Heat

Most farrowing houses are equipped with a heating unit designed to maintain a minimum temperature of at least 60 to 65°F for the sow's comfort, for use in winter farrowing.

Raising the air temperature for the microenvironment of the newborns is important to prevent chilling. Artificial heat usually must be provided, especially for pigs farrowed in the northern United States. Furthermore, providing an area of supplemental heat is necessary to maintain baby pigs in their thermoneutral zone. This heat zone should be 85 to 95°F for the first few days, then decreased to 70 to 80°F. The desired temperature range for the piglets can be obtained by observing them. If they are huddled together, the temperature is too low; if they are avoiding the supplemental heat source, it is too warm.

A major factor adversely affecting piglet survival is its difficulty in maintaining body temperature because of the high ratio between body surface area and size, sparse hair covering, and limited body fat. Even at thermoneutral temperatures, substantial amounts of metabolic energy are required to maintain body functions. A few days after birth, age-related changes occur that markedly improve the thermostability of the piglet.

Sow and Litter

The care and management given the sow and litter should be designed to get the pigs off to a good start. As is true of other young livestock, young pigs make more rapid and efficient gains than older hogs. Strict sanitation and intelligent feeding are especially important for the well-being of the young pig. This and other management practices pertinent to the well-being of young pigs—including adjusting litter size, removal of the needle teeth, ear notching, castrating, and sport—are covered in Chapter 15.

High-producing sows weaning 8 to 11 pigs per litter have much higher nutritional requirements than the sows weaning less than 8 pigs per litter because of the high-nutritional demands of milk production. Additionally, sows farrowing fewer litters per year have more time to recover from heavy lactation weight losses. The observation that they are farrowing less frequently may be the result of excessive lactation weight loss.

High-producing sows should be on full-feed after farrowing. This does not mean that the sow be provided a large quantity of feed the first day, but she should be given whatever amount she is willing to eat. The first day this may be only 1 or 2 handfuls. Continue to provide all sows nursing 7 or more pigs all of the diet they will consume. The goal is to maximize feed intake to minimize the amount of body condition the sow must convert to milk. Excess weight loss will result in delayed estrus and possibly reduce the size of the next litter.

Throughout the lactation period, sows should be fed liberally with feeds that will stimulate milk production. The most essential ingredients fed in the sow's diet during this period are an ample amount of protein (amino acids), vitamins, minerals, and water (see Chapters 6 and 7).

NORMAL BREEDING SEASON AND TIME OF FARROWING

Swine are seasonally polyestrus, meaning that they will breed any time of the year, but as in other farm animals the conception rate is much higher during those seasons when the temperature is moderate. Most producers with environmentally regulated buildings would like to have uniform pig production numbers throughout the year. The suggested breeding coefficients to determine the number of females to breed each month to ensure 1 litter will be farrowed are listed in Table 12.6. Conception rates are

TABLE 12.6   Suggested Coefficients to Determine the Number of Females to Be Bred Each Month

Month	Coefficient[1]
January	1.25
February	1.28
March	1.35
April	1.43
May	1.52
June	1.64
July	1.69
August	1.70
September	1.52
October	1.35
November	1.30
December	1.25

[1]Number to breed each month = number of farrowing stalls × coefficient.

Source: Pork Industry Handbook, *PIH-8, 1995.*

much lower in the United States in the warm months of June, July, and August. However, even in the cool months at least 25% more sows should be exposed to the boar than are expected to farrow.

ARTIFICIAL INSEMINATION[1]

Artificial insemination (AI) is the deposition of spermatozoa in the female genitalia by artificial rather than by natural means. Swine AI is widely used around the world in countries having large swine populations. In 2002, it was estimated that in the United States 55% of the producers used AI and 75% of the litters born were the result of artificial insemination.

The advantages of AI are as follows:

1. It makes it possible to use genetically superior boars more widely than would be possible with natural service.
2. It provides a way to bring in new genetic material with a minimum of disease risk.
3. Many commercial boar studs are now available offering semen from a wide variety of breeds and sire lines. These studs are designed to efficiently and safely handle boars and semen.
4. Fewer boars are needed on the farm. For example, a mature boar should not naturally mate more than 2 females per day. However AI makes

it possible to breed 10 or more sows from one ejaculation. Often farms using AI have only a boar used to stimulate the expression of estrus in the female.
5. There is less risk of injury to the sows, the boar, and the people handling them when AI is used.

Two common misconceptions relative to AI in swine are as follows:

1. The farrowing rate and litter size will be lower with AI than with natural service. In reality, if AI is performed properly, reproductive performance will be equal to the level achieved with natural service.
2. AI matings require more labor than natural matings. In reality, using AI requires less time than is needed to hand-mate sows, which requires the movement of the boar and sows to a breeding pen and observing the mating.

Many AI boar studs operate worldwide and have added top-quality boars to their genetic pool. Also, there is an increasing number of strictly swine service AI organizations in the United States. Generally, swine service businesses sell mostly fresh semen (but some sell frozen semen), provide custom processing and on-site AI training, sell AI supplies, and sell semen throughout the United States and abroad.

Selecting AI Boars

When choosing boars, purebred breeders may select from within their herds or buy new sires from outside their herds. AI gives them a third choice: They may purchase semen from outside the herd, rather than purchase boars. Purchased boars or semen from outside sources may be obtained from other independent purebred breeders or from company seedstock producers. EPDs can be used when selecting boars or semen from outside herds. (See the Chapter 5 section on selection indexes [STAGES, EBV, EPD, and BLUP].

Commercial pork producers rarely have the opportunity to select boars from within their own herd because of the inefficiencies of attempting to maintain purebred lines within a commercial herd. With the ready availability of commercial semen, this inefficiency is costly to overall herd production.

In addition to genetic value, when selecting boars they should be evaluated on those traits that allow them to produce sperm and mate (see Chapter 5, section "Selecting Boars").

AI Laboratory

Where AI is planned, a modest, but adequate, AI laboratory should be arranged. Preferably, it should

---

[1] Material from the following sources was adapted for this section: *Pork Industry Handbook,* PIH 136 and PIH 137, Purdue University, West Lafayette, IN, 1998; *Artificial Insemination* by D. Levis in 1995 Proceedings Nebraska Whole Hog Days, University of Nebraska, Lincoln, NE; and *The Swine AI Book* by Billy Flowers et al., North Carolina State University, Raleigh, NC, 1994.

be adjacent to the semen collection pen, with hand-delivery access between them by a sliding window. The laboratory should be equipped for (1) preparing semen collection equipment, (2) examining semen, (3) preparing semen diluents, (4) diluting semen, (5) storing semen, and (6) cleaning and storing equipment.

Semen Collection Area

A separate semen collection area should be incorporated into the design of commercial boar studs and on most farm studs. The semen collection pen (Figure 12.11) should have at least 2 or 3 of the perimeter walls constructed of 2-in. diameter galvanized pipe. The pipe should be 36 to 42 in. in height and placed at 11- to 12-in. intervals on center, thus a 9- to 10-in. space is provided between pipes. These perimeter pipe walls are a safety feature that allows the handler to enter or exit the collection area without opening a gate or scaling a wall, but the boar is confined within the pen area.

The collection pen and the surrounding area should be void of distractions that may divert the focus of the boar away from the collection dummy. It may be useful to position the dummy within the pen to limit boar movement around the dummy and to aid the handler or collector in directing the boar to the dummy. This can be done by placing the dummy in the corner of the pen or attaching it to a

wall. The recommended width of the collection pen is 6 to 8 ft and the recommended length is 8 to 9 ft. When using a diagonal escape corner(s) on one side of the pen, a width of 8 ft is recommended. Always keep a hurdle (stock panel) nearby when handling boars. A smaller collection pen is helpful when training young boars to mount a dummy.

The collection pen should be equipped with a dummy sow securely fastened to the floor or a pen partition. Swine producers can construct their own dummy or purchase a dummy from a company that specializes in AI equipment (Figure 12.12).

Training Boars

Boars generally show an interest in mounting stationary objects. Therefore, an estrous female is not required when attempting to collect semen. Basic requirements for a good mounting dummy include appropriate height for mounting and straddling of the boar's forequarters, structural stability, and durability. Good footing around the dummy is essential to aid the boar in mounting, thrusting, and the semen collection process. Rubber matting material with openings is a popular choice because it provides for good footing, resiliency to constant use, nonabsorbency, and ease of cleaning between uses.

Training boars for collection of semen requires patience. As soon as the young boar starts to rant, or at approximately 7 to 8 months of age, his training to mount the dummy sow should start. The following techniques have been used to train a boar to mount a dummy sow: (1) introduce the new boar to the dummy sow immediately after another boar has worked on it; (2) pour urine, semen, or preputial fluid from a mature boar on the rear of the dummy; (3) sexually stimulate the boar by exposing him to a strange boar or gilt adjacent to the collection area,

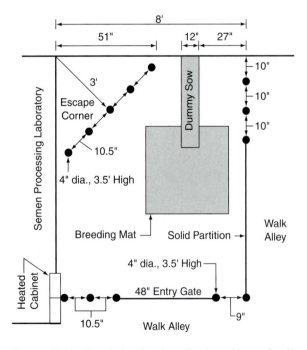

Figure 12.11 Pen design for the collection of boars for AI. Note the escape areas all around the pen for worker safety. *(Courtesy,* Pork Industry Handbook, *Purdue University, West Lafayette, IN, 1998)*

Figure 12.12 A dummy for the boar to mount in a collection pen. Note the pen division of vertical posts allowing the worker to escape the pen quickly. *(Courtesy, Palmer Holden, Iowa State University, Ames, IA)*

then have the strange boar or gilt exit in a manner that leads the new boar to the dummy; (4) allow the boar to mount, but not breed, an estrous female near the dummy; then, have the female exit leading the boar to the dummy; (5) collect from the boar when he is mounted on an estrous female standing near the dummy; or (6) allow the boar to mount an in-heat female standing adjacent to the dummy, ejaculate for about 1 minute, then lift him off the female and place him on the dummy.

Collecting Semen

Following are minimum contamination techniques for preparing and collecting semen.

1. Periodically trim hair from the preputial opening.
2. If needed, clean preputial opening and surrounding area with a single-use disposable wipe (i.e., diaper wipe).
3. Aggressively evacuate preputial fluids from the prepuce manually prior to grasping the penis for semen collection.
4. Have the semen collector wear disposable vinyl gloves or use a hand disinfectant between boars to minimize contamination of semen and reduce risk of cross-contamination.
5. After the boar mounts the dummy his penis will emerge. Grasp the penis with the gloved hand. Hold it perpendicular to the boar (to minimize the chance of preputial fluids running down the penis and directed into the semen collection vessel. Remember that the semen collector's hand must simulate the sow's cervix and that the penis, which forms a rigid spiral, locks into it. Apply firm pressure on the first and second ridges of the spiral portion of the penis; do not loosen grip until ejaculation is finished. Once the lock is formed, the boar will be begin to ejaculate.
6. Three different fractions of semen will be ejaculated in the following order:
   a. The presperm fraction, which should not be collected because it contains few sperm. Allow the first few jets of an ejaculate to fall on the ground rather than into the semen collection vessel.
   b. The sperm-rich fraction, which is opaque and milky.
   c. The final gel fraction, which should not be collected.

A prewarmed (100°F) insulated thermos or styrofoam cup is a convenient and economical collection vessel. The gel fraction should be filtered out of the ejaculate during collection using gauze or a mesh filter that has been placed over the mouth of the thermos or cup. Separation of the gel from the fluid during ejaculation is important because the gel coagulates into a semisolid mass that interferes with harvest of the spermatozoa, semen evaluation, and processing.

Actual time of ejaculation can vary considerably. A minimum of 5 to 9 minutes is usually necessary for a boar to complete ejaculation. Normally, a boar will ejaculate 150 to 250 ml, but the volume can sometimes exceed 400 ml, depending on such things as boar age, size, collection technique, and collection frequency.

Collection will take 5 to 20 minutes. Do not let go of the boar's penis until the boar has finished ejaculating. For adequate semen volume, sperm concentration, and doses per ejaculate, a collection frequency of 48 to 72 hours is recommended.

Chemicals (latex gloves, water, soap residues, alcohol, etc.), light (sun, ultraviolet), and temperature (hot or cold) are detrimental to sperm cells and should be avoided. As a general rule, anything that may come into contact with boar semen should be clean and dry. Single use, disposable products are preferred to minimize the risk of exposure to sperm-killing compounds and to eliminate the chance of cross-contamination between boars. When collecting and handling semen, it is important that semen only come into contact with materials/extenders that are at similar temperatures with the semen. Drastic temperature fluctuation is detrimental to sperm quality. Gently stir or swirl semen, but do not shake.

Assessing Semen Quality

Good quality semen is essential to obtaining satisfactory fertility rates. Standard tests currently used to evaluate boar semen quality include sperm motility, morphology, and concentration. When used individually, these standard tests have limited usefulness in actually determining the fertilizing potential of an ejaculate. These tests do, however, have the ability to identify ejaculates of overtly poor quality. Minimum semen quality values for fresh, unextended boar semen used for AI are indicated in Table 12.7.

Routine examination of AI boar semen quality is very important because its impact on herd reproductive efficiency is increased many fold when compared to natural mating. This examination is insurance against a reproductive catastrophe. The costs from using poor quality semen become quite high when considering its effect on herd farrowing rate, litter size, nonproductive days, and inventory of sows and gilts. A record of semen quality should be kept on each boar.

Semen should be evaluated promptly after collection. Routinely, the first few ejaculates of a new boar should be examined, followed by an examination

TABLE 12.7   Minimum Values of Fresh, Unextended Boar Semen for Use in AI

Semen variable	Value
Appearance	Milky to creamy consistency
Color	Gray-white to white in color
Total sperm numbers	$> 15 \times 10^9$ sperm/ejaculate
Gross motility (unextended)	$\geq 70\%$
Abnormal morphology	$\geq 20\%$[1]
-cytoplasmic droplets[2]	$< 15\%$

[1] The 20% maximum includes cytoplasmic droplets.
[2] Includes both proximal and distal cytoplasmic droplets.
Source: Althouse, G. C. Compend Contin Educ Pract Vet 19(3):400–404, 1997.

monthly thereafter and additional examinations if there are fertility problems. Semen should be evaluated on the following bases:

Volume—Volume of semen can be measured by pouring it into a measuring beaker. Alternatively, the semen can be weighed and the volume computed: 1 g = 1 ml.

Motility—Good motility indicates good viability of semen. Visual assessment of the percentage of motile sperm by light microscopy is still the preferred method. Accuracy of this technique largely depends on the technician's experience and natural ability. Sample preparation (i.e., dilution rate, type of diluent, temperature) must be standardized to reduce laboratory error and variation between examinations. To estimate motility, a small drop of semen is placed on a warmed (98.6°F or 37°C) microscope slide overlaid with a coverslip.

When viewed under a microscope, the sample should be thin enough to visualize individual sperm motility. If individual spermatozoa cannot be seen, a small drop of extender (same temperature as the semen) can be dropped on the sample before overlaying with a coverslip. Sperm motility is then estimated to the nearest 5% by viewing groups of sperm in at least 4 different fields on the slide at 200 or 400X; these readings are then averaged. Only ejaculates with at least 70% gross motility should be used for further processing. This is especially important because sperm motility and viability normally decrease during storage.

Morphology—Morphology is a measure of abnormal sperm, including those with coiled tails, crooked or bent tails, tails without heads, heads without tails, giant heads, and many other ab-

normalities. Several stains are commercially available and are essential for examining sperm morphology using dry mounted slides. Stains accentuate the outline of the sperm when using a light microscope, allowing for easier visualization by the observer. Higher resolution and more expensive phase contrast or differential interference contrast microscopes have internal components that generate their own contrast, allowing wet mount samples to be used for morphological estimation.

A minimum of 100 sperm are then assessed and categorized into 1 of 3 categories: (1) normal sperm (Figure 12.13), (2) sperm with abnormal heads, and (3) sperm with abnormal tails (including cytoplasmic droplets). If a large number of sperm in an ejaculate are abnormal, it indicates that a disruption of some type occurred during the development or maturation of the sperm or that the semen was improperly handled. Ejaculates of the general boar population usually exhibit less than 20% abnormal sperm. Therefore, ejaculates accepted for AI use usually contain greater than 80% normal sperm cells.

Sperm concentration—The most common way of estimating sperm concentration in gel-free boar semen is by measuring the degree of sample opacity. Sample opacity is estimated most commonly using a photometer, an instrument that measures the percentage transmittance or absorbency of light through a sample. In boar semen, sample opacity depends on the number of sperm cells and other ejaculate components that interfere with the movement of light through the sample.

Boar semen is normally too opaque for light to pass readily through it. Therefore, a

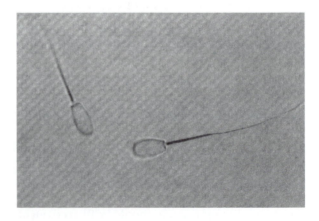

Figure 12.13 Appearance of normal sperm. Note the intact heads and straight tails. *(Courtesy, Pork Industry Handbook, Purdue University, West Lafayette, IN)*

small sample is usually diluted into an isotonic solution before taking a measurement. The photometric measurement is then converted into sperm numbers/ml either internally by the photometer or by the producer using a conversion chart, which accompanies the instrument. For this photometric measurement to be relatively accurate, it is necessary that the instrument be calibrated specifically for boar semen.

Another method of directly determining sperm concentration in boar semen involves using a counting chamber (e.g., hemacytometer). The surface of the counting chamber is etched to outline a defined surface area. After diluting a portion of semen to a 1:200 ratio, a very small portion of this mixture is transferred onto the counting chamber. Avoid overfilling! After allowing 5 minutes for sperm to settle onto the surface of the chamber, the number of sperm are counted within the defined surface area using a microscope at 200 to 400X.

A minimum of 5 large (80 small) squares are counted in the center grid on each side of the hemacytometer (Figure 12.14). Only sperm heads touching the top and left lines of the large square are included in the count, whereas those touching the bottom or right lines are not counted. Tails touching any of the lines are not counted. The two counts are then averaged. If the two counts vary more than 10% from each other, prepare the hemacytometer again and recount the two sides.

The number of sperm cells (N) are then determined by averaging all the counts. This number is then inserted into the formula supplied by the distributor of the counting chamber to determine the number of sperm cells per ml of

semen. The time and tediousness involved with hemacytometric counts make them impractical for most AI laboratories. Thus, photometric analysis remains the most commonly used technique for determining sperm concentration per ml of gel-free ejaculate. Unless sperm numbers are known, doses should be limited to 6 to 10 per collection.

Semen Processing

Processing semen is an exacting process, varying from defined procedures can easily disable or kill the sperm. The following items are essential to producing several quality doses from 1 ejaculate.

Semen extender—When buying powdered extenders in bulk, they should be broken down and repackaged in tightly sealed containers that will make the desired volume of liquid extender. If not mixed in the powdered extender, preservative antibiotics should be added the day the powdered extender is reconstituted with water. The type of extender to use depends on whether the semen is to be used immediately or stored for several days. Purchased extenders should have production dates, be kept in a frost-free refrigerator, and be used within 6 months of purchase.

Water quality—Water quality can have a negative effect on semen viability and fertility. Specific guidelines for water quality to use with boar semen are still being investigated. Types I and II water are currently recommended for use to reconstitute boar semen extender. Type I water is of the highest purity level and is produced using a combination of deionization with distillation, filtration, and reverse osmosis. Type II water is produced by double distillation.

Preparing extender—Extender powder is reconstituted with Type I or II water and incubated at 37°C (98.6°F) in a water bath for a minimum of 1 hour to allow for temperature and pH equilibration. To prevent contamination, it is best to prepare liquid extender in a plastic, single-use, disposable bag.

Extending semen—Total numbers of sperm per dose of semen tend to range from 2 to 6 billion (sperm concentration of 25 to 80×10^6 cells/ml). A dose of semen should contain at least 60 ml and no more than 120 ml total volume, 65 to 85 ml being the most common volumes for a dose of extended porcine semen. The final dilution rate of sperm into extender should depend on initial ejaculate quality, extender type, and anticipated duration of storage time. Some facilities

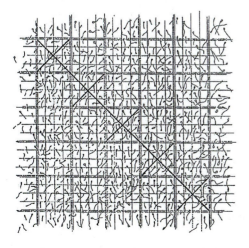

Figure 12.14 Hemacytometer for counting sperm concentration. *(Courtesy,* Pork Industry Handbook, *Purdue University, West Lafayette, IN)*

TABLE 12.8　Suggested Intervals for Inseminating Weaned Sows When Detecting Estrus Once or Twice per Day

| | *Estrous Detection* | | | |
| | *Once per Day* | | *Twice per Day* | |
Wean-to-Estrus Interval	*2 Matings*	*3 Matings*	*2 Matings*	*3 Matings*
3–5 days		A.M.—Day 1		P.M.—Day 1
		A.M. & P.M.—Day 2		A.M. & P.M.—Day 2
6+ days	A.M.—Day 1	A.M. & P.M.—Day 1	P.M.—Day 1	A.M—Day 1
	A.M.—Day 2	A.M.—Day 2	A.M.—Day 2	A.M. & P.M.—Day 2
Returns		A.M. & P.M.—Day 1		P.M.—Day 1
		A.M.—Day 2		A.M. & P.M.—Day 2

Source: Pork Industry Handbook, *PIH-137, Purdue University, West Lafayette, IN, 1998.*

employ an arbitrary extension ratio of 1 part semen (sperm-rich fraction) to 7 to 11 parts extender when storing and using semen within 24 to 72 hours.

If boar semen is to be extended by the volume ratio method, a conservative dilution of 1 part semen (whole ejaculate) to 4 parts extender should be followed, with the extended product used within 24 hours of extension. Problems that can occur when using the volume ratio method are as follows: (1) semen is underdiluted, allowing for exhaustion of available energy substrates and buffers over a shorter period of time, and (2) semen is overdiluted, potentially causing reduced sperm viability and fertility. In addition, the optimum number of doses of semen is not obtained; therefore, an economic and genetic loss occurs because the use of sperm cells is not maximized.

The freshly collected semen and extender should be at similar temperatures for mixing. The mixing of semen and extender can be accomplished by adding either semen into the extender or vice versa. Dilute the semen with extender using either a 1-step (add all of the calculated volume of extender at one time) or 2-step (add one-half the calculated volume of extender to semen, allow to equilibrate for 5 to 10 minutes, then add the remaining extender to achieve final volume) technique. Because the 1-step process is easier and less time consuming, it is the method preferred by many laboratories.

Inseminating Sows and Gilts

The actual process of AI is relatively easy because the anatomy of the female's vagina and cervix guides the inseminating catheter to the cervix without the aid of sight or some manipulation to ensure the passage into the cervix.

The major factor in achieving maximum farrowing rate and litter size is to inseminate females at the proper time (Table 12.8). Although the technique is simple, being certain it is carried out correctly is difficult. When more frequent heat detections are conducted, it is more likely that insemination will occur at the proper time. The best heat check is not to allow boar-to-sow contact for 1 hour before actual time of heat checking. This separation causes the females quickly to exhibit a strong immobilization response when encountering a boar. Where possible, the latter is best accomplished by driving a boar through the passageway in front of the sows, thereby providing nose-to-nose contact.

Ovulation in swine usually occurs about 12 hours before the end of the estrous period, or about 35 to 40 hours after the onset of estrus, and lasts for 1 to 3 hours. The onset and duration of ovulation are quite variable.

Confining the in-heat female to a small pen with a boar nearby, preferably with nose-to-nose or fence-line contact, and applying hand pressure to the back of the female to bring about an immobile stance, or gently rubbing a nervous sow's flank, is all that is necessary for females to stand during insemination.

Several types of insemination catheters are available. Both the rubber catheter, shaped like the boar's penis, and a variety of disposable catheters will work for AI. The major advantage to the disposable ones is that they do not need to be cleaned, eliminating the risk of contamination. Artificial insemination follows these general steps:

1. Wipe vulva with paper towel.
2. Insert the tip of the catheter into the vulva with the tip pointing upward to prevent entrance into the bladder orifice (Figure 12.15).
3. Slide the catheter along the top of the vagina until firm resistance of the cervix is felt, usually about 8 to 10 in.

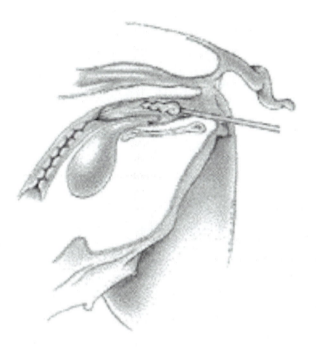

Figure 12.15 Proper insertion of an insemination catheter into the vagina and cervix of a sow. Note the upward pointing position to avoid entry into the bladder. *(Courtesy,* Pork Industry Handbook, *Purdue University, West Lafayette, IN)*

4. Rotate catheter counterclockwise to lock into the cervix.
5. When the catheter is in position, the semen container is attached and semen is squeezed slowly into the catheter. The rate of semen deposition should be slow enough to prevent backflow but otherwise accomplished as quickly as possible. If backflow occurs the catheter should be repositioned in the reproductive tract. If backflow still continues, slow the rate of semen deposition. It normally takes 4 to 10 minutes to complete the insemination.
6. After semen is deposited, rotate the catheter clockwise and slowly remove.
7. Reinseminate the female in 18 to 24 hours.

Cleaning and Storing Equipment

In order to minimize breakdown in sanitation, disposable equipment is recommended, including plastic catheters, plastic insemination bottles, and plastic bags in collection vessels.

If rubber catheters are used, they should be scrubbed with hot water to remove debris, rinsed thoroughly with cold water, and boiled in distilled water for 10 minutes. When removed from the sterilizer, shake them to remove excess water and place them spiral-end up in a drying cabinet. After they are dry, store them in a sealed plastic storage bag.

Frozen Semen

Frozen semen is available in both pellets and macro tubes. But boar semen does not freeze well. Frozen semen can be stored longer than fresh (liquid) semen. However, in comparison with fresh semen, frozen semen (1) produces fewer doses, (2) lowers farrowing rates an average of 30 to 40%, and (3) reduces litter size by as much as 1 pig.

Before use, frozen semen must be thawed, which should be according to supplier's directions, and it should be used immediately after thawing.

Summary

AI requires greater managerial input, a minimum amount of specialized equipment, and specialized training. However, by following a few precautions, litter size and conception rates will be equal to natural service, and the genetic level of the entire herd will be improved.

All purebred swine associations accept registration of AI sired litters. If semen from a boar not owned by the breeder is used, an AI certificate may be required.

QUESTIONS FOR STUDY AND DISCUSSION

1. Diagram and label the reproductive organs of the boar, and briefly describe the function of each organ.
2. Diagram and label the reproductive organs of the sow, and briefly describe the function of each organ.
3. Differentiate between hand-mating, pen-mating, and artificial insemination. List some advantages and disadvantages of each.
4. Describe the estrous cycle of the sow.

5. In order to synchronize ovulation and insemination, when should sows be bred with relation to the heat period?
6. What are the signs of heat?
7. Sows will often come in heat during the first few days after farrowing. If bred at this time, why do they usually fail to conceive?
8. When should sows be bred after farrowing?
9. From 5 to 30% of the fertilized eggs do not develop normally, resulting in embryonic mortality

or death; and from 10 to 20% of the live pigs farrowed die within the first 7 to 10 days of life. Discuss the possible causes and economics of this situation.

10. In your opinion is pregnancy diagnosis a valuable asset to a swine operation? Defend your answer.

11. List the five economically important traits that are usually used in measuring sow productivity. Discuss the importance of each of them.

12. Discuss each of the following aspects of the care and management of boars: (a) purchase and induction of boars, (b) test-mating of boars, (c) housing and feeding, (d) ranting problem, (e) age and service of the boar, (f) maximizing fertility, and (g) keeping and using adequate records.

13. How often should a mature boar be used when hand-mating is practiced? What guidelines would you follow if pen-mating is practiced?

14. Outline some practices that you believe help ensure the survival of a maximum number of pigs at farrowing and the first few days afterward.

15. Describe a sow that shows signs of farrowing and describe the actual birth process.

16. Why must an area of supplemental heat be provided for piglets?

17. What advantages could accrue from the practical and extensive use of artificial insemination in swine?

18. Will AI result in the farrowing rate and litter size being lower than natural service? Will AI require more labor than natural service?

19. How would you go about purchasing a boar or semen for AI in your herd?

20. Discuss training boars for AI.

21. List and discuss bases for evaluating boar semen.

22. List the steps to follow in using either the rubber catheter or plastic catheter in inseminating sows.

SELECTED REFERENCES

Animal Reproduction—Principles and Practices, A. M. Sorensen, Jr., McGraw-Hill Book Co., New York, NY, 1979

Applied Animal Reproduction, H. J. Bearden and J. W. Fuquay, Reston Publishing Co., Inc., Reston, VA, 1996

Managing Swine Reproduction, L. H. Thompson, University of Illinois, Urbana-Champaign, IL, 1981,

Moody, N. W, V. C. Speer and V. W. Hays, 1966, Effects of supplemental milk on growth and survival of suckling pigs, *Journal of Animal Science* 25:1250 (abstract).

Pig Production, J. McGlone and W. Pond, Delmar Learning, Clifton Park, NY, 2003

Pork Industry Handbook, Cooperative Extension Service, Purdue University, West Lafayette, IN, 1998

Reproduction in Farm Animals, Ed. by E. S. E. Hafez, Lea & Febiger, Philadelphia, PA, 1993

Stockman's Handbook, The, 7th ed., M. E. Ensminger, Interstate Publishers, Inc., Danville, IL, 1992

13

Swine Behavior and Their Environment

Swine are adaptable to a wide range of environments. Sows are trained to eat in an electronic feeder in a deep-bedded system. *(Courtesy, Palmer Holden, Iowa State University, Ames, IA)*

Objectives

After studying this chapter, you should:

1. Know the normal and abnormal responses of swine to their environment.
2. Understand the social dominance characteristic and its importance.
3. Be able to identify normal and abnormal behavior.
4. Know the responses of pigs to feeding regimens.
5. Know how pigs change to adapt to their environment.
6. Understand the "five freedoms" of animal welfare and how they can be applied to pork production.

All of the social interaction discussion that follows obviously relates to pigs that are maintained in groups, either indoors or outside on pastures or in dirt/concrete pens. Boars and sows housed in individual stalls have limited opportunities for aggression or many of the behavioral traits discussed in this section. This does not mean that they are not affected by their environment, and some discussion pertaining to individually penned animals is included.

Successful stockmen are "students" of animal behavior. For example, they recognize when females are in heat (estrus), or when parturition is imminent. They know the squeal of a piglet in trouble or the low-pitched rhythmical grunting sound a sow uses to call piglets to nurse. A producer should be able to walk through the facilities and spot sick animals by their behavior, thereby beginning early treatment or taking steps to correct some environmental problem. Moreover, indoor production methods of rearing hogs have renewed the interest in and increased the need for understanding animal behavior (Figure 13.1).

Indoor production must, however, embrace far more than ventilation and heating, along with ample feed and water. The producer needs to be concerned more with the natural habitat of animals. Nature ordained that they do more than eat, sleep, and reproduce. For example, studies on the behavior of swine indicate that they spend much of their day in active investigative behavior, primarily rooting and manipulating their movement. When free ranging, pigs may spend 40% of their day resting, 35% investigating novel surroundings, 15% eating, and 10% in other activities.

What happens when pigs are confined in a building on slotted floors? How is the nervous energy dissipated that would normally be used to satisfy the drives for investigating and rooting? Evidently, environmental deficiencies are often manifested by fighting, tail and ear biting, gastric ulcers, or possibly poor maternal care of the young,

What can be done about it? Preventing disorders by docking the tails of pigs to prevent tail biting is not unlike trying to control malaria fever in humans by the use of drugs without getting rid of mosquitoes. Rather, we need to recognize such a disorder for what it is—a warning signal that conditions are not right. Correcting the cause of the disorder is the best solution.

Unfortunately, this is not usually the easiest. Correcting the cause may involve trying to emulate the natural conditions of swine, such as altering space per animal and group size, promoting exercise, and gradually changing diets. Over the long run, selection provides a major answer to correcting behavioral problems; we need to breed swine adapted to producer-made environments.

This chapter presents some of the principles and applications of animal behavior. Those who have grown up around farm animals and dealt with them in practical ways have already accumulated substantial workaday knowledge about animal behavior. Those with nonlivestock backgrounds need to familiarize themselves with the behavior of ani-

Figure 13.1 Producers have altered the environment of swine. This double-wide, double-curtained, totally slatted building can be modified to optimize pig comfort. *(Courtesy, Palmer Holden, Iowa State University, Ames, IA)*

mals, to better feed and care for them and to recognize the early signs of illness. To all, the principles and applications of animal behavior depend on understanding.

ANIMAL BEHAVIOR

Animal behavior is the reaction of animals to certain stimuli, or the manner in which they react to their environment. The individual and comparative study of animal behavior is known as *ethology*. Through the years, behavior has received less attention than the quantity and quality of the pork produced by swine. But modern breeding, feeding, and management techniques have brought renewed interest in behavior, especially as a factor in obtaining optimum production and efficiency. With the restriction, or confinement, of herds, many abnormal behaviors evolved to plague those who raise them, including cannibalism, loss of appetite, stereotyped movements, poor parental care, overaggressiveness, dullness, degenerate sexual behavior, tail biting, and a host of other behavior disorders. Not only has indoor production limited space but it also has interfered with the habitat and social organization to which, through thousands of years of evolution, the species became adapted and best suited. This is because of a genetic time lag. Swine producers altered the environment faster than they altered the genetic makeup.

HOW SWINE BEHAVE

Different species of animals behave differently. Also, some behavioral systems or patterns are better developed in certain species than in others. Moreover, ingestive and sexual behavioral systems have been most extensively studied because of their importance commercially. Nevertheless, most swine exhibit the following general functions or behavioral systems, each of which is discussed:

1. Ingestive behavior (eating and drinking)
2. Eliminative behavior
3. Sexual behavior
4. Maternal behavior (mothering)
5. Agonistic behavior (competitive or combative)
6. Gregarious behavior
7. Investigative behavior
8. Shelter-seeking behavior

1. Ingestive behavior (eating and drinking). Ingestive behavior is characteristic of all animals of all species and all ages. Animals cannot live without feed and water. Moreover, for high production, animals must have aggressive eating habits. They must consume large quantities of feed.

Figure 13.2 Within minutes of birth, piglets began sucking and by 48 hours have established a "teat order." *(Courtesy, Palmer Holden, Iowa State University, Ames, IA)*

The first ingestive behavior trait common to all young mammals is sucking. Within minutes after birth, piglets find the udder and begin to nurse (Figure 13.2). When nursing occurs after parturition is complete, the sow calls the piglets using a low-pitched rhythmical grunt and the aroused piglets approach the sow, squealing in response. However, by its calls, a single hungry piglet may arouse the sow and the rest of the litter to nurse. Within 48 hours of birth, the piglets establish a "teat order"—each piglet suckles a particular teat. This is a form of dominance hierarchy. Anterior teats produce the most milk and are suckled by the most dominant piglets. Once the teat order is established, there is very little aggression. The sow will nurse her piglets about every 40 to 60 minutes, though the frequency is lower at night and declines as the piglets grow older.

Mature pigs possess teeth in the upper and lower jaws; hence, they bite, chew, and swallow feed. By nature, pigs love to root. If given the opportunity, they will stick their noses into the ground and lift forward and upward, moving soil out of the way and exposing earthworms, grubs, and roots. In the Perigord region of France, pigs are trained to use this natural desire to search for truffles—a type of underground mushroom.

2. Eliminative behavior. If given an opportunity, pigs have very tidy dunging habits. They like to keep their bedding area clean and dry. Hence, they usually deposit their feces and urine in a corner of the pen, away from the sleeping quarters. Modern indoor methods of raising pigs in restricted quarters have disturbed their natural eliminative patterns. Some producers, however, have taken advantage of the social and ingestive behavior of swine to train them to eliminate in specific areas. The use of partially or totally slatted floors is employed to prevent the buildup of manure in indoor production.

3. Sexual behavior. Reproduction is the first and most important requisite of swine breeding. Without sufficient numbers of young being born and born alive, the other economic traits are of academic interest only. Thus, it is important that all those who breed swine should have a working knowledge of porcine sexual behavior.

Sexual behavior involves courtship and mating. It is largely controlled by hormones. Sows in heat (estrus) are nervous and active, and their vulvas are reddened and swollen. Natural detection of estrus involves important nose-to-nose contact between the boar and the sow. The sow in heat will stand to be mounted by a boar. When sows do this, they assume a typical mating stance—rigid limbs and cocked ears. This stance is assumed when the stockman applies pressure to the rump; hence, producers use this as a means of detecting estrus (Figure 13.3).

However, in some sows, the assumption of the mating stance may depend on their hearing, smelling, or seeing the boar. Using a boar is the most accurate method of detecting females in heat, in fact, it increases the chance of finding estrous females, particularly gilts, by 30 to 40%.

Courtship of the boar involves eliciting the immobility stance. The boar often nudges the sow or gilt around the head or in the flanks with his head and nose and emits a courting song—a regular series of soft guttural grunts. Sows also respond to the smell of a boar—his pheromones. Chemically, these pheromones are androstenols, which are metabolites of testosterone. Receptive sows assume the mating stance when the boar attempts to mount. Some boars will mount several times before successfully mating, which takes 3 to 20 minutes.

Figure 13.3 Photo of hand pressure on gilt. When a caretaker applies pressure to the back of a female in estrus, she assumes a rigid stance, with the ears often in an erect position. *(Courtesy,* Pork Industry Handbook, *Purdue University, West Lafayette, IN)*

Abnormal sexual behavior is most often noted in males. For example, boars reared in all male groups form stable homosexual relationships. Also some boars will mount inanimate objects, an especially useful behavior when training boars to mount a dummy for semen collection to be used in artificial insemination. Sows often mount each other when one or both are in estrus.

4. Maternal behavior (mothering). Nest building activities occur 1 to 3 days before farrowing. In a field or pasture, sows will choose a depression in the ground and line it with grass, straw, and other material, forming a nest. Farrowing crates and concrete floors prevent nest building, but sows show increased activity the last 24 hours or more before farrowing. They may grind their teeth, bite and root at the rails, and frequently stand up and lie down. Just before parturition, however, the sow quiets down and lies on one side. Sows usually remain on this side, but some sows will get up and down between the birth of piglets, thereby increasing the chances of crushing a piglet.

Many domestic animals lick their newborn, but swine do not. Sows seldom pay attention to their piglets until the last one is born. Then the sow is very protective of her pigs, especially if they squeal. She will go toward an intruder with mouth open and will emit a series of sharp, barking grunts in rapid succession. She continues to mother her pigs until they are weaned, but after 2 to 3 days of separation she loses interest in them. If pigs are left with the sow for 3 or 4 months, she will usually wean them herself. Sows will readily accept pigs from another litter if the transfer is made the first day or two following farrowing. Transferring pigs among sows, in order to even out the size of litters, is a common practice in herds where many sows are farrowing about the same time.

Some nervous females, often gilts, eat their pigs during or immediately after farrowing. If this trait is observed, all pigs, both live and dead, along with the placental membranes should be removed as soon as possible, before the sow has an opportunity to eat them. Usually, such nervous sows calm down following farrowing, after which their pigs may be returned to them and they will express normal protective behavior.

5. Agonistic behavior (competitive or combative). Agonistic behavior includes fighting, flight, and other related reactions associated with conflict. Among all species of farm mammals, males are more likely to fight than females. Nevertheless, females may exhibit fighting behavior under certain conditions. Gestating sows kept in groups develop definite "pecking orders," with the dominant sow or sows always eating first or having the choice location to lay down or nest. Castrated males are usually quite passive, which indicates that hormones, (especially testosterone) are involved in this type of behavior. Thus, farmers have for centuries used castration as a means of producing docile males, particularly swine, cattle, and horses.

Boars penned together from a very young age seldom fight. Perhaps they have already settled their social rank. However, bringing together sexually mature strange boars almost always results in a fight. Sows and barrows will also fight, but they do not exhibit the jaw-clicking and saliva-producing (champing) characteristics of fighting boars. A sow will try to bite, whereas a boar will slash his opponent with his tusks.

When strange boars are first penned together, they smell one another and begin to circle as they "size up" each other. They frequently strut shoulder to shoulder with the hair on their crests bristled, ears cocked, and heads raised in alert, threatening positions. In a serious encounter, the combatants utter deep-throated barking grunts and champ throughout the fight. As the fighting becomes intense, each boar repeatedly thrusts his head and neck sideways and upward, with his jaws open and his teeth bared. If the boars have tusks, wounds are usually inflicted on the shoulders of each other.

Fighting boars await the opportunity to discontinue shoulder contact and to nip at the ears or the neck and front legs. Sometimes, they even charge the side of the opponent with their mouth wide open. Fighting may continue for as long as an hour, or it may end very quickly. In any event, it will continue until the dominant boar is satisfied and the loser retreats, with the winner biting and slashing him as he scampers away.

This fighting behavior presents a serious risk to people working with boars, whether walking in pens, moving the boars, collecting semen, or performing other activities. Always be cautious around boars and assume they may attack you. Breeding pens should have a narrow slot in the fence allowing the producer to escape an attacking boar. The risk of handling boars is a good reason to consider the use of artificial insemination.

Other types of agonistic behavior among pigs include (1) piglets fighting to establish "teat order," (2) aggressive actions around the feeder, and (3) tail biting. The last two forms of agonistic behavior are likely related to high-stocking rates and feeder space.

6. Gregarious behavior. Gregarious behavior refers to the herding instinct. In the wild state, swine roved through the forest in herds. Usually, these wild groups consisted of 5 to 10 sows, under the leadership of a boar. The wild boar, with his large and long head and strong tusks, was a formidable match for most any enemy.

Under domestication, swine retain their gregarious nature. However, humans have altered it a great deal. Today, hogs are usually confined to a very limited area. Feral pigs, however, return to living in herds of less than 10 individuals, but groups of up to 80 animals sometimes form.

7. Investigative behavior. All animals are curious and have a tendency to explore their environment. Investigation takes place through seeing, hearing, smelling, tasting, and touching. Whenever an animal is introduced into a new area, its first reaction is to explore it. Experienced producers recognize that it is important to allow animals time for investigation before attempting to work them, either when they are placed in new quarters or when new animals are introduced into the herd.

Pigs are curious, too (Figure 13.4). When a strange person approaches a herd of hogs, an alarm, or "woof," is sounded and the animals scatter—scampering as fast as they can for a short distance. In the meantime, if the intruder remains stationary, either standing or sitting, the pigs invariably return to investigate by smelling, rooting, and nibbling. When pigs are housed indoors they have little area to investigate.

8. Shelter-seeking behavior. Hogs are very sensitive to extremes of heat and cold; hence, shelter seeking is a very important trait with them. It is particularly important that swine be provided with shade during hot weather so that they may avoid the direct rays of the sun, because they do not possess an adequate cooling mechanism. In hot weather, hogs will wallow in water if given the opportunity.

Hogs pant rapidly when they are hot and sleep stretched out full length so as to expose the maximum body surface to the air; they sleep curled up and huddled together during cold weather, thereby exposing minimal body surface to the air. Baby pigs

Figure 13.4 Pigs like to investigate new things and are intelligent! *(Courtesy, Palmer Holden, Iowa State University, Ames, IA)*

are particularly sensitive to cold and they seek and need additional warmth.

SOCIAL RELATIONSHIPS

Social behavior may be defined as any behavior caused by or affecting another animal, usually of the same species, but also, in some cases, of another species.

Social organization may be defined as an aggregation of individuals into a fairly well integrated and self-consistent group in which the unity is based on the interdependence of the separate organisms and on their responses to one another.

The social structure and infrastructure in the herd maintained in groups are of great practical importance. Some of the ideas on pecking order have had to undergo changes as a result of increased understanding of the social organization within the herd.

It is obvious that there is no simple hierarchical descent, stepladder, or pecking order in numerical progression. The social structure is much more complex. As a general rule, the older an animal is at the time it is introduced into a group, the less integrated it becomes in the group.

Dominance

When put together, unacquainted pigs fight and establish a linear type of dominance. The top ranking pig has precedence in access to feed and is the winner in an agonistic encounter. Also, studies have demonstrated that pigs of a high rank are heavier and larger in stature. However, there may be situations where two pigs occupy the same social rank or where a triangular type social rank exists. Changes in social rank are common in the middle and lowest ranking pigs, but it is uncommon for the top animal to change its position. In fact some studies have shown that the most dominant pig can be removed for as long as a month and still resume its position on returning to the group.

Once the social rank order is established, it results in a peaceful coexistence. Thereafter, when the dominant pig merely threatens, the subordinate pig submits and avoids conflict. Of course, there are some pairs that fight every time they chance to meet. Also, if strange animals are introduced into such a group, social disorganization results in the outbreak of new fighting, as a new social rank order is established.

Several factors influence social rank; among them, (1) age—both young animals and those that are senile rank toward the bottom; (2) early experience—once a subordinate in a particular herd, usu-

Figure 13.5 Group-fed sows require that the feed be spread over a large area to prevent dominant sows from eating more than their share of the feed. *(Courtesy, Palmer Holden, Iowa State University, Ames, IA)*

ally always a subordinate; (3) weight and size; and (4) aggressiveness or timidity.

Social rank becomes important when a group is fed in a restricted space, and it becomes doubly important if limit feeding is practiced as with gestating sows. Under such circumstances, the dominant females crowd the subordinate ones away from the feeding area, with the result that the latter may not receive sufficient nutrition (Figure 13.5). Providing adequate feeding space will usually overcome this problem; interval feeding also has been successful. Interval-fed sows are given a triple level of feed every third day, providing more feed than a dominate sow can consume. Growing-finishing pigs are rarely limit-fed, but if they are, the feed restriction is usually about 90 to 95% of full-feed.

Dominants should be identified and, if possible, grouped together. Of course, they will fight until a new social order is established. In the meantime, both feed efficiency and gains will suffer. But, as a result of removing the dominants, the feed intake of the rest of the animals will be improved, followed by greater feed efficiency and profit. Among the more settled animals, social facilitation will become more evident. After the dominants have been removed, the rest of the animals will settle down into a new hierarchy, but within the limits of their dominance. Their interaction or social facilitation will be far more likely to have a calming effect on this group, to both the economic and practical advantages of the operator.

Dominance and subordination are not inherited as such; these relations are developed by experience. Rather, the capacity to fight (agonistic behavior) is inherited, and, in turn, this determines dominance and subordination. Hence, when aggressiveness has been bred into the herd, such herds never have the same settled appearance and docility that is desired of high-production animals.

Leadership

Leader–follower relationships are important in hogs. The young follow their mothers; hence, they continue to follow their elders.

It is important to distinguish leader–follower relationships from dominance; in the latter, the herd is driven, rather than led. After the dominants have been removed from the herd, the leader–follower phenomenon usually becomes more evident. It is well known that the dominant animal is not necessarily the leader; in fact, it is very rarely the leader. It pays too much attention to other matters of dominance in its relationship within the herd, with the result that it does not have the quality of leadership.

Interspecies Relationships

Social relationships are normally formed between members of the same species. However, they can be developed between two different species. In domestication this tendency is important (1) because it permits several species to be kept together in the same pasture or corral and (2) because of the close relationship between caretakers and animals. Such interspecies relationships can be produced artificially, generally by taking advantage of the maternal instinct of females and using them as foster mothers. For example, cows and bitches (dogs) have raised pigs.

As a general rule, different species should not be housed or penned together. Many diseases can be transmitted from one species to another, some of which may not be fatal to one species, but almost always are fatal when contracted by another. Aujesky's disease (pseudorabies) is an example of a disease in which pigs are the only natural hosts and often only become sick, but when it is contracted by sheep, cattle, goats, cats, or dogs, death results.

COMMUNICATION AMONG SWINE

Communication involves a signal by one pig, which, on being received by another pig, influences its behavior. Communication between pigs may be via sound, smell, or visual displays.

Sound—Sound communication is of special interest because it forms the fundamental basis of human language. The gift of language alone sets humans apart from the rest of the animals and gives them enormous advantages in their adaptation to their environment and in their social organization.

Sound is also an important means of communication among pigs. They use sounds in many ways, among them, (1) feeding, in sounds of hunger by young, or food finding; (2) distress calls, which announce the approach or presence of an enemy; (3) sexual behavior, courting songs, and related fighting; and (4) mother–young interrelations to establish contact and evoke care behavior. These have already been discussed.

Smell—The sense of smell is well developed in pigs. It is their sense of smell that allows pigs to find food when they root. Sows or gilts in heat search out a boar via olfactory cues or pheromones.

Visual displays—Visual signs play some role in pig communication. Intended agonistic behavior and sexual behavior are, in part, recognized by visual displays. For example, boars will make the hair on top of their necks rise up (bristle). This serves to make them look larger and more formidable.

NORMAL SWINE BEHAVIOR

The producer needs to be familiar with behavioral norms of animals in order to detect and treat abnormal situations—especially illness. Many sicknesses are first suspected because of some change in behavior—loss of appetite (anorexia); changes in water consumption; listlessness; labored breathing; posture; reluctance or unusual movement; persistent rubbing; and altered social behavior, such as one animal leaving the herd and going off by itself—these are among the useful diagnostic tools. Some of the signs of good health are as follows:

1. Contentment
2. Alertness
3. Eating with relish
4. Sleek coat with pliable and elastic skin
5. Bright eyes and pink membranes
6. Normal feces and urine
7. Normal temperature (102.0 to 103.6°F), pulse rate (60 to 80 beats per minute), and breathing rate (8 to 13 breaths per minute)

Vision and sleep are also important aspects of swine behavior. The eyes of pigs, like many animals, are on the sides of their heads. This gives them an orbital or panoramic view—to the front, to the side, and to the back—virtually at the same time. Sometimes this type of vision is referred to as rounded or globular. The field of vision directly in front of a pig is binocular, but on the sides and toward the back it is monocular. Vision of this type leads pigs to interpretations different from that of the binocular-type vision of humans.

The resting position of swine varies according to temperature—in the summer, they sleep stretched full length; in cold weather, they sleep curled up. In any event, pigs sleep soundly.

LEARNED BEHAVIOR

Given the proper reinforcement, usually food, pigs can be easily trained or more specifically conditioned. Conditioning is the type of learning in which the pig responds to a certain stimulus. The type that is used in swine production is operant conditioning—the pig learns to operate some aspect of the environment. For example, pigs can be operantly conditioned to push a panel to turn on a heat lamp or obtain a food reward. In practical application, an example is the lifting of the cover of a self-feeder, or drinking from a watering device. When training sows to eat in electronic feeders, the training period can be difficult, often involving a separate area to accustom sows to the equipment.

ABNORMAL SWINE BEHAVIOR

Abnormal behavior of domestic animals is not fully understood. Like human behavior disorders, more study is needed. However, studies of captured wild animals have demonstrated that when the amount and quality, including variability, of the surroundings of an animal are reduced, there is an increased probability that abnormal behavior will develop. Also, it is recognized that confinement of animals indoors or outside in dirt lots creates a lack of space that often leads to unfavorable changes in habitat and social interactions for which the species have become adapted and best suited over thousands of years of evolution.

Abnormal behavior may take many forms—ingestive, eliminative, sexual, maternal, agonistic, bar biting, or investigative (Figure 13.6). Abnormal sexual behavior is particularly distressing because the whole of production depends on the animal's ability to reproduce. Tail biting is a common form of abnormal behavior. It accompanies close confinement, resulting when pigs are prevented from normal behaviors, such as rooting, nibbling, and chewing. Thus far, docking the tails at birth is the best way to prevent tail biting.

Sometimes tail biting will occur even with tail-docked pigs, and producers have tried all sorts of things to stop it. Some have substituted other materials for the pigs to bite, such as rubber tires, chains hung in the pig pen, or mineral blocks. Occasionally, adding animal protein to the diet or

Figure 13.6   Bar biting is a behavior often observed as sows try to relieve boredom. *(Courtesy, Palmer Holden, Iowa State University, Ames, IA)*

0.25% magnesium oxide (MgO) to the diet has temporarily helped. Changes in the weather may precipitate the problem. Ultimately, tail biting with docked pigs is a comfort issue, especially if feeder or waterer space is limited, pens are crowded, or ventilation is inadequate.

APPLIED SWINE BEHAVIOR

The presentation to this point has been for the purpose of understanding. However, knowledge and understanding must be put into practice to be of value. Hence, the producer must make practical barnyard applications of swine behavior.

Breeding for Adaptation

The wide variety of livestock in different parts of the world reflects a continuous process of natural and artificial selection, which has resulted in the survival of animals well adapted to climate and other environmental factors. Adaptations relate to survival of the animals, but they do not necessarily entail maximum productivity of food for humans. This may be one of the more profound statements in swine behavior and production.

Selection of replacements should be from among animals kept under an environment similar to that in which it is expected that their offspring shall perform. Moreover, these animals should demonstrate high productivity in their environment. Therefore, pigs should be bred and selected that adapt quickly to artificial environments—animals that not only survive, but that thrive, under the conditions that are imposed on them. Properly combined heredity and environment complement each other, but when one or the other is disregarded, they may oppose each other.

ANIMAL ENVIRONMENT

Environment may be defined as all the conditions, circumstances, and influences surrounding and affecting the animal's growth, development, and productivity. The most important influences in the environment are the nutrition and shelter.

The branch of science concerned with the relations of living things to their environments and to each other is known as ecology. People achieve environmental control through clothing and air-conditioned homes and cars. In swine, environmental control involves space requirements, light, air temperature, relative humidity, air speed, wet bedding, ammonia buildup, dust, odors, and manure disposal, along with proper feed and water. Control or modification of these factors offers possibilities for improving animal perfor-mance. Although there is still much to be learned about environmental control, the gap between awareness and application is becoming smaller. Research on animal environment has lagged, primarily because it requires the commingling of several disciplines—nutrition, physiology, genetics, engineering, and climatology.

In the present era, pollution control is the first and most important requisite in locating a new swine unit, or in continuing an old one. The location should be such as to avoid (1) affecting the neighbors with odors, insects, and dust; and (2) contamination of surface and underground water. Without knowledge of animal behavior, or without pollution control, no amount of capital, native intelligence, and sweat will make for a successful swine enterprise.

Pollution of the environment is of worldwide concern. Attention is being directed to the following 4 key areas:

1. The proper application of nutrients in manure to the land.
2. The reduction in ammonia emissions from slurry during storage and spreading.
3. The implementation of production techniques that reduce both the initial amount of manure and its content of nitrates and phosphates.
4. The handling and disposal of dead animals.

The following factors are of special importance in any discussion of animal environment.

Feed and Nutrition

The most important influence in the environment is the feed. Swine may be affected by (1) too much or too little feed; (2) diets that are deficient in one or more nutrients; (3) diets that are excessive in nu-trients, primarily protein (amino acids) and phosphorus; (4) an imbalance between certain nutrients; (5) contaminants in the diet; or (6) the physical form of the diet, for example, it may be ground too finely.

Fortunately, with the vast amount of information available on pigs and their nutritional requirements, nutritional diseases and ailments have become increasingly uncommon. Probably the major area of nutritional deficiency is in the lactating sow and, generally speaking, energy intake is more critical than protein. After giving birth, energy requirements increase tremendously because of milk production; hence, a sow suckling young needs approximately double or triple the daily feed allowance than during the pregnancy period. Otherwise, she will suffer a serious loss in weight, and she may fail to come in heat and conceive. The following additional feed–environmental factors are pertinent:

1. **Regularity of feeding.** Pigs are creatures of habit; hence, if they are not on self-feeders, they should be fed at regular times each day, by the clock.
2. **Underfeeding.** Too little feed results in slow and stunted growth of pigs; in loss of weight, poor condition, and excessive fatigue of mature animals; and in poor reproduction, failure of some females to show heat, more services per conception, lowered litter size, and light birth weights.
3. **Overfeeding.** Too much feed is wasteful. Obviously, feeder adjustment is important for self-fed pigs to reduce spillage. Reproduction is lower in overfed, obese breeding animals, and excessive fat gain actually depresses lactation feed intake and milk production.
4. **Deficiency of nutrient(s).** A deficiency of any essential nutrient will lower production and feed efficiency relative to the ratio of the shortage. An animal can only perform up to the adequacy provided by the most limiting nutrient.
5. **Excess of nutrient(s).** Nutrients appearing in large excess can actually reduce the availability of other nutrients. In some cases large excesses of certain nutrients will cause a pig to reduce feed intake because of the dietary imbalance. Also, large excesses may reduce diet palatability.
6. **Some feed ingredients and diets influence milk composition.** The diet fed during gestation and lactation affects some of the nutrients in the milk. Minerals in milk, such as zinc and manganese, can be increased by feeding higher levels, but milk levels of other minerals, such as calcium, phosphorus, and iron, are not

increased by dietary means. Experiments have demonstrated that the fat content of sow's milk can be increased by feeding a high-fat diet. This may benefit the piglets.

Water

Animals can survive for much longer periods without feed than without water. Water is one of the largest constituents in the animal body, ranging from 50% in fat market hogs to almost 80% in newborn pigs. The percentage of water in the body is important. Excesses or deficits of more than a few percentage points of the total body water are incompatible with health, and large deficits of 10 to 20% of the body weight lead to death.

The total water requirement of swine varies primarily with the weather (temperature and humidity), the feed (kind and amount), the age and weight of the animal, and the physiological state of the animal. The need for water increases with increased intakes of protein and salt, and with increased milk production of lactating sows. Water quality is also important, especially with respect to the content of various salts, bacterial contamination, and toxic compounds.

The water content of feed ingredients ranges from about 8 to 10% in dried or air-dry feeds to more than 80% in fresh, green forage. All may be considered when meeting the daily water requirement. Of course, the majority of the water intake should be provided by continuous access to a drinking source.

The frequency of watering wild and feral swine, as well as domestic swine, is determined primarily by temperature and humidity—the higher the temperature and humidity, the more frequent the watering. Under practical conditions, the frequency of watering is best determined by the animals, by allowing them access to clean, fresh water at all times (Figure 13.7).

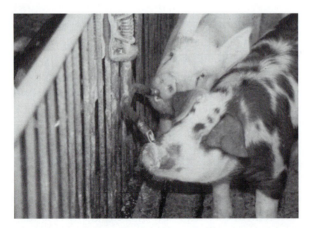

Figure 13.7 Nipple waterers are commonly used for swine. This paired set allows 2 pigs to drink at the same time. *(Courtesy, Palmer Holden, Iowa State University, Ames, IA)*

When pigs are hand-fed, they generally eat all their feed, then drink; when they are self-fed, they alternate between eating and drinking.

Weather

Temperature extremes are easily recognizable by producers and will elicit the obvious responses in the pigs of keeping warm or cool. Humidity levels have a major impact on the effect of temperature. More subtle changes, such as with air pressure, effect the pigs temperament, including irritability, movement, and feed intake.

The maintenance requirements of animals increase as temperature, humidity, and air movements depart from the comfort zone. Likewise, the loss or retention of body heat is affected by these three items.

Weather affects the maintenance requirements of swine. Extreme hot or cold can cause wide fluctuations in animal performance. Variations in weather from year to year create difficulty in making realistic analyses of buildings and management techniques to reduce weather stress. Weather-effected conditions can and should be modified by facilities. Research clearly indicates that pigs raised under extensive conditions (usually pasture or dirt lot) have improved production and feed efficiency when winter shelters and summer shades are provided.

Environmentally regulated buildings attempt to modify the problems imposed by weather changes. With the shift to indoor housing and high-density production operations, building design and environmental control became critical. Animals are more efficient, that is, they produce and perform better, and require less feed, if raised under ideal conditions of temperature, humidity, and ventilation. The fixed cost per head is, however, much higher for environmentally modified facilities. Thus, the decision as to whether indoor production and environmental modification can be justified should be determined by economics. Manure storage and management and pollution control are also considerations.

Buildings for the indoor production of pigs can be roughly divided into two types, modified open front (MOF) and mechanically ventilated, along with some crossover between the two. MOF buildings are designed with curtains or panels on sides that can be opened or closed, usually automatically, depending on the weather conditions. Mechanically ventilated buildings have permanent walls and depend on fans to move outside air through the units for cooling and removing accumulated gases.

Buildings for housing pigs under roof are costly to construct, but they make for the ultimate in animal comfort, health, and efficiency of feed utilization. Also, like any building, they lend themselves to automation, which results in a savings in labor, and, because of minimizing space requirements, they effect a savings in land cost compared to outside lots or pastures. If they malfunction, however, they can cause animals to suffocate and result in large economic losses. Today, under roof is rather common in swine housing.

Before an environmental system can be designed for animals, it is important to know their (1) heat production, (2) moisture production, and (3) space requirements. This information is as pertinent to designing livestock buildings as nutrient requirements are to balancing diets. This information is presented in Chapter 17, Buildings and Equipment.

Regulation of Body Temperature

The comfort zone, optimum temperature, and both upper and lower critical temperatures vary with different species, breeds, ages, body sizes, physiological and production status, acclimatizations, feed consumed (kind and amount), activity, and the opportunity for evaporative cooling. For example, the comfort zone for adult humans is considered to be between 72 and 85°F (22 to 30°C) and the comfort zone of sows after farrowing is 55 to 75°F (12 to 24°C). The comfort zone of newborn pigs is 90 to 95°F (32 to 35°C) for the first few days. Figure 7.3 in Chapter 7 indicates the upper and lower temperatures of the comfort zone as the pigs mature.

Heat production (metabolism) is plotted against environmental temperature in Figure 13.8 to depict the relationship between chemical and physical heat regulation. Note, also, the broad range of accommodation to low (cool) tempera-

tures in contrast to the restricted range of accommodation to high (warm) temperatures. Definitions of terms pertaining to Figure 13.8 follow.

Comfort or thermoneutral zone (B to C) is the range in temperature within which the animal may perform with little discomfort. This is the temperature at which the animal responds most favorably, as determined by optimum production and feed efficiency. In this zone the animal employs physical temperature regulation such as hair coat, layers of fat, sweating or panting, sun, shade, and so on to maintain body temperature

Lower critical temperature (LCT), vertical line A, is the low point of the cold temperature beyond which the animal cannot maintain normal body temperature. When the environmental temperature goes below point A, the chemical regulation mechanism is no longer able to cope with cold and body temperature drops, followed by death.

Chemical temperature regulation maintains body temperature in zone A to B. The French physiologist, Giaja, in 1925 used the term *summit metabolism* (maximum sustained heat production) to indicate the point beyond which a decrease in ambient temperature causes the homeothermic mechanisms to break down, resulting in a decline in both heat production and body temperature and eventually death of the animal.

Upper critical temperature (UCT), vertical line C, is the high point on the range of the comfort zone, beyond which animals are heat stressed and body temperature rises. Feed intake decreases and ceases as the temperature rises beyond line C. Supplementary cooling, such as sprinklers and fans are required to keep body temperature from rising, ultimately causing death.

Swine that consume diets of roughage, higher-fiber grains such as oats, or higher-protein diets produce more heat during digestion; hence, they have critical temperatures that are lower than the same animals fed a high-concentrate, moderate-protein diet. This excess heat of digestion essentially lowers the environmental temperatures corresponding to lines A, B, and C.

Stresses of both high and low temperatures increase with high humidity. The cooling effect of evaporating sprinkled moisture from dripper coolers (or sweating) is minimized and the respired air has less of a cooling effect. As humidity of the air increases, discomfort at any temperature, and feed intake, decrease proportionately.

Air movement (wind) results in body heat being removed at a more rapid rate than when there is no wind. In warm weather, air movement may make the animal more comfortable, but in cold weather it adds to the stress temperature. At low temperatures, the nutrients required to maintain

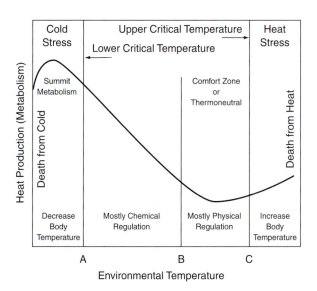

Figure 13.8 Influence of environmental temperature on heat production and body temperature. (*Bioenergetics and Growth, S. Brody. Hafner Publishing Co., NY, 1964, p. 283*)

the body temperature are increased as the wind velocity increases. In addition to the wind, a drafty condition where the wind passes through small openings directly onto some portion or all of the animal body will usually be more detrimental to comfort and nutrient utilization than the wind itself.

Adaptation of Swine to the Environment

Every discipline has developed its own vocabulary. The study of adaptation to the environment is no exception. The following definitions are pertinent to a discussion of this subject:

Adaptation refers to the adjustment of animals to changes in their environment.

Acclimation refers to the short-term (over days or weeks) response of animals to their immediate environment.

Acclimatization refers to evolutionary changes of a species to a changed environment, which may be passed on to succeeding generations.

Habituation is the act or process of making animals familiar with, or accustomed to, a new environment through use or experience.

Species differences in response to environmental factors result primarily from thermoregulatory mechanisms provided by nature, such as type of coat (hair, wool, feathers) and sweat glands. Thus, hogs, which have a light coat of hair, are very sensitive to extremes of heat and cold. However, nature gave cattle an assist through growing more hair for winter and shedding hair for summer, with the result that they can withstand higher and lower temperatures than hogs. The long-haired, shaggy yak of Tibet and the woolly Scotch Highland cattle of Scotland are as cold tolerant as the arctic-dwelling caribou, musk ox, and reindeer.

Facilities

Optimum facility environments can only provide the means for animals to express their full genetic potential of production, but they do not compensate for poor management, health problems, or improper diets. Facilities are discussed in more detail in Chapter 17.

Research has shown that swine are more productive and feed-efficient when raised in an ideal environment. The primary reason for having facilities, therefore, is to modify the environment. Proper barns and other shelters, shades, wallows, sprinklers, insulation, ventilation, heating, cooling, and lighting can be used to approach the desired environment. Also, increasing attention needs to be given to other stress sources such as space require-

ments, and the grouping of animals as affected by class, age, size, and sex.

The principal scientific and practical criteria for decision making relative to the facilities for swine in modern, intensive operations is the productivity and cost of production of animals, which can be achieved only by healthy animals under minimal stress. The investment in environmentally modified facilities is usually balanced against the expected increased returns.

Health

Health is the state of complete well-being, and not merely the absence of disease.

Disease is any departure from the state of health.

Parasites are organisms living in, on, or at the expense of another living organism.

Diseases and parasites (external and internal) are ever-present swine environmental factors. Death takes a tremendous toll. Even greater economic losses result from retarded growth, poor feed efficiency, carcass condemnations, decreases in meat quality, and increases in labor and drug costs. Swine health is discussed in Chapter 14.

Any departure from the signs of good health constitutes a warning of trouble. Most sicknesses are ushered in by one or more signs of poor health—by indicators that tell expert caretakers that all is not well—that tells them that their animals will go off feed tomorrow, and that prompts them to do something about it today.

Typical signs of ill health include inappetence, listlessness, sunken eyes, nervousness, or a humped-up, unusual posture—such as standing with the head down. Feces that are either very hard or watery suggest an upset in the water balance or an intestinal disturbance following infection. Abnormal urine includes repeated attempts to urinate without success or off-colored urine. Abnormal discharges from the nose, mouth, and eyes are also cause for concern.

Additionally, persistent rubbing, hairless spots, dull hair coat, and dry, scruffy hidebound skin are indicators of less-than-normal health. Pale red, or purple mucous membranes lining the eyes and gums; reluctance to move; fever; labored breathing—increased rate and depth; altered social behavior, such as going off alone; and sudden decrease in weight gains are signs a good stockman observes.

Stress

Stress is the nonspecific response of the body to any demand. As used herein, stress indicates an environmental condition that is adverse to an animal's well-being, either external or internal. However, a certain amount of stress is normal, in fact it has been said that "The absence of stress is death." The ability of the animal to successfully respond to the stresses are the important factor.

Swine experience many periods of stress—physical or psychological tension or strain. Stress of any kind affects animals. Among the external forces that may stress swine are previous nutrition, abrupt diet changes, change of water, space, level of production, number of animals together, changing quarters or mates, irregular care, transporting, excitement, presence of strangers, fatigue, illness, management, weaning, temperature, and abrupt weather changes.

In the life of a pig, some stresses are normal, and they may even be beneficial—they can stimulate favorable action on the part of an individual. Thus, we need to differentiate between stress and distress. Distress—not being able to adapt—is responsible for harmful effects. The trick is to manage stress so that it does not become distress and cause damage, and to recognize the warning signals of distress.

The principal criteria used to evaluate, or measure, the well-being or stress of people are increased blood pressure, increased muscle tension, body temperature, rapid heart rate, rapid breathing, and altered endocrine gland function. In the whole scheme, the nervous system and the endocrine system are intimately involved in the response to stress and the effects of stress.

The principal criteria used to evaluate, or measure, the well-being or stress of swine are growth rate or production, efficiency of feed use, efficiency of reproduction, body temperature, pulse rate, breathing rate, mortality, and morbidity. Other signs of swine well-being, any departure from which constitutes a warning signal, are contentment, alertness, eating with relish, sleek coat and pliable and elastic skin, bright eyes and pink eye membranes, and normal feces and urine.

Stress is unavoidable. Wild animals were often subjected to great stress; there were no caretakers to modify their weather; often their range was overgrazed; and malnutrition, predators, diseases, and parasites still take a tremendous toll.

Domestic animals are subjected to different stresses than their wild ancestors, especially to more restricted areas and greater animal density. However, in order to be profitable, their stresses must be minimal.

The porcine stress syndrome (PSS) is a special problem in pigs. Some pigs subjected to the stress of management procedures of handling, crowded transportation, or sudden environmental change adversely react and may even succumb. PSS is discussed in detail in Chapter 4.

POLLUTION

Pollution has been the lead issue since the 1990s in the pork industry. Anything that defiles, desecrates, or makes impure or unclean the surroundings pollutes the environment and can have a detrimental effect on swine health and performance. Thus, gases, odorous vapors, and dust particles from swine wastes (feces and urine) in buildings directly affect the quality of the environment. Muddy lots may also pollute the environment. For healthy and productive pigs, each of these pollutants must be maintained at an acceptable level.

Pollution potential, affecting the environment of both people and swine, increased as the U.S. swine industry moved toward specialization, mechanization, and high animal density. However, these factors are highly management dependent and not necessarily causes of pollution. But these fewer and larger specialized farms are much more visible than when most farms raised livestock.

To operate compatibly within the community, to provide maximum self-protection, and to avoid neighbor complaints and legal actions seeking either monetary damages or court injunctions, swine producers must be aware of some basic information and strategy concerning pollution and apply pollution control measures appropriate to the location.

The most troublesome swine pollutants are odors, dust, manure, muddy lots, pesticides, and nitrates in water. A discussion of each of these follows.

Odor

The hog odor issue heats up in swine producing areas where neighbors and communities complain of offensive odors. Today, many neighbors are former city dwellers who moved to the country for that fabled "fresh air" and have no farm experience. Also only a small percentage of the farms still raise livestock and many are retired farmers who are unhappy with the changes in agricultural structure. Jobs, a market for corn, and smelly pig manure are at the center of a debate that has moved from farms and towns to state capitals. To protect farmers from people moving next door and then complaining about odors, right-to-farm laws have been enacted in 42 states, but have been challenged in the courts in some cases.

Researchers are exploring odor control solutions including (1) composting manure (Figure 13.9), (2) the microbial flora and odors that make up manure and the part each plays in producing odors, (3) slowing down or stopping microbial activity, (4) natural or synthetic covers over lagoons and storage tanks, (5) the effect of swine diets, (6) the evaluation of different products for treating odors in

Figure 13.9 Composting appears to be an environmentally friendly and economical means of processing manure and dead animals with no odor or runoff when properly constructed. *(Courtesy, Palmer Holden, Iowa State University, Ames, IA)*

Figure 13.10 An outdoor gestation lot with no vegetation and heavy rainfall in Great Britain. *(Courtesy, Palmer Holden, Iowa State University, Ames, IA)*

manure or mixing in the feed, and (7) methods and rates of spreading manure on the land.

Dust

Dust may be defined as a mixture of fine, dry pulverized particles of matter resulting from feed processing, dried manure, and/or animal dandruff. Dust is a contributing factor to both swine and human health, especially with respect to respiratory diseases.

High levels of dust particles resulting from automated dry feed handling systems, dandruff and hair from hogs, and dried manure particles can occur inside swine buildings. Manure gases can cling to those dust particles in such a way that inhaling these gas-laden particles is uncomfortable and objectionable. Particulate matter also includes viral, bacterial, and fungal agents from the building environment and carries them into the respiratory system of people and hogs.

Manure

There is no one best manure management system for all situations. But one way or another, science and technology must evolve with ways of utilizing manure, and this must be accomplished without polluting streams or the atmosphere or being offensive to the neighbors. See Chapter 15 for a comprehensive discussion of manure management.

Muddy Lots

Muddy lots often plague livestock producers, especially during spring snow thaw, heavy rainfall, or the winter months. Mud increases scours and other diseases in newborn animals and reduces production

and feed efficiency in older animals. They are very unsightly and runoff from muddy lots can carry a significant amount of manure and other soil particles into waterways (Figure 13.10).

Pesticides

A pesticide is a substance (or mixture of substances) that is used to prevent, destroy, repel, or mitigate any pest, usually insects.

A pest is an organism declared to be a pest under circumstances that make it deleterious to man or the environment.

Pesticides are an integral part of modern agricultural production and contribute greatly to the quality of food, clothing, and forest products we enjoy. Also, they protect our health from disease and vermin. Pesticides have been condemned, however, for polluting the environment and, in some cases, for posing human health hazards. Unfortunately, opinions relative to pesticides tend to become polarized. A report, "The Future Role of Pesticides in U.S. Agriculture" (National Academy Press, 2000) summarized the situation as follows, "Although chemical pesticides safeguard crops and improve farm productivity, they are increasingly feared for their potentially dangerous residues and their effects on ecosystems."

The application of certain restricted-use animal health pesticides and fungicides requires training to purchase and apply the products. Check with your state Department of Natural Resources to find out which may be regulated in your state. Permission or training is not needed for application of products to control lice, mange, or internal parasites, but certain withdrawal periods are required before slaughter.

Nitrates in Water

The 10 ppm drinking water standard for nitrate levels (NO_3–N) was established by the U.S. Environmental Protection Agency. It was set at this level to protect the most nitrate-sensitive segment of the human population—infants under 3 months of age. Pigs, conversely, can drink water containing up to 300 ppm and possibly more (Feedstuffs, 1982).

Nitrogen (N) enters surface water primarily by runoff. Nitrates reach groundwater primarily by leaching. Organic nitrogen (N) or ammoniated (NH_4^+) sources are converted to NO_3–N by a process called nitrification. Primary agricultural sources of nitrogen include commercial fertilizer, animal waste, and breakdown of residual nitrogen in soils and crop residues.

Figure 13.11 A deep-bedded sow gestation barn in Great Britain. This meets the welfare-friendly laws of the country. *(Courtesy, Palmer Holden, Iowa State University, Ames, IA)*

ANIMAL WELFARE AND ANIMAL RIGHTS

Animal caretakers must recognize that the principles and application of animal behavior and environment depend on understanding and recognizing that they should provide as comfortable an environment as feasible for their animals, for both welfare and economic reasons. This requires that attention be paid to environmental factors, including feeding and housing, that influence behavioral welfare as well as physical comfort.

Animal welfare issues tend to increase with urbanization. Fewer and fewer urbanites and people moving to rural nonfarm homes have farm backgrounds. As a result, the animal welfare gap between producers and the nonfarm population widens. Also, both the news media and the legislators are increasingly from urban centers, and nonfarm views will have greater and greater impact in the future of livestock production.

In recent years, the behavior and environment of animals in a restricted space has come under increased scrutiny of animal welfare/animal rights groups. For example, in 1988 Sweden passed the Animal Protection Act, legislation designed (1) to phase out layer cages; (2) to discontinue sow stalls and farrowing crates; (3) to require straw bedding for all hogs (Figure 13.11); (4) to provide outside access for sows and cattle; and (4) to provide various restrictions on castration, dehorning, tail docking and several other restrictions. Existing farms were allowed to continue until the end of 1993. Florida passed a constitutional amendment in 2002 banning the use of gestation stalls in that state.

The National Pork Board in 2003 initiated a Swine Welfare Assurance Program (SWAP) to assist pork producers in evaluating welfare and safety issues on their farms. The program includes trained observers who will visit the farms to advise on farm welfare strengths and weaknesses.

Britain passed the Welfare of Pigs Regulations in 1991. These regulations banned tethers, specified pen sizes, and provided other criteria for the well-being of pigs. Existing farms were allowed to continue with existing conditions until 1999, but new ones were to comply immediately. The European Union (EU) has set minimum welfare standards for all EU countries regarding space, transportation, and housing. Gestation stalls will not be permitted after 2013. Individual countries can and do set restrictions in excess of the EU directives.

The Brambell Commission in the United Kingdom investigated farm animal welfare and its report led to the formation of a Farm Animal Welfare Council (FAWC).[1] The Brambell report (1965) included the well-known "five freedoms."

1. Freedom from hunger and thirst (access to fresh water and diet to maintain full health and vigor).
2. Freedom from discomfort (appropriate environment including shelter and a comfortable resting area).
3. Freedom from pain, injury, or disease (prevent problems or provide rapid diagnosis and treatment).
4. Freedom to express normal behavior (sufficient space, proper facilities, and company).
5. Freedom from fear and distress (ensuring conditions and treatment that avoid mental suffering).

How do various production systems relate to the five freedoms? Is one system the best? Table 13.1

---

[1] http://www.fawc.org.uk/reports/pigs/fawcptoc.htm

TABLE 13.1   Comparison of Gestation Systems and the Five Freedoms

Freedom	Outdoor Group Fed	Indoor (straw individually fed)	Stalls
Nutrition	Variable intake	Controlled	Controlled
Comfort (temp, physical)	Variable hot, cold	Controlled	Controlled
Hygiene	Fair to poor	Fair	Good
Health	Parasites Variable treatments	Enhanced	Enhanced
Pain and injury	Variable	Minimal	Minimal
Behavior	Good, enriched	Good, enriched	Limited, barren
Fear	Variable (fighting and predators)	Acceptable (fighting)	Acceptable

Source: Developed by J. McKean, P. Holden, E. Stevermer, and V. Meyer, Iowa State University, 1993.

compares three gestation management systems—outdoor group fed, housed indoors on straw and individually fed, and gestation stalls—with the five freedoms. When comparing the outdoor group fed, the nutritional freedom is variable because of group fighting and dominance; the comfort is variable depending on current temperatures and the hygiene is probably from fair to poor. Parasites are a health hazard and the ability to treat sows is variable; pain and injury depend on the level of aggression; and the freedom of behavior is good with a very enriched environment and social interaction. The fear level is variable, depending on the amount of fighting and presence of predators.

Gestation stalls, on the other extreme, control the nutrition (individual feeding) and the temperature is regulated. Hygiene is good because they are on slatted floors, immediately separating them from the manure. Health is enhanced because of individual observations and ease of treatment. Pain and injury is minimal because there is no fighting or aggression, but there may be foot injuries from the concrete flooring. The sow's ability to express normal behavior is severely limited in a barren environment. The fear factor is acceptable and depends on pleasant interactions with the stockman.

Animal welfarists are not all of one opinion. Some see many modern practices as unnatural and not conducive to the welfare of animals. In general, they construe animal welfare as the well-being, health, and happiness of animals, and they believe that certain intensive production systems are cruel and should be outlawed. Most livestock producers believe they also are animal welfarists because the well-being of their charges is essential to the success of the farm operation (Figure 13.12).

But the two differ in the means of evaluating welfare. Whereas the livestock producer may view animal performance (satisfactory reproduction or growth) as an adequate measure of welfare, the welfare groups contend the evaluation, additionally, should include evidence of behavioral, physiologi-

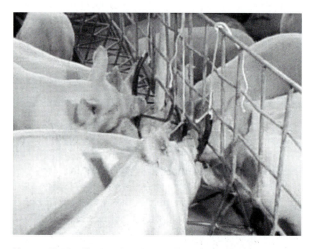

Figure 13.12   Pieces of rubber tubing attached to the pen partitions provide playthings for the pigs to chew on to relieve boredom, reduce aggression, and curb tail biting. *(Courtesy,* Pork Industry Handbook, *Purdue University, West Lafayette, IN)*

cal, and environmental well-being. Livestock producers know animal neglect or abuse leads to lowered productivity and income. They recognize that practices that impact labor and housing may result in physical and social conditions that may alleviate or increase animal problems. Means of reducing stress are needed so that changes in labor and housing costs are not offset by losses in productivity.

The animal rightists go further; they maintain that humans are animals, too, and that all animals should be accorded the same moral protection. They contend that animals have essential physical and behavioral requirements, which, if denied, lead to privation, stress, and suffering; and they conclude that all animals have the right to live. Rightists usually do not consume animal products (vegetarians) or use any animals products, such as leather (vegans) (Figure 13.13). Discussion of the validity of livestock production with animal rights groups is fruitless.

But wild animals are often more severely stressed than domesticated animals. They did not

Figure 13.13 Meat Free Zone poster in Great Britain sponsored by a vegetarian group. *(Courtesy, Palmer Holden, Iowa State University, Ames, IA)*

Figure 13.14 Moorman swinging gestation stalls. Side panels pivot from the front to allow sows to turn around. This provides her some activity but still provides individual feeding and protection from aggression. *(Courtesy, Palmer Holden, Iowa State University, Ames, IA)*

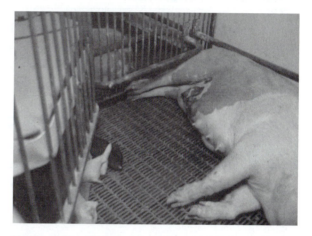

Figure 13.15 "Comfort" farrowing pen where the sow is allowed greater freedom of movement after the piglets are 3 or 4 days old and strong enough to avoid being overlain. *(Courtesy, Palmer Holden, Iowa State University, Ames, IA)*

have caretakers to store feed for winter or to irrigate during droughts; to provide protection against storms, extreme temperatures, and predators; and to control diseases and parasites. Often survival is a grim business.

Animal welfare issues tend to increase with urbanization because the city dwellers have little or no knowledge of livestock production practices. As a result, the animal welfare gap between town and country widens. Consumer concern about animal welfare has contributed to the development of niche markets where the animal products are advertised as being "welfare friendly" or "natural" and often command a higher price for the products.

Animal/Environmentally Friendly Innovations

Scientists and technologists are working ceaselessly to develop new facilities and equipment designed and thought to be animal or environmentally friendly. Examples are the Moorman Gestation Stalls (Figure 13.14) and "comfort" farrowing pens (Figure 13.15). The Moorman gestation stalls pivot from the front so the back widens, allowing the sow to turn around. It does not involve any more space per sow than conventional gestation stalls. The farrowing pens are based on a European system in which the sow is confined to a farrowing stall for a

few days after farrowing and then one side gate of the stall is swung back, allowing a guarded area for the piglets but allowing the sow to turn around and be more active. Like the Moorman gestation stall the "comfort" farrowing pen can be placed in the same space as a standard farrowing stall.

QUESTIONS FOR STUDY AND DISCUSSION

1. Why has increased restricting of space for pigs made for great interest in the subject of animal behavior?
2. Define *animal behavior* and *ethology*.
3. Why has there been a genetic time lag or why have swine producers altered the environ- ment faster than the genetic makeup of animals?
4. Discuss the following behavioral systems as they pertain to swine:
   a. Ingestive behavior
   b. Eliminative behavior

    c. Maternal behavior (mothering)

    d. Sexual behavior

5. How is the dominance hierarchy established when several swine are brought together to form a herd? How is it maintained?

6. Discuss how swine communicate with each other and the importance of each method.

7. List and discuss the significance of one example of the practical application of swine behavior in breeding for adaptation.

8. Discuss how each of the following environmental factors affects swine:

    a. Feed and nutrition

    b. Water

    c. Weather

    d. Facilities

    e. Health

    f. Stress

9. Detail why the following pollutants are especially troublesome in swine operations:

    a. Odor

    b. Dust

    c. Manure

    d. Muddy lots

    e. Nitrates in water

10. Assume that you plan to establish a herd of 1,000 swine. With what environmental laws and regulations must you comply?

11. Define (a) animal welfare and (b) animal rights.

12. What is involved with the National Pork Board's Swine Welfare Assurance Program?

13. Discuss the five freedoms relative to three types of finishing pigs: pasture, hoops, and indoors on concrete slats.

SELECTED REFERENCES

Animal Science, 9th ed., M. E. Ensminger, Interstate Publishers, Inc., Danville, IL, 1991

Animal Welfare, M. C. Appleby and B. O. Hughes, CAB International, New York, NY, 1997

Behavior of Domestic Animals, The, Ed. by E. S. E. Hafez, The Williams & Wilkins Company, Baltimore, MD, 1975

Dictionary of Farm Animal Behavior, 2nd ed., J. F. Hurnik, A. B. Webster, and P. B. Siegel, Iowa State University Press, Ames, IA, 1995

Farm Animal Welfare, B. E. Rollin, Iowa State University Press, Ames, IA, 1995

Farm Animal Well-Being, S. A. Ewing, D. Lay, Jr., and E. von Borell, Prentice Hall, Upper Saddle River, NJ, 1999

Future Role of Pesticides in U.S. Agriculture, The National Academy Press, Washington, D.C., 2000

Pig Production: Biological Principles and Applications, John McGlone and Wilson Pond, Thomson-Delmar Learning, Clifton Park, NY, 2003

Pigs Display Wide Tolerance for Salinity in Water, John Goihl, Feedstuffs, Minnetonka, MN, 1982

Pork Industry Handbook, Cooperative Extension Service, Purdue University, West Lafayette, IN, 2002

Pork Quality Assurance Program, http://www.porkboard.org/PQA/manualHome.asp, National Pork Board, Des Moines, IA

Stockman's Handbook, The, 7th ed., M. E. Ensminger, Interstate Publishers, Inc., Danville, IL, 1992

Swine Welfare Assurance Program, http://www.porkboard.org/SWAPHome/default2.asp, National Pork Board, Des Moines, IA

C H A P T E R

14

Swine Health[1]

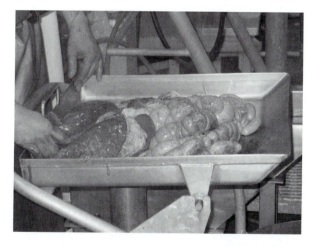

Inspecting the entrails postslaughter for evidence of infection.
(Courtesy, Sherry Hoyer, IPIC, Iowa State University, Ames, IA)

[1] A special appreciation is extended to Dr. James McKean, university professor and extension swine veterinarian, Iowa State University, Ames, IA, for his assistance in preparing this chapter.

Objectives

After studying this chapter, you should:

1. Know the normal vital signs of the pig.
2. Know the types of immunity.
3. Know the FDA classifications of drugs and limits on their usage.
4. Understand the aspects of a herd health program.
5. Know techniques for reducing the spread of diseases.
6. Know what diseases are not present in your country and what is being done to prevent reentry or to eradicate current diseases.
7. Recognize symptoms of some of the diseases most commonly appearing in your area.
8. Be aware of the internal parasites affecting swine in your area.
9. Be able to identify lice and mange infestations.
10. Be aware of the various disinfectants available to swine producers.
11. Know what zoonoses are and why they are of concern.

This chapter presents a combination of scientific and practical information relative to swine health, disease prevention, and parasite control. This should enhance the services of the veterinarian and assist producers in doing a more effective job in controlling diseases with quality information available.

Also, swine producers should be informed relative to the relationship of swine diseases and parasites to other classes of animals and to human health, because many of them are transmissible between species and humans. These diseases are called zoonoses. Accordingly, other classes of animals and humans necessarily are mentioned in the discussion that follows relative to swine diseases and parasites.

NORMAL VITAL SIGNS

The normal temperature, heart rate (pulse rate), and respiration rate (breathing rate) of swine are presented in Figure 14.1. In general, any marked or persistent deviations from these normal values may be considered as a sign of animal ill health or stress.

Swine producers should have an animal thermometer, which is heavier and more rugged than

Swine Vital Signs

Rectal Temperature
102.5°F (101.6–130.6)
39.2°C (38.7–39.8)

Heart Rate
70–120 beats per minute

Resting Respiration
32–58 breaths per minute

Figure 14.1   Swine vital signs. *(Courtesy, Palmer Holden, Iowa State University, Ames, IA)*

the ordinary human thermometer. Also, at the end opposite to the bulb, animal thermometers have an eye to which a 12-in. length of string with a clip is tied. The temperature is measured by inserting the thermometer full length into the rectum, where it should be left a minimum of 3 minutes. The clip on the string is attached to the hair of the animal.

In general, infectious diseases are first recognized with a rise in body temperature, but it must be remembered that body temperature is affected by environmental temperature, exercise, excitement, age, feed, and so on. It is lower in cold weather, in older animals, and at night.

The heart rate (pulse) is felt at the front (anterior) rim of the ear or the underside of the tail. Also, the heart beat can be felt by placing the palm of the hand between the left elbow and the side of the rib cage. It should be noted that the younger, the smaller, and the more nervous the animal, the higher the heart rate. Also, the heart rate increases with exercise, excitement, digestion, and high outside temperature.

The respiration (breathing rate) can be determined by placing the hand on the flank, by observing the rise and fall of the flanks, or, in the winter, by watching the breath condensate coming from the nostrils. Rapid breathing as a result of recent exercise, excitement, hot weather, or poorly ventilated buildings should not be confused with disease. Respiration is accelerated in pain and in febrile (feverish) conditions.

DISEASE CONTROL METHODS

Immunity

No discussion of health and disease can be complete without a brief explanation of immunity. When an animal is immune to a certain disease, it simply means that it is not susceptible to that disease. There are two forms of immunity, natural and acquired.

When immunity to a disease is inherited, it is referred to as a natural immunity. For example, when sheep are exposed to hog cholera they never contract the disease because they have a type of natural immunity referred to as species immunity. Likewise, humans are naturally immune to Texas fever. Algerian sheep are said to be highly resistant to anthrax; this constitutes a type of natural immunity called racial immunity.

The body also has the ability, when properly stimulated by a given organism or toxin, to produce antibodies and/or antitoxins. When an animal has enough antibodies for overcoming particular (disease-producing) organisms, it is said to be immune to that disease. This type of immunity is referred to as acquired.

Acquired immunity or resistance is either active or passive. When the animal is stimulated in such a manner (vaccination or actual disease) as to cause it to produce antibodies, it is said to have acquired active immunity. However, if an animal is injected with the antibodies (or immune bodies) produced by an actively immunized animal, it is referred to as an acquired passive immunity. Such immunity is usually conferred by the injection of blood serum from immunized animals, the serum carrying with it the substances by which the protection is conferred. Passive immunization confers immunity on its injection, but the immunity disappears within 3 to 6 weeks. Young mammals secure passive immunity from the mother's colostrum the first few days following birth.

In active immunity, resistance is not developed until after 1 or 2 weeks, but it is far more lasting because the animal apparently keeps on manufacturing antibodies. It can be said, therefore, that active immunity has a great advantage. There are exceptions, however—for example, the tetanus antitoxin.

Drugs

Pharmaceutical compounds and terms—For regulatory purposes, pharmaceutical compounds are termed as follows, categories with which livestock producers should be familiar:

1. Over-the-counter (OTC) drugs. These are drugs that are available to the general public without prescription. They are sold for general use by the purchaser in accordance with the application on the label.

2. Prescription drugs. These are drugs that are for use by, or on the order of, the veterinarian. These are always identified by the following statement on the label:

"Caution. Federal law restricts this drug to use by or on the order of a licensed veterinarian."

3. Extralabel drugs. The Animal Medicinal Drug Use Clarification Act of 1994 (AMDUCA)[2] allows veterinarians to prescribe extralabel uses of certain approved animal drugs and approved human drugs for animals under certain conditions. Extralabel use refers to the use of an approved drug in a manner that is not in accordance with the approved label directions. The key constraints of AMDUCA are that any extralabel use must be by or on the order of a veterinarian within the context of a veterinarian–client–patient relationship, must not result in violative residues in food-producing animals, and the use must be in

---

[2]*http://www.fda.gov/cvm/index/amducca/amducatoc.htm*

conformance with the implementing regulations published at Title 21 CFR 530 (Code of Federal Regulations).

A list of drugs specifically prohibited from extralabel use (Title 21 CFR Part 530.41) appears in the Code of Federal Regulations. *Note:* No coccidiostats are approved for use in swine. The treatments suggested in this book for the control of coccidiosis are extralabel uses.

This category is not a part of the law or regulations, but it is recognized in discretionary privileges granted to veterinarians. Use of over-the-counter drugs in therapies or dosages not approved by the label constitutes extralabel drug use. Feed additives cannot be used as extralabel products.

4. Veterinary feed directives (VFD).[3] A veterinary feed directive is a written statement that authorizes the client (the owner of the animal or animals or other caretaker) to obtain and use animal feed containing a VFD drug to treat their animals only in accordance with the FDA-approved directions for use. A veterinarian may issue a VFD only if a valid veterinarian–client–patient relationship exists, as defined in Title 21 CFR 530.3(i). The information needed on a VFD is stated in the final rule.

A VFD drug is a drug approved by the FDA for use in animal feeds and which is limited to use under the professional supervision of a licensed veterinarian. No extralabel use of a VFD drug is permitted. Although statutory controls on the distribution and use of VFD drugs are similar to those for prescription animal drugs, the implementing VFD regulations are tailored to the unique circumstances relating to the distribution of animal feeds containing a VFD drug. This rule helps ensure the protection of public health while enabling animal producers to obtain and use needed drugs as efficiently and cost effectively as possible. Appropriate withdrawal times must be set by the veterinarian writing the directive.

Extralabel feed additives cannot be prescribed by veterinarians or mixed into livestock diets by commercial feed manufacturers or on-farm mixers in the United States.

LIFE CYCLE HERD HEALTH

Successful swine production necessitates the application of health-conserving, disease-prevention, and parasite-control measures to the breeding, feeding, and management of the herd. Although the basic principles and objectives of swine health remain the same, their application has changed with the increase in unit size, combined with greater intensification and sophistication. Two other relevant developments have evolved: The growth of large seedstock producers, and the production and sale of feeder pigs for growing-finishing by another hog feeder.

Two management techniques developed that enhanced herd health are early weaning and age segregation (all-in/all-out or AIAO). These two techniques have often been combined under the title of segregated early weaning (SEW) but obviously the AIAO continues with a group of pigs from weaning until they are marketed. A more complete discussion of SEW is found in Chapter 15.

Today, swine enterprises are so diverse and patterns of swine health are so varied that methods of disease prevention and parasite control must be adapted to each operation. Table 14.1 is presented for guidance to assist operators and their veterinarians in developing programs adapted to their individual enterprises. Special problems of each hog operation should be taken into consideration. Also, locale, separation distances, type and size of operation, and government regulations will influence the herd health program. Various biosecurity programs, including early weaning, segregated early weaning (SEW), all-in/all-out (AIAO), specific pathogen-free (SPF), and the McLean County System, are discussed in Chapter 15.

ANIMAL MOVEMENT REGULATIONS

Federal and State Regulations

Certain diseases are so devastating that swine producers cannot protect their herds against invasion. Moreover, where human health is involved, the problem is too important to be entrusted to individual action. In the United States, therefore, certain regulatory activities in animal disease control are under the supervision of various federal and state organizations. Federally, this responsibility is entrusted to the USDA Animal and Plant Health Inspection Service (APHIS).

In addition to the federal interstate regulations, each of the states has requirements for the entry and movement of livestock. Generally, these requirements include compliance with interstate regulations. States usually require a certificate of health or a permit, or both, and additional testing requirements depending on the class of livestock involved. Abide by the federal and state animal health regulations. These laws are for your and the livestock industry's protection.

---

[3] *http://www.fda.gov/cvm/index/updates/vfdfinal.htm*

TABLE 14.1   Herd Health Management Timetable for the Swine Breeding Herd

Time (age)	Vaccines and Parasite Control	Management and Breeding
Gilts/Sows		
6 1/2 months	Deworm, treat for lice and mange, feed fresh manure from boars and sows to cause an infection to which antibodies are produced. Commingle with cull sows, and initiate fence line contact with boars. Vaccinate for leptospirosis, erysipelas, parvovirus, PRRS, and PRV.	Isolate purchased gilts for 60 days. Blood test for important disease(s) not already present in the herd.
7 1/2 months	Repeat vaccinations, except PRRS.	
3 weeks postbreeding		Pregnancy check nonreturns to heat.
35 to 60 days postbreeding		Pregnancy test (36 to 60 days postbreeding).
6 weeks prior to farrowing	*Clostridium* toxoid	
4 to 6 weeks prior to farrowing	*E. coli* bacterin, *Pasteurella* (AR), *Mycoplasma*, TGE, and PRV. Treat for lice and mange.	
2 weeks prior to farrowing	*E. coli* bacterin, *Clostridium*, *Mycoplasma*, TGE, AR.	May include feed additives through lactation to prevent *Clostridium*.
7 to 10 days prior to farrowing	Deworm; treat for lice and mange.	May include feed additives to prevent constipation. Wash sows thoroughly with detergent before entering farrowing house.
Farrowing	Leptospirosis, parvovirus, and erysipelas, PRRS, PRV for sows.	Record litter and sow information.
2 to 5 weeks postfarrow	Treat for lice and mange.	Wean pigs. Provide comfort, sanitation, and adequate diet.
Boars		
4 to 6 months		Select and bring to farm at least 60 days prior to breeding. (Boars are ready for limited use at 8 months of age.) Isolate purchased boars for 60 days. Blood test for important diseases not already present in the herd.
First 30 days following purchase in isolation	Test for brucellosis, leptospirosis, PRRS, parvovirus, *Actinobacillus*, TGE, and PRV. Treat for lice and mange and deworm.	Feed unmedicated feed, and observe for diarrhea, lameness, pneumonia, and ulcers.
Second 30 days following purchase in isolation	Vaccinate for erysipelas, leptospirosis, and parvovirus.	Feed manure from other boars and sows. Commingle with cull gilts, and observe desire and ability to breed. Provide fence line contact with gilts and sows to be bred.
Every 6 months	Revaccinate PRV, leptospirosis, erysipelas, and parvovirus; then, deworm. Treat for lice and mange.	Detusk.
Pigs		
1 day	*Clostridium* antitoxin.	
1 to 3 days	Iron injection (200 mg).	Clip needle teeth. Dock tails. Ear notch. Castrate.
3 to 7 days	Vaccinate for AR and TGE.	
10 to 21 days		Start creep feed; wean if SEW.
3 to 4 weeks	Vaccinate for AR, PRRS, *Mycoplasma*, and *Salmonella choleraesuis*.	Wean.

(*continued*)

TABLE 14.1   Herd Health Management Timetable for the Swine Breeding Herd *(continued)*

Time (age)	Vaccines and Parasite Control	Management and Breeding
Pigs		
Weaning + 10 days	Treat for lice, mange; then deworm.	
Weaning + 20 days	Vaccinate with erysipelas and *Actinobacillus pleuropneumoniae* bacterins.	
10 to 12 weeks	Vaccinate for PRV and revaccinate with erysipelas and *Actinobacillus pleuropneumoniae* bacterins.	Fecal exam for internal parasites.
5 to 6 months	Follow all vaccination withdrawal times prior to slaughter.	Health check 20% or up to 30 hogs from a market group. Follow all feed and injectable antibiotic withdrawal times prior to slaughter.

Source: Adapted from Pork Industry Handbook, *PIH-68, Purdue University, West Lafayette, IN, 1997.*

Figure 14.2   Producers need to be able to identify pigs that are not growing normally or exhibit symptoms of distress, such as this pig with the sore legs and abnormally arched back. *(Courtesy, Palmer Holden, Iowa State University, Ames, IA)*

Vaccinations may or may not be required depending on various state and current disease control efforts. Visual inspection and health certificates are required to be furnished by the seller to the buyer. All animals in interstate commerce with the exception of those going directly to slaughter must be accompanied by an interstate health certificate signed by a veterinarian who has inspected them (Figure 14.2).

Detailed information relative to animal disease control can be obtained from federal and state animal health officials, or from accredited veterinarians in all states. Shippers are urged to obtain such information prior to making interstate shipments of livestock.

Quarantine—Many highly infectious diseases are prevented by quarantine from (1) gaining a foothold in a country or (2) spreading. Quarantine *involves (1) segregating and confining of one* or more animals in the smallest possible area to prevent any direct or indirect contact with animals not so restrained or (2) regulating movement of animals at points of entry. When an infectious disease outbreak occurs, drastic quarantine measures must be imposed to restrict movement out of an area or within areas. The type of quarantine varies from one involving a mere physical examination and movement under proper certification to the complete prohibition against the movement of animals, produce, vehicles, and even human beings.

Indemnities—Where certain animal diseases are involved, the swine producer can obtain financial assistance in eradication programs through federal and state sources. Information relative to indemnities paid to owners by the federal government may be secured from APHIS and for each state by writing to the respective state's departments of agriculture.

Harry S. Truman Animal Import Center

A federal quarantine center opened in 1979 near Key West, Florida, was named the Harry S. Truman Import Center, and was closed in 1998. It operated under the administration of the USDA's Animal and Plant Inspection Service.

The center had capacity for about 400 head of cattle or other animals at one time, for a 5-month quarantine period. This maximum security station enabled American livestock producers to import breeding animals from all parts of the world, while at the same time safeguarding our domestic herds from foreign diseases.

Plum Island Animal Disease Center

The Plum Island Animal Disease Center (PIADC) is responsible for research and diagnosis to protect

Figure 14.3 The Plum Island Animal Disease Center located east of New York City. *(Courtesy, Plum Island Animal Disease Center, Orient, NY)*

the U.S. animal industries and exports from catastrophic economic losses caused by foreign animal disease agents accidentally or deliberately introduced into the United States (Figure 14.3). Certain highly infectious foreign animal diseases, such as foot-and-mouth disease, can be studied only at Plum Island, not on any other part of U.S. soil. The PIADC was transferred from USDA to the Department of Homeland Security in 2003.

The mission at Plum Island is carried out jointly by scientists and support staff of the USDA's Agricultural Research Service and Animal and Plant Health Inspection Service. Together, the scientists work to

1. Develop new strategies to prevent and control foreign or emerging animal disease epidemics.
2. Conduct diagnostic investigations of suspected cases of foreign or emerging animal diseases in the United States, or in countries abroad

through cooperation with animal health international organizations.

3. Test imported animals and animal products to assure they are free of foreign animal disease agents.
4. Assess risks involved in importation of animals and animal products from countries where epidemic foreign animal diseases occur.
5. Produce and maintain materials used in diagnostic tests for foreign animal diseases.
6. Test and evaluate vaccines for foreign animal diseases, and maintain the North American foot-and-mouth disease vaccine bank.
7. Train veterinarians and animal health professionals in the diagnosis and recognition of foreign animal diseases through courses at PIADC and at other domestic and international locations.

DISEASES OF SWINE

Adequate sanitation is the first and most important requirement that must be fulfilled if the swine enterprise is to reduce its exposure to disease organisms. Filthy quarters, feeding floors, and watering places favor the entrance of disease-producing germs into the body of the animal. In addition to maintaining a program of sanitation, the good caretaker is ever alert in observing any deviation from the normal in the functions of the animal—such as loss of appetite, lameness, digestive disturbances, and so on.

A veterinarian should be called on should there be the appearance of serious trouble, but often intelligent aid may be given before a veterinarian can be reached. Moreover, the appearance of serious disorders can be recognized and control measures instituted before the spread of the disease has made much progress. The caretaker who is sufficiently familiar with the nature and causes of the common diseases is in a position to employ sound preventive measures.

This chapter discusses nonnutritional diseases and ailments. Nutritional diseases caused by excesses or deficiencies of nutrients are reviewed in Chapter 6. The most prevalent diseases in the United States are graphed in Figure 14.4.

Actinobacillus pleuropneumonia (APP)

Actinobacillus pleuropneumonia (APP) was formerly called *Haemophilus pleuropneuomia* and is a severe pneumonia striking young pigs (up to 6 months of age). Although pigs infected with APP will often die, the greatest economic losses result from pigs surviving the disease because the chronic lung lesions from the disease create "poor-doers." Average daily gains decrease and feed per unit of gain increases.

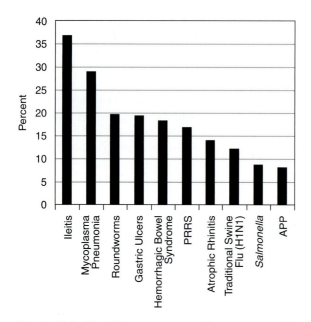

Figure 14.4 The 10 most common diseases reported in growing-finishing swine in the United States in 2000. *(Courtesy, USDA, Washington, DC, http://www.aphis.usda.gov/vs/ceah/cahm/Swine/Swine2000/Swine2Highlights.pdf)*

The distribution of APP is worldwide, causing significant economic losses to the swine industries of such countries as Australia, Canada, Denmark, Germany, Korea, Mexico, Taiwan, South America, Switzerland, and, of course, the United States.

Symptoms and Signs

The incubation period is short, sometimes 8 to 12 hours, and it commonly strikes pigs from 40 lb to market weight. Often the first indication of APP is the sudden death of apparently healthy pigs, though this sudden death follows a stressful period, for example, moving, mixing, or rapid weather change. When observed, healthy pigs may develop labored breathing and die within minutes of a stress. Bleeding from the nose may be observed in some pigs but not all. Pigs less severely affected may demonstrate thumping, a fever of 104 to 107°F, depression, and reluctance to move. Prominent lesions include hemorrhagic, edema-filled lungs. Because the causative organism is spread via aerosol droplets, morbidity reaches 100%, whereas mortality is 20 to 40%, when immediate and effective treatment is instituted.

Cause, Prevention, and Treatment

Often a tentative diagnosis can be made based on the history, age of affected pigs, signs and symptoms, and lung lesions, but a definitive diagnosis may require the culture of the causative bacteria— *Actinobacillus pleuropneumoniae*—from a lung lesion.

Presently, the best prevention seems to be prevention of its spread and introduction into a herd. The major source of infection is the carrier pig that has recovered from APP. New animals should be isolated and tested to determine if they are carriers before being introduced into an uninfected herd. Vaccination with a killed vaccine is effective in protecting susceptible animals and animals exposed to the disease.

Treatment during an acute outbreak consists of maintaining high blood levels of antibiotics, such as procaine penicillin and long-acting oxytetracycline (LA-200), by injection. Following an outbreak, antibiotic sensitivity tests should be used to determine the most effective drug. To reduce death losses, both healthy and sick swine should be treated. The treatment of animals showing clinical signs has not demonstrated uniform success due to the development of lung damage.

African Swine Fever

African swine fever (ASF) was first described in Kenya in 1921. Under natural conditions it affects only swine. It is endemic in Sardinia and parts of Africa, particularly the sub-Saharan countries. It has appeared in other parts of the world, but these are non-ASF-free.

Symptoms and Signs

ASF can resemble classical swine fever (hog cholera) and erysipelas. Clinical symptoms range from peracute (often 100% sudden deaths without previous symptoms) to subclinical. Acute forms are characterized by hemorrhagic lesions in the internal organs and especially the skin of the ears and flank (Figure 14.5). The incubation period is 4 to 19 days depending on the dose and route of infection.

Cause, Prevention, and Treatment

ASF is caused by a virus and maintained by a cycle between wild boars and soft ticks. Wild boars can transmit the disease directly to domesticated pigs. It can be spread between countries through the introduction of uncooked, infected pork at international airports where garbage can be found and fed to pigs.

Prevention is accomplished by preventing the feeding of uncooked garbage from international airports or ports. Controlling pig movement and surveys to identify carrier pigs is an important aspect of prevention. In Africa, reducing contact between domesticated pigs and wild pigs and soft ticks is indicated.

At present there is no treatment or effective vaccine against ASF.

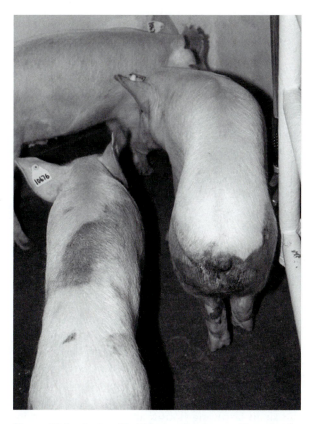

Figure 14.5 A pig with African swine fever with severe internal bleeding occurring through the anus. *(Courtesy, Plum Island Animal Disease Center, Orient, NY)*

Anthrax (Splenic Fever, Charbon)

Anthrax, also referred to as splenic fever or charbon, is an acute infectious disease infecting all warm-blooded animals (and humans). It usually occurs as scattered outbreaks or cases, but hundreds of animals may be involved. Certain geographic regions are known as anthrax districts because of the repeated appearance of the disease. Grazing animals are particularly subject to anthrax, especially when pasturing closely following a drought or on land that has been recently flooded.

Historically, anthrax is of great importance. It is one of the first scourges to be described in ancient and biblical literature; it marks the beginning of modern bacteriology, being described by Koch in 1877; and it is the first disease in which immunization was effected by means of an attenuated culture, Pasteur having immunized animals against anthrax in 1881.

Symptoms and Signs

The mortality rate is usually quite high. It runs a very short course and is characterized by a blood poisoning (septicemia). In swine, the disease is usually evidenced by swelling of the neck region (lymph nodes), which leads to death from suffocation and blood poisoning. It is accompanied by high temperature, loss of appetite, muscular weakness, difficult breathing, depression, and the passage of blood-stained feces.

Cause, Prevention, and Treatment

The disease is identified by a microscopic examination of the blood or lymph nodes in which will be found the typical large, rod-shaped organisms causing anthrax, *Bacillus anthracis*. These organisms can survive in soils and organic material for years in a spore stage, resisting all destructive agents. As a result, anthrax may remain in the soil for extremely long periods. Swine, however, rarely contact a sufficient dosage of anthrax spores from the soil to cause infection. Outbreaks in swine have been traced to contaminated feed, which in most cases contained products of animal origin.

Immunization is not generally practiced on a large scale because swine possess a level of natural resistance sufficient to prevent the disease unless there is heavy exposure to the bacillus. Herds that are infected should be quarantined and withheld from the market until the danger of disease transmission is past. The swine producer should never open the carcass of a dead animal suspected of having died from anthrax; instead, the veterinarian should be summoned at the first sign of an outbreak.

When the presence of anthrax is suspected or proved, all carcasses and contaminated material should be completely burned, or covered with lime and deeply buried, preferably on the spot. This precaution is important because the disease can be spread by dogs, coyotes, buzzards, and other flesh eaters and by flies and other insects.

When an outbreak of anthrax is discovered, state and federal veterinary regulatory officers need to be contacted immediately. All sick animals should be isolated promptly and treated, even though treatment of infected animals is not too satisfactory. Penicillin or the tetracyclines are effective if given early. All exposed healthy animals should be vaccinated; pastures should be rotated; the premises should be quarantined; and a rigid program of sanitation should be initiated. These control measures should be carried out under the supervision of a veterinarian.

Atrophic Rhinitis

Atrophic rhinitis (AR) is widespread throughout the world. Apparently, it affects only swine because it does not seem to be related to atrophic rhinitis in humans. Nationwide, it is estimated that AR is present in market hogs that have at least mild lesions of turbinate atrophy.

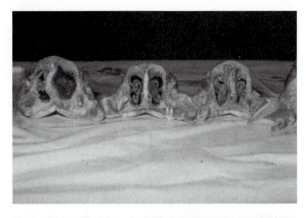

Figure 14.6 The left cross-section depicts a normal turbinate obtained by sawing through the snout of a pig. The center indicates much of the turbinates are atrophied and the right snout has one set of the turbinates completely missing. *(Courtesy, Palmer Holden, Iowa State University, Ames, IA)*

Figure 14.7 A finishing pig with the characteristic twisted snout resulting from turbinate atrophy. *(Courtesy, Palmer Holden, Iowa State University, Ames, IA)*

Symptoms and Signs

Atrophic rhinitis is a transmissible disease that is characterized by rhinitis—inflammation of the mucous membranes of the nose—and wasting away or lack of growth of the turbinate bones of the nose—small, scroll-like structures that warm, moisten, and filter incoming air (Figure 14.6). There are two types of rhinitis, the mild nonprogressive atrophic rhinitis caused by toxigenic *Bordetella bronchiseptica* (Bb) and the progressive atrophic rhinitis caused by toxigenic *Pasteurella multocida* (Pm) alone or in combination with other agents such as Bb.

Persistent sneezing, which becomes more pronounced as the pigs grow older, is the first symptom, sometimes appearing in the first week after birth and more frequently by weaning time. Appetite is only slightly depressed with Bb. The signs tend to ease after a few weeks unless pigs are also infected with Pm. Uncomplicated Bb infections in older pigs produce only mild signs or possibly none.

When also infected with Pm, the most characteristic clinical sign is the deformity of the nose. At 4 to 12 weeks of age, the snout begins to show wrinkles, and it may bulge and thicken. At 8 to 16 weeks of age, the snout and face may twist to one side (Figure 14.7). Nose bleeding is often seen. Infected pigs become rough all over, and make slow and inefficient gains. Actual death may be caused by pneumonia. All ages of swine are susceptible, but the most severe cases develop when pigs are infected at only a few weeks of age. Swabs from the noses of breeding animals can be cultured for the presence of the causative organism.

Cause, Prevention, and Treatment

In the United States, the primary causative agents of atrophic rhinitis are toxigenic strains of *Bordetella bron-* *chiseptica* (Bb) and *Pasteurella multocida* (Pm). *Haemophilus parasuis* may cause mild turbinate atrophy. *Bordetella* is carried in the respiratory tract of numerous mammals including humans, rats, cats, and dogs, though not all strains have equal disease-causing capabilities. Its spread is via infected sow to her litter, the air in farrowing houses and nurseries, and non-swine sources such as cats. A calcium–phosphorus imbalance or a calcium deficiency in growing pigs has been shown to produce similar lesions.

Preventive measures—or at least measures to limit rhinitis—consist of some of the following:

1. Keep the percentage of replacements to a minimum—maintain older breeding stock.
2. Use SPF pigs.
3. Keep farrowing quarters and nursery clean. Use the all-in/all-out system if possible.
4. Isolate bred females in separate lots and never allow contact with any other swine except their offspring until they are culled. Keep individual litters separate until a month after removal of the sow at weaning time. Then, select and isolate new breeding stock from those litters that show no evidence of symptoms, and also react negatively to nasal swabbing tests for *Bordetella*.
5. On a continual basis, cull animals that are visibly affected.
6. Vaccinate, using Bb or Pm bacterins/ toxoids. Most direct benefit is obtained by vaccinating the pigs or in extreme cases a combination of vaccinating the sow twice prefarrowing. Follow the directions of the manufacturer of the product used. Combination Bb/Pm vaccines are also available.

Once a vaccination program is begun, it usually must be continued for many years.

Treatment of rhinitis is based on the use of feed medications containing sulfonamides with tetracyclines. The most effective management control of atrophic rhinitis is all-in/all-out farrowing and nursery with thorough sanitation between groups.

The most effective way to eradicate atrophic rhinitis from a herd is complete depopulation and repopulation with atrophic rhinitis-free swine. Atrophic rhinitis may be eradicated from herds that need to preserve their genetic base (i.e., seedstock herds) by following the SPF program (Caesarean delivery, raised in isolation) or by early weaning. After a repopulation to eliminate the atrophic rhinitis, the following steps help to maintain a rhinitis-free herd:

1. Practice the closed-herd concept.
2. Add boars that have passed 3 consecutive negative nasal swabbing tests, or use SPF boars.
3. Control the cat and rodent population.
4. Although not necessary, nasal swabbing and culturing will assist in monitoring the status of the herd. Routine swabbing does add expense to the operation.

Brucellosis

Brucellosis is an insidious (hidden) disease in which the lesions frequently are not evident. Although the medical term *brucellosis* is used in a collective way to designate the disease caused by the 3 different but closely related *Brucella* organisms, the species names further differentiate the germs as (1) *Brucella abortus*, (2) *B. suis*, and (3) *B. melitensis*.

Brucella abortus is known as Bang's disease (after Professor Bang, noted Danish research worker, who, in 1896, first discovered the organism responsible for bovine brucellosis) or contagious abortion in cattle; *B. suis* causes Traum's disease or infectious abortion in swine; and *B. melitensis* causes Malta fever, or abortion, in goats. The disease is known as Malta fever, Mediterranean fever, undulant fever, or brucellosis in humans. The causative organism is often associated with fistulous withers and poll-evil of horses.

Swine brucellosis control and eradication is important for two reasons: (1) the danger of human infection and (2) the economic loss. The most accurate, and possibly the most sensitive method, of diagnosis of swine brucellosis is isolation of *Brucella* organisms. The U.S. swine population is brucellosis-free except for feral swine or transitional herds with feral swine exposure. Other countries are not free.

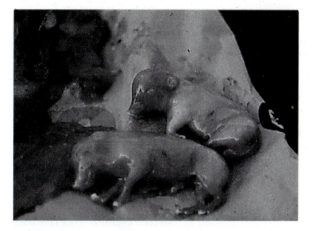

Figure 14.8 Aborted fetuses typically observed with brucellosis infections. *(Courtesy, USDA, Washington, DC)*

Symptoms and Signs

Unfortunately, the symptoms of brucellosis are often rather indefinite. It should be borne in mind that not all animals that abort are infected with brucellosis and that not all animals infected with brucellosis will abort. However, every case of abortion should be regarded with suspicion until proved noninfectious (Figure 14.8).

In swine, abortion and sterility are not as common as in cattle: Infection may cause swollen joints and lameness, and swelling or atrophy of the testes, epididymis, and prostate in the male.

Cause, Prevention, and Treatment

The disease is caused by bacteria called *Brucella suis* in swine, *B. abortus* in cattle, and *B. melitensis* in goats. The *suis* and *melitensis* types are seen in cattle, but the incidence is rare.

Humans are susceptible to all three types of brucellosis. The swine organism causes a more severe disease in human beings than the cattle organism, although not so severe as that induced by the goat type. Fortunately, far fewer people are exposed to the latter simply because of the limited number of goats and the rarity of the disease in goats in the United States. Producers are aware of the possibility that human beings may contract undulant fever from handling infected animals, especially at the time of parturition; from slaughtering operations or handling raw meats from infected animals; or from consuming raw milk or other raw by-products from goats or cows, and eating uncooked meats infected with brucellosis organisms. The simple precautions of pasteurizing milk and cooking meat, however, make these foods safe for human consumption.

The brucella organism is relatively resistant to drying, but is killed by the common disinfectants and by pasteurization. The organism is found in immense numbers in the various tissues of the aborted young and in the discharges and membranes from the aborted animal. It is harbored indefinitely in the udder and may also be found in the sex glands, spleen, liver, kidneys, bloodstream, joints, and lymph nodes.

Brucellosis appears to be commonly acquired through the mouth in feed and water contaminated with the bacteria, or by licking infected animals, contaminated feeders, or other objects to which the bacteria may adhere. There is also evidence that boars frequently transmit the disease through the act of service.

Strict sanitation; the recognition and removal of infected animals through testing programs; isolation at the time of parturition; and the control of animals, feed, and water brought into the premises are the keys to the successful control and eradication of brucellosis. Sound management practices, which include either buying replacement animals that are free of the disease or raising all females, are a necessary adjunct in prevention. Drainage from infected areas should be diverted or fenced off, and visitors (people and animals) should be kept away from animal barns and feedlots. Animals taken to livestock shows and fairs should be isolated on their return and tested 30 days later.

Vaccination procedures in swine have not been successful in brucellosis control programs. Since 1989 all states in the United States have been part of a program for the eradication of brucellosis. The guidelines for this successful eradication program are set forth in the Uniform Methods and Rules for Brucellosis Eradication.[4] Current information regarding the program is available from state veterinarians. In general, herds are established and maintained as validated brucellosis-free through surveillance blood testing. If a herd is infected or suspected of being infected with *B. suis*, 3 plans are recommended:

• **Plan I**—Depopulation. Market the entire herd for slaughter; clean and disinfect houses and equipment; restock premises with animals from validated brucellosis-free herds.

• **Plan II**—For use in salvaging irreplaceable bloodlines. Retain weanling pigs for breeding and market infected adult animals for slaughter as soon as practical; isolate and test replacement gilts before breeding, saving only those that are negative and breeding them to negative boars; retest gilts after farrowing and before moving them from individual farrowing pens, segregate any reactors and retain

only pigs from negative sows for breeding purposes; and repeat this procedure until the herd has 2 consecutive negative tests, at which time the herd is eligible for validation.

• **Plan III**—For use in herds where only a few reactors are found and no clinical symptoms of brucellosis have been noted. Market reactors for slaughter; retest herd at intervals, removing reactors for slaughter, until the entire herd is negative for 2 successive tests. If the herd is not readily freed of infection, abandon this plan in favor of plan I or plan II.

No known medicinal agent is effective in the treatment of brucellosis in any class of farm animals. Therefore, the producer should not waste valuable time and money on so-called "cures."

Classical Swine Fever (Hog Cholera)

Classical swine fever (CSF) is a highly contagious viral disease infecting only swine. Currently Australia, Belgium, Canada, France, Great Britain, Iceland, Ireland, New Zealand, Portugal, the Scandinavian countries, Spain, Switzerland, and the United States are free of the disease. A federal–state hog cholera eradication program was initiated in the United States in 1962 and the last outbreak occurred in 1976. Eradicating hog cholera in the United States cost approximately $140 million.

Symptoms and Signs

The course of the disease may be acute, subacute, chronic, atypical, or inapparent. Acute CSF is caused by a virulent virus and generally results in high morbidity and mortality, whereas the low-virulence infections may go unnoticed. CSF is considered a foreign animal disease for the United States, and a state or federal veterinary official should be contacted if there is any thought that CSF is present.

The symptoms of the acute and subacute appear after an incubation period of 2 to 6 days. The disease is marked by a sudden onset, fever, loss of appetite, and weakness—although some pigs may die without showing any symptoms. Affected animals separate themselves from the herd; show a wobbly, scissorlike gait; and, although refusing feed, may drink much water. The underside of the pigs may show a purplish red coloring (also seen in erysipelas and in other acute febrile diseases). Pigs with acute infection usually die between 10 and 20 days postinfection.

Sometimes there may be chilling, causing the infected animals to bury themselves in the bedding or to pile up. There is constipation alternating with diarrhea, and coughing is often evident. There may be

---

[4] *http://www.aphis.usda.gov/oa/pubs/bruumr.pdf*

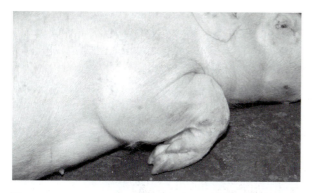

Figure 14.9 A pig infected with classical swine fever. Note the discharge around the eye, lesions on the skin, and reddening of the skin caused by increased blood flow. *(Courtesy, Plum Island Animal Disease Center, Orient, NY)*

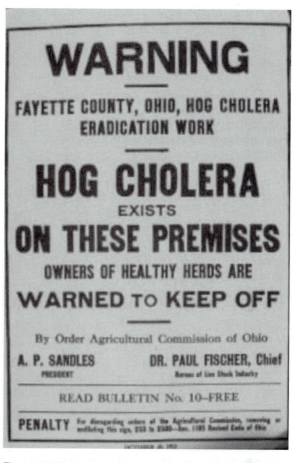

Figure 14.10 A quarantine sign from Fayette County, Ohio, warning of the presence of hog cholera (classical swine fever) on the premises. *(Courtesy, USDA, Washington, DC)*

a discharge from the eyes (Figure 14.9). Often the disease is associated with pneumonia or enteritis. Because this disease is often confused with erysipelas, diagnosis should include a postmortem examination of one or more of the pigs that have recently died, a bacteriological check for the erysipelas bacteria, and perhaps a blood test for white cell count. The Fluorescent Antibody Test (FAT) is widely used by pathologists to diagnose this disease.

Persistent CSF infections are generally caused by strains of reduced virulence and are divided into chronic and late-onset CSF. Initially, chronic CSF resembles acute CSF but the virus spreads more slowly. Late-onset initially is not apparent and symptoms may not appear for several months.

Cause, Prevention, and Treatment

CSF is caused by a virus that belongs to the genus *Pestivirus*. Prevention includes accepting animal imports only from countries free of CSF and only by-products that have been sterilized. If CSF appears in a country free of the disease, all infected herds are destroyed as well as traces to find all possible contacts with the infected herd. No animal movement is permitted and infected farms are intensely disinfected.

In countries where CSF is present, vaccination may be used, often supplementary to destroying infected herds. There is no treatment.

Following is a brief review of the successful eradication program in the United States. The USDA first introduced the simultaneous serum-virus hog cholera vaccination in 1908. This type of immunization checked the widespread hog cholera epidemics of the old days. A negative side effect was that it also kept alive the hog cholera virus (which it contained). It provided sufficient control to allow U.S. swine producers to live with the disease.

In 1951, the USDA approved modified live virus (MLV) preparations of rabbit, swine, or tissue culture

origin consisting of modified (attenuated) live virus capable of producing long-lasting immunity. Unfortunately, they were capable of spreading the infection to susceptible animals. As a result, the USDA banned all interstate shipments of modified hog cholera vaccines after March 1, 1969 (Figure 14.10).

The passage of Public Law 87-209 in 1961 provided for the eradication of hog cholera in the United States. Federal funds were made available in 1962, and the following four-phase program was initiated.

Phase I: Preparation—Education; surveys to determine incidence to improve and standardize diagnostic systems, and to promote disease reporting; and enforcement of garbage cooking or bans on garbage feeding by some states, particularly in the Midwest.

Phase II: Reduction of incidence—Quarantine of infected and exposed pigs, and stopping intrastate movement of infected pigs.

Phase III: Elimination of outbreaks—Depopulation (with indemnity) of infected premises, and increasing the control of biologics used.

Phase IV: Protection against reinfection—Prompt enforcement of all procedures in phase III and a 21-day segregation of all swine imported from outside the state, with restrictions on the importation of pigs from states having endemic hog cholera.

Dysentery, Swine (Bloody Scours, Hemorrhagic Dysentery, Black Scours, and Vibrionic Dysentery)

Swine dysentery, an acute infectious disease, has been reported from coast to coast, but it is most common in the Corn Belt, where the swine population is densest. The disease is primarily spread by movement of pigs or by manure contact. Infected animals are carriers. In untreated herds, morbidity may be up to 90%, and mortality up to 30%.

Symptoms and Signs

The most characteristic symptom of swine dysentery is a profuse bloody diarrhea. Sometimes the feces are black instead of bloody and contain shreds of tissue. Most infected animals go off feed and experience a moderate rise in temperature. Some pigs die suddenly after a couple days of illness, whereas others linger on for 2 weeks or longer. On autopsy or postmortem, the large intestine is found to be inflamed and bloody. The lesions in the large intestine is a differentiating symptom from other enteric pathogens. A whipworm infection can cause similar signs.

Cause, Prevention, and Treatment

Swine dysentery appears to be caused by *Brachyspira hyodysenteriae*, which is ingested with contaminated fecal material. A diagnosis of swine dysentery is confirmed by the isolation and identification of *B. hyodysenteriae* from the lining of the colon or from feces. It can be carried from herd to herd on contaminated boots or vehicle tires.

Eradication of swine dysentery in infected herds requires (1) depopulation, cleanup, disinfection, and repopulation with *B. hyodysenteriae*–free stock; (2) medication (carbadox, lincomycin, and tiamulin); (3) sanitation; and (4) elimination of rodents.

Edema Disease (Enterotoxemia, Gut Edema, Bowel Edema)

Edema disease is an acute, usually fatal disease of young pigs, which usually occurs 1 to 3 weeks after weaning. Often it is found under conditions of excellent management and nutrition. Morbidity is usually low, but case mortality rate is very high—up to 100%. It appears to be increasing in the United States, possibly as a result of earlier weaning.

Symptoms and Signs

The disease usually strikes the most thrifty, rapid-growing pigs. Often, the first indication of an outbreak is the sudden death of one or more pigs in a group. Constipation, inability to eat, swollen eyelids, and a staggering gait may be observed. Affected pigs may display nervous symptoms, such as fits or convulsions. As the disease progresses, the hog becomes completely paralyzed. Death usually follows from a few hours to 2 to 3 days. The term *gut edema* is derived from the common postmortem findings of a jellylike swelling (edema) of the stomach and intestinal submucosa with leakage into the peritoneal cavity.

Cause, Prevention, and Treatment

The disease is caused by certain serotypes of *E. coli* that produce a specific toxin (enterotoxigenic). The recommended prevention and control consists of antibiotics in the feed or water; restricted feeding and small frequent feedings; creep feeding; and high-fiber diets. Presently, autogenous vaccines are available and used to prevent the disease.

If clinical signs have developed, treatment is ineffective. In general, treatment is aimed at reducing the factors believed to predispose an increase of *E. coli* and avoiding sudden changes in management. Overall, the edema disease treatment and management practice that has earned the best rating involves withholding (for 24 hours) or a sharp reduction in the amount of feed for a short but variable length of time, and avoiding stresses.

Enteric Colibacillosis (Baby Pig Scours, Colibacillosis, Scours)

Colibacillosis (scours) may infect piglets soon after birth, older nursing piglets, and weaned piglets. Intestinal disorders can be caused by a variety of bacteria and viruses, and within a herd these infectious agents may cause disorders concurrently or sequentially. Therefore, it is necessary for laboratory tests to identify the agent causing the intestinal disorder. In pigs less than 7 days old, pathogenic (disease-causing) strains of *Escherichia coli* bacteria rapidly proliferate in the small intestine, producing toxins which produce diarrhea. The danger of diarrhea in baby pigs is the death of piglets through dehydration. This disease is often called baby pig scours, scours, or colibacillosis.

Baby pig scours is so acute that even with proper treatment, death and setbacks in performance make it very costly to producers. It is far better to prevent this disease than to be continuously fighting it.

Symptoms and Signs

The symptoms and signs are clear, watery to yellowish-brown pasty diarrhea; dehydration; depression, gauntness, inflamed perineum; and variable death loss—the highest death loss is in pigs from 0 to 4 days old.

Enteric colibacillosis infection in young unweaned pigs must be differentiated from other common infectious causes of diarrhea in pigs of this age group, including TGE virus, rotavirus, and coccidia. A preliminary diagnosis may be made by determining the fecal pH. The pH of the feces of pigs infected with enteric colibacillosis is alkaline, whereas the feces from diarrhea as a result of TGE virus or rotavirus infection are acidic.

Cause, Prevention, and Treatment

Diagnosis requires laboratory tests to be certain that the cause is a pathogenic strain of *E. coli* because other bacteria and viruses can cause diarrhea and not all strains of *E. coli* produce disease.

Prevention of baby pig scours can be viewed as a 3-pronged approach:

1. A good sanitation program reduces the number of *E. coli* bacteria in the environment. In an environment where the ventilation is poor, the humidity high, and it is dirty and wet, large numbers of *E. coli* are present. Moreover, in herds with baby pig scours, affected pigs should be treated and liquid stools covered to limit the spread of the bacteria.

2. An overall program of good management—nutrition and herd health—ensures the delivery of vigorous piglets and satisfactory lactation. Moreover, proper management of the farrowing house to avoid stressful conditions for the piglets, primarily chilling, maintains the resistance of piglets to infections. Also, all-in/all-out farrowing rooms are effective in preventing and controlling baby pig scours.

3. Immunity passed from the sow to the piglet in the colostrum and milk provides protection until piglets start producing their own antibodies at about 10 days of age. Vaccination of the sow late in gestation with the pathogenic strain of *E. coli* promotes the formation of colostral antibodies, which provides the piglets with passive immunity to the *E. coli* until their immune system starts producing antibodies. The veterinarian and the producer need to work together for this portion of the approach to control baby pig scours.

Prevention is better than treatment. However, once *E. coli* diarrhea is diagnosed, those pigs that are affected should be treated with an antibacterial drug known to be effective for the *E. coli* strain in the herd. It is known that *E. coli* is so adaptable that it can rapidly develop a resistance to antibacterial drugs. Besides being a treatment for the infected pigs, administering antibacterial agents lowers the number of pathogenic *E. coli* being shed in the watery feces.

Gilt litters are more likely to have this disease, even in herds where sow litters are unaffected. Immunity is critical and current bacterins are quite effective. *E. coli* is not generally the same as the cause of edema disease; therefore, control of one does not control the other.

Erysipelas, Swine

Erysipelas is an acute or chronic infectious disease of swine, but it has also been reported in sheep, rabbits, and turkeys. When it attacks humans, it is called erysipeloid, which should not be confused with human erysipelas, which is caused by a streptococcal bacteria. When the acute form of erysipelas occurs in swine, death losses are high, ranging from 50 to 75%.

Symptoms and Signs

The disease occurs in 3 forms: acute, subacute, and chronic. Additionally, a subclinical infection can occur with no visible signs and then develop into the chronic form. The symptoms of the acute septicemia form resemble those seen in hog cholera. The infected animals have high fever, and frequently have edema of the nose, ears, and limbs. Edema of the nose usually causes infected animals to breathe with snoring sounds. Also, there may be purplish patches under the belly similar to those described for hog cholera. The characteristic skin lesions called *diamond-skin* lesions appear 2 to 3 days after the onset of clinical signs (Figure 14.11). If severely affected animals do not die, areas of necrotic skin develop, which are dark, dry, and firm; and eventually these areas will slough, particularly on the ears and tail.

In the subacute form, the signs are less severe. Only a few skin lesions may appear, and they may be overlooked. The typical lesions are reddish, rectangular plaques in the skin. If the animal appears visibly sick, it will recover in a shorter time than the acutely affected animal.

The chronic form of erysipelas may follow the acute form, the subacute form, or a subclinical infection with no visible signs. In the chronic form, the heart and/or the joints are usually the areas of localization. The joints of the knees and hocks are most commonly affected, showing enlargement and stiffness—arthritis. Those pigs chronically infected are usually very poor and unthrifty.

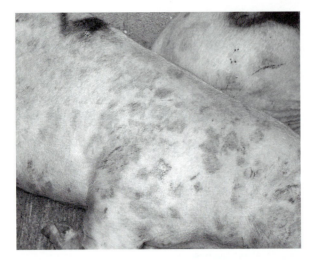

Figure 14.11 Pigs infected with erysipelas and the typical diamond-skin lesions. *(Courtesy, John Carr, Iowa State University, Ames, IA)*

Cause, Prevention, and Treatment

It has been demonstrated that the sole causative agent of erysipelas is *Erysipelothrix rhusiopathiae*, a bacterium. This organism is often found in the tonsils, gall bladders, and intestines of apparently normal pigs, and is common in soils. Infection usually takes place by ingestion. Veterinary aid is necessary for an accurate diagnosis.

Swine erysipelas is prevented by a sound program of herd health management, including active immunization using either an attenuated (living avirulent) vaccine or bacterins (nonliving). Attenuated vaccines may be given orally in the drinking water or via injection. The immunization for erysipelas provides protection for only 3 to 5 months or slightly longer with a double treatment.

When an outbreak occurs, all sick animals should be isolated and the herd examined daily for new cases. Antiserum is available, and it provides immediate passive immunity for about 2 weeks. During a herd outbreak, antiserum may be used to provide protection to suckling pigs not old enough to be vaccinated. Penicillin alone or in conjunction with antiserum provides a satisfactory treatment for acute cases. Tetracyclines, Naxcel™, and tylosin are also effective. Penicillin (and all drug) use must be with proper withdrawal times and usage levels may necessitate a withdrawal change based on weight and proximity to marketing time.

Quarters that housed infected animals should be cleaned and disinfected. Lots or pastures that held infected animals should remain vacant for several months, preferably including the summer season. Animals that are chronically infected should be eliminated from the herd because they can be carriers of the disease.

Foot-and-Mouth Disease (FMD)

Foot-and-mouth disease is an old disease; it was first reported in 1514. It is a highly contagious disease of cloven-footed animals (mainly swine, sheep, and cattle) characterized by the appearance of water blisters in the mouth (and in the snout in the case of hogs), on the skin between and around the claws of the hoof, and on the teats and udder. Fever is another symptom. Humans are mildly susceptible but very rarely infected, whereas the horse is immune.

Unfortunately, one attack does not render the animal permanently immune, but the disease has a tendency to recur perhaps because of the different serotypes and subtypes of the causative virus. The disease is not present in the United States, but there have been at least 9 outbreaks (some authorities claim 10) in this country between 1870 and 1929, each of which was stamped out by the prompt slaughter of every infected and exposed animal. No United States outbreak has occurred since 1929, but the disease is greatly feared. Drastic measures are exercised in preventing the introduction of the disease into the United States. In the case of actual outbreak, eradication of all exposed and infected animals is anticipated.

FMD is constantly present in Europe, Asia, Africa, and South America. However, it has not been reported in New Zealand or Australia. Neither hogs nor uncooked pork products can be imported from any country in which it has been determined that foot-and-mouth disease exists.

In September 1946, an outbreak of FMD appeared in Mexico. From that date until January 1, 1955, the Mexican–U.S. border was closed to imports of virtually all livestock and meat products, except for a 9-month period from September 1, 1952, to May 23, 1953. Likewise, the Canadian–U.S. border was closed from February 1952 to March 1, 1953, because of an outbreak of the disease in Canada in which the virus was transported by infected meat products from Europe. This emphasizes the importance of the ban on the international movement of meat-containing foods.

Symptoms and Signs

The disease is characterized by the formation of blisters (vesicles) and a moderate fever 3 to 6 days following exposure. These blisters are found on the mucous membranes of the tongue, lips, palate, cheeks; on the skin around the claws of the feet; and on the teats and udder. In swine the blisters are usually found on and above the snout also (Figure 14.12, Figure 14.13).

Presence of these vesicles, especially in the mouth of cattle, stimulates a profuse flow of saliva

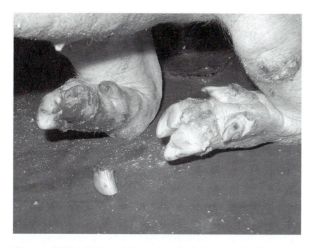

Figure 14.12 Affected feet of a pig infected with foot-and-mouth disease. *(Courtesy, Plum Island Animal Disease Center, Orient, NY)*

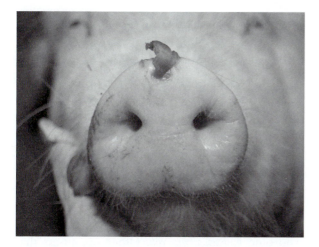

Figure 14.13 Affected snout of a pig infected with foot-and-mouth disease. *(Courtesy, Plum Island Animal Disease Center, Orient, NY)*

that hangs from the lips in strings. Complicating or secondary factors are infected feet, caked udder, abortion, and great loss of weight. The mortality of adult animals is not ordinarily high, but the usefulness and productivity of affected animals is likely to be greatly damaged, thus causing great economic loss.

Cause, Prevention, and Treatment

The infective agent of this disease is a small filterable virus, of which there are at least 7 types, all of which are immunologically distinct from one another. Thus, infection with one does not protect against the others. The virus is present in the fluid and coverings of the blisters, in the blood, meat, milk, saliva, urine, and other secretions of the infected animal. The virus may be excreted in the

urine for more than 200 days following experimental inoculation. The virus can also be spread through infected biological products, such as small pox vaccine and hog cholera virus and serum, and by the cattle fever tick.

Except for the 9 outbreaks mentioned, the disease has been kept out of the United States by extreme precautions, such as quarantine at ports of entry and assistance with eradication in neighboring countries when introduction appears imminent. Neither live animals nor fresh or frozen meats can be imported from any country in which it has been determined that foot-and-mouth disease exists. Meat imports from these countries must be canned or fully cured.

Two methods have been applied in control: the slaughter method and the quarantine procedure. If the existence of the disease is confirmed by diagnosis, the area is immediately placed under strict quarantine; infected and exposed animals are slaughtered and buried, with owners being paid indemnities based on their appraised value. Everything is cleaned and thoroughly disinfected. No attempt is made to treat animals that are known to be infected.

Fortunately, the foot-and-mouth disease virus is quickly destroyed by a solution of the cheap and common chemical, sodium hydroxide (lye). Because quick control action is necessary, state or federal authorities should immediately be notified the very moment the presence of the disease is suspected.

Vaccines containing one or more immunotypes of the virus are manufactured in European countries and in Argentina. The immunity produced from such vaccines lasts for only 3 to 6 months; hence, animals must be vaccinated 3 to 4 times per year. Vaccines have not been used in the outbreaks in the United States because they have not been regarded as favorable to rapid, complete eradication of the infection. A new vaccine against foot-and-mouth disease has been developed by Dr. H. Bachrach and other U.S. Department of Agriculture scientists who have used gene-splicing techniques that allow mass production of a safe and effective supply of the vaccine.

Leptospirosis

Leptospirosis was first observed in humans in 1915–1916, in dogs in 1931, and in cattle in 1934. It has been recognized as an important disease of swine in the United States since about 1952, although it very probably has been present for a much longer period. Humans may contract infections through skin abrasions, when handling or slaughtering infected animals, by swimming in contaminated water, through consuming uncooked foods that are contaminated, or through drinking unpasteurized milk.

Symptoms and Signs

Swine losses from leptospirosis are chiefly in baby pigs that are aborted or born dead or weak, and in the unthriftiness of market hogs. Infected animals may refuse to eat, become mildly depressed, lose weight, and often run a high fever for 2 or 3 days. Occasionally, there may be blood-stained urine. Very frequently, however, the symptoms in growing-finishing pigs are not apparent. Leptospirosis infection is usually confirmed on the basis of herd history, postmortem findings, and blood tests. Leptospirosis should be suspect when reproductive failures are being considered. Chronic infections often result in stillborns, stillbirths, and the birth of weak piglets.

Cause, Prevention, and Treatment

The most common cause of leptospirosis in swine worldwide is *Leptospira pomona*, a corkscrew-shaped bacterium of the spirochete group. It can cause the disease in swine, cattle, sheep, horses, and humans, and it is readily transferred from one species to another.

Leptospirosis organisms enter the body via the skin or mucous membranes, and they are spread by infective urine—direct contact with urine or water contaminated by urine. Control depends on a combination of antibiotic therapy, vaccination, and management. The following preventive measures are recommended:

1. Blood test all animals prior to purchase, isolate for 30 days, and then retest prior to adding them to the herd.
2. Vaccinate with leptospira bacterin, following the manufacturer's directions. Immunity from bacterins is relatively short lived. So, gilts should be vaccinated twice before the first breeding, and sows should be vaccinated at or near each breeding—approximately every 6 months.
3. Keep different classes of livestock separated because leptospirosis can be spread from one species to another.

The tetracycline group of antibiotics is the treatment of choice. If afflicted animals are treated promptly, fairly good results can be expected. To help remove the carrier phase, feed high levels of the tetracycline drugs (400 to 500 g per ton is approved by the FDA in complete feeds for 7 to 14 days). Untreated, swine can remain carriers up to 1 year.

When faced with an outbreak of clinical leptospirosis, streptomycin treatment for both at-risk and infected animals at 25 mg/kg body weight, followed by vaccination is beneficial. Then develop a regular vaccination program.

Mastitis-Metritis-Agalactia Complex (MMA or Agalactia)

Mastitis-metritis-agalactia (MMA) complex is a disease of sows and gilts characterized by an inflammation of the mammary glands (mastitis), and inflammation of the uterus (metritis), and/or a failure to secrete milk (agalactia). Often, however, there is not total lack of milk but rather a reduction in the normal amount of milk (hypogalactia). The death rate in sows is low, but the loss of baby pigs may be high. MMA appears to be a more serious problem in "multiple" and "indoor" farrowing than where portable houses on pasture are used.

Symptoms and Signs

The first signs of the disease usually appear within 3 days after farrowing, although symptoms usually can be seen before farrowing or before the pigs are weaned. A whitish or yellowish discharge of pus appears from the vagina of the infected animal and the temperature rises to 103 to 106°F or higher. The sow goes off feed and stops milking, and the pigs often develop diarrhea. Sometimes the problem is not recognized until the pigs starve to death.

Cause, Prevention, and Treatment

The organism commonly thought to cause the MMA syndrome is *Escherichia coli*. Other organisms that have been associated with the syndrome include *Actinobacillus, Actinomyces, Aerobacter, Citrobacter, Clostridia, Corynebacterium, Enterobacter, Klebsiella, Pseudomonas, Mycoplasma, Proteus, Streptococcus, Staphylococcus,* and *Chlamydia*. It is particularly noteworthy that *E. coli* is also the most common organism isolated as the cause of diarrhea in baby pigs during the first 3 weeks of life. However, the disease is complex, and proof of the cause is often difficult to ascertain.

Prevention revolves around sound management, good nutrition, and sanitation. When sows are confined to stalls or pens, it is especially important to keep them from becoming overweight and constipated, and to reduce stress, particularly near the time of farrowing. The overweight condition can be controlled by limiting the diet and the addition of fiber or chemical laxatives to the diet will prevent constipation. Also, underfeeding may contribute to the problem.

Because many of the organisms causing MMA symptoms are endemic to the farm, infecting females 30 days before breeding with manure from the gestating herd or farrowing house may provide immunity from some of the infective pathogens. The list of suggested treatments is long and not always effective. Among them are the following:

1. Antibiotics to eliminate infections.
2. Oxytocin, a hormone, to cause milk let-down, alleviating the agalactia.
3. Relief or prevention of constipation by feeding a diet high in molasses or fiber or providing morning and evening exercise.
4. Supplemental milk and glucose to keep the baby pigs alive.
5. Oral antibiotics for the baby pigs if diarrhea occurs.
6. Vaccination of sows with a bacterin prior to farrowing; preferably using an autogenous bacterin prepared from the strains of bacteria involved in the herd, or using a stock bacterin.

Mycoplasmal Pneumonia (Enzootic Pneumonia)

Mycoplasmal pneumonia (MP) of pigs is a chronic disease, and one of the world's most important swine diseases, infecting all ages. In the United States, it is the second most reported disease, following ileitis (Figure 14.4), and the feed efficiency of infected pigs may be reduced by 20%.

Symptoms and Signs

Coughing is the first and most characteristic symptom manifested of mycoplasmal pneumonia. It generally begins 10 to 16 days following exposure to diseased animals, and it may persist indefinitely. Coughing is most marked when pigs come out to feed in the morning or following vigorous exercise. Diarrhea usually occurs when pigs first begin to cough, but it lasts only 2 or 3 days. Temperatures are only slightly elevated, and even when they reach 105°F, the pigs do not look sick. In general, infected pigs eat well, but gains are slow and feed utilization poor.

If management is poor, many pigs develop serious pneumonia as a result of secondary bacterial infection of the lungs. Also, severe pneumonia occurs when MP is complicated by large numbers of roundworm larvae passing through the lungs. The age 14 to 26 weeks seems to be the most critical for such secondary lung complications.

Cause, Prevention, and Treatment

The disease is caused by a small coccobacillary organism, *Mycoplasma hyopneumoniae.* Actually, 3 different diseases are caused by different mycoplasma species: mycoplasmal pneumonia, mycoplasmal polyserositis arthritis, and mycoplasmal arthritis.

Because mycoplasmal pneumonia is spread largely by direct contact from infected to normal animals, prevention consists of avoiding the purchase of animals from infected herds. Therefore, strict adherence to all-in/all-out production is the most effective means of controlling the disease.

The disease is continued in infected herds by contact of young pigs with carrier sows. However, older sows tend to rid themselves of the organism. Serotesting older sows and culling positives may allow the establishment of a carrier-free herd. The SPF program has been very effective in controlling MP.

Treatments with lincomycin (200 g/ton of feed for 21 days), tetracyclines, tylosin, or tiamulin reduces the severity of mycoplasmal pneumonia and improves performance.

Pneumonia

Pneumonia is a widespread disease of all animals, including swine. If untreated, 50 to 75% of affected animals die.

Symptoms and Signs

Pneumonia is an inflammation of the lungs. It is ushered in by a chill, followed by elevated temperature. There is quick, shallow respiration, discharge from the nostrils and perhaps the eyes, and a cough. Sick animals stand with their legs wide apart, and there is a drop in milk production, loss of appetite, constipation, crackling noises with breathing, and gasping for breath.

Cause, Prevention, and Treatment

The causes of pneumonia are numerous, including (1) many bacteria, (2) inhalation of water or medicines given by untrained persons as a drench, (3) noxious gases, (4) mycoplasma, (5) viruses, and (6) lungworms. Changeable weather during the spring and fall, and damp or drafty buildings are conducive to pneumonia. Preventive measures include providing good hygienic surroundings and practicing sound husbandry.

Sick animals should be segregated in quiet, clean quarters away from drafts, and given easily digested nutritious feeds. Treatment may include antibiotics or sulfonamides for acute pneumonias or for keeping down secondary bacterial pathogens.

Porcine Parvovirus (PPV)

Losses caused by porcine parvovirus are usually unexpected because the sow or gilt typically appears healthy during gestation. The incidence of parvovirus is high and it is one of the most important infectious agents indicated in SMEDI (discussed later in the chapter). Parvovirus occurs throughout the world.

Symptoms and Signs

Usually, the only clinical sign is maternal reproductive failure such as (1) return to estrus, (2) failure to farrow despite being anestrous, (3) farrowing few pigs per litter, or (4) farrowing a large proportion of mummified fetuses. The observable signs depend on the stage of pregnancy when the sow or gilt is infected. Gilts are especially vulnerable and boars appear to be unaffected.

Cause, Prevention, and Treatment

The disease is caused by a virus of the genus *Parvovirus*, though there is no indication that the porcine parvovirus is the same as the parvoviruses infecting species such as rats, mice, cats, dogs, cows, and sheep.

Diagnosis is made on the basis of differentiating parvovirus from other causes of reproductive failure in swine, and, if available, mummified fetuses can be submitted for immunofluorescence testing.

Prevention consists of naturally infecting or vaccinating gilts with porcine parvovirus before they are bred. Commonly, gilts are exposed to sows or moved to a potentially contaminated area to promote a natural infection and immunity. The effectiveness of this method is, however, uncertain. A vaccination can ensure that gilts develop immunity before breeding. Several inactivated vaccines are on the market. There is no treatment for porcine parvovirus.

Porcine Proliferative Enteropathies (Proliferative Ileitis)

Ileitis, or porcine proliferative enteropathies (PE), was the disease most common on sites with growing-finishing hogs in the United States in the National Animal Health Monitoring System (NAHMS) 2000 report (Figure 14.4), occurring on 36.7% of all sites and on 75% of sites with more than 10,000 pigs and the distribution is worldwide.

Symptoms and Signs

Clinical signs are not dramatic; they are mostly observed in postweaning pigs between 6 and 20 weeks of age. In chronic cases symptoms are minimal and often limited to reduced growth. If diarrhea is present, it usually is not severe and is of normal color. The disease is best detected by maintaining records

on weight gains and feed conversion. Severely affected cases develop necrotic enteritis, often referred to as "garden hose gut."

Cause, Prevention, and Treatment

PE is caused by *Lawsonia intracellularis*. Movement of pigs that involve transport of gilts to new units and mixing of boars and gilts into breeding groups is commonly associated with outbreaks. Extreme weather conditions, such as large changes in temperature or humidity, have also been implicated in outbreaks. The organism is spread through the feces.

Treatment with chlortetracycline (300 ppm), lincomycin (110 ppm), tiamulin (120 ppm), or tylosin (100 ppm) delivered orally via water or feed for 14 days is the preferred treatment.

Porcine Reproductive and Respiratory Syndrome (PRRS)

In 1987, porcine reproductive and respiratory syndrome (PRRS) was first reported in the United States and subsequently in Canada, Japan, and several European countries. When the disease first appeared in the United States, it was designated as the "mystery swine disease" because it could not be readily identified or treated. Because of the cyanosis (a bluish coloration of the skin [especially the ears, vulva, abdomen, and snout] as a result of deficient oxygenation of the blood) of afflicted animals, Europeans called it "blue ear disease." In the mid-1990s, PRRS was considered to be one of the most difficult and costly swine health problems.

Symptoms and Signs

As the name implies, the syndrome consists of two parts: (1) a reproductive component, characterized by late-term abortions (107 to 112 days of pregnancy) accompanied by elevated numbers of stillbirths, mummies, and weak born pigs; and (2) a respiratory disease component, characterized by the piglets in the farrowing and nursery houses exhibiting labored breathing, loss of appetite, gauntness, and rough hair coats. As a result of PRRS infection, piglets are more susceptible to such other infections as bacterial pneumonia and *Streptococcus suis*. Cyanosis of the ears (blue ear disease), vulva, abdomen, and snout may be observed, and afflicted hogs may have a fever (103 to 105°F). Infected herds gradually return to almost normal in 2 to 4 months.

Cause, Prevention, and Treatment

PRRS is caused by a virus first isolated in 1991. Also, researchers have shown that there are different strains of the virus and that the different strains cause diseases that vary in severity.

Prevention and control of PRRS involves (1) determining the status of the herd, (2) segregating age groups better to control viral shedding, (3) isolating incoming replacement boars and gilts a minimum of 30 days, (4) operating on an all-in/all-out basis, (5) depopulating nurseries if necessary, (6) changing feed if current feed is high in mycotoxins, and (7) vaccinating pigs under 16 weeks of age with the newly approved vaccine.

There is no specific treatment of PRRS-afflicted swine. Good nursing and good feed will help. The use of an antibiotic to reduce secondary infection is prudent.

Porcine Stress Syndrome (PSS)

Some pigs are unable to adapt successfully to stressful management practices, and they exhibit a non-pathological disorder that is best described as a syndrome rather than a single abnormal characteristic. It commonly occurs in rapidly growing, heavily muscled pigs, resulting from intensive genetic selection in systems of partial or total indoor production. It is often manifested as sudden unexplained death losses. Furthermore, the syndrome is related to low-quality pork products, or so-called pale, soft, exudative (PSE) pork, when the susceptible pigs survive until slaughter. (See the Chapter 4 section on porcine stress syndrome for details about this genetic disorder.)

Pseudorabies (PR, Aujeszky's Disease, Mad Itch)

Pseudorabies (PR) is not related to rabies, though some of the symptoms and signs are similar. Pseudorabies is an acute, infectious, and frequently fatal disease affecting most species of domestic and wild animals. Humans are resistant to the disease. It was first recognized as an infectious disease of cattle and dogs in Hungary by Dr. Aladár Aujeszky in 1902.

Pseudorabies is an important disease in areas of the world where swine are raised. It was of considerable economic importance in the United States; however, an eradication program was begun in 1989. As of October 2003, 46 states were in stage V PR-free status and the remaining four, Florida, Iowa, Pennsylvania, and Texas, in stage IV-Surveillance (no known infections for 2 years). The disease is also widespread in Europe. It is felt that the pre-eradication growth and severity of the disease may have been a result of rearing more pigs in large indoor production units or by the cessation of the hog cholera (classical swine fever) vaccination.

Symptoms and Signs

Baby pigs may die with few, if any, clinical signs evidenced. However, death is usually preceded by fever which may exceed 105°F, dullness, loss of appetite, vomiting, weakness, incoordination, and convulsions. In pigs less than 3 weeks old, death losses frequently approach 100%. After 3 weeks of age, the signs and death losses decrease.

In adult pigs, the signs may be very mild and include fever, going off feed, salivation, coughing, sneezing, vomiting, diarrhea, constipation, convulsions, itching, middle ear infections, and blindness. Sows infected in middle pregnancy may abort mummified fetuses. Sows infected late in pregnancy often abort or give birth to weak, shaker, or stillborn pigs. Piglets infected prior to birth usually die within 2 days.

In animals other than swine, cattle, sheep, goats, and cats the disease is fatal. Wild animals are also known to be susceptible to PR infection. Diagnosis should be confirmed by a laboratory test, usually performed on serum samples.

Cause, Prevention, and Treatment

PR virus is a member of the group *Herpesviridae*. The main mode of spreading the disease is via direct contact with the nose serving as the primary point of entry for the virus. Nasal discharges and saliva containing the virus contaminate water, bedding, feed, and feeders, as well as the clothes worn by workers.

Vaccines for pseudorabies are available. Modified live virus (MLV), inactivated, and gene-deleted vaccines have been developed for PR control. The use of vaccines has been quite effective in reducing the economic impact of the disease. However, immunized animals generally survive the infection and shed the virus for awhile, thereby acting as carriers. The following control measures are recommended for the protection of herds:

1. Dispose of dead pigs properly (burning or burying).
2. Buy PR-tested breeding stock and isolate them for 30 days.
3. If you raise breeding stock, do not buy feeder pigs.
4. Get feeder pigs from a farm that has not had the disease.
5. Keep visitors away from swine premises.
6. Keep stray dogs, cats, and wildlife off the premises.
7. Keep swine and cattle separate.
8. Isolate show stock for 30 days after the fair is over.

When an outbreak occurs, the premises should be quarantined and the movement of all people strictly controlled. If it is at all possible, sick animals should be separated from unaffected animals.

Rabies

Rabies is an acute infectious disease of all warm-blooded animals and humans. It is characterized by deranged consciousness and paralysis, and terminates fatally.

When a human being is bitten by an animal that is suspected of being rabid, it is important to confine the animal under the observation of a veterinarian until the disease, if it is present, has a chance to develop and run its course. If no recognizable symptoms appear in the animal within a period of 2 weeks after it inflicted the bite, it is usually safe to assume that there was no rabies at the time.

Death occurs within a few days after the symptoms appear, and the animal's brain can be examined for specific evidence of rabies. With this procedure, unless the bite is in the region of the neck or head, there will usually be ample time in which to administer treatment to exposed humans. Because the virus has been found in the saliva of a dog at least 5 days before the appearance of the clinically recognizable symptoms, the bite of a dog should always be considered potentially dangerous until proven otherwise. In any event, when people are bitten or exposed to rabies, they should immediately report to the family doctor. With severe bites, especially those around the head, antiserum is particularly indicated.

Symptoms and Signs

Rabies in swine is not common. The typical course is rapid onset of symptoms, including incoordination, dullness, and later prostration. Affected pigs may be unable to squeal. A paralytic form also exists. Most swine die within 72 hours after developing the clinical signs.

Cause, Prevention, and Treatment

Rabies is caused by a filterable virus that is usually carried into a bite wound by a rabid animal. It is generally transmitted to farm animals by dogs and certain wild animals, such as bats, foxes, skunks, and so on. It is reported that only 3 to 8 cases of rabies occur in swine in the United States annually.

The U.S. Department of Health and Human Services reported in 1989 that of the cases of rabies recorded, only 2.7% were in dogs and 88.4% were in wildlife. With the advent of an improved vaccine for the dog, all dogs should be immunized. Complete eradication would be difficult to achieve because of the reservoir of infection in wild animals.

When swine are bitten or exposed to rabies, see your veterinarian. After the disease is fully developed, there is no known treatment.

Rotavirus

The importance of rotaviruses in pigs is a fairly recent development. In 1976, a rotavirus was first reported as being responsible for diarrhea in pigs. Now it is recognized that the infection is widespread and common in pigs. The mortality rate in piglets varies from 0 to 50%.

Symptoms and Signs

Rotavirus causes diarrhea in piglets. Generally, it strikes 3 to 4 days following weaning. However, neonatal diarrhea from rotavirus is occasionally reported. This diarrhea is characterized by a white or yellow stool that is liquid at first but later becomes creamy and then pasty before returning to normal. Piglets also become depressed, fail to eat, and resist moving. The severity of the disease is influenced by (1) the age of the piglet, (2) stresses such as chilling, (3) concurrent infections with pathogenic *E. coli* or the transmissible gastroenteritis (TGE) virus, and (4) inadequate intake of immune milk. In general, the older the pig, the less severe the outcome, but the severity increases with weaning at any age. In many pigs, however, the infection may occur with no clinical signs.

Cause, Prevention, and Treatment

The diarrhea is caused by a group of viruses from the family *Reoviridae*—a name derived from the wheel-like appearance of the virus through the electron microscope. A rotavirus infection can be confused with TGE. A laboratory diagnosis is necessary to differentiate.

Most swine herds are infected with rotavirus. Orally administered modified live virus vaccines to the sows ensure adequate antibodies in the colostrum and milk. The best methods for the prevention and control of rotaviruses in young pigs appears to be ensuring piglets get adequate colostrum and milk at an early age and providing good sanitation and a comfortable (nonstressful) environment.

All-in/all-out management should be followed and coordinated with cleaning and disinfection between groups. Sows within a room should farrow within a short interval to prevent transmission from older to younger pigs.

No known therapeutic agents are available for the specific treatment of porcine rotavirus infections. General supportive therapy, management procedures, and antibiotics are recommended to minimize mortality as a result of rotavirus and secondary bacterial infections. Electrolyte solutions containing glucose-glycine fed *ad libitum* minimize dehydration and weight loss induced by rotavirus infection.

Salmonellosis

The concern with salmonellosis arises because the causative bacteria are widespread and adaptable; they may cause diarrhea and septicemia in pigs; and they are a source of human salmonellosis. Most outbreaks of the disease have occurred in intensively reared, weaned pigs.

Symptoms and Signs

The signs may result primarily from two forms of the disease: septicemia or diarrhea (enterocolitis). Septicemic salmonellosis occurs mainly in weaned pigs less than 5 months old. It is characterized by restlessness, failure to eat, and a fever of 105 to 107°F. Generally, mortality is high but for some reason, morbidity is low, often less than 10%. The disease is spread via the fecal-oral route, and recovered pigs are carriers and fecal shedders for some time.

Between the ages of weaning and 4 months, outbreaks of enterocolitic salmonellosis may occur. This is characterized by a watery, yellow diarrhea, which rapidly spreads to most pigs in a pen within a few days. Pigs may experience repeated bouts with the disease. Affected pigs have an increased body temperature and decreased food intake. Mortality is low and is the result of several days of diarrhea, which causes dehydration.

Cause, Prevention, and Treatment

Most often the disease is caused by the bacteria *Salmonella choleraesuis* or occasionally *Salmonella typhimurium*. Other species of the *Salmonella* may also be involved in swine diseases. The septicemic form is usually caused by *S. choleraesuis*, whereas diarrhea is caused by *S. typhimurium*.

Prevention of infection of swine with *Salmonella* is not currently possible. The recommended prevention and control of *Salmonella* infection consists of all-in/all-out pig flow and sanitation.

The major sources of *Salmonella* introduction into a herd are (1) the arrival of new animals into a herd and (2) contaminated feed. Quarantining new animals allows time for fecal shedding of *Salmonella* to stop. Other control measures when the disease is not evident include, primarily, keeping stresses to a minimum, cleanliness, and the use of disinfectants.

Often the *Salmonella* bacteria are resistant to a variety of drugs, hence, the choice of medication to control the disease should be based on the susceptibility of the *Salmonella* involved. The use of various antibiotics based on sensitivity is widely advocated. But antibiotics in the feed and water of critically ill animals have limited efficacy.

SMEDI (Stillbirth, Mummification, Embryonic Death, Infertility)

SMEDI refers to a syndrome rather than a disease caused by a specific pathogen. The factors involved result in stillbirths (S), mummified fetuses (M), embryonic deaths (ED), and infertility (I).

Symptoms and Signs

Signs of reproductive failure are anestrus, failure to mate, bleeding at mating, repeat breedings, abortions, mummies, stillbirths, embryonic deaths, infertility, fewer pigs per litter, and pregnant sows failing to farrow.

1. Stillborn piglets are piglets that are alive at the initiation of farrowing, but die intrapartum.
2. Mummies are the remains of fetal tissues after the maternal uterus has removed bodily fluids leaving only the nonabsorbable components of the fetuses, including the partially calcified skeletons.
3. Embryonic deaths are deaths of developing young between the 1st and 114th day of gestation.
4. Infertility in swine refers to animals lacking the capability to reproduce.

Cause, Prevention, and Treatment

Reproductive failures occur in all swine breeding operations, but they are regarded as significant only when reproductive levels fall below the expected norm. These norms vary from operation to operation and are based on such things as percentage of females cycling, conception and farrowing rates, average litter size, and number of pigs produced per sow per year.

Most reproductive problems involve management practices, nutrition, environmental effects, toxicoses, genetics, or disease conditions. The first step in investigating swine reproductive failure is to identify the problem(s), followed by rectifying it. Often exposing gilts to afterbirth or feces from the farrowing house 30 or more days prior to breeding will allow them to develop some immunity. The treatment directed toward reproductive problems often improves the reproductive management and productivity of the herd.

Streptococcal Infections (*Streptococcus suis*)

Streptococcal infections appear to be fairly common in large, intensive hog farms. They are responsible for epidemic outbreaks of meningitis, septicemia, and arthritis.

Symptoms and Signs

In peracute cases, pigs may be found dead with no premonitory signs. Usually, however, signs of meningitis predominate. In most cases a progression of loss of appetite, depression, reddening of the skin, fever, incoordination, paralysis, paddling, opisthotonus (body spasm where the head and heels are bent backward and the body is bowed forward), tremors, and convulsions develop.

Cause, Prevention, and Treatment

The disease is caused by *Streptococcus suis*. Good management, with emphasis on sanitation, and minimizing stress from overcrowding, poor ventilation, and mixing/moving pigs, constitutes the best prevention and control. However, it occurs under both good and poor sanitary management conditions.

Early individual therapy with penicillin and supportive nursing prevent death and may result in complete recovery.

Swine Influenza (Hog Flu)

Influenza in swine resembles influenza in humans, but the effects may be more serious. Swine influenza is an acute respiratory disease. The onset is rapid, and in most cases all swine in an infected herd show symptoms at the same time. The disease is usually first observed in the fall when the weather becomes cold. Swine influenza occurs throughout the United States. However, the mortality seldom exceeds 2% and the principle loss from swine flu is a result of the lingering debility, which results in uneconomical gains.

Symptoms and Signs

The disease makes its appearance suddenly, following an incubation period of less than a week. It is a herd disease and most of the animals show symptoms at the same time. High fever, loss of appetite, a cough, and discharge from the eyes and nose are seen. The animals seem reluctant to move but may sit up like dogs in an attempt to facilitate breathing, which may in some cases be difficult.

Cause, Prevention, and Treatment

Swine influenza is caused by a type A influenza virus, which is similar to the virus that caused the worldwide flu epidemic of 1918 in humans that killed more than 20 million people throughout the world. Indeed, there is substantial evidence that flu was first introduced into swine at that time—a classic example of a human disease transmitted to animals. Moreover, there is evidence of cross infection of humans and swine with type A influenza virus. Swine flu is passed from pig to pig via direct contacts with infectious secretions or exudates, droplet infection, or small particle aerosols. Some investigators have hypothesized that the lungworm, whose intermediate host is the earthworm, serves as a host to the influenza virus, thereby perpetuating and transmitting the virus. The pig then eats earthworms that contain lungworm larvae which may in turn harbor the virus. However, this system is not clearly shown to be necessary for the survival of the virus because some studies have demonstrated that hogs can become carriers.

Standard biosecurity measures to prevent susceptible animals from contacting infected animals is appropriate to prevent infection of the herd. Avoid contact with avian species.

An inactivated-virus vaccine is available. Vaccination is reported to decrease clinical signs and reduce virus shedding in challenged pigs. The only treatment seems to be the provision of warm, dry, clean quarters, minimum diets, and fresh, clean water accessible at all times. Some veterinarians believe that death losses from pneumonia as a result of secondary invading organisms can be lessened by the use of antibiotics or sulfonamides, administered hypodermically or in the drinking water. Expectorants are also often used if respiration is difficult.

Tetanus (Lockjaw)

Tetanus is chiefly a wound infection disease that attacks the nervous system of almost all animals. Also, it is a human disease. In the central states, tetanus frequently infects pigs, lambs, and calves following castration or other wounds.

In the United States, the disease occurs most frequently in the South, where precautions against tetanus are an essential part of the routine treatment of wounds. The disease is worldwide in distribution.

Symptoms and Signs

The incubation period of tetanus varies from 1 to 2 weeks but may be from 1 day to many months. It is usually associated with a wound but may not directly follow an injury. The first noticeable signs of the disease are a stiffened gait and a slightly elevated head. The animal often chews slowly and weakly and swallows awkwardly; hence, the common designation lockjaw. The animal then has uncontrollable spasms or contractions of groups of muscles brought on by the slightest movement or noise, and usually attempts to remain standing throughout the course of the disease. In more than 90% of the cases, however, death ensues, usually because of many factors, but in acute cases respiratory failure is likely the cause.

Cause, Prevention, and Treatment

The disease is caused by the presence of *Clostridium tetani* in tissue in an environment that will permit its growth and toxin production, and the toxin must subsequently reach the central nervous system. This organism is an anaerobe (lives in absence of oxygen), which forms the most hardy spores known. It is a common inhabitant of soil. The organism causes trouble when it gets into a deep skin wound, avoiding that natural defense. In the absence of oxygen, it then grows and liberates the toxin, which apparently passes by diffusion out into the surrounding medium or environment and then spreads to the central nervous system.

The disease can largely be prevented by good sanitation and reducing the probability of wounds by removing sharp objects, performing sanitary castration, and clipping needle teeth. The prognosis in swine is poor and there is little evidence that treatment is beneficial. If valuable animals are wounded, passive immunization with tetanus antitoxin may help. Large doses of long-acting penicillin or tetracyclines may be beneficial if instituted within a few hours of infection. All perceptible wounds should be properly treated, and the animals should be kept quiet and preferably should be placed in a dark, quiet corner free from flies.

Transmissible Gastroenteritis (TGE)

Transmissible gastroenteritis (TGE) is a rapidly spreading disease accompanied by symptoms of inflammation of the stomach and intestine. The incubation period may be only 18 hours. Once the disease strikes, it spreads rapidly; the entire herd may become noticeably affected in 2 to 3 days.

Symptoms and Signs

It is characterized by piglets vomiting, accompanied or rapidly followed by a watery and usually yellowish diarrhea, rapid loss in weight, dehydration, and high morbidity and mortality in pigs under 2 weeks of age. It affects swine of all ages, but the death loss in older swine is low. Death results from starvation, dehydration, and acidosis as a result of changes in the lining of the intestine. Clinical signs in growing-finishing swine and sows are usually limited to inappetence and diarrhea for 1 or a few days.

Cause, Prevention, and Treatment

The disease is caused by a virus that belongs to the group called *coronaviruses*.

Presently, commercial vaccines given to the sow before farrowing are of limited value in preventing the infection or diarrhea of TGE in piglets. The most effective preventive measure consists of exposing the sows to the virus before they farrow—a planned infection using minced intestines of young pigs known to have had TGE. They will then develop resistance or protection, and the pigs will get the antibodies from the colostrum and milk. However, this procedure should be used only where it is inevitable that sows will be exposed at farrowing time and where there is no danger of spreading the disease to neighboring herds. Isolation at farrowing time is also recommended, as is isolation of new additions to the herd. Also, continuous farrowing should be avoided because it tends to perpetuate the disease.

The only treatment for TGE presently available is to alleviate starvation, dehydration, and acidosis. Parenteral treatments are effective but not practical. Oral therapy is not indicated. Keeping fresh water or nutrient solution available and providing a draft-free environment of about 90°F or higher can reduce losses in piglets 3 to 4 days of age. Antibiotics or sulfonamides may minimize secondary bacterial complications.

Tuberculosis

Tuberculosis is a chronic infectious disease of humans and animals. It is characterized by the development of nodules (tubercules) that may calcify or turn into abscesses. The disease spreads very slowly, and affects mainly the lymph nodes. There are three kinds of tubercule bacilli—the human, the bovine, and the avian (bird) types. Swine are subject to all three kinds, and practically every species of animal is subject to one or more of the three kinds.

The percentage incidence of tuberculosis in the United States in pigs has steadily declined since 1922 when 16.38% of slaughtered swine were positive compared to 1995 when the incidence was only 0.21% positives and 0.003% of the carcasses condemned.[5] Unfortunately, that is still close to 200,000 pigs testing positive.

Symptoms and Signs

Tuberculosis may take one or more of several forms. Humans get tuberculosis of the skin (lupus), of the lymph nodes (scrofula), of the bones and joints, of

---

[5] USDA, Statistical Summary. Federal Meat and Poultry Inspection for Calendar Year 1995. 1996.

the lining of the brain (tuberculous meningitis), and of the lungs. For the most part, animals get tuberculosis of the lungs and lymph nodes, although in poultry the liver, spleen, and intestines are chiefly affected. In cows, the udder becomes infected in chronic cases. In swine, infection is most often contracted through ingestion of infected material; hence, the lesions often are in the abdominal cavity.

Many times an infected animal will show no outward physical signs of the disease. There may be a gradual loss of weight and condition and swelling of joints, especially in older animals. If the respiratory system is affected, there may be a chronic cough and labored breathing. Other seats of infection are lymph glands, udder, genitals, central nervous system, and the digestive system. The symptoms are similar, regardless of species.

Cause, Prevention, and Treatment

The causative agent is a rod-shaped organism belonging to the acid-fast group of bacilli—*Mycobacterium tuberculosis, Mycobacterium bovis,* and *Mycobacterium avium.* The disease is usually contracted by eating food or drinking fluids contaminated with tuberculosis material. Hogs may also contract the disease by eating part of a tubercular chicken.

The tuberculin test appears to be the most useful procedure on a herd basis. At present the intradermal injection of tuberculin in the dorsal surface of the ear is recommended for swine. The injection site should be observed for 48 hours. In human beings, the x-ray is usually used for purposes of detecting the presence of the disease.

As a part of the federal–state tuberculosis eradication campaign of 1917, provision was made for indemnity payments on animals slaughtered.

The incidence in swine is affected by their exposure to cattle and poultry and the incidence occurring in these species. It may be transmitted by milk, feces, or uncooked garbage. An effective control program among swine embraces the following procedure: (1) disposing of tubercular swine, cattle, and poultry; (2) applying strict sanitation; and (3) rotating feedlots and pastures.

Preventive treatment for both humans and animals consists of pasteurization of milk and creamery by-products and the removal and supervised slaughter of reactor animals. Also, avoid housing or pasturing swine with chickens.

Vesicular Exanthema

Vesicular exanthema is almost exclusively a disease of swine, but horses are slightly susceptible. Humans are not infected. It first appeared in Buena Park, California, in 1932, where it was mistaken for and treated like foot-and-mouth disease. During the early 1950s, it reached epidemic proportions in the United States and was finally contained by 1956 after costing the federal government about $33 million.

The course of the disease is 1 to 2 weeks, and the mortality is low. Pregnant sows often abort and nursing sows fall off noticeably in milk production.

Symptoms and Signs

The symptoms are similar to foot-and-mouth disease. Small vesicles (like water blisters) appear around the head, particularly on the snout, nose, or lips. Also these blisters appear on the feet where the hair and the horny part of the hoof meet, on the balls of the feet, on the dewclaws, between the toes, and on the udder and teats of nursing sows. Affected animals go lame, have a high temperature, and usually go off feed for 3 to 4 days.

In order to distinguish vesicular exanthema from foot-and-mouth disease, tests may be made with horses and cattle. Also, guinea pigs may be infected with foot-and-mouth disease but are resistant to the virus of vesicular exanthema.

Cause, Prevention, and Treatment

Vesicular exanthema is caused by a virus. Usually symptoms appear 24 to 48 hours after contact with the virus.

No immunizing agents are available. Preventive measures consist of (1) keeping hogs from infested areas and (2) keeping them from consuming feed and water contaminated with the virus. Uncooked garbage that contains pork trimmings or animals from the sea may be the source of the virus. Thus, cooking breaks the chain of infection from feed to susceptible swine, providing the principal means of control.

Good nursing will help, but no treatment is entirely successful. Rigid quarantine of infected areas (applicable to both hogs and pork products) and the destruction of infected herds appear to constitute the best control and eradication. The virus that causes vesicular exanthema is readily killed when exposed to a 2% sodium hydroxide (lye) solution or a 4% sodium carbonate (soda ash) solution.

Because vesicular exanthema is considered an exotic, reportable disease, federal and state authorities should be notified immediately of suspected cases.

PARASITES OF SWINE[6]

Hogs are probably more affected by parasites than any other class of livestock, with the possible exception of sheep. Infection with either internal or external parasites results in unthriftiness, poor development of young pigs, and increased susceptibility to diseases. A total of more than 50 species of worm and protozoan parasites have been reported as being found in swine throughout the world. Fortunately, a number of these species occur only infrequently in the United States and other species, although widespread, are not of major importance under ordinary conditions.

The following discussion on parasites should be helpful in (1) preventing their propagation and (2) causing their destruction through the use of effective anthelmintics (dewormers) or insecticides. Removal of parasites, however, does not ensure against reinfection, thus, if animals are kept on contaminated ground, the relief afforded by treatment may be only temporary. Therefore, the prevention and control of parasites is far more important than any treatment that is prescribed.

In order to initiate and carry out control measures successfully, the swine producer should have a clear understanding of how each parasite develops and where it lives, stage by stage, from the egg to the adult worm. When the life history and habits of the parasite are definitely known, then and only then can plans be made to break the life cycle at some point that will destroy it.

Few dosages for the control of internal parasites are given; instead, users should follow the manufacturer's directions on the label. Also, only a limited number of pesticides are suggested for the control of external parasites because of (1) the diversity of environments and management practices under which they occur, (2) the varying restrictions on the use of pesticides from area to area, and (3) the fact that registered uses of pesticides change from time to time. Information about what is available and registered for use in a specific area can be obtained from the extension agent, extension entomologist, or veterinarian.

At the outset, it should be emphasized that animals fed an adequate diet and kept under sanitary conditions are seldom heavily infected with para-

sites. A discussion of the most damaging internal parasites of hogs follows.

Ascarids (Large Roundworms; *Ascaris suum*)

This is the most common and one of the most injurious parasites of swine.

Distribution and Losses Caused by Ascarids

The roundworm is widespread throughout the world. At one time, it was reported that in the United States 20 to 70% or more of the hogs examined for this parasite were infected. At times, especially in young pigs, the damage inflicted by roundworms may cause death of the animals. Lighter infections result in a stunting of growth and in uneconomical gains.

Life History and Habits

Technically, the parasite is known as *Ascaris suum*. The adult worm is usually yellowish or pinkish in color, 8 to 12 in. long, and almost the size of a lead pencil (Figure 14.14). The life history of the parasite may be briefly described as follows:

1. The female worms lay eggs in the small intestines and these are eliminated with the feces. These eggs are extremely resistant to the usual destructive influences.

2. A small larva develops in the egg and remains there until the egg is swallowed by the pig along with contaminated feed or water. Then it emerges from its shell, bores through the wall of the intestine, and thence gets into the bloodstream—by means of which, after a brief sojourn in the liver, it is carried to the lungs.

3. In the lungs, the larvae break out of the capillaries, enter the windpipe, and migrate to the throat. While in the throat, they are swallowed and

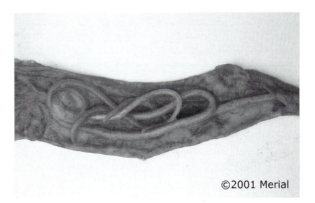

©2001 Merial

Figure 14.14 Adult ascarids in the intestine of a pig. *(Courtesy, Merial, Ltd., Rahway, NJ)*

---

[6] The use of trade names does not imply endorsement, nor is any criticism implied for similar products that are not named. It is recognized that producers and their advisors are generally more familiar with trade names than with generic names.

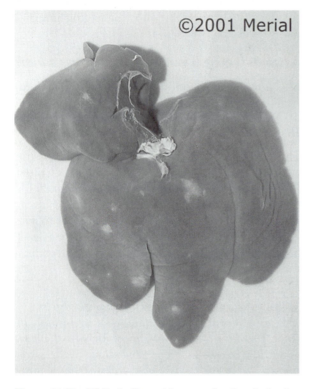

©2001 Merial

Figure 14.15 White "milk spots" on a swine liver indicating damage by migrating large roundworms. *(Courtesy, Merial, Ltd., Rahway, NJ)*

swept to the small intestines where they develop into sexually mature worms, thus completing their life cycle.

Damage Inflicted; Symptoms and Signs of Infected Animals

Principal damage is produced by the migrating larvae which produce liver and lung lesions. These lesions in the liver result in liver condemnations (Figure 14.15) and in the lungs they create conditions favorable to the development of bacterial and viral pneumonias. The evident symptoms are exceedingly variable. Infected young pigs become unthrifty in appearance and stunted in growth. Because of the presence of the larvae in the lungs, coughing and "thumpy" breathing are usually characteristic symptoms. There may be a yellow color to the mucous membrane because of blockage of the bile ducts.

Prevention, Control, and Treatment

Prevention consists of keeping the young pigs away from infection by preventing the recycling of manure containing the eggs (Figure 14.16). For pigs on pasture, the application of the McLean County

Large Roundworms

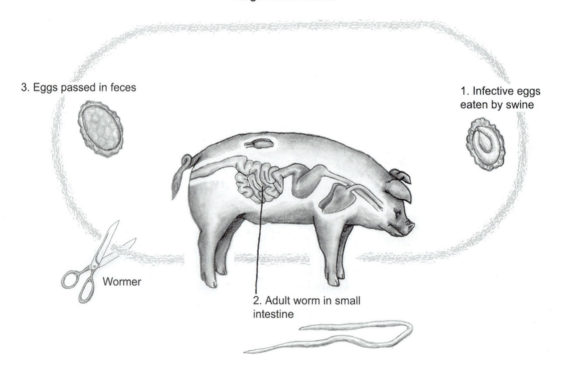

3. Eggs passed in feces

1. Infective eggs eaten by swine

Wormer

2. Adult worm in small intestine

Figure 14.16 A schematic diagram of the life cycle and habits of the large roundworm (*Ascaris suum*). A proper dewormer applied (see scissors) and good sanitation are effective control and preventive measures. *(Courtesy, Nathaniel Klein, Iowa State University, Ames, IA)*

System of Swine Sanitation[7] has proven most effective in protecting pigs from the common roundworm infection. For pigs raised indoors, thorough and frequent cleaning of buildings and floors is the best control program.

Although strict sanitation programs aid in the prevention of worm infections, sows and gilts should be routinely dewormed 10 to 30 days before farrowing. Weanlings, feeder pigs, and boars should also be dewormed on a routine basis.

A variety of anthelmintics, which are effective on ascarids, are available. Most are broad-spectrum products, or effective on a variety of worms. It is recommended that several different anthelmintics be used in rotation (Table 14.2).

Coccidiosis

Coccidiosis, a parasitic disease infecting swine, cattle, sheep, goats, pet stock, and poultry is caused by microscopic protozoan organisms known as coccidia, which live in the cells of the intestinal lining. Each class of domestic livestock harbors its own species of coccidia, thus there is no cross-infection between animals.

Distribution and Losses Caused by Coccidiosis

The distribution of the disease is worldwide. Except in very severe infections, or where a secondary bacterial invasion develops, infested farm animals usually recover. The chief economic loss is in lowered gains.

Life History and Habits

Infected animals may eliminate daily with their droppings thousands of coccidia organisms (in the resistant oocyst stage). Under favorable conditions of temperature and moisture, coccidia sporulate in 3 to 5 days, and each oocyst contains 2 to 4 infective sporocysts. The sporulated oocyst then gains entrance into an animal by being swallowed with contaminated

feed or water. In the host's intestine, the outer membrane of the oocyst, acted on by the digestive juices, ruptures and liberates the eight sporozoites within. Each sporozoite then attacks and penetrates an epithelial cell, ultimately destroying it. While destroying the cell, however, the parasite undergoes sexual multiplication and fertilization with the formation of new oocysts. The parasite (oocyst) is then expelled with the feces and is again in a position to infect a new host.

The coccidia parasite abounds in wet, filthy surroundings; resists freezing and ordinary disinfectants; and can be carried long distances in streams.

Damage Inflicted; Symptoms and Signs of Infected Animals

A severe infection with coccidia produces diarrhea, and the feces may be bloody. The bloody discharge is a result of the destruction of the epithelial cells lining the intestines. Ensuing exposure and rupture of the blood vessels then produces hemorrhage in the intestinal lumen.

In addition to a bloody diarrhea, infected animals usually show pronounced unthriftiness and weakness. It generally occurs, as a recognizable disease, in pigs 7 to 21 days old. Morbidity is usually high, whereas mortality varies, probably because of the number of oocysts ingested.

Prevention, Control, and Treatment

Coccidiosis can be prevented and controlled by (1) practicing all-in/all-out farrowing house management, (2) cleaning thoroughly and disinfecting with 5% bleach (sodium hypochlorate), and (3) using raised farrowing crates with woven wire or expanded metal flooring (Figure 14.17).

The addition of coccidiostats to swine feed is both of little effect and illegal—*it is not approved by the FDA.*

No coccidiostats are approved for use in swine. All treatments presented herein are extralabel uses. Pigs may be treated by (1) giving each pig 2 ml of 9.6% amprolium solution orally for 3 days or (2) giving each pig trimethoprim/sulfa orally. Such pig treatments are labor intensive and of limited efficacy. Swine may develop some immunity to the coccidian *Isospora suis* following infection and recovery.

Kidney Worms (*Stephanurus dentatus*)

Kidney worms are one of the most damaging worm parasites infecting swine in the southern United States with reported infestations up to 60% (Table 14.2). The kidney worm, known as *Stephanurus dentatus*, is a thick-bodied black-and-white worm up to

[7] McLean County System of Swine Sanitation—This system, devised by Drs. B. H. Ransom and H. B. Raffensperger of the U.S. Bureau of Animal Industry, was developed in McLean County, Illinois, as a result of a trial period of 7 years, commencing in 1919. Though this program was worked out chiefly for the purpose of preventing infection of young pigs with the common roundworm, it is equally effective in lessening troubles from other parasites and in disease control, thus making possible cheaper and more profitable pork production. The system involves four simple steps:

1. Clean and disinfect the farrowing quarters.
2. Scrub sows with soap and water before moving into farrowing quarters.
3. Haul the sow and litter to clean pen or pasture.
4. Keep the pigs on clean quarters or pasture.

Coccidiosis

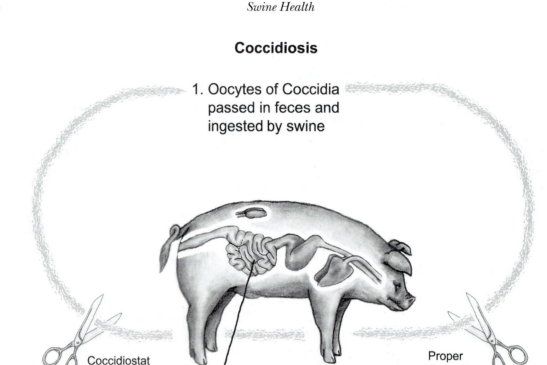

1. Oocytes of Coccidia
passed in feces and
ingested by swine

Coccidiostat

Proper
Sanitation

2. Mature Coccidia organisms
live and reproduce in the
cells of the host's intestines

Figure 14.17　A schematic diagram showing the life cycle of coccidia, the parasite causing coccidiosis in swine. Proper sanitation is the most effective key to preventing the reinfestation of animals. Some cocciodostats approved for other species may be effective, but none are approved for swine. *(Courtesy, Nathaniel Klein, Iowa State University, Ames, IA)*

2 in. long. Though especially harmful to swine, cattle may become infected when running with hogs.

Distribution and Losses Caused by Kidney Worms

The kidney worm is one of the most serious obstacles to profitable swine production in the South. Although it causes initial loss to the producer because of inefficient gains and lowered reproduction, the carcass also is affected, with damage to the liver, kidney, loin, leaf fat, and even the ham. This necessitates severe trimming or even condemnation of the carcass. All such carcass losses are ultimately borne by the swine producer in the form of lowered market prices. In the past, a blanket reduction in price was made on all market hogs coming from certain areas of the South where swine infections with kidney worms are notoriously heavy. This was done in anticipation of the usual damage to the carcass.

Life History and Habits

Adult kidney worms may be found around the kidneys and in cysts in the ureters (tubes leading from the kidneys to the bladder). The mature female worms lay numerous eggs that are discharged with the urine. It has been estimated that, in 1 day, as many as 1 million eggs may be passed in the urine of a moderately infected hog. When eggs fall on moist, shaded soil, a tiny larva hatches from each egg in 1 to 2 days, depending on temperature conditions. In another 3 to 5 days, the larva develops into the infective stage. Hogs then obtain kidney worms by swallowing the infective larvae with contaminated feed and through the skin on ground containing the larvae. Under warm, moist, shaded conditions, the larvae may survive for several months (Figure 14.18).

Kidney worm larvae can also enter the bodies of pigs through the skin, though this is not considered a great source of infection. Regardless of the way in which the larvae enter the body, they get into the blood and migrate to the liver, lungs, and other organs—some of them finally reaching the kidneys. On reaching maturity, which may be 12 to 14 months after the pigs ingest the larvae, the adult female kidney worms begin producing eggs, thus completing the life cycle.

Damage Inflicted; Symptoms and Signs of Infected Animals

There are no definite symptoms ascribable to kidney worm infection. Growth rate is markedly retarded, and health is impaired. Frequently, infected

Kidney Worms

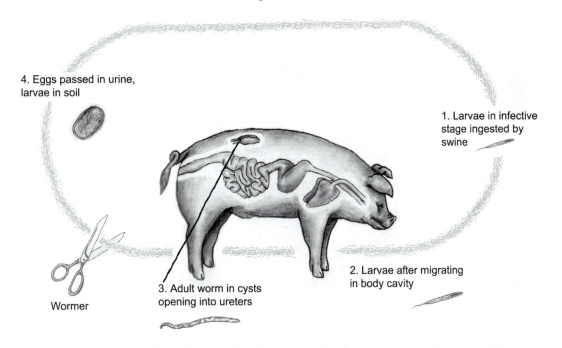

4. Eggs passed in urine, larvae in soil

1. Larvae in infective stage ingested by swine

2. Larvae after migrating in body cavity

3. Adult worm in cysts opening into ureters

Wormer

Figure 14.18 A schematic diagram of the life cycle of the kidney worm (*Stephanurus dentatus*). An approved dewormer and enhanced sanitation reduce exposure to infective larvae. *(Courtesy, Nathaniel Klein, Iowa State University, Ames, IA)*

animals discharge pus in the urine. Most cases are diagnosed at necropsy. Diagnosis is only positively made by microscopically discovering the presence of eggs in the urine.

Primarily, kidney worms damage the liver, ureter, kidneys, and tissues surrounding the kidneys (loin muscle).

Prevention, Control, and Treatment

The "gilt-only" method, first proved at the Coastal Plain Experiment Station, Tifton, Georgia, constitutes an effective prevention and control. With this method, gilts are bred; then, after farrowing and weaning off their first litter, they are sent to slaughter before mature kidney worms develop. This system is based on the fact that it may take the kidney worm as long as a year to reach the egg-laying stage. By following this system for 2 years, the swine producer may completely eliminate kidney worms. Also, several products are effective in controlling kidney worms (Table 14.2).

Lungworms (*Metastrongylus elongatus, M. pudendotectus,* and *M. salmi*)

Lungworms are among the most widespread of the parasites that infect swine. Three species are common in hogs in the United States, namely, *Metastrongylus elongatus, M. pudendotectus,* and *M. salmi.* All lungworms are threadlike in diameter, 1 to 1.5 in. in

length, and white or brownish in color. As the name would indicate, they are found in the bronchi, or air passages, of the lungs. Sheep and cattle also have lungworms.

Distribution and Losses Caused by Lungworms

The lungworm is found in all sections of the United States, but the heaviest infection of swine occurs in the southeastern states. In addition to the usual economic losses and lowered growth and feed efficiency caused by the lungworm, there is evidence to indicate that this parasite may be instrumental in the spread of swine influenza.

Life History and Habits

Female lungworms produce large numbers of thick-shelled eggs, each containing a tiny larva. The eggs are coughed up, swallowed, and eliminated with the feces of the pig. Earthworms, the intermediate hosts, feed on the feces, then swallow the eggs, which hatch in the earthworms' intestines. The larva then develops in the earthworm for about 10 days, after which it is capable of producing an infection. Infection of the pig results from the swallowing of the earthworm, which it usually acquires by rooting and feeding in places where earthworms abound—manure piles; rich, feces-contaminated soil; under trash; and in moist places.

Lungworms

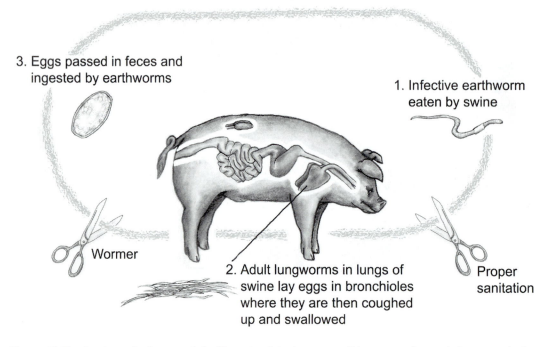

3. Eggs passed in feces and ingested by earthworms

1. Infective earthworm eaten by swine

Wormer

2. Adult lungworms in lungs of swine lay eggs in bronchioles where they are then coughed up and swallowed

Proper sanitation

Figure 14.19   A schematic diagram of the life cycle of the lungworm (*Metastrongylus* spp.). Proper sanitation and preventing contact with earthworms compliment several effective anthelmintics. *(Courtesy, Nathaniel Klein, Iowa State University, Ames, IA)*

After ingestion by the pig, the lungworm larvae are liberated from the earthworm and migrate by way of the lymphatic and blood circulatory systems to the lungs. There they become localized, complete development, and begin to produce eggs within 24 days, thus completing the life cycle (Figure 14.19).

The female lungworm produces an incredibly large number of eggs. It is estimated that as many as 3 million eggs may be eliminated in the droppings of a heavily infected pig in a period of 24 hours. More than 2,000 larvae have been found in a single earthworm collected in a hog lot.

Damage Inflicted; Symptoms and Signs of Infected Animals

Pigs heavily infected with lungworms become unthrifty, stunted, and are subject to spasmodic coughing. Positive diagnosis is made only by fecal examination or by postmortem examination. A cross-section of the lungs exposes the white thread-like worms in the air tubes.

Prevention, Control, and Treatment

Prevention of lungworm infection consists of keeping hogs away from those areas where earthworms are likely to abound. Removal of manure piles and trash and the drainage of low places will help. In brief, the swine producer should provide clean,

dry, well-drained lots—conditions that are not conducive for the intermediate host, the earthworm. Ringing the snout also helps by preventing rooting.

Effective anthelmintics for the control of lungworms are listed in Table 14.2.

Nodular Worms (*Oesophagostomum* spp.)

Nodular worms are most numerous in the southeastern United States. A parasitic survey conducted in North Carolina showed the nodular worm to be second to strongyloides in incidence.

They are called nodular worms because of the nodules or lumps they cause in the large intestine. Four species of *Oesophagostomum* occur in swine, but all of them are slender, whitish to grayish in color, and 0.3 to 0.5 in. in length (Figure 14.20).

Distribution and Losses Caused by Nodular Worms

Nodular worms are widely distributed, but damage is heaviest in the southeastern United States. In addition to the usual lack of thrift that accompanies parasite infections, the intestines of severely infected animals are not suited for either sausage casings or food (chitterlings).

Life History and Habits

The four species of nodular worms affecting swine are *O. dentatum, O. brevicaudum, O. georgianum,* and

O. quadrispinulatum. All have similar life cycles. The adult worms are localized in the large intestine of the host animal. The female worms deposit large numbers of partly developed eggs that become mixed with the intestinal contents and are eliminated with the feces. With favorable conditions of temperature and moisture, a larva emerges from each egg in 1 to 2 days. After another 3 to 6 days of development, the larvae are infective to swine. Pigs then become infected by swallowing the larvae while feeding on contaminated ground or grazing on contaminated pastures.

In the digestive system of the host, the larvae travel to the large intestine where they penetrate into the wall and grow for the next 2 to 3 weeks, forming nodules. They then move into the lumen, or cavity, of the large intestine where they continue development. Within 3 to 7 weeks after ingestion by the pig, the worms are fully grown and have mated and are producing eggs, thus starting a new cycle (Figure 14.21).

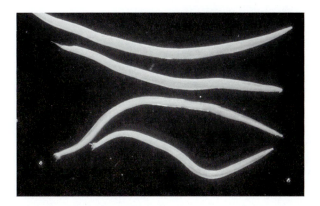

Figure 14.20 Nodular worms are 1 to 2 cm long. (*Oesophagostomum* spp.). *(Courtesy, Merial, Ltd., Rahway, NJ)*

Damage Inflicted; Symptoms and Signs of Infected Animals

No specific symptoms can be attributed to the presence of nodular worms. Weakness, anemia, emaciation, diarrhea, and general unthriftiness have been reported as a result of infection with these parasites.

Prevention, Control, and Treatment

A strict program of swine sanitation, accompanied by a program of regular worming, constitutes a successful and practical preventive measure.

Effective anthelmintics for the control of nodular worms are listed in Table 14.2.

Nodular Worms

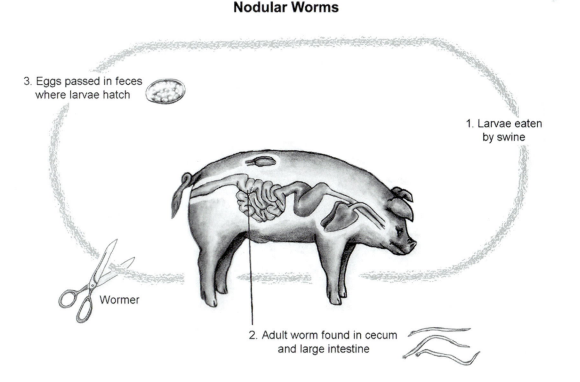

3. Eggs passed in feces where larvae hatch

1. Larvae eaten by swine

Wormer

2. Adult worm found in cecum and large intestine

Figure 14.21 A schematic diagram of the life cycle of the nodular worm (*Oesophagostomum* spp.). A proper dewormer may be used to control and prevent nodular worms. Rigid sanitation, accompanied by pasture rotation (when grazing is involved) compliment a control and prevention program. *(Courtesy, Nathaniel Klein, Iowa State University, Ames, IA)*

Stomach Worms (*Ascarops strongylina, Physocephalus sexalatus, and Hyostrongylus rubidus*)

Three species of small stomach worms infect swine. Two of the species, *Ascarops strongylina* and *Physocephalus sexalatus*, are commonly known as "thick stomach worms." These parasites are reddish in color and nearly 1 in. long in the adult stage (Figure 14.22). The third species, *Hyostrongylus rubidus*, commonly known as the "red, or thin, stom-ach worm," is a small, delicate, slender, reddish worm about 0.2 in. in length.

Distribution and Losses Caused by Stomach Worms

The occurrence of the stomach worm is widespread in hogs throughout the United States. Surveys indi-cate the occurrence in hogs for both the thick stom-ach worm and the red stomach worm is up to 30% (Table 14.2). The move from dirt lots to indoor far-rowing and rearing has greatly reduced their preva-lence in the United States.

Life History and Habits

It has been definitely established that the ordinary dung beetle serves as the intermediate host for the thick stomach worm. The female worm deposits eggs in the stomach of the pig, with each egg con-taining a tiny embryo. Passing with the feces to the outside of the body of the pig, the eggs are eaten by various species of dung beetles, which are the inter-mediate hosts.

After developing for about a month in the bee-tle, the larvae arrive at a stage infective to swine. Hogs feeding on contaminated ground then swal-low the beetles. In the stomach of the pig, the para-sites are liberated from the beetles and make their way into the mucous membrane of the stomach, where they grow to maturity (Figure 14.23).

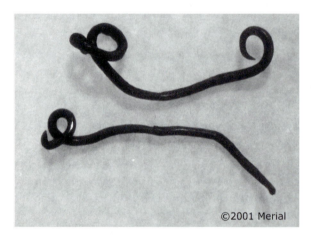

©2001 Merial

Figure 14.22   Thick stomach worms (*Ascarops strongylina*), males. *(Courtesy, Merial, Ltd., Rahway, NJ)*

Thick Stomach Worms

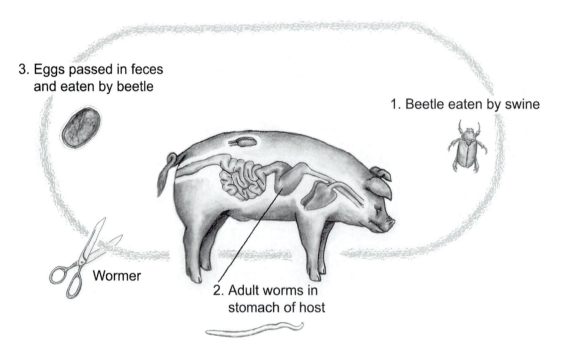

3. Eggs passed in feces and eaten by beetle

1. Beetle eaten by swine

Wormer

2. Adult worms in stomach of host

Figure 14.23   A schematic diagram of the life cycle of the thick stomach worm (*Ascarops strongylina, Physocephalus sexala-tus*). A proper dewormer may be used to control and prevent stomach worms. Good sanitation helps by limiting contact with dung beetles containing the encysted larvae. *(Courtesy, Nathaniel Klein, Iowa State University, Ames, IA)*

The life history of the red stomach worm differs from that of the thick stomach worms in that no intermediate host is necessary, the infections being directly acquired. Eggs are shed in the feces and they hatch in 1 to 2 days. After 7 days the larvae are infective. Swine ingest the stomach worms by feeding in wet areas where they survive, and then they enter the stomach, thus completing their cycle.

Damage Inflicted; Symptoms and Signs of Infected Animals

Because of the burrowing tendency of this parasite, inflammation and gastric ulcers usually follow. In external appearance, the affected animals usually become unthrifty and show a marked loss in appetite.

Prevention, Control, and Treatment

Effective anthelmintics for the control of stomach worms are listed in Table 14.2. Note that different products are required for thick and red, or thin, stomach worms. Carbon disulfide has been effective and the McLean County System of Swine Sanitation should be applied to swine on pastures.

Thorny-headed Worms (*Macracanthorhynchus hirudinaceus*)

The thorny-headed worm is of considerable importance in the southern part of the United States, though its distribution is worldwide. It may be easily distinguished from the common intestinal roundworm by the presence of rows of hooks—a spiny proboscis (snout)—through which it attaches itself to the wall of the small intestine of the host. Thorny-headed worms are milk white to bluish in color and cylindrical to flat in shape, ranging from 2 to 4 in. for males and up to 26 in. long for females (Figure 14.24).

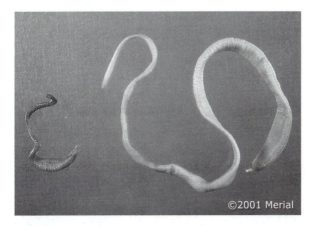

Figure 14.24 Male and female thorny-headed worms (*Macracanthorhynchus hirudinaceus*). *(Courtesy, Merial, Ltd., Rahway, NJ)*

Distribution and Losses Caused by Thorny-Headed Worms

The thorny-headed worm is not a common parasite of hogs grown in the Corn Belt states, but it is of considerable economic importance in the Deep South. In addition to the usual slow growth, inefficient feed utilization, and death losses resulting from other parasites, thorny-headed worms weaken the intestine and make it unfit for sausage casings. This causes a financial loss to the packer that is passed along to the swine producer in the form of lowered meat prices.

Life History and Habits

Adult female thorny-headed worms produce numerous thick-shelled brownish eggs, each containing a fully developed larva. Each female may produce as many as 600,000 eggs per day at the peak of her egg-producing capacity. The eggs, which pass out with the feces, are very resistant to destruction.

Beetle grubs, the larvae of June beetles, or dung beetles, serve as the intermediate hosts. The grubs, feeding on infected manure or contaminated soil, swallow the parasite eggs. The eggs hatch in the bodies of the grubs and in 3 to 6 months develop to a stage that is infective to swine (Figure 14.25).

Pigs rooting in manure or trash piles, rich soil, or low-lying pastures swallow the grubs or beetles. The young thorny-headed worms then escape from the bodies of the grubs or adult beetles through the process of digestion and develop to egg-laying maturity in 3 to 4 months.

Damage Inflicted; Symptoms and Signs of Infected Animals

No special symptoms have been attributed to swine infected with thorny-headed worms, although these parasites are decidedly injurious. Infected swine exhibit the general unthriftiness that is commonly associated with parasites. A heavy infestation may even kill young pigs. On autopsy, a swelling or nodule may be evident at the point of attachment, and the intestinal wall may exhibit great weakness.

Prevention, Control, and Treatment

Prevention consists of keeping pigs from feeding in areas where they might obtain grubs of the June beetle or dung beetle. Thus, sanitation, clean ground, and nose ringing are effective preventive measures. Levamisole will remove thorny-headed worms (Table 14.2).

Thorny-headed Worms

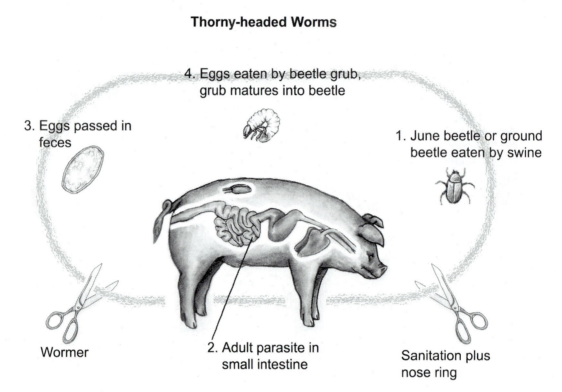

4. Eggs eaten by beetle grub, grub matures into beetle

3. Eggs passed in feces

1. June beetle or ground beetle eaten by swine

Wormer

2. Adult parasite in small intestine

Sanitation plus nose ring

Figure 14.25   A schematic diagram of the life cycle of the thorny-headed worm (*Macracanthorhynchus hirudinaceus*). A proper dewormer is the effective choice for control. Nose ringing of pasture hogs and sanitation are complimentary procedures. *(Courtesy, Nathaniel Klein, Iowa State University, Ames, IA)*

Threadworms *(Strongyloides ransomi)*

The pig is the only known host of *Strongyloides ransomi*. These worms are tiny. The parasitic females are 0.13 to 0.18 in. long, whereas the free-living forms are only about 0.04 in. long.

Distribution and Losses Caused by Threadworms

It is the most important swine parasite in the southern and southeastern states, though it has worldwide distribution.

Life History and Habits

Small, embryonated eggs are passed in the feces and they hatch in 12 to 18 hours. These larvae develop into the infective form, or free-living form, 2 to 3 days after hatching. The infective larvae penetrate the skin and proceed to the lungs via the bloodstream, thence from the alveoli of the lungs to the bronchi, esophagus, stomach, and small intestine, where they become adult parthenogenetic females. Oral ingestion of the infective larvae can also produce infection. Furthermore, baby pigs may obtain the infective larvae from the colostrum and demonstrate an infection as early as 4 days of age. A free-living generation of adult males and females develop into the infective parasitic larvae (Figure 14.26).

Damage Inflicted; Symptoms and Signs of Infected Animals

Diagnosis is best made on autopsy. However, in heavy infestations, it is a serious disease causing scours, anemia, and severe weight loss and death loss, particularly in young pigs. Light infections may show no symptoms.

Prevention, Control, and Treatment

Prevention is aided through (1) a program of strict sanitation; (2) selecting dry, unshaded areas for swine lots; (3) pasture rotation; and (4) a program of frequent deworming, using an effective dewormer of choice. Effective anthelmintics for the control of threadworms are listed in Table 14.2.

Trichinosis *(Trichinella spiralis)*

Trichinosis is a parasitic disease of human beings caused by *Trichinella spiralis*. The main source of the disease is meat from infected pigs or bears eaten raw or improperly cooked. Although the parasite is present in the muscle tissue of swine, it does not induce symptoms in this species. Infections have resulted from the consumption of bear meat in Canada, Alaska, and the northeastern and western states.

Threadworms

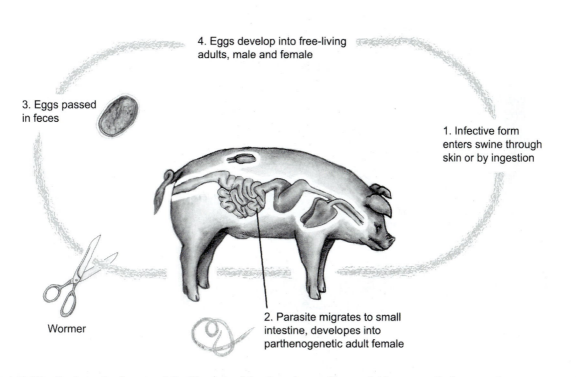

4. Eggs develop into free-living adults, male and female

3. Eggs passed in feces

1. Infective form enters swine through skin or by ingestion

Wormer

2. Parasite migrates to small intestine, develops into parthenogenetic adult female

Figure 14.26 A schematic diagram of the life cycle of the threadworm (*Strongyloides ransomi*). A proper dewormer, complimented by sanitations, is the effective choice for control. *(Courtesy, Nathaniel Klein, Iowa State University, Ames, IA)*

Distribution and Losses Caused by Trichinosis

This disease appears to be worldwide, with the highest incidence occurring in areas where uncooked garbage is fed to hogs. A USDA NAHMS national swine survey conducted in 1995 reported the infection rate in U.S. swine to be 0.013%. Modern swine management systems have virtually eliminated trichinae as a problem in domestic pigs.

Life History and Habits

The adult parasite, a round worm from 0.06 to 0.16 in. in length, lives in the small intestine of humans, hogs, rats, and other animals. The female worms penetrate into the lining of the intestines where they produce numerous young or larvae. Males die soon after mating. The larvae then pass from the wall of the intestine into the lymph system, thence into the bloodstream, and finally into the muscle cells.

In the muscles, the larvae grow until they are about 0.04 in. long, then roll into a characteristic spiral shape, and become surrounded by a capsule. In this environment and stage of development (cysts), these larvae may live for years or until the raw or improperly cooked muscle tissue is eaten by humans or other species of meat eaters. On gaining entrance to the intestines of another host, the worm starts a new life cycle (Figure 14.27).

Damage Inflicted; Symptoms and Signs of Infected Animals

There are no specific symptoms in hogs, even when the parasite is present in the muscle tissue, its usual abode.

The Centers for Disease Control and Prevention reported that cases of human trichinellosis have declined to below 25 annually over the past several years and only a few of these cases have been associated with consumption of pork. The symptoms in humans vary depending on the degree of parasite infection. The disease is usually accompanied by a fever, digestive disturbances, swelling of infected muscles, and severe muscular pain (in the breathing muscles as well as others). A specific skin test, serologic tests, or muscle biopsy may also be applied to aid in the diagnosis in humans. Normally, the symptoms begin about 10 to 14 days after infected pork has been eaten. Infected humans should be under the care of a medical doctor.

Prevention, Control, and Treatment

Essentially, any trichina in pork that is slaughtered is noninfective by the time it reaches the consumer. In processing, much of it is made safe through heating or freezing. Nevertheless, a small amount of infected pork may be sold primarily as ready-to-eat

Trichina

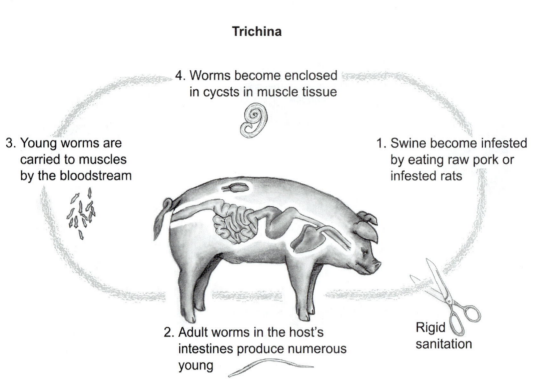

4. Worms become enclosed
in cycsts in muscle tissue

3. Young worms are
carried to muscles
by the bloodstream

1. Swine become infested
by eating raw pork or
infested rats

2. Adult worms in the host's
intestines produce numerous
young

Rigid
sanitation

Figure 14.27   A schematic diagram of the life cycle of trichina (*Trichinella spiralis*). Effective control is obtained through rigid sanitation, including destruction of rats, proper carcass disposal of dead hogs, and cooking all garbage and offal from slaughtering houses. *(Courtesy, Nathaniel Klein, Iowa State University, Ames, IA)*

pork from noninspected facilities, and the only certain protection to humans is to cook pork and pork products to 137°F internal temperature. Thus, prevention consists of the thorough cooking of all pork before it is eaten, either by humans or by swine. *Trichinella* is also destroyed by freezing for a continuous period of not less than 20 days at a temperature of not higher than 5°F.

Microscopic examination of pork is the only way in which to detect the presence of trichinae. Such methods are regarded as impractical, and meat inspection in the United States does not include examination for *Trichinella* (Figure 14.28).

Trichinosis in swine may be prevented by (1) destruction of all rats on the farm, (2) proper carcass disposal of hogs that die on the farm, and (3) cooking of all garbage and offal from slaughtering houses at a temperature of 212°F for 30 minutes. However, as all these preventive measures may not be feasible or practical or carried out with care, the surest and best protection currently available to pork consumers consists of cooking or freezing pork at the temperatures specified. State and federal governments regulate the cooking and feeding of garbage.

Whipworms (*Trichuris suis*)

Whipworms are usually found attached to the walls of the cecum (blind gut) and large intestine of

Figure 14.28   *Trichinella spiralis*, muscle biopsy, 100X. *(Courtesy, Dr. Dietrich Barth, Merial, Ltd., Rahway, NJ)*

swine. They are 1 to 3 in. in length. The worms have a very slender anterior portion and a much enlarged posterior. The anterior resembles the lash of a whip and the posterior resembles the handle, hence, the name whipworm (Figure 14.29).

Distribution and Losses Caused by Whipworms

Whipworms are present in most areas where swine are raised. It appears, however, that the heaviest infection of swine occurs in the southeastern states. There is ample evidence that the whipworm is increasing in the Corn Belt and in the Southeast.

Life History and Habits

The life cycle of the whipworm is simple and direct; that is, no intermediate host is necessary to complete the life cycle. The eggs, which are produced in large numbers, are eliminated with the feces. An infective larva develops in the shell, and swine become infected by swallowing the eggs when feeding on soil where they are present. The eggs hatch in the stomach and intestine, and the larvae make their way to the blind gut, where they grow to maturity in about 10 weeks or longer (Figure 14.30).

Figure 14.29 Whipworms (*Trichuris suis*). *(Courtesy, Merial, Ltd., Rahway, NJ)*

Damage Inflicted; Symptoms and Signs of Infected Animals

Infected animals may develop bloody diarrhea, and in heavy infections the pigs become anemic and dehydrated. In massive infections, growth may be noticeably retarded, and the animals may become weak and finally die. Also, the parasites may cause sufficient damage to enable secondary infections to become established.

Prevention, Control, and Treatment

Clean, well-drained pastures, rotation grazing, and plenty of sunlight are an aid to the prevention and control of whipworms. Effective anthelmintics for the control of whipworms are listed in Table 14.2.

ANTHELMINTICS (DEWORMERS)

Knowing what internal parasites are present within an animal is the first requisite to the choice of the proper drug, or anthelmintic. Because no one drug is appropriate or economical for all conditions, the next requisite is to select the right one, which, when used according to directions, will be most effective and produce a minimum of side effects on the animal treated. Coupled with knowledge of the kind of parasites present, an individual assessment of each animal is necessary.

Whipworms

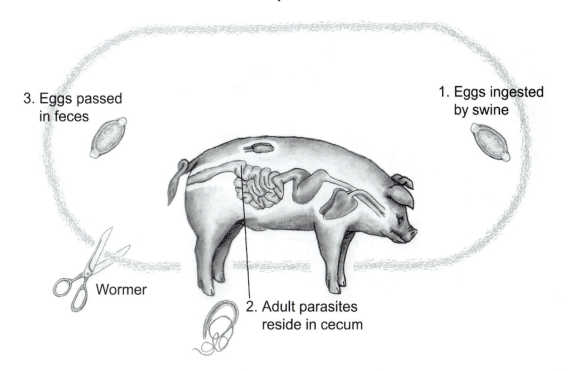

3. Eggs passed in feces

1. Eggs ingested by swine

Wormer

2. Adult parasites reside in cecum

Figure 14.30 A schematic diagram of the life cycle of whipworms (*Trichuris suis*). A proper wormer, complimented by dry, clean lots (or well-drained pastures) is the key to their control. *(Courtesy, Nathaniel Klein, Iowa State University, Ames, IA)*

TABLE 14.2 Approved Products for Removal of Internal Parasites from Pigs[1]

Parasite	Infected Herds	Dichlorvos Atgard	Doramectin	Fenbendazole (Safe-Guard Type "A")	Hygromycin B Hygromix	Ivermectin Ivomec	Levamisole Tramisol	Piperazine Wonder Wormer	Pyrantel Banminth
Withdrawal Period		None	Injection = 24 Days	6 Days	15 Days	Feed = None Injection = 18 Days	72 Hours	21 Days	24 Hours
Kidney worm *Stephanurus dentatus*	Up to 60%		Yes	Yes		Yes	Yes		
Lungworm *Metastrongylus* spp.	Up to 60%		Yes	Yes		Yes	Yes		
Nodular worm *Oesophagostomum* spp.	30–60%	Yes	Yes	Yes	Yes	Yes	Yes	Yes	Yes
Red stomach worm *Hyostrongylus rubidus*	Up to 30%		Yes	Yes		Yes	Yes		
Roundworm *Ascaris suum*	> 60%	Yes	Yes	Yes[2]	Yes	Yes	Yes	Yes	Yes[3]
Thick stomach worm *Ascarops strongylina*	Up to 30%	Yes				Yes			
Thorny-headed worm *Macracanthorhynchus hirudinaceus*	Up to 30%						Yes		
Threadworm *Strongyloides ransomi*	> 60%		Yes			Yes	Yes		
Whipworm *Trichuris suis*	Up to 60%	Yes		Yes	Yes				

[1] Doramectin is an injectable; Ivermectin is available as a feed additive or injectable; others are all feed additives. Read the labels for complete information.

[2] Also effective against immature stages.

[3] Prevents ascarid larval migration by killing the larvae as they hatch from ingested embryonated eggs.

Source: Pork Industry Handbook, *PIH-44, Purdue University, West Lafayette, IN, 2001; and Feed Additive Compendium, 2003.*

334

Among the factors to consider are age, pregnancy, other illnesses and medications, and the method by which the product is to be administered. Some drugs are unnecessarily harsh or expensive for the problem at hand, whereas a safe, inexpensive alternative would be equally suitable.

Each swine producer should, in cooperation with the local veterinarian or other advisor, develop a parasite control program and schedule. It is recommended that several different anthelmintics be used in rotation. Also, a schedule of treatments should be prepared, based on knowledge of the life cycles of the various parasites. Possibly, before an active worming program is begun, the veterinarian should have the feces of pigs in the herd examined for the types of worms present and then choose the anthelmintic. Often, in slotted floor systems, no worm eggs are found.

The recommendations as to the compounds to use for the control of parasites are intended as guidelines and are limited to those approved by the FDA or APHIS (Table 14.2). Furthermore, it is recognized that anthelmintics are constantly being improved, that new ones are becoming available, and that some old ones are banned.

EXTERNAL PARASITES

Blowflies

Blowflies differ from screwworms in that they deposit their eggs in necrotic tissue, whereas screwworms lay their eggs on the edges of wounds. The flies of the blowfly group include a number of species that find their principal breeding ground in necrotic tissue.

Distribution and Losses Caused by Blowflies

Although blowflies are widespread, they present the greatest problem in the Pacific Northwest and in the South and southwestern states. Death losses from blowflies are not excessive, but they cause much discomfort to affected animals and lower production.

Life History and Habits

With the exception of the group known as gray flesh flies, which deposit tiny living maggots instead of eggs, the blowflies have a similar life cycle to the screwworm, except that the cycle is completed in about one-half the time. (See discussion of the screwworm later in this chapter.)

Damage Inflicted; Symptoms and Signs of Infected Animals

The blowfly causes its greatest damage by infesting wounds of living animals. Such damage, which is largely limited to the black blowfly (or woolmaggot fly), is similar to that caused by screwworms. The maggots spread over the body, feeding on the dead skin and exudates, where they produce a severe irritation and destroy the ability of the skin to function. Infested animals rapidly become weak and fevered; and although they recover, they may remain in an unthrifty condition for a long period.

Because blowflies infest fresh or cooked meat, they are often a problem of major importance around packing houses or farm homes.

Prevention, Control, and Treatment

Prevention of blowfly damage consists of eliminating the pest and decreasing the susceptibility of animals to infestation.

Because blowflies breed principally in dead carcasses, the most effective control is effected by promptly destroying all dead animals by burning or deep burial. The use of traps, poisoned baits, and electrified screens is also helpful in reducing trouble from blowflies. Suitable repellents, such as pine tar oil, help prevent the fly from depositing its eggs.

Wounds can be protected from blowflies by the application of Smear 62 or EQ335 (3% lindane and 35% pine oil). After the wounds have been invaded by the larvae, they may be effectively treated with pressurized aerosols containing coumaphos, lindane, or ronnel.

Lice

The louse is a small, flattened, wingless insect parasite of which there are several species. Hog lice, which are the largest species, may range up to 0.25 in. in length (Figure 14.31). The two main types are sucking lice and biting lice. Of the two groups, the sucking lice are the most injurious.

Most species of lice are specific for a particular class of animals; thus, hog lice will not remain on the other farm animals, nor will lice from other animals usually infest hogs. Lice are always more abundant on weak, unthrifty animals and are more troublesome during the winter months than during the rest of the year.

Distribution and Losses Caused by Lice

The presence of lice is almost universal, but the degree of infestation depends largely on the state of animal nutrition and the extent to which the owner will tolerate parasites. The irritation caused by the presence of lice retards growth, gains, and production of milk, and such diseases as swine pox may be transmitted by lice.

©2001 Merial

Figure 14.31   Hog louse. This louse is about 7 mm long. *(Courtesy, Merial, Ltd., Rahway, NJ)*

Life History and Habits

Lice spend their entire life cycle on the host's body. They attach their eggs or "nits" to the hair near the skin where they hatch in about 2 weeks. Two weeks later the young females begin laying eggs, and after reproduction they die on the host. Lice do not survive more than 2 to 3 days when separated from the host. They are found on all parts of the body, but, in particular, areas such as the folds around the neck, jowl, flanks, and inside the legs and ears.

Damage Inflicted; Symptoms and Signs of Infected Animals

Infestation shows up most commonly in winter in ill-nourished and neglected animals. There is intense irritation, restlessness, and loss of condition. Because many lice are blood suckers, they devitalize their host. There may be severe itching and the animal may be seen scratching, rubbing, and gnawing at the skin. The hair may be rough, thin, and lack luster, and scabs may be evident. In pigs, the skin of infested animals becomes thick, cracked, tender, and sore. In some cases, the symptoms may resemble that of mange, and it must be kept in mind that the two may occur simultaneously. With the coming of spring, when the hair sheds and when some animals may go to pasture, lousiness is greatly diminished.

Prevention, Control, and Treatment

Because of the close contact of domesticated animals, especially during the winter months, it is difficult to prevent herds from becoming slightly infested with the pests. Nevertheless, lice can be kept under control. For more effective control, the entire herd must be treated simultaneously, which is especially necessary during the fall months about the time they are placed in winter quarters.

Spraying or dipping during freezing weather should be avoided. It is also desirable to treat the housing and bedding. With a power sprayer, from 100 to 200 lb/sq in. of pressure is adequate for spraying hogs. Dusting is less effective than spraying or dipping, but it may be preferable when only a few animals are to be treated or during the winter months.

The following insecticides are effective in the control of lice when used according to the manufacturer's label directions: amitraz (Takie), diazinon, fenvalerate, lindane, malathion, permethrin, and phosmet (Prolate-Star Bar). Also, ivermectin, available as an injectable or a feed additive, is very effective. Where hog lice are encountered, it is suggested that the producer obtain from local authorities the current recommendations relative to the choice and concentration of the insecticide to use.

Mange

Mange mites produce a specific contagious disease known as mange (or scabies, scab, or itch). These small insectlike parasites, which are almost invisible to the naked eye, constitute a very large group. They attack members of both the plant and animal kingdoms.

They are responsible for the condition known as mange (scabies) in swine, sheep, cattle, and horses. Each species of domesticated animals has its own species of mange mites. The mites from one species of animals cannot live normally and propagate permanently on different species. The disease appears to spread most rapidly among young and poorly nourished animals. Two types of mange infect swine—sarcoptic mange and demodectic mange. Of the two, sarcoptic mange is by far the most prevalent, showing worldwide distribution (Figure 14.32). Demodectic mange is uncommon in swine. Intense itching often accompanies the sarcoptic form because the mites form tunnels in the skin as part of their life cycle.

Distribution and Losses Caused by Mites

Injury from mites is caused by irritation and blood sucking and the formation of scabs and other skin affections. In a severe attack, the skin may be much less valuable for leather. Growth is retarded and animals are unthrifty.

Life History and Habits

The entire life cycle of the sarcoptic mange mite, *Sarcoptes scabiei* var. *suis*, occurs on the body of the

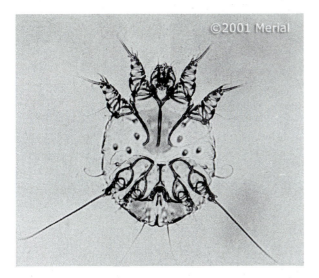

Figure 14.32 Mange mite (*Sarcoptes scabei* var. *suis*). *(Courtesy, Merial, Ltd., Rahway, NJ)*

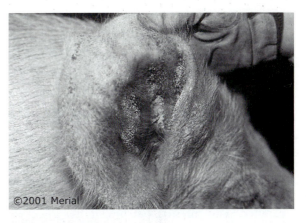

Figure 14.33 Chronic mange with typical crusts. *(Courtesy, Merial, Ltd., Rahway, NJ)*

host pig. After the female mite mates, she lays eggs at the rate of about 1 to 3 per day in tunnels that she carves into the upper two-thirds of the epidermis. In about 5 days the eggs hatch and then mature to adults within the tunnels in about 10 to 15 days. New females then mate near the surface and begin their tunnels. The cycle from egg to egg-laying female requires 10 to 15 days. Mature females die about 1 month after reaching maturity. Generally, mite infestations begin in the inner ear and then spread along the neck and across the body (Figure 14.33).

Mites are more prevalent during the winter months when animals are in close contact with each other and when treatment is more difficult. They spread among the young and among the poorly nourished animals.

Damage Inflicted; Symptoms and Signs of Infected Animals

Infested animals do not eat properly and, as a result, they do not gain at normal rates. Sarcoptic mites cause marked irritation, itching, and scratching, and crusting of the skin which is often accompanied or followed by the formation of a thick, tough, wrinkled skin. Often there are secondary skin infections. The only certain method of diagnosis is to demonstrate the presence of the mites.

Prevention, Control, and Treatment

Prevention consists of avoiding contact with diseased animals or infested premises. In case of an outbreak, the local veterinarian or livestock sanitary officials should be contacted.

Mites can be controlled by spraying animals with suitable insecticides and quarantine of infected herds. Producers should have a routine program for spraying for mange control. For example, gilts and sows should be sprayed 4 to 6 weeks before farrowing and then again 7 to 10 days prior to farrowing. A high-pressure spray (100 to 250 lb/sq in. [psi]) is recommended and repeated treatments are important because eggs in tunnels would not be reached by the first spraying.

The following insecticides are effective in the control of mites when used according to the manufacturer's directions: amitraz (Takie), diazinon, fenvalerate, lindane, malathion, permethrin, and phosmet (Prolate-Star Bar). Also, ivermectin, available as an injectable or feed additive, is very effective. Because new insecticides are constantly being developed, and old ones banned or restricted, when hog mites are encountered, it is suggested that the producer obtain from local authorities the current recommendations relative to the choice and concentration of the insecticide to use. Generally, those products listed for the control of mange are also effective for louse control.

Ringworm

Ringworm, or barn itch, is a contagious disease of the outer layers of skin. It is caused by certain microscopic molds or fungi (*Trichophyton, Achorion,* or *Microsporon*). All animals and humans are susceptible.

Distribution and Losses Caused by Ringworm

Ringworm is widespread throughout the United States. Though it may appear among animals on pasture, it is far more prevalent as a barn disease. It is unsightly, and affected animals may experience considerable discomfort, but the actual economic losses attributed to the disease are not too great.

Life History and Habits

The period of incubation for this disease is about 1 week. The fungi form spores that may live 18 months or longer in barns or elsewhere.

Damage Inflicted; Symptoms and Signs of Infected Animals

Round, scaly areas almost devoid of hair appear mainly in the vicinity of the eyes, ears, side of the neck, or the root of the tail. Crusts may form, and the skin may have a gray, powdery, asbestoslike appearance. The infested patches, if not checked, gradually increase in size. Mild itching usually accompanies the disease.

Prevention, Control, and Treatment

The organisms are spread from animal to animal and through the medium of contaminated fence posts and brushes. Thus, prevention and control consists of disinfecting everything that has been in contact with infested animals. The infected animals should also be isolated. Strict sanitation is essential in the control of ringworm.

The hair should be clipped, the scabs removed, and the area brushed and washed with mild soap. The diseased parts should be painted with tincture of iodine or salicylic acid and alcohol (1 part in 10) every 3 days until cleared up. Animals may be treated orally with nystatin or griseofulvin, or topically with iodine, captan, or salicylic acid.

Screwworms (*Callitroga hominovora*)

Screwworms are parasites that can cause great damage to domestic livestock and other warm-blooded animals. The larvae of this pest enter open wounds of the host animal and feed on the raw flesh. Rare cases of humans being infested with screwworm have been reported. The United States has been free of screwworm since 1966.

In 1972 the Mexican-American Commission for the Eradication of Screwworm was formed at the request of Mexican livestock producers. Working with the commission, USDA-APHIS managed to establish a barrier in Mexico in 1984, but cases remained in Mexico until 1986. The Mexican-American Commission, in cooperation with other commissions formed with each Central American country, has eradicated the screwworm from virtually all of Central America down to the Isthmus of Panama. Today a permanent sterile fly release barrier is maintained in Panama between the Panama Canal and the Colombian border (Figure 14.34).

Figure 14.34 Sterile male screwworm fly marked with a numbered tag to study fly dispersal, behavior, and longevity. *(Photo by Peggy Greb, Courtesy, USDA, Washington, DC)*

Distribution and Losses Caused by Screwworms

Wounds resulting from castrating, docking, dehorning, branding, and shearing afford a breeding ground for this parasite. Add to this the wounds from some types of vegetation, from fighting, and from blood-sucking insects, and ample places for propagation are provided.

Life History and Habits

The primary screwworm fly is bluish-green, with three dark stripes on its back and reddish or orange color below the eyes. The fly generally deposits its eggs in shinglelike masses on the edges or the dry portion of wounds. From 50 to 300 eggs are laid at one time, with a single female being capable of laying about 3,000 eggs in a lifetime. Hatching of the eggs occurs in 11 hours, and the young whitish worms (larvae or maggots) immediately burrow into the living flesh. There they feed and grow for a period of 4 to 7 days, shedding their skin twice during this period.

When these larvae have reached their full growth, they assume a pinkish color, leave the wound, and drop to the ground, where they dig beneath the surface of the soil and undergo a transformation to the hard-skinned, dark brown, motionless pupae. It is during the pupa stage that the maggot changes to the adult fly (Figure 14.35).

After the pupa has been in the soil from 7 to 60 days, the fly emerges from it, works its way to the surface of the ground, and crawls up on some nearby object (bush, weed, etc.) to allow its wings to unfold and otherwise mature. Under favorable conditions, the newly emerged female fly becomes sexually mature and will lay eggs 5 days later. During warm weather, the entire life cycle is usually completed in

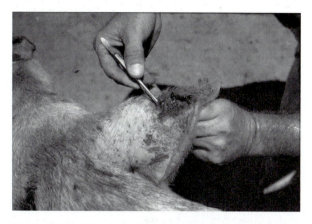

Figure 14.35 Treating the ear of screwworm infested pig by Animal and Plant Health Inspection Service. *(Courtesy, USDA, National Agricultural Library, Washington, DC)*

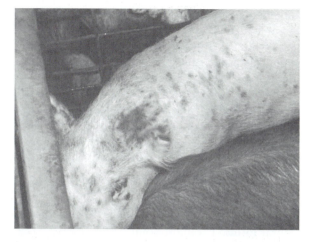

Figure 14.36 A pig with swine pox nodules and vesicles. *(Courtesy, Palmer Holden, Iowa State University, Ames, IA)*

21 days, but under cold, unfavorable conditions the cycle may take as many as 80 days or longer.

Damage Inflicted; Symptoms and Signs of Infected Animals

The injury caused by screwworms is inflicted chiefly by the maggots. Early symptoms in infected animals are loss of appetite and condition, and listlessness. Unless proper treatment is administered, the great destruction of many tissues kills the host in a few days.

Prevention, Control, and Treatment

Prevention in infested areas consists mainly of keeping animal wounds to a minimum and of protecting those that do materialize. The sterile male screwworms released into the environment were the major factor in eradicating them from North America.

When maggots (larvae) are found in an animal, they should be removed and sent to the proper authorities for identification.

Wounds can be protected from screwworm flies by the application of a wound dressing of Smear 62 (containing diphenylamine, benzol, turkey red oil, and lampblack) or EQ335 (containing 3% lindane and 35% pine oil). After wounds are invaded by larvae, they may be effectively treated with pressurized aerosols containing coumaphos, lindane, or ronnel.

Swine Pox

Swine pox is widely distributed in the Midwest, where it is an important disease of young pigs.

Symptoms and Signs

Swine pox is characterized by small red spots, which appear over large parts of the body especially on the ears, neck, undersurface of the body, and inside the thighs. These spots grow rapidly and reach the size of a dime (Figure 14.36). A hard nodule develops in the center of each. Several days later, small, pea-size vesicles (blisters) develop; at first these contain a clear fluid, but later the contents become puslike. Soon, these blisters dry up, leaving dark brown scabs, which fall off. Preceding these skin changes, some animals show fever, chills, and refusal to eat. Very few pigs die.

Cause, Prevention, and Treatment

There are two types of swine pox, each caused by a different type of virus (swine pox virus and vaccinic virus). Swine that recover from the disease are immune to further attacks from the specific type of virus that caused the disease but not to attacks from the other virus. One type of virus is related to that causing pox in various other species of animals, but the other is not.

The disease appears to be transmitted primarily by lice, and less often by other insects and by contact. Therefore, lice control and general sanitation are the best preventive measures. Vaccination is possible but not recommended for two reasons: (1) The disease has no great economic importance and (2) the use of a live vaccine would introduce more virus in an infected environment.

No treatment for swine pox is known. Good management and nursing will help. Also, good, clean quarters prevent secondary infections of the skin lesions.

PESTICIDES

Pesticides are chemicals used to control pests. They are an important part of livestock management and should be applied judiciously, carefully following manufacturer's label directions for use, hazards, and withdrawal times before slaughter. Agricultural chemicals are as vital to the health of animals as modern medicines are to the health of people. Producers depend on chemicals—insecticides, herbicides, fungicides, and similar materials—to control the pests that attack their animals or damage their feed crops, and, when necessary, they use biologics for the prevention or treatment of animal diseases.

Forms of Pesticides

Pesticides for use on animals may be purchased in several forms. The most common are emulsifiable concentrates, dusts, wettable powders, and oil solutions. When treating animals, be sure to use only pesticide formulations that are prepared specifically for livestock.

> **Emulsifiable concentrates (EC)**—Emulsifiable concentrates, which are probably the most common type of formulation, are solutions of pesticides in petroleum oils or other solvents. An emulsifier has been added so that the solution will mix well with water. On occasion, usually after extended storage, an EC may separate into its various parts; in that case, it should be discarded. An emulsion may also separate if it is allowed to stand after the concentrate has been added to the water; periodic agitation will help prevent this.
>
> **Dusts**—Dusts are applied directly to animals in the dry form and cannot be used as sprays.
>
> **Wettable powders**—Wettable powders are also dry, but the addition of a dispersing and wetting agent allows them to be suspended in water for application. Continuous agitation of the mixture is important when treating with wettable powders.
>
> **Oil solutions**—Oil solutions are pesticides dissolved in oil; no emulsifier is added. These materials are usually ready for use and should not be added to water.

Precautions on the Use of Pesticides

Certain basic precautions must be observed when pesticides are to be used because, when used im-

properly, they can be injurious to humans, domestic animals, wildlife, and beneficial insects. If used properly, pesticides should give satisfactory control and should not leave residues that exceed the tolerances established for any specific chemical. To avoid excessive residues in pork products, follow the label directions carefully with respect to dosage levels and chemical safety precautions. Remember that swine producers are responsible for residues in their own animals and products as well as for problems caused by drift from their property. Other pertinent information follows:

> **Selecting pesticides**—Always select the formulation and the pesticide labeled for the purpose for which it is to be used.
>
> **Storing pesticides**—Always store pesticides in original containers. Never transfer them to unlabeled containers or to feed or beverage containers. Store pesticides in a dry place out of reach of children, animals, or unauthorized persons.
>
> **Disposing of empty containers and unused pesticides**—Properly and promptly dispose of all empty pesticide containers. Do not reuse. Break and bury glass containers and chop holes in, crush, and bury metal containers to prevent their reuse. Check with local authorities to determine specific procedures for disposal of used containers.
>
> **Mixing and handling**—Mix and prepare pesticides in the open or in a well-ventilated place. Wear rubber gloves and clean, dry clothing (a respirator device may be necessary with some products). If any pesticide is spilled on you or your clothing, wash with soap and water immediately and change clothing. Avoid prolonged inhalation. Do not smoke, eat, or drink when mixing and handling pesticides.
>
> **Applying**—Use only amounts recommended. Apply at the correct time to avoid unlawful residues. Avoid treating pigs younger than that specified on the label. Avoid re-treating more often than label restrictions specify. Avoid drift on nearby crops, pastures, livestock, or other nontarget areas. Avoid prolonged contact with all pesticides. Do not eat, drink, or smoke until all operations have ceased and hands and face are thoroughly

washed. Change and launder clothing after each day's work.

In case of an emergency—If you accidentally swallow a pesticide, call a doctor or in the United States call the National Poison Center toll-free number, 1-800-222-1222. Do not induce vomiting until consulting with a physician as aspiration of some chemicals into the lungs can cause more problems.

Withdrawal—After spraying, dipping, or dusting your animals with pesticides, observe the prescribed number of days or interval between the last treatment and slaughter. Refer to the container label for this information.

DISINFECTANTS

A disinfectant is an agent, such as heat, radiation, or a chemical, that destroys, neutralizes, or inhibits the growth of disease-carrying microorganisms. The high concentration of swine and the continuous use of buildings often result in a microbiological buildup not limited to bacteria. As disease-producing organisms—viruses, bacteria, fungi, and parasite eggs—accumulate in the environment, disease problems can become more severe and be transmitted to each succeeding group of animals raised on the same premises. Under these circumstances, cleaning and disinfection become extremely important in breaking the life cycle. Also, in the case of a disease outbreak, the premises must be disinfected.

Under ordinary conditions, proper cleaning of buildings removes most of the microorganisms, along with the filth, thus eliminating the necessity of disinfection. Effective disinfection depends on five things:

1. Thorough removal of all manure and washing before application.
2. The phenol coefficient of the disinfectant, which indicates the killing strength of a disinfectant as compared to phenol (carbolic acid). It is determined by a standard laboratory test in which the typhoid fever germ is used as the test organism.
3. The dilution at which the disinfectant is used.
4. The temperature; most disinfectants are much more effective if applied hot.
5. Thoroughness of application and time of exposure.

Disinfection must in all cases be preceded by a very thorough cleaning because organic matter serves to protect disease germs and interferes with the activity of the disinfecting agent.

Sunlight possesses disinfecting properties, but it is variable and superficial in its action. Heat and some of the chemical disinfectants are more effective. The application of heat by steam, hot water, burning, or boiling is an effective method of disinfection. In many cases, however, it may not be practical.

A good disinfectant should (1) have the power to kill disease-producing organisms, (2) remain stable in the presence of organic matter (feces, hair, soil), (3) dissolve readily in water and remain in solution, (4) be nontoxic to animals and humans, (5) penetrate organic matter rapidly, (6) remove dirt and grease, and (7) be economical to use.

The number of available disinfectants is large because there is no ideal universally applicable disinfectant. Table 14.3 summarizes the limitations, usefulness, and strength of some common disinfectants. When using a disinfectant, always read and follow the manufacturer's directions.

POISONS

Poisonous plants and feeds should be avoided. The swine producer should know the poisonous plants common to the area, and avoid them. Also, the following poisons should be avoided: ergot, mycotoxins, pitch, excessive copper, fluorine, lead, mercury, nitrates, and selenium. Because poisons are generally eaten, a more detailed discussion of potentially poisonous elements is provided in Chapter 8, Table 8.12, Poisons and Toxins.

ZOONOSES

Zoonoses are diseases that may spread between animals and humans or between species of animals. Extra care should be employed when working around animals that were exposed to these diseases or are actually infected and exhibiting symptoms (Table 14.4).

TABLE 14.3   Disinfectant Guide[1]

Kind of Disinfectant	Usefulness	Strength	Limitations and Comments
Alcohols (ethyl-ethanol, isopropyl, methanol)	Primarily as skin disinfectant and for emergency purposes on instruments.	70% alcohol—the content usually found in rubbing alcohol.	They are too costly for general disinfection. They are ineffective against bacterial spores.
Boric acid[2]	As a wash for eyes, and other sensitive parts of the body.	1 oz in 1 pt water (about 6% solution).	It is a weak antiseptic. It may cause harm to the nervous system if absorbed into the body in large amounts. For this and other reasons, antibiotic solutions and saline solutions are fast replacing it.
Chlorines (sodium hypochlorate, chlomine-T, chlorine dioxide)	Used for equipment and as deodorants. They will kill all kinds of bacteria, fungi, and viruses, providing the concentration is sufficiently high.	Generally used at about 200 ppm for equipment and as a deodorant.	They are corrosive to metals and neutralized by organic materials.
Cresols (many commercial products available)	A generally reliable class of disinfectant. Effective against brucellosis, swine erysipelas, and tuberculosis.	Cresol is usually used as a 2 to 4% solution (1 cup/2 gal of water makes a 4% solution).	Cannot be used where odor may be absorbed, and, therefore, not suited for use around milk and meat.
Formaldehyde (gaseous disinfectant)	Formaldehyde will kill anthrax spores, TB organisms, and animal viruses in a 1 to 2% solution. It is often used to disinfect buildings following a disease outbreak. A 1 to 2% solution may be used as a footbath to control foot rot.	As a liquid disinfectant, it is usually used as a 1 to 2% solution. As a gaseous disinfectant (fumigant), use 1.5 lb of potassium permanganate plus 3 pt of formaldehyde. Also, gas may be released by heating paraformaldehyde.	It has a disagreeable odor, destroys living tissue, and can be extremely poisonous. The bacterial effectiveness of the gas depends on having the proper relative humidity (above 75%) and temperature (above 65°F and preferably near 80°F).
Heat (by steam, hot water, burning, or boiling)	In the burning of rubbish or articles of little value, and in disposing of infected body discharges. The steam "Jenny" is effective for disinfection if properly employed, particularly if used in conjunction with a phenolic germicide.	10 minutes exposure to boiling water is usually sufficient.	Exposure to boiling water will destroy all ordinary disease germs but sometimes fails to kill the spores of such diseases as anthrax and tetanus. Moist heat is preferred to dry heat, and steam under pressure is the most effective. Heat may be impractical or too expensive.
Iodine[2] (tincture)	Extensively used as a skin disinfectant, for minor cuts and bruises.	Generally used as tincture of iodine, either 2% or 7%.	Never with a bandage. Clean skin before applying the iodine. It is corrosive to metals.
Iodophors (tamed iodine)	Primarily used for dairy utensils. Effective against all bacteria (both gram-negative and gram-positive), fungi, and moist viruses.	Usually used as disinfectants at concentrations of 50 to 75 ppm titratable iodine, and as sanitizers at levels of 12.5 to 25 ppm. At 12.5 ppm titratable iodine, they can be used as an antiseptic in drinking water.	They are inhibited in their activity by organic matter. They are quite expensive.
Lime (quicklime, burnt lime, calcium oxide)	As a deodorant when sprinkled on manure and animal discharges, or as a disinfectant when sprinkled on the floor, or used as a newly made "milk of lime," or as a whitewash.	Use as a dust, as "milk of lime," or as a whitewash, but use fresh.	Not effective against anthrax or tetanus spores. Wear goggles, when adding water to quicklime.

Table 14.3   Disinfectant Guide *(continued)*

Kind of Disinfectant	Usefulness	Strength	Limitations and Comments

Lysol (the brand name of a product of cresol plus soap)	For disinfecting surgical instruments used in castrating and tattooing.
	    Useful as a skin disinfectant before surgery, and for use on the hands before castrating.
		0.5 to 2.0%	Has a disagreeable odor.
			    Does not mix well with hard water.
			    Less costly than phenol.

Phenols (carbolic acid):
1. Phenolics—coal tar derivatives
2. Synthetic phenols
	They are ideal general-purpose disinfectants. Effective and inexpensive.
	    They are very resistant to the inhibiting effects of organic residue; hence, they are suitable for barn disinfection, and foot- and wheel-dip-baths.
		Both phenolics (coal tar) and synthetic phenols vary widely in efficacy from one compound to another. Note and follow manufacturer's directions.
		    Generally used in a 5% solution.
			They are corrosive, and they are toxic to animals and humans.
			    Ineffective on fungi and viruses.

Quaternary ammonium compounds (QAC)	Very water-soluble, ultrarapid kill rate, effective deodorizing properties, and moderately priced.
	    Good detergent characteristics and harmless to skin.
		Follow manufacturer's directions.	They can corrode metal.
			    Not very potent in combating viruses.
			    Adversely affected by organic matter.

Sal soda	It may be used in place of lye against foot-and-mouth disease.	10.5% solution (13.5 oz/1 gal water).	
Sal soda and soda ash (or sodium carbonate)	They may be used in place of lye against foot-and-mouth disease.	4% solution (1 lb/3 gal water). Most effective in hot solution.	Commonly used as cleansing agents, but have disinfectant properties, especially when used as a hot solution.
Soap	Its power to kill germs is very limited. Greatest usefulness is in cleansing and dissolving coatings from various surfaces, including the skin, prior to application of a good disinfectant.	As commercially prepared.	Although indispensable to sanitizing surfaces, soaps should not be used as disinfectants. They are not routinely effective against staphylococci and the organisms which cause diarrheal disease are resistant.

[1] For metric conversions, refer to the Appendix.

[2] Sometimes loosely classified as a disinfectant but actually an antiseptic and practically useful only on living tissue.

Source: Courtesy of Pearson Education.

TABLE 14.4   Zoonoses of Swine That Affect Humans[1]

Disease	Infectious Agent	Primary Host(s)	Diseases in Animals	Diseases in People
Anthrax (LA)	*Bacillus anthracis*	Cattle, horses, sheep, goats, swine, dogs, cats	Sudden death, systemic disease, GI affliction	Malignant pustule, gastroenteritis, pneumonitis
Brucellosis (LA)	*Brucella spp.*	Cattle, goats, swine, sheep, horses, mules, dogs, cats, fowl, deer, buffalo, rabbits	Abortion, lameness, mastitis, granulomas, abscesses	Fever, malaise, lymphadenopathy, bacteremia, splenomegaly, osteomyelitis
Clostridium perfringes type A enteritis (DOS)	*Clostridium perfringens* type A	Swine	Diarrhea	Food poisoning
Cryptosporidium	*Cryptosporidium spp.*	Farm animals, cats, dogs	No apparent disease; enteritis, respiratory disease	Enteritis, dysentery
Cysticercosis (LA, DOS)	*Taenia solium* (*Cysticercus cellulosae*)	Swine	No apparent or known disease	Abdominal pain, diarrhea, weight loss, cysticercosis, life threatening
Erysipelas (LA, DOS)	*Erysipelothrix rhusiopathiae*	Swine, fowl, sheep, fish	Sudden death, "diamond skin" lesions, polyarthritis, septicemia, endocarditis	Skin lesions
Japanese B encephalitis (DOS)	*Flavivirus spp.*	Horses, cattle, sheep, goats, swine	Stillbirths and mummified fetuses	Encephalitis and abortion (spread primarily by mosquitoes)
Pastuerellosis (LA)	*Pasteurella multocida*	Fowl, cattle, sheep, swine, goats, horses, mice, rats, rabbits	Hemorrhagic septicemia, pneumonia	Tonsilitis, rhinitis, sinusitis, pleuritis, appendicitis, septicemia
Ringworm (DOS)	*Microsporum nanum* and *Trichophyton verrucosum*	Swine	Concentrical lesions	Concentrical lesions
Salmonellosis (LA)	*Salmonella spp.* (nontyphoidal)	Fowl, swine, cattle, sheep, horses, dogs, cats, rodents, reptiles, birds, cattle	No apparent disease; enteritis, septicemia, puerperal fever	Gastroenteritis, focal infection, septicemia
Streptococcus	*Streptococcus suis*	Swine	Meningitis, dead pigs, anorexia, paralysis, paddling	Septicemia, meningitis, endocarditis
Swine influenza (DOS)	Type A influenza viruses	Mammals and birds	Anorexia, prostration	Influenza
Swine vesicular disease	Swine vesicular disease virus	Swine		
Toxoplasmosis (LA)	*Toxoplasma gondii*	Cats, dogs, sheep, cattle, swine	Stillbirth, congenital defects, CNS lesions	Stillbirth, congenital defects, retinochoroiditis, encephalitis
Trichinosis (LA, DOS)	*Trichinella spiralis*	Most mammals (swine, bears, walruses)	No apparent or known disease	Trichinosis, chemosis, conjunctivitis, myositis, skin rash
Tuberculosis	*Mycobacterium bovis*	Cattle, horses, swine, cats, dogs	Pulmonary, lymph nodes, udder, GI tuberculosis, spondylitis	Primarily GI, lymph node, bone, pulmonary tuberculosis
Vesicular stomatitis (LA)	*Vesicular stomatitis virus*	Cattle, swine, horses	Ulceration of oral mucosa, feet, and teats	Fever, chills, headache

[1] LA = Los Angeles County, Department of Health Services, Los Angeles, CA, and (DOS) = *Diseases of Swine*, 8th ed., Iowa State University Press, Ames, IA.

QUESTIONS FOR STUDY AND DISCUSSION

1. What is the normal temperature, pulse rate, and breathing rate of hogs? How would you determine each?
2. How can federal and state regulatory officials be of assistance to the individual swine producer?
3. Define the following pharmaceutical terms: (a) over-the-counter drugs, (b) prescription drug, (c) extralabel drugs, and (d) veterinary feed directives.
4. Explain the difference between natural immunity, active immunity, and passive immunity.
5. Select a specific farm (either your own or one with which you are familiar) and outline (in 1, 2, 3 order) a program of life cycle herd health.
6. Give the (a) symptoms and signs, and (b) cause, prevention, and treatment of each of the following swine diseases:
   a. Atrophic rhinitis
   b. Enteric colibacillosis (baby pig scours)
   c. Porcine parvovirus
   d. Porcine reproductive and respiratory syndrome (PRRS)
   e. Porcine stress syndrome (PSS)
   f. Transmissible gastroenteritis (TGE)
7. Is it likely that, in the United States, a person will contract trichinosis? Justify your answer.
8. Explain why, even with antibiotics, drugs, and vaccines, a program of sanitation is important to herd health.
9. Assume that an internal parasite (you name it) has become troublesome in your herd. What steps would you take to meet the situation (list in 1, 2, 3 order; be specific)?
10. Why would you expect to see less of certain parasitic infestations with indoor housing on slotted flooring?
11. Recommend a treatment for lice and mites.
12. How are screwworms controlled?
13. Producers have been criticized for their use of chemicals. To eliminate reasons for the criticism, what precautions should producers take when using a pesticide in their herd health program?
14. List three zoonoses that may be of particular concern for transmission between pigs and humans, and explain your concerns.

SELECTED REFERENCES

Animal Health: A Layman's Guide to Disease Control, J. K. Baker and W. J. Greer, Interstate Publishers, Inc., Danville, IL, 1992

Diseases of Swine, 8th ed., Ed. by B. E. Straw et al., The Iowa State University Press, Ames, IA, 1999

Merck Veterinary Manual, The, 8th ed., Merck & Co., Inc., Rahway, NJ, 2003, http://www.merckvetmanual.com/mvm/index.jsp

Pork Industry Handbook, Cooperative Extension Service, Purdue University, West Lafayette, IN, 2002

Safeguarding Animal Health, USDA, APHIS, Veterinary Services, http://www.aphis.usda.gov/vs

Stockman's Handbook, The, 7th ed., M. E. Ensminger, Interstate Publishers, Inc., Danville, IL, 1992

3

Management Skills

15

Swine Production and Management

The room temperature should be measured at the level of the pigs, not at human eye level. *(Courtesy, Palmer Holden, Iowa State University, Ames, IA)*

Objectives

After studying this chapter, you should:

1. Know the advantages and disadvantages of indoor versus outdoor production.
2. Understand the conditions that allowed producers to move from single to year-round farrowings.
3. Know the management factors involved in handling herd boars.
4. Be able to estimate the number of replacement females needed.
5. Know the factors involved in observing estrus and successful mating strategies.
6. Know the factors that can increase litter size weaned.
7. Understand the various types of pork production systems and why a producer may choose one over the other.
8. Know how to ear notch, cut needle teeth, dock tails, and castrate pigs.
9. Know what factors to consider when determining the age to wean pigs.
10. Know the advantages and disadvantages of bedded production systems.
11. Know the methods of manure collection, transfer, storage, and land application.
12. Know the risks involved with storing liquid manure, especially pertaining to gases.
13. Understand the factors involved in determining the amount of fertilizer to apply to land for crop or pasture production.
14. Be aware of the environmental issue involving pork production.

Swine management is the art of caring for and handling swine. It gives point and purpose to everything else. It is essential to the success and profitability of a swine operation.

MANAGER'S ROLE

Managers of a swine operation must fulfill many roles in order to be successful; among them, planner, organizer, leader, controller, and facilitator of changes. These primary functions are shown in Figure 15.1. Good stockmen are not necessarily good managers. Even though they are adept at managing pigs, they may not have the ability to manage employees or the swine enterprise.

The development of sound management ability is a gradual process incorporating both formal training and practical experience. By itself, formal management education is insufficient in developing sound management abilities. Similarly, practical management experience in the absence of formal training in the use of management skills can hamper success.

Managerial skills need to be developed and rewarded. If the managers are also the owners, they should seek outside advice either formally or as a member of a producer group to compare their skills and operation to those of their peers. Nonowner managers should have their responsibilities clearly defined and periodically evaluated.

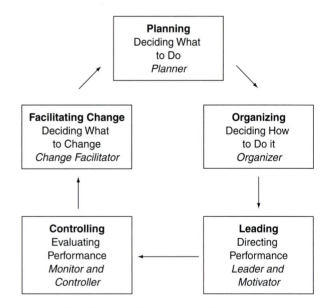

Figure 15.1 The role of the manager. Many swine operators become more involved in labor and fail to realize the value of good planning and execution of a farm plan, both in the short term and the long term. *(Courtesy of Pearson Education)*

SWINE MANAGEMENT SYSTEMS AND PRACTICES

Swine management systems and practices vary widely among areas and individual swine producers. They are influenced by size of operations; available land and its value for alternate uses; markets; available capital, labor, and feed; and the individual whims of producers.

Indoor versus Outdoor Intensive Swine Production

About 1955, the indoor confinement system started evolving, although for some time it had been recognized that hogs are better adapted to restricted rearing than other classes of four-footed farm animals. By 1960, it was estimated that 3 to 5% of the market hogs in the United States were produced indoors. In 1980, a survey revealed that 81% of U.S. swine producers put sows in stalls at farrowing time, 66% provided indoor production for nurseries, 70% confined finishing pigs, but only 24% confined the sow herd.[1] This shift to indoor production went hand-in-hand with (1) greater knowledge of nutrition, (2) improved disease prevention and parasite control, (3) improvements in buildings and equipment, and (4) the growth of large-scale, highly specialized operations.

Contrary to the opinion held by some, this does not mean that the use of pastures for hog production is antiquated. Rather, there are at least two alternatives plus many permutations for the management of swine, and the able manager will choose wisely between them or use the best of each system.

In the early 1990s, intensive outdoor production began to be popular in Britain and northern Europe and a strong market demand exists for pork from pigs with access to pasture or "free-range pork." Several European countries as well as the European Union have passed laws stipulating the pigs access to pasture and bedding.

A comparison of indoor versus outdoor intensive pork production at Texas Tech University compared the two systems. The relative data is presented in Table 15.1. The indoor facility featured totally indoor breeding, farrowing, and nursery units. The outdoor unit was located on a 34-acre site. Gilts were bred in radial-designed paddocks (shaped like the spokes of a wagon wheel with a central hub and triangle-shaped paddocks). Then gilts were penned together in the breeding-gestation paddocks. Bred gilts remained together until they were moved into a farrowing hut, each equipped with a fender (guard rail) 12 in. high. The fender remained up for 3 days after farrowing. Wheat straw was used for bedding. Farrowing paddocks were operated on an all-in/all-out basis.

Gilts and sows were bred to farrow every 3 weeks, year around. Pigs were weaned at 28 days. Diets were identical, except the outdoor diet was cubed. Pertinent results follow:

1. The total number born were identical but the outdoor gestated females farrowed about 0.7 more

TABLE 15.1   Effects of Farrowing Environment on Sow and Litter Performance

Measure	Indoor	Outdoor	P-Value[1]
Number litters	148	210	—
Piglets born/litter	10.99	11.01	0.97
Born live/litter	9.91	10.59	0.08
Found dead/litter	1.47	0.38	0.0001
Birth weight, lb alive	4.14	4.25	0.17
Litter birth wt, lb	43.0	44.9	0.22
Pigs weaned/litter	8.6	7.5	0.007
Preweaning mortality, %	12.3	27.5	0.0001
Weaning wt, lb/pig	14.3	15.3	0.106
Weaning wt/litter, lb	120.5	114.6	0.30

[1] P-value for environment effect. Values less than 0.05 indicate significant differences between the two systems. The smaller the P-value, the more statistically significant are the differences.
Source: Adapted from R. I. Nicholson, et al. 1995, Texas Journal of Agriculture and Natural Resources, *8:19–26. Data in table are least squares means averaged over the first and second parities.*

live piglets. The differences are not statistically significant but do suggest a trend toward more live pigs.

2. Indoor gilts weaned 8.6 pigs per litter, compared with 7.5 pigs for the outdoor gilts, and had only 12.3% preweaning mortality compared to 27.5% for outdoor litters. Crushing was the primary cause of preweaning piglet death in the outdoor farrowing huts.

3. The outdoor pigs weighed more than the indoor pigs at weaning (28 days of age) and at the end of the nursery phase (9 weeks of age). The advantage in weaning weight may be partially because of the fewer number of piglets nursing each sow.

4. Outdoor system required more labor. Sometimes, the caretakers had to work longer and do more heavy lifting (the huts) with the outdoor group. Also, they were exposed to all kinds of weather.

Both production systems can be competitive. Capital requirements are greater for indoor production but productivity is slightly higher. A market is emerging for outdoor reared pigs and appears to be commanding a premium market price. This will make outdoor pigs more competitive than if the specialty market does not exist. For producers not sure of their long-term interest in pork production, the low-investment outdoor facilities are advantageous.

In the mid-1990s, outdoor swine units were common in northern Europe. The warm climate of the southern part of the United States is suited to year-around outdoor intensive production. In addition to a dry climate, for success of outdoor, intensive production it is important that animal density be kept low so that disease incidence is low and manure buildup is not a problem.

Many changes have occurred since the early days of outdoor pork production in the United States. The pork industry has moved from the low-investment, low-intensity system of the 1950s to the modern high-investment, high-intensity indoor production systems in the 1980s. Many improvements in production accompanied the move indoors. Some of these changes can be adapted and applied to improve intensive outdoor production. (See the Chapter 11 introduction and the section, "Gestating and Outdoor Farrowing.")

Gilts versus Older Sows

Controlled experiments and practical observations, in which gilt litters have been compared with sows on their second or later litters, bear out the following facts: (1) gilts have fewer pigs in their litters and (2) the pigs from gilts are slightly lighter at birth and also tend to make slower gains.

Despite these disadvantages, gilts have certain advantages, especially for the commercial pork producer. Their chief superiority lies in the fact that they continue to grow and increase in value while in reproduction. Although overweight gilts marketed after their first litter bring less on the market than prime barrows, from this standpoint alone they generally return a handsome profit when the price of pork is sufficiently favorable. Additionally, the income from the sale of animals used for breeding may receive favorable tax treatment. Typically, about 40% or more of the sows will be replaced annually or about 20% of each farrowing group.

It is not recommended that the purebred breeder rely only on gilts. Proven sows that are regular producers of large litters with a heavy weaning weight and that are good mothers and producers of progeny of the right type should be retained in the herd as long as they are fertile. Although genetic improvement can be increased with turnover of the females, much of the gain will be accomplished by using superior sires.

Year-Round Farrowing

Historically, producers farrowed gilt litters in the spring and then rebred some of them for another litter in the fall. This resulted in two peak farrowings that had major impacts on the market price. In the 1950s the packing industry started "Spring Market Hog" shows to encourage fall farrowings. Today probably more than 95% of the pigs are farrowed in year-round farrowing systems.

Multiple farrowing involves scheduling so the litters arrive in a greater number of farrowing periods throughout the year. Large operations and environmentally regulated buildings have been designed for year-round production. There is noth-ing mysterious about year-round farrowing. It does, however, entail planning and close attention to management details. Among the factors favorable to year-round farrowing are the following:

1. A more stable hog market, with fewer high- and low-market receipts and price fluctuations.
2. A uniform workload for the swine producer in a planned year-round program.
3. Better use of existing buildings and equipment, for example, the farrowing house can accommodate more litters 12 or more times per year instead of 1 or 2.
4. A more sustained flow of hogs to market, which, from the standpoint of the packer, is desirable because (a) it makes for more complete use of labor and plant capacity, and (b) it enables the processor more nearly to meet the demands of the retailer and consumer. Also, the producer's income is distributed throughout the year.
5. A steady supply of pork for retail sale.
6. Avoidance of sharp price rises, which the consumer dislikes.

Among the factors unfavorable to year-round farrowing are the following:

1. The swine enterprise requires competent help be available throughout the year.
2. The possibility of a disease outbreak may be increased and more difficult to stop because of the possible buildup of pathogenic organisms. With year-round farrowing, it becomes more difficult to provide a break in production to clean and sanitize the facilities.
3. Farrowing in seasons of extreme cold or heat increases building and equipment costs.
4. It requires more management expertise including critical skills such as tightly managing breeding periods and maintaining a continuous gilt pool to optimize the use of farrowing facilities.

No one expects the seasonal pattern of hog production to be completely eliminated, but, because of the several recognized advantages of year-round farrowing to both the processor and the producer, it will continue to lessen some of the market gluts of the past. (See the subsequent section headed "Farrowing Management.")

Managing Herd Boars

Herd boars influence the swine breeding program in two important ways: (1) They provide a source of genetic improvement and (2) they affect the farrowing rate and litter size. In addition, replacement boars are a potential source for the production of disease in the herd.

Note: The feeding of boars is discussed in Chapter 7.

Boars should be purchased 45 to 60 days before they are needed for breeding. This allows for 30 days of isolation, testing for diseases of concern, and then acclimating them to microorganisms present in the herd. They need to be test-mated or evaluated for breeding soundness. Boars should be ready for use when they are about 8 months of age (Figure 15.2).

• **Transporting newly purchased boars**—Many seedstock suppliers offer a delivery service to their customers. Proper care in transporting boars ensures maximum animal performance by minimizing stresses, injuries, and disease.

• **Quarantine of new boars**—Newly purchased boars should be isolated in clean, comfortable facilities for a minimum of 30 and preferably 60 days. Following isolation, new additions should be tested for brucellosis, pseudorabies, transmissible gastroenteritis (TGE), and such other diseases as the herd veterinarian recommends. In the meantime, the isolation period should be continued until all health tests are received and evaluated.

• **Feeding boars in isolation**—During the isolation period, boars should initially be fed a diet similar to that fed by the seller. This reduces stress associated with relocation. Gradually, the diet should be changed to match that of the buyer. Because most diets in the United States are based on corn and soybean meal, it is likely that both farms are feeding similar diets. Depending on the diet, age, condition, and housing, boars should be given a balanced diet with about 0.75% lysine. Usually the sow lactation diet will be adequate for feeding boars. Most gestation diets are deficient in lysine for

boars. Large hog operations or boar studs may have enough boars to provide a special boar diet.

• **Test-mating and semen evaluation of boars**—Boars should be test-mated at 7 to 8 months of age. Records indicate that about 10% of new boars have fertility problems. Use an in-estrus (heat) gilt for test-mating, and assist the boar if necessary. If possible, also collect and evaluate a semen sample.

• **Age and service of boar**—The recommended maximum number of services per boar, by age, is given in Table 15.2 and may be used as a guide in determining boar power.

Individual mating systems provide an opportunity to control the use of boars and the opportunity to observe and record matings, but more labor and higher-cost facilities are usually required. Individual mating systems are essential for proper management for a weekly farrowing or for all-in/all-out production.

Pen-mating requires less labor and lower-cost facilities, but the resulting farrowing rate may be lower than for sows which are individually mated. When pen-mating is used, consider dividing weaned sows in two or more groups and rotating boars on a 12- to 24-hour interval.

• **Boar culling and replacement**—Farrowing rate and litter size are generally best when boars 9 to 20 months of age are used. However, as long as boars remain structurally sound and are aggressive breeders, fertility is generally maintained until they are 3 years of age or more. Most boars are culled for reasons of lameness, lack of aggressiveness, and size, rather than because of poor semen quality. Most well-managed herds replace up to 50% of their boars annually. This is especially true for farms that raise their own replacement gilts to prevent sire–daughter matings as the gilts enter the breeding herd.

• **Breeding area**—Provide a breeding area with good footing. Avoid wet, slippery floors. Provide for worker safety when working with boars.

• **Boar housing**—When using an individual mating system, boars are generally penned separately in individual crates about 28 in. wide and 7 ft long, or

Figure 15.2 A Duroc herd boar on an Iowa purebred breeder's farm. *(Courtesy, Palmer Holden, Iowa State University, Ames, IA)*

TABLE 15.2 Recommended Maximum Number of Services per Boar, by Age

Boar	Individual Mating System Maximum Matings		Pen-Mating System[1] Boar-to-Sow Ratio 7- to 10-Day Breeding Period
	Daily	*Weekly*	
Young (8 to 12 months)	1	5	1.2 to 4
Mature (more than 12 months)	2	7	1.3 to 5

[1] Assumes that all sows are weaned on the same day.

in pens 6 ft by 8 ft. Individual housing of boars eliminates fighting, riding, and competition for feed.

Groups of boars used in a pen-mating system should be penned together when they are removed from a sow group.

• **Keep boars cool during summer**—Boars subjected to temperature more than 85°F will have reduced semen quality or even temporary sterility, resulting in lowered farrowing rate and smaller litter size.

Heat stress may be determined by monitoring the respiration rate of boars. The normal respiration rate for a boar is 25 to 35 breaths per minute. During heat stress, respiration rate may increase to 75 to 100 breaths per minute. Respiration rate can be determined by watching the movement of the rib cage; it expands and contracts with each single breath.

A group of boars used in pen-mating can be kept cool by the use of a sprinkler or mister/fogger installed under a shade built over a sand or concrete floor. For boars housed in environmentally regulated buildings, consideration should be given to the use of thermostatically controlled drippers or evaporative coolers.

• **Routine boar management**—Some routine good boar management procedures include (1) observing boars daily for abnormal behavior such as lack of appetite, listlessness, or lameness; (2) having a high–low thermometer in the boar housing facility and in the breeding area; (3) vaccinating boars for reproductive diseases such as erysipelas, leptospirosis, and parvovirus; (4) treating for mange and lice; and (5) cutting the tusks of boars and trimming preputial hairs around the sheath, every 6 to 8 months.

• **Artificial insemination**—The vast majority of swine breeding farms now use some or all artificial insemination for their boar needs. In this case boars are often used only to stimulate the appearance of estrus in the female or to breed gilts in which estrus is often more difficult to detect.

Managing Sows and Gilts

In order to achieve maximum reproductive efficiency, a high level of management must be expended on the breeding herd. Good management of sows and gilts in the herd will pay handsome dividends by increasing the number of live pigs farrowed and marketed.

Note: The feeding of sows and gilts is discussed in Chapter 7.

Gilt Management

Producers should select a certain percentage (such as 20%) of the higher-indexing gilts from each con-

temporary group. These females will go into the gilt pool as candidates for breeding. Within the pool, gilts showing estrus during an assigned breeding period can be bred to serve as replacements for cull sows in the group.

Gilts should be selected from family lines that have superior productivity, primarily mothering ability and secondarily superior carcass traits. A good indication of the gilt's ability to reproduce normally is by having their first estrus cycle by 4 to 6 months of age.

The general recommendations regarding age at first breeding are to breed gilts during the third or later estrous period and wait until they reach 280 to 300 lb body weight. Gilts bred earlier often suffer from poorer performance when attempting remating after the first litter is weaned. It is recommended that gilts not reaching puberty or the desired breeding weight by 7 months of age be culled from the breeding herd.

Anestrous conditions (absence of standing heat) may be the result of one or more of the following conditions:

1. Faulty heat detection.
2. Hot weather stress.
3. Silent heat (ovulation with no visible sign of heat).
4. Sickness.
5. Nutritional deficiency(s) (especially lack of protein or energy).
6. Obesity (too fat).
7. Social stress.
8. Abnormal ovaries.
9. Pregnancy.

Commercial gilts should be selected from the largest, healthiest litters (based on farrowing and weaning data) as replacements on the basis of their ability to come into heat at an early age and conceive within three heat periods after their first exposure to a boar. An easy visual way to identify potential replacements requires ear notching the week of birth in one ear of piglets with the desired genetic background. Then at weaning punch a hole in an ear of female pigs from superior litters.

At selection several months later it is necessary only to observe gilts with holes in their ears. Cull those with unsound feet and legs or unsound underlines. Then select from the largest ones within a farrowing period as they will be the fastest-growing gilts from the best sows (litters), traits they will transmit to their offspring.

Managing the Gilt Pool

The gilt pool refers to a group of females selected as potential brood sows to fill vacancies in the sow herd. Empty farrowing crates indicate a lack of planning and poor use of the gilt pool.

Under normal conditions, about 15 to 30% of the weaned sows from each farrowing are culled. During problem breeding periods (hot weather, disease, or infertile boars), the number of replacement gilts needed increases. If 20 sows are weaned, it is likely at least 4 will need to be replaced. To have 4 gilts cycle during a 7-day breeding period, at least 12 gilts need to be available in the gilt pool. So, approximately 3 replacement gilts are needed in the pool for each sow in a farrowing group that is culled.

The number of gilts that should be selected from each contemporary group will depend on the sow culling policy and the length of the breeding interval. Assuming sows will only be kept for four litters and 20% of the sows are culled after each farrowing, the estimated number of gilts to select from each contemporary group can be calculated using the following equation:

$$\text{Number gilts to select} = \left(\frac{0.38 \times \text{No. females to breed}}{90\%} \right) \div \left(\frac{\text{Breeding interval, days}}{21} \right)$$

Where 0.38 = coefficient based on the producer selling all sows after their fourth litter and culling 20% of the them between farrowing and breeding; 90% = proportion of selected gilts that are cyclic and will mate (use an appropriate value for your herd); and 21 = the length of the estrous cycle.

An example illustrates the use of this equation. Assume 28 females are to be bred during a 14-day period for each group. About 25 sows (90%) farrow in each group. The number of gilts that should be selected from each contemporary group is

$$\text{Number gilts} = \left(\frac{0.38 \times 28}{90} \right) \div \left(\frac{14}{21} \right) = 17.7 \text{ (or 18) gilts}$$

Thus, the 18 top-indexing gilts from each contemporary group should be selected. If 18 top animals are selected from 100 gilts that may be tested within this contemporary group, the selection rate will be 18%. Obviously, additional gilts will be needed if some animals are culled because of health, reproductive, or feet and leg problems. Producers should strive for a gilt selection rate of less than 25% in each contemporary group. (National Swine Improvement Federation, Fact Sheet #15, 1992).

Sow Management

During gestation, gilts should be fed to gain about 90 lb and sows should gain about 70 lb. This provides for approximately 30 to 35 lb of newborn piglets, some increase in mammary tissue, and 30 lb of "products of conception" or afterbirth. The additional gain will be an increase in net body gain. Gilts need to continue to grow but mature sows only need to maintain their nonpregnant body weight.

Depending on each producer's program, lactation may typically last 14 to 28 days. Adequate records of individual performance during all phases of the reproductive cycle will be of benefit in upgrading the herd and making it more profitable.

Selection of sows that cycle within 4 to 7 days postweaning is essential in maintaining breeding management schedules. If a sow fails to conceive within 30 days postweaning, she should be culled. With each 21-day delay in conception, the sow must produce 1 to 2 extra pigs just to pay for the additional labor and feed. Likewise, if cycling gilts do not conceive after three estrous cycles, they should be culled so as not to increase the number of "hard breeders" in future generations.

When adequate nursery facilities are available, weaning at 2 to 4 weeks of age is recommended so that the sows can be returned to production as soon as possible. When adequate facilities, diet, and management are available, sows may be weaned at 10 to 14 days of lactation. Usually, sows weaned at 17 days or later have normal subsequent litter performance without jeopardizing their ability to grow efficiently. Sows weaned at 14 days or less may have a delayed cycle. Sows requiring more than 8 or 10 days to recycle should not be bred on that estrus because the farrowing rate and litter size will likely be reduced. Waiting until the next cycle 21 days later results in normal performance.

Sows should be retained on the basis of their ability to cycle within 7 days postweaning and their ability to wean a large, healthy litter. They do not have to be bred on the first estrus, depending on the producer's management plans, but should cycle.

Usually, lactating sows will secrete more calcium and phosphorus into the milk than they can absorb from the diet. Heavy-milking sows and lactation lengths longer than 21 days are at greatest risk. To prevent sow breakdown (posterior paralysis), the gestation and lactation diets should both be properly fortified with minerals. In addition, dry pens with plenty of space and good footing should be provided so as to prevent slippage. The use of individual stalls will reduce the incidence of injuries.

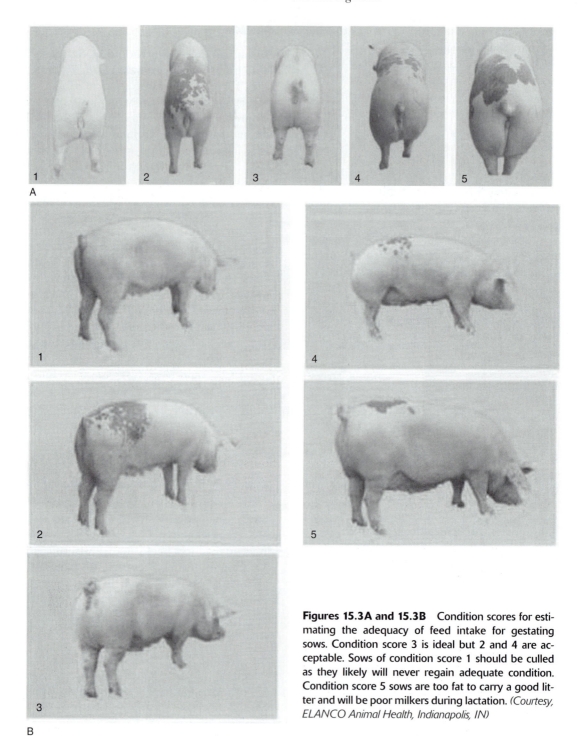

Figures 15.3A and 15.3B Condition scores for estimating the adequacy of feed intake for gestating sows. Condition score 3 is ideal but 2 and 4 are acceptable. Sows of condition score 1 should be culled as they likely will never regain adequate condition. Condition score 5 sows are too fat to carry a good litter and will be poor milkers during lactation. *(Courtesy, ELANCO Animal Health, Indianapolis, IN)*

Sow Condition Scoring

Sow condition should be evaluated periodically to prevent loss of body weight, to maintain an adequate level of backfat, to evaluate the nutritional adequacy of the diet, to enhance longevity in the herd, and to decrease problems associated with farrowing and rebreeding (Figures 15.3A & B). Sows should be scored approximately 2 weeks after weaning, at midgestation, 2 weeks prior to farrowing, and again about 2 weeks after farrowing or near weaning.

Sows in thin condition at weaning usually will have estrus occur later than the desired 4 to 7 days postweaning. Sows cycling after 7 to 10 days should not be bred that estrus and but should be put on a higher plane of nutrition (5 to 7 lb/day) and gain weight before breeding. This will assure maximum ovulation rate by the following estrus.

Ideally, a sow at mating should be condition 2.5 to 3.0 and advanced to condition 3 by midgestation, and then maintain condition 3 until farrowing. The condition of the sows will be affected by the quantity

TABLE 15.3   Body Condition Evaluation

Score	Condition	Backfat	Appearance
1	Emaciated	<0.5 in. (13 mm)	Hips and backbones are prominently seen
2	Thin	0.6 in. (15 mm)	Hips and backbone can be easily felt without applying palm pressure
3	Ideal	0.75 in. (19 mm)	Hips and backbone can be felt only with firm palm pressure
4	Fat	0.9 in. (23 mm)	Hips and backbone cannot be felt
5	Obese	>1.0 in. (25mm)	Hips and backbone are heavily covered with fat

Source: Courtesy, ELANCO Animal Health, Indianapolis, IN.

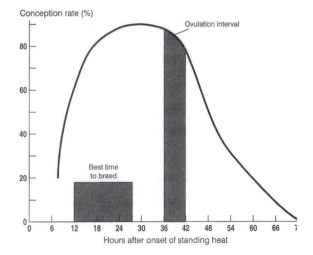

Figure 15.4   The effect of time of insemination on conception rate in swine. *(Courtesy of Pearson Education)*

and quality of the diet, the genetics of the sows, and the attention provided by the employees. The same person should score the herd each time to maintain consistency of the observations. Specific backfat depths and appearances are described in Table 15.3.

Heat Detection/Ovulation/Conception

Heat or estrus is the time that the female accepts the male for mating. Heat detection is more effective in the presence of a sexually mature boar (by placing a boar in the pen with the female, or by nose-to-nose fence line contact with females). The caretaker can confirm heat by applying back pressure to each female in the presence of a boar. Most sows and gilts in heat will respond by standing stiffly and attempting to stiffen their ears (making them erect). If they do not stand solidly and pop their ears, they are not in standing estrus.

High temperatures (above 85°F) will delay or prevent the occurrence of heat, reduce ovulation rate, and increase early embryonic death. Sows in a pen or pasture may be kept cool by the use of a sprinkler installed under a shade built over a sand or concrete floor. For sows housed in environmentally regulated buildings, consideration should be given to the use of thermostatically controlled drippers or evaporative cooling.

Achieving a high conception rate and good litter size necessitates getting sperm into the female's reproductive tract at the time when pregnancy rate and litter size will be maximized. Regardless of the method of breeding, (i.e., pen-mating, hand-mating, or artificial insemination), an adequate number of live sperm must be in the reproductive tract a few hours before ovulation occurs or conception rate and litter size will be reduced. The effect on conception rate of breeding at various times relative to the time of ovulation is presented in Figure 15.4.

When heat lasts 48 hours a female will ovulate 8 to 12 hours before the end of standing heat or 36 to

40 hours after its onset. When mating occurs too early or too late, conception rate and litter size drop dramatically. Gilts will have a shorter estrous cycle than sows.

With once-a-day heat detection and insemination, breed the females each day they will stand. With twice-a-day detection, breed every 12 to 24 hours after they are first detected in heat. Because the exact time of ovulation is not known, multiple matings will enhance both conception rate and litter size.

Producers using unobserved (pen-) mating must have adequate boar power. If sows or gilts are being mated that are not synchronized, allow one mature boar (older than 1 year of age) per 10 sows to be bred over a 21-day period. Decrease that ratio to 4 to 6 sows for each young boar (less than 1 year old).

When sows are weaned in groups (synchronized estrus) allow a sow-to-boar ratio of 4 sows per mature boars and 2 sows per young boars (Table 15.2). Reducing the breeding period in a pen-mating system to no more than 7 to 10 days postweaning simplifies baby pig management at farrowing, but there is a high probability that only 80 to 90% of the sows will cycle within that period.

Pen-mating is not recommended because of poor boar usage, not knowing which sows were bred, and not knowing when they were bred. This can result in under- or overusage of the farrowing stalls.

Double-mating—A boost in conception rate and litter size can be obtained by using more than one boar on each female (double-mating). This is easily accomplished with hand-mating or AI as the matings are controlled. When pen breeding, different boars should be rotated in at least once every day and rested periodically.

Figure 15.5 Prolific crossbred sows in a group lactation, deep-bedded system. *(Courtesy, Palmer Holden, Iowa State University, Ames, IA)*

Pregnancy Detection

Ultrasonic pregnancy diagnosis is routinely used and identifies fluid in the uterus. With ultrasound detectors, pregnancy at 30 to 45 days after breeding can be determined with 90 to 95% accuracy. Often before 30 days insufficient uterine fluid has accumulated for detection and after 60 days accuracy drops rapidly as the growing solid mass of the fetuses begins to displace the fluid. Another instrument, the Doppler, detects fetal heartbeats and may be used over a longer period. Several ultrasound instruments can be used for both pregnancy detection and measuring backfat thickness and loin eye size.

Farrowing Management

Regardless of the frequency of farrowing groups entering the farrowing facility and the breeding system being used, producers should strive to farrow all females in one farrowing room within 1 week. The all-in/all-out management practice can be achieved with wise use of the gilt pool and group weaning.

Farrowing sows should be provided only limited access to feed the day of farrowing but should have access to fresh water. Feed the postfarrowing sow the amount of feed that she will clean up in one day. The day of farrowing this may only be a couple handfuls, but if she is hungry, feed should be available. Then increase the feed daily, providing all of the feed she will consume. However, musty, stale feed should not be allowed to accumulate in the feeder as it will depress intake. The most limiting nutrient for lactating sows is energy and all reasonable efforts should be undertaken to encourage feed intake (Figure 15.5).

Methods of encouraging maximum energy (feed) intake include feeding at least twice per day, requiring the sow to get up at feeding time, minimizing or eliminating added fiber to the diet, keep-

ing the temperature of the lactation area at 65 to 70°F, increasing the feed by 50% for 6 to 8 days prefarrowing and, as a last resort, adding fat to the diet.

The following practices all have been reported at sometime as increasing the litter size weaned by about one pig.

1. Being present when sows farrow. This saves the piglets that get caught in the afterbirth or crawl to a cold corner of the pen rather than to the sow's udder for nursing.
2. Using a farrowing stall.
3. Providing supplementary heat for the piglets. This prevents chilling and keeps the piglets from laying next to the sow for warmth, often resulting in crushing as the sow gets up or lies down.
4. Flushing gilts prior to breeding. Flushing increases the number of eggs shed. Decrease the feed intake to 5 lb postmating.
5. Keeping sows instead of gilts.
6. Breeding twice at each estrus. This ensures that mating will occur closer to ovulation.
7. Limiting energy intake during gestation. This supports greater lactation feed intake.
8. Medicating the sow diets from a few days prefarrowing, through lactation, and until breeding. This reduces bacteria in the sow's feces (reducing piglet exposure to pathogens) and enhances conception rate and litter size.
9. Treating scours in piglets early.
10. Medicating the pig starter with broad spectrum antibiotics.

(Also see the Chapter 7 section headed "Defining Nutrient Requirements of the Breeding Herd.")

TYPES OF PRODUCTION SYSTEMS AND THEIR MANAGEMENT

Commercial swine producers may specialize as follows along with several variations:

1. Farrow-to-finish production.
2. Farrow-to-feeder pig production.
3. Feeder pig-to-finish production.

The type of production chosen depends on the interest, experience, and labor and capital availability of the producer, as well as the equipment, facilities, feed supply, and the markets. Specializing in only a segment of the production scheme allows the producer to develop more refined management skills and facilities for a particular stage as opposed to being an "expert" in all phases of production. Different methods also allow the operator to manage labor requirements to mesh with other farming or off-farm occupations.

TABLE 15.4 Expected Feed Inputs

Feed Inputs	Percentage of Total Feed
Farrow-to-finish (includes all feed)	100%
Farrow-to-wean (includes gestation, lactation, boar feed)	14%
Farrow-to-feeder pigs (includes above plus starter feed)	22%
Wean-to-finish (includes starter and finishing feed)	86%
Feeder pig-to-finish (includes finish feed only)	78%

Source: Adapted from Life Cycle Swine Nutrition, *Iowa State University, Ames, Iowa, 1996.*

Figure 15.6 Sows and litters in farrowing crates. Crates help keep sows from accidentally injuring their pigs. *(Courtesy, Palmer Holden, Iowa State University, Ames, IA)*

For grain-hog farmers, grain production capacity can be a factor in deciding the type of production system to use. Table 15.4 lists the percentage of the total feed required for each of the production systems mentioned. While farrow-to-wean systems require only 14% of the total feed needed to produce a slaughter hog, the wean-to-finish system consumes 86% of the feed. Today, many producers raise only hogs and do not have a grain production operation. For them, the cost of purchasing feed may be a deciding factor.

Farrow-to-Finish Production

Farrow-to-finish producers breed sows, farrow them, and produce pigs to weaning, then finish them for market. For success, farrow-to-finish operators must have expertise in all phases of swine production. This has been the historic type of pork production and still remains the most profitable production method. The better profit results from the fact that high-labor areas of pork production (farrowing and nursery) are better rewarded, and there is not a loss in efficiency by transferring ownership one or more times and the stress of moving pigs to different farm settings.

Weaner and Feeder Pig Production

Several important technological developments that occurred in the 1950s and early 1960s ushered in multiphase production of hogs. Among these developments were (1) improved disease control measures, including specific pathogen-free (SPF) herds; (2) intensive production—which increased specialization in breeding, in farrowing (Figure 15.6), nursery, and in finishing; (3) early weaning; (4) multisite production; and (5) increased mechanization.

Farrow-to-feeder pig production is best suited for producers who have a surplus of labor available and a

limited feed supply. Also, more capital is required because farrowing houses and nurseries are the expensive facilities used in pork production. This system is the first of a two-phase production system in which the pigs change ownership or finishing sites.

Farrow-to-wean production is a subset of farrow–feeder pig production and a phenomenon that began in the 1990s as pork production moved to specialization, particularly with the development of large integrated pork production companies. Weaned pig producers maintain the breeding, gestation, farrowing, and lactation phases. Piglets are usually weaned at 14 to 21 days and are sold either privately or on an open market, on a 10 lb basis and 50 to 54% lean (USDA). These pigs rarely, if ever, appear in an auction market but are sold on a price range either based on the seller delivering the pigs or the buyer coming to the sow farm to claim them.

Many weaned pig producers have developed long-term contracts with buyers, and others maintain ownership of the pigs but may pay another entity to feed the weaner pigs to slaughter weight. Weaner pig producers have no nursery facilities and likely purchase little or no feed for the piglets.

Historically, producing feeder pigs involved the production and sale of pigs often weighing 40 to 60 lb for finishing on other farms. Feeder pigs are sold by private treaty, contracts, or actually may go through an auction market to determine their value, and as with the sale of weaned pigs, (Figure 15.7) the USDA grades assume 50 to 54% lean value.

Following are some of the advantages of raising, producing, and selling weaned pigs or feeder pigs as opposed to finishing pigs:

1. They provide an opportunity to efficiently use a maximum amount of labor because about two-thirds of the labor for raising hogs occurs by the time pigs reach weaning age.

Figure 15.7 Nursery pigs on plastic-coated metal flooring. *(Courtesy, Palmer Holden, Iowa State University, Ames, IA)*

2. They require less grain (and feed) per unit of product sold.
3. They produce less manure.
4. These systems allow a rapid turnover in the number of pigs that can be handled each year. The farrowing schedule can be planned to provide frequent sales for consistent income.

The main disadvantage to feeder pig production is that producers must depend on those engaged in finishing operations for a market for their pigs. Often the weaned and feeder pigs are priced on a percentage of the future's price for slaughter hogs.

Knowledge of the following are pertinent to successful weaner and feeder pig production:

1. **High-level of management skills.** Because 10 to 15% of the pigs born die before weaning, the importance of good management during this critical farrowing and lactation period is obvious.
2. **Dependable markets.** Most weaner and feeder pig producers are not equipped to feed hogs from weaning to market. Modern systems of year-round farrowing require pigs be marketed on schedule to make room for subsequent farrowing and weaning groups. Thus, if dependable markets are not available, feeder pig producers can find themselves in the unenviable position of having unsold pigs on their hands and more of them on the way.
3. **Variations in feeder pig prices.** Feeder pig prices are more erratic than slaughter hog prices. They move upward and downward more than slaughter hog prices, and they do so more rapidly. This is attributed to the fact that there is not a well-organized nationwide system of marketing weaned and feeder pigs where dependable price information is available.

4. **Methods of marketing weaner and feeder pigs.** There are three methods of marketing feeder pigs:
   a. **Contract arrangements.** This is the sale of pigs directly from producers to finishers on a multiyear contract basis. Usually the price is based on a future value of market hogs at the time these pigs would reach market weight. There usually are price adjustments for pigs outside a predefined weight and quality range.
   b. **Competitive organized markets.** The auction method of selling feeder pigs is good for smaller feeder pig producers. Well-conducted auctions provide the advantages of (1) uniform lots in type, color, and size; (2) selling a large number of pigs quickly and at reasonable cost; (3) buyers who are interested in various sizes, qualities, and numbers; and (4) open competition. Health considerations are the major disadvantage of these markets because pigs from many sources come into contact with each other.
   c. **Private treaty.** Private treaty refers to the direct negotiation between producer and buyer for each group of pigs that is sold. Often, newspaper ads or feed suppliers bring the two parties together and then a per head price is established after visual inspection. The major disadvantage to this system is its uncertainty.

Note: The feeding of weaned and feeder pigs is discussed in Chapter 7.

Weaner and Feeder Pig Finishing

Producers that finish pigs buy either the weaned or feeder pigs described previously then feed them to market weight, and sell them. Finishing production is the lowest management and labor phase in a swine operation. Most feeding and manure handling is mechanized and death losses are low, usually 3% or less. Pigs in this stage are more resistant to diseases as well as environmental changes. The quality of the facilities is not as critical as it is for farrowing and nursery stages (Figure 15.8).

Producers with large grain supplies and limited labor and facilities can purchase feeder pigs, then finish, and market them, providing a good means of marketing their grains. Profits will depend on the weaner or feeder pigs' costs, feed costs, efficiency of feed conversion, death losses, slaughter hog price, and labor and capital investment costs. Feeder pigs that are in good condition and, preferably, purchased from a single reliable source, and handled as

Figure 15.8 Feeder pigs on a slotted floor through which the feces and urine drop to a pit below. *(Courtesy, Palmer Holden, Iowa State University, Ames, IA)*

Figure 15.9 Modern finishing building with double-curtain sides and totally slatted floor. *(Courtesy, Palmer Holden, Iowa State University, Ames, IA)*

described next, will have the best chance of getting off to a good start with minimal death losses:

1. Abide by the health regulations of your state.
2. Avoid unnecessary handling, watch the pigs closely during the first 2 weeks, and work with your veterinarian to maintain optimum herd health.
3. Ideally, newly arrived feeder pigs should not have contact with any other pigs on the site (Figure 15.9). Feeder pigs that are weaned early and segregated from other groups of pigs maintain a high-health status and outperform older weaned pigs that are commingled with other groups.
4. Avoid contact between feeder pigs and breeding stock. Feeder pigs carry bacteria and viruses from which they may be immune but which can be spread to other hogs on the farm. And conversely, pigs currently on site have other microorganisms to which the feeder pigs have not been exposed.

5. Wear different outer clothing and boots when working with two sets of pigs to avoid spreading disease. Wash your hands before handling pigs in different phases. Do chores with the most susceptible pigs first, for example, do farrowing chores before going to the nursery or the finisher units.
6. Feed a fresh palatable diet adequate in all nutrients. If feed intake is lower than anticipated increase the nutritional concentrations.
7. Develop a herd health plan with a veterinarian. Items to consider include the following:
   a. Adding 1 lb of copper sulfate ($CuSO_4$) per 1,000 lb of feed (250 ppm copper) or 3.75 lb zinc oxide (ZnO) per 1,000 lb (~3,000 ppm zinc) may be helpful in the nursery diet. Feeding high levels of copper will increase the manure copper concentration and may be of concern if the manure is spread on land that may be grazed by sheep, which are very sensitive to copper.
   b. Use a high antibiotic level in the feed for the first week (200 ppm or more) and then gradually decrease it. If pigs are sick and not eating, put an antibiotic in the water. Some antibiotics may give the water an off-flavor and reduce water intake.
   c. Control of lice and mange. Treatment may be oral, sprayed on, or injected depending on the product used. Do not spray pigs sooner than 5 days after arrival or when weather is cold or damp.
8. Sort pigs by sex if adequate pens and feeding systems exist. Barrows will reach market weight more quickly than gilts and require lower levels of amino acids in the diet. Also try to sort by size because the bigger pigs will crowd the smaller ones away from feed and water.
9. When signs of trouble appear, call your veterinarian.
10. See Table 15.5 for recommended environmental needs such as temperature, floor space, and feeder space, and number of waterers.

Note: The feeding of pigs is discussed in Chapter 7.

Split-Sex Feeding

There are well-defined sex differences in the performance of finishing pigs. This prompts the question: Should barrows and gilts be fed separately?

The answer depends on the number of different age groups in the facility, the number of pigs in

TABLE 15.5   Guidelines to Nursery Conditions for Successful Early Weaning[1]

	Weaning Age (weeks)			
Guideline	2	3	4	5
Minimum pig weight, lb	9	12	15	21
Temperature at pig level, °F	85	83	81	79
Floor space per pig, ft[2]	2	2	2	3
Maximum pigs per linear foot of feeding space	5	5	5	5
Maximum pigs per nipple waterer[3]	8	8	8	8
Maximum group size	10	10	15	25

[1] See the Appendix for conversion of U.S. customary to metric.
[2] The figures given are for partial or total slats. On solid floors, increase requirement by 50%.
[3] Where bowls are used instead of nipples, there should be 1 bowl for each 12 pigs. Provide at least 2 waterers per pen to avoid water stoppages.

each age group, the number of pens, the number of different diets currently being fed, and the number of bulk bins and feed delivery lines available. Separate diets are not needed before the pigs are 60 or 70 lb. At that time the amino acid requirements start to differ. However, pigs of different weights also have different requirements and thus there is not only the need to provide different diets for barrows and gilts but also diets for pigs of different weights in the same finishing facility.

If all pigs in the finishing barn are the same age and weight, put the gilts in pens on one side of the alley or set of feeders and the barrows on another. Slow growers may need to be sorted off and fed separately or culled from the herd. Split-sex feeding of barrows and gilts will necessitate 1 or 2 additional grower diets and 1 or 2 additional finisher diets. In this situation (all pigs the same age) one bulk bin would be required for the barrows and one for the gilts. If there is more than one age group in the barn, at least two bulk bins would be required for each sex, for example, one for the heavy barrows and one for the light barrows.

MANAGEMENT TO PREVENT DRUG RESIDUES IN PORK

Feed additives have been widely used in swine diets for more than 50 years. As a group, they are effective in improving the rate and efficiency of gains and in reducing mortality and morbidity. Certain feed additives require a withdrawal period prior to slaughter in order to ensure that residues do not occur in carcasses. The common swine feed additives and the appropriate withdrawal times are listed in Chapter 8, Table 8.2.

The feed additives that have caused the greatest residue problem and received the most attention in recent years are the sulfonamides. The term, *sulfonamide*, includes sulfamethazine, sulfathiazole, and other sulfonamide drugs. However, any of the products in Table 8.2 have the potential of leaving residues if withdrawal times are not followed or approved levels are exceeded.

The National Pork Board introduced the Pork Quality Assurance (PQA) program in 1989 as a three-level management education program. The PQA program emphasizes good management practices in the handling and use of animal health products, and encourages producers to review their approaches to their herds' health programs. The PQA good production practices (GPPs) are as follows:

1. Identify and track all treated animals. Often it is advisable to place them in an isolated "sick" pen to prevent recycling of feed additives that have withdrawal times via manure and urine from treated animals to adjoining pens that may have pigs nearing market weight.
2. Maintain medication and treatment records.
3. Properly store, label, and account for all drug products and medicated feeds.
4. Use a valid veterinarian–client–patient relationship as the basis for medication decision making.
5. Educate all employees and family members on proper administration techniques.
6. Use drug residue tests when appropriate.
7. Establish and implement an efficient and effective herd health management plan.
8. Provide proper swine care.
9. Follow appropriate on-farm feed and commercial feed processor procedures. This includes good feed mixing practices, including sequencing batches to prevent feed additive residues from contaminating diets of pigs near market. Also clean out or flush feed mixing, conveying, and feeding equipment to reduce drug carryover into finishing feeds.
10. Complete the Pork Quality Assurance checklist every year and the education card every 2 years.

SWINE SKILLS

In order to be successful and to get satisfaction from producing pigs, the operator should be able to perform the skills that follow.

Cutting Navel Cords

The navel cord should be cut on newborn pigs if they are processed before the navel cord has dried.

It should be cut off leaving about 2 in. and dipped in a disinfectant such as iodine solution. If the cord has dried, still clip it. If it has fallen off do not bother disinfecting the navel area (Figure 15.10).

Removing the Needle Teeth

Newborn pigs have eight small, tusklike teeth (so-called needle teeth), two on each side of both the upper and lower jaws. These are of no benefit to the pig and most swine producers prefer to cut them off soon after birth. The needle teeth are very sharp and are often the cause of pain or injury to the sow, particularly if the udder is tender. Moreover, the pigs may bite or scratch each other, and infection may result.

This operation may be done with a small pair of side cutters or wire cutters (Figure 15.11). Care should be taken to avoid injury to the jaw or gums

because injuries may provide an opening for bacteria. For this reason only the tips of needle teeth should be clipped—about one-half of each tooth.

Some research has suggested that teeth do not always need to be clipped. However, if facial scarring occurs it needs to become a routine procedure (Figure 15.12).

Tail Docking

Trimming or "docking" baby pigs' tails is practical for confinement operations, and it is mandatory in some graded feeder pig sales. Tail docking seems to be the best method of preventing, or at least reducing, tail biting, and preventing infections that usually accompany open wounds (Figure 15.13).

The tail should be clipped off leaving about 0.75 to 1.0 in. of the tail. Either sterilized wire cutters or an electric cauterizing blade can be used. Most times the procedure is performed when needle teeth are

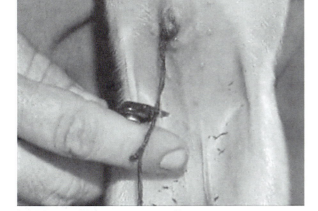

Figure 15.10 Docking the navel on a newborn piglet. *(Courtesy,* Pork Industry Handbook, *Purdue, University, West Lafayette, IN)*

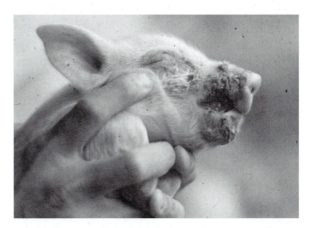

Figure 15.12 Facial lacerations as a result of baby pigs fighting with sharp needle teeth. *(Courtesy, Palmer Holden, Iowa State University, Ames, IA)*

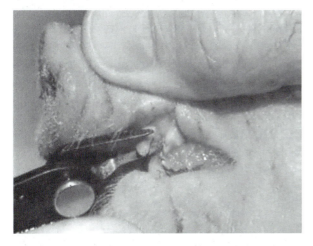

Figure 15.11 Clipping the needle teeth is usually done the first day of the piglet's life. *(Courtesy,* Pork Industry Handbook, *Purdue University, West Lafayette, IN)*

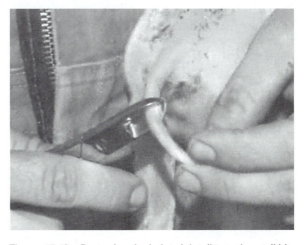

Figure 15.13 Removing the baby pig's tail to reduce tail biting. Leave about a 1-in. stump. *(Courtesy,* Pork Industry Handbook, *Purdue University, West Lafayette, IN)*

clipped or iron injections are given. A wound protection spray or dip may be applied to the tail stump.

Injections

Shortly after birth a pig receives its first injection of iron dextran to prevent nutritional anemia. Proper injection involves using the right-sized needle and the best site for the injection (Figure 15.14). The length of the needle depends on whether the injection is to be subcutaneous (under the skin) or intramuscular. For piglets, a 20-gauge needle works for thin liquids whereas an 18-gauge needle is best for thick liquids. Most sow injections should be given with 16- to 14-gauge needles. The heavier gauge needles reduce the risks of needle breakage.

Needles should be clean and sharp. Intramuscular injections should always be given in the neck muscle. Do not administer any injections in the ham as staining or potential abscesses will damage a very valuable cut of pork. Pigs should be adequately restrained during injections. Use separate needles for removing the material from the vial and for the actual injection to prevent contaminating the vial of product. Also, when injecting iron, it is recommended to use clean needles between litters.

Identifying Swine

The most common method of identifying swine consists of ear notching the litters (Figure 15.15). Pigs are generally marked at the same time that the needle teeth are removed and the tail docked. Purebred breeders find it necessary to employ a system of marking to help identify the parentage of the in-

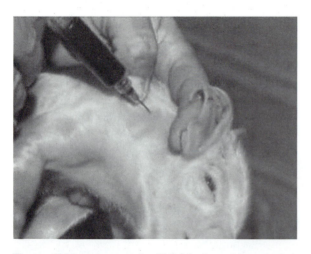

Figure 15.14 Intramuscular (IM) injection of iron dextran into a baby pig. IM injections should be given in the low value neck muscle in case staining or an abscess occurs. *(Courtesy, Pork Industry Handbook, Purdue University, West Lafayette, IN)*

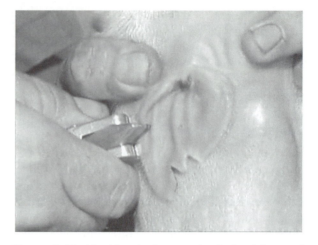

Figure 15.15 Notching the litter number in the right ear of a newborn piglet. When the final cut is made, this pig will be identified as from litter 33. *(Courtesy, Pork Industry Handbook, Purdue University, West Lafayette, IN)*

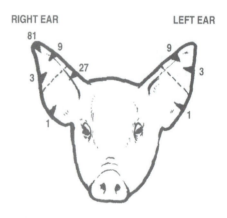

Figure 15.16 Universal Ear Notching System (also known as the 1-3-9-27 system) used, or recommended, by most swine registry associations. The right ear is used to number the litter and the left ear for the individual pig number. Up to 161 litters can be identified before needing to start over. *(Courtesy, Pork Industry Handbook, Purdue University, West Lafayette, IN)*

dividuals for purposes of registration and herd records. Even in the commercial herd, a system of identification is necessary if the gilts are to be selected from the larger and more efficient litters.

The ear notches are usually made with a special V-notcher. Most of the breed associations require the Universal Ear Notching System shown in Figure 15.16. This is the most common notching system, though there are others, including plastic or metal ear tags, tattooing, and branding.

- **Microchips**—Electronic swine tracking systems are being tested in both Europe and the United States. The goal is to assign a number to each farm and each pig. Each animal could carry its own history, for example, reproductive history, health, and medications. Currently, the price is not competitive with other methods and the microchips have a tendency to migrate to different parts of the body from which they were implanted. They must be recoverable at slaughter or inserted into a part of the body that does not end up in human food, for example, inserted into the hind foot. In theory, consumers will be able to identify a retail package of meat to its source.

Castration

Castration is the removal of the testicles from the male and the ovaries from the female. In the United States, female swine usually are not castrated (spayed) and in much of Europe the castration of male pigs is not done.

Male pigs are castrated to maintain the quality of the meat, to prevent uncontrolled breeding, and to prevent the development of the boar odor or flavor that occurs in the cooked meat of a sexually mature boar. Male pigs that are to be castrated should be castrated before 3 days of age. This subjects the pig to minimal stress; in addition, younger pigs are easier to hold and restrain and they bleed very little.

The best way to restrain, or hold, swine that are to be castrated depends on the age and size of the animal and the number of helpers available (Figure 15.17). A very young pig is held with one hand and castrated with the other. A heavier pig may be (1) suspended by its hind legs with the back toward the helper (whose knees are clamped against the pig's ribs, near the shoulders); or (2) held on its back on the top of a table (this requires either [a] a castration crate or rack or [b] two helpers—one grasping the front legs and the other the rear legs). Large boars are usually snared around the upper jaw and behind the tusks, with the free end of the snare tied to a post; then further restraint is applied by either tying all four legs or by hoisting the hind legs, with the animal castrated in either a standing or a lying position. The following method may be used for pigs less than 1 week old:

1. Use a surgical scalpel handle that accepts a hooked blade (Bard Parker No. 3 handle and No. 12 blade).
2. Clean the scrotal area with a mild disinfectant solution.
3. Hold pigs by the hind legs with one hand using the thumb or first finger to push up the testicles until the scrotal skin is tightened.

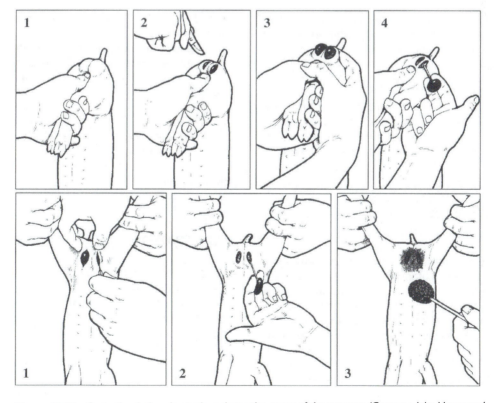

Figure 15.17 Castrating baby pigs early reduces the stress of the surgery. *(Courtesy, John Hammond, Pig Progress)*

4. Push the tip of the hooked blade through the scrotum and direct it forward and upward toward the tail.
5. Repeat procedure over the other testicle. An alternative is to make the cut across the lower part of the scrotum to remove both testicles.
6. Push the exposed testicles out through the incisions.
7. Grasp each testicle separately and pull upward and remove with as much loose tissue as possible.
8. When properly performed, very little bleeding occurs and the incision remains clean.
9. Keep the knife in a disinfectant solution between pigs, and an antiseptic topical powder or spray may be applied to the incision.

Older pigs (more than a week) are castrated by holding them upside down by the hind legs and then pushing the testicles down and cutting under the body between the legs. This method allows for good drainage but it requires two people or a pig holder.

Most swine producers routinely handle this management phase, others may call on the veterinarian to learn the proper techniques; perhaps the most important thing is that it be done at the proper time. Pigs with retained testicles or ruptures (scrotal hernias) should be operated on by a veterinarian.

Boars that are no longer useful in a breeding program may be castrated to remove the boar odor before marketing; such an operation is known as stagging. This should be done with use of an anesthetic. By the time the castration wound has healed (in 3 to 4 weeks), the odor usually disappears enough to allow the stag to be marketed. Most meat from boars or stags will be made into spicy sausages to hide any residual boar odor.

Weaning

The optimum age to wean pigs varies depending on nutritional programs, facilities, environment, health, and management available. The conventional age in the United States is at 3 to 4 weeks. The weaning age varies from herd to herd, according to the facilities available, intensity of operation, and managerial skill of the producer. Generally, pigs can be weaned over a wide age range; however, weaning younger pigs requires more demanding management to be successful. The following guides will reduce the stress at weaning.

1. Wean only pigs weighing 12 lb or more.
2. Split-wean litters, weaning the heavier pigs first and allowing the smaller pigs to nurse 2 or 3 days longer.
3. Provide an initial nursery temperature of 85°F.

4. Group pigs according to size. Split sexing will facilitate later penning for split-sex feeding.
5. Limit pen size to 25. There are some producers using very large groups, but more experience is needed for a recommendation.
6. Limit feed intake for 2 days if postweaning scours are a problem.
7. Treat the drinking water with approved medications or electrolytes if scours develop.
8. Provide 1 feeder hole for 4 to 5 pigs and 1 waterer for each 20 to 25 pigs.

CLIPPING BOAR TUSKS

It is never safe to allow the boar to have long tusks because they may inflict injury on other boars and are hazardous to the caretaker. Above all, such tusks should be removed well in advance of the breeding season when is necessary to handle the boar often. The common procedure in preparation for removing the tusks consists of drawing a strong rope over the upper jaw and tying the other end securely to a post or other object. As the animal pulls back and the mouth opens, the tusks may be cut with a hoof parer (Figure 15.18).

Fortunately, with the fairly common use of artificial insemination, the number of boars on a farm is greatly reduced.

GROUPING HOGS

Proper grouping is important. In addition to the obvious separation of hogs by sex, age, and size, the following grouping practices are generally advocated by successful producers:

1. **Gilts to be retained for breeding herd.** They should be separated from market hogs at 4 to 5 months of age and placed on separate diets.

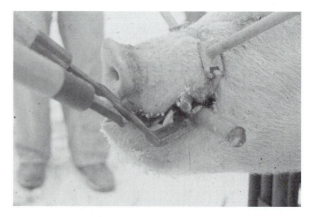

Figure 15.18 Removing the boar's tusks. This is done for the safety of the farm workers as well as for other pigs. *(Courtesy, Maynard Hogberg, Iowa State University, Ames, IA)*

2. Pregnant gilts and sows. They should be kept separate during the gestation period, unless they are self-fed a bulky diet. Gilts and sows should not be penned together because the sows will dominate and possibly injure the smaller gilts.

3. Boars of different ages. Young and mature boars should not be penned together. Boars of the same age or size can be kept together during the off-breeding season. Do not mix boars that were not raised together because the resulting fights may injure them.

4. Pigs of different weights. Finishing pigs of varying weights should not be penned together. It is recommended that the range in weight should not exceed 20% above or below the average.

5. Penning sows and litters together. Piglets should be about 2 weeks old before placing sows and litters together, although small groups may be put together as early as 1 week. The age difference between such litters should not be more than 1 week.

6. Adjusting size of litter. Where possible, the size of litters should be adjusted to the number of functioning teats and nursing ability of the sow. Transferring pigs from sow to sow should be done as early as possible; usually within the first 3 to 4 days after farrowing. If the odor of the pigs is masked, it may be possible to transfer at a later date. It is essential that the piglets all get to consume colostrum.

BIOSECURITY

Sanitation is the foundation of an effective herd health program. It is the use of hygienic measures to promote health and prevent diseases, but sanitation is more than cleaning and disinfecting. It is an attitude. Sanitation begins with the producer who understands diseases and their modes of transmission (Chapter 14). Some of the factors or practices involved in sanitation are as follows:

- **Environmental effects**—Moisture and high humidity favor the survival and transmission of disease-causing organisms. Drying and proper ventilation are important to maintain herd health.
- **Design and construction of facilities**—Buildings and equipment should be durable and easily cleaned. All areas should have proper drainage so water will not stand and invite microbial growth.
- **Vacating facilities**—Combined with thorough cleaning and disinfecting, removing all animals from an area enhances the opportunity to break disease cycles because many disease-causing organisms cannot survive very long outside the pig's body. This includes pasture and dirt lot rotation. For best results, a solid material floor, such as concrete, metal, or plastic, should be empty a week or longer if a disease is present that needs a complete break in

production to eliminate the risk for reinfection. Dirt or pasture may require a complete season or more.

- **Cleaning and disinfecting**—All indoor facilities should be thoroughly cleaned followed by disinfection, including all equipment.
- **Footbaths**—To prevent the spread of disease from one room to another, one production unit to another or from farm to farm, footbaths filled with a suitable disinfectant should be located at entryways. Solutions in the footbaths should be kept fresh and replaced whenever organic material accumulates.
- **Washing sows**—To minimize the exposure of newborn pigs to parasite eggs and other microorganisms, sows should be washed with warm soap or detergent and mild germicidal solutions before farrowing. This should be done immediately before entering the farrowing house. Of course, sows will be placed in a clean and disinfected farrowing room.
- **Carcass and afterbirth disposal**—Dead animals and afterbirth can spread disease, and should be removed immediately by a rendering company, composted (Figure 15.19), incinerated, or properly buried down grade from water sources and covered with a generous amount of quicklime. Remove all dead animals to an off-site collection area. This eliminates the need for rendering trucks to come onto your production site.

Further details regarding diseases and their spread and control are given in Chapter 14.

Early Weaning and Segregated Early Weaning (SEW)

Early weaning *and* segregated early weaning (SEW) *are terms applied to an infectious disease control procedure of which the primary objective is to improve the productiv-*

Figure 15.19 Composting is one of many means of disposing of dead pigs and afterbirth. The carcasses are covered with manure or straw and allowed to decompose. *(Courtesy, Palmer Holden, Iowa State University, Ames, IA)*

ity of the postweaning phases of swine production by preventing the transfer of diseases from sows to litters. Although SEW is a commonly used term, in actuality it needs to be separated into two parts: "early weaning" and "segregation." The procedure is accomplished by (1) weaning piglets at less than 21 days of age, rather than the more conventional 21- to 28-day weaning; and (2) segregating the piglets from their mothers either by transporting them from the farrowing site to another facility, or removing the sows to a centralized breeding/gestating facility and leaving the piglets in the site in which they were born. Thus, the segregation of early weaned pigs prevents the vertical transmission of infectious diseases from sows to their piglets.

Early weaning is critical to the success of the isolation program. Maternal immunity provided by colostrum decreases at different rates for different diseases. Specific dates for weaning to stop the transmission from the sow to the piglets vary depending on the immune status of the sows, the severity of any infections in the sow herd, and the exposure of the piglets to the sows' microorganisms. For example, the organism causing atrophic rhinitis may require a weaning age of 10 days, whereas the transmission by the pseudorabies virus has been prevented by using a 21-day weaning age. The transmission of some disease organisms, such as *Streptococcus suis, Actinobacillosis suis, E. coli*, rotaviruses, and coccidia, are not prevented. Advantages of early weaning include:

1. Heavier pigs at 8 to 9 weeks of age.
2. Lower sow lactation feed costs.
3. Slightly more litters per year.
4. Less weight loss of the sow during lactation and faster rebreeding.
5. Greater flexibility in rebreeding or selling sows.
6. Greater number of sows through the farrowing unit; hence, less total farrowing area space required and lower-facility cost per sow.

Successful early weaning first requires a sound breeding and feeding program of gestating sows to ensure large, healthy pigs at birth. There is a positive relationship between birth weight, survival, and weight at weaning. Good milking sows are essential to get the pigs off to a good start, and positive baby pig and sow management during lactation help ensure strong, uniform, and healthy pigs at weaning.

A good nursery feeding program, beginning with a quality prestarter diet, is essential for very young pigs. Piglets weaned at less than 21 days will not benefit from the labor and expense associated with creep feeding. Expect an increase in labor and facility costs, plus more preventive feed medication is usually recommended. For best results, the guidelines given in Table 15.5 should be observed when planning nursery conditions for early weaned pigs.

Early weaning and segregated early weaning (SEW) are disease control procedures that combine two techniques, early weaning and segregation of the piglets from the sows.

The early weaning and segregation of the piglets prevents the vertical transmission (from sow to her piglets) of infectious disease. The weaned pigs are placed in groups and each group remains segregated by the use of all-in/all-out production. Segregation of age groups of pigs through the use of all-in/all-out production also prevents the horizontal transmission of disease from noncontemporary pigs. Often the term *SEW* is extended from pigs that were weaned early and segregated from the sow herd to include groups of early weaned pigs that are later combined. The use of the term defines the procedures that occur at weaning and not to later phases.

The piglets may be segregated in either of two ways: (1) by transporting the piglets at weaning from the farrowing site to another site or (2) by removing the sows to a centralized breeding/gestating facility and leaving the piglets on the site in which they were born. Additionally, the piglets should be segregated from previously and future-weaned groups of pigs, and commingling of pigs from production units with a different health status should not be allowed. The second part, early weaning, is accomplished by weaning the piglets at less than 21 days of age, usually between 17 and 21 days, rather than at the more conventional 21- to 28-day weaning.

The transfer of colostral immunity from the sow protects the piglets from infection until they are weaned. Infectious organisms circulating in the herd and vaccination of sows stimulate colostral immunity. The shedding of infectious organisms may be decreased by medicating sows. Piglets may also be medicated near weaning to bacterial pathogens.

Weaning age is critical to the success of SEW because the maternal immunity provided by colostrum decreases at different rates for different diseases. For example, vertical transmission of the pseudorabies virus has been prevented by using a 21-day maximum weaning age, whereas the organism causing rhinitis requires a weaning age of 10 days because colostral immunity does not prevent infection. *Streptococcus* spp. may be transmitted within 1 or 2 days of birth so early weaning is not effective for this organism.

The major advantage of SEW production is the increased performance and reduced mortality and morbidity of the hogs. It is conjectured that this advantage is because of the decreased burden of infectious diseases, especially respiratory diseases, stimulating the pigs' immune systems. If the piglets have less need to partition nutrients to the immune system, they can use those nutrients for growth. The decreased pathogenic burden also results in pigs with greater appetites. Moreover, the decreased infectious disease load results in the decreased use of drugs and vaccines, which lowers production costs.

SEW also tends to decrease the backfat and increase loin eye size and percentage of muscle in the carcass. Still another advantage of SEW production is that it makes it possible to maintain the genetic base of the sow herd. For example, if the herd becomes infected with an organism, their offspring can be raised on another site without disposing of the sow herd.

Potential disadvantages of the SEW program are (1) the short lactation period associated with increased weaning-to-service intervals, reduced farrowing rates, and decreased subsequent litter size; (2) the moving of animals from site to site increases expenses and requires coordination; (3) the transportation of very young pigs provides a special challenge, although properly prepared SEW pigs tend to transport with less stress and death loss than typical 8-week-old feeder pigs; and (4) the weaning of piglets at less than 17 days of age requires a modification of nursery equipment, much higher quality and costly diets, and stringent management techniques the first 1 to 2 weeks in the nursery.

All-In/All-Out (AIAO)

All-in/all-out (AIAO) *refers to the movement of pigs in groups through (1) farrowing, (2) nursery, and (3) growing-finishing.*

With all-in/all-out (AIAO) management, pigs are moved in groups through each of the following stages: (1) farrowing, (2) nursery, and (3) growing-finishing. Normally, a barn or room holds 1 week's farrowings, but in AIAO nurseries up to 10 day's production can be housed together and up to 2 or 3 weeks of production can be held in finishing rooms. After each group is moved to the next production phase, the entire barn or room is washed and disinfected before a new group of pigs is moved in. AIAO can benefit any type of operation, including farrowing operations, feeder pig producers, and finishers, large or small.

AIAO is not new. It was first used in farrowing and nursery units to combat scours *(E. coli)* and respiratory problems. Now, AIAO is included with such health sanitary/security procedures as footbaths, separate farm entrances, security fences, bird netting, shower-in/shower-out, controlling traffic, and buying and quarantining breeding stock. AIAO is being used to control a host of swine diseases at farrowing, in the nursery, and in finisher operations.

Purdue University researchers compared the performance and health of growing-finishing pigs in an AIAO system to the conventional (continuous-use) system. Based on 4 complete replications, they reported that the AIAO pigs gained faster (1.72 versus 1.52 lb/day), more efficiently (3.04 versus 3.23 lb feed), and required fewer days to reach market weight of 231 lb (173 versus 185 days) than the conventional pigs.[1]

Specific Pathogen-Free (SPF) Pigs

Specific pathogen-free (SPF) pigs are pigs that are free of certain diseases at birth. Obtaining specific pathogen-free (SPF) pigs is very expensive but is an effective method to replace foundation stock with animals that have not been infected with a specific list of pathogens, including *Mycoplasma hyopneumoniae* (enzootic pneumonia), *Bordetella bronchiseptica* (infectious rhinitis), *Brachyspira hyodysenteriae* (swine dysentery), *Actinobacillus pleuropneumoniae* (porcine pleuropneumonia), pseudorabies (Aujeszky's disease), brucellosis, leptospirosis, and the external parasites lice and mange. The SPF system embraces the following provisions:

1. Obtaining pigs by a surgical process (Caesarean section) 2 to 4 days before normal birth.
2. Rearing pigs in individual isolation until 1 week old, and in groups of 8 to 12 until 4 weeks old.
3. Rearing pigs in groups of 10 to 20 from 4 weeks old to maturity, on farms from which all other swine have been removed and to which no new stock is introduced, and where the producer avoids contact with other swine.
4. Resuming normal birth of SPF pigs on these clean farms.
5. Restocking other "clean" farms—farms that have no swine or only SPF swine, and on which the owner avoids contact with other swine.

Also, pigs may be caught during natural birth in sterile canvas bags, in sterile basins, or on sterile canvas towels. However this method is not recommended because of the variability in farrowing times and possibility of contamination during the birth process.

Hysterectomy means removal of the uterus. With this technique, the sow is killed and the uterus and pigs are taken without passing through the birth canal, eliminating any chance of the pigs becoming infected. Although hysterectomy represents the ultimate in disease control, it has one

---

[1] Cline, T. R. et al, *Journal of Animal Science*, 70(Supp. 1):49, 1992.

major disadvantage. Many laboratories cannot comply with slaughter inspection regulations; hence, they experienced difficulty in marketing the sow's carcass. As a result, this forced commercial laboratories to obtain the pigs by Caesarean section (C-section), with methods developed to keep the newborn pigs separated from the contaminated environment of the sow.

Rearing the SPF piglets is done in one of two ways. One is to allow the piglets to suckle synchronized foster sows on the farm where they will be going. This allows the piglets to get antibodies from the sow and enhances their survival. Do not feed them any other feed until they have suckled or development of the digestive enzymes may be delayed. The second is to rear them artificially in individual cages or germ-free environments. This is risky because the piglets do not get any antibodies from the sow milk.

Another risk incurred with the production of SPF pigs is that, because they have not been exposed to various pathogens, they are much more susceptible to later-occurring infections. Biosecurity is a much more significant item of importance.

Primary SPF herds are those originating from surgically derived stock and maintained in strict isolation. Any new bloodlines added to the herd must also be obtained by surgical means, such as embryo transfer, C-section births from approved labs (to reduce dam to pig pathogen transfer), or artificial insemination. These primary herds are used to supply secondary multiplying herds, which in turn supply breeding stock to commercial swine producers.

Secondary SPF herds are those operated by producers that raise breeding stock for sale to other commercial producers. All breeding stock for secondary herds must only come from primary herds or from procedures used in primary herd production. The secondary herds are monitored for maintenance of SPF status by the accreditation agency.

The National SPF Association in Conrad, Iowa, is responsible for supervising the SPF program and for issuing an accreditation certificate to those who qualify. The association is made up of independent swine seedstock producers involved in certifying and health monitoring of SPF herds. Nebraska also has a state SPF accrediting program.

A diagrammatic outline of the SPF method is presented in Figure 15.20.

McLean County System

One of the early practices of sanitation was the McLean County System of Swine Sanitation developed by Drs. B. H. Ransom and H. B. Raffensperger. The system was developed in McLean County, Illi-

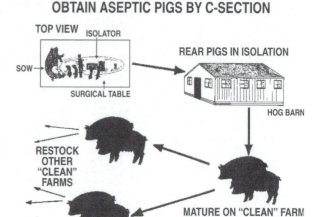

Figure 15.20 A diagrammatic outline for obtaining aseptic pigs by Caesarean section, starting with the surgical removal of the fetuses. *(Courtesy of Pearson Education)*

nois, commencing in 1919. Today, the McLean County System of Swine Sanitation is of historic interest only. But it was effective in reducing swine parasites and diseases by application of the following four simple steps: (1) cleaning and disinfecting the farrowing quarters, (2) washing the sow before placing her in the farrowing quarters, (3) hauling the sow and pigs to clean pasture, and (4) keeping the pigs on clean pasture until they were at least 4 months old.

MANAGEMENT PRACTICES AND ANIMAL WELFARE

Animal caretakers must recognize that the principles and application of animal behavior and environment depend on understanding and recognizing that they should provide as comfortable an environment as feasible for their animals, for both welfare and economic reasons. This requires that attention be paid to environmental factors, including feeding and housing, that influence the behavioral welfare as well as the physical comfort of the animals.

Animal welfare issues tend to increase with urbanization. Fewer and fewer urbanites and people moving to rural nonfarm homes have farm backgrounds. As a result, the animal welfare gap between producers and the nonfarm population widens. Also, both the news media and the legislators are increasingly from urban centers and nonfarm views will have greater and greater impact in the future of livestock production.

The National Pork Board in 2003 initiated the Swine Welfare Assurance Program (SWAP) to assist pork producers in evaluating welfare and

This orange glazed leg of pork free of drug residues is a goal of pork production. *(Courtesy, National Pork Board, Des Moines, IA)*

safety issues on their farms. The program includes trained observers who visit farms to advise on farm welfare strengths and weaknesses.

The National Pork Board introduced the Pork Quality Assurance (PQA) program in 1989 as a three-level management education program. The PQA program emphasizes good management practices in the handling and use of animal health products, and encourages producers to review their approach to their herds' health programs. The PQA good production practices (GPPs) are as follows:

1. Identify and track all treated animals. Often it is advisable to place them in an isolated "sick" pen to prevent recycling of feed additives that have withdrawal times via manure and urine from treated animals to adjoining pens that may have pigs nearing market weight.
2. Maintain medication and treatment records.
3. Properly store, label, and account for all drug products and medicated feeds.
4. Use a valid veterinarian–client–patient relationship as the basis for medication decision making.
5. Educate all employees and family members on proper administration techniques.
6. Use drug residue tests when appropriate.
7. Establish and implement an efficient and effective herd health management plan.
8. Provide proper swine care.
9. Follow appropriate on-farm feed and commercial feed processor procedures. This includes good feed mixing practices, such as sequencing batches to prevent feed additive residues from contaminating diets of pigs near market. Also clean out or flush feed mixing, conveying, and feeding equipment to reduce drug carryover into finishing feeds.
10. Complete the quality assurance checklist every year and the education card every 2 years.

QUESTIONS FOR STUDY AND DISCUSSION

1. Define management. List and discuss the five primary roles of a successful manager of a large swine operation. How can a young person become a manager?
2. What did the Texas Tech University researchers report as a result of their study of "indoor versus outdoor intensive pork production"? Will swine producers in the warmer areas of the United States return to outdoor production?
3. List considerations as to whether to use boars, artificial insemination, or a combination of both.
4. List the most important management aspects of sows and gilts, and discuss each of them.
5. Describe each of the following production systems, and tell the place of each of them:
   a. Farrow-to-finish production
   b. Farrow-to-wean
   c. Feeder pig production
   d. Finishing production
6. Why are each of the following skills important?
   a. Removing the needle teeth
   b. Marking or identifying
   c. Castrating
   d. Ringing
   e. Tail docking
7. At what age, and how, would you wean pigs? Why?
8. What is the SPF program and its values and shortcomings?
9. Define the all-in/all-out system of swine management. Why and how is it used?
10. What is the McLean County System of Swine Sanitation?
11. List and discuss management practices to prevent feed additive residues in pork.

SELECTED REFERENCES

Animal Welfare, M. C. Appleby and B. O. Hughes, CAB International, New York, NY, 1997

Farm Animal Well-Being, S. A. Ewing, D. C. Lay, Jr., and E. von Borell, Prentice Hall, Upper Saddle River, NJ, 1999

Handbook of Livestock Management, 3rd ed., R. A. Battaglia, Prentice Hall, Upper Saddle River, NJ, 1998

Pork Industry Handbook, Cooperative Extension Service, Purdue University, West Lafayette IN, 2002

Pork Quality Assurance, National Pork Board, Des Moines, IA, 2003, http://www.porkboard.org/PQA/default.asp

Stockman's Handbook, The, 7th ed., M. E. Ensminger, Interstate Publishers, Inc., Danville, IL, 1992

Swine Breeding and Gestation Facilities Handbook, MWPS-43, J. Harmon et al., Midwest Plans Service, Iowa State University, Ames, IA, 1997

Swine Care Handbook, National Pork Board, Des Moines, IA, 2003, http://www.porkboard.org/docs/default.asp

Swine Genetics Handbook, National Swine Improvement Federation, North Carolina State University, Raleigh, NC, http://mark.asci.ncsu.edu/nsif/handbook.htm

Swine Nursery Facilities Handbook, L. D. Jacobson, H. L. Person, and S. H. Pohl, Midwest Plans Service, Iowa State University, Ames, IA, 1997

Swine Production and Nutrition, W. G. Pond and J. H. Maner, AVI Publishing Co., Westport, CT, 1984

16

Manure and Environmental Management

Lagoon on hog operation on a farm in Taylor County, Iowa. The system uses a series of hillside terraces that form constructed wetlands that also use bacteria to purify wastewater from a hog operation. *(Photo by Tim McCabe, Courtesy, USDA-NRCS, Washington, DC)*

Swine management is the art of caring for and handling swine. It gives point and purpose to everything else. It is essential to the success and profitability of a swine operation.

MANURE MANAGEMENT

Manure refers to a mixture of animal excrement (consisting of feces and urine) and bedding.

Sustainable Agriculture

Sustainable agriculture is often described as farming that is ecologically sound and economically viable. It may be high or low input, large scale or small scale, a single crop or a diversified farm, and use either organic or conventional inputs and practices. Obviously, the actual practices will differ from farm to farm.

Many of the practices advocated under sustainable agriculture are not new; they involve such timeless agricultural practices as soil erosion control, the protection of groundwater, the use of legumes as a source of nitrogen, biological insect and weed control, and the use of pastures and home-grown grains as primary feed sources.

Sustainable livestock production systems, to be successful, must be (1) profitable, (2) maintain or enhance the quality of the environment, (3) improve the quality of life for the producer, and (4) be friendly to the animal. Ideally, they would minimize purchased inputs, allow for nutrient recycling, market a value-added product, and have a positive public perception.

The development of improved crops, cropping systems, irrigation, farm management, and marketing make farms more profitable. However, to be sustainable, minimizing purchased inputs requires the production of livestock, both to add value to the crops and to recycle the nutrients in the manure back to the cropland (Figure 16.1). Typically, sustainable farms rely more on biological resources and management than on nonrenewable inputs of energy and chemicals. The foundation of a sustainable farm system is a comprehensive understanding of the land, the farm resources and operations, and potential short- and long-term markets.

Manure management is one of the most important problems and issues confronting swine producers today. It must be removed from the hog house for sanitary reasons, its volume makes it difficult to store for long periods, high labor costs reduce its value as a fertilizer, it is too costly to burn, and it is too bulky to bury.

The quantity, composition, and value of manure produced varies by species, weight, kind and amount of feed, and kind and amount of bedding. The amount of manure produced by various farm animals is presented in Figure 16.2. Among the farm animals, swine produce the most manure per 1,000 lb of live-weight followed by dairy cows.

Also, for the operator, the neighbors, travelers, and city dwellers alike, manure odor and potential health risks are taboo. Producers must thoroughly evaluate the methods for handling manure and choose a system that will complement their overall production system and be environmentally friendly.

If not managed properly, animals may produce or contribute to the increase in the following pollutants in troublesome quantities: manure, gases and odors, dust, and flies or other insects. If manure is not properly controlled, it may pollute water supplies.

Basically, manure is handled either as a solid or as a liquid, or some combination of the two, depending on the farm's facilities and stages of production. Solid manure results from scraping semidry manure from outside concrete surfaces and catching or holding excrement in bedding or by allowing the liquids to run off leaving the solids

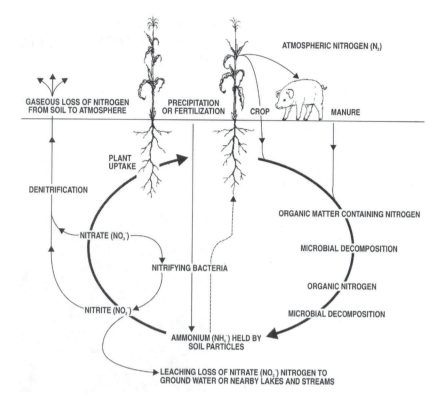

Figure 16.1 A sustainable agriculture enhances the cycle of nature. Manure is applied to soils and decomposed by microbes; complex protein in manure is broken down to release nitrogen as ammonia (NH_4); aerobic microbes convert ammonium nitrogen to nitrite (NO_2), thence to nitrate (NO_3) nitrogen; nitrate is (1) taken up by plants and built back into protein compounds, (2) leached downward when the soil is saturated—contaminating surface and groundwater if excessive nitrogen has been applied to the soil, and/or (3) released into the atmosphere when soils are wet for extended periods of time and the absence of air causes anaerobic microbes to convert the nitrate nitrogen to gaseous form. *(Courtesy of Pearson Education)*

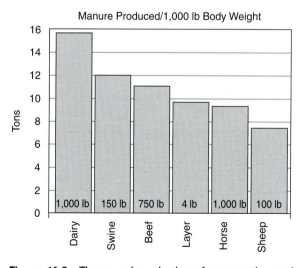

Figure 16.2 The annual production of manure that each species produces per year per 1,000 lb of body weight. *(Midwest Plan Service, MWPS-18, Iowa State University, Ames, IA, 1985)*

behind. Arrangements must be made to control runoff from bedded areas and from outdoor surfaces following a rainfall

Liquids are generally retained in deep pits under slotted floors or drained to outside storage. Outdoor runoff is generally held in settling basins, which retain the solids and the liquid can be pumped off.

Bedding

Bedding or litter is used primarily for the purpose of keeping animals clean and comfortable. But it has the following added values from the standpoint of the manure:

1. It soaks up the urine, which contains about one-half the total plant nutrients of manure.
2. It makes manure easier to handle.
3. It absorbs plant nutrients, fixing both ammonia and potash in relatively insoluble forms that

protects them against losses by leaching. This characteristic of bedding is especially important in peat moss but of little significance with sawdust and shavings.

Often bedding is eliminated, especially in the indoor-type operations, primarily because it reduces costs and makes cleaning easier. Slotted floors of indoor operations are responsible for the elimination of bedding.

Types of Bedding

The kind of bedding material selected should be determined primarily by (1) availability and price, (2) absorptive capacity, (3) cleanliness (no dirt or dust), (4) ease of handling, (5) ease of cleanup and disposal, (6) nonirritability from dust or components causing allergies, (7) texture, and (8) plant nutrient fertility value. In addition, a desirable bedding should not be excessively coarse, should remain in place, and should not be easily kicked aside.

Common bedding materials and their average water absorptive capacity are presented in Table 16.1. In addition to these bedding materials, many other products can be successfully used for this purpose, including various leaves and hulls.

Naturally, the availability and price of various bedding materials varies from area to area and from year to year. Thus, in forested areas shavings and sawdust are readily available, and straws are more plentiful in areas that produce small grains, such as wheat and oats. Corn stalks may be used for bedding for larger pigs in corn-producing areas.

Bedding materials differ considerably in their relative capacities to absorb liquid, and cut straw will absorb more liquid than long straw. But there are disadvantages to chopping; chopped straws do not stay in place and may be dustier. From the standpoint of the value of plant nutrients per unit of air-dry material, peat moss is the most valuable bedding, and wood products (sawdust and shavings) are the least valuable.

The suspicion that sawdust and shavings will hurt the land is unfounded. These products decompose slowly, but this process can be expedited by the addition of nitrogen fertilizers. Also, when plowed under, they increase soil acidity, but the change is both small and temporary.

Bedding Requirements

In most areas, bedding materials are becoming scarcer and higher in price, primarily because geneticists are breeding plants with shorter straws and stalks, fewer grain producers harvest and sell the straw because of its value to the soil, and because of

TABLE 16.1   Water Absorption of Bedding Materials

Material	Water Absorbed by Air-Dry Bedding % (wt:wt)[1]
Barley straw	210
Cocoa shells	270
Corn stover (shredded)	250
Corncobs (crushed or ground)	210
Cottonseed hulls	250
Flax straw	260
Hay (mature, chopped)	300
Leaves (broad leaf)	200
(pine needles)	100
Oat hulls	200
Oat straw (long)	280
(chopped)	375
Paper, newsprint, shredded	340
Peanut hulls	250
Peat moss	1,000
Rye straw	210
Sand	25
Sawdust (top-quality pine)	250
(run-of-the-mill hardwood)	150
Sugarcane bagasse	220
Tree bark (dry, fine)	250
(from tanneries)	400
Vermiculite[2]	350
Wheat straw (long)	220
(chopped)	295
Wood chips (top-quality pine)	300
(run-of-the-mill hardwood)	150
Wood shavings (top-quality pine)	200
(run-of-the-mill hardwood)	150

[1]The weight of the water-saturated material divided by the weight of the dry material. It is an indication of the water-holding ability of the material.
[2]This micalike mineral is mined chiefly in South Carolina and Montana.
Source: Courtesy of Pearson Education.

more competitive uses for some of the materials. The minimum desirable amount of bedding to use is the amount necessary to absorb completely the liquids in manure. Some helpful guides to achieve this end are as follows:

1. Per 24-hour confinement, the minimum bedding requirements of hogs, based on uncut wheat or oat straw, is 0.5 to 1 lb per animal. With other bedding materials, the quantities will vary according to their respective absorptive capacities. Also, more than minimum quantities of bedding may be desirable where cleanliness and comfort of the animal are important. Comfortable animals lie down more and use a higher proportion of the energy of the feed for productive purposes.

2. Under average conditions, about 500 lb of bedding are used for each ton of excrement.

3. Farrowing sows should be bedded lightly with chopped or short material that will not interfere with the movement of the piglets.

Bedding needs vary considerably depending on their use and the season. Winter bedding needs will be greater than for summer. The average amount of different types of bedding needed per pig or per sow in hoop buildings is presented in Table 16.2. Producers may reduce needs and costs as follows:

1. Collect urine separately. When the liquid urine is allowed to run off to a basin or tank, less bedding is required than where the liquid and solid excrement are kept together.

2. Chop bedding. Chopped straw, waste hay, fodder, or cobs will go further and do a better job of keeping animals dry than long materials.

3. Ventilate quarters properly. Proper ventilation lowers the humidity, which increases evaporation and helps keep the bedding dry.

4. Keep feeders and waterers away from sleeping quarters. Animals should be fed and watered in areas removed from their sleeping quarters. With this type of arrangement, they defecate less in the sleeping area.

5. Prevent rain and snow from blowing into the facility.

6. Provide exercise area. Where possible and practical, provide an exercise area, without confining animals to or near their sleeping quarters. This is especially beneficial for producers who turn lactating sows out of the stall for feed and water twice per day.

7. Consider slotted floors. Slotted or wire floors eliminate the need for bedding.

MANURE HANDLING SYSTEMS

Depending on the production system, there are several phases involved in handling manure, each requiring some different equipment and facilities. The phases of manure handling include (1) collection and transfer from the building, (2) storage and treatment, and (3) application and facilities for land application.

Collection and transfer methods include slotted floors over deep or shallow pits, flushing either in open gutters or beneath the slats, floor drains, mechanical scrapers, and power loaders. Transfer of manure includes gravity drains, pumps, agitator, augers, and so on.

Storage and treatment of manure includes deep pits below slotted floors; remote storage slurry, above- or belowground, or in earthen basins; anaer-

TABLE 16.2 Estimated Amount of Bedding Needed for Grow-Finish Pigs and Sows in Hoop Structures

Material	Average, lb/pig	lb/sow
Corn stalks	195	670–1,000
Corn cobs	240	800–1,200
Barley straw (long)	240	800–1,200
Oat straw (long)	180	600–900
Wheat straw (long)	225	750–1,130
Sawdust (hardwood)	335	1,120–1,680
Sawdust (pine)	200	670–1,000
Wood shavings (hardwood)	335	1,120–1,680
Wood shavings (pine)	250	840–1,250

Source: AED 41, 2004; and AED 44, 2004. MidWest Plan Service, Iowa State University, Ames, IA.

obic or aerobic lagoons; oxidation ditches; and solids separation with partial dehydration.

Application requirements include liquid tank spreaders that may inject manure into the soil or broadcast on the surface with or without later incorporation, irrigation, dry manure surface spreading, and, of course, available land for application.

There is no one best manure management system for all situations; rather, it is a matter of designing and using a system that will be most practical for a particular set of conditions. Regardless of the collection, storage, treatment, and application methods, there are many advantages and disadvantages to consider. Some will work better than others for specific circumstances. In many cases, the cost of manure collection, treatment, and application will exceed the value to the user, but it is a cost of operating a swine enterprise.

Often the end use of swine manure dictates the most appropriate management system. But before selecting the system, all variables should be considered. Furthermore, the least expensive method may not meet regulations or be acceptable to the neighbors. But to make an intelligent management selection, the systems should be understood.

Storage Considerations

The storage unit must be leak proof. Its size will depend on the stages of production producing manure, the method under which swine operation is managed, the required length of storage time, and the volume of hogs (Table 16.3). Large storage units provide more flexibility and allow for better scheduling of labor.

Various factors in addition to the volume produced by the pigs must be included. Cleaning swine facilities with high-pressure water may increase the volume by 15 to 20%. Roof or lot drainage will have an unpredictable effect if allowed to run into the storage unit. Diverting runoff will greatly enhance the storage time.

Extra water may have to be added if the manure is to be pumped; anywhere from one-fifth to three-fifths of the storage volume may be needed for extra water. For irrigation, there should be about 95% water and 5% manure. However, for tank transportation extra water should be held to a minimum. The needed storage capacity can be estimated as follows:

Storage capacity = No. of animals
× Daily manure production (Table 16.3)
× Desired storage time (days) + Extra water.

Indoor storage containers should be well ventilated, and all storage containers (indoor and outdoor) should be insulated against freezing. Adding 3 to 4 in. of water to the storage container before it is filled with manure will reduce ammonia losses and odor.

Types

• **Pits, tanks, and holding basins**—Pits under the buildings are usually made from concrete (Figure 16.3) and outside tanks may be concrete, coated steel, or earthen. All of these are holding units to store liquid manure until it is possible to field spread it. Any digesting that takes place is usually incidental.

Under floor pits are the most common method of storing liquid manure. However, to provide a better environment within the building for both the humans and the pigs, more producers are attempting to remove the manure from the barn on a daily or weekly basis to outside storage.

• **Lagoons**—A lagoon is a manure treatment unit that uses microbial breakdown of manure and some bedding. Except in a very dry climate with high evaporation rates, where evaporation exceeds rainfall, excess liquids must be field spread, either with tanks or umbilical hoses connected to irrigators. Most of the nitrogen in lagoon manure is lost through volatilization, and mineral nutrients, such as phosphorus and potassium, settle to the bottom of the lagoon.

Lagoons may be one or two stage. With a one-stage lagoon liquid material is usually spread by a liquid manure wagon or by irrigation. In a two-stage system liquid from the first stage overflows into a second lagoon with much less solids in the liquid. Often liquid from a second stage is recycled back to the hog barn for flushing the manure to the first-stage lagoon. Most lagoons are anaerobic because of the high cost of operating aerators. In some instances small aerators may be used for some aerobic action at the surface to reduce odor (Figure 16.4).

• **Oxidation ditches**—An oxidation ditch is a storage unit, oxygenated to promote aerobic bacterial digestion. The purpose is usually odor control, and the effluent is still a potential pollutant. Usually, the ditch is a circular raceway with one or more paddles operated by electric motors. The cost of adding air via the paddles is quite expensive and this system is rarely used. As with lagoons, much of the nitrogen is lost.

Figure 16.3 A concrete tank used for aboveground storage of liquid manure. Many regulations require a minimum of 8 months capacity so manure will not have to be spread on frozen ground. *(Courtesy, Palmer Holden, Iowa State University, Ames, IA)*

TABLE 16.3 Approximate Daily Manure Production, Without Bedding[1,2]

Swine Stage	Wt (lb)	Total (lb)	Ft³	Gal	Water (%)	Density (lb/ft³)	N (lb)	P₂O₅ (lb)	K₂O (lb)
Nursery	25	2.7	0.04	0.3	89	62	0.02	0.01	0.01
Grow-finish	150	9.5	0.15	1.2	89	63	0.08	0.05	0.04
Gestating	275	7.5	0.12	0.9	91	62	0.05	0.04	0.04
Lactating	375	22.5	0.36	2.7	90	63	0.18	0.13	0.14
Boar	350	7.2	0.12	0.9	91	62	0.05	0.04	0.04

[1] Values are as-produced estimations and do not reflect any treatment. Values do not include bedding. The actual characteristics of manure can vary ±30% from table values. Increase solids and nutrients by 4% for each 1% feed waste more than 5%.
[2] N, P₂O₅, and K₂O are applied forms of nitrogen, phosphorus, and potassium fertilizers.
Source: Midwest Plan Service (MWPS), Publication 18, Section 1, Page 12, 2000.

Figure 16.4 A swine lagoon with an aerator installed to reduce odors. *(Courtesy, Palmer Holden, Iowa State University, Ames, IA)*

• **Settling or debris basins**—A settling or debris basin is a separating and holding unit, usually part of a runoff control system. As liquids enter the basin and slow down, the undissolved solids settle out. The liquids are slowly drained off, leaving the solids to become dry or semidry for removal and field spreading. A settling basin is usually smaller and much shallower than a lagoon. It is intended to dry out. Some method must be employed to retain the liquid runoff so it does not contaminate the environment.

When considering manure storage in a separate tank, lagoon, or in a pit under slotted floors, the storage capacity can be computed as follows:

Storage capacity = Number of animals ×
Daily manure production (Table 16.3)×
Desired storage time (days) + Extra water

Adding 3 or 4 in. of water to a pit initially and after emptying will reduce the loss of ammonia (nitrogen) and odor in the barn. About 20 to 40% of the storage volume may be needed for the extra water depending on water wastage and water used for cleaning. For irrigation, the manure should be about 95% water and 5% solids. Water additions should be kept to a minimum if the manure is to be field spread with a tank wagon.

Most governmental regulations restrict the application of manure to frozen ground. Storage capacity may be specified to provide at least 6 months or longer of storage. Also, extra capacity is needed for storage during times of high rainfall or cropping stages that limit manure application. Cleaning swine facilities with high-pressure water may double the volume of manure.

MANURE GASES

Air quality and potential gaseous emissions from livestock feeding operations should be a concern for all producers, both for their well-being and their neighbors. *Animals and people have been asphyxiated or died as a result of inhaling hydrogen sulfide, carbon dioxide, or gas from methane explosions.*

Emissions come from three primary sources, manure storage, livestock housing, and land application of manure. Of the many gases emitted, ammonia, carbon dioxide, and methane are considered contributors to global climate change and acid rain.

When manure and urine are stored and undergo anaerobic digestion, dangerous and disagreeable gases are produced. The ones of primary concern are ammonia, carbon dioxide, hydrogen sulfide, and methane. Several have undesirable odors and/or possible human and animal toxicity, some are explosive, and some promote corrosion of equipment (Table 16.4).

Most gas problems occur when manure is agitated or when ventilation fans fail. No one should enter a storage tank, unless (1) the space over the liquid is first ventilated with a fan, (2) another person is standing by controlling a line attached to the person entering the tank to give assistance, and (3) they are wearing self-contained breathing equipment of the kind used for fire fighting or scuba diving. Immediately leave any area suggestive of the presence of gas, such as dizziness, or sick or dead animals.

Do not attempt to rescue someone who is unconscious in a manure storage structure. Contact an emergency rescue crew immediately. Multiple deaths have occurred because of unprepared rescue attempts.

It is important that maximum building ventilation be provided when agitating or pumping manure from a pit because large quantities of gas are released. Also, an alarm system (loud bell) to warn of power failures in tightly enclosed buildings is essential because there can be a rapid buildup of gases when forced ventilation ceases.

Odors and emissions can be reduced by providing adequate separation distances between livestock facilities and residences, landscaping, and prompt removal of manure from confinement sites. Adding 2 to 3 in. of water to shallow pits or recharging deep pits with some water after emptying reduces the volatilization of ammonia. Aerating lagoons is effective but very costly. Covering manure storage facilities is very effective in reducing odors.

TABLE 16.4   Properties of Various Manure Gases

Gas	Weight (Air = 1)	Physiologic Effect	Other Properties	Comments
Ammonia (NH_3)	0.67	Irritant	Strong odor, corrosive	5,000 ppm dangerous level
Carbon dioxide (CO_2)	1.33	Asphyxiant	Odorless, mildly corrosive	Normally only about 0.03% of the volume of air
Hydrogen sulfide (H_2S)	1+	Poison	Rotten-egg odor, corrosive	Fatal in 30 min. when present at 800–1,000 ppm
Methane (CH_4)	0.50	Asphyxiant	Odorless, explosive	Explosive when present at only 5% the volume of air

Source: Courtesy of Pearson Education.

Ammonia

The nitrogen in manure and urea in the urine can be converted to ammonia (NH_3). It has a distinct odor and can be detected by humans at concentrations as low as 5 ppm. It is common in winter months to exceed 25 ppm even under winter ventilation rates. If your eyes burn when entering an enclosed livestock facility, the level is at least 20 ppm. Increasing winter ventilation and partially recharging manure pits with water will reduce ammonia emissions. The maximum acceptable level for ammonia is 25 ppm and 5,000 ppm can be deadly.

Carbon Dioxide

Carbon dioxide is a result of animal respiration. At prolonged high exposures carbon dioxide can be fatal. A level of 5,000 ppm is the maximum acceptable level.

Carbon Monoxide

Although it is not a product of manure decomposition, carbon monoxide is often a hazard in confined spaces. Carbon monoxide (CO) is the result of using combustion equipment, such as catalytic heaters in farrowing and nursery rooms, or gas operated power washers in confined areas. At prolonged high exposures, carbon monoxide can be fatal. The maximum acceptable level is 50 ppm.

Hydrogen Sulfide

Hydrogen sulfide (H_2S) is also produced in anaerobic conditions with the microbial decomposition of sulfur-containing organic matter in manure. It is the most dangerous gas found in livestock operations. It is heavier than air and accumulates in underground pits or other low, unventilated areas. At levels of less than 1 ppm it smells like rotten eggs. At higher levels it paralyzes the sense of smell. The maximum acceptable level is 10 ppm and levels of less than 1,000 ppm will cause death.

Methane

Methane (CH_4) is produced by microbial breakdown of organic matter under anaerobic conditions, and warm temperatures accelerate the production. It is odorless and tasteless and, therefore, is difficult to detect. It is lighter than air and will accumulate at the top of unvented areas such as capped pits. The maximum acceptable level is 1,000 ppm, and at concentrations of 50,000 ppm or more it can explode. Prohibit all sparks or flames in areas near manure storage facilities.

FERTILIZER SOURCE

China has kept its soils productive for thousands of years, primarily through the use of night soil (human manure) and every other kind of manure applied to the land in primitive, but effective, fashion. A familiar Chinese saying is "The more pigs, the more manure; and the more manure, the more grain." Indeed, manure is very precious in China; it is carefully conserved and added to the land. Manure is used as a way in which to increase yields of farmland already under cultivation.

Historically, manure has been used as a fertilizer, and it is expected that manure will continue to be used primarily as a fertilizer for many years to come. Also, historically, manure was called waste, and land application referred to as "disposal." It is now well established that manure is an excellent substitute for commercial sources of nitrogen, phosphorus, and potassium plus it adds a large portion of organic material to the soil.

A 5-year comparison of commercial nitrogen (N) and hog manure is presented in Figure 16.5. Soil and yield analyses estimated that 150 lb of added N would be required. The treatments included no added nitrogen, 150 lb of commercial nitrogen, 150 lb of nitrogen from hog manure, or 300 lb of nitrogen from hog manure. Obviously, adding no N resulted in the poorest yields. Adding 150 lb of commercial N boosted yields to 140 bu/acre and

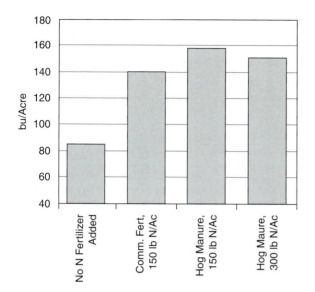

Figure 16.5 Value of swine manure as a fertilizer for corn; 5-year study at Iowa State University, Ames, IA. *(Courtesy, Iowa State University College of Agriculture, Ames, IA)*

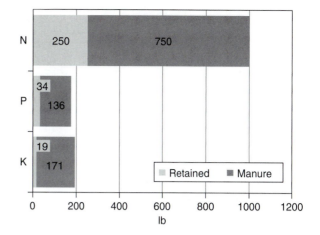

Figure 16.6 The pounds of nitrogen (N), phosphorus (P), and potassium (K) retained and excreted from 1,000 bushels (56,000 lb) of corn. *(Courtesy, Iowa State University College of Agriculture, Ames, IA)*

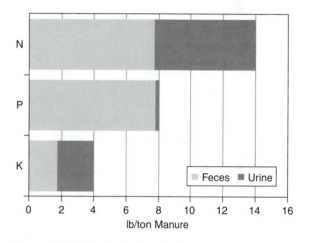

Figure 16.7 The distribution of nitrogen (as N), phosphorus (as P_2O_5), and potassium (as K_2O) between feces and urine. A large portion of the nitrogen and potassium are lost if the urine is not retained. *(Courtesy, Pork Industry Handbook, Purdue University, West Lafayette, IN).*

150 lb from hog manure provided an added yield to 158 bu/acre. Doubling the hog manure N to 300 lb provided no value over the 150 lb.

Rapid incorporation of manure minimizes nitrogen loss to the air and hastens biological activity to release nutrients for plant use. Injecting, chiseling, or knifing liquids beneath the soil surface will minimize odors and nutrient losses to the air and/or runoff. Disking or chiseling after surface application will also reduce odors generated from surface spreading.

Timing manure application can help minimize complaints. Select dry days and pay attention to wind direction with respect to neighboring residences. Direct injection or rapid incorporation should be accomplished where neighbors live close to fields receiving manure. Good communication, sincere response to complaints, and other good-neighbor techniques can be as important as following good waste management practices.

Many factors affect the composition of manure, for example, diet, method of collection and storage, bedding, and added water. The time and method of application affect the composition of the useful nitrogen applied to the land. In general terms, about 75% of the nitrogen (N), 80% of the phosphorus (P), and 85% of the potassium (K) contained in swine feeds are returned as manure (Figure 16.6). In addition, about 40% of the organic matter in feeds is excreted as manure. As a rule of thumb, it is commonly estimated that 80% of the total nutrients in feeds are excreted by animals.

Urine comprises 40% of the total weight of the excrement of swine and contains nearly 50% of

the nitrogen, 4% of the phosphorus, and 55% of the potassium of average manure—roughly one-half the total nutrient content of manure (see Figure 16.7). Also, the plant nutrients in the liquid portion of manure are more readily available to plants than those in the solid portion. It is, therefore, important to conserve the urine to maximize fertilizer value.

The actual monetary value of manure can be based on (1) increased crop yields and (2) the equivalent cost of a like amount of commercial fertilizer. Numerous experiments and practical observations have shown the measurable value of manure in increased crop yields.

On the average, the fertilizer content of liquid manure produced per pig sold from a farrow-to-finish operation is about 7 lb of nitrogen, 6 lb phosphorus (P_2O_5), and 6 lb potassium (K_2O). On the

TABLE 16.5 Average Total Nitrogen (N), Ammonium Nitrogen (NH$_4$+), Phosphate (P$_2$O$_5$), and Potash Content (K$_2$O) of Manure at the Time of Land Application

	Solid Manure (lb/ton)				Liquid Manure (lb/1000gal)				Pit Lagoon (lb/1000gal)[1]			
Stage	*N*[2]	*NH$_4$+*[3]	*P$_2$O$_5$*[4]	*K$_2$O*[5]	*N*[2]	*NH$_4$+*[3]	*P$_2$O$_5$*[4]	*K$_2$O*[5]	*N*[2]	*NH$_4$+*[3]	*P$_2$O$_5$*[4]	*K$_2$O*[5]
Farrow	9	4	6	4	15	7.5	12	11	3	2.8	1.5	1.5
Nursery	13	5	8	4	25	14	19	22	4	3.5	3	3
Grow-Finish	16	6	9	5	33	19	26	25	5	4.5	3	4
Breed-Gest	9	5	7	5	25	12	25	24	3.5	3.2	3.5	4

[1] Includes feedlot runoff water and is sized as follows: single-cell lagoon, 2 ft^3/lb animal weight; two-cell lagoon-cell 1, 1–2 ft^3/lb animal weight, and cell 2, 1 ft^3/lb animal weight.
[2] Ammonium-N plus organic N, which is slow releasing.
[3] Ammonium-N, which is available to the plant during the growing season.
[4] To convert to elemental P, multiply by 0.44.
[5] To convert to elemental K, multiply by 0.83.
Source: Adapted from Pork Industry Handbook, *PIH-25, 1996.*

assumption that nitrogen retails at 20¢, phosphorus at 30¢, and potassium at 10¢ per lb, the manure value per pig sold is worth about $3.80. If 20 pigs are marketed per sow per year the fertilizer value is about $76 per sow. It has additional value for the organic matter that almost all soils need, and which cannot be purchased in a sack or tank.

Further guidelines relative to the production of manure nitrogen, phosphorus, and potassium from different stages of production are given in Table 16.5. Livestock producers sometimes fail to recognize the value of this manure because (1) it is produced regardless of whether it is wanted and (2) it is available without cost.

Also, it is noteworthy that because of the slower availability of its nitrogen and its contribution to the soil humus, manure produces rather lasting benefits, which may continue for many years. Approximately one-half the nitrogen in manure is available to and effective on the crops in the immediate cycle of the rotation to which the application is made. Of the remainder, about one-half in turn, is taken up by the crops in the second year; one-half more in the third year, and so on. Likewise, the continuous use of manure through several rounds of a rotation builds up a backlog bringing additional benefits, and a measurable climb in crop yields.

Figure 16.8 depicts the common types of production areas (pasture, open lot with shelter, and indoor) and various processes that may be involved before the manure is applied to the land.

PUBLIC POLICY ISSUES

Zoning

In recent years, some of the concern over the problem of pollution of the environment (air, water, and soil) and its effect on human health has stemmed

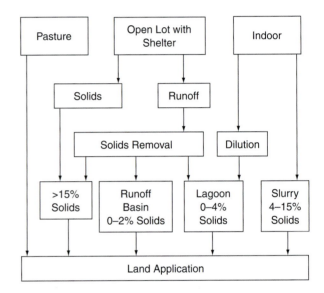

Figure 16.8 Alternatives for swine manure relative to the types of production systems. Ultimately, the end usage is land application for crops or pastures. *(Courtesy of Pearson Education)*

from animals kept in the suburbs, with the neighbors protesting on the basis that the animals make for more flies, dust, and odors. This has resulted in zoning. Thus, persons who desire to keep an animal(s) in their backyards on a plot of land within city limits, should first study the zoning ordinance to determine if there are any restrictions against it.

Pollution Laws and Regulations

The U.S. Environmental Protection Agency (EPA) has set regulations regarding the need for comprehensive nutrient management plans (CNMP). Producers with capacity of more than 1,000 animal units need to develop and file CNMPs. EPA rules do not differentiate between confined or open lots, applying the rules equally (Table 16.6).

TABLE 16.6   Summary of Environmental Protection Agency Regulations

Feedlots with 1,000 or More Animal Units[1]	*Feedlots with Less Than 1,000 but with 300 or More Animal Units[2]*	*Feedlots with Less Than 300 Animal Units*
Permit required for all feedlots with discharges[3] of pollutants.	Permit required if feedlot 1. Discharges pollutants through an unnatural conveyance, or 2. Discharges pollutants into waters passing through or coming into direct contact with animals in the confined area. Feedlots subject to case-by-case designation requiring an individual permit only after on-site inspection and notice to the owner or operator.	No permit required unless 1. Feedlot discharges pollutants through an unnatural conveyance, or 2. Feedlot discharges pollutants into waters passing through or coming into direct contact with the animals in the confined area, and 3. After on-site inspection, written notice is transmitted to the owner or operator.

[1] More than 1,000 feeder or slaughter cattle, 700 mature dairy cows (milked or dry), 2,500 swine weighing more than 55 lb (24.9 kg), 500 horses, 10,000 sheep or lambs, 55,000 turkeys, 100,000 laying hens or broilers with continuous overflow watering, 30,000 laying hens or broilers with liquid manure handling, 5,000 ducks, or any combination of these animals adding up to 1,000 animal units.

[2] More than 300 feeder or slaughter cattle, 200 mature diary cows (milked or dry), 750 swine weighing more than 55 lb (24.9 kg), 150 horses, 3,000 sheep, 16,500 turkeys, 30,000 laying hens or broilers with continuous overflow watering, 9,000 laying hens or broilers with liquid manure handling, 1,500 ducks, or any combination of these animals adding up to 300 animal units.

[3] Feedlot not subject to requirement to obtain permit if discharge occurs only in the event of a 25-year, 24-hour storm event.

Source: U.S. Environmental Protection Agency, Washington, DC.

Animal units are computed as follows: multiply number of slaughter and feeder cattle by 1.0; the number of mature dairy cattle by 1.4; the number of swine weighing more than 55 lb by 0.4; the number of sheep by 0.1; and the number of horses by 2.0 (Table 16.6, footnote 1). For swine 1,000 animal units equals 2,500 pigs weighing more than 55 lb or more than 10,000 nursery pigs.

States can be more restrictive. Iowa, for example, requires the filing of manure management plans (MMP) for 500 animal units instead of the EPA's 1,000. Thus, in Iowa, MMPs must be filed for operations with capacity for more than 2,500 pigs weighing more than 55 lb or 5,000 nursery pigs weighing less than 55 lb. An added requirement is that any farm using earthen storage (lagoons or unlined pits in the ground) must file an MMP regardless of animal capacity.

Land Requirements

With heavy animal concentration in one location, the question is being asked, "How much manure can be applied without depressing crop yield, creating salt problems in the soil, creating nitrate problems in feed, contributing excess nitrate to groundwater or surface streams, or violating state regulations?"

Environmental regulations differ in limiting the rate of manure application. Most states in the United States restrict the amount of manure nitrogen that can be applied to the land to be no more than the amount taken up by the crops grown on the land. This is not a severe restriction for combi-

nation grain and pork producers. However, many states have set in motion regulations that will restrict future applications to a phosphorus balance, a rule that will more than double the amount of land required. It is expected the EPA will implement similar requirements nationwide.

An example of the quantities of swine manure that can be applied on corn cropland is in Table 16.7. Information for both 100 sows, farrow to finish, including all of their offspring, and for 1,000 feeder pigs from 40 to 250 lb is provided. If manure is applied to balance the corn requirements at various yields, note that a large quantity of excess phosphorus will result. For example, with a corn yield of 150 bushels/acre, the manure from the production of 100 sows will require 78 acres for land application. This will result in an excess of 73 lb of phosphorus applied per acre.

What if environmental regulations require a phosphorus instead of a nitrogen balance? The manure from these same 100 sows will require 182 acres of corn to balance the phosphorus application with the corn uptake, more than doubling the land base needed. Producers in most states must develop manure application plans, depicting how and where the manure will be applied, before they can construct new swine facilities. It will be wise to base those plans on balancing phosphorus rather than nitrogen.

Concerns with higher rates annually include nitrate and phosphorus buildup. Excess nitrate from manure will move with water from the land and can pollute streams and groundwater, resulting in toxic levels for human and animal babies that must be removed from the water before it is consumed. Excess phosphorus stays with the soil and only enters waterways with soil erosion. However, high levels of phosphorus in rivers and lakes stimulate plant growth that consumes the oxygen fish need to survive. The maximum rate at which manure can be applied will vary according to current soil nitrogen and phosphorus levels, soil type, slope of the land, and rainfall.

Precautions

Avoiding water pollution and objectionable odors from manure requires planning. Following are some precautions that should be observed when applying manure.

   1. Separation distances for land application of manure from water sources, rivers, neighbors, public

use areas, and so on vary considerably among states. Some of Iowa's requirements are summarized in Table 16.8. Before planning any manure application program visit with local authorities regarding requirements. For example, the distances for Iowa apply only to confinement feeding operations, not to manure from open feedlots or that handled as dry manure.

   2. Test manure and soil to determine appropriate application rates. Different parts of a field may need different amounts.

   3. Limit application of manure nutrients to crop needs. Excess application results in buildup of soil nutrients and may be illegal.

   4. Do not spread on frozen ground because of the risk of runoff.

   5. Distribute the manure as uniformly as possible on the area to be covered.

TABLE 16.7   Acres Required for Manure Application[1]

	Corn Yield		
	150 bu	*175 bu*	*200 bu*
100 Sows, Farrow to Finish			
Nitrogen balance	78 acres	64 acres	57 acres
Excess P, lb/acre	73 lb	93 lb	103 lb
Phosphorus balance	182 acres	156 acres	136 acres
1,000 Feeder Pigs			
Nitrogen balance	110 acres	90 acres	80 acres
Excess P, lb/acre	71 lb	90 lb	100 lb
Phosphorus balance	240 acres	205 acres	180 acres

[1] Assumes continuous corn, building is idle 15 days/year.
Source: Adapted from MicroComputerSoftware, MCS-18, Iowa State University, Ames, IA, 1997.

TABLE 16.8   Required Separation Distances by Type of Manure and Method of Manure Application

	Dry Manure			Liquid Manure (except irrigated)		
	Surface Application				Surface Application	
	Incorporate within 24 hr	*Not incorporated*	*Direct injection*	*Incorporate within 24 hr*	*Not incorporated*
Buildings or public use areas	0 Incorporate same date	0	0	0 Incorporate same date	750 ft[1]
Sinkholes, wells, wetlands, and so on	0	200 ft (50 ft with buffer)	0	0	200 ft (50 ft with buffer)
High-quality water resource	0	800 ft (50 ft with buffer)	0	0	800 ft (50 ft with buffer)
Unplugged drainage wells or surface inlets	0	200	0	0	200

[1] The 750-ft separation distance applies only to liquid manure from confinement feeding operations or from feeding operations with less than 500 animal units (1,250 sows and grow-finish pigs or 5,000 nursery pigs).
Source: Iowa Department of Natural Resources, 113; Rev. Oct. 20, 2002; effective March 1, 2003.

6. Restrict surface application of manure just prior to rain.

7. Minimize odor problems by

    a. Incorporating (preferably by plowing, disking, or injecting) manure into the soil as quickly as possible after application. This will maximize nutrient conservation, reduce odors, and minimize runoff. Many regulations require incorporation within 24 hours (Figure 16.9).

    b. Spreading early in the day as the air is warming up, rather than late in the day when the air is cooling.

    c. Spreading when the wind is not blowing toward populated areas.

8. Limit one-time application volumes of liquid manure to the amount of moisture needed to bring the soil up to field capacity. Heavy applications can result in pooling and runoff.

9. Take care when applying manure to fields with subsurface drainage, especially when the soil is dry and cracked. Consider tilling the soil prior to application if cracks are present.

MANURE AS A FEED SOURCE

Recycled animal waste is a processed feed product for livestock derived from livestock manure or a mixture of manure and litter. Animal wastes contain significant percentages of protein, fiber, and essential minerals and have been deliberately incorporated into animal diets for their nutrient properties for more than 50 years. The most common processing method involves removing the coarse solids with a liquid–solids separation device and feeding the solids to ruminants after some disinfection stage. Another method is the mixing of swine manure and high-cellulose crop residues, such as cornstalks, to produce manure silage for ruminants.

Various risks are involved, including the spread of diseases, recycling of antibiotics that may be restricted in the consuming species and resulting withdrawal periods, and the issue of convincing consumers that food from animals fed manure is a quality product.

The Food and Drug Administration has revoked its previous policy on feeding manure and leaves the regulation to the states because manure feeding is largely an intrastate activity. The FDA does retain the power to regulate manure on an ad hoc basis if a particular instance involves a health hazard that the involved state(s) is unable to control. The Association of American Feed Control Officials (2003) defines dried swine waste offered for sale to have a maximum of 15% moisture, minimum of 20% crude protein, not more than 35% crude fiber (including bedding materials), and not more than 20% ash.

MANURE AS AN ENERGY SOURCE

Manure may also serve as a source of energy, which, of course, is not new. The pioneers burned dried bison dung, which they dubbed "buffalo chips," to heat their sod homes and to cook food over. Methane from manure has been used for power in Asian and European farms (Figure 16.10). Although the costs of constructing plants to produce energy from manure on a large-scale basis may be high, several hog producers have built on-farm methane generators. This is more efficient in

Figure 16.9 Injecting or immediately cultivating liquid manure into the soil essentially eliminates the odor and reduces nitrogen lost from volatilization to 5% or less compared to 25% lost with surface application. *(Courtesy, Palmer Holden, Iowa State University, Ames, IA)*

Figure 16.10 Methane storage bags on a hog farm in the Philippine Islands. *(Courtesy, Palmer Holden, Iowa State University, Ames, IA)*

warm climates because with cold temperatures most or all of the methane produced is required to keep the slurry at a temperature that allows the microorganisms to grow and produce methane. It is possible to run electrical generators with the methane produced.

Methane, of course, is usable like natural gas. Sanitary engineers have long known that a family of bacteria produces methane when they ferment organic material under strictly anaerobic conditions. However, it should be added that, because of capital and technical resources needed for any volume, the production of methane by anaerobic digestion will likely be limited to municipal or corporate industries. There is still the potential for methane production to be economically feasible for small livestock operations. Additionally, processing the manure through a digester and producing methane eliminates much of the odor from the manure. Some of the factors that determine the economical feasibility of obtaining methane gas from hog manure include the following:

1. Efficient on-site digested slurry and use of the methane gas.
2. Efficient collection and management of high-quality (8 to 10% solids) manure.
3. Low-cost digester.
4. Large manure tonnage to achieve economics of scale.

CURRENT MANURE MANAGEMENT ISSUES

Manure spills from hog operations in U.S. hog farms have focused attention on animal manure. The two biggest concerns will be (1) nutrient management, primarily nitrogen and phosphorus, but possibly copper, zinc, and selenium may also limit manure application in the future; and (2) odor control. In preparation for more activities by issue groups, swine producers need to control nutrient and odor pollution, and keep informed of local, state, and national legislation.

The U.S. swine industry has taken the lead in minimizing the problem. The National Pork Board has initiated an Environmental Assurance Program (EAP) to provide pork producers practical, proactive educational information and to enable them to identify and address the key management issues affecting the environmental quality of their operations.

The EAP is a comprehensive science-based program designed to help producers address environmental concerns. It is delivered at the local level by volunteers selected by state pork producer associations and trained by the NPB. The backbone of the educational program focuses on nutrient management, air and water quality, federal and state regulations, and neighbor relations. For more information contact the NPB at http://www.porkenvironment.org/.

QUESTIONS FOR STUDY AND DISCUSSION

1. Is manure important in sustainable agriculture?
2. How would you handle manure without disturbing the neighbors and optimizing the fertilizer value?
3. List the pros and cons of each of the most frequently used ways of handling manure.
4. "Manure management will be an issue in the decades to come." Do you agree or disagree with this statement?
5. In modern swine production units, how is manure handled and what equipment is required?

SELECTED REFERENCES

Common Manure Handling Systems, Ag 101, Environmental Protection Agency, Washington, DC, http://www.epa.gov/agriculture/ag101/porkmanure.html

Confined Animal Feeding Operations (CAFO) Fact Sheets, National Pork Board, Des Moines, IA, 2003, http://www.porkboard.org/cafo.asp

Iowa Manure Management Action Group, Iowa State University, Ames, IA, http://extension.agron.iastate.edu/immag/

Iowa State University Air Quality Publications, Iowa State University, Ames, IA, http://www.extension.iastate.edu/airquality/pubs.html

Managing Manure Nutrients at Concentrated Animal Feeding Operations Draft Guidance, Environmental Protection Agency, Washington, DC, http://www.epa.gov/ost/guide/cafo/pdf/PNPGuide.PDF

Pork Industry Handbook, Cooperative Extension Service, Purdue University, West Lafayette, IN, 2002

Swine Production and Environmental Stewardship, Environmental Protection Agency, Washington, DC, http://www.epa.gov/agriculture/swine.pdf

17

Buildings and Equipment

SELF-FEEDER. HALE'S "BODY OF HUSBANDRY", p.196
LONDON, 1766.

A 1766 self-feeder. Note the feeder appeared to serve all species. *(Hale's "Body of Husbandry," p. 196, London, 1766)*

Objectives

After studying this chapter, you should:

1. Know the stages of production and the type of housing that is most suitable for each stage.
2. Know the types of finishing facilities and the reasons why a producer may select each type.
3. Understand the importance of the heat production and vapor production of pigs and why these are important factors in the design of swine facilities.
4. Know the types of ventilation and why each may be selected for different stages of production.
5. Understand the variety of gestation housing and feeding arrangements and why each may be chosen.
6. Know the types of flooring materials available and where each has the greatest benefit.
7. Know the animal space required designing hog facilities.

Until the mid-1960s, pastures, drylots, and portable houses were an important part of most U.S. swine production systems. About this period of time, specialized commercial pork operations evolved. With their advent, came recognition that a good environment for swine is essential for maximum comfort and optimal performance. With their advent also came larger units, indoor production, and specialized swine buildings for specialized systems.

SPECIALIZED HOUSING

Today, swine producers have a multitude of choices from which they may select the system, along with the buildings and equipment, to match their management, labor, capital, and environmental needs. Any of their choices can be engineered and designed to provide the maximum comfort and optimal performance of the animals. A brief description of each of the most common specialized swine buildings in environmentally regulated or modified operations follows.

• **Breeding and gestation facilities**—Usually, the breeding herd (gilts, sows, and boars) is the last stage to be moved inside. These are mature animals and can better withstand the rigors of environmental changes. Once the sow is bred, she is very resistant to any environmental factors that would interrupt the pregnancy. Also they are not being raised with growth rate and efficiency as major goals that are greatly affected by the environment as with nursery and finishing pigs.

• **Farrowing house**—Central farrowing houses are generally environmentally modified. In comparison with portable houses on pasture, central

farrowing houses require more capital and a higher level of management, but the labor requirements are reduced and the comfort and safety of the sow and pigs are enhanced, especially under very hot or very cold conditions. Most of the floors are slotted, and both central and zone heat is supplied. Sows and litters are usually kept here until weaning from 14 to 35 days postfarrowing. Some producers still use solid floors with bedding and turn the sows out for feed and water twice per day.

• **Nursery**—At weaning, pigs are usually moved to a nursery, where they remain until they weigh 40 to 60 lb. This facility is kept dry, warm, and free of drafts. Slotted floors are used to help separate the pigs from their waste and to keep the floors dry, thereby improving sanitation and aiding in disease control. Raising the slatted flooring, often called raised decks, enhances nursery pig comfort by raising them off the cold floor and into a slightly warmer environment.

• **Finishing facilities**—Pigs are usually moved from the nursery when they weigh from 40 to 60 lb. In a farrow-to-finish operation, they are moved to a nearby finishing facility owned by the same farm. In an operation that produces feeder pigs, they are marketed when moved from the nursery. Pigs are generally considered to be in the growing phase until they reach 120 lb and in the finishing phase from 120 lb to market weight of 240 to 270 lb. The four types of swine buildings used for finishing pigs are (1) open front with an outside concrete apron, (2) modified open front, (3) double curtain, and (4) totally enclosed and artificially ventilated.

An open-front building with an outside concrete apron is the lowest cost of the four types of units (Figure 17.1). But the comfort and performance of the finishing pigs that they house may be affected by weather, either very hot or cold conditions. Some of these buildings are modified hog barns with outside pens, whereas others are commercially made with designed pens. They require some type of manure control facilities because the floor is solid concrete. This is especially important to catch the runoff after rainfalls. This control may be either a pit or settling basin. Environmental rules require that manure cannot flow into areas that may contaminate waterways.

The modified open front is so named because the sides may be opened for summer ventilation but closed to allow only minimum ventilation during winter. This type of building is naturally ventilated, alleviating any need for mechanical ventilation. Historically, these buildings had curtains on the front and wooden baffle doors on the back. The curtains are either adjusted by hand or automatically. The back doors are adjusted manually (Figure 17.2).

Double-curtain buildings are the latest design of modified open-front buildings (Figure 17.3). They are usually placed perpendicular to prevailing winds and have automatically controlled curtains on both sidewalls. Double-curtain buildings are primarily naturally ventilated to maintain proper temperatures and provide fresh air. However, the side curtains can be closed in hot windless weather, and the building can be ventilated with large fans in one end, called tunnel ventilation.

The totally enclosed and environmentally regulated building is the most costly of the four, but it provides the greatest control over temperature, humidity, and insects. It has high electrical demands because fans are used to ventilate the building continuously (Figure 17.4).

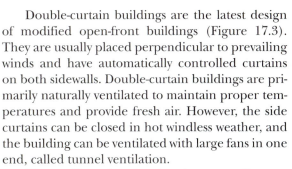

Figure 17.1 A typical open-front building with a shelter for the pigs and a large concrete exercise area with self-feeders. *(Courtesy, Palmer Holden, Iowa State University, Ames, IA)*

Figure 17.3 The double-sided curtain building also relies primarily on natural ventilation. Unlike the modified open-front building, this has totally slatted floors, with pens on both sides of a central alley. *(Courtesy, Palmer Holden, Iowa State University, Ames, IA)*

Figure 17.2 A typical modified open-front building. These units have floors that are usually 25 to 30% slatted at the front of the pens. Feeders are placed on the solid portion of the floor and waterers over the pit to encourage dunging in the slatted area. *(Courtesy, Palmer Holden, Iowa State University, Ames, IA)*

Figure 17.4 This farrowing house is a totally enclosed building that relies on fans for continuous ventilation. The floors are usually totally slatted. These are most commonly used for farrowing and nursery facilities because they provide the operator with better environmental control. *(Courtesy, Palmer Holden, Iowa State University, Ames, IA)*

Typically, the order of movement to indoor production follows the order of the effect of the environment and the age or weight of the pig. The first phase to move indoors is usually farrowing followed by nursery, finishing, and finally the breeding herd. Nursing pigs are most susceptible to the environment and the breeding herd is least affected.

In all walks of life, people expect a return on their investments, and swine producers are people. Thus, the cost associated with an empty farrowing crate or finishing pen must be recovered by the production from those crates or spaces that are occupied. All buildings must be used at or near capacity for maximum return on investment.

Hand-in-hand with indoor operations, waste disposal and animal welfare have become issues with swine producers. To cope with these problems, swine producers and researchers have worked to evolve new building and equipment designs to alleviate these problems. In the meantime, outdoor production has taken on a new name and a new look called *outdoor intensive swine production* (Chapter 15).

ENVIRONMENTAL CONTROL

Most species are able to control their responses to the environment to a large extent. But in terms of coping with heat or cold, pigs are poorly equipped. They possess very little hair for protection against the cold. Piglets, with essentially no hair coat, a large ratio of surface area to body weight, and no fat reserves, are very sensitive to cold. High temperatures will reduce performance and lower fertility. In environmentally regulated rearing, pigs must rely on their caretakers to modify their environments. Pigs respond quickly to environmental stresses, and often their responses are counterproductive.

The primary reason for having swine buildings is to modify the environment. Properly designed barns and other shelters, shades, drippers, insulation, ventilation, heating, and air conditioning can be used to approach the environment that producers desire. Naturally, the investment in environmentally modified facilities must be balanced against the expected increased returns, and there is a point beyond which further expenditures for environmental control will not increase returns sufficiently to justify the added cost. This point of diminishing returns differs from one section of the country to another, among animals of different ages, and among operators. For example, higher expenditures for environmental control can be justified for farrowing units than for gestating sow facilities. Also, labor and feed costs will enter into the picture.

Normal metabolic processes of the body result in the production of heat and water vapor. These two products are important considerations when designing swine buildings.

Heat Production of Swine

Heat produced by swine varies with age, body weight, diet, activity, ambient temperature, and humidity at high temperatures. Table 17.1 gives both "total heat production" and "sensible heat production." Total heat production includes both sensible heat and latent heat combined. Latent heat refers to the energy involved in a change of state and cannot be measured with a thermometer; evaporation of water and respired moisture from the lungs are examples. Sensible heat is that portion of the total heat, measurable with a thermometer, that can be used for warming air, compensating for building losses, and so on.

Note that heat production is highly dependent on the environmental temperature. Finishing hogs in a 35°F environment produce about 40% more total heat than pigs in a 70°F environment (860 vs. 610 Btu/hr).

Vapor Production of Swine

Pigs give off moisture during normal respiration, and the higher the temperature, the greater the

TABLE 17.1   Heat Production of Swine

Heat Source	Unit		Heat Production (Btu/hr)[1]			Heat Production (kcal/hr)[1]		
			Temperature (°F)	Total	Sensible	Temperature (°C)	Total	Sensible
	(lb)	(kg)						
Sow and litter (3 weeks after farrowing)	400	181.4	—	2,000	1,000	—	504.0	252.0
Finishing hog	200	90.7	35	860	740	2	216.7	186.5
			70	610	435	21	153.7	109.6

[1]One Btu (British thermal unit) is the amount of heat required to raise the temperature of 1 lb of water 1°F, while 1 kcal (kilocalorie) is the amount of heat required to raise the temperature of 1 kg of water 1°C.
Source: Adapted by the authors from Agricultural Engineers Yearbook, *St. Joseph, MI, ASAE Data Sheet D-249.2, p. 424.*

moisture. This moisture needs to be removed from buildings through ventilation systems. Most building designers determine the amount of winter ventilation by the need for moisture removal. Also, moisture removal in the winter is lower than that in the summer; hence, less air is needed. However, lack of heat makes moisture removal more difficult in the wintertime because cold air cannot hold as much moisture as warm air. Figure 17.5 gives the information necessary to determine the approximate amount of moisture to be removed.

Because ventilation also involves a transfer of heat, it is important to conserve heat in the building to maintain desired temperatures and reduce the need for supplemental heat. In a well-insulated building, mature animals may produce sufficient heat to provide a desirable balance between heat and moisture, but young animals will usually require supplemental heat in cold outdoor environments. The major requirement of summer ventilation is temperature reduction, which requires moving more air than in the winter.

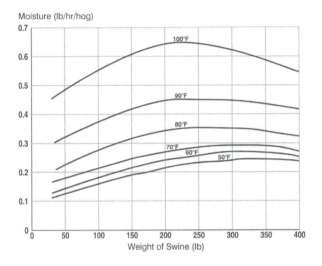

Figure 17.5 The amount of moisture (or vapor) produced per hour per hog by different weight pigs at various temperatures. Essentially, all of this is moisture exhaled from the lungs. As the temperature increases so does the respiration rate, with the hogs exhaling more moisture. Much of this moisture needs to be removed from the building. *(Courtesy of Pearson Education)*

Recommended Environmental Conditions

The comfort of animals (or humans) is a function of temperature, humidity, and air movement. Likewise, the heat loss from animals is a function of these three items. Additionally, when considering buildings, there are certain general requisites that should always be considered; among them, reasonable construction and maintenance costs, reduced labor, and utility value.

Temperature, humidity, and ventilation recommendations for different classes and ages of swine are given in Table 17.2. This table will be helpful in obtaining a satisfactory environment in buildings, which requires careful planning and design. Features of buildings that require emphasis include temperature and humidity control, insulation, vapor barrier, heat, ventilation, and automation.

Humidity influences are tied closely to temperature effects. The relative humidity should be within the range of 60 to 85%. The most desirable temperature and humidity levels for different classes and ages of hogs are given in Table 17.2.

TABLE 17.2   Recommended Environmental Conditions for Swine

Hog Unit	Temperature				Acceptable Humidity	Commonly Used Ventilation Rates[1]			
	Comfort Zone		Optimum			Cold[2]		Hot	
	(°F)	(°C)	(°F)	(°C)	(%)	(cfm)	(m³/min)	(cfm)	(m³/min)
Sow and litter	60–70	15–20	65	17	60–85	20	0.56	500	14.0
Newborn pigs (brooder area)	85–95	29–35	90	32	60–85	—	—	—	—
Finishing hogs									
12–30 lb (5–8 kg)	70–75	21–24	75	24	60–85	2	0.056	25	0.7
30–75 lb (8–34 kg)	65–70	18–21	65	18	60–85	3	0.08	35	1.0
75–150 lb (34–68 kg)	45–75	7–16	60	16	60–85	7	0.7	75	2.0
150–250 lb (68–114 kg)	45–75	7–16	60	16	60–85	10	0.20	120	3.4
Breeding herd									
Sow 325 lb (148 kg)	45–75	7–16	60	16	60–85	12	0.34	150	4.2
Boar 400 lb (182 kg)	45–75	7–16	60	16	60–85	14	0.39	300	8.4

[1]Generally two different ventilating systems are provided: one for winter and an additional one for summer. In practice, in many buildings, added summer ventilation is provided by opening (a) barn doors and (b) high-up hinged walls.
[2]Provide approximately one-fourth the winter rate continuously for moisture removal.
Source: Midwest Plans Service, MWPS-8. Iowa State University, Ames, IA, 1983.

The optimal temperature ranges for growing pigs are presented in Figure 17.6. Young pigs may be provided with zone heating to maintain temperature without heating the entire room. Measure temperatures at pig levels.

During cold weather, the necessary added warmth should be provided through properly constructed buildings and artificial means (heated buildings, brooders, and so on), and summer cooling should be enhanced by shades, wallows, and sprinklers.

Rate of gain and feed efficiency are lowered when swine must endure temperatures appreciably below or above the comfort zone. Feed consumption, growth rate, and feed efficiency of finishing pigs housed at temperatures range from cold stress to heat stress (Table 17.3). Cold pigs eat more and gain less, whereas heat stressed pigs eat less and gain less. From the standpoint of promoting maximum efficiency, therefore, it is desirable to provide animal housing, shades, wallows, and so on that eliminate extreme changes in environmental temperature.

Insulation

Insulation refers to materials that have a high resistance to the flow of heat. Such materials are commonly used in the walls and ceiling of hog houses. Proper insulation maintains a more uniform temperature—cooler houses in the summer and warmer houses in the winter—and permits a substantial fuel saving in artificially heated or air conditioned houses.

Adequate amounts of insulation should be installed during initial construction because it is easier and cheaper to install at that time. Although it will be more difficult and costly, existing inadequately insulated buildings should be better insulated. Spray-on insulation is available for application to the inside or outside of buildings. The energy savings will be long term.

The insulation value, or the R value, of some construction materials is shown in Figure 17.7. Polystyrene foam, polyurethane foam, and mineral wool provide the greatest insulation with the least thickness. But the insulating values of the other materials used in wall and ceiling construction must also be considered.

The amount of insulation desired depends on a number of factors, including climate. In mild and moderate climates, the walls should have R9 to R12 insulation values, and the ceilings about R12 to R16. In cold climates, the walls should be at least R14 and

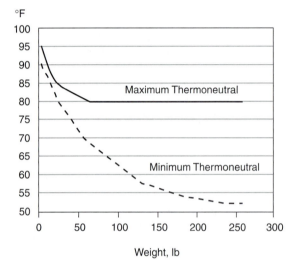

Figure 17.6 The thermoneutral temperature for pigs from birth to market. The earliest stages are the most critical with the narrowest temperature range. As the pig matures, it becomes much abler to adapt to wider temperature variations. *(Courtesy of Pearson Education)*

TABLE 17.3 **Effect of Temperature on Intake, Growth Rate, and Efficiency of Energy Conversion for Finishing Pigs**

Temperaure		Caloric Intake[1]	Growth Rate		Caloric Value of Gain[2]	Caloric Efficiency[3]
°F	(°C)	(kcal/day)	(lb/day)	(kg/day)	(kcal)	(%)
32	(0)	15,377	1.2	0.54	2,991	19.4
41	(5)	11,404	1.2	0.53	2,936	25.7
50	(10)	10,616	1.8	0.80	4,432	41.7
59	(15)	9,554	1.7	0.77	4,376	45.8
68	(20)	9,766	1.9	0.85	4,709	48.2
77	(25)	7,976	1.6	0.72	3,988	50.1
86	(30)	6,703	1.0	0.45	2,493	37.1
95	(35)	4,579	0.7	0.31	1,717	37.4

[1]Digestible energy.
[2]The caloric value for each gram of gain was estimated at 5.54 kcal.
[3]Calculated from the division of the caloric value of the gain by the caloric intake.
Source: Adapted from Ames, D. R. "Thermal Environment Affects Livestock Performance," Bioscience, 30, p. 457.

the ceilings R23. As the cost of heating goes up, more insulation will be justified in fuel savings.

Vapor Barrier

There is much moisture in hog houses; it comes from waterers, wet bedding, the respiration of the animals, and from the feces and urine. When the amount of water vapor in the house is greater than that in the outside air, the vapor will tend to move from inside to outside. The moisture enters the wall and moves outward, condensing when it reaches a cold enough area. Condensed water in the wall greatly reduces the value of the insulation and may damage the wall.

Because warm air holds more water vapor than cold air, the movement of vapor is most pronounced during the winter months. The effective way to combat this problem in a hog house is to use a vapor barrier with the insulation. The barrier should be placed on the warm side or inside of the house. Common vapor barriers are 4-mil (0.004 inches) plastic film and some of the asphalt-impregnated building papers.

Heat

In many hog-producing areas, buildings need supplemental heat. The amount of heat produced by hogs of different weights and their supplemental heat needs are presented in Table 17.4. Even with insulation, about 60% of this body heat is lost by ventilation to remove moisture; without insulation the heat loss is more.

Ventilation

Ventilation refers to the changing of air. Its purpose is (1) replacement of foul air with fresh air, (2) removal of moisture, (3) removal of odors, and (4) removal of excess heat in hot weather. Hog houses should be well ventilated, but care must be taken to avoid direct drafts and coldness. Good

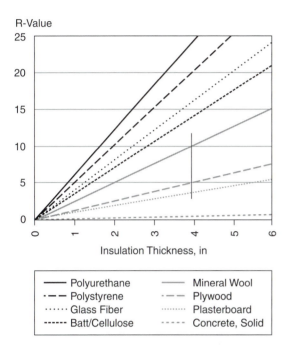

Figure 17.7 The relationship between insulation thickness and its R-value for several materials. Four inches of mineral wool have an R-value of 10. *(Midwest Plans Service, MWPS-8, Iowa State University, Ames, IA, 1983)*

TABLE 17.4   The Heat Production of Swine and Their Supplemental Heat Needs in Cold and Mild Climates[1]

Hog Unit	Building Temperature		Approximate Heat Production[2]	Supplemental Heat Needs			
				Slotted Floor		Bedded-Scraped Floor	
				Cold	Mild	Cold	Mild
	(°F)	(°C)		(Btu/hr)			
Sow and litter							
400 lb (181 kg)	60–80	16–27	2,100	1,500	1,000	2,000[3]	1,400
Pigs							
20–40 lb (9–18 kg)	70	21	330	275[3]	125[3]	300[3]	150[3]
40–100 lb (18–45 kg)	45–75	7–24	410	250	100	500	200
100–150 lb (45–68 kg)	45–75	7–24	500	250	100	500	200
150–210 lb (68–95 kg)	45–75	7–24	600	250	100	500	200
Sow or boar, limit fed							
200–250 lb (91–113 kg)	45–75	7–24	680	250	100	500	200
250–300 lb (113–136 kg)	45–75	7–24	760	250	100	500	200
300–500 lb (136–227 kg)	45–75	7–24	980	250	100	500	200

[1]Mild climates are seldom colder than 0°F (−18°C), whereas cold climates are often colder than 0°F (−18°C).

[2]At 60°F (16°C). Depending on the room temperature, about two-thirds of this total heat production is sensible heat that is available for warming air, building heat losses, and evaporating water.

[3]In addition to this amount, it is necessary to provide brooder heat for small pigs.

Source: Courtesy of Pearson Education.

swine house ventilation saves feed and helps optimize production.

All swine building ventilation is based on using the difference in static pressure inside and outside the building to bring in fresh air and exhaust foul air. Five factors are essential for good ventilation:

1. Fresh air moving into the hog house (inlets).
2. Insulation to keep the house temperatures warm.
3. Supplemental heat in the winter in cold areas.
4. Vapor barrier.
5. Removal of moist air (outlets). The importance of moist air removal becomes evident when it is realized that, through breathing,
    a. Each sow and litter adds about 1 gal of water per day, and
    b. 50 head of 125 lb pigs give off about 1 gal of moisture per hour.

In most hog houses, easily controlled electric fans do the best job of putting air where it is needed. Fans are rated in cubic feet per minute (cfm) of air they move. By selecting and using two or three fans, each of a different size, the needed variable rate, or flexibility, of ventilation can be achieved.

Provision should be made for one or more of the following, should there be a power or equipment failure in a closed building:

1. A battery-powered alarm that sounds in the home of the caretaker if power goes off.
2. A standby generator to supply power.
3. Solenoid or other electrically closed doors that open if power goes off.
4. Manually opened doors and windows.
5. A place to move the hogs to outside if the building cannot be opened sufficiently.

The amount of ventilation air required depends on the (1) inside-of-house temperature desired, (2) outside air temperature, (3) relative humidity, (4) number and size of animals in the building, and (5) amount of insulation. Because size of hog is a factor and because the comfort zone of baby pigs and finishing hogs differs, it is obvious that at least two ventilation tables are needed: one for the farrowing house and the other for the finishing house. Also, ventilation needs are different in the winter than in the summer. Recommended ventilation rates are presented in Table 17.2.

A common misconception is that heating costs will be reduced when the inside temperature is dropped by lowering the thermostat setting. Losses through the structure will be less, but the ventilation loss will be much greater.

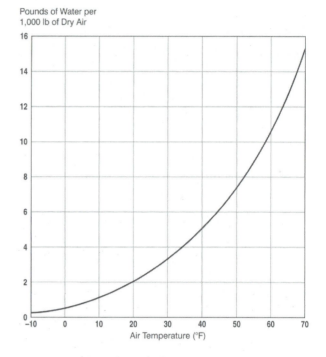

Figure 17.8 The influence of air temperature on its water-holding capacity. The water-holding capacity of air increases with rising temperature. *(Courtesy of Pearson Education)*

Incoming cold air must be heated in order to increase its moisture-holding capacity (Figure 17.8). Cold air has little capacity for picking up moisture. In general, the moisture-holding capacity doubles as the air temperature is raised 20°F.

Natural Ventilation

Swine producers continue their interest in naturally ventilated buildings over mechanically ventilated buildings, primarily because of their significantly lower construction and operating costs. Because only curtains or doors are used to regulate temperature, the costs of heavy insulation, tight-fitting doors and windows, and a mechanical ventilation system are averted. Of course, buildings that, for management reasons, must be maintained at temperatures above winter levels are not suited to natural ventilations, for example, farrowing barns and nurseries.

Naturally ventilated buildings are successfully used for finishing and gestation units. Naturally ventilated buildings are mainly a shell to protect animals from rain and snow, and to protect the building's contents. Most will have insulated ceilings to prevent the condensation of moisture and water dripping on the pigs. When a naturally ventilated barn is full of pigs, winter inside temperatures will rarely, if ever, get below freezing. Only if a significant number of pigs are sold, will the risk of frozen waterers exist.

Naturally ventilated buildings have a continuous opening at the high point (normally the ridge) of the building for air exhaust and continuous openings or inlets along the long sidewalls of the building for fresh air. The minimum area of these openings for winter ventilation can be estimated based on air movement needs, but experience is the best guideline to have sufficient fresh air to remove the moisture the pigs exhale. Air entering along the sidewalls (normally under the eaves) of the building is warmed by the heat from the animals in the building and picks up moisture as it rises toward the ridge. The continuous open ridge allows this warm, moist air to escape, completing the air exchange process.

During the warm weather, the building should serve mainly to keep rain out and to act as a sunshade. Large, continuous openings in the sidewalls allow summer breezes to blow through the building. During winter, the sides are nearly closed and adjusted to provide minimum winter ventilation to remove moisture. These include the modified open-front and double-curtain units.

Forced Air Ventilation

The forced air ventilation system is used in totally enclosed buildings that require heating, such as those used for farrowing or nurseries. Only a few finishing buildings use this system because the need for heating is low for older pigs. The three types of forced ventilation are pressure, exhaust, and tunnel ventilation.

With *pressure ventilation*, fresh air is forced in and stagnant air is allowed to flow out of the building through vents.

With *exhaust ventilation*, stagnant air is forced out of the building by fans and fresh air is sucked into the building through appropriately spaced openings because of the negative pressure caused by the exhaust fans.

With *tunnel ventilation*, air is pushed from one end of the building through the entire building and moist air is exhausted on the other end. These buildings usually have curtain sides that are closed in very warm weather and form the tunnel. Often four or more 48-in. fans force the air into the building.

DESIGNS AND SPECIFICATIONS

No attempt is made here to present detailed buildings and equipment designs and specifications. It is desired merely to convey suggestions regarding some of the desirable features of buildings and equipment in use in various parts of the United States. For detailed plans and specifications for a particular locality, the swine producer should (1) study successful

TABLE 17.5   Approximate Swine Building Component Costs

Building	Cost
Environmentally regulated facility, slatted floor, and so on	
Farrowing facility	\$2,200–\$2,500/crate
Nursery facility	\$110–\$125/pig (3 ft^2/pig)
Finishing facility	\$150–\$200/pig (7.5–8 ft^2/pig)
Crated gestation	\$550–\$625/crate
Hoop facility (\$5–\$6/ft^2 including water, gates, and electricity)	
Finishing facility	\$50–\$90/pig (10–15 ft^2/pig)
Gestation	\$120–\$160/sow (24 to 27 ft^2/sow)

Source: Dr. Jay Harmon, Agricultural and Biosystems Engineering, Iowa State University, Ames, IA, 2003.

buildings and equipment on neighboring hog farms; (2) consult the extension agents, vocational agriculture instructors, or commercial building and equipment suppliers; and/or (3) write Midwest Plans Service, Iowa State University, Ames, IA 50011-3080 (*http://ww.mwpshq.org/*). Nevertheless, the following items should be considered when developing building plans:

1. Be sure drainage is away from the well, farm home, and other buildings. Pasture or feedlot drainage should not enter waterways leading to streams or lakes.

2. Always leave room for more buildings to be located properly with respect to feed storage and roads.

3. Be sure facilities are located properly for efficient snow control. Locate facilities downwind from homes to minimize odor problems.

4. Plan for the necessary movement of animals, operators, and equipment.

5. Be certain of the accessibility of water and electricity.

A primary consideration in constructing livestock facilities is the cost of building. Examples of various costs are provided in Table 17.5. For example, a 20-stall farrowing building would cost from \$44,000 to \$50,000.

BUILDING SYSTEMS

Swine move through a sequence of production steps; as a result, one building does not work alone but it is part of a system. A complete farrow-to-finish system generally has one or more buildings devoted to each of the following phases (with the number of buildings determined by the size of the buildings and the number of hogs):

1. Farrowing building.
2. Nursery unit.
3. Finishing facility.
4. Breeding/gestation and herd boar facility.

In a feeder pig system, the feeder pigs are marketed at the end of the nursery period, either when they are less than 21 days of age (early weaned) or at 40 to 60 lb. Hence, nursery and finishing facilities are not needed. In a finishing system, feeder pigs are purchased; hence, only finishing facilities are needed.

Farrowing Buildings

Farrowing buildings are for the care and protection of sows and baby pigs during farrowing and until weaning. The design and construction of the farrowing building should be such as to provide optimum environmental control, minimum pig losses, maximum labor efficiency, and satisfactory manure handling. To this end, the following features are being incorporated in modern farrowing houses:

1. **Temperature.** Sows eat and milk best in an environment of 60 to 65°F which is too cold for newborn piglets. For the first 3 days, the brooder temperature for newborn pigs should be maintained at 90 to 95°F with a 2° drop per day to about 70°F. Generally, maintaining the room temperature between 65 and 75°F will prove successful while providing supplemental heat for the piglets. Piglet comfort can be observed by noting whether they are huddling (chilled) or avoiding the heated area (too warm).
2. **Floors.** Slotted floors greatly improve sanitation and are almost universally used in farrowing houses. Usually, slotted flooring is used in the entire crate although some have solid flooring under the sow. Slotted floors save more labor in farrowing than in any other phase of production. Solid floors with bedding are occasionally used in older buildings. They have a higher labor requirement but likely will allow the sow to wean as many pigs as with a slotted floor. It is important to slope the solid floors to drain urine and spilled water out of the stall.

Winter environment is more critical inside a slotted floor farrowing house than in a house with a solid floor. More environmental control is required with slotted floors than with solid floors because there are drafts under slotted floors. Coated metal, concrete, triangular bars, or woven wire slats are generally used. If slats are used, they generally should be about 3 in. wide with 3/8-in. spacing. The slots behind the sow should be 1 in. for ease of manure to fall through. These 1- in. slots need to be covered during farrowing.

3. **Farrowing crates (or stalls).** A variety of designs and sizes of farrowing crates or stalls are available (Figure 17.9). Those with a pit under the entire crate are popular. A typical farrowing crate is 5 ft wide and 7 ft long. The width for the sow is usually 24 in., and a creep or brooder area about 18 to 24 in. wide is provided for the baby pigs on both sides. Alleys should be available for easy access of sows to and from crates, and of the operator to the brooder area.

Nurseries

A swine nursery unit is for weaned pigs that are generally between 14 to 35 days old in the United States. They stay in this unit until they usually weigh 40 to 60 lb (Figure 17.10). The following considerations should be given to nursery units:

Figure 17.9   A farrowing room with two rows of crates with access from both the front and rear of the crates. Catalytic heaters hang from the ceiling to provide supplemental heat. This room has just been cleaned and disinfected in preparation for the next farrowing group. *(Courtesy, Palmer Holden, Iowa State University, Ames, IA)*

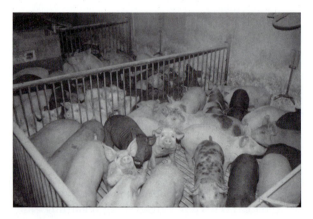

Figure 17.10   Nursery pens with metal panels, plastic slats, and about 20 pigs per 8 ft by 10 ft pen, allowing 4 ft² per pig. One fence line feeder serves two pens. *(Courtesy, Palmer Holden, Iowa State University, Ames, IA)*

1. Temperature. The initial temperature in the nursery should be maintained at 80 to 90°F for the comfort of the piglets. Gradually cool it down to about 70°F as the pigs adapt.

2. Floors. Slotted flooring is recommended. Floors may be either partially or totally slotted. In pig nurseries, coated metal, triangular bars, plastic, or woven wire are desirable because of their cleaning characteristics. Concrete slats are colder and are not used as often as other materials.

3. Decking. Decking refers to the housing of weaned pigs in small pens in single, double, or triple tiers and raised above the concrete level, providing a warmer environment. Single decks are called flat or raised decks. Decking was developed for early-weaned programs, reduced operating costs, better environmental control, and increased stocking density. Multiple decks made better use of the room space but now rarely are used because of sanitation concerns and the difficulty of the producer to observe the pigs.

Raised decking got the pigs off the cold, wet floors. But they were not practical. Producers did not object to lifting 12- to 15-lb pigs into a deck, but they objected to lifting out 40- to 60-lb pigs (Figure 17.11).

4. Feeders. If the nutritious and costly baby pig diet ends up in the manure pit or crammed in inaccessible corners of the feeder, growth, feed efficiency, and health will suffer. Following are recommendations for nursery feeding systems:

a. The feeding space should allow the pig to adopt its normal eating stance and carry out its normal eating motions. Allow 1 feeder space for every 2 pigs less than 30 lb or 1 space for every 3 pigs more than 30 lb.

b. The feeder should prevent feed buildup in the trough yet not restrict intake.

c. The feeder should be easily adjustable, limiting the feed available to cover about half of the feeding trough area.

d. The design should allow the pig to swallow meal with its mouth in or above the feed trough.

e. The feed should be accessible to the pig at or near the floor level.

f. Pigs should not be able to climb into, or become trapped by, the feeder.

Modular Nurseries

Modular nurseries are self-contained buildings that are constructed by the manufacturing plant before being delivered to the farm (Figure 17.12). The producer is generally responsible for preparing the delivery site and installing connections for water, sewer, and electricity. Among the advantages of modular over conventional facilities built on site is speedy delivery. Several modular companies guarantee delivery within 6 weeks or less from the time a contract is signed. Stick-built may take several months. Additionally, modular's second strong selling point is that they are portable—they can be moved easily and quickly.

Finishing Facilities

Finishing is often divided into two phases: growing, considered to be that span in a pig's life from about 40 or 60 to 120 lb, and finishing, the gain from about 120 to 260 lb, or market weight. The temperature needs of finishing pigs are not as critical as those of younger pigs. They are usually kept in full-size finishing pens designed to accommodate both growing and finishing stages.

Occasionally, producers may have separate growing and finishing stages, particularly on farms where the production is not sufficient to fill an entire barn.

Figure 17.11 A nursery with a partial double deck. The higher floor level of the deck provides a warmer environment for smaller pigs. The floor is completely slotted with Tri-bar slats. *(Courtesy, Palmer Holden, Iowa State University, Ames, IA)*

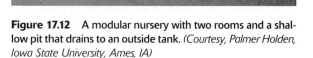

Figure 17.12 A modular nursery with two rooms and a shallow pit that drains to an outside tank. *(Courtesy, Palmer Holden, Iowa State University, Ames, IA)*

In this case often half of the building will be a grower and the other half a finisher. There will be separate ventilation controls to provide a warmer environment for the growing pigs as well as separate bulk feed supply bins with differently formulated diets.

Space requirements vary according to pig size and type of pen floor—bedded, solid, or slotted. The finishing of pigs can be successfully accomplished in an open-front building or in environmentally regulated facilities. When solid floors are used, they should be sloped one-half to 1 in. per foot to aid manure handling and cleanliness. Indoor grower-finishers have floors that are at least one-third slotted, but totally slotted floors with appropriate manure storage and disposal are most common.

A variety of building styles and floor plans are available for the range of designs, from totally enclosed, to modified open front, to open front–outside apron. Floors may be totally or partially slotted with appropriate manure storage and disposal.

Gestation Facilities

A variety of housing facilities are available for gestating sows. A brief discussion of each follows.

1. Open-front houses. Some producers provide about 15 sq ft per gestating sow in open-front houses (cold housing), which they divide into pen areas that will hold 10, 20, or 30 sows. The sows require bedding, and manure is handled as a solid. Feeders or feed stalls are placed outside the shelter on a concrete floor (Figure 17.13).

2. Enclosed houses. Other producers are building completely enclosed (warm) buildings for gestating sows, which allow year-round climate moderation. Such buildings either use individual stalls or are di-

Figure 17.13 Gestating sows in an open-front facility with long concrete runs outside of the shelter. Note the divided feeding troughs along both sides of the pen to provide plenty of feeding space. *(Courtesy, Palmer Holden, Iowa State University, Ames, IA)*

vided into pens holding 10 to 30 sows each. As in open housing the grouped sows are either fed as a group or fed individually.

The inside temperature should be maintained above 50°F, which can usually be achieved in cold weather if the barn is full of sows. Usually, some supplemental heat is required. Some gestation barns use curtain sides. If the walls are solid, they should be insulated, have vapor barriers installed in the walls and ceilings, and be equipped with ventilating fans. During the summer, large ventilator doors in the walls should be opened.

In totally confined gestation sow housing, totally or partially slotted floors are common, no bedding is used, and manure is handled as a liquid. The major disadvantage of indoor housing for gestating sows is the high initial investment cost in buildings and equipment.

Total indoor production of gestating sows has the following advantages:

1. Less labor. The labor requirements are reduced in indoor production, primarily in handling, cleaning, and feeding.

2. More efficient use of land. Hog pastures may be used for crop production, thereby (a) saving fencing and (b) making for a higher return from the land.

3. Better environmental control. Indoor production provides for better control of mud, dust, flies, and manure.

4. Improved parasite control. Indoor production provides improved control of internal and external parasites.

5. Better supervision. Indoor production provides for better supervision of the herd at breeding time as well as ease of observing the pigs.

6. Improved caretaker comfort. Indoor production provides improved operator comfort and convenience, especially in cold or rainy weather.

7. Facilitates individual feeding. The practice of keeping gestating sows indoors and individually feeding is common in most major hog-producing areas. Advantages of individual feeding include discouraging boss sows and fighting, controlling of dosages of feed additives, and providing facilities (stalls) for artificial insemination and other individual sow care needs.

Feeding Arrangements

Following are types of individual feeding arrangements that are being used:

- **Individual stalls.** With individual stalls, one sow is kept per pen during the gestation period. This system

Figure 17.14 Individual gestation stalls, each with its own feeding tube. The floors are typically totally slotted with concrete slats. *(Courtesy, Palmer Holden, Iowa State University, Ames, IA)*

prevents all fighting and provides individual feed supply and attention. The stalls are usually 24 in. wide and the length may vary depending on the size of the animal (Figure 17.14).

• **Tie stalls.** With the tie stall arrangement, sows are tied in their individual stalls with a strap. A tie-ring for the tether is centered in the floor under each sow's neck, about 8 in. behind the feed trough. The use of tie stalls is minimal.

Many European countries have banned individual and tie stalls, opting for what they consider to be more "welfare friendly" loose housing with feeding systems such as loose feeding stalls and electronic feeders. Research information is not consistent as to the best type of system based on the performance of the sow and longevity in the herd.

• **Loose feeding stalls.** Loose feeding stalls are generally about 20 in. wide and 8 ft long. Usually, the back end is open, although a gate may be used to shut the sows in until all are done eating. This method is usually used when sows are housed in groups but individual feeding is still encouraged.

• **Electronic feeders.** The electronic feeder system is also used with group-housed sows. Each sow has an individual electronic device attached to a collar or ear tag that is recognized by the electronic feeder. The feeder then provides the appropriate programmed level of feed to that sow. Electronic feeders are costly and each feeder can accommodate up to about 40 sows.

Not all indoor-raised gestating sows are individually fed. Many continue to be group fed, either in a separate feeding area or in their regular pens.

Boar Housing

When used in an individual mating system, boars are generally penned separately in individual crates about 28 in. wide and 7 ft long, or in pens 6 ft by 8 ft. Individual housing of boars eliminates fighting, riding, and competition for feed. It is recommended that boars be housed in a separate air space from the sows to be bred. Bringing sows into boar air space facilitates the detection of estrus. Groups of boars used in a pen-mating system should be penned together when they are removed from a sow group.

The development of commercial boar studs providing semen is a highly specialized business, and the design and requirements of boar studs is not covered in this book.

INDOOR VERSUS OUTDOOR INTENSIVE SWINE PRODUCTION

Many changes in U.S. swine production have occurred since the 1960s pasture system with portable houses. Beginning in the late 1950s, the industry moved from the low-investment, low-intensity system to the high-investment, high-intensity system.

In the early 1990s, England led the way with a pasture and portable house system to which was given a new name: *outdoor intensive swine production*. Throughout Europe, indoor swine production was being attacked by animal welfare activists.

But neither European nor U.S. swine producers are returning to the old pasture and portable house system. Instead, the outdoor intensive swine production renewing in the 1990s adapted and applied many of the changes that evolved with specialized confinement systems.

The subject of indoor versus outdoor intensive swine production is covered in Chapter 11, Forages and Pasture Production.

FLOORS

Floors are a concern when considering buildings for swine. Although there are many variations, there are basically two types: solid floor and slotted floor.

The question as to what type of floor to use in a building depends on how management intends to handle the manure. If manure is to be handled as a solid or semisolid product, the floor will usually be solid. If manure is to be handled as a liquid, the floor will be partially or totally slotted.

Solid Floors

Solid floors should slope toward alleys and drains. Annoying puddles accumulate when floors do not have a proper slope or slope in the wrong direction. A two-slope floor is recommended if waterers are installed in the stall or pen. Alleys should slope 1/2 in. per ft cross slope to form a crown or 1/4 in. per ft to drains. Floors should slope 1/2 to 3/4 in. per ft without bedding and 1/4 to 1/2 in. per ft with bedding. Usually, the waters are placed at the lower end of the solid floor pen to prevent the entire pen from becoming wet.

Slotted Floors

Five types of slotted floors in swine buildings are illustrated in Figures 17.15 to 17.19.

Slotted floors are floors with slots through which the feces and urine pass to a storage area below or nearby. Such floors are not new; they started in the poultry industry many years ago and have seen increasing usage by pork producers since the 1960s. Slotted floor buildings should be warm, well ventilated, and the floors flat.

The main advantages of slotted floors are (1) facilitating automation and saving labor; (2) reducing or eliminating bedding; (3) facilitating manure handling; (4) reducing the space needed per animal; (5) requiring less land; (6) increasing sanitation; (7) reducing mud, dust, odor, and fly problems; and (8) reducing pollution by collecting the manure.

The chief disadvantages of slotted floors are (1) higher initial cost than solid floors (no pit is required), (2) less flexibility in the use of the building, (3) spilled feed will be lost through the slots, (4) pigs raised on slotted floors resist being driven over a solid floor, and (5) environmental conditions become more critical.

Partial versus Total Slats

Studies have generally shown no difference in performance when comparing similar buildings with partially slotted floors to those with totally slotted floors, except possibly during the winter months in cold climates when heat may be marginal.

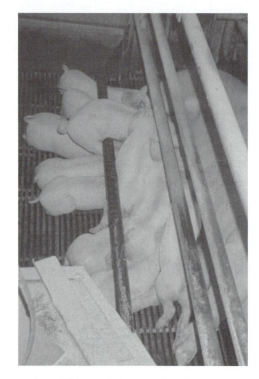

Figure 17.16 Tri-bar or T-bar slats in a farrowing stall. This style of flooring allows for little or no manure accumulation on the slats. Chilling of pigs is a bigger risk because of the small amount of surface area. *(Courtesy, Palmer Holden, Iowa State University, Ames, IA)*

Figure 17.15 Concrete slats in nursery facility. The slot should be at least 1 in. wide and the slat width 4 to 6 in. Nursery slats wider than 4 in. tend to become dirtier than narrower slats. *(Courtesy, Palmer Holden, Iowa State University, Ames, IA)*

Figure 17.17 Wire floors are gaining in popularity. These nursery pigs are slightly chilled as evidenced by their huddling together. Woven wire floors are very easy to clean. Woven wire is often used in farrowing stalls, but the gauge of the wire needs to be strong enough to maintain its rigidity over time. *(Courtesy, Palmer Holden, Iowa State University, Ames, IA)*

Figure 17.18 Coated wire is wire flooring that has been covered with plastic or hard rubber. It is easy to clean, and warmer and more comfortable than wire floors. It also is more expensive. Coated wire needs adequate support to maintain its rigidity with heavier pigs. *(Courtesy, Palmer Holden, Iowa State University, Ames, IA)*

Figure 17.19 Plastic flooring is used primarily in nurseries. It is comfortable and easy to clean. The white color improves the lighting in the room. It requires a lot of under support to maintain its rigidity. It is not strong enough for finishing. *(Courtesy, Palmer Holden, Iowa State University, Ames, IA)*

The initial floor cost increases slightly as the amount of slotted area increases. Despite this fact, most new finishing facilities are totally slotted because they want to avoid the messy pens of partially slotted floors and because of the increased pit capacity for manure storage. Systems with partial slats that store manure in underslat pits will need to be emptied more often because of limited manure storage capacity.

Floor Materials and Design

Durability, ease of cleaning, pig comfort, and cost should be considered when selecting slotted flooring. Slotted flooring material may be made from a variety of materials, which are summarized in Table 17.6 along with pertinent information regarding the expected life and advantages and disadvantages.

Slat width and spacing (slot width) are governed by size of hogs and cleaning efficiency. Narrow slots are usually more effective in farrowing and nursery units because piglets have small feet and can get their hooves caught in slots between 3/8 and 1 in. Slots narrower than 3/8 in. are too small for their feet to get into, and those wider than 1 in. allows them to pull their feet out without getting stuck.

Spacing narrow slats too wide can cause injury to the feet and legs of finishing hogs. However, wide slats and narrow openings result in floors that are not completely self-cleaning. In most phases of production, slats placed so that the openings are parallel to the long dimension of the pen or crate appear to favor pig comfort and mobility. Table 17.7 provides recommendations for slot width.

A host of flooring designs and materials is available for constructing slotted floors. Practically all of these materials and designs have their strengths and weaknesses. When choosing a slotted floor, the first consideration should be given to where it will be installed—farrowing, nursery, or the finishing area. Then, consideration should be given to the following: pig comfort, durability and longevity, traction, cost, cleanability, and quality.

SPACE REQUIREMENTS

Animals

One of the first, and frequently one of the most difficult, problems confronting the producer who wishes to construct a building or item of equipment is that of arriving at the proper size or dimensions. Table 17.8 contains suggested figures that may prove helpful. In general, less space than indicated may jeopardize the health and well-being of the pigs; whereas, more space may make the buildings and equipment more expensive than necessary.

Feed and Bedding

The space requirements for feed storage for the swine enterprise vary so widely that it is difficult to provide a suggested method of calculating space requirements (Table 17.9). The amount of feed to be stored depends primarily on (1) length of pasture season, (2) method of feeding and management, (3) kind of feed, (4) climate, and (5) the proportion of feeds produced on the farm in comparison with those purchased. Normally, the storage capacity should be sufficient to handle all feed grain grown on the farm and to hold purchased supplies for about 30 days.

TABLE 17.6   Materials for Slotted Flooring

Material	Expected Life	Advantages	Disadvantages	Comments
Cast iron	20 years	High cost.	Cast iron pulls heat from the animals.	Avoid sharp edges on cast iron slats.
Concrete slats	20 years	Long life. May be homemade.	Quality control difficult when poured on site.	Should have flat, smooth surface and slightly pencil-rounded edge.
Metal slats (steel or aluminum)	4–8 years	Easily cleaned.	High cost. Aluminum degrades if in contact with other metals, shortening expected life.	Longer life from stainless steel than plain steel. Porcelain coated steel slats available. Some steel slats are perforated.
Plastic-coated metal	3–5 years	Comfortable. Has diamond-shaped holes.	May become slick. Requires adequate support. High cost.	Generally constructed of welded wire or expanded metal, then coated with a plastic substance. Performs well in farrowing and nursery facilities.
Plastic slats	8–10 years	Warmer than concrete and metal. Easy to clean.	May become slick. Requires adequate support.	Modular design is easy to install.
Triangular and T-bar metal slats	5–10 years	Durable, 50% open. Easy to clean.	High cost.	Generally fabricated from steel. Performs well in farrowing and nursery facilities.
Wood slats	2–4 years	Low initial cost.	Difficult to maintain spacing. Not too durable.	Make from hardwood, such as oak.
Woven and welded wire	3-gauge: 5–10 years; 5/16-gauge: 7–15 years	Cleaning ease, low cost.	Must be supported rather thoroughly with a frame.	Galvanized or coated with plastic. Performs well in farrowing and nursery facilities.

Source: Courtesy of Pearson Education.

TABLE 17.7   Recommended Slot Widths

Production Stage	Slot Width		Comments
	(in)	(mm)	
Farrowing	3/8 for baby pigs	9	In a farrowing crate, use larger slot width behind sow and smaller slot width elsewhere. Cover openings behind sow with plywood, sheet metal, or mesh during farrowing.
Nursery (20–60 lb or 9–27 kg)	3/8	9	Spacing depends on width of slat; wider slat needs wider spacing.
Finishing (60 lb or 27 kg to market)	3/4–1.0	19–25	When narrower slats of material other than concrete are used, narrower openings are used.
Gestating sows; boars	3/4–1.1	9–27.9	Slats which are 5–8 in. (13–20 cm) wide need the 1-in. (2.5-cm) opening.

Source: Courtesy of Pearson Education.

Bedding may or may not be stored under cover. Sometimes poled framed sheds or an inexpensive cover of plastic is used for bedding protection. This information may be helpful to the operator desiring to compute the building space required for a specific operation. Table 17.9 also provides a convenient means of estimating the amount of feed or bedding in storage.

SWINE EQUIPMENT

The successful hog producer must have adequate equipment with which to provide feed, water, shelter, and care for the animals. Suitable equipment saves labor and prevents the loss of many baby pigs and even the loss of older animals. It should not be overlooked.

Certain features are desirable in swine equipment. The equipment should be convenient and economical, and, because hogs are subject to many diseases and parasites, it should be constructed for easy cleaning and disinfection. Equipment should be useful, durable, and yet economical.

There are many types and designs of the various pieces of swine equipment, and some producers will introduce their own ideas to fit the material available to local conditions. It is expected that certain adaptations should be made in designs to meet individual conditions.

Creep Area

Almost no creep feed is consumed before 2 weeks of age and very little by 3 weeks of age unless the sow is a poor milk producer. Therefore, with early weaning, many swine producers no longer provide creep feed. Often sprinkling some dry feed on a solid portion of the farrowing stall will accustom the pigs to feed.

A creep area should be provided with feeders for piglets with group lactation. The creep area should be large enough to accommodate the number of pigs intended along with the feeders. There should be enough openings in the fence through which the little pigs can pass. They should be wide enough for the pigs, but narrow enough to keep sows out. The openings should be sufficiently high so the pigs do not have to lower their backs when passing through; otherwise, there is a tendency to produce objectionable sway backs.

Heat Lamps and Brooders

Newborn pigs are quite comfortable at temperatures of 90 to 95°F, shiver when standing alone at 70°F, and are quite cold at 60°F. Recent estimates indicate that swine producers can save an average of 1.5 more pigs per litter by using supplementary heat to keep baby pigs warm and to keep them away from being laid on by the sow.

Brooders, or enclosed heated areas, of various designs are used. Where sows are farrowed in open farrowing pens, a triangular-shaped brooder is usually secured in one corner of the pen (Figure 17.20). Heat may be supplied by infrared-heat lamps, by electric hovers, or by gas-fired radiant heaters. Heat lamps should be of hard Pyrex glass (red filter type) that resists breakage when splashed with water. Use heavy porcelain sockets, a metal hood, and suspend the lamp by a chain. Never suspend the heat lamp by an electrical cord. Adjusting or moving the lamp may weaken the electrical connections and cause an electrical short. A 250-watt lamp should be 24 in.

Figure 17.20 Lactating sows in farrowing crates with hovers. Note the wire flooring under the sows and the heated area for the piglets. (*Courtesy, Palmer Holden, Iowa State University, Ames, IA*)

above the floor of the brooder area. A smaller bulb can be used as the pigs get older.

Any type of heating device can cause a fire, so heaters and lamps should be installed carefully. Clean heating devices regularly to remove dust that may interfere with the proper functioning of the heater. Radiant heating pads and other baby pig heating devices are safer than heat lamps.

Farrowing Crates (Stalls)

Adequate facilities for farrowing are important because one-third or more of all death losses before weaning result from overlaying or crushing. A variety of farrowing crates, stalls, or tethering arrangements of different designs and sizes are available.

The most exciting new systems of gestation and lactation housing preserve the advantages of individual living places, while permitting the sow much more freedom of movement than the traditional crates, stalls, or tethering. The new installations are referred to as turn-around, comfort stalls, or freedom stalls (Figure 17.21).

It is noteworthy that the United Kingdom's agricultural ministry decreed that stall and tethering systems be removed effective in 1998 and that the rest of the European Community countries remove tethers by 1999. These rules were not based on research on pig welfare, but rather on the human's perception of welfare.

(Also see the Chapter 15 section headed "Farrowing Management.")

Loading Chutes

Loading chutes are essential on most hog farms. Chutes make possible loading hogs without injury

TABLE 17.8   Space Requirements of Buildings and Equipment for Swine[1]

| Age and Size of Animal | Pasture or Feeding Floor | | Swine Buildings | | | | | Shades | |
	Good Pasture	Paved Feeding Floor in Addition to Sleeping Space	Inside Sleeping Area Per Animal[3]	Ceiling Height[4]	Pen Gate Height	Hog Door Height	Hog Door Width	Shade Per Pig	Shade Height
	(pigs/acre)	(sq ft)	(sq ft)	(ft)	(in.)	(in.)	(in.)	(sq ft)	(ft)
Gestation									
Gilts	10–12	15–20	15–17	7–8	36	36	24	17	4–6
Mature sows	8–10	8–10	18–20	7–8	36	36	24	20	4–6
Sows with pigs									
Gilts	6–8	48	48	7–8	36	36	24	20	4–6
Mature sows	6–8	64	64	7–8	36	36	24	30	4–6
Herd boars	1/4 acre/boar	15–20	15–20	7–8	48	36	24	15–20	4–6
Finish									
Weaning[7] to 75 lb	50–100	6–8[8]	5–6[9]	7–8	30	36	24	4	4–6
75 lb–125 lb	50–100	7–9[8]	6–7[9]	7–8	33	36	24	6	4–6
125 lb–market	50–100	8–10[8]	8–10[9]	7–8	36	36	24	6	4–6

[1]With the following information, these requirements may be converted to metric equivalents: 1 sq ft = 0.093 m²; 1 ft = 0.305 m; 1 in. = 25.4 mm; 1 acre = 0.405 ha; 1 lb = 0.45 kg.

[2]The drinking water should not fall below 35 to 40°F (2 to 4°C) during winter.

[3]When using slotted floors, only half as much floor space per hog is needed as when using conventional solid floors.

[4]Ceiling heights in excess of 7 to 8 ft create cold hog houses in cold climates.

[5]For example, a 6-ft feeder open on both sides has 12 linear ft of feeding space.

[6]With creep provided for pigs in addition.

[7]For early weaning space requirements, see Chapter 15, Table 15–5.

[8]The larger area when fed from troughs; the smaller area is adequate where self-feeders are used.

[9]The larger area in the summertime.

Source: Courtesy of Pearson Education.

and with the least amount of effort. They may be either fixed or portable. Although slightly more expensive to build, a stair-step loading chute is preferable to the ramp-and-cleat type; hogs find the steps easier to ascend, and will go up more willingly and with less chance of injury. Ramp-type chutes are still the most prevalent because of their lower initial cost.

The chute should be fairly narrow so animals cannot turn around, usually about 22 in. wide. Care should be taken so that no cleats, nails, or other projections injure the hog. A portable chute is convenient because hogs can be loaded from any lot or pasture (Figure 17.22). The chute should be durably constructed. Many farms now have portable pens with hydraulic lifters that allow the trailer to be lowered to ground level for loading the pigs and then raised to truck box level for unloading.

Feeders

The design of the self-feeder is extremely important. An adequate self-feeder should be constructed so that

1. It will be sturdy and durable.
2. The feed will not clog. The feeder should have means for agitating the feed to keep it from bridging and adjustable throats for controlling the rate of flow.
3. Waste will be reduced to a minimum.
4. The feed will be protected from wind, rain, birds, and rodents.
5. It is large enough to hold several days' feed. Some new feeders, either wet-dry or tube feeders, have minimal storage capacity and depend

| Feeding Equipment | | | | Watering Equipment[2] | | | |
| Self-Feeder Space Drylot | Pasture | Percentage of Feeder Space for Protein Supplement Pasture | Drylot | Feed Trough Space for Hand-Feeding[5] | Water Trough Space for Hand-Feeding | Automatic Watering Cups (two openings considered 2 cups) | Comments |
(pigs/ linear ft)[5]	pigs/ (linear ft)[5]	(%)	(%)	(linear ft/ pig)	(linear ft/ pig)		
2	3	15	10–15	1.5	1.5	1 cup/12 gilts	When alfalfa hay is fed
3	4	15	10–15	2	2	1 cup/10 sows	in rack, allow 4 sows/ linear ft.
1[6]	1[6]	15	10–15	1.5[6]	2	1 cup/4 sows	For pig creep: provide a
1[6]	1[6]	15	10–15	1.5[6]	2	1 cup/4 sows	minimum of 1 ft of
1	1	15	10–15	2	2	1 cup/2 boars	feeder space per 5 pigs; lip of the feeder trough no more than 4 in. above the floor; maximum of 40 pigs per creep.
4	4–5	25	20–25	0.75	0.75	1 cup/20 pigs	When salt or mineral is
3	3–4	20	15–20	1.0	1.0	1 cup/20 pigs	fed free choice, provide
3	3–4	15	10–15	1.25	1.25	1 cup/20 pigs	3 linear ft of mineral box space or 3 self-feeder holes per 100 pigs.

Figure 17.21 A bedded version of the freedom farrowing stall. The sow is confined to the left side of the pen until the piglets are 3 or 4 days old. Then both the sow and piglets have access to the entire pen. Some freedom stalls use totally slotted floors but still fit in the standard 5 ft by 7 ft pen area. *(Courtesy, Palmer Holden, Iowa State University, Ames, IA)*

on being filled automatically several times per day (Figure 17.23).

Many satisfactory designs of self-feeders are available, including both commercial and home-made equipment.

When they are located outside, self-feeders should always be placed on concrete platforms; otherwise, mud holes soon appear around the base in wet weather, causing feed wastage and the risk of tipping as they become undermined. Also, they should be located where they can be filled from over a fence so that entering the pen is not necessary.

The hog trough, a universally used piece of feedlot equipment, is often used in sick pens or where only a few pigs are temporarily kept. A good trough should be easy to clean and should be constructed so that hogs cannot lie in it or otherwise contaminate the feed. The little pig trough is similar in construction and is useful for small creep-fed pigs.

Shade

Hogs require shade during the hot months of the year. Heat-stressed pigs eat less feed, gain less weight, and the gains are less efficient (see Table 17.3). When the temperature gets about 10°F above thermoneutral, the pigs essentially stop eating (about 90°F for a finishing pig). Fortunately, it usually cools at night and the pigs will consume feed at that time.

TABLE 17.9   Storage Space Requirements for Feed and Bedding[1]

Feed or Bedding	lb/ft³	ft³/ton	lb/bu of grain	ft³/bu
Hay—Straw				
1. Loose				
Alfalfa	4.4–4.0	450–500		
Nonlegume	4.4–3.3	450–600		
Straw	3.0–2.0	670–1,000		
2. Baled				
Alfalfa	10.0–6.0	200–330		
Nonlegume	8.0–6.0	250–330		
Straw	5.0–4.0	400–500		
3. Chopped				
Alfalfa	7.0–5.5	285–360		
Nonlegume	6.7–5.0	300–400		
Straw	8.0–5.7	250–350		
Silage				
Corn or sorghum in tower silos	40	50		
Corn or sorghum in trench silos	35	57		
Grain				
Corn, shelled[2]	45	45	56	1.25
Corn, ear	28	72	70	2.50
Corn, shelled, ground	38		48	
Barley	39	51	48	1.25
Barley, ground	28		37	
Oats	26	77	32	1.25
Oats, ground	18	106	23	
Rye	45	44	56	1.25
Rye, ground	38		48	
Sorghum	45	44	56	1.25
Wheat	48	42	60	1.25
Wheat, ground	43	46	50	
Mill Feed				
Bran	13	154		
Middlings	25	80		
Linseed meal	23	88		
Cottonseed meal	38	53		
Soybean meal	42	43		
Alfalfa meal	15	134		
Miscellaneous				
Soybeans	48	56		
Salt, fine	50	40		
Pellets, mixed feed	37	58		
Shavings, baled	20	100		

[1]For metric conversion, refer to the appendix.
[2]These values are for corn with 15% moisture. Whole kernel corn that is 30% moisture weighs 51 lb/cu ft shelled and 36 lb/cu ft ground or 68 lb/bu shelled and 90 lb/bu ground.
Source: Courtesy of Pearson Education.

In many cases, sleeping sheds may double for shade purposes. But it is often desirable to provide additional protection from the sun.

Sow Wash

As part of a herd health program, sows should be washed before being placed in farrowing rooms. Areas for washing sows are usually located at the exit of the gestation area or near the entrance to the farrowing house. A variety of designs will work

as an area for washing sows, but it is a good idea to include both a hot and a cold water source and a floor drain.

Vaccinating and Castrating Rack

A vaccinating and castrating rack makes for convenience in treating young pigs. This is a simple rack, made like a common sawbuck with a V-shaped trough added. A rope or strap is attached at one or both ends of the trough to hold the pigs securely in

Figure 17.22 A portable stair-step loading chute. Pigs prefer the stair-step design compared to the ramp with cleats across the floor. *(Courtesy, Department of Agricultural and Biosystems Engineering, Iowa State University, Ames, IA)*

Figure 17.23 A tube feeder with two feed delivery pipes into a stainless steel trough with nipple waterers at each end of the trough. The tube is slid up or down to adjust the feed level. If the water level in the trough becomes too high, it could restrict feed intake. *(Courtesy, Palmer Holden, Iowa State University, Ames, IA)*

place. Also, some commercial holders are available that hold piglets upside down by their hind legs.

Wallows; Sprinklers

Hogs have very few sweat glands; thus they cannot cool themselves by perspiration. In the warm climates, therefore, mature breeding animals and finishing hogs kept outdoors need a wallow during the hot, summer months. Instead of permitting an unsanitary mud wallow, swine producers may install hog

wallows. This equipment, which may be either movable or fixed, will keep the animals cool and clean and make for faster and more economical gains.

The wallow should be in close proximity to shade, but in no case should shade be built directly over the wallow. Such an arrangement will cause the hogs to lie in the water all day. The size of wallow to build depends on the number and size of animals. Up to 50 finishing pigs can be accommodated per 100 sq ft of wallow provided shade or shelter is nearby.

Sprinklers are often used indoors for cooling. They should be intermittent rather than constant because the cooling accomplished by wetting the pigs is followed by evaporation of the water. They should be located over the manure collection area of the pens. They are usually set to begin 80 to 85°F. Outdoor sprinklers should be limited to concrete lots or sandy soils where mud holes do not develop.

Provide 1 nozzle per 20 to 30 hogs, the number usually housed per pen. The nozzles should be 4 to 6 ft from the floor or ground and about 8 ft apart. They should be controlled by a timer and spray 2 minutes out of 10 or 1 minute out of 30. Sprinkling will require from 0.4 to 0.6 gallons per hour per sprinkler. It is important to provide an inline filter or settling tank to prevent clogging of the nozzles.

Water

Swine will consume 2 to 3 units of water per unit of dry feed or about 1/4 to 1/3 of a gallon per pound of feed. The higher the temperature, the greater the water consumption. It is preferable that swine have access to automatic waterers, with cool, clean water available at all times. Otherwise, they should be hand-watered at least twice daily. During winter, the drinking water should not be permitted to fall below 40°F. The estimated water needs for various classes of swine are in Table 17.10.

For running water to pastures and other fields, plastic pipe is recommended because it is easily installed either on top of the ground or in a trench. By locating the waterer in a line fence, two pastures can be supplied from one unit.

FENCES

Good fences (1) maintain farm boundaries, (2) make livestock operations possible, (3) reduce losses to both animals and crops, (4) increase land values, (5) promote better relationships between neighbors, (6) lessen accidents from animals getting on roads, and (7) add to the attractiveness and distinctiveness of the premises.

TABLE 17.10   Water Requirements of Swine

Class of Swine	Water Consumption[1] (gal/head/day)
Gestating sow	6
Sow and litter	8
Nursery pig (15–40 lb)	1
Growing pig (40–120 lb)	3
Finishing pig (120–250 lb)	4
Boar	8

[1]Water requirements are highly variable because of temperature, water wastage, and feed intake.
Source: Swine Housing and Equipment Handbook,
Midwest Plans Service, Iowa State University, Ames, IA, 1983.

The discussion is limited primarily to wire fencing, although such materials as rails, poles, boards, stone, hedge, pipe, and concrete have a place and are used under certain circumstances. Also, where there is a heavy concentration of animals, such as on feeding floors, there is need for a more rigid type of fencing material than wire. Moreover, certain fencing materials have more artistic appeal than others, and this is an especially important consideration for the purebred establishment.

Wire Options

The kind of wire to purchase should be determined primarily by the class of animals to be confined. The following additional points are pertinent in the selection of wire:

1. **Styles of woven wire.** The standard styles of woven wire fences are designated by numbers such as 958, *1155*, 849, *1047*, 741, *939*, *832*, and *726* (the figures in italics are most common). The first one or two digits represent the number of line (horizontal) wires; the last two the height in inches (i.e., 832 has 8 horizontal wires and is 32 in. in height). Each style can be obtained in either (a) 12-in. spacing of stays (or mesh) or (b) 6-in. spacing of stays.

2. **Mesh.** Generally, a close-spaced fence with stay or vertical wires 6 in. apart (6-in. mesh) will give better service than a wide-spaced (12-in. mesh) fence. However, some fence manufacturers believe that a 12-in. spacing with a No. 9 wire is superior to a 6-in. spacing with No. 11 filler wire (about the same amount of material is involved in each case).

3. **Weight of wire.** A fence made of heavier weight wires will usually last longer and prove cheaper than one made of light wires. Heavier or larger size wire is designated by a smaller gauge number. Thus, 9-gauge wire is heavier and larger than 11 gauge. Woven wire fencing comes in 9, 11, 12 1/2, and 16 gauges—which refer to the gauges of the wires other than the top and bottom line wires. Heavy barbed wire is 12 1/2 gauge. But there is a lighter, high-tensile barbed wire available in 14 and 16 gauge.

Heavier or larger wire than normal should be used in those areas subject to (a) salty air from the ocean, (b) smoke from close proximity industries that may give off chemical fumes into the atmosphere, (c) rapid temperature changes, or (d) overflow or flood. Also, heavier or larger wire than normal should be used in fencing (a) small areas, (b) where a dense concentration of animals is involved, and (c) where animals have already learned to get out.

4. **Styles of barbed wire.** Styles of barbed wire differ in the shape and number of the points of the barb, and the spacing of the barbs on the line wires. The 2-point barbs are commonly spaced 4 in. apart, whereas 4-point barbs are generally spaced 5 in. apart. Because any style is satisfactory, selection is a matter of personal preference.

5. **Standard size rolls or spools.** Woven wire comes in 20- and 40-rod rolls; barbed wire in 80-rod spools.

6. **Wire coating.** The kind and amount of coating on wire definitely affects its lasting qualities. Galvanized coating is most commonly used to protect wire from corrosion. Coatings are specified as Class I, Class II, and Class III. The higher the class number, the greater the coating thickness and performance.

Posts

Three kinds of material are commonly used for fence posts: wood, metal, and concrete. The selection of the particular kind of posts should be determined by (1) the availability and cost of each, (2) the length of service desired (posts should last as long as the fencing material attached to them, or the maintenance cost may be too high), (3) the kind and amount of livestock to be confined, and (4) the cost of installation.

Electric Fences

Where a temporary enclosure is desired or where existing fences need bolstering from roguish or pushy animals, it may be desirable to install an electric fence, which can be done at minimum cost. The following points are pertinent in the construction of an electric fence:

1. **Safety.** When an electric fence is installed and used, (a) necessary safety precautions against accidents to both persons and animals should be taken and (b) the hog producer should first check into the applicability of state regulations relative to the

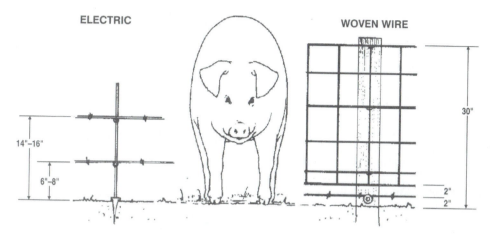

Figure 17.24 Woven wire fences for hogs may be 30 or 36 in. high with a strand of barbed wire at the bottom to prevent rooting. Electric fences with two strands of barbed wire may be used for temporary fencing. If producers spread some feed along the fence, they can make the pigs aware of the fence as they sample the feed. *(Courtesy of Pearson Education)*

installation and use of electric fences. *Remember that an electric fence can be dangerous.* Fence controllers should be purchased from a reliable manufacturer; homemade controllers may be dangerous.

2. Wire height. As a rule of thumb, the correct wire height for an electric fence is about three-fourths the height of the animal; with two wires provided for swine—one wire should be 6 to 8 in. from the ground, and the other 14 to 16 in. from the ground (Figure 17.24).

3. Posts. Either plastic or steel posts may be used for electric fencing. Corner posts should be as firmly set and well braced as required for any nonelectric fence so as to stand the pull necessary to stretch the wire tight. Line posts (a) need only be heavy enough to support the wire and withstand the elements and (b) may be spaced 25 to 40 ft apart for hogs.

4. Wire. Barbed wire is recommended because the barbs will penetrate the hair of animals and touch the skin, but smooth wire can be used satisfactorily. Rusty wire should never be used because rust is an insulator.

5. Insulators. Wire should be fastened to the posts by insulators and should not come into direct contact with posts, weeds, or the ground. Inexpensive solid glass, porcelain, or plastic insulators should be used, rather than old rubber or necks of bottles. Plastic or fiberglass posts do not require insulators.

6. Charger. The charger should be safe and effective (purchase one made by a reputable manufacturer). There are five types of chargers: (a) *the battery charger,* which uses a 6-volt hot shot battery; (b) *the inductive discharge system,* in which the current is fed to an interrupter device called a circuit breaker or chopper that energizes a current limiting transformer; (c) *the capacitor discharge system,* in which the power line is rectified to direct current and the current is stowed in the capacitor; (d) *the continuous current type,* in which a transformer regulates the flow of current from the power line to the fence; and (e) *the solar (sun) charger.*

7. Grounding. One lead from the controller should be grounded to a pipe driven into the moist earth. *An electric fence should never be grounded to a water pipe because it could carry lightning directly to connecting buildings.* A lightning arrestor should be installed on the ground wire.

Note: Do not string an electric fence across a gate to be used by the pigs. Use wood or metal gates. If pigs get shocked by an electrified gate, it will be very difficult to get them through the gate even though it is wide open.

QUESTIONS FOR STUDY AND DISCUSSION

1. In the late 1950s, specialized pork producers evolved. With their advent, came indoor production, replacing much of the old system of pastures, drylots, and portable houses. Why did this happen?

2. Why are hogs more sensitive to extremes in temperatures—either hot or cold—than other farm animals? What is the most desirable temperature for each class of hogs? What can be done to modify (a) winter and (b) summer temperatures?

3. Why is knowledge of the heat production of hogs important?

4. How are the effects of cold stress and heat stress responded to by hogs?

5. The major requirement of winter ventilation is moisture removal, whereas the major requirement of summer ventilation is temperature control. Why is there a difference?

6. What is natural ventilation? How is natural ventilation accomplished?

7. A complete farrow-to-finish confinement system generally has one or more buildings devoted to each of following phases: (a) farrowing, (b) nursery, (c) finishing, (d) breeding/gestation and herd boars. Describe the types of buildings used for each of these phases.

8. Why is there renewed interest in outdoor swine production, now referred to as *outdoor intensive swine production*?

9. List and discuss the advantages and disadvantages of the various materials for slotted floors.

10. Describe and justify the type of flooring you would use in a nursery.

11. In planning to construct new buildings and equipment for swine, what factors and measurements for buildings and equipment should be considered?

12. Why should a swine producer be knowledgeable about the storage space requirements for feed and bedding?

13. Make a critical study of your own swine buildings and equipment, or those on a hog farm with which you are familiar, and determine their (a) desirable and (b) undesirable features.

14. Is a turn-around farrowing crate necessary? Why or why not?

15. What features would you look for in a self-feeder?

16. Why are shades and wallows important for swine on pasture?

17. Describe the construction of an electric fence for swine.

SELECTED REFERENCES

Alternative Systems for Farrowing in Cold Weather, AED-47, 2004

Hoop Structures for Gestating Swine, AED-44, 2004

Hoop Structures for Grow-Finish Swine, AED-41, 2004

Livestock Environment, Ed. by E. Collins and C. Boon, American Society of Agricultural Engineers, 1993

Midwest Plan Service, Iowa State University, Ames, IA. http://www.mwpshq.org/

Natural Ventilating Systems for Livestock Housing, MWPS-33, 1989

Pork Facts 2002–2003, National Pork Board

Pork Industry Handbook, Cooperative Extension Service, Purdue University, West Lafayette, IN, 2002

Swine Breeding and Gestation Facilities Handbook, MWPS-43, 1983

Swine Farrowing Handbook: Housing and Equipment, MWPS-40, 1992

Swine Housing and Equipment Handbook, MWPS-8, 1983

Swine Nursery Facilities Handbook, MWPS-41, 1997

*Swine Wean-to-Finish Building*s, AED-46, 2000

18

Fitting and Showing Swine

4-H girls showing pigs at the Montgomery County Fair. *(Photo by Bill Tarpenning, Courtesy, USDA, Washington, DC)*

Objectives

After studying this chapter, you should:

1. Understand the influence (both positive and negative) of the show-ring on the development of types of pigs.
2. Know why some shows require performance as well as conformation.
3. Know why different feeding programs may be employed.
4. Know how to prepare a pig for exhibition.
5. Know why oiling, painting, and powdering show hogs is usually prohibited.
6. Be aware of the health requirements for exhibiting swine.

Through the years show-ring fashions in swine have fluctuated more radically than those in any other class of livestock. Fancy breed points and decisions made by judges are but passing fads when they conflict with the primary objective of producing swine, which is pork over the counter.

Comply with the rules of the show. Today, major shows do not permit the use of oil, paint, powder, or other dressing on market swine. Also, the use of unapproved drugs (drugs not approved by the FDA or the USDA) is forbidden. Exhibitors violating these rules have been barred from showing, sometimes for life.

ADVANTAGES OF SHOWING

Though not all exhibitors share equally in the advantages that may accrue from showing hogs, in general the following reasons may be advanced for exhibiting swine:

1. It serves as an available medium for molding breed type.
2. It gives breeders an opportunity to observe the impartial appraisal by a competent judge of their entries in comparison with others.
3. It offers an opportunity to study the progress being made within other breeds and classes of livestock.
4. It provides an excellent medium for advertising.
5. It gives breeders an opportunity to exchange ideas, thus serving as an educational event.
6. It offers an opportunity to sell a limited number of breeding animals.
7. It sets sale values for the animals back home, with values being based on the sale of show animals.
8. It provides a medium for 4-H and FFA members to become educated in one of the long traditions of the swine industry (Figure 18.1).
9. It should teach ethics, good sports competition, and professionalism.

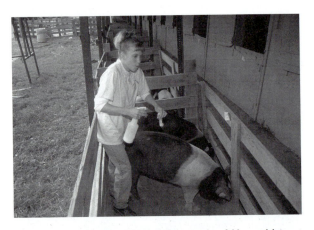

Figure 18.1 A 4-H girl readying a pair of Hampshires at the Prince George's County Fair, Maryland. *(Photo by Bill Tarpenning, Courtesy, USDA, Washington, DC)*

SELECTING SHOW ANIMALS

The first and most important assignment in preparation for the show is the selection of the prospective show animals. Unless the exhibitor has had considerable experience and is a good judge of hogs, it is well to secure the assistance and advice of a competent judge, breeder, or club leader when selecting the animals to be developed for the show-ring.

Many hog shows now combine visual live appraisal with growth rate and carcass merit. Selections need to be made as far in advance of the show as is possible. In fact, the show-ring date is usually kept in mind at the time matings are made or feeder pigs are purchased for a show. In general, all breeding animals intended for show, except those in the junior pig classes, should be selected at least 4 to 6 months in advance, thus allowing ample time for developing, fitting, and training.

Because some pigs may not develop properly, it is advisable to select larger numbers than are intended to show. In this manner, those animals that fail to grow rapidly and develop as expected may be culled and a stronger group for exhibition assembled.

Exhibitors may show (1) breeding animals, either boars or females, or (2) market hogs, either barrows or gilts. When possible and when the classifications are available, for purebred exhibitors it is desirable to show in both groups. In recent years, great emphasis has been placed on market hog shows, and winning in a strong market hog show is a fine accomplishment for the producer and breed/strain that produced the champion. Especially when a carcass evaluation is part of the contest, the type of winning pigs generally reflects consumer demands.

Exhibitors can enhance their chances of winning and of securing sufficient premium money to cover expenses by filling as many of the individual classes and groups as possible, and this applies to both breeding animals and market hogs. Provided that the animals are good, it costs little more and is usually good economy to have a sizable and well-balanced show herd.

Type

The animals in modern shows must possess adequate growth rate or size for age and be of lean meat type. They should be trim about the head and neck; the back should be moderately and evenly arched, and of adequate width; the loin should be wide and strong; the hams should be deep, thick, slightly bulging, and muscled well down to the hocks; the shoulders should be well laid in and smooth; the sides should be long, deep, and smooth; the legs should be set well apart, straight, and squarely on the corners; and the pasterns should be strong. With this special meat type, there should be a proper balance and blending of all parts, and the animals should be stylish and showy. An alert, active walk is a decided asset. Sex character, breed type, and adherence to distinctive breed characteristics are important for boars and sows.

In market hog shows special emphasis should be placed on trimness of middle and quality throughout as well as on the other characteristics mentioned. The ideal market pigs when ready for the show must possess superior conformation, finish, and quality. There should be the maximum development in the high-priced cuts and a minimum of fatness.

Group classes (pen of three, truckload, and so on) in both breeding and market classes should be as uniform as possible because group placings are determined by the merits and uniformity of the entire group rather than on the basis of 1 or 2 outstanding individuals therein.

Further information relative to breed characteristics and other factors of importance in making selections is available in Chapters 3, 4, and 5.

Show Classifications

Classifications for breeding animals are based on age. It is advantageous to select them as old as possible within the respective age classifications. Because the oldest should be heaviest in the class, they will show to the best possible advantage. Many shows include weight gain per day of age and either paint that value on the pig's rump or provide it to the judge on a sheet of paper. Barrow classifications are usually based on weight rather than age and many include a measure of growth rate. Some production hog shows will divide classes based on the starting weight of pigs that were weighed and entered 3 to 4 months before the actual show.

Many shows, in addition to the on-foot classes, are adding classes for market hogs that are judged both on-foot and in carcass (Figure 18.2). Placings are based on measures including backfat thickness, size of loin area, and rate of lean gain. Estimates of carcass quality including color and marbling of the loin eye muscle.

The weight divisions of different market hog shows vary considerably, usually specifying a minimum and maximum show weight for events where growth rate is not part of the contest. The total of the pigs entered are then divided into classes of approximately equal numbers based on weight.

Besides the breeding swine and market hog classifications, a few fairs may have classes for groups such as (1) produce of one sow, (2) get (offspring) of sire, (3) premier exhibitor, and (4) premier sire.

The swine classifications of four major livestock shows are presented in Table 18.1. As noted, shows vary greatly in classifications.

Breeding

Animals selected for the breeding shows must have a breed pedigree to be eligible. Ensuring that the animals are of proven ancestry adds assurance to the likelihood of producing offspring that will be similar in appearance to the parents. Breeding animals should show the distinctive breed characteristics. Purebred shows usually include market hog classes of both purebreds as well as for those sired by the respective breeds.

FEEDING AND HANDLING FOR THE SHOW

All animals intended for show purposes must be placed in proper condition, a process requiring great attention to details.

Figure 18.2 Catalogs for the Open Class, 4-H, and FFA shows at the 2003 Iowa State Fair. *(Courtesy, Palmer Holden, Iowa State University, Ames, IA)*

Managing the Pig before the Show

Prospective show animals should have quality quarters on the farm, and, above all, they should be kept cool, clean, and free from parasites. A certain amount of exercise is necessary to promote good circulation and to increase the thrift and vigor of the animal. Exercise tends to stimulate the appetite, making for greater feed consumption; keeps the animals sound on their feet and legs; and promotes firmness of fleshing and trimness of middle. Mature boars, sows, and market hogs may need to be walked, thus forcing exercise. During warm weather, it is best to exercise animals in the early morning or late evening, avoiding unnecessary handling in the heat of the day.

Segregate the Boars and Sows

More individual attention can be given, especially in the matter of feeding, if show hogs are handled in small groups. With junior pigs, the boars should be separated from the gilts when the pigs are 4 months old. At the time young show boars begin to rant, usually about 4 to 7 months of age, it is desirable that they be placed in isolated lots, preferably where they cannot see or hear other hogs. For the most part, show boars that have been raised to-

gether may be fed together but boars that come from different pens need to be penned separately at the show. Several gilts of similar age or weight can be penned together.

Suggested Diets

Any of the diets discussed in Chapter 7 for the respective classes and ages of swine are suitable for use in fitting show animals of similar stages of development. In general, however, instead of self-feeding most experienced exhibitors believe that they can get superior bloom and condition by either (1) hand-feeding or (2) using a combination of hand-feeding and self-feeding (hand-feeding twice daily and allowing free access to a self-feeder). When hand-feeding, mixing the diet with skimmed milk, buttermilk, or condensed buttermilk and feeding the entire diet in a form will enhance feed intake. This should also enhance daily gain.

Adding milk to a diet properly balanced does result in a higher protein content than necessary. In general, when skimmed milk or buttermilk is used, the dietary protein may be reduced by 1 or 2 percentage points without harm to the animal.

In fitting show hogs, it may be necessary to decrease or discontinue slop-feeding 2 to 4 weeks be-

TABLE 18.1 Swine Show Classifications of Four Major Livestock Shows

Show	Breeds	Open Class	Youth Classes	Youth Carcass
Houston, Texas, Livestock Show and Rodeo, 2003	Berkshire Chester White Crossbred Duroc Hampshire Poland China Spotted Yorkshire	None	Only 660 barrows will come to show. Market barrows must weigh 230–270 lb. Limited to TX 4-H or FFA. Allowed 1 market barrow and 2 breeding gilts. **Barrow Awards** Grand Champion Reserve Champion Breed Champion Reserve Breed 1st to 10th plus "All remaining placing barrows" Breeding gilts Farrow between July and Sept. 1 for Feb./March show **Awards** Supreme Champion Reserve Supreme Breed Champions Reserve Breed Champions 1st to 5th placings Showmanship, Junior, and Senior	1st and 2nd place barrows in each class will be in the carcass contest and awards for the top 10 **Carcass Standards** 152–214 lb 1. Carcass meat quality based on 2002 "Composition and Quality Assessment Procedures" of National Pork Board 2. Minimum of 47.0% fat-free lean 3. Cryptorchids are disqualified 4. Carcasses requiring more than 1% trim are disqualified
Iowa State Fair, Des Moines, IA, Aug. 2003	Berkshire Chester White Duroc Hampshire Landrace (youth only) Poland China Spotted Yorkshire Other breeds (youth only) Commercial gilt (youth)	**Breeding** Boars and gilts farrowed Dec.–March **Breed Awards** Champion Reserve Champion Class 1st to 15th Premier Sire Premier Exhibitor **Market Hogs** **Derby Show** Grand and Reserve Champions, Live and Carcass Class awards depend on number exhibited **Hawkeye** Purebred and crossbred classes, 230–280 lb Grand Champion Reserve Champion Breed and Cross Champion Reserve Breed and Cross Truckload Purebred and crossbred classes, 230–280 lb, 6 pigs, maximum of 3 gilts Grand Champion Reserve Champion 5 placings/class	**4-H Classes** Breeding gilts (Feb. and March) Purebred and Commercial Purple, blue, red, white, and premiums **Market Swine** Purebred and Crossbred classes (must have one derby pig/2 market swine) 225–280 lb Overall Champion Overall Reserve Champion Class Grand Champions Class Reserve Champions Ribbons and premiums **Derby Swine** (initial wt taken in April) Classes for barrows, gilts, and purebreds if more than 5 entered) **Live Awards** Grand Champion Reserve Champion (Grand and Reserve compete for overall in Market Swine) Ribbons and premiums **Showmanship** Three age divisions **FFA Classes** Similar to 4-H except purebred boars are also shown	**Derby Hogs** Top 100 pigs in lean gain/day are eligible for the carcass contest. NPPC quality guidelines are used and placings are based on lean gain/day. **Carcass Awards** Champion Reserve Champion 3rd to 10th placings

(continued)

TABLE 18.1   Swine Show Classifications of Four Major Livestock Shows (*continued*)

Show	Breeds	Open Class	Youth Classes	Youth Carcass
National Western Stock Show, Rodeo and Horse Show, Denver, CO, Jan. 2003	Crossbred Duroc Hampshire Yorkshire Other purebreds	None	**Awards** Grand Champion Reserve Champion Breed Champion Reserve Breed Champion Breed/class 1st to 11th placings	
North American International, Louisville, KY, 2002	Crossbred Duroc Hampshire Landrace Yorkshire	None	**Awards** Breed Champion Reserve Breed Champion Overall Grand Champion Overall Reserve Grand Champion	

fore the show to avoid paunchiness and a lower dressing percentage.

Feeding Guidelines

The general principles and practices of swine feeding are discussed in Chapters 7 and 8. The feeding of show hogs differs from the feeding of the rest of the herd primarily in that the former are fed for maximum development and bloom, with less attention being paid to economy from the standpoint of both feed and labor. It is to be noted, however, that it is a major disadvantage to have fat hogs for modern shows, thus rendering possible harm to breeding animals and making market hogs too fat. Rather, a firm, smooth finish is desired.

The most successful producers have worked out systems of their own as a result of years of practical experience and close observation. The beginner may well emulate their methods. Some rules of feeding show hogs as practiced by experienced producers are as follows:

1.   Type of show. Does the evaluation include only the visual appearance of the pig or will performance and carcass merit be part of the contest? The type of show will define the method of fitting the pigs.

2.   Practice economy, but avoid false economy. Although the diet should be as economical as possible, it must be remembered that proper condition is the primary objective, even at somewhat additional expense. Perhaps the most common mistake made in the fitting diet, especially of breeding animals, is the heavy feeding of corn or other grains.

3.   Hand-feeding versus self-feeding. Market hogs are more often self-fed than breeding animals. Many exhibitors prefer mixing the grain diet with

highly palatable skimmed milk, buttermilk, or condensed buttermilk, and feeding the entire diet as a liquid.

4.   Feed a balanced diet. A balanced diet will be more economical and will result in better growth and finish. Experienced producers usually prefer a diet that is on the narrow side (high in proteins). For assistance in selecting a diet, the reader is referred to Chapter 7.

5.   The diet must not be too bulky. Hogs cannot handle a great amount of bulk. Feeding too much bulk will reduce the growth rate, cause the animal to become paunchy, and lower the dressing percentage, a condition severely criticized in a show animal.

6.   Feed regularly. Animals intended for showing should be fed with exacting regularity. In the early part of the feeding period, 2 feedings per day may be adequate. Later, the animals should be fed 3 to 5 times a day, especially when rapidly growing pigs are the goal (Figure 18.3).

7.   Avoid sudden changes. Sudden changes in either the kind or amount of feed are apt to cause digestive disturbances. Any necessary changes should be gradual.

8.   Additives. Consider the use of feed additives to maintain and enhance the pigs' performance. A new additive, Paylean™ (Ractopamine·HCl), markedly increases growth rate and muscling and reduces backfat.

EQUIPMENT FOR FITTING AND SHOWING HOGS

In the fitting, training, and grooming operations, the essential equipment consists of brushes, clipper, rasp, sharp knife, soap, canes or whips, and hurdles. In loading for the fair, all of these items should be

Figure 18.3 Growing pigs on a well-adjusted feeder.
(Courtesy, Palmer Holden, Iowa State University, Ames, IA)

included in the show box. In addition, most exhibitors take water buckets, light troughs, a sprinkling can, a fork and a broom, oil and powder if permitted by the show rules, a saw, hammer, hatchet, a few nails, some rope, flashlight, blankets or a sleeping bag, and a limited and permissible supply of feed and bedding.

The show box, in which all smaller equipment is stored, is usually of durable wood construction, freshly painted, with a neat sign on the top or front giving the name and address of the exhibitor. The feed taken to the show should be identical to the diet that was fed at home.

TRAINING AND GROOMING

Most show-rings are cursed with the presence of too many squealing, scampering, and unmanageable pigs. Such animals make an adverse impression on ringside spectators, annoy other exhibitors, and fail to catch the eye of the judge. When competition is keen, there may be several well-bred and beautifully fitted individuals of the right type in each class, with the result that the winner will be selected by a very narrow margin. Under such circumstances, proper training, grooming, and showing is often a deciding factor.

Training the Pig for Exhibition

Proper training of the pig requires time, patience, and persistence (Figure 18.4). Such schooling makes it possible for the judge to see the animal to the best advantage. Some exhibitors prefer to use the whip, whereas others use the conventional producer's cane. The pig should be trained to respond to either one or the other (cane or whip) but not to both. With mature boars, it is good protection to use a hand hurdle. However, the use of a hurdle in showing young animals is usually an admission of either a mean disposition or a lack of training.

Long before the show, the exhibitor should study the individual animal, arriving at a decision as to the most advantageous pose. Then the animal should be trained to perfection so that it will execute the proper pose at the desired time. It is important that the pig be trained to stop when necessary, perhaps when the whip or cane is placed gently in front of its face. The direction of walking should be easily guided by merely placing the cane or whip alongside the animal's head.

The exhibitor should avoid making a pet of the pig. Such an animal may display a nasty disposition when placed in the show-ring, or it may slouch down when anyone comes near, including the judge. Also, animals should be trained to stand squarely on their feet and to keep their backs up and heads down.

Trimming the Feet

In order that the animal may stand squarely and walk properly and that the pasterns appear straight and strong, the toes and dew claws should be trimmed regularly. Moreover, long toes and dew claws are unsightly in appearance. Trimming can best be done with the animal lying on its side. Usually this position can best be acquired by merely stroking the animal's belly. With this procedure, tying and possible resulting injury is unnecessary.

The practice of standing the animal on a hard surface and cutting off the ends of the toes with a hammer and chisel gives only temporary relief and is not recommended. The bottoms of the toes should be trimmed. Proper trimming can best be accomplished by using a small rasp and a knife.

The outside toes of the rear feet grow faster than the inside toes. In extreme cases this condition may result in crooked hind legs. This condition may be corrected by trimming the outside toes more frequently.

The toes should be trimmed regularly, with some animals as often as every 6 weeks. Too much trimming at any one time, however, may result in lameness. For this reason, it is not advisable to work on the feet within 2 weeks before the show. In no case should trimming be so severe as to draw blood.

The dew claws should be cut back and dressed down neatly, making the pasterns appear short and straight.

Removing Tusks

Boars more than 1 year of age usually have tusks of considerable size. These should be removed a month or 2 before the show. To remove the tusks, tie the animal to a post with a strong rope, draw up the upper jaw, and use a hoof parer (see Chapter 15).

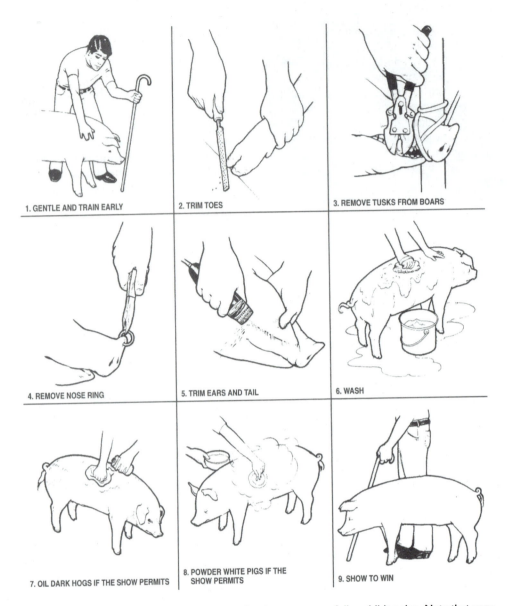

1. GENTLE AND TRAIN EARLY

2. TRIM TOES

3. REMOVE TUSKS FROM BOARS

4. REMOVE NOSE RING

5. TRIM EARS AND TAIL

6. WASH

7. OIL DARK HOGS IF THE SHOW PERMITS

8. POWDER WHITE PIGS IF THE SHOW PERMITS

9. SHOW TO WIN

Figure 18.4 Some essential tasks that need to be done to successfully exhibit swine. Note that very few shows still permit oiling or powdering. *(Courtesy of Pearson Education)*

Clipping

Clipping is usually done with hand clippers. It should be done a few days before the show. Many successful exhibitors prefer to use the clippers twice, about 2 weeks before the show and a second time the day before entering the ring. The usual practice is to clip the ears, both inside and outside, and to clip the tail. Clipping of the tail should begin above the switch and extend up to the tail head. The exact amount of switch to leave will vary somewhat with each individual, requiring judgment on the part of the exhibitor.

It may also help to remove long hairs from about the head and jowl. The udder of a gilt will often show to better advantage if some of the hairs on the belly are carefully removed. In all trimming, it is important that the hair be gradually tapered off so that the clipped and unclipped areas blend nicely. Remember to clip the pigs before arriving at the show as some prohibit clipping on the show grounds.

Washing

An occasional washing and brushing keeps the animal clean and makes the skin smooth and mellow. It also assists in shedding the coats of older animals. Lukewarm water, plenty of soap, and a fairly stiff brush will do the job. After the skin and hair have been dampened thoroughly, rub the hair with soap until suds form and then work this lather into the hide with the hands and brush. Clean all parts of the body thoroughly, using care not to get water into the ears. A small amount of bluing in the water is helpful in bleaching out stained spots on white hogs. Following washing, the animal should be rinsed off to remove all traces of soap from the hair and skin. The animal should then be placed in a clean pen to dry.

Oiling

Many hog shows forbid oiling pigs for several reasons. Pigs with too much oil get other animals and exhibitors oily. If a carcass contest is involved, oiled pigs often fail to dehair properly in the slaughtering process and have to be skinned, disqualifying them from competition.

Oiling softens the skin and hair and gives bloom to the coat. It is most important that oil be used sparingly and that it be evenly distributed over the body. For best results a light application of oil should be given the night before the show, using an oiled cloth or brush. Then give the animal a thorough grooming with a brush and a woolen rag just before entering the ring. When the pig enters the ring, surplus oil should not be in evidence. Any clear, light vegetable oil is satisfactory for oiling purposes. Paraffin oil is very satisfactory when mixed with small proportions of rubbing alcohol.

Oiling mature boars and sows that are in high condition may cause them to overheat more easily during warm weather. Under such conditions, some experienced exhibitors use no oil but sprinkle water over the animals before entering the ring, and then follow with the brush.

Feeding a diet containing 10 to 15% linseed meal for 30 days before the show will provide an excellent skin and hair condition.

Powdering

Many shows have eliminated powdering, and one should check the rules carefully before using powder. White hogs or white spots on dark hogs benefit from powdering with talcum powder or corn starch. The animal should first be thoroughly washed and allowed to dry. The powder should be dusted on before entering the ring. In order to be most effective, it should be evenly distributed.

MAKING FAIR ENTRIES

Well in advance of the show, the exhibitor should request that the fair manager or secretary forward the exhibition rules, premium list, and necessary entry blanks. Usually, entries close from 2 to 4 weeks prior to the opening date of the show. Production-oriented, or derby shows, require that pigs be nominated and weighed in 3 to 4 months before the show. All rules and regulations of the show should be studied carefully and followed to the letter—including requirements relative to entrance, registration certificates, vaccinations, health certificates, pen fees, exhibition and helper's tickets, and other matters pertaining to the show.

Most entry blanks for purebred hogs call for the following information: breed, sex, name and registration number of the entry, date farrowed, name and registration number of the sire and dam, a description of markings (such as ear notches), and the class in which the animal is to be shown. Registration certificates usually must be presented at the show for purebred classes. Entries should be made in all individual and group classes but not in breed specials or championship classes. It is not necessary to specify the identity of individuals constituting herds or groups because, when the exhibitor has a choice, the winnings in the individual class will largely determine this.

Providing Health Certificates

All fairs require that swine must be accompanied by a health certificate when they enter the exposition grounds. Although the stipulations vary, perhaps the following provisions of the State of Iowa are typical:

2004 HEALTH REQUIREMENTS FOR THE EXHIBITION OF LIVESTOCK, POULTRY AND BIRDS AT STATE FAIR AND DISTRICT SHOWS
(http://www.ipic.iastate.edu/prv/healthexhibit04.pdf)

SECTION 1 – GENERAL

A. All animals, poultry, and birds intended for exhibition within the State of Iowa will be considered under quarantine and not eligible for showing until the owner or agent presents a CERTIFICATE OF VETERINARY INSPECTION, stating the animals, poultry or birds are apparently free from symptoms of infectious or communicable diseases as determined on clinical inspection by an accredited veterinarian within 30 days (14 days for sheep) prior to date of entry to exhibition grounds, or

B. in certain classes, if the division superintendent has made prior arrangements with the official fair veterinarian to have all animals and/or birds inspected on arrival and prior to exhibition.

ANY EVIDENCE OF WARTS, RINGWORM, FOOT ROT, PINK EYE, DRAINING ABSCESSES, OR ANY OTHER CONTAGIOUS DISEASE WILL ELIMINATE THE ANIMAL FROM THE SHOW.

CONSULT YOUR STATE AND/OR COUNTY FAIR BOOK FOR ADDITIONAL OR SUPPLEMENTAL REGULATIONS.

SECTION 4 – SWINE

General

1. All swine must be individually identified on a Certificate of Veterinary Inspection and originate from herds or areas not under Quarantine. Plastic tags issued by 4-H officials can be substituted for an official metal test tag, when an additional identification (ear notch) is also recorded on the test chart and Certificate of Veterinary Inspection. All identification is to be recorded on the pseudorabies test chart and the Certificate of Veterinary Inspection.

Brucellosis

1. Native Iowa Swine—No brucellosis test required for exhibition purposes.
2. Swine from Out of State—All breeding swine six months of age and older must either:
   a. Originate from a Brucellosis Class "Free" state; or
   b. Originate from a brucellosis validated herd with herd certification number and date of last test listed on the Certificate of Veterinary Inspection; or
   c. Have a negative brucellosis test conducted within 60 days prior to show and confirmed by a state-federal laboratory.

Aujeszky's Disease (Pseudorabies)— All Swine

1. Native Iowa swine. Native Iowa swine originating from a Stage 4 or lower status county must present a test record and Certificate of Veterinary Inspection that indicate that each swine has had a negative test for pseudorabies within 30 days prior to the show (individual show regulations may have more restrictive time restrictions), regardless of the status of the herd, and that lists the individual official identification. Native Iowa swine originating from a Stage 5 status county must present a Certificate of Veterinary Inspection listing individual official identification. No pseudorabies testing requirements will be necessary for native Iowa swine originating from Stage 5 counties. Electronic identification will not be considered official identification for exhibition purposes.

2. Swine originating outside Iowa. All exhibitors must present a test record and Certificate of Veterinary Inspection that indicate that each swine has had a negative test for pseudorabies within 30 days prior to the show (individual show regulations may have more restrictive time restrictions), regardless of the status of the herd, and that lists the individual official identification. Electronic identification will not be considered official identification for exhibition purposes.

3. Swine that return from an exhibition to the home herd or that are moved to a purchaser's herd following an exhibition or consignment sale must be isolated and retested negative for pseudorabies not less than 30 and not more than 60 days after reaching their destination. (Code of Iowa 166D.13(2).)

2004 HEALTH REQUIREMENTS FOR EXHIBITION OF LIVESTOCK, POULTRY, AND BIRDS AT A COUNTY 4H/FFA FAIR
(http://www.agriculture.state.ia.us/2004exhibition2.htm)

ANY EVIDENCE OF WARTS, RINGWORM, FOOT ROT, PINK EYE, DRAINING ABSCESSES OR ANY OTHER CONTAGIOUS OR INFECTIOUS CONDITION WILL ELIMINATE THE ANIMAL FROM THE SHOW.

No individual Certificate of Veterinary Inspection will be required on animals or poultry exhibited at County 4-H/FFA FAIR, but the animals must be inspected when unloaded or shortly thereafter by an accredited veterinarian. Each show must have an official veterinarian.

Quarantined animals or animals from quarantined herds cannot be exhibited.

Swine exhibitors at county fairs that do not require a Certificate of Veterinary Inspection, must sign and present an owner affidavit that the animals being exhibited did not originate from a quarantined herd and to the best of their knowledge, swine dysentery has not been in evidence in their herd for the past 12 months.

All swine exhibited must be accompanied by a record of a negative pseudorabies test, the test having been performed within 30 days prior to show, for swine originating from a Stage 4 or lower status county, subject to 64.35(2). No pseudorabies testing is required for swine originating from a Stage 5 county.

Swine returning from an exhibition to its home herd or moved to a purchaser's herd, following an exhibition or consignment sale, must be isolated and retested negative for pseudorabies not less than 30 days and not more than 60 days after reaching the swine's destination. (Code of Iowa 166D.13(2))

Exceptions:

a. No testing is required for swine at an exhibition that involves only market classes, provided all swine are consigned directly to a slaughter establishment from the exhibition. The site that the swine originate from must have a current monitored status in order for the swine to be transported to the fairgrounds (statistical testing completed within the last twelve months or originate from a site in Stage III or higher area). Swine leaving the exhibition from a market class must be consigned and moved direct to a slaughtering establishment.

b. If counties have a split show and the breeding animals are exhibited and returned home before the market classes arrive, it will not be necessary to have a test record on the animals showing in the market classes; however, market class animals must have a current monitored status in order to be transported to the fairgrounds.

THE DECISION OF THE OFFICIAL SHOW VETERINARIAN WILL BE FINAL

TRANSPORTATION TO THE FAIR

Show hogs should be transported so that they arrive within the limitations imposed by the fair and preferably a minimum of 2 days in advance of showing. Because of the greater convenience and speed, most hauls are made by private truck or trailer. Public conveyances should always be thoroughly disinfected before loading hogs. Show hogs should not be crowded, and those of different age groups and sexes should be separated by suitable partitions. During warm weather, sand should be wetted down before loading and while en route. Also, hogs may be drenched when necessary, but water should never be applied to the backs of hot hogs. Regardless of the method of transportation, animals should be fed lightly (about a half ration) just prior to and during shipping. A heavy fill is likely to result in digestive disturbances and overheating in warm weather.

PEN SPACE, FEEDING, AND MANAGEMENT AT THE FAIR

Most swine pens at fairs or exhibitions range from 6 to 8 ft sq. When it is not too hot and the hogs are used to each other, the following numbers of the

Figure 18.5 Three 4-H members penning their pigs at the Prince George's County Fair, Maryland. *(Photo by Bill Tarpenning, Courtesy, USDA, Washington, DC)*

same sex can be accommodated in one pen: 3 to 5 junior pigs or market hogs, 2 seniors, and 2 each of the older age groups—except that boars older than junior pigs had best be kept in separate pens. Sufficient pens should be obtained to avoid overcrowding, especially during warm weather. It is easier to keep the animals clean when there is ample space.

The advertising value of the exhibit will be enhanced through displaying a neat and attractive sign over the pens, and giving the name and address of the breeder, farm, or ranch. It is also important that the pens and alleys be kept neat, clean, and attractive at all times, impressing spectators, exhibitors, and fair management with your desire for a good show.

Following a day of rest and light feeding after unloading at the show, normal feeding may be resumed provided that it is accompanied by exercise. Usually, the animals are fed twice daily. Most exhibitors prefer to clean the pens thoroughly in the early morning, while the animals are confined with hurdles for feeding or are being taken for exercise. It is important that all animals receive sufficient and regular exercise while they are on the fairgrounds (Figure 18.5).

The final washing may be given a day or 2 after arrival, but the coat should be brushed daily.

SHOWING THE PIG

Expert showmanship cannot be achieved through reading any set of instructions. Each show and each ring will present unusual circumstances. However, there are certain guiding principles that are always adhered to by the most successful exhibitors, including the following:

1. Train the animal well, long before entering the ring.
2. Have the animal carefully groomed and ready to parade before the judge.

3. Dress neatly for the occasion. Some youth shows specify dress. Check the rules.
4. Enter the ring promptly when the class is called.
5. Do not crowd the judge.
6. Be courteous and respect the rights of other exhibitors.
7. Work in close partnership with the animal.
8. Keep your animal in view at all times.
9. Do not allow your hog to fight or bite other animals.
10. Always be showing. Keep one eye on the judge and the other on your pig. The animal may be under the observation of the judge when you least suspect it.
11. Keep calm, confident, and collected. Remember that the nervous exhibitor creates an unfavorable impression.
12. Be a good sport. Win without bragging and lose without squealing (Figure 18.6).

Figure 18.6 4-H member with his blue ribbon Landrace gilt at the Prince George's County Fair, Maryland. *(Photo by Bill Tarpenning, Courtesy, USDA, Washington, DC)*

AFTER THE FAIR IS OVER

Before an exhibitor can leave the show grounds, it is customary to require a signed release from the superintendent of the show. Immediately prior to that time, all of the equipment should be loaded, followed by loading the hogs. The same care and precautions that applied in travel to the show should prevail in the return trip.

Because of the possible disease and parasite hazards resulting from contact with other herds and through transportation facilities, quarantine the show herd for a period of 4 weeks following return from the fair. Consider blood tests for diseases of concern before allowing reentry into the herd.

QUESTIONS FOR STUDY AND DISCUSSION

1. Do you approve or disapprove of the rules of some shows not permitting the use of oil, paint, powder, or other dressing on market swine; and of forbidding the use of drugs not approved by the FDA? Defend your position.
2. Discuss the similarities and the differences in the swine show classifications of the major livestock shows presented in Table 18.1. What do you like or dislike about them?
3. Why or why not should a purebred pork producer, a commercial pork producer, and an FFA or 4-H club member show or not show hogs?
4. Defend either the positive or the negative position of the following statements:
   a. Fitting and showing does not harm animals.
   b. Livestock shows have been a powerful force in swine improvement.
   c. Too much money is spent on livestock shows.
   d. Unless all animals are fitted, groomed, and shown to the same degree of perfection, show-ring winnings are not indicative of the comparative quality of animals.
5. How may livestock shows be changed so that they (a) more nearly reflect consumer preference and (b) enhance swine improvement?
6. It is generally agreed that livestock shows caused the radical shifts in swine types of the past, some of which later proved to be detrimental. How could this have been averted?
7. How does feeding for show differ from commercial feeding for market?
8. Why do the health certificates of most fairs detail the requirements relative to brucellosis and pseudorabies?
9. What advice would you give someone showing swine for the first time?

SELECTED REFERENCES

Pork Industry Handbook, Cooperative Extension Service, Purdue University, West Lafayette, IN, 2002

Stockman's Handbook, The, 7th ed., M. E. Ensminger, Interstate Publishers, Inc., Danville, IL, 1992

Various state 4-H Club Extension Service Offices

4

Business Skills

19

Marketing Hogs

Swine carcasses in a cooler. The carcasses are chilled about 24 hours before cutting. *(Courtesy, Swift and Company, Greeley, CO)*

Livestock marketing embraces those operations beginning with loading animals out on the farm and extending until they are sold to go into processing channels.

Marketing—along with breeding, feeding, and management—is an integral part of the modern livestock production process. It is the end of the line; that part which gives point and purpose to all that has gone before. Market receipts constitute the only source of reimbursement to the producer; market day is the producer's payday—hence, it is the most important single day of the operation. The importance of hog marketing is further attested by the following facts:

1. In the past 10 years (1994–2003), an average 97,608,130 hogs were slaughtered annually in the United States (USDA).
2. In the past 10 years (1992–2001) U.S. farmers received an average 11.8% of their cash farm marketing income of livestock and products from hogs (PorkFacts 2002–2003).
3. Livestock markets establish values of all animals, including those down on the farm. On December 1, 2003, there were 60,040,000 hogs in the United States, with an aggregate value of $4,712 million, or $78 per head (USDA-NASS).

THE CHANGING HOG MARKET

In the good old days, farm produce marketing was relatively simple. On Saturdays, the farmer headed to town and sold to the corner produce store a basket of eggs, a gunny sack of old hens or fryers, and a jar of sour cream. Surplus market hogs or feeder pigs (Figure 19.1) were usually sold to local buyers. The farmer did little figuring as to the best time to sell animals; the chief worry was in growing rather than in selling. The farmer could be successful by knowing how to breed, feed, and manage stock. Today, this is not enough; preconsidered, if not prearranged, marketing plans are essential.

Consumer preference has dictated, and will continue to dictate, changes in market hogs. In recent years, the consumer has been demanding (1) less lard, (2) smaller cuts with less fat and more lean, and (3) a larger proportion of the valuable cuts. These requirements are met by meat-type hogs (Figure 19.2). Further, from the producers' standpoint, there is a savings in feed in marketing meat-type hogs. The feed costs of production of fat are about 2.5 times greater than those for lean.

• **Standard of excellence in hog type**—In order to meet consumer preference, the National Pork Board has provided a description of the ideal market pig (known as Symbol II, see Chapter 4) for the year 2,000—and beyond. It is a pig marketed at 260 lb, with a carcass weight of 195 lb, grown from a ter-

Figure 19.1 Selling a wagon load of feeder pigs about 1900. *(Courtesy, USDA, Washington, DC)*

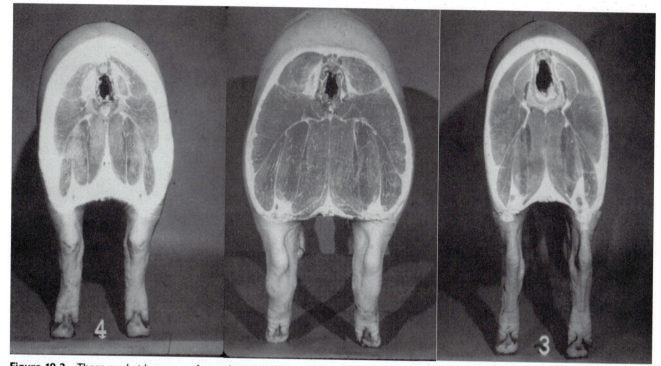

Figure 19.2 These market hogs were frozen in a standing position and then cross-sectioned through the ham. The left pig is an extremely fat pig with very little muscle and a lot of external and internal fat. The center cross-section typifies an ideal pig with a large volume of muscle mass and very little external fat. The right hog is a "meatless wonder," no more fat than the well-muscled pig, but very little lean meat. *(Courtesy, College of Agriculture, Iowa State University, Ames, IA)*

minal crossbreeding program and maternal line capable of weaning 25 pigs per year, free of the halothane or "stress" gene (Chapter 4).

MARKET CHANNELS FOR SLAUGHTER HOGS

Hog producers are confronted with the perplexing problem of determining where and how to market their animals. Usually, there is a choice of market outlets, and the one selected often varies between classes and grades of hogs and among sections of the country. Thus, the method of marketing usually differs between slaughter hogs and feeder pigs, and both of these differ from the marketing of purebreds to be used as breeding animals. Most hogs and pigs are sold through the following channels:

1. Terminal or central markets.
2. Auction markets.
3. Direct to buyer (including country dealers).
4. Carcass grade and yield basis.
5. Slaughter hog pooling.
6. Packer marketing contract.
7. Feeder pig marketing.
8. Segregated early weaned (SEW) pig marketing.
9. Futures markets (see discussion in Chapter 21)

There are other methods of marketing, but the methods listed are the primary ones.

Note: The listing and discussion of market channels for hogs is in the historical order and importance in which they evolved, rather than in the current annual volume of hogs marketed through them.

Terminal or Central Markets

Terminal markets (also referred to as terminals, central markets, public stockyards, and public markets) are livestock trading centers that generally include several commission firms and an independent stockyards company. Up through World War I, the majority of slaughter livestock in the United States was sold through terminal public markets by farmers or local buyers shipping to them. Since then, the importance of these markets has dramatically declined in relation to other outlets.

• **Leading terminal hog markets**—As would be expected, the leading hog markets of the United States were located in or near the Corn Belt—the area of densest hog population. As the truck replaced the railroad and with improved highways, the pork packing plants were moved nearer the areas of hog raising. Thus, during the last half of the

TABLE 19.1   Receipts at Leading Terminal Hog Markets

Market	1984	1994	2000
	(1,000 head)		
National Stockyards, East St. Louis, IL	1,065		
South St. Paul, MN	960	406	203
South St. Joseph, MO	669	451	59
Omaha, NB	837		
Sioux City, IA	1,230		
All others reporting	<u>4,175</u>	<u>1,368</u>	<u>998</u>
U.S. Total[1]	9,139	3,018	1,260
Direct receipts, interior Iowa and southern Minnesota[2]	23,116	28,669	36,504

[1]Number of stockyards varies from 41 to 68.
[2]Data from 14 packing plants and 30 concentration yards.
Source: Agricultural Statistics 1994, *2003, USDA-National Agricultural Statistics Service.*

1990s, local, or interior, packers increased in numbers. In order to meet this added competition, the large packers at more distant points resorted to direct buying and the purchase of interior plants (Table 19.1).

The rank of the major hog markets shifted considerably. One drastic example is the Union Stockyards in Chicago, which was the leading terminal market for years; in 1970, the Union Stockyards stopped receiving hogs and eventually closed. In 1984 the terminal markets received almost 40% of the Corn Belt hogs at 5 major terminals. In 2000 only 2 of the major terminals remained, South St. Paul and South St. Joseph, and the terminals only accounted for about 3.5% of the Corn Belt hogs.

Auction Markets

Auction markets (also referred to as sales barns, livestock auction agencies, community sales, and community auctions) are trading centers where animals are sold by public bidding to the buyer who offers the highest price per hundredweight or per head. Auctions may be owned by individuals, partnerships, corporations, or cooperative associations.

This method of selling livestock in the United States is very old, apparently based on a system from Great Britain where auction sales date back many centuries.

The auction method of selling was used in many of the colonies as a means of disposing of property, imported goods, secondhand household furnishings, farm utensils, and animals. Ac-

cording to available records, the first U.S. public livestock auction sale was held in Ohio in 1836, by the Ohio Company, whose business was importing English cattle.

Livestock auction markets had their greatest growth since 1930, both in numbers established and the extensiveness of the area over which they operated. Their most rapid growth occurred between 1930 and 1937. Today, most auction markets are limited to feeder pig and feeder cattle sales. In the case of pigs, they often are sold only at auction and do not actually appear in the sale ring, but are shipped directly from the seller's farm to the buyer's farm.

Several factors contributed to the growth in auction markets, some of which were the following:

1. The decentralization of markets, the improvement of hard-surfaced roads, and the increased use of trucks to transport livestock.
2. The desire to obtain a competitive bid for the animals being sold.
3. Improvements in the collection and dissemination of market news.
4. The convenience afforded in disposing of small lots of livestock and in purchasing stockers, feeders, and breeding animals.
5. The desire to sell near home.

Prior to the advent of local livestock auctions, small livestock operators had two main market outlets for their animals—(1) shipping them to the nearest terminal public market or (2) selling them to buyers who came to their farms or ranches. Generally, the first method was too expensive because of the transportation distance involved and the greater expense of shipping small lots. The second method put producers at the mercy of buyers because they had no good alternative to taking the price the buyers offered, and often they did not know the value of their animals.

The auction market method of selling is similar to the terminal market in that both markets (1) are an assembly or collection point for livestock being offered for sale, (2) furnish or provide all necessary services associated with the selling activity, (3) are supervised by the federal government in accordance with the provisions of the National Packers and Stockyards Act, and (4) are characterized by buyers purchasing their animals on the basis of visual inspection.

But there are several important differences between terminals and auctions; among them, auction markets (1) are not always terminal with respect to livestock destination; (2) are generally smaller; (3)

are usually single-firm operations; (4) sell by bid, rather than by offer and counteroffer; and (5) are completely open to the public with respect to bidding, and all buyers present have an equal opportunity to bid on all livestock offered for sale, whereas the terminal method is by private treaty (the negotiation is private).

Livestock auctions are really of greatest importance to the small operator. Also, auctions are an important outlet for feeder pigs; thus, a large proportion of the hogs sold through auctions go back to the farm. Auctions are also popular at breeding livestock shows as a means of promoting and selling prize animals. Additionally some purebred breeders will have an annual or semiannual auction of their animals.

Direct Selling

Direct selling, including selling to country dealers, refers to producers selling hogs directly to packers or local dealers, without the support of commission firms, selling agents, buying agents, or brokers. Direct selling does not involve a recognized market. The selling usually takes place at the farm, a nonmarket buying station, or collection yard. Some country buyers purchase livestock at fixed establishments similar to packer-owned country buying points.

Direct selling is similar to terminal market selling with respect to price determination; both are by private treaty and negotiation. But it permits producers to observe and exercise some control over selling while it takes place, whereas consignment to distant terminal markets usually represents an irreversible commitment to sell. Larger and more specialized hog producers feel competent to sell their livestock direct.

Prior to the advent of public stockyards in 1865, country selling accounted for virtually all sales of livestock. Sales of livestock in the country declined with the growth of terminal markets until the latter method reached its peak at the time of World War I. Country selling was accelerated by the large nationwide packers following World War I in order to meet the increased buying competition of the small interior packers.

Improved highways and trucking facilitated the growth of direct selling. Farmers were no longer tied to outlets located at important railroad terminals or river crossings. Livestock could move in any direction. Improved communications, such as the radio and telephone, and an expanded market information service also aided in the development of direct selling of livestock, especially in sales direct to packers.

Today, most slaughter animals are sold directly to the slaughter plants. Many go through a nearby collection point. Only a few pigs are still priced on a live basis; most are sold on a grade and yield basis.

Carcass Grade and Yield Basis

Carcass selling is commonly called grade and yield selling. All hogs should be evaluated by a practical, accurate, and objective measure of carcass composition. Any of several measures may be used, among them, (1) backfat rulers, (2) lean meters, (3) ultrasound, or (4) electrical conductivity. The backfat rulers were once the primary means of estimating carcass lean and the backfat depth has a high correlation with lean. However, it was done by a human measuring the fat on the split carcass, usually at the last rib, as it was easy to locate. Using the ruler incurred the risk of recording errors.

The lean meter advanced technology in two ways. It was probed through the backfat and lean at the 10th or last rib, so it included the depth of the loin muscle in addition to the fat depth. This added some precision to the estimation of carcass lean. Also, the measurement was recorded electronically, minimizing the errors in data collection (Figure 19.3). Ultrasound has the same benefits as the lean meter and has the potential of adding a cross section of the loin muscle. Ultrasound is not commonly used in slaughter houses.

Electrical conductivity is slowly gaining acceptance in packing plants as a more precise measurement of carcass lean and can actually sort carcasses to different processing lines by providing data on the musculature of different parts of the carcass. Its major limitation is that it has difficulty handling

Figure 19.3 Measuring backfat and loin depth with an electronic probe in a Danish slaughter house. *(Courtesy, Palmer Holden, Iowa State University, Ames, IA)*

carcasses that move at the high speeds on current slaughter lines, often more than 900 per hour.

In addition to physical carcass measurements, pH is often recorded as in indication of carcass muscle quality, primarily for PSE. It currently is not being used as part of the price structure but producers marketing pigs that have low carcass pH values are being informed that the quality of their pigs is lacking. (See the Chapter 5 section headed "Mechanical Measures.")

Grade and yield marketing hogs are becoming more common in countries with large hog operations and large central slaughter facilities. The bargaining is in terms of the price to be paid per hundred pounds dressed weight for carcasses that meet certain grade specifications. It is a more accurate and unassailable evaluation of a carcass than attempting to predict the carcass merits by observing the live hog. Even producers still selling on a live basis are, in reality, paid on what the buyer thinks those pigs will do on a carcass basis based on the past history of the seller.

In general, hog producers who market superior animals benefit from selling on the basis of carcass grade and weight, whereas the producers of lower-quality animals usually believe that this method unjustly discriminates against them. In countries where rail grading has been used extensively, there has been an unmistakable improvement in the breeding and feeding of swine.

The factors favorable to selling on the basis of carcass grade and weight may be summarized as follows:

1. It encourages the breeding and feeding of quality hogs.
2. It provides the most unassailable evaluation of the product.
3. It allows the packer to trace losses from condemnations, bruises, soft pork, and diseases to the producer responsible for them.
4. It is the most effective approach to animal improvement.

Note: It is not live animals, but pounds of lean on the rail that counts.

The factors unfavorable to selling on the basis of carcass grade and weight are as follows:

1. From the standpoint of the meat packer, there is less flexibility in the operations.
2. There is delay in returns to the seller. In the United States the Packers and Stockyards Act requires producers be paid within 24 hours of slaughter. Sellers of live pigs are usually paid immediately on conclusion of the transaction.
3. The seller usually assumes the risk and loss of totally or partially condemned carcasses.

• **USDA guidelines**—The U.S. Department of Agriculture guidelines, wherein meat packers buy livestock on the basis of carcass grade, carcass weight, or a combination of the two, are as follows:

1. Make known to the seller, before sale, significant details of the purchase contract.
2. Maintain identity of each seller's livestock and carcass.
3. Maintain sufficient records to substantiate settlement for each purchase transaction.
4. Make payment on the basis of actual carcass weight as it leaves the kill floor.
5. Use hooks, rollers, gambrels, and similar equipment of uniform weight in weighing carcasses of the same species of livestock in each packing plant.
6. Make payment on the basis of USDA carcass grades or furnish the seller with detailed written specifications for any other grades used in determining final payment.
7. Grade carcasses by the close of the second business day following the day of slaughter. In practice, most U.S. pigs are graded as they leave the kill floor before entering the coolers.

It is generally agreed that there is need for a system of marketing that favors payment for (1) pounds of lean or percent of lean in the carcass; and (2) the maximum yield in the four primal cuts: the ham, loin, picnic shoulder, and Boston butt, which account for three-quarters of the value of the entire carcass. Selling on the basis of carcass grade fulfills these needs. Producers need the following information on every hog they sell:

1. Carcass identification.
2. Carcass weight.
3. A quality score.
4. Pounds of lean in each carcass, determined by an objective measure.

Slaughter Hog Pooling

Slaughter hog pooling, consisting of a number of smaller hog producers joining together for the purpose of marketing hogs in truckload lots, is generally prompted because of lack of access to a nearby market. Typically, a person is hired to manage or coordinate the pooling program.

Pooled slaughter hogs are individually identified, usually by tattoos on their backs, to each owner. Then they are commingled for shipment with all hogs on a single truckload or compartmentalized on the truck to prevent stress. In either case, the producers are charged a fee by the marketing association for services rendered.

Pooling slaughter hogs from small producers into truckload lots and timing deliveries to fit buyers' needs increases marketing efficiency and usually increases sale prices because pool managers can offer for sale a larger volume of pigs.

Successful slaughter hog pooling must be customer oriented. Hog producers should understand the needs and wants of different packers. For example, some packers may pay premiums for heavier-than-normal pigs or at least not severely discount heavy hogs when compared to other packers. The National Pork Board several years ago "marketed" various loads of "make-believe" pigs on several marketing programs in the United States. No single packer consistently paid the highest price for each lot. Thus, the manager needs to be able to sort the pooled pigs to maximize the producers' returns.

Packer Marketing Contracts

A hog marketing contract is an agreement between a seller (usually a producer) and a meat packer to sell/buy at a specified future date a specified number of hogs of a specified weight and grade for a certain price. The most common hog contracts are packer marketing contracts. Typically, the following terms are included in a packer marketing contract:

1. Quantity. The quantity to be delivered, with the minimum amount varying anywhere from 5,000 to 40,000 lb (the Chicago Mercantile Exchange contract is for 40,000 lb of lean value hog carcasses).

2. Location and date of delivery. The delivery location and date are usually specified with the time of delivery within a specified time interval, and with provision to change time of delivery by mutual agreement.

3. Acceptable weights and grades. Additionally, there may be provision for premiums and discounts. For example, the packer may require that all hogs be at least 50% lean. Some contracts now price hogs on a carcass grade and yield basis, which rewards better producers.

4. Pricing arrangement. The most popular contract sets a floor and a ceiling for the duration of the contract. This type of contract would be made directly with a packer. For example, the contract might specify a floor of $38 per cwt and a ceiling of $48 cwt. If the cash market stays within the range, the producer gets the going cash market when the hogs are delivered. But if the market drops to $34, the producer and the packer split the difference between the floor set in the contract and the actual cash market; in this case, the producer would get $36. Should hog prices rise above the $48 ceiling, the same rule holds; the producer and the packer split the difference, so hogs sold on a $52 cash market would return $50 to the producer.

Some packer contracts specify fixed ceiling and floor prices, without provision to split the difference. Thus, in the preceding example, the producer would never take less than $38 or get more than $48 regardless of how low or high actual hog prices might go.

Both of these types of contracts may include a formula for recalculating the price if the cost of production changes, especially as feed prices rise or fall. Thus, in the Midwest where corn and soybeans account for about 60% of the cost of production, the contract may call for adjustment in case corn and soybean meal prices go up or down.

5. Duration of contracts. Contracts are usually long term—5 to 7 years—because bankers and other financiers may require long-term marketing arrangements before financing facilities. In some respects, a 5- to 7-year contract is a gamble. The producer is betting hog prices will drift lower in the next few years, whereas the packer is betting prices will gradually move higher.

6. Some financial consultants advise producers against putting their entire hog output under contract. These specialists advise producers to contract enough hogs (a) to be sure a bad market will not ruin them and (b) to keep their bankers satisfied. But they recommend that some hogs (perhaps 40 to 50%) be without a contract in case prices advance. This allows producers some breathing space in case they are hit with unusually large pig losses (for example, an outbreak of TGE) during the contract.

7. Future of packer contracting. Packer market hog contracting now dominates the scene. The increasing number of hogs committed to a packer before delivery are shown in Table 19.2. Currently, more than 80% of market hogs are not priced on a negotiated competitive market. Because all producer-packer contracts base the contract price on the daily market bids, there is a valid concern on how the true value of the daily market is determined when less than 20% of the hogs are competitively purchased.

8. Miscellaneous provisions of packer marketing contracts. Typically, marketing contracts also make provisions for (a) nonacceptable hogs and carcasses, (b) credit arrangement, (c) buyer inspection of the hogs while on the seller's premises, and (d) breach

TABLE 19.2   The Changing Structure of Pork Marketing

	1999	*2000*	*2001*	*2002*
	\multicolumn{4}{c}{*Percentage of U.S. Hogs Sold through Various Pricing Arrangements, January 1999–2002*}			
Hog or meat market formula	44.2	47.2	54.0	44.5
Other market formula	13.2	20.8	21.9	11.8
Other purchase arrangement	4.6	4.6	6.6	8.6
Packer-sold				2.1
Packer-owned				16.4
Negotiated—spot	35.8	25.7	17.3	16.7

Source: 1999–2001 data based on industry survey, University of Missouri and National Pork Board; 2002 data based on USDA Mandatory Reports.

Figure 19.4   A group of feeder pigs on plastic slats. *(Courtesy, Palmer Holden, Iowa State University, Ames, IA)*

of contract. Before signing a contract, producers are admonished to read the stipulations of the contract with care. It may be wise to have an attorney review the terms to completely understand them.

9. Concerns. If packers have 70% or more of their kill under contract, what impact will 5- to 7-year contracts have on the cash market? If a minority of the hogs marketed (e.g., 25%) set the cash market, will they be hogs of lesser quality? Then, would these few hogs of lesser quality be used to calculate prices for the contract hogs?

FEEDER PIG MARKETING

Feeder pig marketing is an important part of the swine industry and farms producing feeder pigs are getting larger. Also, through market price, packers are signaling pork producers that high-quality, consistent, and lean slaughter animals will receive significant premiums. In turn, this places increasing pressure on feeder pig producers to improve the quality, uniformity, and health of feeder pigs (Figure 19.4). The most prevalent marketing methods for feeder pigs are the following:

1. Direct. Direct marketing consists of a feeder pig producer marketing to a feeder pig finisher. Price may either be negotiated or determined by long-term contractual arrangement based on a formula price.

2. Public auctions. Pigs may be sold as delivered or graded and pooled for sale to buyers. Pigs are sold to the highest bidder, with price determined by demand.

Electronic feeder pig auctions have gained increasing popularity in recent years. Feeder pigs offered for sale are scored on the basis of health

programs, feeding programs, and herd management. This information, along with identification of the seller, is made available to prospective buyers via computer modems. Consigned feeder pigs can be reviewed by potential buyers who may participate in an open auction as each pen is offered for sale. No commingling or central collection of the feeder pigs occurs. Pigs move directly from the seller to the buyer. Feeder pigs that are not accurately represented, or that do not meet weight and grade specifications, are either price adjusted at delivery or rejected by the buyer.

Other innovative feeder pig marketing methods include telephone marketing and video marketing. In order to reduce exposure to disease when feeder pigs are sold live in a public auction ring or terminal market, it is predicted that the use of electronic auctions of feeder pigs will increase.

• **Feeder pig price, supply, and demand**—Feeder pig prices are volatile, reacting to numerous economic factors, supply, and demand. Thus, it is noteworthy that, during a 10-year period, feeder pig prices varied from less than $15 per head to more than $50 per head.

Feeder pig demand is influenced by numerous factors (Figure 19.5). Profit expectations of feeder pig buyers is the major pricing factor. Expected market hog prices at the time the feeder pigs will be slaughtered are an important part of this consideration. The lean hog future price for contracts maturing near the expected slaughter hog marketing date serve as a barometer of future cash slaughter hog market price levels. Thus, increases in deferred live hog future prices generally signal increased cash feeder pig prices. Any economic factor that affects market hog prices, including pork production, production of competing meats, meat exports and imports, consumer income, and

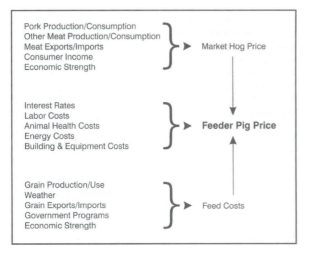

Figure 19.5 Factors affecting feeder pig prices. *(Courtesy, Pork Industry Handbook, Purdue University, West Lafayette, IN)*

the strength of the economy exert indirect influences on feeder pig prices.

The second factor affecting feeder pig demand and prices is feeding costs. Feeding costs are affected both by the cost of the feed itself and the efficiency with which the animal can turn these inputs into salable pork. Increases in feeding costs reduce feeder pig demand driving prices down. As a result, weather affecting feed grain and soybean meal prices can have dramatic influences on feeder pig prices. Characteristics of the pig itself, such as its genetics and health status plus the environment within which the pig will be finished, also affect feeding costs and thereby feeder pig prices.

Other costs affecting feeder pig costs include interest rates, labor costs, routine health costs, energy costs, and buildings and equipment. One method to project the expected influence of changes in feed costs and expected slaughter hog prices on feeder pig prices is to develop feeder pig finishing budgets to determine break-even feeder pig purchase prices for different feeding costs and slaughter hog prices. These may be obtained at low costs from various extension services in the United States. Actual feeder pig prices usually follow break-even prices (perhaps with a profit adjustment).

• **Feeder pig price information**—Daily feeder pig price information is limited in many regions. Producers often must rely on weekly auction quotations, terminal market sales reports, cooperative market reports, releases from agribusiness firms, electronic wire service reports, and periodic government releases. The percentage of feeder pigs being marketed directly from feeder pig producers to hog finishers in private treaties reduces data availability and, therefore, how representative open market

feeder pig price data may be. A partial listing of feeder pig auctions may be found on the Agricultural Marketing Service Web site (*http://www.ams. usda.gov/lsmnpubs/txPFAuction.htm*).

Feeder pig prices usually are quoted on a per head basis or, occasionally, on per hundredweight (cwt) basis. Interpretation of price reports requires understanding the market conditions, location, terms of sale, grade standards, number of pigs represented by price quote, feeder pig quality, and preconditioning and vaccination backgrounds of the pigs.

As an alternative to using a competitive bid, some feeder pig producers use a formula to establish the sale price for their pigs. Making a formula pricing system work is very difficult because of changing price relationships among slaughter hogs, hog futures, feed prices, interest rates, and other production costs. Formula prices will rarely exactly match market prices. When the formula price is below the market price, the pig seller often feels shortchanged. When the formula price is above the market, the pig buyer is likely to feel disadvantaged. For these reasons, formula pricing works best when the feeder pig producer deals with the same buyer(s) over an extended period. The most successful formulas are those based on an established feeder pig market.

• **Guidelines for selling feeder pigs**—Historically, most feeder pig producers have been able to rely on organized public markets as an option for selling some or all of their pigs. For a variety of reasons, the number of organized feeder pig auctions has declined greatly in the past decade and is likely to continue to drop. This means that direct sales to finishers may become the only option available to many feeder pig producers.

Direct and auction sales place importance on lot size, pig uniformity, genetics, and health status. It is clear that feeder pig buyers place a significant premium on large lots of uniform pigs from a single producer. Lot sizes have a direct relationship on price per pig (Figure 19.6). Larger lot sizes, especially 10- and 50-lb feeder pigs, receive higher prices per pig. The smallest lots (less than 250 pigs) receive the lowest price per pig.

The seller should work with the buyer and the buyer's veterinarian to ensure that the health status of the pigs delivered meets the demands of the buyer. Selling feeder pigs to an individual buyer allows the seller to refine and customize a health program to a buyer's needs. Some general guidelines for selling feeder pigs follow:

1. Provide healthy pigs of uniform size and quality. Remove the "bottom-sort" or "tail-enders" from the

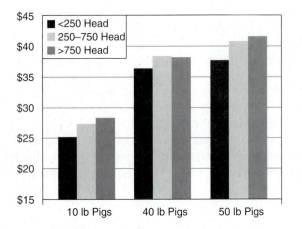

Figure 19.6 Effect of lot size on price paid per head for different weights of feeder pigs for Iowa and Central U.S. Direct Delivered Feeder Pig Data, 2002, 50–54 Percent Lean. *(Courtesy, Bruce Thomas, USDA-Iowa Department of Agriculture, Market News, Des Moines, IA)*

group. All-in/all-out nursery production is critical to overall disease control and helps with uniformity of pigs.

2. Provide pigs within a narrow age spread (2 weeks or less). This helps reduce the transmission of disease organisms, especially respiratory diseases, from older to younger pigs.

3. Dock pigs' tails and castrate male pigs well in advance of delivery so they are healed by the time the buyer receives them. Pigs with physical defects, such as hernias or other blemishes, should not be delivered.

4. Establish parasite control practices in the seller's breeding herd to ensure the delivery of pigs free of internal and external parasites.

5. Market healthy feeder pigs. The parent herd should have no history of *Actinobacillus pleuropneumonia* (APP) or swine dysentery. The pigs should be free of pseudorabies (PRV) or their status should be known. State laws vary on movement of PRV-infected pigs, depending on the status of each state in a national eradication program.

6. Vaccinate pigs against erysipelas. If the parent herd has a history, or if the buyer requests it, atrophic rhinitis and APP vaccinations may be necessary.

7. Supply the herd history and background information to the buyer and the buyer's veterinarian. A health certificate must be provided by the seller's veterinarian.

Some guidelines for selling feeder pigs through a market follow. Most of the guidelines that apply to selling to an individual buyer apply to selling through a public auction market, except that usually the communication between seller and buyer is not practical, so a specific program cannot

be developed. Most markets have general health requirements concerning weight range and physical appearance of pigs. Uniformity of pigs is still recommended but not critical because they will usually be sorted by weight and grade by market personnel. Markets now require that pigs be from PRV-free herds before they will be accepted on the sale premises.

- **Co-op feeder pig production marketing**—As the pork production industry evolves, pressures for volume production of consistent, high-quality pigs are resulting in the reemergence of cooperative feeder pig production. Commonly, several producers pool their resources to develop a feeder pig production site. Each producer may then claim one week's (or some time fraction) production on a rotating basis with excess pigs (if available) contractually sold to third parties. A separate site is obtained for the sow herd and nursery to take advantage of multiple-site health benefits. Pigs usually are priced to cooperative members based on cost. Such an arrangement allows specialization, consistency, and the exploitation of economies of size. This trend is increasing, with more interest in the production of segregated early weaned pigs.
- **Grades of feeder pigs**—The grades of feeder pigs are closely correlated with the standards for slaughter barrows and gilts. The standards on which these grades are based embrace two general value-determining characteristics of feeder pigs—(1) ability to relate the feeder pig to the final, finished quality of the animal, expected to produce market hogs with carcasses containing 50 to 54% lean, and (2) their thriftiness. For example, if a feeder pig is graded U.S. No. 1, it has the potential for developing into a U.S. No. 1 slaughter hog that will produce a U.S. No. 1 carcass containing 50 to 54% lean. Thriftiness indicates the ability of a feeder pig to gain weight rapidly and efficiently (Figure 19.7).

Segregated Early Weaned (SEW) Pig Marketing

Segregated early weaned (SEW) pigs (pigs weaned at less than 3 weeks of age, rather than the more conventional 21- to 28-days of age) are segregated from their mothers and this separation is maintained in the nursery and finisher stages.

Traditionally, feeder pigs have been sold, transported, and mixed at about 30- to 60-lb weight. Removing pigs at less than 3 weeks of age is desirable from a health standpoint; it removes them from the primary source of exposure to infectious organisms (the sow herd) while they still have maternal antibodies (passive immunity) to protect them from

FEEDER PIGS
U.S. GRADES

U.S. NO. 1

U.S. NO. 2

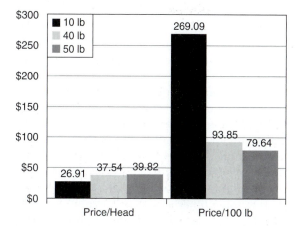

Figure 19.8   The effect of the market weight of feeder pigs on the price per head and per 100 lb of weight. Iowa and Central U.S. Direct Delivered Feeder Pig Data, 2002, 50–54 Percent Lean. *(Courtesy, Bruce Thomas, USDA-IA Department of Agriculture, Market News, Des Moines, IA)*

U.S. NO. 3

U.S. NO. 4

U.S. UTILITY

Figure 19.7   The five market grades of feeder pigs. *(Courtesy, USDA, Washington, DC)*

many disease organisms. A heated, covered method of transportation and a modern, environmentally controlled hot nursery in which to be housed are necessary. This technology may allow buyers to acquire pigs from more than one source without the severe disease consequences that often occur when mixing multiple source 40- to 60-lb feeder pigs.

Often a premium price is paid per pound for 10-lb SEW pigs compared to the traditional 40- to 50-lb feeder pigs (Figure 19.8). Two reasons include the health of the SEW pigs as well as the potential to put on early gains which are very efficient. As expected, 40- and 50-lb feeder pigs cost more than 10-lb pigs. However, when the price is converted to dollars per 100 lb, the price of the 10-lb pig becomes $269/cwt compared to $93 and $79/cwt for the heavier pigs.

In the quest for healthy growing-finishing pigs, more producers are turning to the SEW market. In the mid-1990s, many of these 10-lb pigs were sold on contract from the production units at around $32 a head. Today, the prices are more likely to be based on a potential market price. (See Chapter 15, Swine Production and Management, in the section headed "Early Weaning and Segregated Early Weaning.")

CHOICE OF MARKET OUTLETS

Marketing is dynamic. Changes are inevitable in types of market outlets, market structures, and market services. Some outlets have gained in importance, such as market contracts; others, such as terminal markets, have declined.

The choice of a market outlet represents the seller's evaluation of the most favorable market among the number of alternatives available. No simple and brief statement of criteria can be given as a guide to the choice of the most favorable market channel. Rather, an evaluation is required of the contributions made by alternative markets in terms of available services offered, selling costs, the competitive nature of the pricing process, and, ultimately, the producer's net return. Thus, an accurate appraisal is not simple.

From time to time, producers can be expected to shift from one market to another. Because price changes at different market do not take place simultaneously, or in the same amount, or even in the same direction, one market may be the most advantageous outlet for a particular class and grade of hogs at one time, but another may be more advantageous at some other time. The situation may differ for different classes and kinds of livestock and may vary from one area to another.

Regardless of the channels through which producers market their hogs, in one way or another, they pay or bear, either in the price they receive for the hogs or otherwise, the entire cost of marketing. Because of this, they should never choose a market because of convenience or habit, or because of personal acquaintance with the market and its operator. Rather, the choice should be determined strictly by the net returns from the sale of hogs; effective selling and net returns are more important than selling costs.

SELLING PUREBRED AND SEEDSTOCK HOGS

Selling purebred and seedstock animals is a specialized and, hopefully, scientific business.

Purebred animals are members of a breed that possess a common ancestry and distinctive characteristics and are either registered or eligible for registry. Purebred hogs are also seedstock, but here are differentiated from seedstock suppliers in the context of purebred versus commercially available lines developed by commercial companies.

Seedstock suppliers are competitors of purebred breeders who sell boars and gilts of specialized bloodlines. These lines of breeding, usually called hybrids, originate from crossing two or more breeds, then applying some specialized selection programs. Most of their sales will be by salespeople developing relationships with large herd producers and developing a complete genetic program, supplying all of the sires and often all of the replacement gilts.

Most purebreds are sold from the breeder's farm, usually by private treaty. Prices are based on demand as well as on an index based on the tested performance of the animals. Consignment sales are sponsored by a breed association, local, statewide, or national in character. Purebred auction sales are conducted by highly specialized auc-

tioneers. In addition to being good salespersons, such auctioneers must have a keen knowledge of values and must be familiar with the bloodlines of the breeding stock.

In general, the vast majority of purebred and seedstock boars saved for breeding purposes go into commercial herds. Only the elite sires are retained with the hope of effecting further improvement in the source herds. However, the sale of purebred or seedstock females is fairly well restricted to meeting the requirements for replacement purposes in existing herds or for establishing new herds.

LIVESTOCK MARKET NEWS SERVICES

Accurate market news is essential to the efficient marketing of livestock, both from the standpoint of the buyer and the seller. In the days of trailing, the meager market reports available were largely conveyed by word of mouth. Moreover, the time required to move livestock from the farm or ranch to market was so great that detailed market information would have been of little benefit even if it had been available. With the speed in transportation afforded by trucks, late information on market conditions have become critical.

The Federal Market News Service was initiated by the USDA, beginning in 1916. This service was established for the purpose of providing unbiased and uniformly interpretable market information. Originally, it depended on voluntary cooperation in gathering information. Today, large hog packers are required to report volume and prices daily, including both the range of prices paid as well as a weighted average. Feeder pig auction reports are obtained largely by telephone on scheduled sale days. The Federal Market News Service relies on local and privately owned newspapers, radio stations, and TV stations for disseminating market reports. Because at least a portion of the readers or listeners are interested in this type of information, most local papers and radio stations in agricultural areas are usually glad to serve as media for releasing these reports.

Other important sources of market information include farm and trade magazines. Also, many private market agencies prepare and distribute market information. By means of weekly market newsletters or cards, they commonly emphasize the price, market conditions, and trends of the particular market they serve.

• **Terminology of market reports**—Knowledgeable producers must follow market reports in order to determine the best channels through which to market their hogs, as well as to project future trends

TABLE 19.3   Terms Used in Federal and State Market News Reports[1]

Terms	Definitions
Market	1. A geographic location where a commodity is traded.
	2. The price, or price level, at which a commodity is traded.
	3. To sell.
Market activity	The pace at which sales are being made.
Active	Available supplies (offerings) are readily clearing the market.
Moderate	Available supplies (offerings) are clearing the market at a reasonable rate.
Slow	Available supplies (offerings) are not readily clearing the market.
Inactive	Sales are intermittent with few buyers or sellers.
Price trend	The direction in which prices are moving in relation to trading in the previous reporting period(s).
Higher	The majority of sales are at prices measurably higher than the previous trading session.
Firm	Prices are tending higher, but not measurably so.
Steady	Prices are unchanged from previous trading session.
Weak	Prices are tending low, but not measurably so.
Lower	Prices for most sales are measurably lower than the previous trading session.
Supply/Offering	The quantity of a particular item available for current trading.
Heavy	The volume of supplies is above average for the market being reported.
Moderate	The volume of supplies is average for the market being reported.
Light	The volume of supplies is below average for the market being reported.
Demand	The desire to possess a commodity coupled with the willingness and ability to pay.
Very good	Offerings or supplies are readily absorbed.
Good	Firm confidence on the part of buyers that general market conditions are good. Trading is more active than normal.
Moderate	Average buyer interest and trading.
Light	Demand is below average.
Very light	Few buyers are interested in trading.
Mostly	The majority of sales or volume.
Undertone	Situation or sense of direction in an unsettled market situation.

[1]Glossary of Terms Used in Federal–State Market News Reports, Agricultural Marketing Service, USDA.

in supply and demand so that they plan their programs accordingly. Terms commonly associated with the marketing of commodities are defined in Table 19.3.

PREPARING AND SHIPPING HOGS

Improper handling of hogs immediately prior to and during shipment may result in excessive shrinkage, high death, bruises, crippled pigs, disappointing sales, and dissatisfied buyers (Figure 19.9). Often pork producers who do a superb job of producing hogs, dissipate all the good things that have gone before by doing a poor job of preparing and shipping. Generally speaking, such omissions are a result of lack of know-how, rather than any deliberate attempt to take advantage of anyone. Even if the sale is consummated prior to delivery, negligence at shipping time will make for a dissatisfied customer. Buyers soon learn what to expect from various producers and place their bids accordingly.

In addition to the important specific considerations covered in later sections, the following general considerations should be accorded in preparing hogs for shipment and in transporting them to market:

Figure 19.9 Hog alleys at slaughter plants should have solid panels to facilitate moving hogs and secure footing to prevent slipping. *(Courtesy,* Pork Industry Handbook, *Purdue University, West Lafayette, IN)*

1. Select the best method of transportation. The distance of haul is the greatest single factor for consideration in this regard. All truckers should clean, disinfect, and bed facilities prior to loading, but it is always well that shippers make their own inspection to ensure these matters have been handled to their satisfaction. Generally, hogs are bedded with damp sand in hot weather if the floors are slippery, and in the winter straw is used for warmth. To avoid any misunderstanding, it is recommended that all requests for hauling facilities be either requested or confirmed in writing.

2. Feed and water properly prior to loading. Never ship hogs on an excess fill. Instead, fast them for 10 hours before shipping. If hogs are in transit longer than 10 hours, feed them lightly en route. Although feed is cut off, never withhold water. Access to water is necessary in order to lessen stress in shipment as well as to avoid tissue shrink in long transits and penning.

Hogs too full of feed at the time of loading will scour and urinate excessively. As a result, the floors become dirty and slippery and the animals befoul themselves. Such hogs undergo a heavy shrink and present an unattractive appearance when unloaded.

3. Keep hogs quiet. Prior to and during shipment, hogs should be handled carefully. Hot, excited animals experience more shrinkage and are more apt to be injured or die.

Although loading may be exasperating at times, take it easy; never lose your temper. Avoid hurrying and striking. Never beat an animal with such objects as pipes, sticks, canes, or forks; instead, use (a) a flat, wide canvas slapper with a handle or (b) a broom to guide them.

4. Comply with the requirements for health certificates and permits. When hogs are to be shipped into another state, the shipper should verify and comply with the state regulations relative to health certificates and permits. Usually, the local veterinarian will have this information. Should there be any question about the health regulations, however, the state livestock sanitary board (usually located at the state capital) of the state of destination should be consulted. Knowledge of and compliance with such regulations well in advance of shipment will avoid frustrations and costly delays. Usually, hogs shipped directly to slaughter have minimal or no health certificate requirements, unless they are coming from another country.

5. Use partitions in the truck or car when necessary. When mixed loads (consisting of hogs, cattle, and/or sheep) are placed in the same truck, partition each class off separately. Also, boars, stags, and sows should be properly partitioned.

6. Avoid shipping during extremes in weather. Whenever possible, avoid shipping when the weather is either very hot or very cold. During such times, shrinkage and death losses are higher than normal. During warm weather, avoid transporting hogs during the heat of the day; travel at night or in the evening or early morning. In hot weather, wet the sand.

Preventing Bruises, Crippling, and Death Losses

Losses from bruising, crippling, and death that occur during the marketing process represent a part of the cost of marketing hogs, and indirectly, even though they usually are insured, the producer foots most of the bill. The following precautions are suggested as a means of reducing hog market losses from bruises, crippling, and death:

1. Remove projecting nails, splinters, and broken boards from feeding areas and fences.
2. Keep feedlots free from old machinery, trash, and any obstacles that may bruise.
3. Do not feed wet feeds heavily just prior to loading.
4. Use good loading chutes; not too steep.
5. Bed with sand free from stones, to prevent slipping. Cover sand with straw in cold weather, but do not use straw in hot weather. Wet the sand bedding in summer before loading hogs and while traveling. Wet the pigs down again when necessary.
6. Use partitions in trucks that are not fully loaded to keep animals closer together, and in very long trucks to keep animals from crowding from one location to another.
7. Provide covers for trucks to protect from sun in summer and cold in winter.
8. Always partition mixed loads into separate classes. Partition boars and stags.
9. Remove protruding nails, bolts, and any sharp objects in the truck.
10. Load slowly to prevent crowding against sharp corners and to avoid excitement. Do not overload.
11. Use canvas slappers or panels instead of clubs or canes (Figure 19.10).
12. Drive trucks carefully; slow down on sharp turns, and avoid sudden stops.
13. Inspect load during the trip to prevent trampling of animals that may be down. Get downed animals back on their feet immediately.
14. Back truck slowly and squarely against unloading dock.
15. Unload slowly. Do not drop animals from upper to lower decks; use cleated inclines.
16. If porcine stress syndrome (PSS) is known to be in a herd, exercise extreme care in all stages of getting hogs to market.

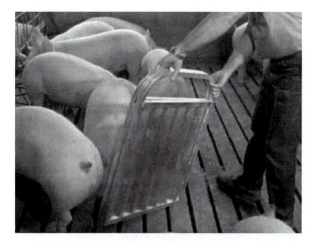

Figure 19.10 Using a light panel, paddle, or broom facilitates moving pigs. Never use electric shockers or whips. *(Courtesy, Pork Industry Handbook, Purdue University, West Lafayette, IN)*

Number of Hogs Per Truck

Overcrowding of market animals causes heavy losses. Sometimes a truck is overloaded in an attempt to effect a saving in hauling charges. More frequently, however, it is simply the result of not knowing space requirements. The suggested number of animals should be tempered by such factors as distance of haul, class of livestock, weather, and road conditions.

Truck beds vary in length. The size of the truck and the class and size of animals determine the number of head that can be loaded in a truck. For comfort in shipping, the truck should be loaded heavily enough so that the animals stand close together, but both underloading and overcrowding are to be avoided (Table 19.4).

Bedding Hogs in Transit

Among the several factors affecting hog losses, perhaps none is more important than proper bedding and footing in transit. Footing, such as sand, is required at all times of the year, to prevent the car or truck floor from becoming wet and slick, thus predisposing animals to injury by slipping and falling. Bedding, such as straw, is recommended for warmth in the shipment of swine during extremely cold weather. During warm weather, the sand should be wet down prior to loading. Recommended kinds and amounts of bedding and footing materials are given in Table 19.5.

Shrinkage in Marketing Hogs

Shrinkage refers to the weight loss encountered from the time animals leave the feedlot until they are weighed over the scales at the market. Thus, if a hog weighed 260 lb at the feedlot and had a market weight of 255 lb, the shrinkage would be 5 lb or 1.9%. Shrink is usually expressed in terms of a percentage. Most of this weight loss is due to excretion, in the form of feces and urine and the moisture in the expired air. With long hauls or time off feed, usually more than 8 or 10 hours, there is some tissue shrinkage, which results from metabolic or breakdown changes. The most important factors affecting shrinkage of hogs are as follows:

1. **Season.** Extremes in temperature, either very hot or very cold weather, result in higher shrinkage. Shrink is at a minimum between 20 and 60°F. When the temperature is above 80°F, hogs in transit should be sprinkled.
2. **Age and weight.** Young pigs shrink proportionally more than older animals because of less body fat and a greater fill in proportion to live-weight.
3. **Overloading or underloading.** Either overloading or underloading always results in abnormally high shrinkage. Shrinkage from underloading may be prevented by using proper partitions.
4. **Rough ride, abnormal feeding, and mixed loads.** All of these factors increase animal discomfort.

TABLE 19.4   Number of Hogs for Safe Loading per Truck

Floor Length		Weight of Hogs								
		100 lb (45 kg)	150 lb (68 kg)	175 lb (79 kg)	200 lb (91 kg)	225 lb (102 kg)	250 lb (113 kg)	300 lb (136 kg)	350 lb (159 kg)	400 lb (181 kg)
(ft)	(m)									
8	2.4	27	21	19	18	16	14	13	11	9
10	3.1	33	26	24	22	20	18	16	14	12
12	3.7	40	31	28	26	24	22	19	17	14
15	4.6	50	39	36	33	30	27	24	21	17
18	5.5	60	47	43	40	36	33	28	25	21
20	6.1	67	52	48	44	40	35	32	28	24
24	7.3	80	62	57	52	48	44	38	34	28

Source: Courtesy of Pearson Education.

TABLE 19.5   Bedding and Footing Material for Transporting Swine[1, 2, 3]

Class of Livestock	Bedding for Moderate or Warm Weather, above 50°F (10°C)	Bedding for Cool or Cold Weather, below 50°F (10°C)
Swine	Sand, 0.5–2 in. (1.3 to 5 cm)	Sand covered with straw

[1]Straw or other suitable bedding (covered over sand) should be used for protection and cushioning breeding stock that are loaded lightly enough to permit their lying down in the truck.
[2]Sand should be clean and medium-fine, and free from brick, stones, coarse gravel, dirt, and dust.
[3]In hot weather, wet sand down before loading. Never apply water to the backs of hot hogs; it may kill them.
Source: Courtesy of Pearson Education.

Mixed loads of strange pigs usually will incur more fighting and bruising.

Most hogs shrink about 1.7% with a 100-mile shipment or less from farm to market, with the majority of the shrinkage occurring during the first few miles from feces and urine.

Air Transportation

With modern communication and transportation, the world is becoming smaller and smaller. Along with this, there is an increasing need for and desire to move breeding animals efficiently between countries separated by great distances. Ancestors to the present-day U.S. hog endured a long sea voyage, which was often a hardship on humans. For many years, ships were considered the only economical method of moving large numbers of animals across the seas, but animals never adapted well to the pitching and tossing of sea travel. In some cases, losses of up to 50% were not uncommon.

During the late 1960s, the concept of air transportation of large numbers of animals was born. Whole planes were adapted for livestock comfort and maximum capacity—literally "flying corrals." With this modern concept, animals can arrive at any destination in the world within hours, thereby minimizing stress and virtually eliminating death losses. Thus, numerous swine and other species are successfully airlifted worldwide.

MARKET CLASSES AND GRADES

Swine market classes and grades differ from those used in cattle and sheep in that (1) there are no age divisions by years (e.g., cattle are classified as yearlings and 2 year olds and over) and (2) rarely are hogs of any kind purchased on the slaughter feeding market for use as breeding animals. The class of market hogs indicates the use to which the animals are best adapted, whereas the grade indicates the degree of perfection within the class.

Factors Determining Market Classes

The market classes of hogs are determined by the following factors: (1) hogs or pigs, (2) use selection, (3) sex class, and (4) weight divisions.

The generally accepted market classes and grades of hogs are summarized in Table 19.6.

Hogs or Pigs

All swine are first divided into two major groups according to age: hogs or pigs. Although actual ages are not observed, the division is made largely by weight in relation to the animal's apparent age. Young animals weighing less than 120 lb (under about 3 months of age) are generally known as pigs, whereas those weighing more than 120 lb are called hogs.

Use Selection

Hogs or pigs are further divided into two subdivisions as slaughter hogs or feeder pigs. Slaughter swine are those that are suitable for immediate slaughter. The demand for lightweight slaughter pigs is greatest during the holiday season when they are wanted for roasting pigs by hotels, clubs, restaurants, steamships, and other consumers. Such pigs weigh from 30 to 60 lb, are dressed shipper style (with the head on), and must produce a plump and well-proportioned carcass. Slaughter hogs (the older animals) are in demand throughout the year.

Feeder pigs include those animals that show ability to grow additional weight. Moreover, because of the greater disease hazard with hogs, this class is under very close federal supervision. Before being released for return to the country, feeder swine must be inspected from a health standpoint, and then either sprayed or dipped as a precautionary measure to prevent the spread of disease germs or parasites.

Sex

The sex class is used only when it affects the usefulness and selling price of animals. In hogs, this subdivision

TABLE 19.6   Market Classes and Quality Grades of Hogs

Hogs or Pigs	Use Selection	Sex Class	Weight Divisions				Commonly Used Grades
			(lb)		(kg)		
Hogs	Slaughter hogs	Barrows and gilts (often called butcher hogs)	120–140 140–160 160–180 180–200 200–220 220–240	240–270 270–300 300–330 330–360 360–400 400 lb up	55–64 64–73 73–82 82–91 91–100 100–109	109–123 123–136 136–150 150–163 163–182 182 kg up	U.S. No.1, U.S. No. 2, U.S. No. 3, U.S. No. 4, U.S. Utility
		Sows (or packing sows)	270–300 300–330 330–360 360–400	400–450 450–500 500–600 600 lb up	123–136 136–150 150–163 163–182	182–204 204–227 227–272 272 kg up	U.S. No. 1, U.S. No. 2, U.S. No. 3, Medium, Cull
		Stags Boars	All weights All weights				Ungraded Ungraded
	Feeder hogs	Barrows and gilts	120–140 140–160 160–180		55–64 64–73 73–82		U.S. No.1, U.S. No. 2, U.S. No. 3, U.S. No. 4, U.S. Utility, Cull
Pigs	Slaughter pigs	Barrows, gilts, and boars	Under 30 30–60		13.6 13.6–27.2		Ungraded
		Barrows and gilts	60–80 80–100 100–120		27.2–36.3 36.3–45.4 45.4–54.5		Ungraded
	Feeder pigs	Barrows and gilts	80–100 100–120		36.3–45.4 45.4–54.5		U.S. No.1, U.S. No. 2, U.S. No. 3, U.S. No. 4, U.S. Utility, Cull

Source: Courtesy, USDA, Washington, DC.

is of less importance than in cattle where steers and heifers are priced differently. Barrows and gilts are always classed together in the case of both slaughter and feeder hogs. This is done because their sex affects their usefulness so little that a price differentiation is not warranted. In addition, because the carcass is not affected, no sex differentiations are made for slaughter pigs less than 60 lb in weight. The terms *barrow, gilt, sow, boar*, and *stag* are used to designate the sex classes of hogs. The definition of each of these terms follows:

• **Barrow**—A male swine that was castrated at an early age—before reaching sexual maturity and before developing the physical characteristics particular to boars.
• **Gilt**—A female swine that has not produced pigs and that has not reached an evident stage of pregnancy.
• **Sow**—A female swine that shows evidence of having produced pigs or that is in an evident stage of pregnancy.
• **Boar**—An uncastrated male swine of any age. Mature boars should always be stagged (castrated) and

fed 3 weeks or longer (until the wound heals) before being sent to market. The market value of boars is very low, primarily because of odor. Much of the boar meat is processed into heavily spiced dried sausages.
• **Stag**—A male swine that was castrated after it had developed the physical characteristics of a mature boar. Because of relatively thick skins, coarse hair, and heavy bones, stags are subject to dockage. They are usually docked 70 lb, but may be docked from 40 to 80 lb, depending on the market.

Boars and stags are usually marketed on a live basis and are purchased at a price that reflects their true value from a meat standpoint. Often smaller collection markets gain reputations as places to sell boars and stags to processors that primarily slaughter only these classes of hogs.

Weight

Occasionally, the terms *light, medium*, and *heavy* are used to indicate approximate weights, but most generally the actual range is weight in pounds and is specified both in trading and in market reporting.

Moreover, hogs are usually grouped according to relatively narrow weight ranges because variations in weight affect (1) the dressing percentage, (2) the weight and desirability of the cuts of meat, and (3) the amount of lard produced (heavier weights produce more lard). Boars and stags are not usually subdivided according to weights.

LEAN PORK MARKETING

The majority of slaughter hogs today are sold grade and yield with a base price that is adjusted for carcass weight and for a certain percentage of lean meat in the carcass. Standard percentage leans usually are from 50 to 54%, with premiums or discounts given for carcasses above or below the standard. The percent lean is estimated by measuring the fat depth and sometimes the size of the loin muscle. Newer ultrasound and other techniques are becoming available.

Optimum carcass weight ranges vary among packers as well as the equations used to predict the percent lean. Most packers use equations that combine the carcass weight with a measure of lean to estimate the percent lean. Kill sheets the producer receives from the packer will list the pigs that fall into each carcass weight range and the percent lean, usually on a "fat-free lean" basis, for each hog as well as for the entire load.

Percent Lean

Percent lean is the amount of lean muscle in a pork carcass. It may be based on muscle with a certain percentage of intramuscular fat or it may be based on a "fat-free" system. Because pork muscle often contains from 3 to 5 percent intramuscular fat, those systems will produce reports with the lean percentages somewhat greater than "fat-free" reports.

Also, percent lean is based on the amount of lean predicted to be in the carcass, not in the live pig. For example, a 250-lb hog dresses 74% and yields 185 lb of carcass. The difference, 65 lb, includes primarily the internal organs and the head. If this carcass contains 50% lean, it has 92.5 lb of muscle and 92.5 lb of skin, separable fat, and bone. Working back, the 250-lb pig produced 92.5 lb of muscle, or about 37% of the live-weight of the pig is actual muscle.

FEDERAL GRADES OF HOGS

Federal grades are rarely used in the hog industry, but they do set the standard on which other methods of pricing pigs are based. USDA grade descriptions provide the uniform terminology for trading nationwide or internationally, using communication technology and accurate formula pricing. Some organizations modify the USDA standards to conform with their own grading systems. USDA slaughter grades include U.S. numbers 1 through 4 and Utility (Figure 19.11). Additional information about USDA grades can be obtained from the Packers and Stockyards division of the USDA.

The market grade for live swine, as for other kinds of livestock, is a specific indication of the degree of excellence within a given class based on conformation, finish, and quality. The two chief factors that serve to place a hog in a specific grade are the visible degrees of fatness and amounts of muscling. Although no official grading of live animals is done by the USDA, market grades do form a basis for uniform reporting of livestock sales. Today very few U.S. slaughter hogs are sold on a live basis, and the federal market reports from packing plants are all on a carcass basis. Hogs still sold on a live basis at terminal markets are priced on live-weight.

It is intended that the grade of live slaughter hogs be correlated with the carcass grade. However, it takes a great deal of experience and study to correlate live animals with the type of carcass they will produce. Tentative standards for grades of pork carcasses and fresh pork cuts were first issued by the USDA in 1931. Subsequently, these standards were revised in 1933, 1952, 1955, and 1968. In January 1985, the standards were further revised, based on backfat thickness of the last rib and muscling.

The current federal market grades of live slaughter barrows and gilts, which were adopted in 1985, are U.S. No. 1, U.S. No. 2, U.S. No. 3, U.S. No. 4, and U.S. Utility (Figure 19.11). The grades may be described as follows:

- **U.S. No. 1**—Slaughter barrows and gilts in this grade will produce carcasses with acceptable lean quality, acceptable belly thickness, and a high percentage of lean cuts (60.4% and over).
- **U.S. No. 2**—Slaughter barrows and gilts in this grade will produce carcasses with acceptable lean quality, acceptable belly thickness, and a slightly lower percentage of lean cuts (57.4 to 60.3%).
- **U.S. No. 3**—Slaughter barrows and gilts in this grade will produce carcasses with acceptable lean quality, acceptable belly thickness, and a slightly lower percentage of the four lean cuts (54.4 to 57.3%).
- **U.S. No. 4**—Slaughter barrows and gilts in this grade will produce carcasses with acceptable lean quality and acceptable belly thickness. However, they are fatter and less muscular and will have a lower carcass yield of the four lean cuts than those in the U.S. No. 3 grade (less than 54.4% of four lean cuts).
- **U.S. Utility**—Barrows and gilts typical of this grade will have a thin covering of fat. The sides are wrinkled and the flanks are shallow and thin. They will produce carcasses with unacceptable lean quality and/or unacceptable belly thickness.

SLAUGHTER SWINE
U.S. GRADES

U.S. NO. 1

U.S. NO. 2

U.S. NO. 3

U.S. NO. 4

U.S. UTILITY

Figure 19.11 The five market grades of slaughter swine. Although no hogs are graded by federal graders, these grades form the basis for other grading systems. *(Courtesy, USDA, Washington, DC)*

The federal grades of slaughter sows, which became effective in 1956, are U.S. No. 1, U.S. No. 2, U.S. No. 3, medium, and cull. The grades are based on differences in yields of lean cuts and fat cuts and differences in quality of pork. Medium-grade sows have a degree of finish less than the minimum required to produce pork of acceptable palatability. Cull-grade sows have a very low degree of finish.

As a rule, slaughter pigs that weigh less than 60 lb are not graded because they have not reached sufficient maturity for variations in their conformation, finish, and quality to affect the market value materially. Slaughter hogs weighing less than 220 lb live are severely discounted at the market because the small size of the lean cuts do not fit the processor's marketing program.

Federal Grades of Pork Carcasses

The grade of a pork carcass may be defined as a measure of its degree of excellence based chiefly on quality of lean and the expected yield of trimmed major wholesale cuts. It is intended that the specifications for each grade shall be sufficiently definite to make for uniform grades throughout the United States and from season to season, and that on-rail grades shall be correlated with on-foot grades.

Because of the relationship between maturity in pork and the acceptability of the prepared meat to the consumer, separate standards have been developed for (1) barrow and gilt carcasses and (2) sow carcasses. Only barrow and gilt carcasses are discussed herein.

The U.S. grades of barrow and gilt carcasses are based on two general considerations: (1) the quality-indicating characteristics of the lean and fat and (2) the expected combined yields of the four primal cuts (ham, loin, picnic shoulder, and Boston butt).

If a carcass qualifies as acceptable in quality of lean and in belly thickness, and is not soft and oily, it is graded U.S. No. 1, 2, 3, or 4, based entirely on projected carcass yields of the four lean cuts. The expected yields of each of the carcass grades in the four lean cuts, based on using the USDA standard cutting and trimming methods, are given in Table 19.7. Carcasses vary in their yields of the four lean cuts because of variations in their degree of fatness and in their degree of muscling (thickness of muscling in relation to skeletal size).

From the standpoint of quality, two general levels are recognized—"acceptable" and "unacceptable." Acceptability is determined by direct observation of the cut surface and is based on considerations of firmness, marbling, and color, along with the use of such indirect indicators as firmness of fat and lean, and feathering between the ribs. The degree of external fat is not considered in evaluating the quality of the lean. Suitability of the belly for bacon (in terms of thickness) is also considered in quality evaluation, as is the softness and oiliness of the carcass. Carcasses that have unacceptable quality of lean, and/or bellies that are too thin, and/or carcasses which are soft and oily are graded U.S. Utility.

The grade of a barrow or gilt carcass is determined on the basis of the following equation:

Carcass grade = (4.0 × Backfat thickness over the last rib, inches) − (1.0 × Muscling score)

To apply this equation, muscling should be scored as follows: thin muscling = 1, average muscling = 2, and thick muscling = 3. Carcasses with thin muscling cannot grade U.S. No. 1. The grade may also be determined by calculating a preliminary grade according to the backfat schedule shown in Table 19.8, and adjusting up or down one grade for thick or thin muscling, respectively. The location of the last rib measurement is shown in Figure 19.12.

The second factor considered in barrow and gilt carcass grading is the degree of muscling. The degree of muscling is determined by a subjective evaluation of the thickness of muscling in relation to skeletal size. Because the total thickness of a carcass is affected by both the amount of fat and the amount of muscle in relation to skeletal size, the fatness must also be considered when degree of

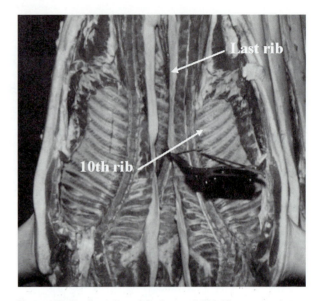

Figure 19.12 Location of the last and 10th ribs on a pork carcass. The last rib can be located on the inside of the carcass. The last rib measurement is taken on the midline of the carcass, measured from the outside of the skin to the bottom of the fat layer. The 10th rib is located by counting the number of ribs from the front of the carcass. A cut is made between the 10th and 11th ribs (shown by saw) and the measurement taken three-fourths of the distance of the loin muscle from the spine side. *(Courtesy, Palmer Holden, Iowa State University, Ames, IA)*

muscling is evaluated. To best evaluate muscling, primary consideration is given to those parts least affected by fatness, such as the ham. In evaluating the ham for degree of muscling, consideration should be given to both the stifle and back views. The size of lumbar lean area and the relative width through the back of the loin and through the center of the ham are also good indications of muscling.

In barrow and gilt carcass grading, three degrees of muscling—thick (superior), average, and thin (inferior)—are considered.

Thus, the on-foot and carcass federal grades of slaughter barrows and gilts are U.S. No. 1, U.S. No. 2, U.S. No. 3, U.S. No. 4, and U.S. Utility.

Unlike meat inspection, government grading is purely voluntary, on a charge basis. Only a very small proportion of U.S. commercial pork production is federally graded because carcasses are cut and trimmed at the packing plant, and sold as trimmed primal and subprimal cuts. Graded carcasses are stamped (with an edible vegetable dye) so that the grade appears on the retail cuts as well as on the carcass and wholesale cuts.

Only negligible quantities of pork carcasses are actually federally graded. Most carcasses are packer graded using a metal ruler on the midline at the last rib (similar to the location of the federal grade), 2 in. off the 10th rib midline with a metal probe, or at the

TABLE 19.7   Expected Yields of Four Lean Cuts Based on Chilled Carcass Weight, by Grade

Preliminary Grade	Yield of Four Lean Cuts
U.S. No. 1	60.4% and over
U.S. No. 2	57.4–60.3%
U.S. No. 3	54.4–57.3%
U.S. No. 4	Less than 54.4%

Source: USDA-AMS, 1985. U.S. Standards for Grades of Pork Carcasses.

TABLE 19.8   Preliminary Carcass Grade Based on Backfat Thickness over the Last Rib

Preliminary Grade	Backfat Thickness Range
U.S. No. 1	Less than 1.00 in.
U.S. No. 2	1.00–1.24 in.
U.S. No. 3	1.25–1.49 in.
U.S. No. 4	1.50 in. and over

Source: USDA-AMS, 1985. U.S. Standards for Grades of Pork Carcasses.

10th rib as shown in Figure 19.12. Packers use these measurements to predict the percentage lean content of the carcass and base their premiums on the percentage of lean rather than the weight of the four primal cuts.

OTHER HOG MARKET TERMS AND FACTORS

In addition to the rather general terms used in designating the different market classes and grades of hogs, the following terms and factors are frequently of importance.

Roasters

Roasters refer to fat, plump, suckling pigs, weighing 30 to 60 lb on foot. These are dressed shipper style (with the head on), and they are not split at the chest or between the hams. When properly roasted and attractively served with the traditional apple in the mouth, roast pig is considered a great delicacy for the holiday season.

Suspects (Governments)

Suspects, or governments, are suspicious animals that federal inspectors tag at the time of the antemortem (before death) inspection to indicate that more careful scrutiny is to be given in the postmortem inspection. If the carcass is deemed unfit for human consumption, it is condemned and sent to the inedible tank.

Cripples

Cripples are hogs that are too lame or injured to walk. Such animals should never reach a market; they should be euthanized on the farm.

Dead Hogs

Dead hogs are those that arrive dead at the market. They have no edible salvage value. These carcasses are sent to rendering for conversion into inedible grease, fertilizer, and so on.

HOG MARKETING CONSIDERATIONS

Enlightened and shrewd marketing practices generally characterize the successful hog enterprise. Among the considerations of importance in marketing hogs are those which follow.

Secular Trends

Secular trends are longtime trends that persist over a period of several cycles. The long-run trend in

U.S. hog numbers from 1925 to 2002 have been upward, as shown in Figure 19.13. Likewise, the longtime trend in the nation's pork production has been upward because of increased hog numbers, along with improved breeding, feeding, and management, which has increased productivity per head and resulted in marketing hogs at younger ages. The two significant depressions in the long-term trend were the Great Depression in the 1930s and World War II in the early 1940s.

Cyclical Movements

Cyclical movements follow a pattern that repeats itself. Hog cycles average about 4 years—2 years of expansion, and 2 years of liquidation (Figure 19.14). Basically, cycles are the response of producers to prices, and prices reflect supply and demand. In the

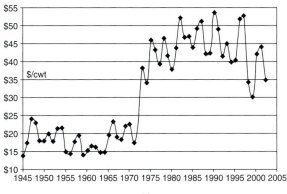

Figure 19.13 Annual commercial U.S. hog slaughter. The long-term trend of the industry is increased production. With the exception of the Great Depression and World War II, the trend has been upward. *(Courtesy, USDA Hog & Pig Reports)*

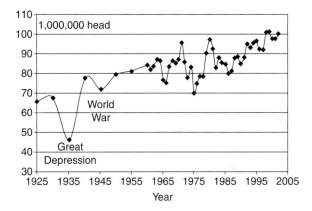

Figure 19.14 The cyclical trend of the annual average price received by U.S. farmers for hogs from 1945 to 2002. The hog price cycle typically lasts from 3 to 4 years. *(Prices Received by Farmers: Historic Prices and Indexes 1908–1992, USDA-ESS and 2003 Agricultural Statistics, USDA-NASS)*

past, the hog–corn ratio served as a barometer; when it was favorable—above 12—an expansion in hog numbers followed; when it was unfavorable—below 12—a cutback followed. Today, the hog–corn ratio is not as much of a factor as formerly. Also, large swine units are inclined to maintain production near optimum levels for the size unit without regard to cyclical movements.

These cycles are a direct reflection of the rapidity with which the numbers of each class of farm animals can be shifted under practical conditions to meet consumer meat demands. Thus, litter-bearing and early-producing swine can be increased in numbers much more rapidly than either sheep or cattle. Normal cycles are disturbed by droughts, wars, general periods of depression or inflation, and federal controls.

Seasonal Variations

Hog prices vary seasonally (within a year) as a result of the variation in market receipts. As would be expected, seasons of high market prices are generally associated with light sales and seasons of low market prices with heavy sales. It must be realized, however, that the normal seasons of high and low prices may be changed by such factors as (1) federal farm programs and controls, (2) business conditions and general price levels, (3) feed supplies and weather conditions, and (4) wars, and so on.

In recent years, seasonal patterns have not been as reliable as they used to be. Year-round farrowing of sows in confinement has made for more uniform marketing throughout the year and reduced seasonality in livestock marketing. Thus, when arriving at livestock forecasts and marketing advice, proper reservation should be exercised in considering seasonal patterns. Historically, planning to increase summer sales has the potential to increase market prices. It is not always wise to plan

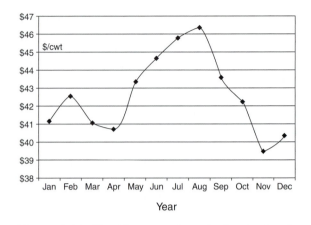

Figure 19.15   Monthly average live price for slaughter barrows and gilts, 1970 to 1999. Prices peak in the summer months of June through August and bottom out in November and December. A slight rebound occurs in February. *(Prices Received by Farmers: Historic Prices and Indexes 1908–1992, USDA-ESS and USDA Red Meat Yearbook)*

production to hit the highest market because added production costs of winter farrowing may offset the gains from higher prices. Nevertheless, a careful study of normal seasonal prices will serve as a useful guide (Figure 19.15).

Dockage

The value of some market animals is low because dressing losses are high or because part of the product is of low quality. Some common dockages on hog markets are as follows:

1. **Piggy sows.** Usually docked 40 lb, but it may range from 0 to 50 lb, depending on the market and stage of pregnancy.
2. **Stags (hogs).** Usually docked 70 lb, but it may range from 40 to 80 lb, depending on the market.

QUESTIONS FOR STUDY AND DISCUSSION

1. Define *livestock marketing.*
1. Why is hog marketing important?
3. In recent years, terminal and auction markets have declined in importance, whereas direct selling, carcass grade and yield selling, and contract selling have increased. Why has this happened?
4. Briefly, describe terminal markets, auction markets, direct selling, carcass grade and yield selling, slaughter hog pooling, packer marketing contracts, feeder pig marketing, and early-weaned pig marketing.
5. Why are only negligible quantities of pork federally graded?

6. What method of marketing do you consider most advantageous for the hogs sold off your home farm (or a farm with which you are familiar)? Justify your choice.
7. How does selling purebred and seedstock hogs differ from selling slaughter hogs?
8. Many producers and marketing specialists express the following concerns if packers have 80% or more of their kill under contract:
   a. If a minority (20% or fewer) of the hogs marketed set the cash market, will they be hogs of lesser quality?
   b. Would these few hogs of lesser quality be used to calculate prices for the contract hogs?

c. Do you believe that the preceding two concerns are justified?

d. Why might these open negotiated pigs be priced higher or lower than expected?

9. Which is the more important to the hog producer: (a) low marketing costs or (b) effective selling and net returns?

10. List and discuss the factors affecting feeder pig prices.

11. What prompted the rise of the segregated early weaned (SEW) pig market?

12. Why are livestock market news services important?

13. Outline, step by step, how you would prepare and ship hogs to market.

14. Discuss practical ways and means of reducing shrinkage in market hogs.

15. List the federal market grades of slaughter barrows and gilts, and briefly give the specifications of each.

16. What are roasters?

17. Discuss practical ways through which the hog producer can take advantage of cyclical trends and seasonal changes.

18. Describe the federal meat inspection of hogs.

19. Why are slaughter weights increasing?

SELECTED REFERENCES

Animal Science, 9th ed., M. E. Ensminger, Interstate Publishers, Inc., Danville, IL, 1991

Chicago Mercantile Exchange, http://www.cme.com/

Meat We Eat, The, J. R. Romans et al., Interstate Publishers, Inc., Danville, IL, 1994

Pork Facts 2002/2003, Staff, National Pork Producers Council, Des Moines, IA, 1995–96

Pork Industry Handbook, Cooperative Extension Service, Purdue University, West Lafayette, IN, 2002

Reports by Commodity, USDA-National Agricultural Statistics Service, Washington, DC, http://usda.mannlib.cornell.edu

Stockman's Handbook, The, 7th ed., M. E. Ensminger, Interstate Publishers, Inc., Danville, IL, 1992

20

Pork and Pork By-Products

Grilled honey-soy pork steaks. Bone-in pork steaks with grilled corn on the cob and red and green peppers. *(Courtesy, National Pork Board, Des Moines, IA)*

Contents

Objectives

After studying this chapter, you should:

1. Have an understanding of the many facets of pork quality.
2. Know how the pork industry has changed to meet the consumer demands for quality.
3. Know the wholesale cuts of pork and what retail products are produced from each.
4. Know why pork is cured and what curing does toward value.
5. Know why pork fits into a balanced human diet.
6. Be able to discuss the risks of cholesterol and nitrate, nitrites, and nitrosamines.
7. Be aware of pork safety issues and industry responses.
8. Know the value of pork by-products to the industry and the consumer.

Pork may be defined as the edible flesh of pigs or hogs, whereas by-products include all products, both edible and inedible, other than the carcass meat. The edible glands and organs are usually classed as by-products, but lard is usually grouped along with pork.

Although this chapter is devoted primarily to the final animal product—meat—it must be remembered that the top grades of this food represent the culmination of years of progressive breeding; the best nutrition; vigilant sanitation and disease prevention; superior care and management; and modern marketing, slaughtering, processing, and distribution. Much effort and years of progress have gone into the production of tasty pork chops and hams. Pigs lead the meat species in the quantity of meat produced per breeding female per year (Figure 20.1). Even though much of this change over time is because of an increase in slaughter weights, the increased productivity of the sow plays an important role.

PORK TO THE CONSUMER: THE ULTIMATE OBJECTIVE

The end product of all breeding, feeding, care and management, marketing, and processing is pork available to the consumer. It is imperative, therefore, that the pork producer, the student, and the swine scientist have a reasonable working knowledge of pork and of the by-products from hog slaughter. Such knowledge will be of value in selecting animals and in determining policies relative to their handling.

The type of animals best adapted to the production of meat over the counter has changed in a changing world. Thus, in the early history of the United States, the very survival of animals often depended on their speed, hardiness, and ability to fight. Moreover, long legs and plenty of bone were important attributes when it came time for animals

to be driven to market (Figure 20.1). The Arkansas Razorback was adapted to these conditions.

With the advent of rail transportation and improved care and feeding methods, the ability of animals to travel and fight diminished in importance. It was then possible, through selection and breeding, to produce meat animals better suited to the needs of more critical consumers. With the development of large cities, artisans and craftspeople and their successors in industry required fewer

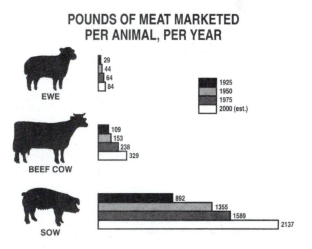

POUNDS OF MEAT MARKETED PER ANIMAL, PER YEAR

EWE
29
44
64
84

1925
1950
1975
2000 (est.)

BEEF COW
109
153
238
329

SOW
892
1355
1589
2137

Live-weight marketed per breeding ewe, beef cow, and sow in 1925, 1950, and 1975. Years of progress and much effort have gone into the efficient production of quality meat. The sow has done her part. Pounds of meat marketed per year and dressing percentages for beef cattle and hogs from *Meat and Poultry Facts, 1994*, pp. 22–23, American Meat Institute, Washington, DC; and for sheep from *Sheep and Goat Science,* 5th ed., p. 216. Year 2000 estimates uses the CAST figures on live-weight marketed per breeding female multiplied by the percentage of edible meat obtained at slaughter from *Meat and Poultry Facts* and *Sheep and Goat Science.* (Food for Animals, *CAST Report No. 82, March 1990, p. 13, Table 2)*

October 31, 1868

Figure 20.1   An 1868 hog drive. The owner rides in the buggy, the herdsman drives the pigs, and the sign says "Chicago, 118 miles." *(Courtesy,* Swine in America, *F. D. Coburn,. 1916)*

calories than those who were engaged in the more arduous tasks of logging, building railroads, and so on. Simultaneously, the American family decreased in size. The demand shifted, therefore, to smaller and less fatty cuts of meats, and, with greater prosperity, high-quality hams, bacons, and chops were in demand. To meet the needs of the consumer, the producer gradually shifted to the breeding and marketing of younger animals with maximum cutout value of the primal cuts. The need was for meat-type hogs.

Consumer demand has exerted a powerful influence on the type of hogs produced. To be sure, it is necessary that such production factors as prolificacy, feed efficiency, rapid gains, size, and so on receive due consideration along with consumer demands. But once these production factors have received due weight, hog producers—whether they be purebred or commercial operators—must remember that quality pork available to the consumer is the ultimate objective.

Now, and in the future, hog producers need to select and feed so as to obtain increased lean meat without excess fat. Production testing programs have been reoriented to give greater emphasis to these consumer demands.

DESIRABLE PORK QUALITIES

The term *pork quality* conveys different messages to different people. To pork processors, it relates primarily to functional properties and color of the muscle. To retailers, it relates to appearance of retail cuts, including fat and bone content, as well as color and juice loss or retention. To consumers, all factors that affect pork-eating satisfaction, safety, convenience, and nutritional value falls within the definition of

pork quality. Pork producers must recognize all these requirements and use the management practices that maximize pork quality for the entire industry.

Data presented in Table 20.1 indicate how producers have adapted production to produce more retail meat cuts and less lard, in keeping with consumer demand. Market hogs from 1950 to 1960 had more than 30 lb of lard produced per animal. From 1965 to date lard production per pig has declined to about 11 lb. The dressing percentage has steadily improved and the amount of retail meat per hog has climbed to more than 150 lb.

Because consumer preference is such an important item in the production of pork, the producer, the packer, and the meat retailer must be familiar with these qualities, which are summarized as follows:

1. **Quality.** The quality of lean pork muscle is based on firmness, texture, marbling, and color.
    a. **Firmness.** Pork muscle should be firm so as to display attractively. Firmness is affected by the kind and amount of fat. For example, pigs fed liberally on peanuts or full-fat soybeans produce a softer pork, those fed on low-fat grains with saturated fats have a firmer fat.
    b. **Texture.** Pork lean that has a fine-grained texture is preferred. Coarse-textured lean is generally indicative of greater animal maturity and less tender meat.
    c. **Marbling.** Marbling contributes to buyer appeal. Feathering (flecks of fat) between the ribs and within the muscles is indicative of marbling. Although consumers prefer to purchase lean meat, a minimum amount of fat in the meat is essential to flavor.
    d. **Color.** Most consumers prefer pork to be a bright reddish lean marbled with flecks of fat with a white fat on the exterior.
2. **Maximum muscling; moderate fat.** Maximum thickness of muscling influences materially the acceptability by the consumer. Also, consumers prefer a uniform cover not to exceed 1/8 inch of firm, white fat on the exterior. The amount of external fat is determined by the trimming procedures of the packer or the retailer (Figure 20.2).
3. **Repeatability.** The consumer wants to be able to secure a standardized product—meat of the same tenderness and other eating qualities as the previous purchase.
4. **Safety.** The United States has one of the safest food supplies in the world. Yet, the Centers for Disease Control and Prevention (2002) estimated food-borne diseases cause 76 million illnesses, 325,000 hospitalizations, and 5,200

TABLE 20.1   Changes in the Production of Lard and Retail Meat

Year	Live-Weight		Dressing Percentage	Dressed Weight		Lard Yield		Retail Meat Yield	
	(lb)	*(kg)*	*(%)*	*(lb)*	*(kg)*	*(lb/hog)*	*(kg/hog)*	*(lb/hog)*	*(kg/hog)*
1950	240	109	68.9	165	75.0	35.4	16.1	137	62.1
1955	237	107	69.5	165	74.7	34.9	15.8	136	61.7
1960	236	107	69.5	164	74.4	32.2	14.6	139	63.0
1965	238	108	70.1	167	75.6	27.9	12.6	147	66.7
1970	240	109	70.3	169	76.5	22.8	10.3	155	70.3
1975	240	109	70.6	169	76.8	14.8	6.7	166	75.3
1980	242	110	71.7	174	78.7	12.8	5.8	172	78.0
1985	245	111	71.4	175	79.5	11.0	5.0	136	61.7
1990	249	113	72.7	181	82.2	11.1	5.0	140.6	64.0
1995	257	117	72.0	185	84.0	11.5	5.2	143.6	65.2
1996	255	116	72.5	185	84.0	11.1	5.0	143.6	65.2
1997	256	116	73.4	188	85.4	10.0	4.5	145.4	66.0
1998	257	117	73.2	188	85.4	10.9	5.0	146.7	66.6
1999	258	117	73.4	190	86.3	11.4	5.2	148.2	67.3
2000	262	119	73.9	193	87.7	11.0	5.0	150.8	68.5
2001	264	120	74.0	195	88.7	10.7	4.9	151.6	68.8

Source: Pork Facts 2002–2003, *Industry Statistics, National Pork Board, Des Moines, IA; USDA-NASS Agricultural Statistics, 2003.*

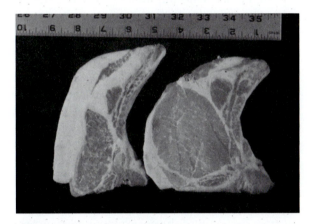

Figure 20.2   Consumers want pork with a minimum of fat and a maximum amount of lean. These two pork chops show the variation, a loin eye with 2.7 in.^{2} loin eye with lots of fat and a 7.1 in.^{2} loin eye. *(Courtesy, Palmer Holden, Iowa State University, Ames, IA)*

deaths in the United States each year. Food-borne diseases are caused by consuming contaminated foods or beverages. In most of these cases, the health problem is a mere inconvenience, but it can be life threatening. (See also the section in Chapter 15 "Preventing Drug Residues.")

If these qualities are not met by pork, other products will meet them. Recognition of this fact is important because competition is keen for space on the shelves of modern retail food outlets.

The pork and other meat commodity organizations continue to conduct more experimental studies on consumer preference and demand. Factors involved in producing higher-quality pork are covered in Chapter 5, Applied Swine Genetics; Chapter 6, Fundamentals of Swine Nutrition; and Chapter 7, Nutrient Requirements and Allowances, Diet Formulation, and Feeding Programs.

PACKER SLAUGHTERING AND DRESSING OF HOGS

Hog slaughtering is unique in that much more pork is cured than is the case with beef or lamb, and pork fat (lard) is not classed as a by-product, although the surplus fats of beef, veal, mutton, and lamb are in the by-product category.

From 1990 to 2002, an average of 99.8% of hogs slaughtered were processed in commercial facilities under federal inspection or in approved state inspected plants (Table 20.2). This amounted to more than 97 million head per year. An average of 154,000 were slaughtered on the farm and the number of farm-slaughtered hogs is steadily declining.

Federal Meat Inspection

The federal government requires supervision of establishments that slaughter, pack, render, and prepare meats and meat products for interstate shipment and foreign export. It is the responsibility of the respective states to have and enforce

TABLE 20.2   Proportion of Hogs Slaughtered Commercially and On-Farm

Year	Commercial Slaughter (1,000 head)	Farm Slaughter (1,000 head)	Total Number Slaughtered (1,000 head)	Commercial Slaughter (%)
1990	85,136	295	85,431	99.7
1995	96,325	210	96,535	99.8
1996	92,394	175	92,569	99.8
1997	91,960	165	92,125	99.8
1998	101,029	165	101,194	99.8
1999	101,544	150	101,694	99.9
2000	97,976	130	98,106	99.9
2001	97,962	120	98,082	99.9
2002	100,263	115	100,378	99.9
9-year average	97,432	154	97,585	99.8

Source: Livestock Slaughter Annual Summary, USDA.

legislation governing the slaughtering, packaging, and handling of meats shipped intrastate. The meat inspection laws do not apply to farm slaughter for home consumption, although all states require inspection if the meat is sold.

The meat inspection service of the USDA was inaugurated and is maintained under the Meat Inspection Act of June 30, 1906. This act was updated and strengthened by the Wholesome Meat Act of December 15, 1967. The latter statute (1) requires that state standards be at least to the levels applied to meat sent across state lines and (2) assures consumers that all meat sold in the United States is inspected either by the federal government or by an equal state program. State regulations must meet federal standards but do not qualify the meat for interstate shipment. The Animal and Plant Health Inspection Service (APHIS) of the USDA is charged with the responsibility of meat inspection.

The purposes of meat inspection are (1) to safeguard the public by eliminating diseased or otherwise unwholesome meat from the food supply, (2) to enforce the sanitary preparation of meat and meat products, (3) to guard against the use of harmful ingredients, and (4) to prevent the use of false or misleading names or statements on labels.

The personnel responsible for carrying out the provisions of the act are of two types: professional or veterinary inspectors who are graduates of accredited veterinary colleges, and nonprofessional food inspectors who are required to pass a Civil Service examination. In brief, the inspections consist of the following two types:

1. **Antemortem (before slaughter) inspection** is made in the pens or as the animals move from the scales after weighing. The inspection is performed to detect evidence of disease or any abnormal condition that would indicate a disease.

Suspects are provided with a metal ear tag bearing the notation "U.S. Suspect No. . ." and are given special postmortem scrutiny. If in the antemortem examination there is definite and conclusive evidence that the animal is not fit for human consumption, it is "condemned," and no further postmortem examination is necessary.

2. **Postmortem (after slaughter) inspection** is made at the time of slaughter and includes a careful examination of the carcass and the viscera (internal organs). All healthy carcasses (no evidence of disease) are stamped "U.S. Inspected and Passed," whereas the inedible carcasses are stamped "U.S. Inspected and Condemned." The latter are sent to the rendering tanks, the products of which are not used for human food.

In addition to the antemortem and postmortem inspections, the government meat inspectors have the power to refuse the application of the mark of inspection to meat products produced in a plant that is not sanitary. All parts of the plant and its equipment must be maintained in a sanitary condition at all times. In addition, plant employees must wear clean, washable garments, and suitable lavatory facilities must be provided for hand washing.

Meat inspection regulations require the condemnation of all or affected portions of carcasses of animals with various disease conditions, including pneumonia, peritonitis, abscesses and pyemia, uremia, tetanus, rabies, anthrax, tuberculosis, various neoplasms (cancers), arthritis, actinobacillosis, and many others.

- **Federally inspected meat plants**—There were 683 plants under federal inspection on January 1, 2003, that slaughtered hogs, with 13 of them accounting for 57% of the total. This compares with 830 slaughter plants in 1994.

Steps in Slaughtering and Dressing Hogs

After purchase, hogs are driven from the holding pens to the packing plant where they are given a shower and are held temporarily in a small pen while awaiting slaughter. In large meat packing plants, a series of chutes, and retainer conveyors move the hogs into position for stunning.

The chain method of slaughtering is used in killing and dressing hogs. In this method, the following steps are carried out in rapid succession:

1. **Rendering insensible.** The hogs are rendered insensible by use of a captive bolt stunner, electric current, carbon dioxide, or gun shot.
2. **Shackling and hoisting.** The hogs are shackled just above the hoof on one hind leg and are then hoisted to an overhead rail.
3. **Sticking.** The sticker sticks the hog just under the point of the breast bone, severing the arteries and veins leading to the heart. The animal is allowed to bleed for a few minutes. In some plants the instrument used for sticking the hog is a vacuum, allowing the blood to be used for edible purposes.
4. **Scalding.** The animals are next placed in a scalding vat for about 4 minutes. By means of automatically controlled steam jets, the temperature of the water in the vats is maintained at about 150°F. The scalding process loosens the hair and scurf.
5. **Dehairing.** After scalding, the carcasses are elevated into a dehairing machine that scrapes them mechanically.
6. **Returning to overhead tracks.** As the carcasses are discharged from the dehairing machine, the gam cords of the hind legs are exposed and gambrel spreaders are inserted in the cords. Then the carcass is again hung from the rail.
7. **Dressing.** A conveyor then moves the carcass slowly along a prescribed course where attendants perform the following tasks:
   a. Washing and singeing.
   b. Removing the head.
   c. Opening the carcass and eviscerating.
   d. Splitting or halving the carcass with an electric saw or a cleaver.
   e. Removing the leaf fat.
   f. Exposing the kidneys for inspection and facing the hams (removing the skin and fat from the inside face or cushion of the ham).
   g. Washing the carcass and then sending it to the coolers where the temperature is held at about 34°F. (Rapid chilling is desirable as it reduces shrinkage and maintains meat quality.)

Packer Versus Shipper Style

The two common styles of dressing hogs in packing plants are packer style and shipper style. The packer style is most common and this system is used almost exclusively when carcasses are to be converted into the primal cuts. In packer-style dressing, the backbone is split full length through the center; the head, without the jowl, is removed; and the kidneys and the leaf fat are removed and the hams faced.

The shipper style is ordinarily limited to light-weight slaughter pigs that are sold as entire carcasses into the wholesale trade. In this style of dressing, the carcass is merely opened from the crotch to the tip of the breastbone; the backbone is left intact; the leaf fat is left in; and the entire head is left attached. Roasting pigs are dressed shipper style and prior to cooling are placed in a trough with front legs doubled back from the knee joints and the hind legs extending straight back from the hams.

Dressing Percentage of Hogs

Dressing percentage is the percentage yield of chilled carcass in relation to the weight of the animal on-foot. For example, a hog that weighed 250 lb on-foot and yielded a carcass weighing 185 lb has a dressing percentage of 74%.

The degree of fatness and the style of dressing are the important factors affecting dressing percentage in hogs. U.S. No. 1 hogs dressed packer style (with head, leaf fat, and kidneys removed) dress about 74%, whereas hogs dressed shipper style (head left on, and leaf fat and kidneys in) dress 4 to 8% higher.

It is generally recognized that fat, lardy-type hogs give a higher dressing percentage than expected with meat-type or bacon-type animals. Because lard frequently sells at a lower price than is paid for hogs on-foot, an excess yield of lard very obviously represents an economic waste of feed in producing the animals and is undesirable from the standpoint of the processor.

Attaching great importance to the projected dressing percentage of hogs is outmoded. Because most packers currently pay for the hogs based on carcass weight, the dressing percentage has little effect in today's marketing systems.

Hogs have a relatively smaller barrel and chest cavity than cattle and sheep. In addition, they are dressed with their skin and shanks on. Consequently, they dress higher than other classes of slaughter animals.

Since 1960 market weights have increased 0.7 lb per year and dressing percentage has increased 4.5 percentage points in total. This indicates that more carcass (meat) is being produced both by increasing live-weights and the quantity of that live-weight that results in carcass.

DISPOSITION OF THE PORK CARCASS

Almost all hog carcasses are cut up at the slaughtering plant and are sold in the form of wholesale and retail cuts. In most parts of the United States, less than 1% of the pork in large packing plants is sold in carcass form. The whole-carcass trade is largely confined to roasting and slaughter pigs.

The handling of pork differs further from that of beef and lamb in that much of it is cured by various methods, rendered into lard, or manufactured into meat products. In general, loins, Boston shoulders, and spareribs are most likely to be sold as fresh cuts. But it must be remembered that practically every pork cut may be cured and, under certain conditions, is cured. Because pork is well adapted to curing, it has a decided advantage over beef and lamb, which are sold almost entirely in the fresh state. The hog market is stabilized to some extent by this factor.

Hog Carcass and Wholesale Cuts

Almost 100% of the carcasses are split in half through the spinal column on the slaughter floor and then moved to cooling rooms. A minimum of 24 hours chilling at temperatures ranging from 33 to 38°F is necessary to remove the animal heat properly and give the carcasses sufficient firmness to make possible a neat job of cutting. After chilling, the carcasses are brought to the cutting floor where they are reduced to the wholesale cuts. Some hot processing is increasing as a means of saving energy.

The method of cutting varies somewhat according to the relative demand for different cuts. Despite some variation, the most common wholesale cuts of pork are leg (ham), side (bacon), loin, picnic shoulder, and the Boston shoulder (butt) indicated in Figure 20.3. The jowls and feet are also part of the carcass. Most of the internal organs and glands are also harvested and these are discussed later in this chapter.

A market hog weighing 250 lb will have about 58% of live-weight in the five primal cuts: the ham, loin, side, Boston butt, and the picnic. Yet because of the relatively higher value per pound of these cuts, they make up about three-fourths of the value of the entire carcass.

Lard

Lard is the fat rendered (melted out) from fresh, fatty pork tissue. It is considered a primary product of hog slaughter and not a by-product. The proportion varies with the type, weight, and finish of the hogs. Lard production per slaughtered hog has decreased sharply in recent years, with the shift away from fat, lardy-type hogs to lean, muscular animals (Table 20.1).

Lard is classified according to the part of the animal from which the fat comes and the method of rendering as follows: kettle-rendered lard, steam-rendered lard, dry-processed-rendered lard, neutral lard, lard substitutes, and lard oil and stearin. Stearin is characterized by its solidity and contributes materially to the hardness of fat.

- **Modern lard**—In an all-out attempt to meet consumer demands, meat packers modernized lard. They (1) discolored it—they made it white as snow (natural lard is pale bluish in color); (2) deodorized it; (3) hydrogenated it—raised the melting point; (4) added an antioxidant—so that it would keep on the shelf; and (5) placed it in a container that would preserve these qualities. But to no avail! The U.S. per capita consumption of lard plummeted from 14.2 lb in 1940 to 2.0 lb in 1999.

PORK RETAIL CUTS AND HOW TO COOK THEM

The common retail cuts of pork are illustrated in Figure 20.4. Although the majority of the pork is cut into the five wholesale cuts in the United States, growing ethnic markets are desiring pork cuts more traditional to their cultures.

It is important that pork be cooked at low temperatures, usually between 300 and 350°F. At these temperatures, it cooks slowly, and, as a result, is juicier, shrinks less, and has a better flavor than when cooked at high temperatures.

For best results, a meat thermometer should be used to test the doneness of roasts (and also for thick steaks and chops). It takes the guesswork out of meat cooking. Allowing a certain number of minutes to the pound is not always accurate, for example, rolled roasts take longer to cook than bone-in roasts.

The thermometer is inserted into the cut of meat so that the end reaches the center of the largest muscle, and so that it is not in contact with fat or bone. Frozen roasts need to be partially thawed before the thermometer is inserted, or a metal skewer or ice pick will have to be employed in order to make a hole in frozen meat.

PURCHASING PORK

A Consumer Guide To Identifying Retail Pork Cuts.

Left: tenderloin
Right: Canadian-style bacon

TENDERLOIN & CANADIAN-STYLE BACON

CHOPS

CHOPS
Upper row (l-r):
sirloin chop, rib chop,
loin chop.
Lower row (l-r):
boneless rib end chop,
boneless center loin
chop, butterfly chop.

ROASTS
Upper row (l-r):
center rib roast
(Rack of Pork),
bone-in sirloin roast.
Middle: boneless
center loin roast.
Lower row (l-r):
boneless rib end roast,
boneless sirloin roast.

RIBS

ROASTS

RIBS Left: country-style ribs.
Right: back ribs.

BLADE BOSTON-STYLE SHOULDER

BLADE BOSTON-STYLE SHOULDER
Upper row (l-r):
bone-in blade roast, boneless blade roast.
Lower row (l-r): ground pork, sausage,
blade steak.

PICNIC SHOULDER

PICNIC SHOULDER
Upper row (l-r): smoked picnic, arm
picnic roast.
Lower row: smoked hocks.

SIDE Top: spareribs.
Bottom: slab bacon, sliced bacon.

SIDE

LEG

LEG
Upper row (l-r):
bone-in fresh ham,
smoked ham. Lower
row (l-r): leg cutlets,
fresh boneless ham
roast.

Diagram labels: BLADE BOSTON-STYLE SHOULDER, LOIN, ARM PICNIC SHOULDER, SIDE, LEG

pork checkoff
©1997 NATIONAL PORK BOARD

Figure 20.3 The wholesale cuts of pork are shown on the pig diagram and the corresponding retail cuts. Most carcasses leave the packing plant cut into these five sections and are sold in boxes to meat processors or retailers for further cutting. *(Courtesy, National Pork Board, Des Moines, IA)*

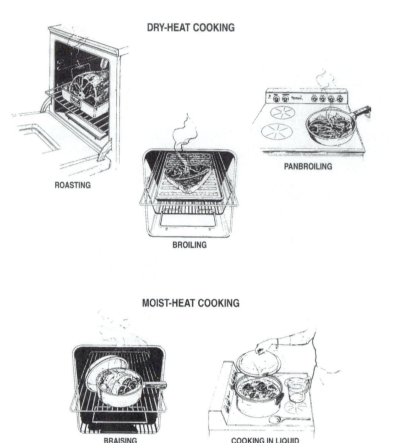

DRY-HEAT COOKING

ROASTING

PANBROILING

BROILING

MOIST-HEAT COOKING

BRAISING

COOKING IN LIQUID

Figure 20.4   Some common methods of pork cookery. More information is available from the National Pork Board on cooking methods, specific recipes, cuts of pork, and temperatures (www.otherwhitemeat.com/default.asp). *(Courtesy, National Pork Board, Des Moines, IA)*

As the oven heat penetrates the meat, the temperature at the center of it gradually rises and registers on the thermometer. Although most meat can be cooked as desired—rare, medium, or well done—pork should always be cooked medium, 160°F, or well done, 170°F, for fresh pork and 155°F for cured pork.

The methods used in meat cookery depend on the nature of the cut to which it is applied. In general, the types of meat cookery may be summarized as follows:

1. **Dry-heat cooking.** Dry-heat cooking is used in preparing the more tender cuts, those that contain little connective tissue. This method of cooking consists of surrounding the meat by dry air in the oven or under the broiler. The common methods of cooking by dry heat are (a) roasting, (b) oven broiling, and (c) pan broiling (see Figure 20.4).

2. **Moist-heat cooking.** Moist-heat cooking is generally used in preparing the less tender cuts containing more connective tissues. These require moist heat to soften them and make them tender. In this type of cooking the meat is surrounded by hot liquid or steam. The common methods of moist-heat cooking are (a) braising and (b) cooking in water or stewing (see Figure 20.4).

3. **Frying.** When a small amount of fat is added before cooking, or allowed to accumulate during cooking, the method is called pan frying. This is suitable for preparing comparatively thin pieces of tender meat, or those pieces made tender by pounding, scoring, cubing, or grinding, or for preparing leftover meat. When meat is cooked immersed in fat, it is called deep-fat frying. This method of cooking is sometimes used for preparing brains, liver, and leftover meat. Usually the meat is coated with eggs and crumbs or a batter, or dredged with flour or cornmeal.

4. **Microwave cooking.** Meat cookery researchers do not recommend cooking fresh pork in the microwave oven. The microwave is, however,

Figure 20.5 Supermarket bacon and sausage in the Mapledale Giant in Dale City, Virginia. *(Photo by Ken Hammond, Courtesy, USDA, Washington, DC)*

very acceptable for cooking cured pork products or reheating cooked fresh pork cuts.

CURING PORK

In the United States, meat curing is largely confined to pork, primarily because of the keeping qualities and palatability of cured pork products. Considerable beef is corned or dried, and some lamb and veal are cured, but none of these is of such magnitude as cured pork (Figure 20.5).

Meat is cured with salt, sugar, and certain curing adjuncts (ascorbate, erythrobate, and so on); with sodium nitrite or sodium nitrate (the latter is used only in certain products); and with smoke. Reasons for each of the most common curing additives are as follows:

- **Salt**—Sodium chloride is added to cured meats as preservative and for palatability. The addition of salt makes it possible to transport certain perishable meats through often lengthy and complicated distribution systems. Also, in products such as wieners, bologna, and canned ham, salt solubilizes myosin, the major meat protein, and causes the meat particles to hold together; the meat can then be sliced without falling apart.
- **Sugar**—Sugar, sucrose, dextrose, or invert sugar counteracts the harshness of salt, enhances the flavor, and lowers the pH of the cure.
- **Sodium nitrite ($NaNO_2$), sodium nitrate ($NaNO_3$), and potassium nitrate or saltpeter (KNO_3)**—Nitrite and nitrate contribute to (1) the prevention of *Clostridium botulinum* spores in or on meat (*C. botulinum* bacteria produce the botulin toxin that causes botulism, a deadly form of food poisoning); (2) the development of the characteris-

tic flavor and pink color of cured meats; (3) the prevention of "warmed-over flavor" in reheated products; and (4) the prevention of rancidity. Of all the effects of nitrite and nitrate, the antibotulism effect is by far the most important.

Nitrate is considered essential in cured meats because it performs several important functions, including (1) preventing botulism, (2) retarding liquid oxidation, (3) giving cured meats their characteristic cured flavors, and (4) imparting the characteristic cured pink color.

Nitrite usage is permitted only in the production of country-cured hams and in some sausages that are cured for several weeks.

- **Smoking**—Smoking produces the distinctive smoked-meat flavor that consumers demand in certain meats. Many meat packers and processors now use either (1) "liquid smoke," made from natural wood smoke treated so as to remove certain components; or (2) synthetic smoke made by mixing pure chemical compounds found in natural smoke.
- **Phosphate (PO_4)**—Phosphate is added to increase the water-binding capacity of meat and its use increases yields up to 10%. Also, it enhances the juiciness of the cooked product. Federal regulations restrict the amount of phosphates to (1) not more than 0.5% in the finished product and (2) not more than 5% in the pickle solution based on 10% pumping pickle. The addition of phosphate must be declared on the label.

Farm meat curing other than freezing is largely confined to pork, primarily because of the keeping qualities and palatability of cured pork products. The secret of pork curing is to use good sound meat, the correct curing method and formula, clean containers, and to be fortunate enough to have cool curing weather.

The primary meat curing ingredients are salt, sugar, and saltpeter (potassium nitrate). A combination of 7 lb of salt and 3 lb of white or brown sugar is a basic mixture. In addition to salt or salt and sugar, commercial cures frequently contain spices and flavorings to impart characteristic flavor, appearance, and aroma.

NUTRITIONAL QUALITIES

Perhaps most people eat pork simply because they like it. They derive a rich enjoyment and satisfaction from the flavor and the variety of preparation methods (Figure 20.6).

But pork is far more than simply a very tempting and delicious food. Nutritionally, it contains certain essentials of an adequate diet: high-quality

Figure 20.6 Raisin-apricot glazed ham is a great winter holiday treat. *(Courtesy, National Pork Board, Des Moines, IA)*

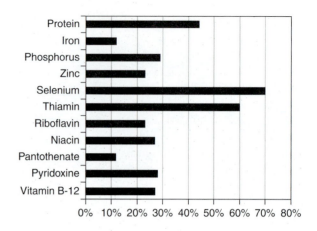

Figure 20.7 Nutritional value of pork, in percentage of the recommended daily allowance (RDA). Based on a 3-oz (85-g) serving of a composite of fresh, trimmed retail cuts (leg, loin, and shoulder), separable lean only, cooked. *(Courtesy, USDA-ARS, 2003. USDA National Nutrient Database for Standard Reference, Release 16 [www.nal.usda.gov/fnic/foodcomp] and RDA for a 19- to 50-year-old male [www.iom.edu/project. asp?id=4574], 2003)*

proteins, minerals, and vitamins. This is important because how we live and how long we live are determined in large part by our diet. Effective pork promotion necessitates full knowledge of the nutritive qualities of meats, the pertinent facts of which follow:

1. Proteins. The word *protein* is derived from the Greek word *proteios*, meaning *primary*. Protein is needed for growth. Fortunately, meat contains the proper quantity and quality of protein for the building and repair of body tissues. On a fresh basis, pork contains 15 to 20% protein. Also, it contains all of the amino acids, or building blocks, that are necessary for the making of new tissue. Pork is a high-quality protein—a complete protein—because it contains all the essential amino acids. A 3-oz serving of pork supplies 44% of the daily protein requirement of a 19- to 50-year-old male (Figure 20.7).

2. Calories. Pork is a good source of energy and its energy value dependent largely upon the amount of fat it contains.

3. Minerals. Minerals are necessary to build and maintain the body skeleton and tissues and to regulate body functions. Pork is a rich source of several minerals but is especially good as a source of iron, phosphorus, zinc, and selenium. Phosphorus com-

bines with calcium in building the bones and teeth. Phosphorus also enters into the structure of every body cell, helps to maintain the alkalinity of the blood, is involved in the output of nervous energy, and has other important functions.

Iron is necessary for the formation of blood, and its presence protects against nutritional anemia. It is a constituent of the hemoglobin or red pigment of the red blood cells. Thus, it helps to carry the life-giving oxygen to every part of the body. Iron from meat, such as pork, is absorbed most readily by the body, and it also increases the absorption of iron from vegetable sources.

4. Vitamins. As early as 1500 B.C., the Egyptians and Chinese discovered that eating liver would improve one's vision in dim light. We now know that liver furnishes vitamin A, a very important factor for night vision. In fact, medical authorities recognize that night blindness, glare blindness, and poor vision in dim light are all common signs pointing to the fact that the persons so affected are not getting enough vitamin A in their diets.

Meat is one of the richest sources of the important B group of vitamins, especially thiamin, riboflavin, niacin, B_6, and vitamin B_{12} (Figure 20.8). Pork is the leading dietary source of thiamin, containing three times as much as any other food.

These B vitamins are now being used to fortify certain foods and are indispensable in our daily diet. They are necessary for energy metabolism, synthesis of new tissue (growth), nervous function, and many other functions. A marked deficiency of thiamin causes beriberi. Niacin prevents and cures the disease pellagra. Indeed, one of the reasons for the rapid decline in B vitamin deficiencies in the United States may well be the increased amount of meat and other B vitamin–containing foods in the daily diet.

5. Digestibility. Finally, in considering the nutritive qualities of meats, it should be noted that pork is highly digestible. About 97% of meat proteins and 96% of meat fats are digested. The statement often is heard that "pork is hard to digest." This is not true. Pork, in common with all meats, is well utilized by the body.

Pork is playing an important part in the nutrition of the United States.

CONSUMER HEALTH ISSUES

Much has been written and spoken linking the consumption of meat, including pork, to certain diseases. The primary concerns of consumers about pork pertain to (1) fats and cholesterol, (2) nitrites and nitrosamines, and (3) safety and quality assurance. Each of these concerns is discussed in a separate section that follows.

Cholesterol

In 1953, Dr. Ancel Keys of the University of Minnesota first reported a positive correlation between the consumption of animal fat (which is high in cholesterol) and the occurrence of atherosclerosis (heart disease) in humans. Subsequently, other studies correlated high blood levels of cholesterol with increased incidence of atherosclerosis in humans.

Because cholesterol is present in many animal and food products, including pork and lard, it was inevitable that they would be incriminated as causes of heart disease. It is important, therefore, that all members of the pork team—producers, processors, and retailers—along with consumers, are aware of the cholesterol issue.

From 1963 to 1990, hogs slimmed down so much that, on the average, fresh pork was 76% lower in fat and 53% lower in calories (Table 20.3).

TABLE 20.3   Improved Leanness of 3-oz Serving of Broiled Pork Loin

Year	Grams of Fat	Calories
1963	29.6	351
1983	11.7	202
1990	6.9	165
2003	8.2	165

Source: U.S. Department of Agriculture, Agricultural Research Service.

Around 1990 it was believed that much of the pork was getting to be devoid of marbling and flavor was lacking. Since then the fat content has been raised to about 8.2 g of fat in a 3-oz serving.

- **About cholesterol**—Cholesterol is found in all body tissues, especially the brain and spinal cord. Further, cholesterol is an essential ingredient for certain biochemical processes, including the production of sex hormones in humans.

Levels and kinds of cholesterol and the associated risks are presented in Table 20.4. Many factors such as (1) age, (2) time of day, (3) physical condition, (4) stress, (5) genetic background, and (6) laboratory expertise may all affect the levels. Therefore, routine screening is essential if you are in an "at-risk" category. Following is an explanation of these categories.

Total Blood Cholesterol

Desirable—Your heart attack risk is relatively low, unless you have other risk factors. Even with a low risk, it is still smart to eat foods low in saturated fat and cholesterol, and also get plenty of physical activity. Have your cholesterol levels measured every 5 years or more often if you are a man over 45 or a woman over 55.

Borderline high risk—About a third of all U.S. adults are in the borderline group; almost half of adults have total cholesterol levels below 200 mg/dL. You should also lower your intake of foods high in saturated fat and cholesterol

TABLE 20.4   Risk Levels of Blood Cholesterol, LDL Cholesterol, and HDL Cholesterol

Total Blood Cholesterol	Level (mg/dL)
Desirable	Less than 200
Borderline high risk	200–239
High risk	240 and over
LDL Cholesterol	*Level (mg/dL)*
Optimal	Less than 100
Near optimal/above optimal	100–129
Borderline high	130–159
High	160–189
Very High	190 and above
HDL Cholesterol	*Level (mg/dL)*
Average man	40–50
Average woman	50–60
Low	Less than 40

Source: American Heart Association, 2002; http://www.american-heart.org/presenter.jhtml?identifier=183

to reduce your blood cholesterol level to below 200 mg/dL. Even if your total cholesterol is between 200 and 239 mg/dL, you may not be at high risk for a heart attack. Some people, such as women before menopause and young, active men who have no other risk factors, may have high HDL cholesterol and desirable LDL levels.

High risk—Your risk of heart attack and stroke is greater. In general, people who have a total cholesterol level of 240 mg/dL have twice the risk of heart attack as people whose cholesterol level is 200 mg/dL.

LDL Cholesterol

Your LDL (low density lipoprotein) cholesterol level greatly affects your risk of heart attack and of stroke. The lower your LDL cholesterol, the lower your risk. In fact, it is a better gauge of risk than total blood cholesterol.

Your doctor may prescribe a diet low in saturated fat and cholesterol, regular exercise and a weight management program if you are overweight. If you cannot lower your cholesterol with these efforts, medications may also be prescribed to lower your LDL cholesterol.

HDL Cholesterol

HDL (high density lipoprotein) cholesterol levels are beneficial. If your HDL cholesterol is low, less than 40 mg/dL, you are at high risk for heart disease. Smoking, being overweight, and being sedentary can all result in lower HDL cholesterol. If you have low HDL cholesterol, you can help raise it by (1) not smoking, (2) losing weight (or maintaining a healthy weight), and being physically active for at least 30 to 60 minutes a day on most or all days of the week.

Cholesterol Ratio

Knowing your total blood cholesterol level is an important first step in determining your risk for heart disease. However, a critical second step is knowing your HDL or "good" cholesterol level. The cholesterol ratio is obtained by dividing the HDL cholesterol level into the total cholesterol. For example, if a person has a total cholesterol of 200 mg/dL and an HDL cholesterol level of 50 mg/dL, the ratio would be stated as 4:1. The goal is to keep the ratio below 5:1; the optimum ratio is 3.5:1.

Cholesterol in the body arises from two sources: (1) that from the diet, or exogenous cholesterol; and (2) that manufactured in the body, or endogenous cholesterol. Cholesterol in the blood reflects the overall cholesterol metabolism, that derived from both sources. Pertinent facts about the metabolism of cholesterol in the human body follow.

1. Digestion and absorption. The average individual consumes between 500 and 800 mg of cholesterol each day. Dietary fat aids the absorption of cholesterol. At high levels, a person absorbs less than 10%, and the remainder leaves the body via the feces. About 2 to 4 hours after eating, the blood cholesterol is indistinguishable between that consumed and that synthesized in the body.

2. Synthesis. The liver is the major site of synthesis, but most tissues, except possibly the brain, are capable of synthesizing cholesterol. Each day the body manufactures 1,000 to 2,000 mg of cholesterol. However, on a day-to-day basis the synthesis and metabolism of cholesterol is controlled by such factors as (1) fasting; (2) caloric intake; (3) cholesterol intake; (4) bile acids; (5) hormones, primarily the thyroid hormones and estrogen; and (6) disorders such as diabetes, gallstones, and hereditary high blood cholesterol—hypercholesterolemia. Control of cholesterol synthesis by cholesterol intake is important because this means that when intake is high then synthesis is low and vice versa.

3. Functions. Cholesterol is vital to the body. Its primary importance concerns tissues, bile acids, and hormones.

4. Excretion. Removal from the body occurs primarily via the conversion of cholesterol to bile acids. About 0.8 mg of cholesterol is degraded daily by this method. Also, a minor amount is converted to the previously mentioned hormones. Additionally, some cholesterol is never digested, and hence, excreted via the feces, particularly when intake is high.

The following seven changes should be made in the diets of persons having high blood cholesterol:

1. **Eat less high-fat food.** The intake of total fat should constitute less than 30% of the calories. *Note: Eating less total fat is an effective way to eat less saturated fat and fewer calories.*
2. **Eat less saturated fat.** The intake of saturated fat should constitute less than 10% of the calories. *Note: Saturated fats are found primarily in animal products. But a few vegetable fats and many commercially processed foods also contain saturated fat. Read labels carefully.*
3. **Substitute unsaturated fats for saturated fat.** Unsaturated fats (polyunsaturated and monounsaturated fats) should be substituted for saturated fat to the extent practical. *Note: Unsaturated fats lower blood cholesterol levels when substituted for saturated fats.*

TABLE 20.5 Cholesterol Content of Some Common Foods

Food	Cholesterol (mg/100 g)
Beef, chuck, 1/4 in. fat, cooked	104
Beef, frankfurter,	53
Beef, ground, 90% lean, broiled	85
Beef, liver, braised	396
Beef, sirloin, trimmed to 0 in. fat, roasted	83
Butter, salted	215
Cheese, cottage, 2% milkfat	8
Cheese, cheddar	105
Cheese, Swiss	92
Chicken, breast, roasted	85
Chicken, broiler, roasted	107
Chicken fillet sandwich, fast food	33
Egg, whole, chicken, hard-boiled	424
Egg yolks, fresh, chicken	1,234
Fish, catfish, cooked, dry heat	64
Fish, cod, cooked, dry heat	55
Fish, salmon, pink, cooked, dry heat	67
Ice cream, 11% fat	44
Ice cream, soft serve, 2.6% fat	12
Margarine	0
Milk, 1% fat	4
Milk, 2% fat	8
Milk, whole, 3.7% fat	14
Pork, ham, cured, cooked	57
Pork, liver, braised	355
Pork, loin, cooked	85
Shrimp, cooked, moist heat	195
Turkey, breast, roasted	83
Turkey, whole, roasted	105
Veal, leg, cooked	134
Yogurt, plain, low fat	6

Source: U.S. Department of Agriculture, Agricultural Research Service, 2003. USDA National Nutrient Database for Standard; Reference, Release 16; http://www.nal.usda.gov/fnic/foodcomp

4. **Eat less high-cholesterol food.** The intake of cholesterol should be less than 300 mg/day. Dietary cholesterol can raise the blood cholesterol level. Therefore, it is important to eat less food that is high in cholesterol (Table 20.5).

   Note: There is very little cholesterol in low-fat dairy foods, such as skim milk or yogurt, and no cholesterol in food from plants, such as fruits, vegetables, vegetable oils, grains, cereals, nuts, and seeds.

5. **Substitute complex carbohydrates for saturated fats.** Breads, pasta, rice, cereal, dried peas and beans, fruits, and vegetables are good sources of complex carbohydrates (starch and fiber). They are excellent substitutes for foods that are high in saturated fat and cholesterol.

   Note: Foods that are high in complex carbohydrates, if eaten plain, are low in saturated fat and cholesterol as well as being good sources of minerals, vitamins, and fiber.

6. **Maintain a desirable weight.** People who are overweight frequently have higher blood cholesterol levels than people of desirable weight.

   Note: To achieve or maintain a desirable weight, caloric intake must not exceed the number of calories the body burns.

7. **Eat foods that are high in soluble fiber.** Among such foods is oat bran.

   Note: Basically, there are two types of fibers: water soluble fiber (such as oat bran) and nonsoluble fiber (such as wheat bran). Only soluble fiber is effective in lowering cholesterol. Soluble fiber dissolves in water.

All of the preceding indicate that the development of atherosclerosis is not simply a matter of eating too much cholesterol. Moreover, these recommendations encompass accepted measures of good health—cessation of smoking, normal blood pressure, ideal weight, exercise, control of stress, and awareness of family history. Each of the recommendations complements the others and contributes to decreasing the risks of atherosclerosis and heart disease.

Nitrates/Nitrites and Nitrosamines

Generations of consumers have been accustomed to the bright pink color of cured meat imparted by nitrates and nitrites. However, they are considered essential in cured pork because they perform several important functions, the most important of which is their antibotulism effect (Figure 20.8).

Both nitrate and nitrite are allowed in meats in specified, limited levels under the Meat Inspection Act, but in fact nitrate is not used by the industry with the possible exception of country-cured hams and in some sausages that are cured for several weeks. Additionally, ascorbates are now added because they accelerate and improve the curing process and also inhibit the formation of carcinogenic nitrosamines.

Controversy over the use of nitrates/nitrites, and their production of nitrosamines, stems primarily from nitrosamines producing cancer in test animals. Pork producers, processors, and retailers, along with consumers, need to know the facts about nitrates/nitrites and nitrosamines.

- **About nitrates/nitrites and nitrosamines**—Nitrate refers to the chemical union of 1 nitrogen (N) and 3 oxygen (O) atoms, or NO_3, whereas nitrite refers to the chemical union of 1 nitrogen (N) and 2 oxygen (O) atoms, or NO_2. Of prime concern in foods are sodium nitrate (Chile saltpeter), potassium nitrate (saltpeter), sodium nitrite, and potassium nitrite.

- **Occurrence and exposure**—Nitrates and nitrites are common chemicals in our environment whether they come from natural sources, food additives, or drinking water.

Figure 20.8 Polish sausages, one of many products produced with nitrites, on a bed of cabbage. *(Courtesy, National Pork Board, Des Moines, IA)*

TABLE 20.6 Estimated Average Daily Ingestion of Nitrate and Nitrite per Person in the United States

Source	Nitrate (mg)	Nitrite (mg)
Vegetables	86.1	0.20
Cured meats[1]	9.4	2.38
Bread	2.0	0.02
Fruits, juices	1.4	0.00
Water	0.7	0.00
Milk and products	0.2	0.00
Total	99.8	2.60
Saliva[2]	30.0	8.62

[1]Because of added ascorbate and decreased levels of nitrite used, the residual nitrite content of modern cured meats is about one-fifth of that typically found 20 years ago. Nitrate is no longer used by the meat industry. Therefore, the value for nitrate in cured meats could be near zero and the nitrite about 0.50 mg (*Food Technology,* 51:2, pp. 53–55, 1997).
[2]Not included in the total because the amount of nitrite produced by bacteria in the mouth depends directly on the amount of nitrate ingested.
Source: Nitrates: An Environmental Assessment, *1978, National Academy of Sciences, p. 437, Table 9.1.*

1. **Naturally occurring.** Most green vegetables contain nitrates. The level of nitrates in vegetables depends on (1) species; (2) variety; (3) plant part; (4) stage of plant maturity; (5) soil condition, such as deficiencies of potassium, phosphorus, and calcium or excesses of soil nitrogen; and (6) environmental factors, such as drought, high temperature, time of day, and shade. Regardless of the variation of nitrate content in plants, vegetables are the major source of nitrate ingestion, followed by saliva (Table 20.6).

Those vegetables that are most apt to contain high levels of nitrates include beets, spinach, radishes, and lettuce. Other natural sources of nitrate are negligible.

2. **Food additives.** Both nitrates and nitrites are used as food additives, mainly in meat and meat products, according to the guidelines presented in Table 20.7. Their use in meat has been the subject of much publicity, though the origin of their use to cure meat is lost in antiquity. The role of nitrates in meats is not clear, though it is believed that they provide a reservoir source of nitrite because microorganisms convert nitrate to nitrite. It is the nitrite that decomposes to nitric oxide, NO, and reacts with heme pigments to form nitrosomyoglobin, giving meats their red color. Taste panel studies on bacon, ham, hot dogs, and other products have demonstrated that a definite preference is shown for the taste of those products containing nitrites.

In addition, nitrites retard rancidity, but more importantly, they inhibit microbial growth, especially *Clostridium botulinum.* Hence, cured meats provide a source of ingested nitrates and nitrites. However, as a source of nitrate, cured meats are minor compared to vegetables. The major dietary source of nitrites is cured meats, but this is small when compared to that produced by the bacteria in the mouth and swallowed with saliva. Currently, there is no other protection from botulism as effective as nitrites. So, for those wishing to eliminate nitrites as possible carcinogens, there is a risk—eliminate the nitrites and increase botulism poisoning, a proven danger.

3. **Other sources.** Aside from very unusual circumstances, other sources of exposure to nitrates and nitrites are relatively minor. Nitrate concentrations in groundwater used for drinking range from several hundred micrograms per liter to a few milligrams per liter. Nitrates are generally higher in groundwater than in surface because water because plants remove the nitrogen from surface water.

• **Dangers of nitrates and nitrites**—These chemicals may present a hazard to people through two routes. First, under certain circumstances, nitrates and nitrites can be directly toxic. Second, nitrates and nitrites contribute to the formation of cancer-causing nitrosamine. Third, there are many non-food exposures to nitrosamines.

TABLE 20.7   Federal Nitrate and Nitrite Allowances in Meat

Meat Preparation	Level Allowed	
	Sodium or Potassium Nitrate	*Sodium or Potassium Nitrite*
Finished product	200 ppm or 91 mg/lb (maximum)	200 ppm or 91 mg/lb (maximum)
Dry cure	3.5 oz/100 lb or 991 mg/lb	1.0 oz/100 lb or 283 mg/lb
Chopped meat	2.75 oz/100 lb or 778 mg/lb	0.25 oz/100 lb or 71 mg/lb

Source: USDA Food Safety and Inspection Service, Washington, DC.

1. Toxicity. Our knowledge of the toxic effects of nitrates and nitrites is derived from its long use in medicine, accidental ingestion, and ingestion by animals. Overall, poisoning by nitrates is uncommon. An accidental ingestion of 8 to 15 g causes severe gastroenteritis, blood in the urine and stool, weakness, collapse, and possibly death. Fortunately, nitrate is rapidly excreted from the adult body in the urine, and the formation of methemoglobin generally is not part of the toxic action of nitrates.

Almost all cases of nitrate-induced methemoglobinemia in the United States have resulted from the ingestion of infant formula made with water from a private well containing an extremely high nitrate level. Overall, it is comforting to note that several hundred million pounds of beets and spinach—nitrate-containing vegetables—are eaten yearly without injury.

2. The cancer question. Without doubt, the greatest concern of people is the involvement of nitrates and nitrites in directly causing cancer, or in indirectly producing compounds known as nitrosamines. Because nitrosamines are definitely accepted as carcinogens in test animals, a majority of the furor around nitrates and nitrites stems from this fact.

It cannot be stated that any human cancer has been positively attributed to nitrosamines. However, some nitrosamines have caused cancer in every laboratory animal species tested.

3. Exposure to nitrosamines. When nitrosamines are mentioned, foods are the first items that come to mind. However, there are numerous other sources of nitrosamines, including cosmetics, lotions, and shampoos containing N-nitrosodiethanolamine (NDELA, carcinogenic in the rat). Other preformed nitrosamines have been found in tobacco and tobacco smoke. Hence, human exposure can result from breathing or eating preformed nitrosamines, or by applying them to the skin.

On an individual basis, after carefully considering the issue, it is the old question of benefit versus risk. The risk of botulism in cured meats in the absence of nitrite is both real and dangerous, whereas the risk of cancer from low levels of nitrosamines or nitrites remains uncertain. Thus, the risk of botulism is considered greater than the risk of nitrates, so even though little or no nitrate is used, it is allowed. Furthermore, no acceptable alternative is as effective as nitrite in preventing botulism. Nevertheless, nitrite should be reduced in all products to the extent protection against botulism is not compromised.

Food Safety

America's food supply is the safest in the world! Nevertheless, there is need for constant vigilance and improvement, especially in animal products, which are subject to all the hazards of other foods (spoilage, pesticides, toxicities), plus being capable of transmitting, or serving as passive carriers of, certain diseases to humans.

In colonial times, the livestock producer slaughtered animals and processed meats, milked the cows, and gathered the eggs, then delivered the products door to door to nonfarm customers. If the products were not acceptable (spoiled meat, sour milk, cracked eggs), the matter was resolved quickly and on the spot, or the producer lost a customer. Today, the public expects the livestock team—farmers, processors, and retailers—to provide wholesome and safe products free from disease agents, toxic substances, and pesticide and drug residues.

Uptake of pesticides by animals, leading to residues in animal products, can result either from direct application of pesticides to animals or from animals ingesting feeds carrying pesticide residues. Drug residues are caused by (1) producers failing to withdraw drugs from livestock far enough in advance of marketing products; (2) contaminated feed storage, mixing, and handling equipment; or (3) the wastes (feces and urine) of treated animals coming in contact with untreated animals. *Reading and following directions on labels is key to safe pesticide and drug use.*

Dr. C. Everett Koop, former U.S. Surgeon General, said, "People who are worried about pesticides fail to recognize that cancer rates in the United States have dropped remarkably over the past 40 years. During this period of time, stomach cancer has dropped more than 75%, and rectal cancer has dropped more than 65%. The only cancer going up today is cigarette-induced lung cancer.

"The same hysteria that sometimes accompanies food safety issues also has affected how Americans look at diet and health. However, one major issue—cholesterol—is waning, as consumers realize how much it has been oversold. The cholesterol bubble has been pricked and is slowly deflating. While cholesterol is a risk factor in coronary heart disease, scientists consider other risk factors, such as smoking, hypertension, and heredity, to be much more significant than cholesterol. Because cholesterol is manufactured in the body naturally, diet does not have the direct relationship to blood levels of cholesterol that many misled laymen assume."[1]

Because the welfare of the nation depend on the health of its people, animal (and other) products are carefully monitored by various government agencies to assure consumers that they are wholesome and safe; and because of recognizing the importance of consumers in the safety of their food, the private sector may do additional testing. The agencies most responsible for this important work are

1. The U.S. Department of Health and Human Services, including the Centers for Disease Control and Prevention (CDC), the Food and Drug Administration (FDA), and the National Institutes of Health (NIH).
2. The U.S. Department of Agriculture, including the Agricultural Research Service (ARS), the Animal and Plant Health Inspection Service (APHIS), the Food Safety and Inspection Service (FSIS), and the Cooperative State Research, Education, and the Extension Service (CSREES).
3. State and local government agencies.
4. International organizations engaged in health or nutrition activities, including the World Health Organization (WHO) and Food and Agriculture Organization (FAO).

Government and Industry Programs

In addition to nutritional quality, consumers are concerned about the safety of their food. Food safety and improved meat inspection became public issues in the 1990s. Several pork safety proposals followed:

1. **Irradiation.** The World Health Organization (WHO), American Medical Association (AMA), and the International Atomic Energy Agency have approved the safety of the irradiation process. Yet, consumers are concerned about something they as-

sociate with a deadly force. Currently, some of the ground beef produced in the Midwest is being irradiated by a company in Iowa. Also, many spices, meals for astronauts and the military, and hospital supplies are irradiated.

2. **Hazard Analysis, Critical Control Point (HACCP).** The HACCP plan identifies hazards in food processing, then monitors crucial points. Problems are fixed as they occur as the entire processing and production of a product is monitored. It is a way of processing safer food. Most, if not all, major and local meat processors have undertaken HACCP training.

3. **Rapid microbial test.** The rapid microbial test, which adapts technology being used in the pharmaceutical and beer industries, takes 5 minutes and can be used in commercial meat plants. It provides a means of verifying that meat and poultry plants are operating under appropriate microbiological controls.

4. **Traceback.** Traceback involves a livestock identification system that makes it possible to trace animals back to their origin in order to locate the source of contamination. Some countries have mandatory identification requirements that allow them to trace animals from birth to slaughter.

5. **Trichinosis prevention.** Trichinosis is caused by a microscopic parasite, *Trichinella spiralis.* The Centers for Disease Control and Prevention (2001) reported that cases of human trichinellosis have declined to less than 25 annually over the past several years. Only a few of these cases have been traced and associated with consumption of pork. A USDA, National Animal Health Monitoring System national swine survey conducted in 1995 reported the infection rate in United States swine to be 0.013%. Modern swine management systems have virtually eliminated trichinae as a problem in domestic pigs.
Note: Trichinella is destroyed by cooking pork to 140°F (60°C) internal temperature for 1 minute, or by freezing for 4 days at 14°F (−10°C). (See the Chapter 14 section "Trichinosis [Trichinella spiralis].")

6. **Food safe labeling.** Effective July 6, 1994, the food labeling requirement became effective. It mandates safe cooking and handling labels for all uncooked meat and poultry products. The labels note that some food products may contain bacteria and can cause illness if mishandled or not cooked properly, and instruct consumers to keep raw meat and poultry refrigerated or frozen. Also, the labels warn that raw meat and poultry should be thawed only in a refrigerator or microwave, kept separate from other foods, cooked thoroughly, and refrigerated immediately or discarded.

7. **The National Pork Board's Pork Quality Assurance (PQA) program.** Pork producers are re-

Figure 20.9 Pork Quality Assurance, commonly known as PQA, assists producers in the production of safe pork, free of feed additive residues. Many U.S. pork slaughter plants require all of their pig suppliers to be PQA-certified. *(Courtesy, National Pork Board, Des Moines, IA)*

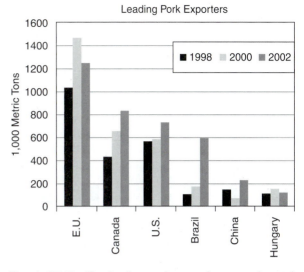

Figure 20.10 The leading pork exporting countries and their changes in exports. The 2002 data are preliminary. *(World Markets and Trade, U.S. Foreign Agricultural Service, 2003)*

quired to follow the FDA's published Compliance Policy Guide (CPG) 7125.37, "Proper Drug Use and Residue Avoidance by Non-veterinarians." In 1989 pork producers created the Pork Quality Assurance program to help producers conform with the FDA policy and to meet consumer demands for quality and safety. Since then the PQA materials have been revised as needed to provide timely, accurate information for producing safe, wholesome pork (Figure 20.9<). The current revision includes the U.S. government's rules and regulations for pork producers and explains how the National Pork Board incorporates these and other scientific principles into the PQA program. The good production practices (GPPs) are outlined in Chapter 15 in the section "Preventing Drug Residues."

DETERMINING RETAIL PORK PRICES

During those periods when pork is high in price, especially the choicest cuts, there is a tendency on the part of the consumer to blame any or all of the following: (1) the producer, (2) the packer, (3) the meat retailer, (4) the government, and these four may blame each other.

Who or what is to blame for high meat prices? If good public relations are to be maintained, it is imperative that each member of the meat team—the producer, the packer, and the meat retailer—be fully armed with documented facts and figures with which to answer such questions and to refute such criticisms. Also, the consumer should know the truth of the situation.

Pork prices are determined by the laws of supply and demand; that is, the price of meat largely depends on what the consumers as a group are able and willing to pay for the available supply.

International pork production also affects pork prices. The leading pork exporting countries or regions include the the European Union, Canada, the United States, and Brazil (Figure 20.10). The remarkable growth in Brazilian production is indicative of the potential of South America to become a major pork producer. As production costs increase and decrease in these areas, the world market will be affected.

Available Supply

Because pork is a perishable product, the supply of this food is very dependent on the number and weight of hogs available for slaughter at a given time. In turn, the number of market animals is largely governed by the relative profitability of the swine enterprise in comparison with other agricultural pursuits. That is to say, swine producers, like other good businesspeople, generally do those things that are most profitable to them. Thus, a short supply of market animals at any given time usually reflects unfavorable and unprofitable production factors that existed some months earlier and that caused curtailment of breeding and feeding operations.

Historically, when short pork supplies exist, pork prices rise, and the market price on slaughter hogs usually advances, making hog production more profitable. But, unfortunately, swine breeding and feeding operations cannot be turned on and off like a spigot.

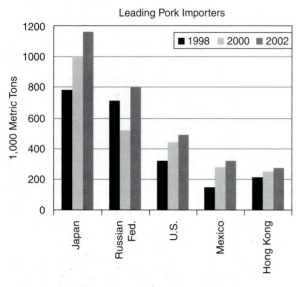

Figure 20.11 The major pork importing countries and the changes since 1998. The 2002 data are preliminary. *(World Markets and Trade, U.S. Foreign Agricultural Service, 2003)*

History also shows that if hog prices remain high and feed abundant, producers will step up their breeding and feeding operations as fast as they can within the limitations imposed by nature, only to discover when market time arrives that too many other producers have done likewise. Overproduction, disappointingly low prices, and curtailment in breeding and feeding operations are the result.

Nevertheless, the operations of livestock producers do respond to market prices, bringing about so-called cycles. Thus, the intervals of high production, or cycles, in hogs—which are litter bearing, breed at an early age, have a short gestation period, and go to market at an early age—occur every 3 to 5 years.

Demand for Pork

The demand for pork is primarily determined by buying power and competition from other products. Stated in simple terms, demand is determined by the spending money available and the competitive bidding of millions of homemakers who are the chief home purchasers of meats. On a nationwide basis, a high buying power and great demand for meats exist when most people are employed and wages are high.

Worldwide, as personal incomes increase demand for meat increases also. The major pork importing countries are shown in Figure 20.11. Note that the United States is both a leading importer and exporter of pork. The majority of the imports are high-value products, such as hams. The exports are generally lower-value products.

Also, it is generally recognized that in boom periods—periods of high personal income—meat

purchases are affected in three ways: (1) more total meat is desired; (2) there is a greater demand for the choicest cuts; and (3) because of the increased money available and shorter working hours, there is a desire for more leisure time, which in turn increases the demand for those meat cuts or products that require a minimum of time in preparation (such as pork chops and hams). In other words, during periods of high buying power, not only do people want more meats but also they compete for the choicer and more easily prepared cuts of meats.

Because of the operation of the old law of supply and demand, when the choicer and more easily prepared cuts of pork are in increased demand, they advance proportionately more in price than the cheaper cuts. This results in a great spread in prices, with some pork cuts very much higher than others. Thus, whereas pork chops may be selling for 4 or 5 times the cost per pound of the live animal, less demanded cuts may be priced at less than half the cost of the more popular cuts. This is so because a market must be obtained for all the cuts.

But the novice may wonder why these choice cuts are so scarce, even though people are able and willing to pay premiums for them. The answer is simple: Hogs are not all pork, and pork is not all chops (Table 20.8). Besides, nature does not make many choice cuts or top grades, regardless of price; a hog is born with only 2 hams and a limited number of pork chops. It is important, therefore, that those who produce and slaughter animals and those who purchase wholesale or retail cuts know the approximate (1) percentage yield of chilled carcass in relation to the weight of the animal on-foot and (2) yield of different retail cuts.

For example, the average hog weighing 250 lb live will only yield about 139.8 lb of retail pork cuts and 44.4 lb of other products (the balance consists of internal organs and so on). Thus, only about 55 to 60% of a live hog can be sold as retail cuts of pork. In other words, the price of pork at retail would have to be nearly double the live cost even if there were no processing and marketing charges at all. In addition, the higher-priced cuts make up only a small part of the carcass. Thus, this 139.8 lb will cut out only about 10.7 lb of boneless loin. The other cuts retail at lower prices than do these choice cuts; also, there are bones, fat, and cutting losses that must be considered.

Thus, when the national income is exceedingly high, there is a demand for the choicest but limited cuts of pork from the very top grades. This is certain to make for high prices because the supply of such cuts is limited, but the demand is great. Under these conditions, if prices did not move up to balance the supply with demand, there would be a marked shortage of the desired cuts at the retail counter.

TABLE 20.8   Hogs Are Not All Pork, and Pork Is Not All Chops

	Retail Pork (lb)	Other Products (lb)	Carcass Total (lb)	Percentage
Ham				
Cured ham	25.5			
Fresh ham	2.3			
Trimmings	5.8			
Skin, fat, bone		11.4		
Total	33.6	11.4	45.0	24.4%
Loin				
Backribs	3.2			
Boneless loin	10.7			
Country style ribs	7.6			
Sirloin roast	5.7			
Tenderloin	1.6			
Trimmings	1.6			
Fat and bone		3.4		
Total	30.4	3.4	33.8	18.3%
Side				
Cured bacon	19.0			
Spareribs	5.8			
Trimmings	9.1			
Fat		1.0		
Total	33.9	1.0	34.9	18.9%
Boston butt				
Blade steaks	4.4			
Blade roast	7.8			
Trimmings	1.7			
Fat		0.8		
Total	13.9	0.8	14.7	8.0%
Picnic				
Boneless picnic meat	12.6			
Skin, fat, bone		4.0		
Total	12.6	4.0	16.6	9.0%
Miscellaneous				
Jowls, feet, tail				
Neckbones and so on	15.4			
Fat, skin, bone		22.0		
Shrink and loss		1.8		
Total	15.4	23.8	39.2	21.3%
Total	**139.8**	**44.4**	**184.2**	100.0%

Source: Pork Facts 2002–2003, National Pork Board, Des Moines, IA.

BY-PRODUCTS FROM HOG SLAUGHTER

The meat or flesh of hogs is the primary object of slaughtering. The numerous other products are obtained incidentally. Thus, all products other than the carcass meat and lard are designated as by-products, even though many of them are wholesome and highly nutritious articles of the human diet. Yet it must be realized that on slaughter live hogs yield an average of 40 to 45% of products other than retail cuts of pork. When meat packers buy hogs, they buy far more than the 55 to 60% of a hog that becomes the retail cuts of meat that are eventually obtained from the carcass.

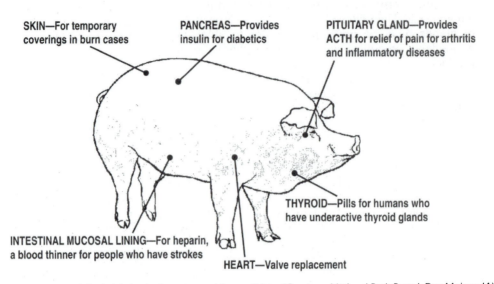

SKIN—For temporary
coverings in burn cases

PANCREAS—Provides
insulin for diabetics

PITUITARY GLAND—Provides
ACTH for relief of pain for arthritis
and inflammatory diseases

THYROID—Pills for humans who
have underactive thyroid glands

INTESTINAL MUCOSAL LINING—For heparin,
a blood thinner for people who have strokes

HEART—Valve replacement

Figure 20.12   **The parts of the hog's body that are used in medicine.** *(Courtesy, National Pork Board, Des Moines, IA)*

In the early days of the meat-packing industry, the only salvaged animal by-products were hides, wool, tallow, and tongue. The remainder of the offal was usually carted away and dumped into the river, burned, or buried. In some instances, packers even paid for having the offal taken away. In due time, factories for the manufacture of glue, fertilizer, soap, buttons, and numerous other by-products sprang up in the vicinity of the packing plants. Some factories were company owned; others were independent industries. Soon much of the former waste material was being converted into materials of value.

Naturally, the relative value of carcass meat and by-products varies both according to the class of livestock and from year to year. The longtime trend for by-product values has been downward relative to the value of the live animal, largely because of technological progress in competitive products derived from nonanimal sources.

The complete utilization of by-products is one of the chief reasons large packers are able to compete so successfully with local butchers. Were it not for this conversion of waste material into salable form, the price of meat would be higher than under existing conditions.

This chapter does not describe all of the by-products obtained from hog slaughter. Rather, a few of the more important ones are listed and briefly discussed (see Figure 20.12).

1. Skins. For the most part pigskins in the United States are sold along with the wholesale cuts, bellies, loins, hams, and so on. These skins are primarily used in gelatin production because they are scalded in the slaughter process, making them unsuitable for tanning into leather. The gelatin is used for coating pills and making capsules. Also, a porcine collagen product has been developed for stimulating clotting during surgery.

Because of its similarity to human skin, specially selected and treated hog skins are used in treating massive burns in humans, injuries that have removed large areas of skin, and in healing persistent skin ulcers. It is shaved and split to 0.008 to 0.020 in., cleansed, and sanitized. It can be applied directly to the injured areas.

Several packing plants are removing the whole skin rather than scalding, which provides a large piece of pigskin for tanning. Pigskin is used for wallets, handbags, shoes, clothing, sporting goods, upholstery, saddles, gloves, book bindings, and razor strops.

2. Fats. Inedible pork fat, known as grease, is rendered and used in animals feeds. In addition to energy value, fats reduce dust in feed processing, improve the color and texture of feed, enhance palatability, increase pelleting efficiency, and reduce machinery wear in the production of animals feeds. It includes renderings from soft tissues, bones, dead pigs, condemned products, and other sources not acceptable for human consumption.

Fatty acids obtained from animal fats through a process referred to as "splitting" are used in the manufacture of a host of products.

Lard oil, made from white grease, is used for making a high-grade lubricant, which is used on delicate, running machine parts.

During the mid-twentieth century, soap making declined significantly, primarily because of the increased use of phosphate-based detergents, powders, and liquids. But soap is biodegradable, whereas phosphate-based detergents are not. So, fat-based cleaning products with detergentlike traits, effective in hard and cold water, have been developed and are in use in various parts of the world. Environmental concerns and fat utilization research have reinstated fat-based materials in the cleaning market.

3. Variety meats. The edible by-products include the brains, heart, kidneys, liver, lungs, stomach, ears, pork skins, snouts, weasand (esophagus), intestines, and many others. These are used both for direct human consumption or in sausage manufacturing. Many are sold over the counter as variety meats or fancy meats. They represent about 3.25% of the live-weight of a hog; the inedible by-products represent about 2.5% of the live-weight of a hog.

Some edible hog by-products that are known as meat and not meat by-products, but which come under the "variety meats" classification, are head and cheek meat, tongue, heart, tail, and feet.

4. Intestines and bladders. Intestines and bladders are used as containers for sausage, lard, cheese, snuff, and putty.

5. Blood. Blood albumen from pigs is used in human blood Rh-factor typing and to make amino acids that are part of parenteral solutions for nourishing certain types of surgical patients. Fetal pig plasma is important in the manufacture of vaccines and tissue culture media. *Note:* Fetal blood does not contain any antibodies, so it is not likely to stimulate immune reactions.

Thrombin, a blood protein, helps create significant blood coagulation. Plasmin, a hog blood enzyme that has the unique ability to digest fibrin in blood clots, is used to treat patients who have suffered heart attacks. Hog blood is also used in cancer research, microbiological media, and cell cultures.

6. Hair. Hog bristles for making brushes were formerly imported from China but are now produced in the United States in increasing quantities. Proper length bristles are found over the shoulder and back of the hog. The fine hair of most U.S. hogs is not suitable for brush making; it is processed and curled for upholstering purposes.

7. Heart. Hog heart valves, specially treated and preserved, are surgically implanted in humans to replace heart valves weakened by disease or injury. Since the first surgery in 1971, thousands of heart valves have been successfully implanted in human recipients of all ages.

8. Meat scraps and muscle tissue. After the grease is removed from meat scraps and muscle tissue, they are made into meat meal or tankage as protein supplements for livestock feed.

9. Bones. The bones and cartilage are converted into stock feed (bone meal), fertilizer, glue, crochet

needles, dice, knife handles, buttons, toothbrush handles, and numerous other articles.

10. Endocrine glands. Various endocrine glands of the body, including the thyroid, parathyroid, pituitary, pineal, adrenals, and pancreas, are used in the manufacture of numerous pharmaceutical preparations. Hog pancreas glands are an important source of insulin used to treat diabetics. Hog insulin is especially important because its chemical structure most nearly resembles that of humans.

Proper preparation of glands requires quick chilling and skillful handling. Moreover, a very large number of glands must be collected in order to obtain any appreciable amount of most of these pharmaceutical products. At one time the endocrine glands were the only source of many pharmaceutical preparations. Now, many of these products are made synthetically.

11. Collagen. The collagen of the connective tissues—sinews, lips, head, knuckles, feet, and bones—is made into glue and gelatin. The most important uses for glues are in the woodworking industry. Gelatin is used in canning hams and other large cuts, and in baking, ice cream making, capsules for medicine, coating for pills, photography, and culture media for bacteria. Also, a porcine collagen product has been developed for stimulating clotting during surgery.

12. Contents of the stomach. Contents of the stomach are used in making fertilizer.

13. Drugs and pharmaceuticals. Hogs are a source of nearly 40 drugs and pharmaceuticals presented in Table 20.9

14. Industrial and consumer products. Hogs make a very significant contribution to industrial and consumer products. Hog by-products are sources of chemical used in the manufacture of a wide range of products which cannot be duplicated by synthesis. A list of many of the products are in Table 20.10.

Thus, in a modern packing plant, there is no waste; literally speaking, "everything but the squeal" is saved. These by-products benefit the human race in many ways. Moreover, their use makes it possible to slaughter and process pork at a lower cost. But scientists are continually striving to find new and better uses for packing house by-products in an effort to increase their value.

PORK PROMOTION

In 1954, a voluntary producers organization known as the National Swine Growers Council was formed to develop meat-type hogs and pork-specific promotion. In 1966 a producer poll favored a voluntary check-off that was initiated in 1967 in six Illinois counties of 5¢ on market hogs and 3¢ on feeder pigs

TABLE 20.9   Drugs and Pharmaceuticals derived from pigs.

Adrenal Glands	**Intestines**	**Skin**	**Pineal Gland**
Corticosteroids	Enterogastrone	Porcine burn dressings	Melatonin
Corrtisone	Heparin	Gelatin	**Pituitary Gland**
Epinephrine	Secretin	**Spleen**	ACTH—
Norepinephrine	**Liver**	Splenin fluid	_adrenocorticotropic
Blood	Desiccated liver	**Stomach**	_hormone
Blood fibrin	**Ovaries**	Intrinsic factor	ADH—antidiuretic
Fetal pig plasma	Progesterone	Mucin	_hormone
Plasmin	Relaxin	Pepsin	Oxytocin
Brain	**Pancreas Gland**	**Thyroid Gland**	Prolactin
Cholesterol	Insulin	Thyroxin	TSH—thyroid
Hypothalamus	Glucagon	Calcitonin	_stimulating hormone
Gall Bladder	Lipase	Thyroglobulin	
Chenodeoxycholic acid	Pancreatin		
Heart	Trypsin		
Heart valves	Chymotrypsin		

Source: PorkFacts 2002–2003, *National Pork Board, Des Moines, Iowa.*

TABLE 20.10   Industrial and Consumer Products Derived from Pigs

Blood	**Fatty Acids & Glycerine, (continued)**
Sticking agent	Fiber softeners
Leather treating agents	Floor waxes
Plywood adhesive	Insecticides
Protein source in feeds	Insulation
Fabric printing & dyeing	Linoleum
Bones & Skin	Lubricants
Glue	Matches
Pigskin garments, gloves, & shoes	Nitroglycerine
Bones, Dried	Oil polishes
Bone China	Paper sizing
Buttons	Phonograph records
Bone Meal	Plasticizers
Fertilizer	Plastics
Glass	Printing rollers
Mineral source in feed	Putty
Porcelain enamel	Rubber
Water filters	Water-proofing agents
Brains	Weed killers
Cholesterol	**Gall Stones**
Fatty Acids & Glycerine	Ornaments
Antifreeze	**Hair**
Cellophane	Artist brushes
Cement	Insulation
Chalk	Upholstery
Cosmetics	**Meat Scraps**
Crayons	Commercial feeds
	Feed for pets

Source: PorkFacts 2002–2003, *National Pork Board, Des Moines, Iowa.*

to fund a national pork promotion program. A national program, "Nickels for Profit" was launched in 1968. It was raised to a dime in 1977. The council was renamed the National Pork Producers Council (NPPC). They lobbied hard for congressional approval to activate a market deduction on livestock to fund product promotion. They succeeded in getting amendments to the Packers and Stockyards Act that permitted voluntary check-off on each pig sold.

In December 1985, Congress approved a 100% National Legislative Check-off, with the stated purpose of providing funds for pork promotion and research to enhance the pork producers' opportunity for profit. In 1986, U.S. pork producers voted overwhelming approval (77% of those who voted) for a mandatory 25¢ check-off fee on every $100 value of hog sold in the United States. Imported pork is included at a separate rate determined by U.S. market prices. The check-off climbed as high as 45¢/$100 value and was reduced to 40¢ in 2002. At the writing of this text the validity of a mandatory check-off is being litigated.

The National Pork Board (NPB) was formed as a separate entity from the NPPC in 2002 to administer the check-off funds and educational, research, and promotion programs. The NPB moved to Des Moines and the NPPC moved their headquarters to Washington, DC, where they could function more effectively as a lobby group for pork interests.

QUESTIONS FOR STUDY AND DISCUSSION

1. Discuss the significance of the message conveyed by the Figure on pg. 449, the amount of meat marketed per animal by different species.
2. What different messages does the term *pork quality* convey?
3. List the qualities that consumers desire in pork.
4. What is the dressing percentage of a hog that weighed 265 lb on-foot and yielded a carcass weighing 192 lb?
5. Outline the hog slaughtering process.
6. Describe the packer and shipper styles of dressing.
7. Sketch and label the wholesale cuts of a pork carcass.
8. In the United States, meat curing is largely confined to pork, with little beef and lamb cured. Why?
9. Discuss the nutritive qualities of pork.
10. Discuss consumer health concerns relative to pork fats and cholesterol.
11. Discuss consumers health concerns relative to nitrates/nitrites and nitrosamines.
12. How are pork retail prices determined by supply and demand?
13. Table 20.8 shows that, on the average, only 139.8 lb of retail pork is obtained from a 250-lb live hog. What accounts for the rest of the original live-weight of 250 lb?
14. What are the major pork exporting and importing countries? Why?
15. Why does the farmer get such a small share of the consumer's dollar?
16. List and discuss 10 important packinghouse by-products from hog slaughter.
17. Trace and discuss the history of pork promotion.

SELECTED REFERENCES

Agricultural Statistics, 2002 (updated annually), U.S. Department of Agriculture, Washington, DC

Guide to Identifying Meat Cuts, The, American Meat Science Association, Savoy, IL, 1998

Meat and Poultry Facts, American Meat Institute, Washington, DC, 1994

Meat We Eat, The, J. R. Romans et al., Interstate Publishers, Inc., Danville, IL, 1994

Pork Facts 2002–2003, National Pork Board, Des Moines, IA

Pork Industry Handbook, Cooperative Extension Service, Purdue University, West Lafayette, IN, 2002

Pork, The Other White Meat, National Pork Board, Des Moines, IA, 2003, http://www.otherwhitemeat.com/default.asp

Pork Quality Assurance Manual, National Pork Board, Des Moines, IA, 2003, http://www.porkboard.org/PQA/manualHome.asp

Statistical Abstracts of the United States, U.S. Department of Commerce, Washington, DC, 2000

USDA National Nutrient Database for Standard Reference, Release 16, U.S. Department of Agriculture, Agricultural Research Service, 2003, Nutrient Data Laboratory home page, http://www.nal.usda.gov/fnic/foodcomp

21

Business Aspects of Swine Production

The Number of U.S. Hog Farms Decreased from
2.39 Million in 1954 to 75,350 in 2002

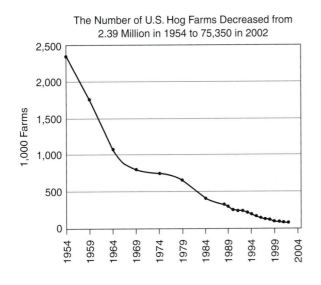

...and the Number of Hogs per
Farm Increased 30-Fold

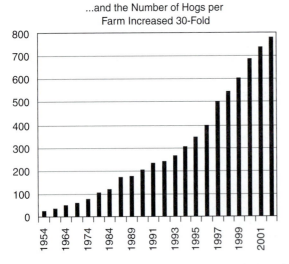

Over the past 50 years the number of U.S. farms with hogs has decreased dramatically along with a commensurate increase in the number of hogs on each farm. The net result is very little change in the total number of hogs marketed per year, usually about 100,000,000. *(Courtesy, USDA, Washington, DC)*

Objectives

After studying this chapter, you should:

1. Know the relative feed and nonfeed costs of producing pigs in the different production systems.
2. Understand what the hog cycle is and why it occurs.
3. Know the various business associations and how they benefit different types of swine enterprises.
4. Know the types of contracts available to pork producers.
5. Know what lean hog futures are.

The great changes that occurred in the U.S. swine industry in the 1980s and 1990s resulted from a shift of our labor-based economy to a knowledge- and capital-based economy. More of the traditional hog farms, such as farrow to finish, are being replaced with producers owning only a segment of the production. This group is being replaced by contract producers who own only the facilities but not the hogs. They are paid per head per day or per head space in the building per year.

Labor is decreasing and automation dominates the swine production industry. As the industry changes, the returns to labor will shrink and the returns to knowledge and capital will continue to grow and grow.

The size of the operation depends on several factors, one of them being the need for family income and how many pigs marketed per year are needed to provide that profit margin. But bigger isn't always better! An Iowa State University evaluation several years ago noted that the cost of producing 100 lb of gain on farrow-to-finish operations decreased

until about 2,500 head were marketed and then the costs remained relatively stable. Additionally, the bigger the unit, the more critical the environmental controls.

COST AND PROFIT/LOSS IN SWINE PRODUCTION

With constant changes in the swine industry, pork producers must continually strive to improve their enterprises. They need to identify their strengths and weaknesses and then determine the opportunities for, and the threats to, their individual swine enterprise.

To analyze the costs and returns from swine production, annual data from 2001 and 2002 for the following three types of swine operations are presented: farrow-to-feeder pigs (Table 21.1a), farrow-to-finish pigs (Table 21.1b), and feeder pigs-to-market (Table 21.1c). The production costs on returns are summarized monthly by Iowa State University Extension Economics.

TABLE 21.1a Estimated Returns from Farrowing and Marketing 50-lb Feeder Pigs[1]

Year	Feed ($/litter)	Nonfeed ($/litter)	Total ($/litter)	Cost/50-lb Pig ($/pig)	Sale Price ($/pig)	Profit/Pig ($/pig)
2001	106.46	239.24	345.69	39.28	51.85	12.57
2002	105.87	239.26	345.13	39.22	43.66	4.44

[1] Assumes 8.8 pigs weaned per litter and no change in value of the sow.
Source: Adapted from Estimated Returns for Farrowing and Finishing Hogs or Producing Feeder Pigs in Iowa, *M-1284, John Lawrence, Iowa State University, Ames, IA.*

TABLE 21.1b Estimated Returns from Farrowing and Marketing 260-lb Slaughter Hogs[1]

Year	Cost/50-Lb Pig[2] ($/pig)	Feed ($/pig)	Nonfeed ($/pig)	Total ($/pig)	Mkt Price ($/cwt)	Profit/Pig ($/pig)
2001	39.28	39.01	20.49	98.78	44.02	15.67
2002	39.22	40.65	20.54	100.41	33.73	−12.71

[1] Assumes 8.8 pigs weaned per litter and no change in value of the sow.
[2] Value from Table 21.1a.
Source: Adapted from Estimated Returns for Farrowing and Finishing Hogs or Producing Feeder Pigs in Iowa, *M-1284, John Lawrence, Iowa State University, Ames, IA.*

TABLE 21.1c Estimated Returns from Purchasing 50-lb Feeder Pigs and Marketing as 260-lb Slaughter Hogs

Year	Pig Cost (50-lb) ($/pig)	Feed ($/pig)	Nonfeed ($/pig)	Total ($/pig)	Mkt Price ($/cwt)	Profit/Loss ($/pig)
2001	51.94	39.01	25.92	116.86	44.02	−2.41
2002	43.66	40.65	25.37	109.68	33.73	−21.98

Source: Adapted from Estimated Returns for Finishing Feeder Pigs in Iowa, *M-1284, John Lawrence, Iowa State University, Ames, IA.*

Although not consistent, generally the farrow-to-finish operation has reaped the greatest return per pig sold. This obviously was not true for 2002 as a marked downswing in prices occurred changing the $4.44 per pig profit in 2002 to a $12.71 loss when that same pig was marketed 4 months later. The major advantage in farrow-to-finish is the producer retains ownership and does not incur added expenses from marketing or purchasing feeder pigs. However the producer needs to be aware of costs and returns always as the $4.44 profit can turn into an $12.71 loss 4 months later (Tables 21.1a and 21.1b).

Neither 2001 nor 2002 were good years for purchasing and feeding a 50-lb hog to slaughter weight, both years resulting in losses. Additional information is available in the Selected References at the end of the chapter.

HOG CYCLE–LONGER AND LESS ABRUPT

Concentrating the hog industry in fewer producers is making a difference in the price cycle. Formerly, it averaged about 4 years (see the Chapter 19 section headed "Cyclical Movements"). Another occurring change is a decrease in the peaks and valleys of the hog cycle (Figure 21.1). The general trend in annual hog slaughter continues to increase but the cyclical changes are not as wide as they were in the late 1970s and early 1980s. The decreases are also moderating, compared to the years prior to 1985.

PORK PRODUCTION SYSTEMS

Many different techniques and facilities are being used in hog production. Basically, there are the following three systems of production: (1) farrow-to-finish, (2) farrow-to-feeder pig production, and (3) feeder pig finishing to market. Within each system there are the following variations: high-investment, high-intensity; low-investment, low-intensity; and indoor versus outdoor production.

Farrow-to-finish production, farrow-to-wean production, feeder pig production, and feeder pig finishing systems are discussed in Chapter 15. Each system requires unique facilities, records, and management skills. Successful pork producers may function in more than one system, but the trend is toward specialization of efforts.

BUSINESS ORGANIZATIONS[1]

The success of a contemporary swine enterprise is very dependent on the type of business organization. No one type of organization is superior under all circumstances; rather, each situation must be considered individually. The size of the operation, the family situation, the enterprises, the objective—all these and more are important in determining the best way in which to organize the swine enterprise.

Six types of business organizations are commonly found among swine enterprises: (1) proprietorship (individual), (2) partnership (general partnership), (3) corporations, (4) contracts, (5) co-ops, and (6) networking.

Among the factors that should be considered when deciding which business form best fits a given set of circumstances are the following:

1. Which type of organization is most likely to be looked on favorably from the standpoint of more credit and capital?

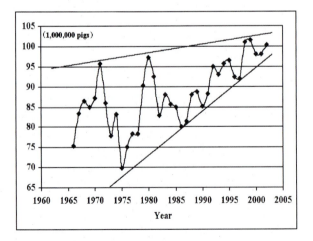

Figure 21.1 The variability of the annual commercial U.S. hog slaughter continues to decrease as the number of swine farms decreases and their size increases. *(Courtesy, USDA, Hog and Pig Reports)*

---

[1] This section contains information adapted from USDA, Economic Research Service, Government Publication AIB768.

2. How much capital will be required of each individual involved?
3. Are there tax advantages to be gained from the business organization?
4. Is expansion of the business feasible and facilitated?
5. Which type of organization reduces risks and liability most?
6. Which type of organization can be terminated most easily and readily?
7. Which type of ownership provides for the most continuity and ease of transfer?
8. What costs for legal and accounting fees are involved in setting up the organization and in the preparation of the annual reports required by law?
9. Who will manage the business?

Farmers use a variety of business arrangements. Although most are sole proprietorships, they may have many formal or informal linkages to other firms. These linkages include arrangements to procure inputs (leasing land and machinery, custom work, hired or contract labor, forward pricing inputs), as well as marketing and production contracts. The extent of these linkages varies across the types of farms and by commodity.

Farmers use a combination of forms of business organization and arrangements to structure a firm that meets personal, business, and household goals. Readily recognizable forms of business organization include sole proprietorships, partnerships, and corporate businesses. In addition to these more traditional forms of organization, farms are also organized using other forms of organization, such as limited liability companies as well as trusts and cooperatives (Table 21.2).

Moving beyond forms of business organization, farmers may choose to use a wide range of formal and informal business arrangements to gain access to technology, markets, equity capital, or other inputs important to achieving business or other goals. Commonly used arrangements include marketing and production contracts, joint ventures, strategic alliances, leases, and a variety of agreements and licenses.

Farmers may use any or all of these business arrangements in various combinations. A farm may have a marketing contract with an elevator company to market grain, a production contract with a meat processor to produce livestock, an arrangement with a farming neighbor to share equipment purchases or use, and an agreement with a relative to jointly rent land from another neighbor. Any formal or informal business arrangement can be used in conjunction with any of the various forms of business organization.

TABLE 21.2 Organizational Choices and Business Arrangements for Farm Operators

Farmers' Choices Of

Business Organization	Business Arrangements
Proprietorships	Independent producer
Modified proprietorships	Contract producer
Partnerships	Subcontract producer
Corporations	Strategic alliances
Family	Franchise agreements
Nonfamily	Licensing
Cooperatives, estates,	Alliances or joint
and trusts	ventures
Limited liability companies	Leasing

The complexity of today's farm business structure suggests that a farm's form of business organization alone is not sufficient to assess the extent of business linkages or the degree to which production or market integration may exist.

The 1998 Agricultural Resource Management Study (USDA, Economic Research Service) confirms that the sole proprietorship is the most prevalent form of business ownership (Table 21.3). Other traditional forms of business organization include partnerships (general and limited), corporations (farm and nonfarm), and cooperatives, estates, and trusts. In any given year, partnerships typically account for 5 to 6% of farms. Some persons or entities associate to produce commodities or services but are not legally organized as partnerships. These informal associations or alliances are typically considered to be proprietorships, but more than one household shares in the asset base and income of the farm.

Corporations, cooperatives, and other forms of business organization account for the remaining 4 to 5% of farms. Although considerable attention is paid to potential expansion in the number of large nonfarm corporations and their share of production, the predominant form of corporate farm in terms of numbers of farms has traditionally been, and continues to be, the family corporation.

Choice of form of business organization varies across the farm types. Important factors include simplicity of forming the business, control over business and financial decisions, business continuity, owner liability, business and personal tax liability, estate transfer issues, and access to capital. More complex forms of organization, such as partnerships and corporations, are typically associated with larger farms where decisions and business issues regularly extend beyond the production and disposition of agricultural commodities to include other key concerns.

TABLE 21.3 Business Organization of Farms, by Type of Farm, 1998

	Occupation—	Small Family Farms[1] Occupation Farming[2]		Family Farms			
Item	Nonfarming	Low Sales	High Sales	Large[1]	Very large[1]	Nonfamily[3]	All farms
Total farms/operators	1,275,527	422,205	171,469	91,939	61,273	42,296	2,064,709
Percentage of farms	62%	20.4%	8.3%	4.5%	3%	2%	100%
Business organization							
Sole proprietorship	29%	92.4%	85.4%	68.7%	59.1%	31%	90.4%
Partnership	5%	5.2%	8.2%	11.9%	23.9%	2.5%	5.3%
Corporation or cooperative	0%	2.4%	6.4%	19.5%	17.1%	66.6%	4.3%
Family corporation	0%	2.4%	6.4%	19.5%	17.1%	16.8%	3.3%
Nonfamily corporation or other[4]	na	na	na	na	na	49.8%	1%

[1] Small family farms have sales less than $250,000; large family farms, between $250,000 and $499,999; and very large family farms have sales of $500,000 or more.

[2] Small farms that report farming as their major occupation are divided into low sales (less than $100,000) and high sales (between $100,000 and $249,999).

[3] Nonfamily farms include nonfamily corporations or cooperatives, as well as farms operated by a hired manager.

[4] "Other" includes estates, trusts, and cooperatives. "na" equals data "not applicable."

Source: USDA Economic Research Service, Washington, DC.

Proprietorship (Individual)

A sole proprietorship is a business that is owned and operated by one individual. This is the most common type of business organization in U.S. farming—90% of U.S. farms are individually or family owned. Proprietorships are generally the simplest form of business organization to use and understand. In effect, individuals (or married couples) establish a business and operate as a proprietorship, unless they take steps to operate using another form of organization.

The individual or couple forming the business hold managerial control, are liable for debt and business decisions, and receive income produced by the business. Other forms of business organization are more complex as a result of shared management, differences in tax treatment and liability, distribution of income among multiple owners, and legal documents needed to establish life for the organization that may transcend that of the owners.

Most hog enterprises are operated as sole proprietorships, not necessarily because this is the best type of organization, but with no effort to form some other type of organization it naturally results. The partnership, the corporation, contracts, co-ops, and networking, which require special planning and effort to bring about, are well suited to the operation of large swine establishments.

The advantages of a sole proprietorship include ease of formation, direct control over the business, relative freedom from government regulations, allocation of all profits to the owner, and tax deduction of business loans.

In comparison with other forms of organization, the sole proprietorship has three major limita-

tions: (1) unlimited personal liability; (2) difficulty acquiring new capital for expansion; and (3) lack of continuity to keep the present business viable, with the result that it usually goes out of existence with the passing of the owner.

Partnership (General Partnership)

A partnership is an association of two or more persons who, as co-owners, operate the business. About 5% of U.S. farms are partnerships.

The basic idea of two or more persons joining together to carry out a business venture can be traced back to the syndicates that were used in major trading centers in western Europe in the Middle Ages. Many of the early efforts to colonize the New World were also partnerships, or "companies," which provided venture capital, ships, provisions, and trade goods to induce settlement of large land grants.

Most swine partnerships involve family members who have pooled land, machinery, working capital, and often their labor and management to operate a larger business than would be possible if each member limited his or her operation to his or her own resources. It is a good way in which to bring a son or daughter, who is usually short on capital, into the business, yet keep the parent in active participation. Although there are financial risks to each member of such a partnership, and potential conflicts in management decisions, the existence of family ties may minimize such problems.

In order for a partnership to be successful, the enterprise must be sufficiently large to use the abilities and skills of the partners and to compensate them adequately in keeping with their contribution to the business. Hopefully, all partners can contribute complementary skills to the enterprise.

A partnership has the following advantages:

1. Ease of formation, low cost of organizing, and little government control. These are very real advantages.

2. Combining resources. A partnership often increases returns from the operation as a result of combining resources. For example, one partner may contribute labor and management skills, whereas another may provide the capital. Under such an arrangement, it is very important that the partners agree on the value of each person's contribution to the business and that this be clearly spelled out in the partnership agreement.

3. Equitable management. Unless otherwise agreed on, all partners have equal rights, regardless of financial interest. Any limitations, such as voting rights proportionate to investment, should be a written part of the agreement.

4. Tax savings. A partnership does not pay any tax on its income, but it must file an information return. The tax is paid as part of the individual tax returns on the respective partners, usually at lower tax rates.

5. Flexibility. Usually, the partnership does not need outside approval to change its structure or operation—the vote of the partners suffices.

Partnerships may have the following disadvantages:

1. Liability for debts and obligations of the partnership. In a partnership, each partner is liable for all the debts and obligations of the partnership.

2. Uncertainty of length of agreement. A partnership ceases with the death or withdrawal of any partner, unless the agreement provides for continuation by the remaining partners.

3. Difficulty of determining value of partner's interest. Because a partner owns a share of every individual item involved in the partnership, it is often very difficult to judge value. This tends to make transfer of a partnership difficult. This disadvantage may be reduced by determining market values regularly.

4. Limitations on management effectiveness. Limitations on management effectiveness may be a result of personal differences among partners, and the responsibility of each partner for the acts of the other partners.

In sum, a partnership or general partnership is characterized by (1) management of the business being shared by the partners and (2) each partner

Figure 21.2 Good records are essential to evaluating business arrangements and opportunities. *(Courtesy, Palmer Holden, Iowa State University, Ames, IA)*

being responsible for the activities and liabilities of all the partners, in addition to self-activities within the partnership (Figure 21.2).

Limited Partnership

A limited partnership is an arrangement in which two or more parties supply the capital, but only one partner is involved in the management. This is a special type of partnership with one or more "general partners" and one or more "limited partners."

The limited partnership avoids many of the problems inherent in a general partnership and has become the chief legal device for attracting outside investor capital into farm ventures. Although this device has been widely used in the oil and gas industry, and for acquiring income-producing urban real estate for a number of years, its application to agricultural ventures on a national scale is quite new. As the term implies, the financial liability of each partner is limited to each partner's original investment, and the partnership does not require, and in fact prohibits, direct involvement of the limited partners in management. In many ways, a limited partner is in a similar position as a stockholder in a corporation.

A limited partnership must have at least one general partner who is responsible for managing the business and who is fully liable for all obligations. The advantages of a limited partnership are as follows:

1. It facilitates bringing in outside capital.
2. It need not dissolve with the loss of a partner.
3. Interests may be sold or transferred.
4. The business is taxed as a partnership.
5. Liability is limited.
6. It may be used as a tax shelter.

The disadvantages of a limited partnership are as follows:

1. The general partner has unlimited liability.
2. The limited partners have no voice in management.

Corporation

A corporation is a device for carrying out a pork enterprise as an entity entirely distinct from the persons who are interested in and control it. Each state authorizes the existence of corporations. As long as the corporation complies with the provisions of the law, it continues to exist—regardless of changes in its membership.

Until about 1960, few farms and ranches were operated as corporations. In recent years, however, there has been increased interest in the use of corporations for the conducting of farm and ranch business. Even so, only about 4.0% of all U.S. family farms use the corporate structure. In contrast, about 50% of the nonfamily farms are incorporated.

From an operational standpoint, a corporation possesses many of the privileges and responsibilities of a real person. It can own property; it can hire labor; it can sue and be sued; and it pays taxes.

Separation of ownership and management is a unique feature of corporations. The owners' interest in a corporation is represented by shares of stock. The shareholders elect the board of directors, who, in turn, elect the officers. The officers are responsible for the day-to-day operation of the business. Of course, in a close family corporation, shareholders, directors, and officers can be the same persons.

The major advantages of a corporate structure are as follows:

1. It provides continuity despite the death of a stockholder.
2. It facilitates transfer of ownership.
3. It limits the liability of shareholders to the value of their stock.
4. It may make for some savings in income taxes.

The major disadvantages of a corporation are as follows:

1. It is restricted to doing only what is specified in its charter.
2. It must register in each state.
3. It must comply with stipulated regulations that involve considerable paperwork and expense.
4. It is subject to the hazard of higher taxes.
5. It is possible to lose control.

Family-Owned (Privately Owned) Corporation

One type of corporation is the family-owned (privately owned). About 17 to 19% of the large and very large family corporations use this business organization (Table 21.3). It enjoys most of the advantages of its generally larger outside investor counterpart, with few of the disadvantages. The chief advantages of the family-owned corporation over a partnership arrangement are as follows:

1. **It alleviates unlimited liability.** For this reason, a lawsuit cannot destroy the entire business and all the individual partners with it.
2. **It facilitates estate planning and ownership transfer.** It makes it possible to handle the estate and keep the business in the family and going if one of the partners should die. Each of the heirs can be given shares of stock—which are easy to sell or transfer and can be used as collateral to borrow money—while leaving the management of the enterprise to those heirs interested in operating it, or even to outsiders.

Tax-Option Corporation (Subchapter S Corporation)

Instead of paying a corporate tax, a corporation with no more than 35 stockholders may elect to be taxed as a partnership, with the income or losses passed directly to the shareholders, each of whom pays taxes on his or her share of the profits. This special type of corporation is variously referred to as *a tax-option corporation, subchapter S corporation,* pseudocorporation, or elective corporation.

For income tax purposes, the owners of a tax-option corporation are taxed as if they were in a partnership. That is, income earned by the corporation passes through the corporation to the personal income tax returns of the individual shareholders. Thus, the corporation does not pay any income tax. Instead, the shareholders pay tax on their share of corporate income at their individual tax rate; and the shareholders report their share of long-term capital gains and receive their deductions from the corporation. Although each shareholder's portion of any corporate losses from current operations is deducted from his or her personal return, capital losses incurred by the corporation cannot be passed on to the shareholders.

Thus, there are some very real advantages to be gained from a subchapter S or tax-option corporation. However, in order to qualify as a subchapter S corporation, the following requisites must be met:

1. There cannot be more than 35 stockholders.
2. All stockholders must agree to be taxed as a partnership.

3. Nonresident aliens cannot own stock.
4. There can be only one class of stock.
5. Not more than 20% of the gross receipts of the corporation can be from royalties, rents, dividends, interest, or annuities plus gains from the sale or exchange of stock and securities; and not more than 80% of the gross receipts can be from sources outside the United States.

Advantages of Limited Partnerships and Corporations

In addition to the advantages peculiar to limited partnerships and corporations, and covered under each, limited partnerships and corporations have the following advantages over individual ownership in the acquisition of capital:

1. They make it possible for several producers to pool their resources and develop an economically sized operation, which might be too large for any one of them to finance individually.
2. They make it possible for persons outside agriculture to invest through the purchase of shares of stock in the business.
3. They can generally borrow money more easily because the strength of the loan is not dependent on the financial and management capability of one person.
4. They give assurance that the business will continue, even if one of the owners should die or decide to sell out.
5. They provide built-in management, with continuity, and, generally speaking, they attract very able management.

Thus, those engaged in the livestock business can and do use either of these two business organizations—a limited partnership or a corporation—to develop and maintain an economically sound operation. Actually, no one type of business organization is best suited for all purposes. Rather, each case must be analyzed, with the assistance of qualified specialists, to determine whether there is an advantage to using one of these types of organizations, and, if so, which organization is best suited to the proposed business.

Contracts[2]

A contract is an agreement between two or more persons to do or refrain from doing certain things.

---

[2] This section contains information adapted from the *Pork Industry Handbook,* PIH 6, "Producing and Marketing Hogs under Contract," Purdue University, West Lafayette, IN, 1992; and "Guide to Contracting," National Pork Board, Des Moines, IA, 2000.

Forward pricing (marketing) contracts for market hogs have been available from most major meat packers for a number of years. They are the most commonly used marketing contracts in the industry.

Under a hog marketing contract, the packer buys a known quantity and quality of a commodity from a farm for a negotiated price. The feeder owns the commodity while it is being produced and receives a price reflecting the value of the commodity. More than 85% of the U.S. hogs sold are produced under some form of a marketing contract (see Chapter 19, Table 19.2 on the changing structure of pork marketing).

Through contracting, producers are able to achieve more stable returns, trading the possibility of large profits for the assurance of a more reliable return. Many producers enter into contracts because they either lack the capital or they do not wish to tie up a large amount of capital in hog production. Following is an overview of the most common contracts in the pork industry.

Slaughter Hog Marketing Contracts

The forward sale contract of market hogs is a contract between a buyer (normally a meat packer or a marketing agent) and a seller (normally a producer), where the producer agrees to sell, at a future date, a specified number of hogs to a buyer for a certain price. A variety of hog contract pricing methods are presented in Table 21.4.

A producer uses a forward sale contract to reduce the risk of price fluctuations and to lock in an acceptable selling price. Although the forward sale contract allows the producer to lock in a particular selling price, it may cause the producer to miss out on greater profits if prices rise. Thus, the decision to contract must be based on each producer's willingness or ability to bear the risk of price uncertainty. Some producers may be forced to contract because of a lack of diversification, indebtedness, or at the request of creditors, whereas other more financially stable or diversified producers may be in

TABLE 21.4 Hog Contract Pricing Methods

Pricing Method	1997	1999	2000
Formula cash based	39.1%	44.2%	47.2%
Formula futures based	2.9%	3.4%	8.5%
Feed price based	5.3%	9.8%	12.3%
Window contract	3.1%	4.6%	4.6%

Source: G. Grimes and S. Meyer, National Pork Producers Council, Des Moines, IA, 2000.

a better position to withstand the risk of price movements.

Formula pricing involves establishing a transaction price using a formula that relies on an external reference price. With a formula cash–based contract the seller agrees to deliver a specified number of hogs to a buyer at a future date and the buyer guarantees the seller a minimum price (the floor price) for the hogs. Usually, the seller receives the higher of the floor price or market price at delivery minus a discount. The discount compensates the buyer for the costs (options premiums and other variable costs associated with the contract) of providing the guaranteed minimum price. This is by far the most popular pricing method.

The price based on the futures market accounted for 8.5% of the methods used in 2000. It obviously is based on what the futures price will be at the time of delivery. Feed price–based contracts allow the base price to vary depending on a defined market for corn and soybean meal. As the feed prices increase, so does the price received by the producer.

The window contract assures the producer of the market price within a defined window with adjustments for markets above or below the window. For example a packer may offer a window price of $40 to $50 per cwt of carcass. Thus, if the market price is within that range on the day the hogs are delivered, the producer receives that base bid for the hogs, plus or minus any discounts for weight or grade.

If the price goes above or below the window prices, the packer and producer split the price. For example if the market price goes to $60, the producer receives half of the difference, or a market price of $55. If the price drops to $30, the producer receives $35.

Some packers maintain a ledger account. In this account the packer records the "losses" or "profits" incurred when the market price is outside of the window. After a defined time period, for example 5 years, the ledger is balanced. If the packer has a positive balance in the ledger, a payment is made to the producer for that amount. Likewise, if there is a negative balance, the producer makes a payment to the packer.

The size of the contract varies among packers and is usually based on so many pounds of lean value hog carcasses. Terms typically found in a forward contract include the following:

• The date and location of delivery. The delivery date may normally be changed by mutual agreement. The seller may have the option of selecting the delivery date within a specified time interval.
• Acceptable weights and grades, including provisions for premiums and discounts.

• Provisions for nondeliverable hogs and unacceptable carcasses. The buyer normally deducts from the seller's receipts for unacceptable hogs and carcasses.
• Provisions outlining the credit requirements of the seller and inspection of the hogs by the buyer. The buyer may request to inspect the hogs while on the seller's premises.
• A provision dealing with breach of contract. Typically, the seller is liable for all losses incurred by the buyer when the seller is in breach of contract.

The producer retains all production risks, other than the selling price, under a fixed price forward sale contract.

Price risk may be reduced by hedging in the futures market. A marketing contract may be preferred to hedging for the following reasons:

• Marketing contracts can typically be written for smaller sizes than the 40,000-lb Chicago Mercantile Exchange contract.
• A fixed price marketing contract locks in the delivered price. A futures contract hedge locks in the futures price, but the local price differential (basis) will vary.
• A marketing contract does not require an initial margin or additional margin calls be paid should the price increase after the contract is signed.
• A marketing contract is typically made with a local marketing agent or packer rather than dealing with the Chicago Mercantile Exchange. However, unlike a futures contract, the producer is required to deliver the hogs to fulfill a marketing contract.

These contracts reduce only the risk of hog price fluctuations. The producer must still bear the other risks associated with hog production.

Feeder Pig Marketing Contracts

Typically, feeder pig marketing contracts are between a marketing agency, often a cooperative or feed supplier, and a pig producer, where the marketing agency agrees to market the pigs for the producer in exchange for a fee. In addition many feeder pig producers have developed long-term contracts with feeder pig finishers.

Producers are essentially hiring marketing expertise to enhance their market prices and minimize the time and effort of locating buyers for their pigs. A marketing contract might contain the following provisions:

• The producer agrees to market all pigs through the marketing agency.

• The marketing agency prescribes specific management practices to be followed by the producer. These may relate to the weight at which the pigs are to be marketed, health status of the animals, parasite treatment, and immunizations.

• Many larger marketing agencies will provide technical assistance to the producer.

Production Contracts

There is a new era of livestock production across the United States. Traditional patterns of livestock production are giving way to new care and feeding arrangements. Many of these arrangements involve livestock production contracts.

A livestock production contract can be defined as an agreement under which a producer agrees to feed and care for livestock owned by a contractor in return for a payment. Production contracts should be distinguished from marketing agreements, cash forward contracts, and futures contracts, which involve the sale of livestock produced and owned by the producer.

Production contracts have become relatively common in the production of select commodities, such as broilers, swine, and processing vegetables. Farms frequently enter into two types of contracts. A production contract is a legal agreement between a farm operator (contractee or agent) and another person or firm (contractor or principal) to produce a specific type, quantity, and quality of agricultural commodity. The contractor usually owns the commodity being produced and the farmer receives a service fee for the facilities and labor provided.

There is increasing interest in contracting hog production by the farmer, in part because of the difficulty many producers have obtaining adequate financing. Contracting also is being used to coordinate pork production from genetics and nutrition to the retail meat counter.

Production contracts for market hog finishing were used by 9.5% of the U.S. producers and more than 3% of farrowing and nursery units in 2001 (Table 21.5). Approximately 18% of U.S. producers were raising pigs under some type of production contract.

Of those with finishing contracts, most produced more than 2,500 head per year, including 17.1% of the producers who were finishing more than 10,000 pigs annually (Table 21.6).

To expand more rapidly their own production, many larger producers use contract production as a way to hold down risk and capital required.

Investors, feed dealers, farmers, and others often are interested in producing hogs but are unwilling or unable to provide the necessary labor,

TABLE 21.5 Stages of Contract Hog Production in the United States[1]

Stage	Percentage
Finishing	9.5%
Farrowing	3.5%
Nursery	3.3%
Wean to finish	1.3%
Seedstock	0.7%
Total	18.3%

[1] Percents may reflect multiple answers.
Source: 2002 National Hog Farmer Reader Profile, 83 responses.

TABLE 21.6 Number of Hogs Finished under Contract in 2001 in the United States

No. Finished	Percentage
500–999 head	5.7%
1,000–2,499 head	17.1%
2,500–4,999 head	20.0%
5,000–9,999 head	40.0%
10,000 head or more	17.1%
Total	100.0%

Source: 2002 National Hog Farmer Reader Profile, 35 responses.

facilities, and equipment. Therefore, they search out producers who are willing to furnish the labor and equipment in exchange for a fixed wage or share of the profits. The resulting contracts, between owner and producer, vary considerably in form and responsibility of each party involved. These contracting arrangements are attractive to young or financially strapped producers and would-be producers who do not have the capital to invest in a herd, and for producers with underutilized facilities.

FARROW-TO-FINISH CONTRACTS. Although base-payment plus bonus contracts are offered in some regions, many farrow-to-finish contracts are on a percentage basis to reflect the relative inputs supplied by each person or form. Three farrow-to-finish options include variable returns based to a degree on the amount of risk assumed.

Option 1: The producer supplies facilities, labor, veterinary care, utilities, and insurance for an appropriate percentage of gross sales based on input costs. The feed retailer supplies feed, standard feed medications, and receives a predetermined percentage of returns. The capital partner and breeding stock supplier get another percentage. The management firm receives a percentage for supplying computerized records services and management consultation.

Option 2: The current hog inventory is purchased outright by a limited partnership and it will supply sow replacements. The producer supplies facilities, labor, utilities, veterinary costs, repairs, and manure removal. The feed retailer provides feed and standard feed medications. A management agency supplies production and marketing guidance. Each of the contract participants receives a percentage of the proceeds when the hogs are marketed. The remaining percentage is split between the limited partnership and the general partner for managing the partnership.

Option 3: The contractor provides breeding stock, feed, and a prescribed system of management. The producer provides facilities, labor, utilities, insurance, and manure removal. The producer receives fees per head or per lb of hogs marketed, plus possibly additional compensation for farrowing and feeding efficiency.

FEEDER PIG PRODUCTION CONTRACTS. Feeder pig production contracts come in several forms.

Option 1: The producer provides everything but the breeding stock and bids a price for which he or she is willing to produce a feeder pig, based on production criteria such as pigs weaned per litter, and so on, with discounts and bonuses based on target levels, such as number weaned or pigs per sow per year. Most of the production risk is borne by the producer.

Option 2: A contractor provides breeding stock, feed, management assistance, and supervision, and pays the feeder pig producer a flat fee for each pig. This fee varies according to pig weight and current production costs. In this example most of the risk falls on the person providing breeding stock, feed, and management.

Option 3: The contractor provides breeding stock, feed, facilities, and veterinary costs. The producer provides labor, utilities, maintenance, and manure handling. A fee for each pig produced and a monthly fee for each sow and boar maintained is paid to the producer. This option fits owners who no longer want to be actively involved in production but have a good manager with limited cash willing to take over the operation.

Option 4: A shared revenue program with revenues divided in proportion to inputs provided. One example would be where the producer supplying facilities, veterinary care, utilities, labor, and insurance would receive a negotiated percentage of gross sales in return for his or her share of production costs for each pig sold. The feed dealer would receive a certain percentage based on his or her share of the total inputs. The remaining percentage would go to the breeding stock supplier and the management firm that supplies computerized records and consultations. Negotiated percentage shares should be based on inputs provided and risks borne by each participant.

FINISHING CONTRACTS. There are three basic types of hog finishing contracts offered, each with variations on payments and resources provided.

Option 1: A fixed payment contract guarantees the producer a fixed payment per head as well as bonuses and discounts based on performance. Under a fixed payment contract for finishing hogs, the producer normally provides the building and equipment, labor, utilities, and the necessary insurance. The contractor supplies the pigs, feed, veterinary services and medication, and transportation. The contractor usually provides a prescribed management system and supervises its conduct. The contractor, as the owner of the hogs, does the marketing.

The producer often receives an incoming payment based on the weight of the feeder pigs when they come into the producer's facilities. For example, $5 for a 30-lb pig and $4 for a 40-lb pig. The remainder of the producer's payment is made when the hogs are sold. The method of calculating base payment varies by contract. Some contracts offer a fixed dollar per head regardless of the weight gained. Other contracts pay a fixed amount per pound of gain based on pay-weights in and out of the facility. Others pay a fixed amount per head per day spent in the facility.

Most contracts contain bonuses for keeping death loss low and improved feed efficiency, as well as penalties for high death losses and unmarketable animals. Producers should have control over factors that impact their bonuses and penalties, for example, the right of refusal on obviously unhealthy pigs, or to negotiate a more lenient bonus schedule for multiple-source pigs. Contract payment methods typically range from a low base payment with high incentive bonuses to a high base with relatively low bonuses.

Option 2: Directed feeding by a cooperative or feed dealer that contracts with a producer to finish-out hogs. The contractor's objective when entering into a directed feeding contract is to increase feed sales.

The contractor provides the feed and some management assistance, and typically directs the

Figure 21.3 A modern wean-to-finish building designed to feed pigs from weaning to slaughter with a totally slotted floor. *(Courtesy, Palmer Holden, Iowa State University, Ames, IA)*

feeding program. The contracting firm often will purchase the feeder pigs, in which case profits from the sale of the hogs are shared as discussed later, or it will help the producer obtain financing to purchase the pigs (Figure 21.3). The producer agrees to purchase all feed and related services from the contractor and is responsible for all costs of production. The producer receives all proceeds from the sale of the hogs minus any outstanding balance owed to the contractor.

Option 3: In a profit-sharing contract, the producer and contracting firm divide the profit in proportion to the share of the inputs provided by each party. Typically, the producer provides the facilities, labor, utilities, and insurance for his or her portion of the profit. The contracting firm normally purchases the pigs and is responsible for all feed, the veterinary services, transportation, and marketing expenses.

Over the duration of the contract, the contractor's costs are charged to an account. This account balance is then subtracted from the sale proceeds to determine the profit. The contracting firm often will use its own feed and provide management assistance. The producer is normally guaranteed a minimum amount per head as long as death loss is below a set percentage. For instance, depending on contract terms, the producer may receive $5/head if death loss is 3% or less and $3/head if death loss is more than 5%. The producer receives this payment regardless of whether a profit is made. The contractor's return depends on the profit made on the sale of the hogs and the gain received from the markup on feed, pigs, and supplies provided.

Breeding Stock Leasing

The popularity of breeding stock leases has declined in recent years and presently they are seldom

used. Many contractors were dissatisfied with the care of the breeding herd and sometimes were unable to collect their payments from producers. Many producers were paying more than was a reasonable return to the contractor. One lease involves a payment-in-kind for the use of breeding stock. This lease is particularly attractive to producers with limited capital but ample feed, facilities, and labor to produce hogs. The producer pays all production costs and pays the breeding stock owner, for example, one market weight hog per litter.

Characteristics of a Good Contract

The relationships between producers and contractors are generally more complex and interdependent for production contracts than for marketing agreements. Hence, production contracts need to be evaluated with special care. When considering contract production, contractors and producers need to evaluate each contract on its own merit.

Each party should look for a contract that best fits their operation and management capabilities, and both parties must know their cost of production to make an informed decision. Simply signing a contract will not necessarily improve efficiency or ensure a profit. It is doubtful that producers will receive a bonus for feed efficiency better than 2.9 lb feed per lb gain if they have been only achieving 3.5 on their own, for example.

Also, carefully scrutinize the examples used to demonstrate cash flow or producer returns. Unless otherwise stated these are only examples and not guarantees. Producers should consider the impact on cash flow and debt repayment if payments are less than projected. Is there a guarantee of contract length if new facilities or other major capital expenditures are required to obtain the contract? Most contracts guarantee a stated number of turns (groups of hogs) or are in force for a stated length of time. Few, if any, guarantee the number of hogs that will be put through the facility in a set time, say 1 year. Facilities that sit idle during an unprofitable period in the hog cycle may profit the contractor, but disrupt the producer's debt repayment schedule.

Livestock Contract Checklist[3]

A. Consult Experts. Before committing yourself to this contractual obligation, be absolutely sure you understand the entire document.
    1. Attorneys. If you do not fully and completely understand the legal consequences of the contract or the legal, financial, or tax consequences

---

[3] Adapted from Livestock Production Contract Checklist, Iowa Attorney General, Des Moines, IA, 2000 (*www.iowaattorneygeneral.org/ working_for_farmers/farm_brochures.html*).

of the contract, then you should consult an attorney or financial expert.

2. Other producers. Talk to other producers who have had experience with contracts. They may be a good source of advice.

B. Facility Requirements

1. Can you have livestock other than the contractor's livestock in the facilities or on the farm?

2. Construction specifications/modernization. Who provides the specifications for construction? Is the facility standard for the industry? Could the facility be used for other purposes? Are you required to pay for future modernization or upgrades in the facility or its equipment?

3. Who is responsible for obtaining governmental permits and/or county zoning approval? What happens if the facility is not approved?

4. Who has access to the facility? If the contractor or others have access, then do they have to give you advance notice?

5. Is the duration of the contract adequate to recover your investment in the facility and equipment? Is there a guarantee of minimum occupancy for the facility?

C. Operational Issues

1. Is there a set schedule for livestock deliveries? Do you have a guaranteed minimum or maximum occupancy rate for your facilities? Who bears the risk of death loss of livestock while trucked in or out?

2. Who is responsible for providing feed and guaranteeing feed quality? Who is responsible if feed conversions are below expectations? If marketing is delayed and feed efficiency declines, are you compensated for the extra feed costs?

3. Livestock health. Can you reject livestock you think are sick? Who bears death loss risk because of failure of equipment or extreme weather conditions? Who bears the costs of poor performance because of unhealthy or low-quality livestock? Can you renegotiate compensation terms? Who determines and pays for programs for scheduled or unscheduled health care? What are your responsibilities for cleaning or disinfecting facilities between turns of livestock?

4. Who is responsible for manure management? Who responds to complaints, lawsuits, or alleged violations of law, involving odor, dust, water quality, or other types of nuisance? Who is ultimately liable for damages, penalties, or legal expenses from complaints, lawsuits, or enforcement actions? Who owns the manure?

5. Labor and management/record keeping. Who provides labor and management to raise the livestock? Who sets and "judges" husbandry practices? Are you or your employees required

Figure 21.4 An on-farm incinerator for disposing of dead pigs. *(Courtesy, Palmer Holden, Iowa State University, Ames, IA)*

to have special skills or training? What production records are you required to maintain?

6. Insurance and other costs. Who pays for liability and casualty insurance? Is there coverage for suffocation of livestock as a result of equipment failure? Who is responsible for utilities? Who is responsible for dead animal removal (Figure 21.4), dust control, or weed control?

D. Payment

1. Liens. The feeder has a statutory lien on the hogs, but this lien is usually subject to all prior liens of record. This means that the owner's secured creditors can remove the hogs without paying the grower. Various states have different laws regarding liens, check with your attorney. It is possible for the grower to receive a first lien on the hogs if the owner and his or her creditors are willing to give the grower a lien subordination or the state allows it.

2. Payment terms. On what basis are you being paid? Will the schedule of payments satisfy your cash flow and are there penalties for late payments? Will the last payment be made before the livestock leave your facilities? Will your lender's name be on the check?

3. If production incentive payments (based on factors such as death loss, feed conversion, or rate of gain) are involved, then exactly what do you have to do to receive the incentive payments? How are the incentives calculated and when are the payments made?

4. Do you know your costs of production to determine the profitability of the contract? If you do not have cost of production records, then you should consult with the extension service or others to arrive at estimated production costs.

5. If you have concerns about getting paid, will the contractor provide you with a financial statement or with a list of producers with whom the contractor has raised pigs?

6. Your credentials. If the contractor has questions about your ability to perform the contract, are you willing and able to release a financial statement and the names of individuals who will verify your financial stability and management abilities?

7. Parent company responsibility. If the contractor is a subsidiary company, does the contract make the parent company responsible for payment if the contractor defaults?

E. Legal Issues

1. Does the contract require forms of dispute resolution, such as mediation or arbitration?

2. Under what conditions can the contractor terminate the contract? Are there objective standards or is it at the discretion of the contractor? Under what conditions can you terminate the contract?

3. Does the contract excuse nonperformance caused by "Acts of God," meaning occurrences out of human control?

4. Under what conditions can the contract be renewed? Are there standards for renewal, or is it up to the contractor?

5. What legal relationship does the contract establish between you and the contractor? Is it a landlord–tenant relationship, employer–employee relationship, independent contractor, partnership, joint venture, agency? The legal status of the relationship not only affects your rights and responsibilities under the contract, but it also has important tax consequences.

6. Approval of contract by others/assignment. Do other parties have to approve the contract, such as your landlord, your lender, your spouse? Can the contract be assigned or transferred by you or by the contractor to others, such as a lender?

7. If the contractor is from another state, does the contract specify the state law that governs? Does the contract set a venue (location) for any lawsuit that might be filed? Does the contract permit renegotiation or nullification of the contract if the laws governing production contracts are changed?

8. Put it in writing. You should not rely on oral agreements or interpretations of the contract. Reduce all understandings or modifications to writing.

F. Neighbors/Goals

1. Will raising livestock under this contract affect your relationship with your neighbors? Have you talked with your neighbors about your plans?

2. How does this contract fit into your long-term goals for your farm, your family, your community?

The key to feeding or producing hogs under contract is finding the type of contract that will allow each individual to profit most from his or her skills, resources, and ability to bear risk associated with hog production. This strength may be record keeping, producing with a low mortality rate, or an ability to maximize herd feed efficiency. Whatever the case, producers should make certain that the contract will reward them appropriately for what they do best.

Once the best contract type has been found, the written contract itself should be carefully read and understood. Although a well-written contract is essential to successful contract production, it is also important that both parties are professional and willing to work out any problems that arise. A contract can never be so complete that every possible problem is anticipated. Individuals interested in contract production should check laws regarding contracting in their state.

Cooperatives (Co-ops)

A cooperative is a business formed by a group of people to get certain services for themselves more effectively or more economically than they can get individually. These people own, finance, and operate the business for their mutual benefit. Often by working together through such a cooperative business, member–owners obtain services not otherwise available to them. Co-ops are engaged in every facet of pork production, including supplies, marketing, processing, and retailing. Like all businesses, their success depends on management.

Networking

Networking is two or more people working together to achieve individual and group goals that would be more difficult to obtain alone. It is a banding together of small and independent pork producers to compete with the megapork producers. Actually, networking is a new name for an old practice. A few years ago, we would have called it cooperating, but some people do not want co-ops raising hogs, so another term evolved.

Networking is not unique to the pork industry. The term was born in the 1960s in the business world, and it flourished through the 1980s. In the business world, the term has many definitions as it does in the pork industry. For example, automakers may not own a glass factory to manufacture their car windows; instead, they may network with a glass company to make their windows.

Although networking possibilities in the pork industry are unlimited, three categories of networks appear to have merit for pork producers: (1) information, (2) marketing, and (3) production.

• **Information networks**—Information networks are generally informal groups—for example, producers with the same record system—who share information with, and learn from, each other. Some groups charge a membership fee, which may be used to bring in consultants and speakers to address the group, or to cover all or part of the cost of a study-tour of swine production in other areas.

• **Marketing networks**—Marketing networks require producer commitment relative to the time, number of hogs, and quality of hogs that they will market in the future. Usually, a better price is secured because of both volume and access to additional markets. Marketing networks include input purchasing as well as hog sales.

• **Production networks**—Production networks may require joint ownership of facilities or hogs. Joint sow ownership, gilt multipliers, and congregate nurseries are all examples of production networks. Although joint ownership in centralized production facilities/hogs is typical, it is not a requisite. Networks in which one producer farrows, a second operates a nursery, and third finishes the hogs, all in their own facilities, is possible.

The key to success of any network is the people involved and their commitment to making it work. The benefits of successful networking are as follows:

1. **Capturing proven technology,** which an individual may not be able to do. For example, by joining together, they may be able to afford multisite technology; one may farrow, a second may operate the nursery, and a third may finish the hogs. Multisites improve hog health, feed efficiency, and rate of gain.

2. **Capturing real economics in volume sales and purchases;** producers are able to sell hogs at a higher price and buy supplies at a lower price.

3. **Improving product quality and market access.** Uniform, high-quality hogs in sufficient volume will gain market access.

4. **Utilizing production, marketing, and information systems.** A "system" merely implies that the entire pork channel is considered when production and marketing decisions are made, and that actions are taken to improve product quality and efficiency where possible.

The limitations of networking are as follows:

1. **Commitment of people.** Successful networking depends on the commitment of people.

2. **Joint responsibility.** Networking depends on the performance of its individual members. Some producers may not wish to assume responsibility for their fellow members.

3. **Formal business procedures.** Tighter control and more formal business procedures are necessary as transactions become more complex. Also, networking requires increased communication between all people involved.

4. **Loss of markets and suppliers.** Typically, networking involves direct negotiation and a likely contract agreement with the supplier and the buyer. This may eliminate local suppliers and buyers.

VERTICAL INTEGRATION

Vertical integration in pork refers to a structure in which, through alliances (complete ownership, joint ownership, or contract), an individual or company controls the product through two or more phases; for example, a meat packer owns hogs from birth through slaughter.

The U.S. poultry industry led the way in vertical integration and today is essentially 100% integrated, followed closely by the turkey industry, which is more than 90% integrated. Currently, the pork industry is following suit. Most vertically integrated hog enterprises are packers that also farrow and finish the hogs they slaughter.

More than 86% of pigs are prepriced before slaughter, but the packer control of their genetics, feed, and management is only loosely scrutinized by packers. In 2002, within this 86%, at least 1.45 million sows were owned by companies that have slaughter facilities, accounting for about 30% of the U.S. hog slaughter. In these companies, everything from facilities, genetics, feed, and management is controlled.

FUTURES MARKETS

Futures trading is a well-accepted, century-old procedure used in many commodities for protecting profits, stabilizing prices, and smoothing out the flow of merchandise (Figure 21.5). For example, it has long been an integral part of the grain industry; grain elevators, flour millers, feed manufacturers, and others have used it to protect themselves against losses that result from price fluctuations. Many of these processors prefer to forego the possibility of making high speculative profits in favor of earning normal margins or service charges through efficient operation of their businesses. They look to futures markets to provide (1) an insurance medium in the marketing field and (2) the facilities and machinery for underwriting price risks.

The unique characteristic of futures markets is that trading is carried out in terms of contracts to deliver or to take delivery, rather than on the immediate transfer of the physical commodity. In practice, however, very few contracts are held until the delivery date. The vast majority of them are canceled by offsetting transactions made before the delivery date.

The biggest uncertainty in the hog business is the selling price of market hogs. Producers know their costs of producing pigs, and, from their past records, they are usually able to project with reasonable accuracy the costs of production to market time. But unless they contract ahead, they have no assurance of what the hogs will bring when they are ready to market. Moreover, there is little flexibility in market time because excess finish is costly and unwanted by consumers.

Most producers contract their pigs for future delivery to slaughter without the medium of a futures contract. They contract to sell and deliver to a packer a certain number and kind of pigs at an agreed-on price and place. Hence, the risk of loss is reduced from a decrease in price after the contract is shifted to the buyer; and, by the same token, the seller foregoes the possibility of a price rise. However, unlike futures trading, contracting requires actual delivery of the pigs.

Lean Hog Futures

The Chicago Mercantile Exchange (CME) Live Hogs futures and options contracts have undergone considerable improvements, including getting a new name—Lean Hogs—that became effective in February 1997. Along with the new name, new specifications make this contract a viable hedging tool for U.S. producers and packers. The new cash settlement feature also makes this a viable hedging tool for international producers and pork importers/exporters. It replaces and upgrades the live hog futures and options that began trading on the CME in 1966.

The Lean Hog contract is for 40,000 lb of lean value, the estimated meat produced from about 220 hogs. There shall be no delivery of hogs in settlement of this contract. All contracts open as of the termination of trading shall be cash settled based on the CME Lean Hog Index for the 2-day period ending on the day on which trading terminates.

The CME Lean Hog Index is a 2-day weighted average of cash prices. Beginning with the April 2003 contract, the cash prices used in the index will be the average net prices at the average percent lean for slaughtered hogs. These prices are available in USDA Report LM_HG201-National Daily Direct Hog Prior Day Report-Slaughtered Swine. (*http://www.ams.usda.gov/mnreports/lm_hg201.txt*)

The index is based on a sample of transactions for packer base weight and base cost hogs. The sample is a 2-day, 3-area weighted average price per lb for packer base weight hogs, 51 to 52% lean with 0.80 to 0.99 in. of backfat at the last rib or equivalent, and 170 to 191 lb dressed weight.

Each packer reports this base cost and average net price, the number of hogs slaughtered and the average carcass weight to the USDA. The USDA then calculates an average base cost and net price, weighted by the number of hogs slaughtered and the average carcass weight. This weighting allows each price to be represented in proportion to the number of hogs sold so that the index better represents production patterns and prices received by producers (Figure 21.6).

On the day after slaughter, the USDA releases the weighted average base cost and average net price, the total number of hogs slaughtered and the average carcass weight. Because the information is reported the day following the actual slaughter, the index itself has a lag.

Figure 21.6 Graded pork carcasses in the cooler waiting for cutting into primal cuts. (*Courtesy, Swift and Company, Greeley, CO*)

Figure 21.5 Trading pit at the Chicago Mercantile Exchange. (*Courtesy, Chicago Mercantile Exchange, Chicago, IL*)

When the contract expires, the cash settlement price must, by definition, be equal to the futures price. This is expected to reduce, substantially, futures price variability at contract expiration because it will be explicitly linked to the cash market. This makes it much easier for producers to hedge, because the basis risk should be much lower near contract expiration.

This system also removes the incentive to get out of the futures position before the time of delivery. Liquidation of futures positions prior to final contract expiration can lead to highly variable end-of-contract price movements.

The primary difference between the new lean hog contract and the old live hog contract are as follows:

- **The Lean Hog Futures Contract is still 40,000 lb, but that amount is on a carcass weight basis**—So, whereas about 167 hogs at 240 lb were necessary to fill a Live Hog Futures Contract, it will now require about 216 hogs at 250 lb (40,000 lb ÷ 185 lb carcass) to fill the contract on a carcass weight basis. Thus, it will take more animals to fill the 40,000 lb requirement. The number of hogs is still important, even without delivery, because to form a hedge you will still need to match futures quantities with spot quantities as always. It may be tempting for producers to mismatch the two with no delivery requirement hanging over their heads, but that would constitute increasing risk.
- **The hogs to be slaughtered must also meet the specifications of 51 to 52% lean, 0.80 to 0.99 in. of backfat at the last rib, and a dressed weight of 170 to 191 lb**—These are new because the old contract was based on live animals' measurements.
- **The daily price limit is still $2.00/cwt**—This represents a lower percentage of the total value of the contract than on a live-weight equivalent.

Buying or Selling Lean Hog Futures

Here is how to buy or sell hog futures:

1. Have good and accurate records of costs.
2. Contact a brokerage house that holds a membership in the commodity exchange.
3. Open up a trading account with the broker by signing an agreement authorizing the broker to execute trades.
4. Deposit with the broker the necessary performance money for each contract desired. The broker will then maintain a separate account for the producer. The contract commission fee is due when the contract is fulfilled by either delivery or offsetting purchase or sale of another contract.

Futures Market Terms

Futures markets have a jargon and sign language all their own. It is not necessary that producers dealing in futures master many of them, but it will facilitate matters if they have at least a working knowledge of the following terms:

- **Basis**—The difference between the spot or cash price and the futures price of the same or a related commodity. The local cash market price minus the price of the nearby futures contract.
- **Basis movement**—The change in the basis within a particular period of time in one location, or from location to location at the same time.
- **Hedgers**—A person or firm who uses the futures market to offset price risk when intending to sell or buy the actual commodity. Persons who cannot afford risks, and who try to increase their normal margins through buying and selling of futures contracts.
- **Limit order**—Usually equivalent to "price." An order in which the customer sets a limit on price or other condition, as contrasted with the trading floor definition of a market order, which implies that the order should be filled as soon as possible.
- **Long position**—A position in which the trader has bought a futures contract that does not offset a previously established short position.
- **Offset**—Taking a second futures position or an option on futures positions opposite to the initial or opening position. Selling if one has bought, or buying if one has sold.
- **Pit**—The platform on the trading floor of an exchange where traders and brokers stand while executing futures trades.
- **Premium**—The amount agreed on between the buyer and seller for the purchase or sale of a futures option. The buyer pays the premium and the seller receives the premium. The excess of one futures contract price over that of another or over the cash market price. The price of an option. The price is made up of the intrinsic value (in the money) and time/volatility value (out of the money).
- **Short position**—The sale of a futures contract in anticipation of a later cash market sale. Used to eliminate or reduce the possible decline in value of ownership of an approximately equal amount of the cash financial instrument or physical commodity.
- **Speculators**—Investors and traders who want to profit from price changes. Speculators accept the price risks and rewards that hedgers wish to avoid. Futures markets provide the forum in which speculators can buy or sell contracts quickly and can exit their positions just as quickly to react to market changes.

WORKERS' COMPENSATION

Workers' compensation laws, now in full force in all 50 states, cover on-the-job injuries and protect disabled workers regardless of whether their disabilities are temporary or permanent. Although broad differences exist among the individual states in their workers' compensation laws, principally in their benefit provisions, all statutes follow a definite pattern as to employment covered, benefits, insurance, and so on.

Workers' compensation is a program designed to provide employees with assured payment for medical expenses or lost income as a result of injury on the job. Whenever an employment-related injury results in death, compensation benefits are generally paid to the worker's surviving dependents.

Generally, all employment is covered by workers' compensation, although a few states provide exemptions for farm labor, or exempt farm employers with fewer than 10 full-time employees, for example. Farm employers in these states, however, may elect workers' compensation protection. Livestock producers in these states may wish to consider coverage as a financial protection strategy because under workers' compensation, the upper limits for settlement of lawsuits are set by state laws.

This government-required employee benefit is costly for livestock producers. Costs vary among insurance companies because of dividends paid, surcharges, minimum premiums, and competitive pricing. Some companies, as a matter of policy, will not write workers' compensation in agricultural industries. Some states have a quasi-government provider of workers' compensation to assure availability of coverage for small businesses and high-risk industries. For information, contact an insurance agent experienced in marketing workers' compensation and liability insurance.

MANAGEMENT

Four major ingredients are essential to success in the hog business: (1) good healthy hogs, (2) good feeding, (3) good records, and (4) good management. Management gives point and purpose to everything else. The skill of the manager materially affects how well hogs are bought and sold, the health of the animals, the results of the diets, the stress of the hogs, the rate of gain and feed efficiency, the performance of labor, the public relations of the establishment, and even the expression of the genetic potential of the hogs. Indeed, a manager can make or break a pork enterprise, a fact often overlooked in the present era with the accent on scientific findings and automation. Management is still the key to success.

In manufacturing and commerce, the importance and scarcity of top managers are generally recognized and reflected in the salaries paid to persons in such positions. Unfortunately, agriculture in this area as a whole has lagged, and too many owners still subscribe to the philosophy that the way to make money out of the pork business is to hire a manager cheap, with the result that they usually get what they pay for—a "cheap" manager.

Traits of a Good Manager

There are established bases for evaluating many articles of trade, including hogs and grain. They are graded according to well-defined standards. Additionally, we chemically analyze feeds and conduct feeding trials. But no such standard or system of evaluation has evolved for managers, despite its acknowledged importance.

A Hog Manager Checklist is presented in Table 21.7 that (1) employers may find useful when selecting or evaluating a manager, (2) managers may apply to themselves for self-improvement purposes,

TABLE 21.7   Hog Manager Checklist

- Character
  Absolute sincerity, honesty, integrity, and loyalty; ethical.

- Industrious
  Work, work, work; enthusiasm, initiative, and aggressiveness.

- Ability
  Hog know-how and experience, business acumen—including ability to arrive at the financial aspects systematically and to convert this information into sound and timely management decisions; knowledge of how to automate and cut costs; common sense; organized; growth potential.

- Plans
  Sets goals, prepares organization chart and job description, plans work, and works plans.

- Analytical
  Identifies the problem, determines pros and cons, then comes to a decision

- Courage
  Accepts responsibility to innovate and to keep on keeping ahead.

- Promptness and dependability
  A self-starter and reliable. Follows through on assignments.

- Leadership
  Stimulates employees and delegates responsibility.

- Personality
  Cheerful; not a complainer.

Source: Courtesy of Pearson Education.

and (3) students may use for guidance as they prepare for managerial positions. No attempt has been made to assign a percentage score to each trait because this will vary among swine establishments. Rather, it is hoped that this checklist will serve as a useful guide to the traits of a good manager and to what the employer wants.

Organization Chart and Job Description

It is important that workers know to whom they are responsible and for what they are responsible; and the bigger and the more complex the operation, the more important this becomes. This should be written down in an organization chart along with job descriptions.

Providing Incentives

Large farms must rely on hired labor, all or in part. Good help—the kind that everyone wants—is hard to come by; it is scarce, in strong demand, and difficult to keep. And the farm labor situation is going to become more difficult in the years ahead. There is need for a system that will (1) give a big assist in getting and holding top-quality help and (2) cut costs and boost profits. An incentive basis that makes employees partners in profits is the answer.

Many manufacturers have long had an incentive basis. Laborers may receive bonuses based on piecework or quotas (number of units, pounds produced). Also, most factory workers get overtime pay and have group insurance and retirement plans. A few industries have true profit-sharing arrangements based on net profits as such, a specified percentage of which is divided among employees. No two systems are alike. Yet, each is designed to pay more for labor, provided labor improves production and efficiency. In this way, both owners and laborers benefit from better performance.

Family-owned and family-operated farms have built-in incentive bases; there is pride of ownership, and all members of the family are fully aware that they prosper as the business prospers.

Many different incentive plans can be, and are, used. There is no best one for all operations. The incentive basis chosen should be tailored to fit the specific operation, with consideration given to kind and size of operation, extent of owner's supervision, present and projected productivity levels, mechanization, and other factors.

QUESTIONS FOR STUDY AND DISCUSSION

1. The great changes that occurred in the U.S. swine industry in the 1980s and 1990s resulted from a shift from our labor-based economy to a knowledge- and capital-based economy. Why did this occur?
2. Estimated costs and profit/loss in swine production are presented in Tables 21.1a, 21.1b, and 21.1c. How would you use the data presented in these tables to select the type of pork enterprise you would favor: (a) farrow to finish, (b) farrow to feeder pig, or (c) feeder pig to market?
3. Figure 21.1 shows that hog cycles are becoming longer and less abrupt. What has caused this change?
4. Describe briefly each of the following pork production systems: (a) farrow to finish, (b) farrow to feeder pig, and (c) finish to market.
5. Describe briefly each of the following types of business organizations: (a) proprietorship (individual), (b) partnership (general partnership), (c) corporations, (d) contracts, (e) co-ops, and (f) networking.
6. What has caused the great increase in contract hog production in recent years?
7. What is vertical integration? What are the resulting benefits and risks for the pork industry? Justify your answer.
8. If you were becoming a pork producer, would you get involved in lean hog futures trading? Justify your answer.
9. List the traits of a good manager. How should a graduate fresh out of college train to become the manager of a megahog operation?

SELECTED REFERENCES

Estimated Livestock Returns, J. D. Lawrence, Department of Economics, Iowa State University, Ames, IA, 2003, http://www.econ.iastate.edu/faculty/lawrence/EstRet/Index.html

Livestock Enterprise Budgets for Iowa—2003, FM 1815 revised, May, G. et al, Iowa State University, Ames, IA 2003, http://www.extension.iastate.edu/Publications/FM1815.pdf

Livestock Production Contract Checklist, Iowa Attorney General Marketing Contracts, http://www.iowaattorneygeneral. org/working_for_farmers/brochures/livestock_production.html

Pork Industry Handbook, Cooperative Extension Service, Purdue University, West Lafayette, IN, 2002

5

Reference Information

22

Feed Composition Tables

Harvesting corn husks on a farm by the Hudson River.
(Photo by, John Collier, Courtesy, USDA, Washington, DC, 1941)

Contents

Objectives

After studying this chapter you should:

1. Know the basic factors that are used when describing feedstuffs.
2. Understand the importance of moisture content when describing feedstuffs.
3. Know the nutrient analyses of ingredients used in swine diets are routinely listed on an as-fed basis, with the dry matter content provided as part of the analysis.

Both nutritionists and swine producers should have access to accurate and up-to-date nutrient compositions of feedstuffs to formulate diets for optimum production and net returns. The ultimate goal of feedstuff analysis, and the reason for feed composition tables is to predict the productive response of animals when they are fed diets of a given composition. The monumental work of Lorin Harris and others from Utah State University provides the basis for the information along with more recent composition tables from the National Research Council's *Nutrient Requirements of Swine* and *Animal Feed Resources Information System* (*http://www.fao.org/ag/aga/agap/frg/afris/index_en.htm*). Additional feed and composition information was obtained from experimental reports, industry composition publications, and other reliable sources.

FEED NAMES

Ideally, a feed name should conjure up the same meaning to all those who use it, and it should provide helpful information. This was the guiding philosophy of the authors when choosing the names given in the feed composition tables. Genus and species—Latin names—are also included. To facilitate worldwide usage, the international feed number of each feed is given. To the extent possible, consideration was also given to source (or parent material), variety or kind, processing, and grade.

MOISTURE CONTENT OF FEEDS

It is necessary to know the moisture content of feeds in diet formulation and buying. Usually, the composition of a feed is expressed according to one or more of the following bases:

1. As-fed (wet, fresh). This refers to feed as normally fed to animals. It may range from 0 to 100% dry matter. Most grains are approximately 85 to 90% dry matter, liquid milk by-products less than 10%, and forages in between.

2. Air-dry (approximately 90% dry matter). This refers to feed that is dried by means of natural air movement, usually in the open. It may be either an actual or an assumed dry matter content; the latter is approximately 90%. Most feeds are fed in an air-dry state.

3. Moisture-free (oven-dry, 100% dry matter). This refers to a sample of feed that has been dried in an oven at 221°F (105°C) until all the moisture has been removed.

Swine diets are routinely prepared on an as-fed basis. Therefore, feed tables in this book have nu-

trient compositions presented on an as-fed basis with the dry matter content noted.

PERTINENT INFORMATION ABOUT DATA

The following information is pertinent to the feed composition tables presented in this chapter.

• **Variations in composition**—Feeds vary in composition. Thus, actual analyses of feedstuffs should be obtained and used wherever possible, especially where a large amount of feed from one source is involved. Many times, however, either it is impossible to determine actual compositions or there is insufficient time to obtain such analyses. Under such circumstances, tabulated data may be the only information available.

• **Feed composition change**—Feed compositions change over a period of time, primarily because of (1) the introduction of new varieties, (2) climatic changes, and (3) modifications in the manufacturing process from which by-products evolve.

• **Available nutrients**—The availability of the nutrients is a function of the feed's chemical composition and the ability of the animal to derive useful nutrient value from the feed. The latter relates to the digestibility, or availability, of the nutrients in the feed. Thus, soft coal and shelled corn may have the same gross energy value in a bomb calorimeter but markedly different useful energy values when consumed by an animal. Biological tests of feeds are more laborious and costly to determine than chemical analyses, but they are much more accurate in predicting the response of animals to a feed.

• **Where information is not available**—Where information is not available or reasonable estimates could not be made, no values are shown. Users may estimate missing values by comparing similar feeds but should realize the accuracy may be quite variable. Many of the net energy values were calculated based on a formula from the National Research Council's *Nutrient Requirements of Swine* (1998).

Energy

Digestible energy (DE) probably is the determining factor controlling voluntary feed intake. It does not take into account energy lost in the urine.

Metabolizable energy (ME) is between 94 and 97% of DE with an average of 96%. It represents that portion of gross energy not lost in feces, urine, and gas (mainly methane). It does not take into account the energy lost as heat, commonly called heat increment.

Net energy is the difference between ME and the heat increment (HI). HI is not used for productive purposes but can be used to maintain body temperature. Net energy is the best indicator of the energy available to the animal for maintenance and production.

Protein

Crude protein values are given. Crude protein represents Kjeldahl nitrogen value times 6.25 most proteins contain about 16% nitrogen (100/16).

Fiber

Three values for fiber are given in Table 22.1: neutral detergent fiber (NDF or cell walls), acid detergent fiber (ADF or cell contents), and crude fiber.

Crude fiber is declining as a measure of low digestible material in the more fibrous feeds. However U.S. feed tags are still required to report this value. The newer method of forage analysis, developed by Van Soest and associates of the U.S. Department of Agriculture, separates feed dry matter into two fractions, one of low digestibility (cell walls) and the other of high digestibility (cell contents).

Cell wall or neutral detergent fiber (NDF). This is the insoluble fraction resulting from boiling a 0.5 to 1.0 g sample of the feed in a neutral detergent solution (3% sodium lauryl sulfate buffered to a pH of 7.0) for 1 hour, then filtering. It contains cellulose, hemicellulose, silica, some protein, and lignin. Cell wall, or NDF, components are of low digestibility and entirely dependent on the microorganisms of the digestive tract for any digestion that they undergo; hence, they are essentially not digested by nonruminants. This fraction affects the volume that will occupy the digestive tract, a principal factor limiting the amount of feed consumed, especially if high-fiber feeds, such forages, are a major component of the diet.

Acid detergent fiber (ADF). This involves boiling a 1.0 g sample of air-dry material in a specially prepared acid detergent solution for 1 hour, then filtering. The insolubles, or residue, consist primarily of cellulose, lignin, and variable amounts of silica. ADF is the best predictor of forage digestible dry matter and digestible energy.

Crude fiber (CF). This is the residue that remains after boiling a feed in a weak acid, and then in a weak alkali, in an attempt to imitate the process that occurs in the digestive tract. The procedure assumes that carbohydrates, which are readily dissolved, will also be readily digested by animals and those not soluble under such conditions are not readily digested. Unfortunately, the treatment dissolves much of the lignin, a nondigestible component. Hence, crude fiber is only an approximation of the indigestible material in feedstuffs. Nevertheless, it is a rough indicator of the energy value of feeds.

The soluble fraction—the cell contents—consists of sugars, starch, fructosans, pectin, protein, nonprotein nitrogen, lipids, water, soluble minerals, and vitamins. This portion is highly digestible (about 98%) by both ruminants and nonruminants.

Minerals

A certain level of minerals are integral to the plant, but variation of minerals is largely determined by the mineral content of the soil on which the feeds are grown. Calcium, phosphorus, iodine, and selenium are well-known examples of soil nutrient–plant nutrient relationships. Phosphorus availability is provided where information is available. The levels of all trace minerals, with the possible exceptions of copper, iodine, iron, manganese, selenium, and zinc are adequate in feedstuffs to meet the pigs' requirements. These six trace minerals are routinely added to swine diets.

Vitamins

Many vitamins found in feedstuffs meet the requirements for pigs and do not need to be supplemented. The fat soluble vitamins (A, D, E and K) and the B vitamins (niacin, pantothenic acid, riboflavin, and B_{12}) are routinely supplemented in swine diets. Biotin, folic acid, and choline may be supplemented under various conditions but are not routinely supplemented.

Where carotene has been converted to vitamin A, the conversion rate of the rat has been used as the standard value, with 1 mg of β-carotene equal to 1,667 IU of vitamin A. Generally speaking, it is unwise to rely on harvested feeds as a source of carotene (vitamin A value), unless pigs are being fed fresh forage or that of a good green color and not more than a year old. Vitamin A is a very inexpensive vitamin, and it is better to include it in a vitamin mix than to rely on forages as a consistent source.

Shelling corn for storage in the "ever-normal granary." *(Photo by, Arthur Rothstein, Courtesy, USDA, Washington, DC, 1939)*

TABLE 22.1 Macronutrient Composition of Feed Ingredients Used for Swine (as-fed basis)[1]

No.	Feed	International Feed Number	Dry Matter (%)	Digestible Energy (kcal/kg)	Metabolizable Energy (kcal/kg)	Net Energy (kcal/kg)[2]	Crude Protein (%)	Ether Extract (%)	Linoleic Acid (%)	Neutral Detergent Fiber (%)	Acid Detergent Fiber (%)	Crude Fiber (%)
	Alfalfa (Lucerne) *Medicago sativa*											
1	-hay, sun-cured, all analyses	1-00-078	90	1,421	1,153	99	15.9	2.3	—	42.9	32.6	27.2
2	-hay, sun-cured	1-08-331	91	—	—	—	14.0	2.1	—	—	37.5	30.3
3	-leaves, meal, dehydrated	1-00-137	—	—	—	—	20.0	—	—	—	—	—
4	-leaves, sun-cured, ground	1-00-246	—	—	—	—	20.5	—	—	—	—	—
5	-meal, dehy, 15% CP (crude protein)	1-00-022	90	—	1,291	259	15.4	2.2	—	46	33.8	26.2
6	-meal, dehy, 17% CP	1-00-023	92	1,830	1,650	910	17.0	2.6	0.35	41.2	30.2	24.4
7	-meal, dehy, 20% CP	1-00-024	92	2,095	1,885	1,290	19.6	3.3	0.44	38.8	26.4	20.4
8	-meal, dehy, 22% CP	1-07-851	93	—	—	—	—	—	—	—	—	—
	Animal											
9	-animal plasma, spray dried	—	91	—	—	—	78.0	2.0	—	—	—	—
10	-blood cells, spray dried	—	92	—	—	—	92.0	1.5	—	—	—	—
11	-blood meal, conventional	5-00-380	92	2,850	2,350	1,950	77.1	1.6	0.09	13.6	1.8	—
12	-blood meal, flash dried	5-26-006	92	2,300	1,950	1,385	87.6	1.6	—	—	—	—
13	-blood meal, spray or ring dried	5-00-381	93	3,370	2,945	2,070	88.8	1.3	0.17	—	—	1.4
14	-fat, animal	4-00-376	99	8,800	7,900	—	—	—	—	—	—	—
15	-fat, swine (lard)	4-04-790	99	8,285	7,950	5,100	—	15.8	—	—	—	—
16	-liver meal, dehy	5-00-389	93	—	—	—	66.7	—	—	—	—	1.4
17	-meat, meal, rendered	5-00-385	94	2,695	2,595	2,175	54.0	12.0	0.80	31.6	8.3	2.4
18	-meat with blood, meal, tankage rendered	5-00-386	92	2,305	2,248	—	59.5	9.0	—	—	—	2.2
19	-meat with blood with bone, meal, tankage rendered	5-00-387	92	2,962	2,618	—	50.2	10.5	—	—	—	2.4
20	-meat meal with bone rendered	5-00-388	93	2,440	2,225	1,355	51.5	10.9	0.72	32.5	5.6	2.0
21	-tallow, beef	4-08-127	99	8,000	7,680	4,925	—	—	—	—	—	—
	Animal-poultry (see poultry)											
	Bakery waste											
22	-dried bakery product	4-00-466	91	3,940	3,700	2,415	10.8	11.3	5.70	2.0	1.3	1.5
	Barley *Hordeum vulgare*											
23	-grain	4-00-549	88	3,052	2,743	1,772	12.2	1.9	—	—	—	5.0
24	-grain, Pacific coast	4-07-939	90	3,156	2,577	—	9.6	1.7	—	—	—	6.3
25	-grain, two row	4-00-572	89	3,050	2,910	2,340	11.3	1.9	0.88	18.0	6.2	04
26	-grain, six row	4-00-574	89	3,050	2,910	2,310	10.5	1.9	0.91	18.6	7.0	05
27	-grain, hulless	4-00-552	88	3,360	3,320	2,650	14.9	2.1	1.14	10.1	2.2	06
28	-malt sprouts, dehy	5-00-545	92	1,538	1,422	916	26.1	1.3	—	—	—	14.5
	Bean *Phaseolus vulgaris*											
29	-kidney seeds	5-00-600	89	—	—	—	22.0	1.3	—	—	—	4.2

No.	Feed	IFN										
30	-navy seeds	5–00–623	90	—	1,323	933	22.8	1.4	—	—	—	4.4
31	-pinto seeds	5–00–624	90	—	—	—	22.6	1.3	—	—	—	4.0
	Beet, sugar *Beta vulgaris, saccharifera*											
32	-molasses, > 48% invert sugar, > 79.5° Brix	4–00–668	78	—	2,025	1,532	6.0	0.1	—	—	—	—
33	-pulp, dried	4–00–669	91	2,865	2,495	1,860	8.6	0.8	—	42.4	24.3	17.9
	Blood (see animal)											
	Brewers' grains											
34	-dehy (brewers' dried grains)	5–02–141	92	2,100	1,960	1,630	26.5	7.3	3.14	48.7	21.9	14.4
	Broomcorn (see millet, proso)											
	Buckwheat *Fagopyrum* spp											
35	-grain	4–00–994	88	2,825	2,640	1,620	11.1	2.4	0.53	17.8	14.3	10.3
	Canola (rapeseed) *Brassica* spp											
36	-meal, solv extr (solvent extracted)	5–06–145	90	2,885	2,640	1,610	35.6	3.5	0.42	21.2	17.2	9.3
37	-meal, prepressed, solv extr, 40% CP	5–08–135	92	—	—	—	40.5	1.1	—	—	—	—
	Cassava *Manihot* spp											
38	-meal (Tapioca or Manioc)	4–01–152	88	3,385	3,330	2,330	3.3	0.5	—	7.7	4.6	4.5
	Cattle *Bos taurus*											
39	-buttermilk, condensed	5–01–159	29	—	1,084	913	10.7	2.3	—	—	—	0.1
40	-buttermilk, dehy	5–01–160	92	2,280	2,066	1,432	31.5	4.7	—	—	—	0.4
41	-casein, dried	5–01–162	91	4,135	3,535	2,555	88.7	0.8	0.03	—	—	—
42	-cheese rind	5–01–163	36	—	—	—	46.2	20.0	—	—	—	0.2
43	-milk, dehy, cow's	5–01–167	95	4,817	4,270	2,749	25.1	26.4	—	—	—	0.2
44	-milk, fresh, cow's	5–01–168	13	712	644	577	3.6	3.7	—	—	—	91
45	-milk, skimmed, fresh, cow's	5–01–170	9	390	347	220	3.1	0.1	0.01	—	—	0.3
46	-milk, skimmed, dehy, cow's	5–01–175	96	3,980	3,715	2,360	34.6	0.9	0.01	—	—	0.3
47	-whey, dehy	4–01–182	96	3,335	3,190	2,215	12.1	0.9	—	—	—	0.2
48	-whey, fresh	4–08–134	7	—	—	—	0.9	0.3	—	—	—	0.2
49	-whey, low lactose, dried	4–01–186	96	3,045	2,910	2,030	17.6	1.1	0.04	—	—	—
50	-whey, permeate, dried	—	96	3,435	3,300	2,260	3.8	0.2	—	—	—	—
	Chufa *Cyperus esculentus*											
51	-roots	4–08–374	27	930	878	760	2.1	1.8	—	—	—	2.0
	Citrus *Citrus* spp											
52	-pulp without fines, dehy (dried citrus pulp)	4–01–237	90	3,059	2,423	1,310	6.3	3.4	—	21	21.0	11.6
53	-syrup, molasses	4–01–241	68	2,369	2,235	1,580	4.8	0.2	—	—	—	—
	Clover, ladino											
54	-hay, sun-cured	1–01–378	90	—	—	—	19.1	2.4	—	32	28.5	18.5
	Coconut *Cocos nucifera*											
55	-meal, mechanically extracted (copra meal)	5–01–572	93	3,308	3,393	1,946	21.2	6.5	—	—	—	11.5
56	-meal, solvent extr (copra meal)	5–01–573	92	3,010	2,565	1,695	21.9	3.0	0.03	51.3	25.5	13.6
	Corn *Zea mays*											
57	-distillers dried grains	5–02–842	94	3,100	2,715	1,170	24.8	7.9	4.46	40.4	17.5	11.3

(continued)

TABLE 22.1 Macronutrient Composition of Feed Ingredients Used for Swine (as-fed basis) *(continued)*

No.	Feed	International Feed Number	Dry Matter (%)	Digestible Energy (kcal/kg)	Metabolizable Energy (kcal/kg)	Net Energy (kcal/kg)[2]	Crude Protein (%)	Ether Extract (%)	Linoleic Acid (%)	Neutral Detergent Fiber (%)	Acid Detergent Fiber (%)	Crude Fiber (%)
58	-distillers dried grains with solubles	5–02–843	93	3,200	2,820	2,065	27.7	8.4	2.15	34.6	16.3	—
	-distillers dried grains with solubles,	5–02–844	89	3,528	3,196	1,906	26.8	9.7	—	37.4	14.9	6.3
	Midwest ethanol plants											
59	-distillers dried solubles	5–02–844	92	3,325	2,945	2,250	26.7	9.1	5.36	24.8	7.5	5.2
60	-ears, ground (corn and cob meal)	4–02–849	86	3,108	2,676	1,636	7.8	3.2	—	—	—	8.3
61	-gluten feed, (gluten with bran)	5–02–903	90	2,990	2,605	1,740	21.5	3.0	1.43	33.3	10.7	8.7
62	-gluten, meal	5–02–900	91	—	3,190	—	43.1	2.2	—	34	8.2	4.5
63	-gluten meal, 60% CP	5–28–242	90	4,225	3,830	2,550	60.2	2.9	1.17	8.7	4.6	—
64	-grain	4–02–935	89	3,525	3,420	2,395	8.3	3.9	1.92	9.6	2.8	2.1
65	-grain, flaked	4–02–859	89	3,744	3,509	2,360	10.0	2.0	—	—	—	0.6
66	-grain, grade 2, 54 lb/bu or 695 g/1	4–02–931	89	3,620	3,397	2,250	8.7	3.9	—	8	5.3	2.2
67	-grits by-product (hominy feed)	4–03–011	90	3,355	3,210	2,260	10.3	6.7	2.97	28.5	8.1	—
68	-hominy feed	4–02–887	90	3,349	3,132	1,995	11.0	7.4	—	—	—	5.0
69	-grain, opaque-2 (high lysine)	4–11–445	87	3,274	3,070	2,050	9.6	4.0	—	—	—	2.6
	Corn, dent white *Zea mays indentata*											
70	-grain	4–02–928	91	—	—	—	10.9	—	—	—	—	—
	Cotton *Gossypium* spp											
71	-meal, mech extr, 36% CP	5–01–625	92	—	—	—	38.9	—	—	—	—	—
72	-meal, mech extr, 41% CP	5–01–617	92	2,945	2,690	1,870	42.4	6.1	3.15	25.7	18.0	11
73	-meal, prepressed, solv extr, 41% CP	5–07–872	90	2,575	2,315	1,325	41.4	1.5	0.51	28.4	19.4	13.6
	Emmer *Triticum dicoccum*											
74	-grain	4–01–830	91	3,095	2,890	1,715	11.7	2.0	—	—	—	9.7
	Fababean (broadbean)											
75	-seeds	5–09–262	87	3,245	3,045	2,000	25.4	1.4	0.62	13.7	9.7	—
	Fat (see animal)											
	Feather meal (see poultry)											
	Fish											
76	-meal, mechanically extracted	5–01–977	92	—	—	—	64.5	—	—	—	—	0.4
77	-solubles, condensed	5–01–969	51	1,910	1,625	995	32.7	5.6	—	—	—	—
78	-solubles, dehy	5–01–971	92	3,310	3,045	1,770	64.2	7.4	0.12	—	—	1.6
	Anchovy *Engraulis ringen*											
79	-Anchovy meal, mech extr	5–01–985	92	3,230	2,695	1,695	64.6	7.9	0.27	—	—	1.0
	Fish, herring											
80	-Herring meal, mech extr	5–02–000	93	3,960	3,260	2,020	68.1	9.2	0.15	—	—	1.8
	Menhaden *Brevoortia tyrannus*											
81	-meal, mech extr	5–02–009	92	3,770	3,360	2,335	62.3	9.4	0.12	—	—	0.8
	Fish, sardine *Clupea* spp, *Sardinops* spp											

#	Feed	Feed No.										
82	-meal, mech extr	5-02-015	93	2,942	2,527	1,415	65.1	5.0	—	—	—	1.0
83	-solubles, condensed	5-02-014	50	2,051	1,723	1,265	29.5	—	9.7	—	—	0
	Fish, white families Gadidae, Lophiidae, and Rafidae											
84	-meal, mech extr	5-02-025	91	3,395	2,810	2,020	63.3	4.8	0.08	—	—	0.7
	Flax Linum usitatissimum											
85	-meal, mech extr, 33% CP (linseed meal)	5-02-045	91	3,309	2,791	1,680	34.3	4.8	—	23.0	15.4	8.9
86	-meal, solv extr, 33% CP (linseed meal)	5-02-048	90	3,060	2,710	1,840	33.6	1.8	0.36	23.9	15.0	
	Garbage											
87	-hotel and restaurant, boiled, wet	4-07-865	26	1,212	1,124	940	4.3	5.8	—	—	—	0.7
88	-hotel and restaurant, cooked, dehy	4-07-879	54	2,680	2,477	1,705	9.5	14.6	—	—	—	1.6
	Lentil Lens spp											
89	-seeds	5-02-506	89	3,540	3,450	2,205	24.4	1.3	0.41	10.1	5.4	
	Lupin (sweet white) Lupinus spp											
90	-seeds	5-27-717	89	3,450	3,305	2,130	34.9	9.2	1.62	20.3	16.7	
	Maize (see corn)											
	Meat (see animal)											
	Milk (see cattle)											
	Millet, proso Setaria spp											
91	-grain	4-03-120	90	3,020	2,950	2,095	11.1	3.5	1.92	15.8	13.8	6.8
	Milo (see sorghum)											
	Molasses and syrup (see source plant)											
	Oats Avena sativa											
92	-cereal by-product (feeding oat meal, oat middlings)	4-03-303	91	3,470	3,218	2,090	14.7	6.5	—	—	—	3.6
93	-grain, all analyses	4-03-309	89	2,770	2,710	1,760	11.5	4.7	1.62	27.0	13.5	4.4
94	-grain, Pacific Coast	4-07-999	91	3,026	2,452	1,270	9.2	4.9	—	—	—	14.5
95	-grain, naked	4-25-101	86	3,480	3,410	2,160	17.1	6.5	2.52	9.9	3.7	2.4
96	-groats	4-03-331	90	3,690	3,465	2,310	13.9	6.2	2.40	—	—	11.3
	Pea Pisum spp											
97	-seeds	5-03-600	89	3,435	3,210	2,195	22.8	1.2	0.47	12.7	7.2	5.6
	Peanut (groundnut) Arachis hypogaea											
98	-kernels, hulls added, meal, solv extr	5-03-656	93	—	—	—	47.4	—	—	—	—	—
99	-kernels, meal, mech extr, 45% CP (peanut meal)	5-03-649	92	3,895	3,560	2,280	43.2	6.5	1.73	14.6	9.1	7.9
100	-kernels, meal, solv extr, 47% CP (peanut meal)	5-03-650	92	3,415	3,245	2,170	49.1	1.2	0.30	16.2	12.2	9.7
	Potato Solarium tuberosum											
101	-pulp, dehy	4-03-775	88	3,399	3,207	2,025	6.2	0.1	—	—	—	8.8
102	-protein concentrate	5-25-392	91	4,140	3,880	2,040	73.8	1.7	—	—	1.8	

(continued)

TABLE 22.1 Macronutrient Composition of Feed Ingredients Used for Swine (as-fed basis) *(continued)*

No.	Feed	International Feed Number	Dry Matter (%)	Digestible Energy (kcal/kg)	Metabolizable Energy (kcal/kg)	Net Energy (kcal/kg)[2]	Crude Protein (%)	Ether Extract (%)	Linoleic Acid (%)	Neutral Detergent Fiber (%)	Acid Detergent Fiber (%)	Crude Fiber (%)
103	-tubers, boiled	4-03-784	22	741	711	698	2.4	0.1	—	—	—	0.6
104	-tubers, dehydrated	4-07-850	91	3,344	3,210	2,145	7.9	0.5	—	—	—	2.0
	Poultry											
105	-by-product meal, rendered	5-03-798	93	3,000	2,860	1,945	64.1	12.6	2.54	—	—	2.3
106	-fat, poultry	4-09-319	99	8,520	8,180	5,230	—	99.1	19.5	—	—	—
107	-feathers meal, hydrolyzed	5-03-795	93	2,990	2,485	2,250	84.5	4.6	0.83	—	—	1.2
	Rape (see canola)											
	Rice *Oryza sativa*											
108	-bran with germ (rice bran)	4-03-928	90	3,100	2,850	2,040	13.3	13.0	4.12	23.7	13.9	11.6
109	-grain, ground (ground rough rice)	4-03-938	89	3,005	2,535	1,520	8.4	1.7	0.28	—	—	9.1
110	-grain, polished + broken (brewers' rice)	4-03-932	89	3,565	3,350	2,295	7.9	1.0	—	12.2	3.1	—
111	-groats, polished (polished rice)	4-03-942	89	3,529	3,329	2,265	7.3	0.4	—	—	—	0.4
112	-polishings	4-03-943	90	3,770	3,350	2,070	13.0	13.7	3.58	—	4.0	3.2
	Rye *Secale cereale*											
113	-distillers grains, dehy	5-04-023	92	—	—	—	21.6	7.2	—	—	—	2.2
114	-distillers grains with solubles, dehy	5-04-024	91	—	—	—	27.4	4.1	—	—	—	12.3
115	-grain	4-04-047	88	3,270	3,060	2,300	11.8	1.6	0.76	12.3	4.6	8.1
	Safflower *Carthamus tinctorius*											
116	-meal, solvent extracted	5-04-110	92	2,840	2,170	870	23.4	1.4	0.84	55.9	38.8	32.3
117	-meal without hulls, mech extr	5-08-499	90	3,352	2,908	1,741	42.1	6.7	—	—	—	8.5
118	-meal without hulls, solv extr	5-07-959	92	3,055	2,910	1,585	42.5	1.3	0.74	25.9	18.0	8.5
119	-seeds, whole	4-07-958	93	2,346	1,995	743	18.2	—	—	—	37.2	23.6
	Sesame *Sesamum indicum*											
120	-meal, mech extr	5-04-220	93	3,350	3,035	2,090	42.6	7.5	3.07	18.0	13.2	15.0
	Sorghum *Sorghum vulgare*											
121	-gluten meal	5-04-388	90	—	—	—	44.1	2.7	—	—	—	6.4
122	-gluten with bran (Gluten feed)	5-08-089	89	3,565	3,220	2,024	23.1	3.4	—	—	—	6.5
123	-grain	4-20-893	89	3,380	3,340	2,255	9.2	2.9	1.13	18.0	8.3	2.3
124	-grain, Feterita (USA)	4-04-369	90	—	—	—	12.4	2.8	—	—	—	1.6
125	-grain, Hegari (Thailand)	4-04-398	89	—	—	—	10.0	0.7	—	—	—	1.6
126	-grain, Kaffir (USA)	4-04-428	89	3,567	3,117	2,080	11.0	2.8	—	—	—	1.6
	Soybean *Glycine max*											
127	-meal, mech extr, 41% CP	5-04-600	90	3,378	2,731	1,715	43.8	4.7	—	—	—	5.8
128	-meal, solv extr	5-04-604	89	3,490	3,180	1,935	43.8	1.5	0.69	13.3	9.4	3.6
129	-meal without hulls, solv extr	5-04-612	90	3,685	3,380	2,020	47.5	3.0	0.60	8.9	5.4	5.4
130	-protein concentrate	5-32-183	90	4,100	3,500	2,000	64.0	3.0	—	—	—	—
131	-protein isolate	5-24-811	92	4,150	3,560	2,000	85.8	0.6	—	—	—	—

#	Feed name	Int'l feed no.										
132	-seeds, heat processed	5-04-597	90	4,140	3,690	35.2	2,880	9.13	18.0	13.9	8.0	5.1
133	-seeds, whole	5-04-610	91	4,048	3,533	39.3	2,215	—	17.5	—	8.2	5.3
134	**Spelt *Triticum spelta***											
	-grain	4-04-651	90	3,074	2,869	12.0	1,840	—	1.9	—	—	9.1
	Sugarcane *Saccharum officinarum*											
135	-sugarcane, molasses, dehydrated	4-04-695	90	2,935	2,368	8.4	—	—	0.9	—	—	—
136	-sugarcane, molasses (blackstrap) >46% invert sugar, >79.5° Brix	4-04-696	75	2,085	2,002	3.9	957	—	0.1	—	—	4.5
	Sunflower *Helianthus* spp											
137	-meal, solvent extracted	5-09-340	90	2,010	1,830	26.8	1,230	0.98	1.3	42.4	30.3	—
138	-meal without hulls, solv extr	5-04-739	93	2,840	2,735	42.2	1,635	1.07	2.9	27.8	18.4	11.7
	Sweet potato *Ipomoea batata*											
139	-pulp, dehy	4-08-535	90	3,351	3,198	2.5	2,250	9.1	0.3	—	—	11.0
140	-tubers	4-04-788	31	1,195	1,134	1.7	680	9.1	0.4	—	—	9.6
	Tallow (see animal)											
	Triticale *Triticale hexaloide*											
141	-grain	4-20-362	90	3,320	3,180	12.5	2,420	0.71	1.8	12.7	3.8	2.9
	Wheat *Triticum aestivum*											
142	-bran	4-05-190	89	2,420	2,275	15.7	1,400	1.80	4.0	42.1	13.0	10.0
143	-distillers grains, dehy	5-05-193	93	—	—	31.5	—	—	6.7	—	—	10.4
144	-germ, ground (wheat germ meal)	5-05-218	88	3,527	2,860	24.5	1,870	—	8.4	—	4.4	3.1
145	-grain	4-05-211	88	3,157	3,258	14.9	2,155	—	1.8	—	3.3	2.5
146	-grain, hard red spring	4-05-258	88	3,400	3,250	14.1	2,150	—	2.0	—	11.0	2.5
147	-grain, hard red winter	4-05-268	88	3,365	3,210	13.5	2,225	0.93	2.0	13.5	4.0	2.6
148	-grain, soft red winter	4-05-294	88	3,450	3,305	11.5	2,400	—	1.9	—	—	2.3
149	-grain, soft white winter	4-05-337	89	3,400	3,285	11.8	2,375	0.83	2.1	12.0	3.7	2.3
150	-grain, soft white winter, Pacific Coast	4-08-555	89	3,382	3,379	10.1	2,230	—	1.8	—	—	2.3
151	-middlings, < 9.5% fiber	4-05-205	89	3,075	3,025	15.9	1,560	1.74	4.2	35.6	10.7	7.8
152	-mill run, < 9.5% fiber	4-05-206	90	3,014	2,866	15.6	1,710	—	4.1	—	—	8.2
153	-red dog, < 4% fiber	4-05-203	88	3,140	2,925	15.3	2,090	—	3.3	18.7	4.3	2.9
154	-shorts, < 7% fiber	4-05-201	88	2,985	2,820	16.0	2,120	1.90	4.6	28.4	8.6	2.4
	Wheat, durum *Triticum durum*											
155	-grain	4-05-224	87	3,273	3,037	13.8	2,030	—	1.8	—	—	2.2
	Whey (see cattle)											
	Yeast *Saccharomyces cerevisiae*											
156	-brewers' dried yeast	7-05-527	93	3,325	3,025	45.9	2,075	0.04	1.7	4.0	3.0	3.2
157	-irradiated dried yeast	7-05-529	94	—	—		—	—	32.4	—	—	6.2
	Yeast, torula *Candida utilis* (formerly *Torulopsis utilis*)											
158	-torula dried yeast	7-05-534	93	3,110	2,765	46.4	1,985	0.05	2.4	—	4.0	2.5

[1]Dashes indicate that no data were available.

[2]Many net energy values are derived using National Research Council *Nutrient Requirements of Swine* (1998) formula 1-12.

501

TABLE 22.2 Mineral Composition of Feed Ingredients Used for Swine (as-fed basis)[1]

No.	Feed	International Feed Number	Calcium (Ca, %)	Phosphorus (P, %)	Phosphorus Available (AP, %)	Sodium (Na, %)	Chlorine (Cl, %)	Potassium (K, %)	Magnesium (Mg, %)	Sulfur (S, %)	Copper (Cu, mg/kg)	Iodine (I, mg/kg)	Iron (Fe, mg/kg)	Manganese (Mn, mg/kg)	Selenium (Se, mg/kg)	Zinc (Zn, mg/kg)
	Alfalfa (lucerne)															
	Medicago sativa															
1	-hay, sun-cured, all analyses	1-00-078	1.38	0.20	—	0.14	0.31	1.98	0.30	0.25	12	—	180	27	—	15
2	-hay, sun-cured	1-08-331	—	—	—	—	—	—	—	—	—	—	—	—	—	—
3	-leaves, meal, dehydrated	1-00-137	—	—	—	—	—	—	—	—	—	—	—	—	—	—
4	-leaves, sun-cured, ground	1-00-246	—	—	—	—	—	—	—	—	—	—	—	—	—	—
5	-meal, dehy, 15% CP (crude protein)	1-00-022	—	—	—	—	—	—	—	—	—	—	—	—	—	—
6	-meal, dehy, 17% CP	1-00-023	0.26	0.26	0.26	0.09	0.47	2.30	0.23	0.29	10	0.15	333	32	0.34	24
7	-meal, dehy, 20% CP	1-00-024	0.28	0.28	0.28	0.09	0.47	2.40	0.36	0.26	11	0.13	346	42	0.29	21
8	-meal, dehy, 22% CP	1-07-851	—	—	—	—	—	—	—	—	—	—	—	—	—	—
	Animal															
9	-animal plasma, spray dried	—	0.15	1.71	—	3.02	1.50	0.20	0.34	—	—	—	55	—	—	—
10	-blood cells, spray dried	—	0.02	0.37	—	0.58	1.40	0.62	0.11	0.48	—	—	2,700	—	—	—
11	-blood meal, conventional	5-00-380	0.37	0.27	—	0.50	0.30	0.11	0.21	0.45	11	—	1,922	6	0.58	38
12	-blood meal, flash dried	5-26-006	0.21	0.21	—	0.29	0.38	0.14	—	—	6	—	2,341	10	—	16
13	-blood meal, spray or ring dried	5-00-381	0.41	0.30	0.28	0.44	0.25	0.15	0.11	0.47	8	—	2,919	6	—	30
14	-fat, animal	4-00-376	—	—	—	—	—	—	—	—	—	—	—	—	—	—
15	-fat, swine (lard)	4-04-790	—	—	—	—	—	—	—	—	—	—	—	—	—	—
16	-liver meal, dehy	5-00-389	0.56	1.27	—	—	—	—	—	—	90	—	630	9	—	—
17	-meat, meal, rendered	5-00-385	7.69	3.88	—	0.80	0.97	0.57	0.35	0.45	10	—	440	10	0.37	94
18	-meat with blood, meal, tankage rendered	5-00-386	5.80	2.99	—	1.68	1.73	0.57	0.34	0.70	39	—	2,100	19	—	—
19	-meat with blood with bone, meal, tankage rendered	5-00-387	6.97	4.59	—	1.71	—	0.57	—	0.26	40	—	—	19	0.26	—
20	-meat meal with bone rendered	5-00-388	9.99	4.98	4.50	0.63	0.69	0.65	0.41	0.38	11	1.31	606	17	0.31	96
21	-tallow, beef	4-08-127	—	—	—	—	—	—	—	—	—	—	—	—	—	—
	Bakery waste															
22	-dried bakery product	4-00-466	0.14	0.32	—	1.14	1.48	0.39	0.24	0.02	5	—	50	65	—	15

No.	Feed	IFN														
	Barley *Hordeum vulgare*															
23	-grain	4-00-549	0.04	0.33	—	0.18	0.03	0.40	0.14	0.15	8	0.04	80	16	0.17	45
24	-grain, Pacific Coast	4-07-939	0.05	0.34	—	0.15	0.02	0.53	0.12	0.15	8	—	100	16	0.10	15
25	-grain, two row	4-00-572	0.06	0.35	—	0.12	0.04	0.45	0.14	0.15	7	—	78	18	0.19	25
26	-grain, six row	4-00-574	0.06	0.36	0.11	0.15	0.02	0.47	0.12	0.15	8	—	88	16	0.10	15
27	-grain, hulless	4-00-552	0.04	0.45	—	0.10	0.02	0.44	0.12	0.12	5	—	56	16	—	27
28	-malt sprouts, dehydrated	5-00-545	0.21	0.72	—	0.36	1.35	0.21	0.18	0.79	—	—	—	31	—	—
	Bean *Phaseolus vulgaris*															
29	-kidney seeds	5-00-600	0.11	0.40	—	—	0.01	0.98	—	—	—	—	70	—	—	—
30	-navy seeds	5-00-623	0.13	0.52	—	0.04	0.05	1.26	0.17	0.23	10	—	100	21	—	—
31	-pinto seeds	5-00-624	0.13	0.46	—	—	—	—	—	—	—	—	—	—	—	—
	Beet, sugar *Beta vulgaris, saccharifera*															
32	-molasses, > 48% invert sugar, > 79.5° Brix	4-00-668	0.12	0.02	—	1.50	1.10	4.63	0.19	0.47	17	—	70	4	—	—
33	-pulp, dried	4-00-669	0.70	0.10	—	0.10	0.20	0.61	0.22	0.31	11	—	411	46	0.09	12
	Blood (see animal)															
	Brewers grains															
34	-dehy (brewers dried grains)	5-02-141	0.30	0.53	—	0.15	0.26	0.08	0.16	0.31	21	0.06	250	38	0.70	62
	Broomcorn (see millet, proso)															
	Buckwheat, common *Fagopyrum sagittatum*															
35	-grain	4-00-994	0.10	0.33	—	0.05	0.05	0.41	0.09	0.14	10	—	44	34	0.18	9
	Canola (rapeseed) *Brassica spp*															
36	-meal, solvent extracted	5-06-145	0.63	1.01	0.21	0.11	0.07	1.22	0.51	0.85	6	—	142	49	1.10	69
37	-meal, prepressed, solv extr, 40% CP	5-08-135	0.66	0.93	—	—	—	—	—	—	—	—	—	—	—	—
	Cassava *Manihot spp*															
38	-meal (Tapioca or Manioc)	4-01-152	0.22	0.13	—	0.07	0.03	0.49	0.11	0.50	4	—	18	28	0.10	10
	Cattle *Bos taurus*															
39	-buttermilk, condensed	5-01-159	0.44	0.26	0.25	0.12	0.31	0.23	0.19	0.03	—	—	—	—	—	—
40	-buttermilk, dehydrated	5-01-160	1.31	0.93	0.90	0.47	0.80	0.89	0.48	0.08	—	—	10	3	—	—
41	-casein, dried	5-01-162	0.61	0.82	0.80	0.04	0.01	0.01	0.01	0.60	4	—	14	4	0.16	30
42	-cheese rind	5-01-163	0.99	0.57	0.55	0.61	0.82	0.28	0.02	—	—	—	—	—	—	—
43	-milk, dehy, cow's	5-01-167	1.29	1.02	0.97	0.90	0.35	1.56	0.12	0.32	12	—	50	2	0.12	41
44	-milk, fresh, cow's	5-01-168	0.12	0.10	0.10	0.20	0.05	0.14	—	—	0	—	—	—	—	—
45	-milk, skimmed, fresh, cow's	5-01-170	0.12	0.09	0.09	—	—	0.14	0.01	0.03	1	—	10	0	—	—
46	-milk, skimmed, dehy, cow's	5-01-175	1.31	1.00	0.97	1.00	0.48	1.60	0.12	0.32	5	—	8	2	0.12	42

(continued)

TABLE 22.2 Mineral Composition of Feed Ingredients Used for Swine (as-fed basis) (continued)

No.	Feed	International Feed Number	Calcium (Ca, %)	Phosphorus (P, %)	Phosphorus (AP, %)	Sodium (Na, %)	Chlorine (Cl, %)	Available Potassium (K, %)	Magnesium (Mg, %)	Sulfur (S, %)	Copper (Cu, mg/kg)	Iodine (I, mg/kg)	Iron (Fe, mg/kg)	Manganese (Mn, mg/kg)	Selenium (Se, mg/kg)	Zinc (Zn, mg/kg)
47	-whey, dehy	4-01-182	0.75	0.72	0.70	0.94	1.40	1.96	0.13	0.72	13	—	130	3	0.12	10
48	-whey, fresh	4-08-134	0.05	0.05	0.05	—	—	0.19	—	—	—	—	20	0	—	—
49	-whey, low lactose, dried	4-01-186	2.00	1.37	1.33	1.85	3.43	4.68	0.25	1.59	3	—	85	8	0.06	11
50	-whey, permeate, dried	—	0.86	0.66	0.64	1.00	2.23	2.10	—	—	—	—	—	—	—	—
	Chufa *Cyperus esculentus*															
51	-roots	4-08-374	0.01	0.07	—	—	—	0.14	—	—	—	—	—	—	—	—
	Citrus *Citrus* spp															
52	-pulp without fines, dehy (dried citrus pulp)	4-01-237	—	—	—	—	—	—	—	—	—	—	—	—	—	—
53	-syrup, molasses	4-01-241	1.09	0.10	—	0.27	0.07	0.09	0.14	—	73	—	340	26	—	93
	Clover, ladino															
54	-hay, sun-cured	1-01-378	—	—	—	—	—	—	—	—	—	—	—	—	—	—
	Coconut *Cocos nucifera*															
55	-meal, mech extr (copra meal)	5-01-572	0.20	0.62	—	0.04	—	1.54	0.31	0.34	14	—	1,320	65	—	49
56	-meal, solv extr (copra meal)	5-01-573	0.16	0.58	—	0.04	0.37	1.83	0.31	0.31	25	—	486	69	—	49
	Corn *Zea mays*															
57	-distillers dried grains	5-02-842	0.10	0.40	—	0.09	0.08	0.17	0.25	0.43	45	0.04	220	22	0.40	55
58	-distillers dried grains with solubles	5-02-843	0.20	0.77	0.59	0.25	0.20	0.84	0.19	0.30	57	—	257	24	0.39	80
	-distillers dried grains with solubles, Midwest ethanol plants		0.05	0.79	0.71	0.41	—	0.83	0.29	0.41	5	—	106	14	—	86
59	-distillers dried solubles	5-02-844	0.29	1.03	—	0.26	0.25	1.50	0.64	0.37	83	0.11	560	74	0.33	85
60	-ears, ground (corn and cob meal)	4-02-849	0.07	0.23	—	0.04	—	0.45	0.12	0.19	6	0.02	80	24	0.074	15
61	-gluten feed, (gluten with bran)	5-02-903	0.22	0.83	0.49	0.15	0.22	0.98	0.33	0.22	48	0.07	460	24	0.27	70
62	-gluten, meal	5-02-900	0.15	0.47	—	0.08	0.07	0.03	0.05	—	28	—	390	7.3	1.01	—
63	-gluten meal, 60% CP	5-28-242	0.05	0.44	0.06	0.02	0.06	0.18	0.08	0.43	26	—	282	4	1.00	33
64	-grain	4-02-935	0.03	0.28	0.04	0.02	0.05	0.33	0.12	0.13	3	—	29	7	0.07	18
65	-grain, flaked	4-02-859	—	—	—	—	—	—	—	—	—	—	—	—	—	—
66	-grain, grade 2, 54 lb/bu or 695 g/l	4-02-931	0.02	0.30	—	0.01	0.04	0.28	—	—	—	—	—	5	—	10
67	-grits by-product (hominy feed)	4-03-011	0.05	0.43	0.06	0.08	0.07	0.61	0.24	0.03	13	—	67	15	0.10	30

No.	Feed	Ref. No.														
68	-hominy feed	4-02-887	0.05	0.53	—	0.08	0.05	0.54	0.23	0.03	14	—	70	14	14	—
69	-grain, opaque-2 (high lysine)	4-11-445	0.02	0.19	—	—	—	—	—	—	—	—	—	—	—	—
	Corn, dent white Zea mays indentata															
70	-grain	4-02-928	—	—	—	—	—	—	—	—	—	—	—	—	—	—
	Cotton Gossypium spp															
71	-meal, mech extr, 36% CP	5-01-625	—	—	—	—	—	—	—	—	—	—	—	—	—	—
72	-meal, mech extr, 41% CP	5-01-617	0.23	1.03	—	0.04	0.04	1.34	0.52	0.40	19	—	160	23	0.90	64
73	-meal, prepressed, solv extr, 41% CP	5-07-872	0.19	1.06	0.01	0.04	0.05	1.40	0.50	0.31	18	—	184	20	0.80	70
	Emmer Triticum dicoccum															
74	-grain	4-01-830	0.05	0.36	—	—	—	0.47	—	—	31	—	60	78	—	—
	Fababean (broadbean)															
75	seeds	5-09-262	0.11	0.48	—	0.03	0.07	1.20	0.15	0.29	11		75	15	0.02	42
	Fat (see animal)															
	Feather meal (see poultry)															
	Fish															
76	-meal, mechanically extracted	5-01-977	—	—	—	—	—	—	—	—	—	—	—	—	—	—
77	-solubles, condensed	5-01-969	0.22	0.59	—	0.21	2.70	1.61	0.02	0.12	45	1.10	160	14	2.00	38
78	-solubles, dehydrated	5-01-971	0.55	1.25	—	0.37	6.29	2.03	0.30	0.40	35	—	300	50	2.20	76
	Anchovy Engraulis ringen															
79	-Anchovy meal, mech extr	5-01-985	3.93	2.55	—	0.88	1.02	0.75	0.24	0.77	9	0.86	220	10	1.36	103
	Fish, herring															
80	-herring meal, mech extr	5-02-000	2.40	1.76	—	0.61	1.12	1.01	0.18	0.69	6	—	181	8	1.93	132
	Menhaden Brevoortia tyrannus															
81	-meal, mech extr	5-02-009	5.21	3.04	2.85	0.40	0.55	0.70	0.16	0.45	11	1.09	440	37	2.10	147
	Fish, sardine Clupea spp, Sardinops spp															
82	-meal, mech extr	5-02-015	4.60	2.68	—	0.18	0.41	0.28	0.10	—	20	—	300	23	1.77	—
83	-solubles, condensed	5-02-014	0.14	0.83	—	0.18	0.28	0.18	—	—	—	—	—	25	—	—
	Fish, white families Gadidae, Lophiidae, and Rafidae															
84	-meal, mechanically extracted	5-02-025	6.65	3.59	—	0.78	1.28	0.85	0.18	0.48	6	—	299	12	1.62	90
	Flax Linum usitatissimum															
85	-meal, mech extr, 33% CP (linseed meal)	5-02-045	0.41	0.87	—	0.11	0.04	1.22	0.58	0.37	26	0.60	176	38	0.80	33
86	-meal, solv extr. 33% CP, (linseed meal),	5-02-048	0.39	0.83	—	0.13	0.06	1.26	0.54	0.39	22	—	270	41	0.63	66

(continued)

TABLE 22.2 Mineral Composition of Feed Ingredients Used for Swine (as-fed basis) (continued)

No.	Feed	International Feed Number	Calcium (Ca, %)	Phosphorus (P, %)	Available Phosphorus (AP, %)	Sodium (Na, %)	Chlorine (Cl, %)	Potassium (K, %)	Magnesium (Mg, %)	Sulfur (S, %)	Copper (Cu, mg/kg)	Iodine (I, mg/kg)	Iron (Fe, mg/kg)	Manganese (Mn, mg/kg)	Selenium (Se, mg/kg)	Zinc (Zn, mg/kg)
	Garbage															
87	-hotel and restaurant, boiled, wet	4–07–865	0.11	0.07	—	—	—	—	—	—	—	—	—	—	—	—
88	-hotel and restaurant, cooked, dehy	4–07–879	0.32	0.22	—	—	—	—	0.17	—	10	—	130	5	—	—
	Lentil *Lens* **spp**															
89	-seeds	5–02–506	0.10	0.38	—	0.02	0.03	0.89	0.12	0.20	10	—	85	13	0.10	25
	Lupin (sweet white) **Lupinus spp**															
90	-seeds	5–27–717	0.22	0.51	—	0.02	0.03	1.10	0.19	0.24	6	—	54	1,390	0.07	32
	Maize (see corn) **Meat (see animal)** **Milk (see cattle)** **Millet, proso** *Setaria* **spp**															
91	-grain	4–03–120	0.03	0.31	—	0.04	0.03	0.43	0.16	0.14	26	—	71	30	0.70	18
	Milo (see sorghum) **Molasses and syrup (see source plant)** **Oats** *Avena sativa*															
92	-cereal by-product (feeding oat meal, oat middlings)	4–03–303	0.07	0.44	—	0.09	0.05	0.53	0.16	0.26	4	—	300	43	—	140
93	-grain, all analyses	4–03–309	0.07	0.31	0.07	0.08	0.10	0.42	0.16	0.21	6	0.09	85	43	0.30	38
94	-grain, Pacific Coast	4–07–999	0.10	0.31	—	—	—	—	—	0.21	—	—	—	—	0.07	—
95	-grain, naked	4–25–101	0.08	0.38	—	0.02	0.11	0.36	0.12	0.14	4	—	58	37	0.09	34
96	-groats	4–03–331	0.08	0.41	0.05	0.05	0.09	0.38	0.11	0.20	6	—	49	32	—	—
	Pea *Pisum* **spp**															
97	-seeds	5–03–600	0.11	0.39	—	0.02	0.11	0.36	0.12	0.14	4	—	58	37	0.09	34
	Peanut (groundnut) *Arachis hypogaea*															
98	-kernels, hulls added, meal, solv extr	5–03–656	—	—	—	—	—	—	—	—	—	—	—	—	—	—
99	-kernels, meal, mech extr, 45% CP (Peanut meal)	5–03–649	0.17	0.59	—	0.06	0.03	1.20	0.33	0.29	15	—	285	39	0.28	47
100	-kernels, meal, solv extr, 47% CP (Peanut meal)	5–03–650	0.22	0.65	0.08	0.07	0.04	1.25	0.31	0.30	15	0.06	260	40	0.21	41

	Feed	Code															
	Potato *Solarium tuberosum*																
101	-pulp, dehy	4-03-775	0.09	0.25	—	—	—	—	—	—	—	—	—	—	—	—
102	-protein concentrate	5-25-392	0.17	0.19	—	0.03	0.20	0.80	0.05	0.23	13	—	40	5	1.00	25
103	-tubers, boiled	4-03-784	—	—	—	—	—	—	—	—	—	—	—	—	—	—
104	-tubers, dehy	4-07-850	0.06	0.19	—	0.01	0.36	1.99	—	—	—	—	—	2	—	2
	Poultry																
105	-by-product meal, rendered	5-03-798	4.46	2.41	—	0.49	0.49	0.53	0.18	0.52	10	3.07	442	9	0.88	94
106	-fat, poultry	4-09-319	—	—	—	—	—	—	—	—	—	—	—	—	—	—
107	-feathers meal, hydrolyzed	5-03-795	0.33	0.50	0.15	0.34	0.26	0.19	0.20	1.39	10	0.04	76	10	0.69	111
	Rape (see canola)																
	Rice *Oryza sativa*																
108	-bran with germ (rice bran)	4-03-928	0.07	1.61	0.40	0.03	0.07	1.56	0.90	0.18	9	—	190	228	0.40	30
109	-grain, ground (ground rough rice)	4-03-938	0.06	0.43	—	0.04	0.08	0.52	0.23	0.05	6	0.04	100	92	—	13
110	-grain, polished + broken (brewers' rice)	4-03-932	0.04	0.18	—	0.04	0.07	0.13	0.11	0.06	21	—	18	12	0.27	17
111	-groats, polished (polished rice)	4-03-942	0.02	0.11	—	0.02	0.04	0.10	0.02	0.08	3	—	10	11	—	2.0
112	-polishings	4-03-943	0.09	1.18	—	0.06	0.11	1.11	0.65	0.17	6	0.06	160	12	—	26
	Rye *Secale cereale*																
113	-distillers grains, dehy	5-04-023	0.15	0.48	—	0.17	0.05	0.07	0.17	0.44	—		—	18	—	—
114	-distillers grains with solubles, dehy	5-04-024	—	—	—	—	—	—	—	—	—		—	—	—	—
115	-grain	4-04-047	0.06	0.33	—	0.02	0.03	0.48	0.12	0.15	7		60	58	0.38	31
	Safflower *Carthamus tinctorius*																
116	-meal, solvent extracted	5-04-110	0.34	0.75	—	0.05	0.08	0.76	0.35	0.13	10	—	495	18	—	41
117	-meal without hulls, mech extr	5-08-499	0.32	0.59	—	—	—	—	—	—	—	—	—	—	—	—
118	meal without hulls, solv extr	5-07-959	0.37	1.31	—	0.04	0.16	1.00	1.02	0.20	9	—	484	39	—	33
119	-seeds, whole	4-07-958	—	—	—	—	—	—	—	—	—	—	—	—	—	—
	Sesame *Sesamum indicum*																
120	-meal, mech extr	5-04-220	1.90	1.22	—	0.04	0.07	1.10	0.54	0.56	34	—	93	53	0.21	100
	Sorghum *Sorghum vulgare*																
121	-gluten meal	5-04-388	—	—	—	—	—	—	—	—	—	—	—	—	—	—
122	-gluten with bran (gluten feed)	5-08-089	—	—	—	—	—	—	—	—	—	—	—	—	—	—
123	-grain	4-20-893	0.03	0.29	0.06	0.01	0.09	0.35	0.15	0.08	5	—	45	15	0.20	15
124	-grain, Feterita (USA)	4-04-369	—	—	—	—	—	—	—	—	—	—	—	—	—	—
125	-grain, Hegari (Thailand)	4-04-398	—	—	—	—	—	—	—	—	—	—	—	—	—	—

(*continued*)

507

TABLE 22.2 Mineral Composition of Feed Ingredients Used for Swine (as-fed basis) (continued)

No.	Feed	International Feed Number	Calcium (Ca, %)	Phosphorus (P, %)	Available Phosphorus (AP, %)	Sodium (Na, %)	Chlorine (Cl, %)	Available Potassium (K, %)	Magnesium (Mg, %)	Sulfur (S, %)	Copper (Cu, mg/kg)	Iodine (I, mg/kg)	Iron (Fe, mg/kg)	Manganese (Mn, mg/kg)	Selenium (Se, mg/kg)	Zinc (Zn, mg/kg)
126	-grain, Kaffir (USA)	4-04-428	0.03	0.31	—	0.05	0.10	0.33	0.15	0.16	7	—	60	16	0.80	13
	Soybean Glycine max															
127	-meal, mech extr, 41% CP	5-04-600	0.27	0.63	—	0.24	0.07	1.71	0.25	0.33	18	—	160	32	—	—
128	-meal, solv extr	5-04-604	0.32	0.65	0.20	0.01	0.05	1.96	0.27	0.43	20	—	202	29	0.32	50
129	-meal without hulls, solv extr	5-04-612	0.34	0.69	0.16	0.02	0.05	2.14	0.30	0.44	20	—	176	36	0.27	55
130	-protein concentrate	5-32-183	0.35	0.81	—	0.05	—	2.20	0.32	—	13	—	110	—	—	30
131	-protein isolate	5-24-811	0.15	0.65	—	0.07	0.02	0.27	0.08	0.71	14	—	137	5	0.14	34
132	-seeds, heat processed	5-04-597	0.25	0.59	—	0.03	0.03	1.70	0.28	0.30	16	—	80	30	0.11	39
133	-seeds, whole	5-04-610	0.25	0.60	—	0.12	0.03	1.61	0.28	0.22	16	—	80	30	—	—
	Spelt Triticum spelta															
134	-grain	4-04-651	0.12	0.38	—	—	—	—	—	—	—	—	—	—	—	—
	Sugarcane Saccharum officinarum															
135	-sugarcane, molasses, dehy	4-04-695	0.79	0.26	—	0.18	—	3.31	0.39	0.41	65	—	210	46	—	—
136	-sugarcane, molasses (blackstrap) > 46% invert sugar, > 79.5° Brix	4-04-696	0.78	0.09	—	0.17	2.78	2.85	0.35	0.35	60	1.58	190	43	—	22
	Sunflower Helianthus spp															
137	-meal, solv extr	5-09-340	0.36	0.86	0.02	0.02	0.10	1.07	0.68	0.30	26	—	254	41	0.50	66
138	meal without hulls, solv extr	5-04-739	0.37	1.01	—	0.04	0.13	1.27	0.75	0.38	25	—	200	35	0.32	98
	Sweet potato Ipomoea batata															
139	-pulp, dehy	4-08-535	—	—	—	—	—	—	—	—	—	—	—	—	—	—
140	-tubers	4-04-788	0.03	0.05	—	0.02	0.02	0.31	0.05	0.04	1.3	—	20	3	—	—
	Tallow (see animal)															
	Triticale Triticale hexaloide															
141	-grain	4-20-362	0.05	0.33	0.15	0.03	0.03	0.46	0.10	0.15	8	—	31	43	—	32
	Wheat Triticum aestivum															
142	-bran	4-05-190	0.16	1.20	0.35	0.04	0.07	1.26	0.52	0.22	14	0.06	170	113	0.51	100
143	-distillers grains, dehy	5-05-193	0.11	0.58	—	—	—	—	—	—	—	—	—	15	—	—
144	-germ, ground (wheat germ meal)	5-05-218	0.05	0.91	—	0.02	0.08	0.97	0.24	0.24	9	—	50	133	0.34	119

145	-grain	4-05-211	0.03	0.38	—	0.03	0.07	0.36	0.15	0.16	6	0.09	60	36	0.22	44
146	-grain, hard red spring	4-05-258	0.05	0.36	—	0.02	0.09	0.41	0.16	0.17	7	—	64	42	0.30	43
147	-grain, hard red winter	4-05-268	0.06	0.37	0.19	0.01	0.06	0.49	0.13	0.15	6	—	39	34	0.33	40
148	-grain, soft red winter	4-05-294	0.04	0.39	0.20	0.01	0.08	0.46	0.11	0.16	8	—	32	38	0.28	47
149	-grain, soft white winter	4-05-337	0.05	0.35	—	0.01	0.07	0.44	0.15	0.18	7	—	60	37	0.26	28
150	-grain, soft white winter, **Pacific Coast**	4-08-555	0.09	0.30	—	0.05	—	0.40	—	—	10	—	100	50	—	13
151	-middlings, < 9.5% fiber	4-05-205	0.12	0.93	0.38	0.05	0.04	1.06	0.41	0.17	10	0.11	84	100	0.72	92
152	-mill run, < 9.5% fiber	4-05-206	0.15	1.03	—	0.22	—	1.28	0.51	—	19	—	100	102	—	—
153	-red dog, < 4% fiber	4-05-203	0.07	0.57	—	0.04	0.10	0.63	0.16	0.24	6	—	46	55	0.30	65
154	-shorts, < 7% fiber	4-05-201	0.09	0.84	—	0.02	0.04	1.06	0.25	0.20	12	—	100	89	0.75	100
	Wheat, durum *Triticum durum*															
155	-grain	4-05-224	0.08	0.35	—	—	—	0.44	0.14	—	7	—	40	28	0.88	32
	Whey (see cattle)															
	Yeast *Saccharamyces cerevisiae*															
156	-brewers' dried yeast	7-05-527	0.16	1.44	—	0.10	0.12	1.80	0.23	0.40	33	—	215	8	1.00	49
157	-irradiated dried yeast	7-05-529	0.78	1.42	—	—	—	—	—	—	—	—	—	—	—	—
	Yeast, torula *Candida utilis* (formerly *Torulopsis utilis*)															
158	-torula dried yeast	7-05-534	0.58	1.52	—	0.07	0.12	1.94	0.20	0.55	17	—	222	13	0.02	99

[1]Dashes indicate that no data were available.

TABLE 22.3 Vitamin Composition of Feed Ingredients Used for Swine (as-fed basis)[1]

No.	Feed	International Feed Number	A (IU/kg)	β-Carotene (mg/kg)	E (mg/kg)	K (mg/kg)	Biotin (mg/kg)	Choline (mg/kg)	Folic Acid (mg/kg)	Niacin[2] (mg/kg)	Pantothenic Acid (mg/kg)	Riboflavin B2 (mg/kg)	Thiamin B1 (mg/kg)	Pyridoxine B6 (mg/kg)	Vitamin B12 (ug/kg)
Alfalfa (Lucerne) Medicago sativa															
1	-hay, sun-cured, all analyses	1-00-078	68,400	—	83.0	15.89	0.20	—	3.07	38	28.6	12.0	2.7	5.7	2
2	-hay, sun-cured	1-08-331	201,500	—	124.3	8.63	0.30	1,405	4.50	39	29.4	13.2	3.4	8.0	—
3	-leaves, meal, dehydrated	1-00-137	—	—	—	—	—	—	—	—	—	—	—	—	—
4	-leaves, sun-cured, ground	1-00-246	—	—	—	—	—	—	—	—	—	—	—	—	—
5	-meal, dehy, 15% CP (crude protein)	1-00-022	—	—	—	—	—	—	—	—	—	—	—	—	—
6	-meal, dehy, 17% CP	1-00-023	25,250	94.6	49.8	—	0.54	1,401	4.36	38	29.0	13.6	3.4	6.5	0
7	-meal, dehy, 20% CP	1-00-024	25,250	94.6	49.8	14.53	0.54	1,419	4.36	45	34.0	15.2	5.8	8.0	0
8	-meal, dehy, 22% CP	1-07-851	—	—	—	—	—	—	—	—	—	—	—	—	—
Animal															
9	-animal plasma, spray dried	—	—	—	—	—	—	—	—	—	—	—	—	—	—
10	-blood cells, spray dried	—	—	—	—	—	—	—	—	—	—	—	—	—	—
11	-blood meal, conventional	5-00-380	—	—	1.0	—	0.03	852	0.10	31	2.0	2.4	0.4	4.4	44
12	-blood meal, flash dried	5-26-006	—	—	1.0	—	0.08	781	0.10	23	1.0	1.4	1.0	4.4	44
13	-blood meal, spray or ring dried	5-00-381	—	—	1.0	—	0.28	485	0.40	23	3.7	3.2	0.3	4.4	—
14	-fat, animal	4-00-376	—	—	22.8	—	—	—	—	—	—	—	—	—	—
15	-fat, swine (lard)	4-04-790	—	—	—	—	—	—	—	—	—	—	—	—	—
16	-liver meal, dehy	5-00-389	—	—	—	—	0.02	1,142	5.59	206	29.3	36.4	0.2	—	503
17	-meat, meal, rendered	5-00-385	—	—	1.0	—	0.13	2,046	0.37	58	8.4	5.3	0.2	3.9	64
18	-meat with blood, meal, rendered	5-00-386	—	—	1.2	—	0.08	2,077	0.50	57	5.0	4.7	0.6	2.4	80
19	-meat with blood with bone, meal, tankage rendered	5-00-387	—	—	0.8	—	0.07	2,150	0.57	46	3.7	3.6	0.2	—	82
20	-meat meal with bone rendered	5-00-388	—	—	1.6	—	0.08	1,996	0.41	49	4.1	4.7	0.4	4.6	90
21	-tallow, beef	4-08-127	—	—	—	—	—	—	—	—	—	—	—	—	—
Animal-poultry (see poultry)															
Bakery waste															
22	-dried bakery product	4-00-466	1,120	4.2	—	—	0.07	923	0.20	26	8.3	1.4	2.9	4.3	0
Barley Hordeum vulgare															
23	-grain	4-00-549	4,700	—	15.8	—	0.14	903	0.55	85	8.2	1.6	4.4	6.5	—
24	-grain, Pacific Coast	4-07-939	—	—	21.1	—	0.15	1,003	0.51	48	7.1	1.6	4.3	2.9	—
25	-grain, two row	4-00-572	1,100	4.1	7.4	—	0.14	1,034	0.31	55	8.0	1.8	4.5	5.0	0
26	-grain, six row	4-00-574	1,100	4.1	7.4	—	0.15	1,034	0.40	48	7.0	1.6	4.0	2.9	0
27	-grain, hulless	4-00-552	—	—	6.0	—	0.07	—	0.62	48	6.8	1.8	4.3	5.6	0

(continued)

No.	Feed	IFN													
28	-malt sprouts, dehy	5-00-545	—	—	20.6	—	—	1,576	0.20	52	8.6	6.7	4.9	—	—
	Bean *Phaseolus vulgaris*														
29	-kidney seeds	5-00-600	—	—	—	—	—	—	—	25	—	2.1	5.7	—	—
30	-navy seeds	5-00-623	—	—	1.0	—	0.11	1,675	1.30	25	2.4	2.0	6.4	0.3	—
31	-pinto seeds	5-00-624	—	—	—	—	—	—	—	22	2.2	3.1	8.6	—	—
	Beet, sugar *Beta vulgaris, saccharifera*														
32	-molasses, > 48% invert sugar, > 79.5° Brix	4-00-668	—	—	—	—	—	829	—	41	4.5	2.3	—	—	—
33	-pulp, dried	4-00-669	2,830	10.6	13.2	—	—	818	—	18	1.3	0.7	0.4	1.9	0
	Blood (see animal)														
	Brewers grains														
34	-dehy (brewers dried grains)	5-02-141	53	0.2	—	—	0.24	1,723	7.10	43	8.0	1.4	0.6	0.7	0
	Broomcorn (see millet, proso)														
	Buckwheat *Fagopyrum sag/ttatum*														
35	grain	4-00-994	—	—	—	—	0.06	440	0.64	19	12.0	5.5	4.0	3.0	0
	Canola (Rapeseed) *Brassica spp*														
36	-meal, solvent extracted	5-06-145	—	—	13.4	—	0.98	6,700	0.83	160	9.5	5.8	5.2	7.2	0
37	-meal, prepressed, solv extr, 40% CP	5-08-135	—	—	—	—	—	—	—	—	—	—	—	—	—
	Cassava *Manihot spp*														
38	-meal (Tapioca or Manioc)	4-01-152	—	—	0.2	—	0.05	—	—	3	0.3	0.8	1.6	0.7	0
	Cattle *Bos taurus*														
39	-buttermilk, condensed	5-01-159	25,300	—	—	—	0.29	1,665	0.40	—	—	12.4	—	—	—
40	-buttermilk, dehydrated	5-01-160	—	—	—	—	0.04	—	—	9	36	31.0	3.4	2.4	19
41	-casein, dried	5-01-162	—	—	—	—	—	205	0.51	1	2.7	1.5	0.4	0.4	—
42	-cheese rind	5-01-163	—	—	—	—	—	—	—	—	—	—	—	—	—
43	-milk, dehy, cow's	5-01-167	1,500	—	—	—	—	—	—	—	—	1.8	0.4	—	—
44	-milk, fresh, cow's	5-01-168	—	—	—	—	—	—	—	1	2.9	1.9	0.4	—	—
45	-milk, skimmed, fresh, cow's	5-01-170	—	—	—	—	—	—	—	1	3.3	19.1	3.7	—	—
46	-milk, skimmed, dehy, cow's	5-01-175	—	—	—	—	0.25	1,393	0.47	12	36.4	27.1	4.1	4.1	36
47	-whey, dehy	4-01-182	—	—	—	—	0.27	1,820	0.85	10	47.0	0.8	0.3	4.0	23
48	-whey, fresh	4-08-134	—	—	—	—	—	—	—	1	5.4	37.2	5.7	—	—
49	-whey, low lactose, dried	4-01-186	—	—	—	—	0.27	3,571	0.69	19	69.0	—	—	4.4	25
50	-whey, permeate, dried	—	—	—	—	—	—	—	—	—	—	—	—	—	—
	Chufa *Cyperus esculentus*														
51	-roots	4-08-374	—	—	—	—	—	—	—	—	—	—	—	—	—
	Citrus *Citrus spp*														
52	-pulp without fines, dehy (Dried citrus pulp)	4-01-237	—	—	—	—	—	—	—	—	—	—	—	—	—
53	-syrup, molasses	4-01-241	—	—	—	—	—	—	—	27	12.6	6.2	—	—	—
	Clover, ladino														

TABLE 22.3 Vitamin Composition of Feed Ingredients Used for Swine (as-fed basis) (continued)

No.	Feed	International Feed Number	A (IU/kg)	β-Carotene (mg/kg)	E (mg/kg)	K (mg/kg)	Biotin (mg/kg)	Choline (mg/kg)	Folic Acid (mg/kg)	Niacin[2] (mg/kg)	Pantothenic Acid (mg/kg)	Riboflavin B2 (mg/kg)	Thiamin B1 (mg/kg)	Pyridoxine B6 (mg/kg)	Vitamin B12 (ug/kg)
54	-hay, sun-cured	1-01-378	—	—	—	—	—	—	—	—	—	—	—	—	—
	Coconut Cocos nucifera														
55	-meal, mech extr (Copra meal)	5-01-572	—	—	—	—	—	1,015	1.39	25	6.2	3.2	0.8	—	—
56	-meal, solv extr (Copra meal)	5-01-573	—	—	7.7	—	0.25	1,089	0.30	28	6.5	3.5	0.7	4.4	—
	Corn Zea mays														
57	-distillers dried grains	5-02-842	5,200	—	—	—	—	1,261	0.88	38	11.8	5.3	1.7	4.4	—
58	-distillers dried grains with solubles	5-02-843	800	3.0	12.9	—	0.49	1,180	0.90	37	11.7	5.2	1.7	4.4	0
	-distillers dried grains with solubles, Midwest ethanol plants														
59	-distillers dried solubles	5-02-844	934	3.5	—	—	0.78	2,637	0.90	75	14.0	8.6	2.9	8.0	0
60	-ears, ground (corn and cob meal)	4-02-849	1,300	—	—	—	1.66	4,842	1.10	116	21.0	17.0	6.9	8.8	3
61	-gluten feed, (gluten with bran)	5-02-903	267	1.0	8.5	—	0.14	1,518	0.28	66	17.0	2.4	2.0	13.0	0
62	-gluten, meal	5-02-900	27,200	—	33.9	—	—	368	0.34	50	9.9	1.4	0.2	7.9	—
63	-gluten meal, 60% CP	5-28-242	—	—	6.7	—	0.15	330	0.13	55	3.5	2.2	0.3	6.9	0
64	-grain	4-02-935	213	0.8	8.3	—	0.06	620	0.15	24	6.0	1.2	3.5	5.0	0
65	-grain, flaked	4-02-859	—	—	0.8	—	—	—	—	—	—	—	—	—	—
66	-grain, grade 2, 54 lb/bu or 695 g/l	4-02-931	2,900	—	22.0	—	0.06	620	0.36	24	4.8	1.3	3.5	7.0	—
67	-grits by-product (hominy feed)	4-03-011	2,403	9.0	6.5	—	0.13	1,155	0.21	47	8.2	2.1	8.1	11.0	0
68	-hominy feed	4-02-887	15,300	—	—	—	0.13	993	0.28	47	7.5	2.1	7.9	10.9	—
69	-grain, opaque-2 (high lysine)	4-11-445	—	—	—	—	—	500	—	19	4.5	1.0	—	—	—
	Corn, dent white Zea mays indentata														
70	-grain	4-02-928	—	—	—	—	—	—	—	—	—	—	—	—	—
	Cotton Gossypium spp														
71	-meal, mech extr, 36% CP	5-01-625	—	—	—	—	—	—	—	—	—	—	—	—	—
72	-meal, mech extr, 41% CP	5-01-617	534	0.2	35.0	—	0.30	2,753	1.65	38	10.0	5.1	6.4	5.3	0
73	-meal, prepressed, solv, 41% CP	5-07-872	534	0.2	14.0	—	0.30	2,933	1.65	40	12.0	5.9	7.0	5.1	0
	Emmer Triticum dicoccum														
74	-grain	4-01-830	—	—	—	—	—	—	—	—	—	—	—	—	—
	Fababean (broadbean)														
75	-seeds	5-09-262	—	—	0.8	—	0.09	1,670	—	26	3.0	2.9	5.5	—	0
	Fat (see animal)														
	Feather meal (see poultry)														
	Fish														

#	Feed name	Code														
76	-meal, mech extr	5-01-977	—	—	—	—	—	—	—	—	—	—	—	—	—
77	-solubles, condensed	5-01-969	2,200	—	—	—	0.18	3,519	0.02	169	35.0	14.6	5.5	12.2	347
78	-solubles, dehy	5-01-971	—	—	—	—	0.26	5,507	0.60	271	55.0	15.6	7.4	23.8	401
	Anchovy Engraulis ringen														
79	-Anchovy meal, mech extr	5-01-985	—	—	5.0	—	0.13	4,408	0.37	100	15.0	7.1	0.3	4.0	280
	Fish, herring														
80	-Herring meal, mech extr	5-02-000	—	—	15.0	—	0.13	5,306	0.37	93	17.0	9.9	0.4	4.8	403
	Menhaden Brevoortia tyrannus														
81	-meal, mech extr	5-02-009	—	—	5.0	—	0.13	3,056	0.37	55	9.0	4.9	0.5	4.0	143
	Fish, sardine Clupea spp, Sardinops spp														
82	-meal, mechanically extracted	5-02-015	—	—	—	—	0.10	3,272	—	75	11.0	5.4	0.3	—	237
83	-solubles, condensed	5-02-014	—	—	—	—	0.13	3,009	—	356	41.2	16.8	4.0	—	1,041
	Fish, white families Gadidae, Lophiidae, and Rafidae														
84	-meal, mech extr	5-02-025	—	—	5.0	—	0.13	3,099	0.37	59	9.9	9.1	1.7	5.9	90
	Flax Linum usitatissimum														
85	-meal, mech extr; 33% CP (linseed meal)	5-02-045	—	—	—	—	—	—	—	—	—	—	—	—	—
86	-meal, solv extr (linseed meal),	5-02-048	53	0.2	2.0	—	0.41	1,512	1.30	33	14.7	2.9	7.5	6.0	0
	Garbage														
87	-hotel and restaurant, boiled, wet	4-07-865	—	—	—	—	—	—	—	—	—	—	—	—	—
88	-hotel and restaurant, cooked, dehy	4-07-879	—	—	—	—	—	—	—	—	—	—	—	—	—
	Lentil Lens spp														
89	-seeds	5-02-506	267	1.0	0.0	—	0.13	—	0.70	22	14.9	2.4	3.9	5.5	0
	Lupin (sweet white) Lupinus spp														
90	-seeds	5-27-717	—	—	7.5	—	0.05	—	—	—	—	—	—	—	—
	Maize (see corn)														
	Meat (see animal)														
	Milk (see cattle)														
	Millet, proso Setaria spp														
91	-grain	4-03-120	—	—	—	—	0.16	440	0.23	23	11.0	3.8	7.3	5.8	0
	Milo (see sorghum)														
	Molasses and syrup (see source plant)														
	Oats Avena sativa														
92	-cereal by-product (feeding oat meal, oat middlings)	4-03-303	—	—	24.0	—	0.22	1,148	0.51	21	18.0	1.8	7.0	—	—
93	-grain, all analyses	4-03-309	988	3.7	7.8	—	0.24	946	0.30	19	13.0	1.7	6.0	2.0	0
94	-grain, Pacific Coast	4-07-999	—	—	20.2	—	—	918	—	14	11.7	1.2	—	—	—

(continued)

TABLE 22.3 Vitamin Composition of Feed Ingredients Used for Swine (as-fed basis) (continued)

No.	Feed	International Feed Number	A (IU/kg)	β-Carotene (mg/kg)	E (mg/kg)	K (mg/kg)	Biotin (mg/kg)	Choline (mg/kg)	Folic Acid (mg/kg)	Niacin[2] (mg/kg)	Pantothenic Acid (mg/kg)	Riboflavin B_2 (mg/kg)	Thiamin B_1 (mg/kg)	Pyridoxine B_6 (mg/kg)	Vitamin B_{12} (ug/kg)
95	-grain, naked	4-25-101	—	—	2.0	—	0.12	1,240	0.50	20c	7.1	1.3	5.2	9.6	0
96	-groats	4-03-331	—	—	—	—	0.20	1,139	0.50	14	13.4	1.5	6.5	1.1	0
	Pea Pisum spp														
97	-seeds	5-03-600	267	1.0	0.2	—	0.15	547	0.20	31	18.7	1.8	4.6	1.0	0
	Peanut (groundnut) Arachis hypogaea														
98	-kernels, hulls added, meal, solv extr	5-03-656	—	—	—	—	—	—	—	—	—	—	—	—	—
99	-kernels, meal, mech extr, 45% CP (Peanut meal)	5-03-649	400	—	2.7	—	0.35	1,848	0.70	166	47.0	5.2	7.1	7.4	0
100	-kernels, meal, solv extr, 47% CP (Peanut meal)	5-03-650	—	—	2.7	—	0.39	1,854	0.50	170	53.0	7.0	5.7	6.0	0
	Potato Solarium tuberosum														
101	-pulp residue dehy	4-03-775	—	—	—	—	—	—	—	—	—	—	—	—	—
102	-protein concentrate	5-25-392	—	—	—	—	—	—	—	—	—	—	—	—	—
103	-tubers, boiled	4-03-784	—	—	—	—	—	—	—	—	—	—	—	—	—
104	-tubers, dehy	4-07-850	—	—	—	—	0.10	2,620	0.6	33	20.0	0.7	—	14.1	—
	Poultry														
105	-by-product meal, rendered	5-03-798	—	—	—	—	0.09	6,029	0.50	47	11.1	10.5	0.2	4.4	—
106	-fat, poultry	4-09-319	—	—	7.8	—	—	—	—	—	—	—	—	—	—
107	-feathers meal, hydrolyzed	5-03-795	—	—	7.3	—	0.13	891	0.20	21	10.0	2.1	0.1	3.0	78
	Rape (see canola)														
	Rice Oryza sativa														
108	-bran with germ (rice bran)	4-03-928	—	—	9.7	—	0.35	1,135	2.20	293	23.0	2.5	22.5	26.0	0
109	-grain, ground (ground rough rice)	4-03-938	—	—	9.9	—	0.08	925	0.36	35	7.0	1.0	2.9	4.4	—
110	-grain, polished + broken (brewers' rice)	4-03-932	—	—	2.0	—	0.08	1,003	0.20	25	3.3	0.4	1.4	28.0	0
111	-groats, polished (polished rice)	4-03-942	—	—	3.6	—	—	904	0.15	16	3.6	0.5	0.7	0.4	—
112	-polishings	4-03-943	27	0.1	61.0	—	0.37	1,237	0.20	520	47.0	1.8	19.8	27.6	0
	Rye Secale cereale														
113	-distillers grains, dehy	5-04-023	—	—	—	—	—	—	—	17	5.3	3.3	1.3	—	—
114	-distillers grains with solubles, dehy	5-04-024	—	—	—	—	—	—	—	63	17.5	8.2	3.1	—	—
115	-grain	4-04-047	—	—	9.0	—	0.08	419	0.60	19	8.0	1.6	3.6	2.6	0

	Safflower *Carthamus tinctorius*													
116	-meal, solv extr	5-04-110	—	—	16.0	—	1.03	820	11	33.9	2.3	4.6	12.0	0
117	-meal without hulls, mech extr	5-08-499	—	—	—	—	—	2,541	22	88.0	4.1	4.5	11.3	—
118	-meal without hulls, solv extr	5-07-959	—	—	16.0	—	1.03	3,248	22	39.1	2.4	4.5	11.3	0
119	-seeds, whole	4-07-958	—	—	—	—	—	—	—	—	—	—	—	—
	Sesame *Sesamum indicum*													
120	-meal, mech extr	5-04-220	53	0.2	1.0	—	0.24	1,536	30	6.0	3.6	2.8	12.5	0
	Sorghum *Sorghum vulgare*													
121	-gluten meal	5-04-388	—	—	—	—	—	—	—	—	—	—	—	—
122	-gluten with bran (gluten feed)	5-08-089	—	—	—	—	—	—	—	—	—	—	—	—
123	-grain	4-20-893	—	—	5.0	—	0.26	668	41[c]	12.4	1.3	3.0	5.2	0
124	-grain, Feterita (USA)	4-04-369	—	—	—	—	—	—	—	—	—	—	—	—
125	-grain, Hegari (Thailand)	4-04-398	—	—	—	—	—	—	—	—	—	—	—	—
126	-grain, Kaffir (USA)	4-04-428	600	—	—	—	0.24	436	38	11.9	1.3	3.8	6.7	—
	Soybean *Glycine max*													
127	-meal, mech extr, 41% CP	5-04-600	300	—	6.6	—	0.30	2,673	30	14.9	3.5	4.0	—	0
128	-meal, solv extr	5-04-604	53	0.2	2.3	—	0.27	2,794	34	16.0	2.9	4.5	6.0	0
129	-meal without hulls, solv extr	5-04-612	53	0.2	2.3	—	0.26	2,731	22	15.0	3.1	3.2	6.4	0
130	-protein concentrate	5-32-183	—	—	—	—	0.30	2	6	4.2	1.7[3]	0.3[3]	5.4[3]	0
131	-protein isolate[3]	5-24-811	—	—	18.1	—	0.24	2,307	22	15.0	2.6	11.0	10.8	0
132	-seeds, heat processed	5-04-597	507	1.9	—	—	0.38	2,898	22	15.8	2.9	11.1	—	—
133	-seeds, whole	5-04-610	1,500	—	—	—	—	—	22	—	—	—	—	—
	Spelt *Triticum pelta*													
134	-grain	4-04-651	—	—	—	—	—	—	48	—	—	—	—	—
	Sugarcane *Saccharum officinarum*													
135	-sugarcane, molasses, dehy	4-04-695	—	—	5.1	—	—	772	35	37.5	3.3	0.9	—	—
136	-sugarcane, molasses (blackstrap) >46% invert sugar, > 79.5° Brix	4-04-696	—	—	5.0	—	0.71	744	41	39.2	2.9	0.9	6.50	—
	Sunflower *Helianthus* spp													
137	-meal, solv extr	5-09-340	—	—	9.1	—	1.40	3,791	264	29.9	3.0	3.0	11.1	0
138	-meal without hulls, solv extr	5-04-739	—	—	9.1	—	1.45	3,150	220	24.0	3.6	3.5	13.7	0
	Sweet potato *Ipomoea batata*													
139	-pulp, dehy	4-08-535	—	—	—	—	—	—	—	—	0.6	1.1	—	—
140	-tubers	4-04-788	222,800	—	—	—	—	—	6	—	—	—	—	—
	Tallow (see animal)													
	Triticale *Triticale hexaloide*													
141	-grain	4-20-362	—	—	1.7	—	—	462	—	—	0.4	—	—	—
	Wheat *Triticum aestivum*													
142	-bran	4-05-190	267	1.0	16.5	—	0.36	1,232	186	31.0	4.6	8.0	12.0	0
143	-distillers grains, dehy	5-05-193	1,800	—	—	—	—	—	56	8.1	3.7	2.0	—	—
144	-germ, ground (wheat germ meal)	5-05-218	—	—	141.1	—	0.22	3,056	72	20.9	6.1	22.	11.3	—

(continued)

TABLE 22.3 Vitamin Composition of Feed Ingredients Used for Swine (as-fed basis) (continued)

No.	Feed	Inter-national Feed Number	A (IU/kg)	β-Carotene (mg/kg)	E (mg/kg)	K (mg/kg)	Biotin (mg/kg)	Choline (mg/kg)	Folic Acid (mg/kg)	Niacin[2] (mg/kg)	Panto-thenic Acid (mg/kg)	Ribo-flavin B₂ (mg/kg)	Thia-min B₁ (mg/kg)	Pyri-doxine B₆ (mg/kg)	Vita-min B₁₂ (ug/kg)
145	-grain	4-05-211	16,900	—	13.7	—	0.10	1,005	0.40	57	9.7	1.4	4.2	5.0	1
146	-grain, hard red spring	4-05-258	16,900	—	—	—	0.11	1,026	0.44	56	12.5	1.3	5.1	3.6	0
147	-rain, hard red winter	4-05-268	107	0.4	11.6	—	0.11	778	0.22	48	9.9	1.4	4.5	3.4	0
148	-grain, soft red winter	4-05-294	—	—	—	—	0.11	1,092	0.35	48	9.9	1.4	4.5	2.2	0
149	-grain, soft white winter	4-05-337	107	0.4	11.6	—	0.11	1,002	0.22	57	11.0	1.3	4.3	4.0	0
150	-grain, soft white winter, **Pacific Coast**	4-08-555	—	—	13.1	—	—	872	—	50	9.8	0.9	5.0	—	—
151	-middlings, < 9.5% fiber	4-05-205	800	3.0	20.1	—	0.33	1,187	0.76	72	15.6	1.8	16.5	9.0	0
152	-mill run, < 9.5% fiber	4-05-206	—	—	—	—	—	989	—	111	13.2	1.6	15.2	—	—
153	-red dog, < 4% fiber	4-05-203	—	—	—	—	0.11	1,534	0.80	42	13.3	2.2	22.8	4.6	0
154	-shorts, < 7% fiber	4-05-201	—	—	—	—	0.24	1,170	1.40	107	22.3	3.3	18.1	7.2	0
	Wheat, durum *Triticum durum*														
155	-grain	4-05-224	—	—	—	—	—	—	0.38	52	8.8	1.0	4.6	2.98	—
	Whey (see cattle)														
	Yeast *Saccharamyces cerevisiae*														
156	-brewers' dried yeast	7-05-527	—	—	10.0	—	0.63	3,984	9.90	448	109	37.0	91.8	42.8	1
157	-irradiated dried yeast	7-05-529	—	—	—	—	—	—	—	—	—	18.5	—	—	—
	Yeast, torula *Candida utilis* (formerly *Torulopsis utilis*)														
158	-torula dried yeast	7-05-534	—	—	—	—	0.58	2,881	22.4	492	84.2	49.9	6.2	36.3	—

[1] Dashes indicate that no data were available.
[2] The niacin in corn, oats, sorghum, and wheat grain is totally unavailable. The bioavailability of niacin in most by-products from these grains is probably also low.
[3] Riboflavin, thiamin, and vitamin B₆ in soybean protein isolate are totally unavailable.

TABLE 22.4 Amino Acid Composition of Feed Ingredients Used for Swine (as-fed basis)[1]

No.	Feed	International Feed Number	Dry Matter (%)	Crude Protein (%)	Arginine (%)	Histidine (%)	Isoleucine (%)	Leucine (%)	Lysine (%)	Methionine (%)	Cystine (%)	Phenylalanine (%)	Tyrosine (%)	Threonine (%)	Tryptophan (%)	Valine (%)
	Alfalfa (lucerne) *Medicago sativa*															
1	-hay, sun-cured	1-00-078	90	15.9	0.64	0.27	0.74	1.15	0.77	0.16	0.21	0.69	0.41	0.67	0.22	0.70
2	-hay, sun-cured	1-08-331	90	14.0	0.68	0.24	0.60	1.26	0.69	0.20	0.24	0.69	0.57	0.61	0.36	0.77
3	-leaves, meal, dehydrated	1-00-137	92	20.0	0.96	0.42	1.00	1.53	1.08	0.30	—	0.99	—	0.90	0.41	1.10
4	-leaves, sun-cured, ground	1-00-246	92	20.5	1.20	0.37	0.92	1.38	1.01	0.37	0.37	0.92	—	0.74	0.46	1.01
5	-meal, dehy, 15% CP (crude protein)	1-00-022	91	15.4	0.59	0.26	0.64	1.03	0.60	0.22	0.24	0.62	0.41	0.55	0.39	0.72
6	-meal dehydrated, 17% CP	1-00-023	92	17.0	0.71	0.37	0.68	1.21	0.74	0.25	0.18	0.84	0.55	0.70	0.24	0.86
7	-meal dehydrated, 20% CP	1-00-024	92	19.6	0.91	0.38	0.89	1.40	0.90	0.34	0.26	0.93	0.60	0.82	0.35	1.05
8	-meal, dehy, 22% CP	1-07-851	93	22.0	0.99	0.44	1.07	1.60	1.00	0.34	0.34	1.13	0.65	0.98	0.48	1.29
	Animal															
9	-animal plasma, spray dried	—	92	78.0	4.55	2.55	2.71	7.61	6.84	0.75	2.63	4.42	3.53	4.72	1.36	4.94
10	-blood cells, spray dried	—	92	92.0	3.77	6.99	0.49	12.70	8.51	0.81	0.61	6.69	2.14	3.38	1.37	8.50
11	-blood meal, conventional	5-00-380	92	77.1	3.34	5.06	0.91	10.99	7.04	0.99	1.09	5.34	2.29	4.05	1.08	7.05
12	-blood meal, flash dehydrated	5-26-006	92	87.6	3.37	4.57	0.88	11.48	7.56	0.95	1.20	6.41	2.32	4.07	1.06	8.03
13	-blood meal, spray or ring dried	5-00-381	93	88.8	3.69	5.30	1.03	10.81	7.45	0.99	1.04	5.81	2.71	3.78	1.48	7.03
14	-fat, animal	4-00-376	100	—	0	0	0	0	0	0	0	0	0	0	0	0
15	-fat, swine (lard)	4-04-790	100	—	0	0	0	0	0	0	0	0	0	0	0	0
16	-liver meal, dehy	5-00-389	93	66.5	4.11	1.50	3.36	5.41	4.81	1.30	0.90	2.91	1.70	2.61	0.60	4.21
17	-meat meal rendered	5-00-385	94	54.0	3.60	1.14	1.60	3.84	3.07	0.80	0.60	2.17	1.40	1.97	0.35	2.66
18	-meat with blood, meal, tankage, rendered	5-00-386	92	59.5	3.59	1.90	1.90	5.09	3.73	0.73	0.46	2.43	—	2.39	0.72	3.75
19	-meat with blood, with bone, meal, tankage rendered	5-00-387	92	50.4	3.09	1.75	1.86	5.23	3.30	0.69	0.30	2.27	—	2.17	0.62	3.40
20	-meat meal with bone rendered	5-00-388	93	51.5	3.45	0.91	1.34	2.98	2.51	0.68	0.50	1.62	1.07	1.59	0.28	2.04
21	-tallow, beef	4-08-127	100	—	0	0	0	0	0	0	0	0	0	0	0	0
	Animal-poultry (see poultry)															
	Bakery waste															
22	-dried bakery product	4-00-466	91	10.8	0.46	0.24	0.38	0.80	0.27	0.18	0.23	0.50	0.36	0.33	0.10	0.46
	Barley *Hordeum vulgare*															
23	-grain	4-00-549	88	12.2	0.53	0.25	0.47	0.80	0.42	0.15	0.23	0.60	0.32	0.38	0.15	0.60
24	-grain, Pacific Coast	4-07-939	90	9.6	0.44	0.20	0.41	0.60	0.25	0.14	0.20	0.47	0.31	0.30	0.13	0.47
25	-grain, two row	4-00-572	89	11.3	0.54	0.25	0.39	0.77	0.41	0.20	0.28	0.55	0.29	0.35	0.11	0.52
26	-grain, six row	4-00-574	89	10.5	0.48	0.22	0.37	0.68	0.36	0.17	0.20	0.49	0.32	0.34	0.13	0.49
27	-grain, hulless	4-00-552	88	14.9	0.56	0.23	0.41	0.77	0.44	0.16	0.24	0.61	0.40	0.40	0.13	0.55
28	-malt sprouts, dehy	5-00-545	92	26.1	1.11	0.53	1.09	1.63	1.22	0.33	0.23	0.91	—	1.01	0.41	1.45

(continued)

517

TABLE 22.4 Amino Acid Composition of Feed Ingredients Used for Swine (as-fed basis) (continued)

No.	Feed	International Feed Number	Dry Matter (%)	Crude Protein (%)	Arginine (%)	Histidine (%)	Isoleucine (%)	Leucine (%)	Lysine (%)	Methionine (%)	Cystine (%)	Phenylalanine (%)	Tyrosine (%)	Threonine (%)	Tryptophan (%)	Valine (%)
	Bean *Phaseolus vulgaris*															
29	-kidney seeds	5–00–600	—	—	—	—	—	—	—	—	—	—	—	—	—	—
30	-navy seeds	5–00–623	90	22.7	1.19	—	—	—	1.29	0.25	0.23	—	—	—	0.24	—
31	-pinto seeds	5–00–624	90	22.7	1.55	0.64	1.14	1.11	1.60	0.26	—	1.20	—	1.09	0.32	1.23
	Beet, sugar *Beta vulgaris saccharilera*															
32	-molasses, > 48% invert sugar, >79.5° Brix	4–00–668	74	4.4	—	—	—	—	—	—	—	—	—	—	—	—
33	-pulp, dried	4–00–669	91	8.6	0.32	0.23	0.31	0.53	0.52	0.07	0.06	0.30	0.40	0.38	0.10	0.45
	Blood (see animal)															
	Brewers' grains															
34	-brewers' dried grains	5–02–141	92	26.5	1.53	0.53	1.02	2.08	1.08	0.45	0.49	1.22	0.88	0.95	0.26	1.26
	Broomcorn (see millet, proso)															
	Buckwheat *Fagopyrunm* spp															
35	-grain	4–00–994	88	11.1	0.92	0.25	0.40	0.64	0.57	0.19	0.23	0.45	0.31	0.41	0.17	0.56
	Canola (rapeseed) *Brassica* spp															
36	-meal, solvent extracted	5–06–145	90	35.6	2.21	0.96	1.43	2.58	2.08	0.74	0.91	1.43	1.13	1.59	0.45	1.82
37	-meal, prepressed, solv extr, 40% CP	5–08–135	92	40.5	2.23	1.09	1.46	2.71	2.15	0.77	—	1.54	0.85	1.70	0.49	1.94
	Cassava *Manihot* spp															
38	-meal (Tapioca or Manioc)	4–01–152	88	3.3	0.18	0.08	0.11	0.19	0.12	0.04	0.05	0.15	0.04	0.11	0.04	0.14
	Cattle *Bos taurus*															
39	-buttermilk, condensed	5–01–159	29	10.8	—	—	—	—	0.78	—	—	—	—	—	0.12	—
40	-buttermilk, dehy	5–01–160	92	31.5	1.08	0.85	2.37	3.20	2.28	0.71	0.39	1.47	1.01	1.52	0.49	2.56
41	-casein, dried	5–01–162	91	88.7	3.26	2.82	4.66	8.79	7.35	2.70	0.41	4.79	4.77	3.98	1.14	6.10
42	-cheese rind	5–01–163	—	46.2	—	—	—	—	—	—	—	—	—	—	—	—
43	-milk, dehy, cow's	5–01–167	96	25.1	0.92	0.72	1.33	2.56	2.25	0.61	—	1.33	1.33	1.02	0.41	1.74
44	-milk, fresh, cow's	5–01–168	13	3.4	0.14	0.10	0.32	0.26	0.26	0.07	—	0.17	—	0.17	0.05	0.26
45	-milk, skimmed, fresh, cow's	5–01–170	9	3.4	—	—	—	—	0.30	—	—	—	—	—	0.05	—
46	-milk, skim, dried	5–01–175	96	34.6	1.24	1.05	1.87	3.67	2.86	0.92	0.30	1.78	1.87	1.62	0.51	2.33
47	-whey, dried	4–01–182	96	12.1	0.26	0.23	0.62	1.08	0.90	0.17	0.25	0.36	0.25	0.72	0.18	0.60
48	-whey, fresh	4–08–134	7	0.9	—	—	—	—	0.07	—	—	—	—	—	0.01	—
49	-whey, low lactose, dried	4–01–186	96	17.6	0.53	0.33	1.16	1.61	1.51	0.39	0.46	0.63	0.52	1.17	0.31	1.15
50	-whey, permeate, dried	—	96	3.8	0.06	0.05	0.17	0.22	0.18	0.03	0.04	0.06	—	0.14	0.03	0.13
	Chufa *Cyperus esculentus*															
51	-roots	4–08–374	—	—	—	—	—	—	—	—	—	—	—	—	—	—

No.	Feed	Ref. No.															
52	**Citrus** *Citrus* **spp** -pulp without fines, dehy (dried citrus pulp)	4-01-237	90	6.3	0.23	—	—	—	0.20	0.09	0.11	—	—	—	0.06	—
53	-syrup, molasses	4-01-241	—	—	—	—	—	—	—	—	—	—	—	—	—	—
54	**Clover, ladino** -hay, sun-cured	1-01-378	90	19.1	0.99	0.45	1.08	1.89	1.08	0.27	0.36	1.08	0.63	1.17	0.45	1.17
55	**Coconut** *Cocos nucifera* -meal, mech extr (Copra meal)	5-01-572	93	21.2	2.30	—	—	—	0.54	0.33	0.20	—	—	—	0.20	—
56	-meal, solv extr (Copra meal)	5-01-573	92	21.9	2.38	0.39	0.75	1.36	0.58	0.35	0.29	0.84	0.58	0.67	0.19	1.07
57	**Corn, yellow** *Zea mays* -distillers dried grains	5-02-842	94	24.8	0.90	0.63	0.95	2.63	0.74	0.43	0.28	0.99	0.82	0.62	0.20	1.24
58	-distillers dried grains with solubles	5-02-843	93	27.7	1.13	0.69	1.03	2.57	0.62	0.50	0.52	1.34	0.83	0.94	0.25	1.30
	-distillers dried grains with solubles, Midwest ethanol plants		89	30.2	1.07	0.68	1.00	3.16	0.75	0.49	—	1.31	—	1.00	0.22	1.33
59	-distillers dried solubles	5-02-844	92	26.7	0.90	0.66	1.21	2.25	0.82	0.51	0.46	1.38	0.80	1.03	0.23	1.50
60	-ears, ground (corn and cob meal)	4-02-849	86	7.8	0.36	0.16	0.34	0.85	0.17	0.14	0.13	0.39	0.32	0.32	0.07	0.31
61	-gluten feed	5-02-903	90	21.5	1.04	0.67	0.66	1.96	0.63	0.35	0.46	0.76	0.58	0.74	0.07	1.01
62	-gluten meal	5-02-900	91	43.3	1.42	0.97	2.24	7.43	0.83	1.07	0.65	2.82	1.01	1.43	0.21	2.24
63	-gluten meal, 60% CP	5-28-242	90	60.2	1.93	1.28	2.48	10.19	1.02	1.43	1.09	3.84	3.25	2.08	0.31	2.79
64	-grain	4-02-935	89	8.3	0.37	0.23	0.28	0.99	0.26	0.17	0.19	0.39	0.25	0.29	0.06	0.39
65	-grain, flaked	4-02-859	89	9.9	0.44	0.28	0.34	1.24	0.25	0.15	0.25	0.44	0.39	0.35	—	0.47
66	-grain, grade 2, 54 lb/bu or 695 g/1	4-02-931	89	8.7	0.50	0.20	0.40	1.10	0.20	0.13	0.13	0.50	0.44	0.40	0.09	0.38
67	-grits by-product (Hominy Feed)	4-03-011	90	10.3	0.56	0.28	0.36	0.98	0.38	0.18	0.18	0.43	0.40	0.40	0.10	0.52
68	-hominy feed	4-02-887	90	10.9	0.45	0.20	0.40	0.84	0.40	0.14	0.18	0.35	0.50	0.40	0.10	0.50
69	-grain, opaque-2 (High lysine)	4-11-445	87	9.6	0.60	0.33	0.32	0.96	0.39	0.16	0.21	0.37	0.38	0.32	0.10	0.44
70	**Corn, dent white** *Zea mays indentata* -grain	4-02-928	91	10.9	0.27	0.18	0.45	0.91	0.27	0.09	0.09	0.36	0.45	0.36	0.09	0.36
71	**Cotton** *Gossypium* **spp** -meal, mech extr; 36% CP	5-01-625	92	38.9	3.56	0.91	1.32	—	1.22	0.55	0.79	1.88	—	1.12	0.46	2.84
72	-meal, mech extr; 41% CP	5-01-617	92	42.4	4.26	1.11	1.29	2.45	1.65	0.67	0.69	1.97	1.23	1.34	0.54	1.76
73	-meal, prepressed, solv extr, 41% CP	5-07-872	90	41.4	4.59	1.10	1.33	—	1.71	0.52	0.64	2.22	—	1.32	0.47	1.88
74	**Emmer** *Triticum dicoccum* -grain	4-01-830	91	11.7	0.46	0.20	0.42	0.67	0.29	0.16	—	0.46	—	0.38	0.12	0.47
75	**Fababean (broadbean)** -seeds	5-09-262	87	25.4	2.28	0.67	1.03	1.89	1.62	0.20	0.32	1.03	0.87	0.89	0.22	1.14
	Fat (see animal)															
	Feather meal (see poultry)															
	Fish															
76	-meal, mech extr	5-01-977	92	64.5	3.88	1.54	3.68	4.98	5.87	1.79	0.70	2.69	1.89	2.69	0.76	3.43
77	-solubles, condensed	5-01-969	51	32.7	1.61	1.56	1.06	1.86	1.73	0.50	0.30	0.93	0.40	0.86	0.31	1.16
78	-solubles, dried	5-01-971	92	64.2	2.67	1.23	1.56	2.68	2.84	0.98	0.49	1.22	0.62	1.40	0.34	1.94

(continued)

TABLE 22.4 Amino Acid Composition of Feed Ingredients Used for Swine (as-fed basis) (continued)

No.	Feed	International Feed Number	Dry Matter (%)	Crude Protein (%)	Arginine (%)	Histidine (%)	Isoleucine (%)	Leucine (%)	Lysine (%)	Methionine (%)	Cystine (%)	Phenylalanine (%)	Tyrosine (%)	Threonine (%)	Tryptophan (%)	Valine (%)
	Anchovy *Engraulis ringen*															
79	-Anchovy meal, mech extr	5-01-985	92	64.6	3.68	1.56	3.06	5.00	5.11	1.95	0.61	2.66	2.15	2.82	0.76	3.51
	Fish, herring															
80	-Herring meal, mech extr	5-02-000	93	68.1	4.01	1.52	2.91	5.20	5.46	2.04	0.66	2.75	2.18	3.02	0.74	3.46
	Menhaden *Brevoortia tyrannus*															
81	-Menhaden meal, mech extr	5-02-009	92	62.9	3.66	1.78	2.57	4.54	4.81	1.77	0.57	2.51	2.04	2.64	0.66	3.03
	Fish, sardine *Clupea* spp. *Sardinops* spp															
82	-Sardine meal, mech extr	5-02-015	93	65.3	2.70	1.80	3.34	—	5.91	2.01	0.80	2.00	—	2.60	0.50	4.10
83	-solubles, condensed	5-02-014	50	29.5	1.50	2.00	0.90	1.60	1.60	0.90	0.20	0.80	—	0.80	0.10	1.00
	Fish, white Families *Gadidae*, *Lophiidae, and Rafidae*															
84	-White meal, mech extr	5-02-025	91	63.3	4.04	1.34	2.61	4.39	4.51	1.76	0.68	2.32	2.03	2.60	0.66	3.06
	Flax (linseed) *Linum usitatissimum*															
85	-meal, mech extr, 33% CP (linseed meal)	5-02-045	91	34.3	2.72	0.64	1.76	1.88	1.19	0.54	0.58	1.41	0.89	1.12	0.50	1.57
86	-meal, solv extr, 33% CP (linseed meal)	5-02-048	90	33.6	2.97	0.68	1.56	2.06	1.24	0.59	0.59	1.57	1.03	1.26	0.52	1.74
	Garbage															
87	-hotel and restaurant, boiled, wet	4-07-865	26	4.3	—	—	—	—	—	—	—	—	—	—	—	—
88	-hotel and restaurant, cooked, dehy	4-07-879	54	9.5	—	—	—	—	—	—	—	—	—	—	—	—
	Lentil *Lens* spp															
89	-seeds	5-02-506	89	24.4	2.05	0.78	1.00	1.84	1.71	0.18	0.27	1.29	0.70	0.84	0.21	1.27
	Lupin (sweet white) *Lupinus* spp															
90	-seeds	5-27-717	89	34.9	3.38	0.77	1.40	2.43	1.54	0.27	0.51	1.22	1.35	1.20	0.26	1.29
	Maize (see corn)															
	Meat (see animal)															
	Milk (see cattle)															
	Millet, proso *Setaria* spp															
91	-grain	4-03-120	90	11.1	0.41	0.20	0.46	1.24	0.23	0.31	0.18	0.56	0.31	0.40	0.16	0.57
	Milo (see sorghum)															
	Molasses and syrup (see source plant)															
	Oats *Avena sativa*															
92	-cereal by-product < 4% fiber (feeding oatmeal, oat middlings)	4-03-303	91	14.6	0.88	0.30	0.54	1.08	0.48	0.21	0.25	0.70	0.75	0.49	0.20	0.75
93	-grain, all analyses	4-03-309	89	11.5	0.87	0.31	0.48	0.92	0.40	0.22	0.36	0.65	0.41	0.44	0.14	0.66
94	-grain, Pacific Coast	4-07-999	91	9.2	0.60	0.15	0.37	—	0.33	0.13	0.17	0.42	—	0.28	0.12	0.48

No.	Ingredient	Ref. Code														
95	-grain, naked	4-25-101	86	17.1	0.77	0.26	0.48	0.86	0.47	0.19	0.32	0.60	0.42	0.40	0.16	0.63
96	-groats	4-03-331	90	13.9	0.85	0.24	0.55	0.98	0.48	0.20	0.22	0.66	0.51	0.44	0.18	0.72
	Pea *Pisum* spp															
97	-seeds	5-03-600	89	22.8	1.87	0.54	0.86	1.51	1.50	0.21	0.31	0.98	0.71	0.78	0.19	0.98
	Peanut (groundnut) *Arachis hypogaea*															
98	-kernels, hulls added, meal, solv extr	5-03-656	93	47.4	5.49	1.22	2.03	3.80	1.83	0.44	0.71	2.74	1.85	1.52	0.50	2.84
99	-kernels, meal, mech extr, 45% CP (peanut meal)	5-03-649	92	43.2	4.79	1.01	1.41	2.77	1.48	0.50	0.60	2.02	1.74	1.16	0.41	1.70
100	-kernels, meal, solv extr, 47% CP (peanut meal)	5-03-650	92	49.1	5.09	1.06	1.78	2.83	1.66	0.52	0.69	2.35	1.80	1.27	0.48	1.98
	Potato *Solanum tuberosum*															
101	-pulp, dehy	4-03-775	88	6.2	0.02	0.01	0.23	0.42	0.22	0.14	0.14	0.27	0.23	0.26	—	0.30
102	-protein concentrate	5-25-392	91	73.8	3.80	1.71	4.09	7.61	5.83	1.68	1.20	4.89	4.27	4.30	1.02	4.89
103	-tubers, boiled	4-03-784	22	2.4	0.11	0.04	0.06	0.10	0.12	0.03	0.03	0.08	0.07	0.07	—	0.11
104	-tubers dehy	4-07-850	—	—	—	—	—	—	—	—	—	—	—	—	—	—
	Poultry															
105	-by-product meal, rendered	5-03-798	93	64.1	3.94	1.25	2.01	3.89	3.32	1.11	0.65	2.26	1.56	2.18	0.48	2.51
106	-fat, poultry	4-09-319	99	0	—	—	—	—	—	—	—	—	—	—	—	—
107	-feathers meal, hydrolyzed	5-03-795	93	84.5	5.62	0.93	3.86	6.79	2.08	0.61	4.13	4.01	2.41	3.82	0.54	5.88
	Rape (see canola)															
	Rice *Oryza sativa*															
108	-bran with germ (rice bran)	4-03-928	90	13.3	1.00	0.34	0.44	0.92	0.57	0.26	0.27	0.56	0.40	0.48	0.14	0.68
109	-grain, ground (ground rough rice)	4-03-938	89	8.4	0.88	0.19	0.39	0.72	0.34	0.17	0.14	0.47	0.70	0.31	0.11	0.56
110	-grain, polished + broken (brewers' rice)	4-03-932	89	7.9	0.52	0.18	0.34	0.67	0.30	0.18	0.11	0.39	0.38	0.26	0.10	0.49
111	-groats, polished (polished rice)	4-03-942	89	7.2	0.44	0.18	0.45	0.71	0.28	0.25	0.09	0.53	0.62	0.36	0.09	0.53
112	-polishings	4-03-943	90	12.1	0.63	0.19	0.36	0.65	0.53	0.21	0.14	0.39	0.42	0.35	0.10	0.73
	Rye *Secale cereale*															
113	-distillers grains, dehy	5-04-023	—	—	—	—	—	—	—	—	—	—	—	—	—	—
114	-distillers grains with solubles, dehy	5-04-024	91	27.2	1.00	0.70	1.50	2.10	1.00	0.40	—	1.30	0.50	1.10	0.30	1.60
115	-grain	4-04-047	88	11.8	0.50	0.24	0.37	0.64	0.38	0.17	0.19	0.50	0.26	0.32	0.12	0.51
	Safflower *Carthamus tinctorius*															
116	-meal, solvent extracted	5-04-110	92	23.4	2.04	0.59	0.67	1.52	0.74	0.34	0.38	1.07	0.77	0.65	0.33	1.18
117	-meal without hulls, mech extr	5-08-499	90	42.1	4.48	—	—	—	1.29	0.68	0.67	—	—	0.79	0.60	—
118	-meal without hulls, solv extr	5-07-959	92	42.5	3.59	1.07	1.69	2.57	1.17	0.66	0.69	2.00	1.08	1.28	0.54	2.33
119	-seeds, whole	4-07-958	93	18.2	1.60	0.48	0.80	1.20	0.60	0.33	0.35	1.00	—	0.64	0.28	1.00
	Sesame *Sesamum indicum*															
120	-meal, mech extr	5-04-220	93	42.6	4.86	0.98	1.47	2.74	1.01	1.15	0.82	1.77	1.52	1.44	0.54	1.85
	Sorghum *Sorghum bicolor*															
121	-gluten meal	5-04-388	90	44.1	1.27	0.99	2.42	8.00	0.73	0.73	0.80	2.73	—	1.47	0.44	2.50

(continued)

TABLE 22.4 Amino Acid Composition of Feed Ingredients Used for Swine (as-fed basis) (continued)

No.	Feed	International Feed Number	Dry Matter (%)	Crude Protein (%)	Arginine (%)	Histidine (%)	Isoleucine (%)	Leucine (%)	Lysine (%)	Methionine (%)	Cystine (%)	Phenylalanine (%)	Tyrosine (%)	Threonine (%)	Tryptophan (%)	Valine (%)
122	-gluten with bran (gluten feed)	5-08-089	89	23.1	0.90	0.60	1.00	2.50	0.70	0.40	0.20	1.00	0.90	0.80	0.20	1.30
123	-grain	4-20-893	88	9.2	0.38	0.23	0.37	1.21	0.22	0.17	0.17	0.49	0.35	0.31	0.10	0.46
124	-grain, Feterita (USA)	4-04-369	90	12.4	0.46	0.26	0.58	1.78	0.20	0.18	—	0.67	—	0.46	0.17	0.67
125	-grain, Hegari (Thailand)	4-04-398	89	10.0	0.29	0.18	0.47	1.40	0.17	0.11	0.16	0.54	—	0.36	0.11	0.55
126	-grain, Kaffir (USA)	4-04-428	89	11.0	0.37	0.27	0.55	1.62	0.26	0.19	—	0.63	—	0.45	0.16	0.61
	Soybean Glycine max															
127	-meal, mech extr, 41% CP	5-04-600	90	43.8	2.90	1.11	2.83	3.64	2.81	0.71	0.63	2.12	1.42	1.72	0.59	2.23
128	-meal, solv extr	5-04-604	89	43.8	3.23	1.17	1.99	3.42	2.83	0.61	0.70	2.18	1.69	1.73	0.61	2.06
129	-meal without hulls, solv extr	5-04-612	90	47.5	3.48	1.28	2.16	3.66	3.02	0.67	0.74	2.39	1.82	1.85	0.65	2.27
130	-protein concentrate	5-32-183	90	64.0	5.79	1.80	3.30	5.30	4.20	0.90	1.00	3.40	2.50	2.80	0.90	3.40
131	-protein isolate	5-24-811	92	85.8	6.87	2.25	4.25	6.64	5.26	1.01	1.19	4.34	3.10	3.17	1.08	4.21
132	-seeds, heat processed	5-04-597	90	35.2	2.60	0.96	1.61	2.75	2.22	0.53	0.55	1.83	1.32	1.41	0.48	1.68
133	-seeds, whole	5-04-610	91	39.4	7.01	1.93	1.81	3.67	1.47	0.22	0.37	2.90	2.17	1.48	—	2.29
	Spelt Triticum spelta															
134	-grain	4-04-651	90	12.0	0.45	0.18	0.36	0.63	0.27	0.18	—	0.45	—	0.36	0.09	0.45
	Sugarcane Saccharum officinarum															
135	-sugarcane, molasses, dehy	4-04-695	—	—	—	—	—	—	—	—	—	—	—	—	—	—
136	-sugarcane, molasses (blackstrap) > 46% invert sugar, > 79.5° Brix	4-04-696	—	—	—	—	—	—	—	—	—	—	—	—	—	—
	Sunflower Helianthus annuus															
137	-meal, solv extr	5-09-340	90	26.8	2.38	0.66	1.29	1.86	1.01	0.59	0.48	1.23	0.76	1.04	0.38	1.49
138	-meal without hulls, solv extr	5-04-739	93	42.2	2.93	0.92	1.44	2.31	1.20	0.82	0.66	1.66	1.03	1.33	0.44	1.74
	Sweet potato Ipomoea batata															
139	-pulp, dehydrated	4-08-535	—	—	—	—	—	—	—	—	—	—	—	—	—	—
140	-tubers	4-04-788	—	—	—	—	—	—	—	—	—	—	—	—	—	—
	Tallow (see animal)															
	Triticale Triticale hexaloide															
141	-grain	4-20-362	90	12.5	0.57	0.26	0.39	0.76	0.39	0.20	0.26	0.49	0.32	0.36	0.14	0.51
	Wheat Triticum aestivum															
142	-bran	4-05-190	89	15.7	1.07	0.44	0.49	0.98	0.64	0.25	0.33	0.62	0.43	0.52	0.22	0.72
143	-distillers grains, dehy	5-05-193	93	31.6	1.10	0.80	2.01	1.71	0.70	—	—	1.71	0.50	0.90	—	1.71
144	-germ, ground (wheat germ meal)	5-05-218	88	24.5	1.88	0.65	0.88	1.56	1.54	0.44	0.47	0.94	0.73	0.97	0.30	1.17
145	-grain	4-05-211	88	14.9	0.58	0.28	0.47	0.87	0.37	0.18	0.31	0.61	0.41	0.38	0.16	0.56
146	-grain, hard red spring	4-05-258	88	14.1	0.67	0.34	0.47	0.93	0.38	0.23	0.30	0.67	0.40	0.41	0.16	0.61
147	-grain, hard red winter	4-05-268	88	13.5	0.60	0.32	0.41	0.86	0.34	0.20	0.29	0.60	0.38	0.37	0.15	0.54
148	-grain, soft red winter	4-05-294	88	11.5	0.50	0.20	0.45	0.90	0.38	0.22	0.27	0.63	0.37	0.39	0.26	0.57

149	-grain, soft white winter	4-05-337	89	11.8	0.55	0.27	0.44	0.79	0.33	0.20	0.28	0.55	0.36	0.35	0.15	0.53
150	-grain, soft white winter, Pacific Coast	4-08-555	89	10.1	0.45	0.20	0.40	0.59	0.30	0.14	0.24	0.42	0.36	0.28	0.12	0.41
151	-middlings, < 9.5% fiber	4-05-205	89	15.9	0.97	0.44	0.53	1.06	0.57	0.26	0.32	0.70	0.29	0.51	0.20	0.75
152	-mill run, < 9.5% fiber	4-05-206	90	15.6	0.90	0.40	0.70	1.20	0.50	0.40	0.20	—	0.50	0.50	0.20	0.80
153	-red dog, < 4% fiber	4-05-203	88	15.3	0.96	0.41	0.55	1.06	0.59	0.23	0.37	0.66	0.46	0.50	0.10	0.72
154	-shorts, < 7% fiber	4-05-201	88	16.0	1.07	0.43	0.58	1.02	0.70	0.25	0.28	0.70	0.51	0.57	0.22	0.87
	Wheat, durum Triticum durum															
155	-grain	4-05-224	87	13.8	0.66	0.34	0.58	1.04	0.38	0.18	—	0.80	0.39	0.42	—	0.69
	Whey (see cattle)															
	Yest Saccharomyces spp															
156	-brewers' dried yeast	7-05-527	93	45.9	2.20	1.09	2.15	3.13	3.22	0.74	0.50	1.83	1.55	2.20	0.56	2.39
157	-irridiated dried yeast	7-05-529	94	48.1	2.46	1.00	2.94	3.56	3.70	1.00	—	2.77	—	2.41	0.73	3.06
	Yeast, torula Candida utilis (formerly Torulopsis utilis)															
158	-torula dried yeast	7-05-534	93	46.4	2.48	1.09	2.50	3.32	3.47	0.69	0.50	2.33	1.65	2.30	0.51	2.60

[1] Dashes indicate that no data were available.

TABLE 22.5 Mineral Composition of Macromineral Sources Used for Swine (as-fed basis)[1]

No.	Feed	International Feed Number	Dry Matter (%)	Calcium (Ca, %)	Phosphorus (%)	Available Phosphorus (%)	Sodium (Na, %)	Chlorine (Cl, %)	Potassium (K, %)	Magnesium (Mg, %)	Sulfur (S, %)	Flourine (Fl, mg/kg)	Copper (Cu, mg/kg)	Iron (Fe, mg/kg)	Manganese (Mn, mg/kg)	Zinc (Zn, mg/kg)
1	Ammonium-polyphosphate solution	6-08-042	60	0.1	14.5	—	—	—	—	—	—	—	—	—	—	—
2	Bone meal	6-00-397	95	25.9	12.4	—	—	—	0.20	—	—	—	—	—	—	—
3	Bone meal, steamed	6-00-400	97	29.8	12.5	11.2	0.04	—	0.14	0.30	2.40	—	11	850	300	126
4	Bone, black, spent	6-00-404	90	27.1	12.7	—	—	—	0.14	0.53	—	—	—	—	—	—
5	Bone, charcoal	6-00-402	90	27.1	12.7	—	—	0.02	0.08	0.53	—	—	—	—	—	—
6	Calcium carbonate, $CaCO_3$	6-01-069	99	35.8	0	—	0.08	0.00	0.16	1.61	0.08	—	24	600	200	—
7	Calcium phosphate, monocalcium, $CaH_4(PO_4)_2 \cdot H_2O$	6-26-334	100	17.0	21.1	21.1	0.20	0.47	0.15	0.09	0.80	—	80	7,500	100	220
8	Calcium phosphate, dicalcium $CaHPO_4 \cdot 2H_2O$ and $CaHPO_4$	6-01-080	96	21.3	18.7	18.7	0.18	—	—	0.80	0.80	1,800	—	7,900	1,400	—
9	Calcium phosphate, tricalcium	6-01-084	100	38.0	18.0	15.1	—	—	—	—	—	—	—	—	—	—
10	Calcium sulfate anhydrous	6-01-087	85	22.0	0	—	—	—	—	2.21	20.01	—	—	—	—	—
11	Calcium sulfate, dihydrate (gypsum)	6-01-090		21.8	0	—	—	—	—	0.48	16.19	—	—	1,171	—	—
	Dicalcium phosphate (see calcium phosphate, dicalcium)															
12	Limestone, dolomite (limestone, magnesium)	6-02-633	99	22.0	0	—	—	0.12	0.36	9.87	—	—	—	760	—	—
13	Limestone, ground	6-02-632	99	38.0	0	—	0.06	0.02	0.11	2.06	0.04	—	—	3,500	200	—
14	Magnesium carbonate	6-02-754	81	0.0	0	—	—	—	—	30.20	—	—	—	—	100	—
15	Magnesium oxide	6-02-756	100	1.7	0	—	0.50	0.01	0.02	55.00	0.10	200	—	—	—	—
16	Magnesium sulfate (Epsom Salts)	6-02-758	49	0.0	0	—	0.00	0.01	0.00	9.60	13.04	—	—	1,060	—	—
17	Oyster shell, ground (flour)	6-03-481	99	37.6	0	—	0.21	0.01	0.10	0.30	—	—	—	2,840	133	—
18	Phosphate, diammonium	6-00-370	97	0.5	20.0	—	0.04	—	0.01	0.45	2.50	2,100	91	12,000	400	342
19	Phosphate, monoammonium	6-09-338	97	0.3	23.0	21.9	0.20	—	0.16	0.75	1.50	2,500	80	4,100	100	300
20	Phosphate, rock, Curaçao	6-05-586	100	35.1	14.2	7.1	0.20	—	—	0.80	—	5,500	—	3,500	—	—
21	Phosphate, rock, defluorinated	6-01-780	100	32.0	18.0	16.2	3.27	—	0.10	0.29	0.13	1,800	22	8,400[2]	500	—
22	Phosphate, rock, raw	6-03-945	100	35.0	13.0	—	—	—	—	—	—	35,000	—	—	—	—
23	Phosphate, rock, low fluorine	6-03-946	100	36.0	14.0	—	—	—	—	—	—	—	—	—	—	—
24	Phosphate, rock, soft (colloidal clay)	6-03-947	100	16.1	9.0	36.0	0.10	—	—	0.38	—	15,000	—	19,200	1,000	—
25	Phosphoric acid, feed grade	6-03-707	75	0.1	23.7	23.7	0.01	—	0.01	—	0.05	3,100	—	20	—	—
26	Potassium and magnesium sulfate	6-06-177	100	0	—	—	0.76	1.25	18.45	11.58	21.97	—	—	100	20	—

#	Feedstuff	Number																	
27	Potassium chloride	6–03–755	100	0	—	1.00	46.93	51.37	0.23	0.32	—	—	—	—	600	10	—	—
28	Sodium phosphate, dibasic	6–04–286	100	—	21.1	21.1	31.04	—	—	—	—	—	—	—	—	10	—	—
29	Sodium phosphate, monobasic, anhydrous, NaH_2PO_4	6–04–288	87	0.1	24.9	24.9	18.65	0.02	0.01	0.01	—	—	—	—	10	—	—
30	Sodium tripolyphosphate	6–08–076	96	—	24.0	24.9	28.80	—	—	—	—	—	—	40	—	—	—

[1]Dashes indicate that no data were available.
[2]The iron in defluorinated phosphate is about 65% as available as that in ferrous sulfate.

TABLE 22.6 Mineral Composition of Trace Mineral Sources (dry matter basis)[1,2]

Feed	Chemical Formula	Mineral (%)	Relative Bioavailability (%)	Sodium (Na, %)	Chlorine (Cl, %)	Sulfur (S, %)	Comments
Copper							
Cupric carbonate (1H$_2$O)	CuCO$_3$·Cu(OH)$_2$·H$_2$O	50	60–100				Dark green crystals
Cupric chloride, tribasic	Cu$_2$(OH)$_3$Cl	58	100		41.64		Green crystals
Cupric iodide	CuI	33.36	100				Black powder or granules
Cupric oxide	CuO	79.88	0–10				
Cupric sulfate (5H$_2$O)	CuSO$_4$·5H$_2$O	25.45	100			12.84	Blue or ultramarine crystals
Cupric sulfate (anhydrous)	CuSO$_4$	39.8	100			40.09	
Iodine							
Calcium iodate	Ca(IO$_3$)$_2$	65.08	100				Stable source
Calcium periodate	Ca$_5$(IO$_5$)$_2$	39.28	90				
Cupric iodide	CuI	66.6	100				
Ethylenediamine dihydroiodide, EDDI (organic iodide)	C$_2$H$_8$N$_2$·2HI	80.3	100				White
Potassium iodide	KI	76.4	100	0			Used in iodized salt (0.01%)
Sodium iodide	NaI	84.6	—	15.33			
Iron							
Ferric chloride	FeCl$_3$·6H$_2$O	20.66	44–100				
Ferric oxide	Fe$_2$O$_3$	69.94	0				Red—used as coloring pigment; do not use as iron supplement
Ferrous carbonate	FeCO$_3$	48.2	15–80				Beige
Ferrous chloride (6 H$_2$O)	FeCl$_3$·6H$_2$O	20.6	40–100		39.35		Reddish-brown
Ferrous fumarate	C$_4$H$_2$FeO$_4$	32.87	100				
Ferrous oxide	FeO	77.8	—				Black powder
Ferrous sulfate (1 H$_2$O)	FeSO$_4$·H$_2$O	32.8	100			18.87	Green to brown crystals
Ferrous sulfate (7 H$_2$O)	FeSO$_4$·7H$_2$O	20.08	100			11.53	Greenish crystals
Manganese							
Manganese sulfate	MnSO$_4$·7H$_2$O	22.8	100				Rose-colored crystals
Manganous carbonate	MnCO$_3$	47.8	30–100				Rose-colored crystals
Manganous chloride (4 H$_2$O)	MnCl$_2$·4H$_2$O	27.7	100		19.0		
Manganous dioxide	MnO$_2$	63.1	35–95				
Manganous oxide	MnO	77.4	70				Green to brown powder
Manganous sulfate	MnSO$_4$·H$_2$O	29.5	100			30.8	White to cream powder
Selenium							
Sodium selenate (10 H$_2$O)	Na$_2$SeO$_4$·10H$_2$O	21.4	100	12.13			White crystals
Sodium selenite	Na$_2$SeO$_3$	45	100	26.00			White to light pink crystals

Zinc

Zinc carbonate	$Zn \cdot CO_3$	52.1	100		White crystals
Zinc chloride	$ZnCl_2$	47.97	100	52.0	
Zinc oxide	ZnO	80.3	50–80		Greyish powder
Zinc sulfate (1 H_2O)	$ZnSO_4 \cdot H_2O$	36.4	100	17.86	White crystals
Zinc sulfate (7 H_2O)	$ZnSO_4 \cdot 7H_2O$	22.7	100	11.15	

[1]Adapted from Iowa State University *Life Cycle Swine Nutrition*, 1996 Iowa State University, Ames, IA; *Kansas Swine Nutrition Guide* Kansas State University, Manhattan, KS; NRC *Nutrient Requirements of Swine*, National Academy Science, Washington 1998; and University of Nebraska and South Dakota State University *Swine Nutrition Guide*, University of Nebraska, Lincoln, NE and South Dakota State University, Brookings, SD, 1995.

[2]Dashes indicate that no data were available.

SELECTED REFERENCES

Animal Feed Resources Information System (AFRIS), http://www.fao.org/ag/aga/agap/frg/afris/default.htm

Nutrient Requirements of Swine, 7th rev., National Academy Press, National Research Council Washington, DC, 1998, http://www.nap.edu/catalog/6016.html

United States–Canadian Tables of Feed Composition: Nutritional Data for United States and Canadian Feeds, 3rd rev., National Academy Press, Washington, DC, 1982, http://books.nap.edu/books/0309032458/html/ R1.html

APPENDIX

Contents	Page

The appendix is essential to the completeness of *Swine Science.* It provides useful supplemental information about (1) animal units, (2) conversions of weights and measures, (3) some uses of weights and measures, (4) swine magazines, (5) breed registry associations, and (6) poison information centers.

ANIMAL UNITS

An animal unit is a common animal denominator based on feed consumption. It is assumed that one mature cow represents an animal unit. Then, the comparative (to a mature cow) feed consumption of other age groups or classes of animals determines the proportion of an animal unit that they represent. For example, it is generally estimated that the ration of one mature cow will feed five hogs raised to 200 lb. For this reason, the animal unit on this class and age of animals is 0.2. Table A.1 gives the animal units for different classes and ages of livestock.

WEIGHTS AND MEASURES

Weights and measures are standards employed in arriving at weights, quantities, and volumes. Even among primitive people, such standards were necessary, and with the growing complexity of life, they become of greater and greater importance.

Weights and measures form one of the most important parts of modern agriculture. This section contains pertinent information relative to the most common standards used by U.S. pork producers (Figure A.1).

TABLE A.1   Animal Units

Animal Type	Animal Units
Cattle, excluding mature dairy or veal cattle	1.0
Veal	1.0
Mature dairy cattle	1.4
Swine greater than 55 lb	0.4
Swine less than 55 lb	0.1
Chickens	0.01
Turkeys	0.018
Ducks	0.2
Horses	2.0
Sheep or lambs	0.1

Source: Adapted from the Federal Register, *Vol. 66, No. 9, Page 2962, Proposed Rules, January 12, 2001.*

Metric System

The United States and a few other countries use standards that belong to the *customary,* or English, system of measurement. This system evolved in England from older measurement standards, beginning about the year 1200. All other countries—including England—now use a system of measurements called the *metric system,* which was created in France in the 1790s. The metric system is used for all scientific research as well as growing use in the United States. Hence, everyone should have a working knowledge of it.

The basic metric units are the *meter* (length/distance), the *gram* (weight), and the *liter* (capacity). The units are then expanded in multiples of 10 or

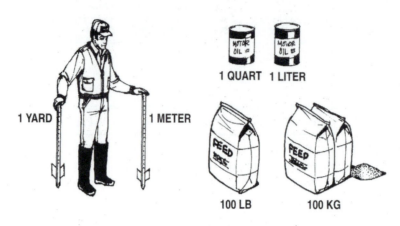

Figure A.1 Metric versus U.S. customary; length, weight, and volume comparisons. *(Courtesy of Pearson Education)*

made smaller by 1/10. The prefixes, which are used in the same way with all basic metric units, follow:

"milli-" = 1/1,000

"centi-" = 1/100

"deci-" = 1/10

"deca-" = 10

"hecto-" = 100

"kilo-" = 1,000

The following tables facilitate conversion from metric units to U.S. customary, and vice versa:

Table A.2, Weight-Unit Conversion Factors

Table A.3, Weight Equivalents

Table A.4, Weights and Measures

    Length

    Surface or Area

    Volume

    Weight

    Weights and Measures per Unit

Temperature

One centigrade (C) degree is 1/100 the difference between the temperature of melting ice and that of water boiling at standard atmospheric pressure. One centigrade degree equals 1.8°F.

    One Fahrenheit (F) degree is 1/180 of the difference between the temperature of melting ice and that of water boiling at standard atmospheric pressure. One Fahrenheit degree equals 0.556°C (Figure A.2).

TABLE A.2 Weight–Unit Conversion Factors

Units Given	Units Wanted	For Conversion Multiply By
lb	g	453.6
lb	kg	0.4536
oz	g	28.35
kg	lb	2.2046
kg	mg	1,000,000
kg	g	1,000
g	mg	1,000
g	ug	1,000,000
mg	ug	1,000
mg/g	mg/lb	453.6
mg/kg	mg/lb	0.4536
ug/kg	ug/lb	0.4536
Mcal	kcal	1,000
kcal/kg	kcal/lb	0.4536
kcal/lb	kcal/kg	2.2046
ppm	ug/g	1
ppm	mg/kg	1
ppm	mg/lb	0.4536
mg/kg	%	0.0001
ppm	%	0.0001
mg/g	%	0.1
g/kg	%	0.1

Source: Courtesy of Pearson Education.

TABLE A.3 Weight Equivalents

1 lb	= 453.6 g	= 0.4536 kg	= 16 oz
1 oz	= 28.35 g		
1 kg	= 1,000 g	= 2.2046 lb	
1 g	= 1,000 mg		
1 mg	= 1,000 ug	= 0.001 g	
1 ug	= 0.001 mg	= 0.000001 g	
1 ug/g	= 1 mg/kg	ppm	

Source: Courtesy of Pearson Education.

TABLE A.4   Weights and Measures (metric and U.S. customary)

Length

Unit	Is Equal To	
Metric System	Metric	(U.S. Customary)
1 millimicron (mu)	0.000000001 m	0.000000039 in.
1 micron (u)	0.000001 m	0.000039 in.
1 millimeter (mm)	0.001 m	0.0394 in.
1 centimeter (cm)	0.01 m	0.3937 in.
1 decimeter (dm)	0.1 m	3.937 in.
1 meter (m)	1 m	39.37 in.; 3.281 ft; 1.094 yd
1 hectometer (hm)	100 m	328 ft, 1 in.; 19.8338 rod
1 kilometer (km)	1,000 m	3,280 ft, 10 in.; 0.621 mi

U.S. Customary	U.S. Customary	(Metric)
1 inch (in.)		25 mm; 2.54 cm
1 hand[1]	4 in.	
1 foot (ft)	12 in.	30.48 cm; 0.305 m
1 yard (yd)	3 ft	0.914 m
1 fathom[2] (fath)	6.08 ft	1.829 m
1 rod (rd), pole, or perch	16.5 ft; 5.5 yd	5.029 m
1 furlong (fur)	220 yd; 40 rd	201.168 m
1 mile (mi)	5,280 ft; 1,760 yd; 320 rd; 8 fur	1,609.35 m; 1,609 km
1 knot or nautical mile	6,080 ft; 1.15 land mi	
1 league (land)	3 miles (land)	
1 league (nautical)	3 miles (nautical)	

[1]Used in measuring height of horses.
[2]Used in measuring depth at sea.

Conversions

To Change	To	Multiply By
inches	centimeters	2.54
feet	meters	0.305
meters	inches	39.73
miles	kilometers	1.609
kilometers	miles	0.621

(To make opposite conversion, divide by the number given instead of multiplying).

Surface or Area

Unit	Is Equal To	
Metric System	Metric	(U.S. Customary)
1 square millimeter (mm^2)	0.000001 m^2	0.00155 in.^{2}
1 square centimeter (cm^2)	0.0001 m^2	0.155 in.^{2}
1 square decimeter (dm^2)	0.01 m^2	15.50 in.^{2}
1 square meter (m^2)	1 centare (ca)	1,550 in.^{2}; 10.76 ft^2; 1.196 yd^2
1 acre (a)	100 m^2	119.6 yd^2
1 hectare (ha)	10,000 m^2	2.47 acres
1 square kilometer (km^2)	1,000,000 m^2	247.1 acres; 0.386 mi^2

TABLE A.4 **Weights and Measures (metric and U.S. customary)** *(continued)*

U.S. Customary	U.S. Customary	(Metric)
1 square inch (in.^{2})	1 in. × 1 in.	6.452 cm^2
1 square foot (ft^2)	144 in.^{2}	0.093 m^2
1 square yard (yd^2)	1,296 in.^{2}; 9 ft^2	0.836 m^2
1 square rod (rod^2)	272.25 ft^2; 30.25 yd^2	25.29 m^2
1 rood	40 rod^2	10.117 acres
1 acre	43,560 ft^2; 4,840 yd^2; 160 rd^2; 4 roods	4,046.87 m^2; 0.405 ha
1 square mile (mi^2)	640 acres	2.59 km^2; 259 ha
1 township	36 sections; 6 mi^2	

Conversions

To Change	To	Multiply By
square inches	square centimeters	6.452
square centimeters	square inches	0.155
square yards	square meters	0.836
square meters	square yards	1.196

(To make opposite conversion, divide by the number given instead of multiplying.)

Volume

Unit Metric System Liquid and Dry	Is Equal To (U.S. Customary)		
	Metric	(Liquid)	(Dry)
1 milliliter (ml)	0.001 liter	0.271 dram (fl)	0.061 in.^{3}
1 centiliter (cl)	0.01 liter	0.338 oz (fl)	0.610 in.^{3}
1 deciliter (dl)	0.1 liter	3.38 oz (fl)	
1 liter (l)	1,000 cubic centimeters	1.057 qt; 0.2642 gal (fl)	0.908 qt
1 hectoliter (hl)	100 liter	26,418 gal	2.838 bu
1 kiloliter (kl)	1,000 liter	264.18 gal	1,308 yd^3

Volume

Unit U.S. Customary Liquid	Is Equal To			
	U.S. Customary	(Ounces)	(Cubic Inches)	(Metric)
1 teaspoon (t)	60 drops	0.1666		5 ml
1 dessert spoon	2 t			
1 tablespoon (T)	3t	0.5		15 ml
1 fl oz		1	1.805	29.57 ml
1 gill (gi)	0.5 c	4	7.22	118.29 ml
1 cup (c)	16 T	8	14.44	236.58 ml; 0.24 l
1 pint (pt)	2 cups	16	28.88	0.47 l
1 quart (qt)	2 pt	32	57.75	0.95 l
1 gallon (gal)	4 qt	8.34 lb	231	3.79 l
1 barrel (bbl)	31.5 gal			
1 hogshead (hhd)	2 bbl			
Dry	U.S. Customary	(Ounces)	(Cubic Inches)	(Metric)
1 pint (pt)	0.5 qt		33.6	0.55 l
1 quart (qt)	2 pt		67.20	1.10 l
1 peck (pk)	8 qt		537.61	8.81 l
1 bushel (bu)	4 pk		2,150.42	35.24 l

TABLE A.4   Weights and Measures (metric and U.S. customary) *(continued)*

Solid

Metric System	(Metric)	(U.S. Customary)
1 cubic millimeter (mm³)	0.001 cc	
1 cubic centimeter (cc)	1,000 mm³	0.061 in.³
1 cubic decimeter (dm³)	1,000 cc	61.023 in.³
1 cubic meter (m³)	1,000 dm³	35.315 ft³; 1.308 yd³

U.S. Customary	(U.S. Customary)	(Metric)
1 cubic inch (in.³)		16.387 cc
1 board foot (fbm)	144 in.³	2,359.8 cc
1 cubic foot (ft³)	1,728 in.³	0.028 m³
1 cubic yard (yd³)	27 ft³	0.765 m³
1 cord	128 ft³	3.625 m³

Conversions

To Change	To	Multiply By
ounces (fluid)	cubic centimeters	29.57
cubic centimeters	ounces (fluid)	0.034
quarts	liters	0.946
liters	quarts	1.057
cubic inches	cubic centimeters	16.387
cubic centimeters	cubic inches	0.061
cubic yards	cubic meters	0.765

(To make opposite conversion, divide by the number given instead of multiplying.)

Weight

Unit Metric System	Is Equal To Metric	(U.S. Customary)
1 microgram (mcg)	0.001 mg	
1 milligram (mg)	0.001 g	0.015432356 grain
1 centigram (cg)	0.01 g	0.15432356 grain
1 decigram (dg)	0.1 g	1.5432 grains
1 gram (g)	1,000 mg	0.03527396 oz
1 decagram (dcg)	10 g	5.643833 dr
1 hectogram (hg)	100 g	3.527396 oz
1 kilogram (kg)	1,000 g	35.274 oz; 2.2046223 lb
1 ton	1,000 kg	2,204.6 lb; 1.102 tons (short); 0.984 ton (long)

U.S. Customary	U.S. Customary	(Metric)
1 grain	0.037 dr	64.798918 mg; 0.064798918 g
1 dram (dr)	0.063 oz	1.771845 g
1 ounce (oz)	16 dr	28.349527 g
1 pound (lb)	16 oz	453.5924 g; 0.4536 kg
1 hundredweight (cwt)	100 lb	
1 ton (short)	2,000 lb	907.18486 kg; 0.907 (metric) ton
1 ton (long)	2,200 lb	1,016.05 kg; 1.016 (metric) ton
1 part per million (ppm)	1 mcg/g; 1 mg/l; 1 mg/kg; 0.0001%;	0.4535924 mg/lb; 0.907 g/ton; 0.00013 oz/gal
1 percent (%) (1/100)	10,000 ppm; 10 g/l	1.28 oz/gal; 8.0 lb/100 g

TABLE A.4 Weights and Measures (metric and U.S. customary) *(continued)*

Conversions

To Change	To	Multiply By
grams	milligrams	64.799
ounces (dry)	grams	28.35
pounds (dry)	kilograms	0.4535924
kilograms	pounds	2.2046223
milligrams/pound	parts/million	2.2046223
parts/million	grams/ton	0.90718486
grams/ton	parts/million	1.1
milligrams/pound	grams/ton	2
grams/ton	milligrams/pound	0.5
grams/pound	grams/ton	2,000
grams/ton	grams/pound	0.0005
grams/ton	pounds/ton	0.0022
pounds/ton	grams/ton	453.5924
grams/ton	percent	0.00011
percent	grams/ton	9,072
parts/million	percent	move decimal four places to left

(To make opposite conversion, divide by the number given instead of multiplying.)

Weights and Measures Per Unit

Unit	Is Equal To
Volume per Unit Area	
1 liter/hectare	0.107 gal/acre
1 gallon/acre	9.354 liter/ha
Weight per Unit Area	
1 kilogram/cm^2	14.22 lb/in.^{2}
1 kilogram/hectare	0.892 lb/acre
1 pound/square inch	0.0703 kg/cm^2
1 pound/acre	1.121 kg/ha
Area per Unit Weight	
1 square centimeter/kilogram	0.0703 in^2/lb
1 square inch/pound	14.22 cm^2/kg

Source: Courtesy of Pearson Education.

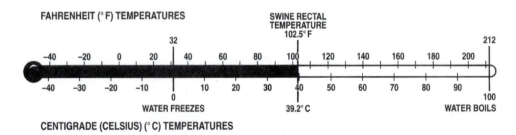

Figure A.2 Fahrenheit–centigrade (Celsius) scale for direct conversion and reading. *(Courtesy of Pearson Education)*

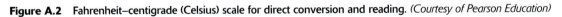

To Change	To	Do This
Degrees centigrade	Degrees Fahrenheit	Multiply by 9/5 and add 32
Degrees Fahrenheit	Degrees centigrade	Subtract 32, then multiply by 5/9

Weights and Measures of Common Feeds

In calculating diets and mixing concentrates, it is usually necessary to use weights rather than measures. However, in practical feeding operations, it is often more convenient for the producer to measure

TABLE A.5 Weights and Measures of Common Feeds

Feed	Approximate Weight[1] lb/quart	lb/bushel
Alfalfa meal	0.6	19
Barley	1.5	48
Beet pulp (dried)	0.6	19
Brewers' grain (dried)	0.6	19
Buckwheat	1.6	50
Buckwheat bran	1.0	29
Corn, husked ear	—	70
Corn, cracked	1.6	50
Corn, shelled	1.8	56
Corn meal	1.6	50
Corn-and-cob meal	1.4	45
Cottonseed meal	1.5	48
Cowpeas	1.9	60
Distillers' grain (dried)	0.6	19
Fish meal	1.0	35
Gluten feed	1.3	42
Linseed meal (old process)	1.1	35
Linseed meal (new process)	0.9	29
Meat scrap	1.3	42
Milo (grain sorghum)	1.7	56
Molasses feed	0.8	26
Oats	1.0	32
Oats, ground	0.7	22
Oat middlings	1.5	48
Peanut meal	1.0	32
Rice bran	0.8	26
Rye	1.7	56
Sorghum (grain)	1.7	56
Soybeans	1.7	60
Tankage	1.6	51
Velvet beans, shelled	1.8	60
Wheat	1.9	60
Wheat bran	0.5	16
Wheat middlings, standard	0.8	26
Wheat screenings	1.0	32

[1]To convert to metric, refer to Table A.4.
Source: Courtesy of Pearson Education.

the concentrates. Table A.5 serves as a guide in feeding by measure.

Estimating Weight of Hogs

Hog weights can be estimated by measuring the heart girth in inches using a cloth measuring tape. Place the tape directly behind the front legs, wrapped snugly around the heart girth and read directly behind the shoulders. Use the figures in Table A.6. The weight will be +/− 10 lb of the pig's weight. Take three separate measurements and use the average to obtain a more precise girth measurement. Pigs should be measured while on full feed and water.

TABLE A.6 Estimating Pig Weight

Heart Girth, in.	Pig Weight, lb
25	49
26	59
27	69
28	79
29	89
30	99
31	110
32	120
33	130
34	140
35	150
36	160
37	171
38	181
39	191
40	201
41	211
42	221
43	232
44	242
45	252
46	262
47	272
48	282

Source: Swine Update, Vol. 25:1, 2003, Kansas State University, Manhattan, KS.

SWINE MAGAZINES

Livestock magazines publish news items and informative articles of special interest to swine caretakers. Also, many of them employ field representatives whose chief duty it is to assist in the buying and selling of animals.

In the compilation of the list presented (see Table A.7), no attempt was made to list the general livestock magazines of which there are numerous outstanding ones. Only those magazines with major emphasis on swine are included.

BREED REGISTRY ASSOCIATIONS

A breed registry association consists of a group of breeders banded together for the purposes of (1) recording the lineage of their animals, (2) protecting the purity of the breed, (3) encouraging further improvement of the breed, and (4) promoting the interest of the breed. A list of the swine breed registry associations is given in Table A.8.

TABLE A.7 Livestock Industry Magazines

National Hog Farmer	*Pork Checkoff Report*
PRIMEDIA Business Magazines and Media, Inc.	National Pork Board
9800 Metcalf Avenue	P.O. Box 9114
Overland Park, KS 66212-2215	Des Moines, IA 50306
http://www.nationalhogfarmer.com	
	Pork Magazine
Pig Progress	10901 West 84th Terrace
International Agri- and Horticulture	Lenexa, KS 66214
P.O. Box 4, 7000 BA Doetinchem	*http://www.porkmag.com*
The Netherlands	
http://www.AgriWorld.nl	*Swine Practitioners*
	Livestock Division, Vance Publishing Corp.
	10901 West 84th Terrace
	Lenexa, KS 66214

Source: Courtesy of Pearson Education.

TABLE A.8 Breed Registry Associations

American Berkshire Association	National Hereford Hog Record Association
P.O. Box 2436	Route 1, Box 37
West Lafayette, IN 47906	Flandreau, SD 57028
(317) 497-3618	(605) 997-2116
Breed Publication: *The Berkshire News*	
http://www.americanberkshire.com	National Spotted Swine Records
	Member of CPS (Certified Pedigreed Swine)
American Landrace Association	P.O. Box 9758
Member of the National Swine Registry	Peoria, IL 61612
P.O. Box 2417	(309) 693-1804
West Lafayette, IN 47996-2417	Breed Publication: *Breeders Digest Magazine*
(765) 463-3594	*http://www.cpsswine.com*
Breed Publication: *Seedstock Edge*	
http://www.nationalswine.com	Poland China Record Association
	Member of CPS (Certified Pedigreed Swine)
American Yorkshire Club	P.O. Box 9758
Member of the National Swine Registry	Peoria, IL 61612
P.O. Box 2417	(309) 691-6301
West Lafayette, IN 47996-2417	Breed Publication: *Breeders Digest Magazine*
(765) 463-3594	*http://www.cpsswine.com*
Breed Publication: *Seedstock Edge*	
http://www.nationalswine.com	Tamworth Swine Association
	621 N CR 850 W
Chester White Swine Record Association	Greencastle, IN 46135
Member of CPS (Certified Pedigreed Swine)	(765) 653-4913
P.O. Box 9758	Breed Publication: *The Tamworth News*
Peoria, IL 61612	
(309) 691-0151	United Duroc Swine Registry
Breed Publication: *Breeders Digest Magazine*	Member of the National Swine Registry
http://www.cpsswine.com	P.O. Box 2417
	West Lafayette, IN 47996-2417
Hampshire Swine Registry	(765) 463-3594
Member of the National Swine Registry	Breed Publication: *Seedstock Edge*
P.O. Box 2417	*http://www.nationalswine.com*
West Lafayette, IN 47996-2417	
(765) 463-3594	
Breed Publication: *Seedstock Edge*	
http://www.nationalswine.com	

Source: Courtesy of Pearson Education.

CODE OF FAIR PRACTICES (NATIONAL ASSOCIATION OF SWINE RECORDS)

Buyers of purebred, registered boars and gilts buy them to be breeders. Many factors may affect an animal's breeding capabilities. Many of these are not visible at the time of purchase. Some problems may be the result of management before the sale. Some may be the result of handling and management by the buyer after the purchase. Some may be hereditary. Because of this, adjustments should be a shared responsibility. All adjustments involve matters between buyers and sellers.

Standard Warranty

All purebred, registered hogs more than 5 months of age (not used for breeding under 7 months of age) sold as breeding animals for breeding purposes are sold with a warranty that they are capable of and will breed. If and when any said animal does prove to be a nonbreeder, the seller shall make an adjustment to the satisfaction of the buyer, provided the buyer informs the seller of the situation within 90 days after purchase. In all purebred transactions, the registration certificate is an integral part of the transaction and shall be delivered to the buyer, properly transferred on the association records, at the expense of the seller. Following are some suggestions considered as generally acceptable within the industry for remediating a problem situation. Other adjustments may be made if they are satisfactory to both the buyer and seller.

Boars Failing to Serve or Settle Sows

1. Refund the difference between the purchase price and the market value as shown by a sales receipt, if the boar is sold on the market.
2. Make a replacement of another boar satisfactory to the buyer.
3. Give the buyer credit (amount to be agreed on by both the buyer and the seller) on the purchase of another animal or animals.

Gilts Sold as Open

A. If proven to be bred
   1. Refund the purchase price on return to the seller.
   2. Refund the difference, if any, between price paid for gilt and value of a commercial bred sow.
B. If proven to be a nonbreeder
   1. Make a replacement of another gilt satisfactory to the buyer.
   2. Refund the difference between the purchase price and the market value of the gilt as shown by a sales receipt, if the gilt is sold at market.
   3. Give the buyer credit (amount to be agreed on by both the buyer and seller) on the purchase of another gilt in the future.

Bred Sows

Bred sows are expected to be bred to a designated boar on the date of service. When proven otherwise:

1. Replace the sow with another, satisfactory to the buyer.
2. Refund the difference between the purchase price and the market value of the sow as shown by a sales receipt, if the sow is sold at market.
3. If the buyer desires to keep the sow, refund one-half the difference between the purchase price and the market value of the sow at the time of purchase.

POISON INFORMATION CENTERS

With the large number of chemical sprays, dusts, and gases now on the market for use in agriculture, accidents may arise because operators are careless with their use. Also, there is always the hazard that a child may eat or drink something that may be harmful. Centers have been established in various parts of the United States where doctors can obtain prompt and up-to-date information on treatment of such cases, if desired.

The *National Capital Poison Center* is associated with the George Washington University Medical Center, 3201 New Mexico Avenues, NW, Suite 310, Washington, DC 20016. It is open 24 hours a day, every day of the week. The *hot line* number is 1-800-222-1222. The toxicology group is staffed to answer questions about known or suspected cases of poisoning or chemical contaminations involving any species of animal.

GLOSSARY OF SWINE TERMS

The mark of distinction of good swine producers is that they "speak the language"—they use the correct terms and know what they mean. Even though swine terms are spoken commonly by people in the business, often they are confusing to the newcomer.

Many terms that are defined or explained elsewhere in this book are not repeated in this Glossary. Thus, if a particular term is not listed herein, the reader should look in the Index or in the particular chapter and section where it is discussed.

A

Abattoir. A slaughterhouse.

Ablactation. The act of weaning.

Abortion. The expulsion or loss of fetuses before the completion of pregnancy, and before they are able to survive.

Abscess. Localized collection of pus.

Acclimation. Short-term response of animals to their immediate environment.

Acclimatization. Complex of processes of becoming accustomed to a new climate or other environmental conditions.

Acquired immunity. The immunity that an animal generates during its lifetime, either passively or actively.

Active immunity. Immunity created when an animal recovers from natural infection or is vaccinated.

Acute. Referring to a disease that has a rapid onset, short course, and pronounced signs.

Ad libitum. Free-choice access to feed.

Adaptation. Adjustment of an organism to a new or changing environment.

Additive. An ingredient or substance added to a basic feed mix, usually in small quantities, for the purpose of fortifying it with certain nutrients, stimulants, and/or medicines.

Adipose tissue. Fatty tissue.

Adrenal gland. One of the endocrine (ductless) glands of the body, located near the kidney. It secretes hormones needed to use nutrients.

Aerobe. In the presence of air. The term is usually applied to microorganisms that require oxygen to live and reproduce.

A-frame house. A portable hog house shaped like an "A," and used by a sow and her litter on pasture.

Afterbirth. The placenta and allied membranes associated with the fetus that are expelled from the uterus at farrowing.

Agalactia. Failure to secrete milk following parturition.

Agonistic behavior. Combat or fighting behavior.

AI. Abbreviation for artificial insemination.

Air-dry (approximately 90% dry matter). Feed that is dried by means of natural air movement, usually in the open. It may be either an actual or an assumed dry matter content; the latter is approximately 90%. Most feeds are fed in the air-dry state.

Albumen. One of the important proteins of blood, milk, eggs, and other substances.

Alkali. A soluble salt or a mixture of soluble salts present in some soils of arid or semiarid regions in quantities detrimental to ordinary agriculture.

Allelomimetic behavior. Doing the same thing.

All-in, all-out (AIAO). System in which pigs are moved in groups through each of the following stages: (1) farrowing, (2) nursery, and (3) growing-finishing. After each group is moved, the facilities are washed and disinfected. AIAO is used to control a host of swine diseases.

Ambient temperature. The prevailing or surrounding temperature.

American Feed Industry Association, Inc. (AFIA). Nationwide organization of feed manufacturers banded together to (1) improve the quality and promote the use of commercial feeds, (2) encourage high standards on the part of its members, and (3) protect the best interests of the feed manufacturer and the producer in legislative programs. Their address is 1501 Wilson Blvd., Suite 1100, Arlington, VA 22209. (*http://www.afia.org/*)

Amino acids. Nitrogen-containing compounds that constitute the "building blocks" or units from which more complex proteins are formed. They contain both an amino (NH_2) group and a carboxyl (COOH) group.

Anabolism. The conversion of simple substances into more complex substances by living cells (constructive metabolism).

Anaerobe. A microorganism that normally does not require air or free oxygen to live and reproduce.

Animal behavior. The reaction of animals to certain stimuli, or the manner in which they react to their environments.

Animal protein. Protein derived from meat-packing or rendering plants, surplus milk or milk products, and marine sources. It includes proteins from meat, milk, poultry, eggs, fish, and their products.

Animal rights. Concept that maintains that humans are animals, too; and that all animals should be accorded the same protection, including the right to live.

Animal welfare. The well-being, health, and happiness of animals.

Anorexia. A lack or loss of appetite for food.

Anoxia. Lack of oxygen in the blood or tissues. This condition may result from various types of anemia, reduction in the flow of blood to tissues, or lack of oxygen in the air at high altitudes.

Antemortem inspection. The before-slaughter inspection of animals.

Anthelmintic (dewormer). A drug that kills or expels worms.

Antibiotic. A compound synthesized by living organisms, such as bacteria or molds, which inhibits the growth of other bacteria or molds.

Antibody. A protein substance (modified type of blood-serum globulin) developed or synthesized by lymphoid tissue of the body in response to an antigenic stimulus. Each antigen elicits production of a specific antibody. In disease defense, the animal must have an encounter with the pathogen (antigen) before a specific antibody is developed in its blood.

Antigen. A high-molecular-weight substance (usually protein) that, when foreign to the bloodstream of an animal, stimulates formation of a specific antibody and reacts specifically in vivo or in vitro with its homologous antibody.

Antioxidant. A compound that prevents oxidative rancidity of polyunsaturated fats. Antioxidants are used to prevent rancidity in feeds and foods.

Antiseptic. A chemical substance that prevents the growth and development of microorganisms.

Appetite. The immediate desire to eat when food is present. Loss of appetite in an animal is usually caused by illness or stress.

Arch. The convex curvature of the top line of swine.

Arthritis. Inflammation of a joint.

Artificial insemination (AI). The introduction of semen into the female reproductive tract by a technician, using a pipette.

As-fed basis. The feed as normally fed to animals. It may range from 0 to 100% dry matter.

Ash. The mineral matter of a feed. The residue that remains after complete incineration of the organic matter.

Assay. Determination of (1) the purity or potency of a substance or (2) the amount of any particular constituent of a mixture.

Assimilation. A physiological term referring to the group of processes by which the nutrients in feed are made available to and used by the body; the processes include digestion, absorption, distribution, and metabolism.

Atrophy. A wasting away of a part of the body, usually muscular, induced by injury or disease.

Auction markets. Trading centers where animals are sold by public bidding to the buyer who offers the highest price per hundredweight or per head. Auctions may be owned by individuals, partnerships, corporations, or cooperative associations. They are also referred to as sales barns, livestock auction agencies, community sales, and community auctions.

Autosomes. All chromosomes except the sex chromosomes.

Averge daily gain (ADG). The average daily liveweight increase of an animal.

Avoirdupois weights and measures. Avoirdupois is a French word meaning "to weigh." The old English system of weights and measures is referred to as the avoirdupois system, or U.S. customary weights and measures, to differentiate it from the metric system.

B

Backcross. The mating of a crossbred (F_1) animal to one of the parental breeds.

Bacon. Cured and smoked meat from the side of the hog.

Bacteria. Microscopic, single-cell plants, found in most environments, often referred to as microbes; some are beneficial, others are capable of causing disease.

Bactericide. A product that destroys bacteria.

Bacterin. A suspension of killed bacteria (vaccine) used to increase disease resistance.

Balanced diet. Diet that provides an animal the proper amounts and proportions of all the required nutrients.

Barrow. A male hog whose testicles were removed before reaching breeding age, and before the development of secondary sex characteristics.

Basal metabolic rate (BMR). The measured value of heat produced by an animal during complete rest (but not sleeping) following fasting, when using only enough energy to maintain vital cellular activity, respiration, and circulation. Basal conditions include thermoneutral environment, resting, postabsorptive state (digestive processes are quiescent), consciousness, quiescence, and sexual repose. It is determined in humans 14 to 18 hours after eating and when at absolute rest. It is measured by means of a calorimeter and is expressed in calories per square meter of body surface.

Base mix. A dietary supplement usually containing the minerals, vitamins, and additives needed to supplement the grain and protein portion of the diet.

Battery. A series of pens or cages used to house animals in concentrated confinement rearing systems.

Best linear unbiased prediction (BLUP). Statistical procedure that can be used to analyze swine performance data.

Bioavailability. The availability of a substance to the animal, for example, the availability of different vitamin sources to pigs.

Biological value of a protein. The percentage of the protein of a feed or feed mixture that is usable as a protein by the animal. Thus, the biological value of a protein is a reflection of the kinds and amounts of amino acids available to the animal after digestion. A protein that has a high biological value is said to be of *good quality*.

Biosynthesis. The production of new material in living cells or tissues.

Biotechnology. The use of living organisms or parts of organisms, such as enzymes, to make or modify products.

Blind teat. A small, functionless teat.

Bloom.
- Said of an animal that has beauty and freshness. An animal in bloom has a glossy hair coat and an attractive appearance.
- The condition or time of flowering.

Boar. A male hog, generally used for breeding purposes.

Bomb calorimeter. An instrument used to measure the gross energy content of any material, in which the feed (or other substance) tested is placed and burned in the presence of oxygen.

Brand name. Any word, name, symbol, or device, or any combination of these, often registered as a trademark or name, that identifies a product and distinguishes it from others.

Bred.
- Refers to an animal that is pregnant.
- Sometimes used synonymously with the term *mated*.

Breed.
- Animals that are genetically pure enough to have similar external characteristics of color and conformation, and, when mated

together, produce offspring with the same characteristics.
- The mating of animals.

Breed type. The combination of characteristics that makes an animal better suited for a specific purpose.

Breeding and gestating facilities. Usually, the breeding herd (gestating sows and gilts, and herd boars) is the last group to be moved inside. So, the breeding and gestating facilities may be inside or outside.

British thermal unit (Btu). The amount of energy required to raise 1 lb of water 1°F; equivalent to 252 calories.

Brix. A term commonly used to indicate the sugar (sucrose) content of molasses. It is expressed in degrees and was originally used to indicate the percentage by weight of sugar in sucrose solutions, with each degree Brix being equal to 1% sucrose.

Buffer. A substance in a solution that makes the degree of acidity (hydrogen ion concentration) resistant to change when an acid or a base is added.

Bushel. A unit of capacity equal to 2,150.42 cu in. (approximately 1.25 cu ft).

By-product feeds. The innumerable roughages and concentrates obtained as secondary products from plant and animal processing, and from industrial manufacturing.

C

Cake (presscake). The mass resulting from the pressing of seeds, meat, or fish in order to remove oils, fats, or other liquids.

Calcification. The process by which organic tissue becomes hardened by a deposit of calcium salts.

Caloric. Pertaining to heat or energy.

Calorie (cal). The amount of heat as energy required to raise the temperature of 1 g of water from 14.5 to 15.5°C. This is equivalent to 4.185 J. A kilocalorie (kcal) is 1,000 calories and a megacalorie (Mcal) is 1,000,000 calories.

Calorimeter. An instrument for measuring the amount of energy.

Canadian bacon. The bacon that results from the mild curing and light smoking of pork sirloin muscle.

Canned ham. Cured outside the can and cooked inside to reduce shrinkage. A small amount of gelatin is added to set the juices.

Cannibalism.
- The habit of one animal pecking at or eating on another animal, such as a fowl pecking at

or eating on another fowl or one pig biting the tail of another.

- The eating of young, such as a sow may do after farrowing or a doe rabbit may do if disturbed soon after kindling.

Canola. An improved variety that was developed by Canadian scientists in the 1970s. It is low in glucosinolates and low in erucic acid (a long-chain fatty acid).

Carcass. The dressed body of a meat animal, the usual items of offal having been removed.

Carcass weight. Weight of the carcass of an animal following slaughter, as it hangs on the rail, expressed either as warm (hot) or chilled (cold) carcass weight.

Carcass yield. The carcass weight as a percentage of the live-weight.

Carrier.

- A disease-carrying animal.
- A heterozygote for any trait.
- An edible material to which ingredients are added to facilitate their uniform incorporation into feeds.

Carrying capacity. The number of animal units a property or area will carry on a year-round basis. This includes the land grazed plus the land necessary to produce the winter feed.

Castrate.

- To remove the testicles or ovaries.
- An animal that has had its testicles or ovaries removed.

Catabolism. The conversion or breaking down of complex substances into more simple compounds by living cells (destructive metabolism).

Catch. Term used by livestock caretakers to indicate that conception has taken place following breeding.

Centigrade (C). One centigrade (C) degree is 1/100 of the difference between the temperature of melting ice and that of water boiling at standard atmospheric pressure. One centigrade degree equals 1.8°F. To convert to Fahrenheit, multiply by 9/5 and add 32.

Cereal. A plant in the grass family *(Gramineae)*, the seeds of which are used for human and animal food, for example, maize (corn) and wheat.

Chelated mineral. A mineral that is bound to a compound such as protein or an amino acid that helps to stabilize it.

Chemical temperature regulation. Body temperature as maintained by normal body metabolism, for example, the breakdown of nutrients in the body resulting in the production of heat.

Chemotherapeutics. Compounds similar to antibiotics but that are produced chemically rather than microbiologically.

Cholesterol. A white, fat-soluble substance found in animal fats and oils, bile, blood, brain tissue, nervous tissue, the liver, kidneys, and adrenal glands. It is important in metabolism and is a precursor of certain hormones. Cholesterol is implicated in arteriosclerosis.

Chromosomes. Structures that contain DNA, normally in a double helix.

Chronic. Referring to a disease condition that is continuous and long-lasting.

Clinical. Referring to direct observation.

Close breeding. A form of inbreeding, such as brother to sister or sire to daughter.

Coefficient of digestibility. The percentage value of a food nutrient that is absorbed. For example, if a food contains 10 g of nitrogen and it is found that 9.5 g are absorbed, the digestibility is 95%.

Coenzyme. A substance, usually containing a vitamin, that works with an enzyme (protein mainly) to perform a certain function.

Collagen. A white, papery transparent type of connective tissue that is of protein composition. It forms gelatin when heated with water.

Colostrum. The milk secreted by mammalian females for the first few days following parturition that is high in antibodies and is laxative.

Combustion. The combination of substances with oxygen accompanied by the liberation of heat.

Commercial feeds. Feeds mixed by manufacturers who specialize in the feed business.

Commodity exchange. A place where buyers and sellers meet on an organized market and transact business on paper, without the physical presence of the commodity.

Complete diet. All feedstuffs (forages and grains) combined in one feed. A complete diet fits well into mechanized feeding and the use of computers to formulate least-cost diets.

Complimentary. The advantage of one cross over another cross or over a purebred, resulting from the manner in which two or more traits combine or complement each other.

Concentrate. A broad classification of feedstuffs that are high in energy and low in crude fiber (less than 18%). For convenience, concentrates are often broken down into (1) carbonaceous feeds and (2) nitrogenous feeds.

Condition.

- The state of health, as evidenced by the coat and general appearance.
- The amount of flesh or finish (fat covering).

Conformation. The shape and design of an animal.

Contagious. Transmissible by contact.

Contentment. A stress-free condition exhibited by healthy animals; the cow will stretch on rising, the sheep will stand or lie quietly, the pig will curl its tail, and the horse will look completely unworried when resting.

Contract. An agreement between two or more persons to do or refrain from doing certain things. The most common kinds of hog contracts are
1. Marketing contracts
    a. Marketing contracts for slaughter hogs.
    b. Marketing contracts for feeder pigs.
2. Production contracts
    a. Farrow-to-finish contracts.
    b. Feeder pig production contracts.
    c. Finishing contracts.
3. Breeding stock leasing

Cooked. Heated to alter chemical or physical characteristics or to sterilize.

Cooperative (co-op). Business formed by a group of people to get certain services for themselves more effectively or more economically than they can get individually.

Corporation. Business device that may be used for carrying out a pork enterprise as an entity distinct from the persons who are interested in and control it.
- **Family-owned corporation** A privately owned corporation.
- **Tax-option corporation (subchapter S corporation)** Type of corporation that cannot have more than 35 members. The income or losses are passed back to the shareholders, each of whom pays his or her share of taxes.

Country ham (dry-cured). Ham produced using a dry-cured, slow-smoking, and long-drying process. Country hams are heavily salted and may require soaking and simmering before roasting. They may be called Virginia or Georgia (or another state name) ham, depending on the state of origin.

Cracked. Particle size reduced by combined breaking and crushing action.

Creep. An enclosure or feeder used for supplemental feeding of nursing young, which excludes their dams.

Creep feeding. Providing a supplemental source of feed for nursing animals in an area isolated from the dam.

Cripples. Animals that are too lame or injured to walk.

Crossbreeding. The mating of animals of different breeds.

Crude fat. Material extracted from moisture-free feeds by ether. It consists largely of fats and oils with small amounts of waxes, resins, and coloring matter. In calculating the energy value of a feed, the fat is considered to have 2.25 times as much energy as either nitrogen-free extract or protein.

Cryptorchid. A male that is often sterile because one or both testicles are retained in the abdominal cavity.

Cull. An animal taken out of a herd because it is below herd standards.

Cutting.
- Removing the testicles.
- Separating one or more animals from a herd.

Cyclical movements. Basically, cycles are the response of producers to prices, and prices reflect supply and demand.

D

Decortification.
- Removal of the bark, hull, husk, or shell from a plant, seed, or root.
- Removal of portions of the cortical substance of a structure or organ, as in the brain, kidneys, and lung.

Defecation. The evacuation of fecal material from the rectum.

Deficiency disease. A disease caused by a lack of one or more basic nutrients, such as a vitamin, a mineral, or an amino acid.

Dehydrate. To remove most or all moisture from a substance for the purpose of preservation, primarily through artificial drying.

Deoxyribonucleic acid (DNA). The genetic information source. The DNA molecule has a double-helical structure.

Dermatitis. Inflammation of the skin.

Desiccate. To dry completely.

Dewclaw. The false hoof of some mammals such as deer, cattle, and hogs, consisting of two rudimentary toes.

Dewormer. Compound that helps to control internal or external parasites (anthelmintic).

Dextrose. Form of glucose found naturally in animal and plant tissue and derived synthetically from starch.

Diet. Feed ingredient or mixture of ingredients, including water, that is consumed by animals.

Digestible nutrient. The part of each feed nutrient that is digested or absorbed by the animal.

Digestible protein. That protein of the ingested food protein that is absorbed.

Digestion coefficient (coefficient of digestibility). The difference between the nutrients consumed and the nutrients excreted expressed as a percentage.

Direct selling. Producers selling hogs directly to packers or local dealers, including sale to country dealers, without the support of commission firms, selling agents, buying agents, or brokers.

Disease. Any departure from the state of health.

Disinfectant. A chemical capable of destroying disease-causing microorganisms or parasites.

Diuresis. An increased excretion of urine.

Diuretic. An agent that increases the flow of urine.

DNA. *See deoxyribonucleic acid.*

Dock.
- To cut off the tail.
- To reduce the weight and/or price, for example, piggy sows are usually docked 40 lb and stags 70 lb when marketed.

Dominant. Describes a gene that, when paired with its allele, covers up the phenotypic expression of that gene.

Dressing percentage. The percentage of the live animal that becomes the carcass at slaughter.

Dried. Materials from which water or other liquids have been removed.

Drugs. Substances of mineral, vegetable, or animal origin used in the relief of pain or for the cure of disease.

Dry. Nonlactating female. The dry period is the time between lactations (when a female is not secreting milk).

Dry matter basis. A method of expressing the level of a nutrient contained in a feed on the basis that the material contains no moisture.

Drylot. A relatively small enclosure without vegetation, either (1) with shelter or (2) on an open yard, in which animals may be confined.

Dry-rendered. Residues of animal tissues cooked in open, steam-jacketed vessels until the water has evaporated. Fat is removed by draining and pressing the solid residue.

Dust. A mixture of small particles of different sizes of dry matter.

E

Ear notching. Making slits or perforations in an animal's ears for identification purposes.

Early maturing. Completing sexual development at an early age.

Early weaner pig marketing (SEW pigs). Marketing segregated early weaned (SEW) pigs—pigs weaned at less than 3 weeks of age, rather than the more conventional 21 to 28 days of age; then, segregating the piglets from their mothers.

Early weaning. The practice of weaning young animals earlier than usual; weaning pigs less than 3 weeks of age.

Easy keeper. An animal that grows or fattens rapidly on limited feed.

Ecology. The branch of science concerned with the relations of living things to their environments and to each other.

Edema. Swelling of a part or all of the body due to the accumulation of excess water.

Efficiency of feed conversion. Units of feed per unit of live-weight or meat produced.

Electrolyte. A chemical compound that in solution dissociates by releasing ions. An ion is an atomic particle that carries a positive (+) or a negative (−) charge.

Electronic feeder pig auctions. Feeder pigs offered for sale are scored on the basis of health programs, feeding programs, and herd management. This information, along with the identification of the seller, is made available to prospective buyers via computer modems. Consigned feeder pigs may be reviewed by potential buyers. Buyers participate in an open auction. Pigs move directly from the seller to the buyer, thereby minimizing disease and stress.

Element. One of the 118 known chemical substances that cannot be divided into simpler substances by chemical means.

Emaciated. Excessive loss of flesh.

Endocrine. Pertaining to glands and their secretions that pass directly into the blood or lymph instead of into a duct (secreting internally). Hormones are secreted by endocrine glands.

Endogenous. Originating within the body, for example, hormones and enzymes.

Energy.
- Vigor or power in action.
- Capacity to perform work.

Energy feeds. Feeds that are high in energy and low in fiber (less than 18%), and that generally contain less than 20% protein.

Enteritis. Inflammation of the intestines.

Environment. The sum total of all external conditions that affect the life and performance of a pig.

Ergosterol. A plant sterol that, when activated by ultraviolet rays, becomes vitamin D_2. It is also called provitamin D_2 and ergosterin.

Ergot. A fungus disease of plants.

Essential amino acids. Those amino acids that cannot be made in the body from other substances or that cannot be made in sufficient quantity to supply the animal's needs.

Essential fatty acid. A fatty acid that cannot be synthesized in the body or that cannot be made in sufficient quantities for the body's needs.

Estimated breeding value (EBV). The entire worth of the parent as a source of genetic material.

Estimated progeny difference (EPD). Half the genetic worth of each parent; an estimate of how much of their performance will be passed on to their offspring. EPD is a prediction of the progeny performance of an animal compared to the progeny of an average animal in the population, based on all information currently available.

Estrus. The period when the gilt or sow will accept service by the boar.

Ether extract (EE). Fatty substances of feeds and foods that are soluble in ether.

Euthanasia. Causing a painless death.

Evaporated. Reduced to a denser form; concentrated as by evaporation or distillation.

Excreta. The products of excretion—primarily feces and urine.

Exogenous. Provided from outside of the organism.

Experiment. The word *experiment* is derived from the Latin *experimentum*, meaning proof from experience. It is a procedure used to discover or to demonstrate a fact or general truth.

Extralabel drugs. Use of over-the-counter drugs in therapies or dosages not approved by the labeling constitutes extralabel drug use.

Extrinsic factor. A dietary substance that was formerly thought to interact with the intrinsic factor of the gastric secretion to produce the antianemic factor, now known to be vitamin B_{12}. (Also see intrinsic factor.)

F

Fahrenheit. One Fahrenheit (F) degree is 1/180 of the difference between the temperature of melting ice and that of water boiling at standard atmospheric pressure. One Fahrenheit degree equals 0.556°C.

Farrow. To give birth to piglets.

Farrowing house. Central farrowing houses are generally environmentally controlled.

Farrow-to-finish. A type of hog farm operation that covers all aspects of breeding, farrowing, and raising pigs to slaughter.

Fat. Term frequently used in a general sense to include both fats and oils, or a mixture of the two. Both fats and oils have the same general structure and chemical properties, but they have different physical characteristics. The melting points of most fats are such that they are solid at ordinary room temperatures, whereas oils have lower melting points and are liquids at these temperatures.

Fattening. The deposition of energy in the form of fat within the body tissues.

Fatty acids. The key components of fats (lipids). Their degree of saturation (hydrogenation) and the length of their carbon chains determine many of the physical aspects—melting point and stability—of fats (lipids).

Feces. The excreta discharged from the digestive tract through the anus.

Fecundity. Ability to produce many offspring.

Feed (feedstuff). Any naturally occurring ingredient, or material, fed to animals for the purpose of sustaining them.

Feed additive. An ingredient or a substance added to a feed to improve the rate or efficiency of gain of animals, prevent certain diseases, or preserve feeds.

Feed efficiency. The ratio expressing the number of units of feed required for one unit of production (meat) by an animal. This value is commonly expressed as pounds of feed eaten per pound of gain in body weight.

Feed grain. Any of several grains most commonly used for livestock or poultry feed, such as corn, sorghum, oats, and barley.

Feed out. To feed a pig until it reaches market weight.

Feeder pig production. The production and sale of 30- to 60-lb pigs for growing and finishing on other farms.

Feeder's margin. The difference between the cost per hundredweight of feeder animals and the selling price per hundredweight of the same animals when finished.

Feedlot. A lot or plot of land on which animals are fed or finished for market.

Feedstuff. Any product, of natural or artificial origin, that has nutritional value in the diet when properly prepared.

Feral. Referring to domesticated animals that have reverted back to their original or untamed state.

Fermentation. Chemical changes brought about by enzymes produced by various microorganisms.

Fetus. A young organism in the uterus from the time the organ systems develop until birth.

Fiber content of a feed. The amount of hard-to-digest carbohydrates. Most fiber is made up of cellulose and lignin.

Fill.
- A term designating the fullness of the digestive tract of an animal.
- With market animals, the amount of feed and water consumed on their arrival at the market and prior to selling.

Fines. Finely ground particles present in a feed usually resulting from overgrinding ingredients.

Finishing pigs. The phase in the life cycle of market hogs from approximately 120 lb to market weight.

Fitting. The conditioning of an animal for show or sale; usually involves a combination of special feeding plus exercise and grooming.

Flora. The plant life present. In nutrition, it generally refers to the bacteria present in the digestive tract.

Flushing. The practice of feeding females more generously 1 to 2 weeks before breeding so that they gain in weight from 1.0 to 1.5 lb daily. The beneficial effects attributed to this practice are (1) more eggs (ova) are shed, resulting in more offspring; (2) females come in heat more promptly; and (3) conception is more certain.

Folacin (folic acid/folate). This includes a group of compounds with folic acid activity. Folic acid participates in many enzymatic reactions.

Following cattle. The former practice of allowing feeder pigs to run behind feedlot cattle so they may glean unused grains and other nutrients from the cattle manure.

Food and Drug Administration (FDA). (*http:// www.fda.gov/*) The federal agency in the Department of Health and Human Services that is charged with the responsibility of safeguarding U.S. consumers against injury, unsanitary food, and fraud. It protects industry against unscrupulous competition, and it inspects and analyzes samples and conducts independent research on such things as toxicity (using laboratory animals), disappearance curves for pesticides, and long-range effects of drugs.

Food safety labeling. The food labeling requirement became effective June 6, 1994. It mandates safe cooking and handling labels for all uncooked meat and poultry products.

Fortify. Nutritionally, to add one or more feeds or feedstuffs.

Free choice. Free to eat from a combination of feeds at will.

Freedom stalls. System of gestation housing that preserves the advantages of individual living places but permits more freedom than traditional crates, stalls, or tethering.

Freeze drying. See lyophilization.

Full-feed. The term indicating that animals are being provided as much feed as they will consume safely without going off feed.

Full-sib mating. The mating of a brother to his sister.

Fumigant. A liquid or solid substance that forms vapors that destroy pathogens, insects, and rodents.

Fungi. Plants that contain no chlorophyll, flowers, or leaves, such as molds, mushrooms, toadstools, and yeasts. They may get their nourishment from either dead or living organic matter.

Futures contract. Standardized, legal, binding paper transaction in which the seller promises to make delivery and the buyer promises to take delivery on a specified quantity and type of commodity at a specified location(s) during a specified future month.

Futures trading. The futures market is a way in which to provide (1) an insurance medium in the marketing field and (2) the facilities and machinery for underwriting price risks.

G

Gastrointestinal. Pertaining to the stomach and intestines.

Gene. A segment of DNA that carries inherited information.

Genome. All the inherited information in a cell.

Genotype. An animal's true (genetic) makeup.

Gestation (pregnancy). Time between breeding and farrowing, about 114 days for swine.

Get. The offspring of a male animal—his progeny.

Gilt. Young female, less than 1 year of age, that has not had her first litter.

Girth. The circumference of the body of an animal behind the shoulders.

Glucose. A hexose monosaccharide obtained upon the hydrolysis of starch and certain other carbohydrates. Also called dextrose.

Goitrogenic. Producing or tending to produce goiter.

Gossypol. A toxic yellow pigment found in cottonseed, which is toxic to swine and certain other nonruminants, and which may cause discoloration of egg yolks during cold storage.

Grade.
- An animal having parents that cannot be registered by a breed association.
- A measure of how well an animal or product fulfills the requirements for the class, for example, the federal grades of hogs and their carcasses are a specific indication of the degree of excellence.

Grading up. The continued use of purebred sires of the same breed in a grade herd.

Grain. Seed from cereal plants.

Grind. To reduce to small segments by impact, shearing, or attrition (as in a mill).

Groats. Grain from which the hulls have been removed.

Ground pork. At least 70% lean and no seasonings.

Grow out. To feed animals so that they attain a certain desired amount of growth with little or no fattening.

Growing and finishing facilities. Pigs are usually moved from the nursery to the growing facilities where they remain until they weigh 120 lb, then to the finishing facilities from 120 lb to market weight.

Growing-finishing. The stage of pig production beginning at 30 or 60 pounds body weight until slaughter at 260 to 280 pounds.

Growth. The increase in size of the muscles, bones, internal organs, and other parts of the body.

Growthy. Describes an animal that is large and well developed for its age.

Gruel. A feed prepared by mixing ground ingredients with hot or cold water.

H

Habituation. The act or process of making animals familiar with or accustomed to a new environment through use or experience.

Half-sib mating. The mating of a boar to his half-sister, that is, a gilt with either the same sire or the same dam.

Ham. The thigh of a hog prepared for food, or the hind leg of a swine from the hock upward on the live animal.

Hand-feeding. To provide a certain amount of a diet at regular intervals.

Hand-mating. Controlled breeding with confined boars rather than allowing boars to run loose with groups of unbred sows.

Hard keeper. An animal that is unthrifty and grows or fattens slowly regardless of the quantity or quality of feed.

Hazard analysis and critical control point (HACCP). The HACCP plan identifies hazards in food processing, followed by monitoring those crucial points. Problems are remedied as they occur. It is a way of processing safer food.

Health. The state of complete well-being, not merely the absence of disease.

Heat (estrus). The period when the female will accept service by the boar.

Heat increment (HI). The increase in heat production following consumption of feed when the animal is in a thermoneutral environment.

Heat labile. Unstable to heat.

Hedging. Offsetting transaction in which purchases or sales of a commodity are counterbalanced by sales or purchases of an equivalent quantity of futures contracts in the same commodity.

Hemoglobin. The oxygen-carrying, red-pigmented protein of the red blood cells.

Heritability. A measure of the proportion of phenotypic variation that is based on the additive effects of genes.

Hernia. The protrusion of some of the intestine through an opening in the body wall—commonly called a rupture.

Heterosis (hybrid vigor). Amount the F_1 generation exceeds the P_1 generation for a given trait, or the amount the crossbreds exceed the average of the 2 pure breeds that are crossed to produce the crossbreds.

Heterozygous. Two pairs of a gene that are not the same.

High-lysine corn (opaque-2). Corn that is much higher than normal corn in lysine and tryptophan; hence, it has a better balance of the amino acids for monogastric animals. Also, high-lysine corn is higher in total protein, but lower in leucine than regular corn.

Hog. A large or mature animal of either sex, generally weighing more than 120 lb.

Hog–corn ratio. Relationship determined by dividing the price of live hogs per cwt by the price of corn per bushel.

Hog down. Practice of allowing pigs to "harvest" a crop in the field.

Homogenized. Having particles broken down into evenly distributed globules small enough to remain emulsified for long periods of time.

Homologous chromosomes. Chromosomes having the same size and shape, and containing the genes affecting the same characteristics.

Homozygous, homozygosity. Genes are called homozygous when both pairs of the gene are identical. Homozygosity is the increasing appearance of homozygous genes that result from inbreeding or linebreeding.

Hormone. A body-regulating chemical secreted by an endocrine gland into the bloodstream, then transported to another region within the animal where it elicits a physiological response.

Hull. Outer covering of grain or other seed, especially when dry.

Hydrogenation. The chemical addition of hydrogen to any unsaturated compound.

Hydrolysis. The splitting of a substance into the smaller units by chemically adding water to the material.

Hypertrophied. Having increased in size beyond the normal growth.

Hypervitaminosis. An abnormal condition resulting from the intake of an excess of one or more vitamins.

Hypocalcemia. Below normal concentration of ionic calcium in blood resulting in convulsions, as in tetany or parturient paresis (milk fever).

Hypoglycemia. A reduction in concentration of blood glucose below normal.

Hypomagnesemia. An abnormally low level of magnesium in the blood.

Hypothalamus. A portion of the brain found in the floor of the third ventricle. It regulates body temperature, appetite, hormone release, and other functions.

I

Ideal protein. A protein that provides a perfect pattern of essential and nonessential amino acids in the diet without any excesses or deficiencies.

Immunity. The ability of an animal to resist or overcome an infection to which most members of its species are susceptible.

Immunoglobulins. A family of proteins found in body fluids that have the property of combining with antigens, and, when the antigens are pathogenic, sometimes inactivating them and producing a state of immunity. Also called antibodies.

In vitro. Occurring in an artificial environment, as in a test tube.

In vivo. Occurring in the living body.

In pig. A pregnant sow or a gilt.

Inbreeding. The mating of individuals that are more closely related than average individuals in a population. It increases homozygosity.

Industrialization of hog production. The production of hogs in large specialized facilities staffed with specialized labor.

Inflammation. The reaction of tissue to injury, characterized by redness, swelling, pain, and heat.

Ingest. To eat or take in through the mouth.

Ingesta. Food or drink taken into the stomach.

Ingestion. The taking in of food and drink.

Ingredient. A constituent feed material.

Insulin. A hormone secreted by the pancreas into the blood that regulates sugar (glucose) metabolism.

International unit (IU). A standard unit of potency of a biologic agent (e.g., a vitamin, a hormone, an antibiotic, an antitoxin) as defined by the International Conference for Unification of Formulae. Potency is based on bioassay that produces a particular effect agreed on internationally. Also called a U.S. pharmacopaeia unit. This type of measure is used for the fat-soluble vitamins (such as vitamins A, D, and E) and certain hormones, enzymes, and biologicals (such as vaccines).

Intradermal. Into, or between, the layers of the skin.

Intramuscular. Within the muscle.

Intraperitoneal. Within the peritoneal cavity.

Intravenous. Within the vein or veins.

Intrinsic factor. A chemical substance secreted by the stomach that is necessary for the absorption of vitamin B_{12}. The exact chemical nature of intrinsic factor is not known, but it is thought to be a mucoprotein or mucopolysaccharide. A deficiency of this factor may lead to a deficiency of vitamin B_{12}, and, ultimately, to pernicious anemia. (Also see extrinsic factor.)

Involution. Return of an organ to its normal size and condition after enlargement, as of the uterus after farrowing.

Irradiated yeast. Yeast that has been irradiated. Yeast contains considerable ergosterol, which, when exposed to ultraviolet light, produces vitamin D.

Irradiation. Exposure to ultraviolet light.

J

Joule (J). Proposed international unit (4.184 J = 1 calorie) for expressing mechanical, chemical, or electrical energy, as well as the concept of heat. In the future, energy requirements and feed values will likely be expressed by this unit.

Jowl. Meat from the cheeks of hogs.

K

Kernel. The whole grain of a cereal. The meat of nuts and drupes (single-stoned fruits).

Killed vaccine. A vaccine in which the antigen has been inactivated so that it cannot produce the disease. Toxoids and subunit vaccines also fall in this category.

Kjeldahl. A method of determining the amount of nitrogen in an organic compound. The quantity of nitrogen measured is then multiplied by 6.25 to calculate the protein content of the feed or compound analyzed. The method was developed by a Danish chemist, J. G. C. Kjeldahl, in 1883.

L

Labile. Unstable. Easily destroyed.

Lactation. The period in which an animal is producing milk.

Lactose (milk sugar). A disaccharide found in milk, having the formula $C_{12}H_{22}O_{11}$. It hydrolyzes to glucose and galactose. Commonly known as milk sugar.

Lagoon. A waste management treatment unit—a digester, for the purpose of biochemical breakdown of organic wastes (manure, straw).

Lard. Fat rendered (melted out) from fresh pork tissue.

Larva. The immature form of insects and other small animals.

Laxative. A feed or drug that will induce bowel movements and relieve constipation.

Lean cuts. Ham, loin, Boston butt, and picnic.

Lean meter. A large, precisely wound coil of wire through which the carcass passes.

Limit feeding. Feeding animals less than they would like to eat. Giving sufficient feed to maintain weight and growth, but not enough for their potential production or finishing.

Limited partnership. An arrangement in which two or more parties supply the capital, but only one partner is involved in the management.

Limiting amino acid. The essential amino acid of a protein that shows the greatest percentage deficit in comparison with the amino acids contained in the same quantity of another protein selected as a standard.

Linebreeding.   A form of inbreeding that attempts to concentrate the inheritance of some ancestor in the pedigree.

Linecross.   A cross of two inbred lines.

Lipolysis.   The hydrolysis of fats by enzymes, acids, alkalis, or other means to yield glycerol and fatty acids.

Litter.   The pigs farrowed by a sow at one delivery. Such individuals are called *littermates.*

Live vaccine.   A vaccine in which the live organism has been altered (attenuated) so that it can no longer cause disease but can replicate in the animal and stimulate an immune response.

Live-weight.   Weight of an animal on foot.

Liver abscesses.   Single or multiple abscesses on the liver, observed at slaughter. Usually, the abscess consists of a central mass of necrotic liver surrounded by pus and a wall of connective tissue. At slaughter, those livers affected with abscesses are condemned for human food.

Loin.   That portion of the back between the thorax and pelvis.

Loin eye area.   Cross-section of the pork chop muscle, usually measured between the 10th and 11th ribs.

Lower critical temperature (LCT).   This is the low point of the cold temperature beyond which the animal cannot maintain normal body temperature.

Lumen.   The cavity inside a tubular organ—the lumen of the stomach or intestine.

Lymph.   The slightly yellow, transparent fluid occupying the lymphatic channels of the body.

Lyophilization.   The evaporation of a liquid from a frozen product with the aid of a high vacuum. Also called freeze drying.

M

Macrominerals.   The major minerals—calcium, phosphorus, sodium, chlorine, potassium, magnesium, and sulfur.

Maintenance requirement.   A ration that is adequate to prevent any loss or gain of tissue in the body when there is no production.

Malnutrition.   Any disorder of nutrition. Commonly used to indicate a state of inadequate nutrition.

Maltose.   A disaccharide, also known as malt sugar, having the formula $C_{12}H_{22}O_{11}$. Obtained from the partial hydrolysis of starch. It hydrolyzes to glucose.

Management, swine.   The art of caring for and handling swine.

Mangy.   Infected with a skin disease or parasite so that the skin is dry and scaly.

Manure.   A mixture of animal excrements (consisting of undigested feeds plus certain body wastes) and bedding.

Manure gases.   The common gases produced by the decomposition of manure are methane, ammonia, hydrogen sulfide, and carbon dioxide. When methane and carbon dioxide displace oxygen, people can be killed.

Margin (spread).   The difference between the purchase price and the selling price.

Marker.   A part of DNA that provides for the detection of genetic variation from animal to animal.

Market class.   Animals grouped according to the use to which they will be put, such as slaughter or feeder.

Market grade.   Animals grouped within a market class according to their value. It is a specific indication of the degree of excellence within a given class based on conformation, finish, and quality.

Marketing contract.   An agreement between a seller (usually a producer) and a meat packer to sell/buy at a specified future date a specified number of hogs of a specified weight and grade for a specified price.

Mash.   A mixture of ingredients in meal form.

Mast.   Nuts, such as oak, beech, or chestnut, used as fodder for pigs.

Mastication.   The chewing of feed.

Mastitis.   Inflammation of the mammary gland.

Meal.

- A feed ingredient having a particle size somewhat larger than flour.
- Mixture of concentrate feeds, usually in which all of the ingredients are ground.

Meats.

- Animal tissues used as food.
- The edible parts of nuts and fruits.

Meconium.   Excrement accumulated in the bowels during fetal development.

Medicated feed.   Any feed that contains drug ingredients intended or represented for the cure, mitigation, treatment, or prevention of diseases of animals (other than humans).

Metabolism.   All the changes that take place in the nutrients after they are absorbed from the digestive tract, including (1) the building-up processes in which the absorbed nutrients are used in the formation or repair of body tissues, and (2) the breaking-down processes in which nutrients are oxidized for the production of heat and work.

Microbe.   Same as microorganism.

Microchips, swine.   Mechanisms for electronic swine tracking, the goal of which is to assign a number to each farm and each pig. Ultimately, meat retailers will be able to trace meat to its source. Such chips are now being tested.

Microflora. Microbial life characteristic of a region, such as the bacteria and protozoa populating the rumen.

Microingredient. Any dietary component, such as minerals, vitamins, antibiotics, and drugs, normally measured in mg or mcg per kg, or in parts per million.

Microminerals. Minerals required by animals in small amounts—mg/lb or smaller units. These are also called *trace elements*, or *trace minerals*.

Microorganism. Any organism of microscopic size, applied especially to bacteria and protozoa.

Milk ejection or "let-down." The process, controlled by the hormone oxytocin, in which milk is forced from the alveoli, where it is stored, into the larger ducts and cisterns, where it is available to suckling pigs.

Mill by-product. A secondary product obtained in addition to the principal product in milling practice.

Mineral supplement. A rich source of one or more of the inorganic elements needed to perform certain essential body functions.

Minerals (ash). The inorganic elements of animals and plants, determined by burning off the organic matter and weighing the residue, which is called ash.

Miniature swine. Genetically small pigs. They are good experimental animals for biomedical studies.

Modular nursery. Self-contained portable nurseries that are constructed by the manufacturer before being delivered to the farm.

Moisture. A term used to indicate the water contained in feeds—expressed as a percentage.

Moisture-free (MF, oven-dry, 100% dry matter). Any substance that has been dried in an oven at 221°F (105°C) until all the moisture has been removed.

Molds (fungi). Fungi that are distinguished by the formation of mycelium (a network of filaments or threads) or by spore masses.

Morbidity. A state of sickness or the rate of sickness.

Mortality. Death or death rate.

Mule foot. Swine hoof having the shape of a mule's foot—not halved.

Multiple farrowing. The practice of breeding sows to farrow pigs throughout the year, thereby marketing hogs more frequently.

Mummified fetus. Shriveled or dried fetus that remains in the uterus instead of being expelled or aborted after dying.

Mutation. Sudden variation that results from changes in a gene or genes that is later passed on through inheritance.

Mycotoxins. Toxic metabolites produced by molds during growth. Sometimes present in feed materials.

N

Nano. A prefix meaning one billionth (10^{-9}).

National Pork Board (NPB). The NPB is the largest commodity organization in the United States with an identified membership. Headquartered in Des Moines, Iowa, the board's purpose is to enhance the quality, production, distribution, and sale of pork and pork products.

National Pork Producers Council (NPPC). The NPPC is the pork producers' lobbying voice and advocate to solve problems efficiently for the industry. It is headquartered in Washington, DC.

National Research Council (NRC). A division of the National Academy of Sciences established in 1916 to promote the effective use of scientific and technical resources. Periodically, this private, nonprofit organization of scientists publishes bulletins giving nutrient requirements and allowances of domestic animals, copies of which are available on a charge basis through the National Academy of Sciences, National Research Council, 2101 Constitution Avenue, NW, Washington, DC 20418.

Native defense system. Qualities in a normal, healthy animal that help it fight disease, including the skin and mucous membranes, stomach acid, gut bacteria, enzymes, and types of white blood cells.

Necropsy. An examination of the internal organs of a dead animal to determine the apparent cause of death—an autopsy or a postmortem.

Necrosis. Death of tissue.

Needle teeth. Eight small sharp teeth on the upper and lower corners of a baby pig's mouth.

Neonate. A newborn.

Nephritis. Inflammation of the nephrons of the kidneys.

Networking. Two or more people working together to achieve individual and group goals that would be more difficult to obtain alone. A banding together of small and independent pork producers to compete with the large integrated pork producers.

Nick. The result of a certain mating that produces an animal of high order of excellence, sometimes from mediocre parents.

Nitrate/nitrite. Nitrate refers to the chemical union of 1 nitrogen (N) and 3 oxygen (O) atoms, or NO_3, whereas nitrite refers to the chemical union of 1 nitrogen (N) and 2 oxygen (O) atoms, or NO_2. Of prime concern in foods are sodium nitrate, potassium nitrate, sodium nitrite, and potassium nitrite.

Nitrogen.   A chemical element essential to life. Animals get it from protein feeds; plants get it from the soil; and some bacteria get it directly from the air.

Nitrogen balance.   The nitrogen in the feed intake minus the nitrogen in the feces, minus the nitrogen in the urine.

Nitrogen fixation.   Conversion of free nitrogen of the atmosphere to organic nitrogen compounds by symbiotic or nonsymbiotic microbial activity.

Nitrogen-free extract (NFE).   Principally, a combination of sugars, starches, pentoses, and nonnitrogenous organic acids. The percentage NFE is determined by subtracting the sum of the percentages of moisture, ether extract, crude protein, crude fiber, and ash from 100.

Nitrosamines.   Nitrates and nitrites contribute to the formation of cancer-causing nitrosamines. Foods are not the only source of nitrosamines.

Nonprotein nitrogen (NPN).   Nitrogen that comes from other than a protein source but may be used by a ruminant in the building of protein. NPN sources include compounds such as urea and anhydrous ammonia, which are used in feed formulations for ruminants only.

Nursery.   The building to which pigs are usually moved at weaning. Normally, they remain in the nursery until they weigh 40 to 60 lb.

Nutrient allowances.   Feeding recommendations that allow for variations in feed composition; possible losses during storage and processing; day-to-day and period-to-period differences in needs of animals; age and size of animal; stage of gestation and lactation; the kind and degree of activity; the amount of stress; the system of management; the health, condition, and temperament of the animal; and the kind, quality, and amount of feed—all of which exert a powerful influence in determining nutritive needs. Nutrient allowances are higher than nutrient requirements.

Nutrient requirements.   This refers to meeting the animal's minimum needs to optimize growth, without margins of safety, for maintenance, growth, reproduction, lactation, and work. To meet these nutritive requirements, the different classes of animals must receive sufficient feed to furnish the necessary quantity of energy (carbohydrates and fats), proteins, minerals, and vitamins.

Nutrients.   The chemical substances found in feed materials that can be used, and are necessary, for the maintenance, production, and health of animals. The chief classes of nutrients are carbohydrates, fats, proteins, minerals, vitamins, and water.

Nutrition.   The science encompassing the sum total of processes that have as a terminal objective the provision of nutrients to the component cells of an animal.

Nutritive ratio (NR).   The ratio of digestible protein to other digestible nutrients in a feedstuff or diet. (The NR of shelled corn is about 1:10.)

O

Offal.   All organs or tissues removed from the carcass in slaughtering.

Oil.   Although fats and oils have the same general structure and chemical properties, they have different physical characteristics. The melting points of oils are such that they are liquid at ordinary room temperatures.

Oil crops.   Crops grown primarily for oil, including soybeans, cottonseed, peanuts, canola, flaxseed, sunflower seed, safflower, and castor bean.

Oiling.   The application of oil to the coat of an animal to soften the skin and hair and give the coat a desired gloss. This practice is prohibited in most swine shows.

Optimum temperature.   Temperature at which the animal responds most favorably, as determined by maximum production and feed efficiency.

Option.   A choice; the right, but not the obligation, to buy or sell something, at a specified price on or before a certain expiration date.

Ossification.   The process of bone formation; the calcification of bone with advancing maturity.

Osteitis.   Inflammation of a bone.

Osteomalacia.   A bone disease of adult animals caused by lack of vitamin D, inadequate intake of calcium or phosphorus, or an incorrect dietary ratio of calcium and phosphorus.

Osteoporosis.   Abnormal porosity and fragility of bone as the result of (1) a calcium, phosphorus, or vitamin D deficiency, or (2) an incorrect ratio between the two minerals.

Outcross.   The introduction of genetic material from some outside and unrelated source, but of the same breed, into a herd that is more or less related.

Outdoor intensive swine production.   A term in vogue in Britain in the 1990s. It refers to outdoor production that is modernized by incorporating some of the improvements of confinement and environmentally controlled systems.

Overfeeding.   Excess feeding.

Overfinishing.   Excess finishing or fatness—a wasteful practice.

Over-the-counter (OTC) drugs.   Drugs that may be purchased by the general public for application according to the label.

Oxidation. The combination with oxygen, or the loss of a hydrogen, or the loss of an electron, all of which render an ion more electropositive. The animal combines carbon from feedstuffs with inhaled oxygen to produce carbon dioxide, energy (as ATP), water, and heat.

Oxytocin. The hormone that controls milk letdown.

P

Palatability. Factors sensed by the animal that results in locating and consuming feed: appearance, odor, taste, texture, temperature, and, in some cases, auditory properties of the feed (such as the sound of pigs eating corn). These factors are affected by the physical and chemical nature of the feed.

Pale, soft, and exudative (PSE) pork. Related to the porcine stress syndrome (PSS).

Pantothenic acid. One of the B vitamins. It is a constituent of coenzyme A, which plays an essential role in fat and cholesterol synthesis.

Parasites. Organisms living in, on, or at the expense of another living organism.

Partnership. An association of two or more persons who, as co-owners, operate the business.

Parts per billion (ppb). It equals micrograms per kilogram or microliters per liter.

Parts per million (ppm). It equals milligrams per kilogram or milliliters per liter.

Parturition. The act of giving birth—farrowing.

Passive immunity. Short-lived immunity an animal generates by drinking colostrum, or by receiving hyperimmune serum via injection or orally.

Pathogenic. Disease causing.

Pearled. Dehulled grains that are reduced into smaller and smoother particles by machine brushing, or abrasion.

Pedigree. A written statement giving the record of an animal's ancestry.

Pen-mating. Breeding sows or gilts by releasing one or more boars into a pen with the females. The time of the matings is usually not known.

Per os. Oral administration (by the mouth).

Performance testing. The evaluation of an animal by its own performance.

Pest. An organism declared to be a pest under circumstances that make it deleterious to man or the environment.

Pesticide. Any substance that is used to control pests.

pH. A measure of the acidity or alkalinity of a solution. Values range from 0 (most acid) to 14 (most alkaline), with neutrality at pH 7.

Phase feeding. Refers to changes in the animal's diet (1) to adjust for age and stage of production, (2) to adjust for season of the year and for temperature and climatic changes, (3) to account for differences in body weight and nutrient requirements of different strains of animals, or (4) to adjust one or more nutrients as other nutrients are changed for economic or availability reasons.

Phenotype. The characteristics of an animal that can be seen and/or measured.

Photosynthesis. The process whereby green plants use the energy of the sun to build up complex organic molecules containing energy.

Physiological. Pertaining to the science that deals with the functions of living organisms or their parts.

Physiological fuel values. Units, expressed in calories, used in the United States to measure food energy in human nutrition. Similar to metabolizable energy.

Physiological saline. A salt solution (0.9% NaCl) having the same osmotic pressure as the blood plasma.

Phytin. A form of phosphorus that is poorly utilized.

Picnic. A shoulder cut often cured and smoked like ham, but containing more internal fat and connective tissue.

Pig. A young swine, generally less than 120 lb and less than 4 months of age.

Piggy. A sow that has the appearance of having recently suckled pigs or that is due to farrow soon.

Piglet. A small pig.

Plant proteins. This group includes the common oilseed by-products—soybean meal, cottonseed meal, linseed meal, peanut meal, safflower meal, sunflower seed meal, and canola meal.

Pneumonia. Inflammation of the lungs.

Pollution. Anything that defiles, desecrates, or makes impure or unclean the surrounding environment.

Polyneuritis. Neuritis of several peripheral nerves at the same time, caused by metallic and other poisons, infectious disease, or vitamin deficiency. In people, alcoholism is also a major cause of polyneuritis.

Polyunsaturated fatty acids. Fatty acids having more than one double bond. Linoleic acid, which contains two double bonds, is the primary dietary essential fatty acid of humans.

Porcine. Pertaining to swine.

Porcine reproductive and respiratory syndrome (PRRS). In 1987, this disease was first reported in North America, at which time it was known as "Mystery Disease." Europeans called it "Blue Ear Disease" because of the bluish coloration of the skin.

Porcine stress syndrome (PSS). Susceptibility to PSS is caused by a single autosomal recessive gene. The disease is manifested only in pigs that are homozygous recessive for this gene, which means that they inherited it from both the sire and the dam. It causes some pigs subjected to the stress of management or sudden environmental changes to react adversely and even succumb. Pigs with PSS are usually associated with low-quality or pale, soft, and exudative (PSE) pork. Symptoms are extreme muscling, nervousness, easily frightened, tail tremors, and skin blotching.

Pork. The edible flesh of pigs or hogs.

Pork by-products. All products of the pig except the carcass meat. It includes organs, skin, feet, dissected bones, and similar products.

Pork carcass grade. A measure of the degree of excellence based chiefly on quality of lean, amount of backfat, and expected yield of trimmed major wholesale cuts.

Pork Quality Assurance Program. The NPPC first announced its Pork Quality Assurance Program in 1989. Its purpose is to enhance consumer confidence in the safety of pork and encourage increased pork consumption.

Porker. A young hog (pig).

Postmortem inspection. The inspection made at the time of slaughter.

Postnatal. Occurring after birth.

Postpartum. Occurring after the birth of the offspring, when referring to the sow.

Pot-bellied. Designating any individual that has developed an abnormally large abdomen.

Precursor. A compound that can be used by the body to form another compound, for example, carotene is a precursor of vitamin A.

Prehension. The seizing (grasping) and conveying of feed to the mouth.

Premix. A uniform mixture of one or more microingredients and a carrier, used in the introduction of microingredients into a larger mixture.

Prenatal. Before birth.

Prepotency. The ability of an individual to transmit its own qualities to its offspring.

Prescription drugs. Drugs that are for use as directed by, or on the order of, the veterinarian.

Preservatives. Materials that are available to incorporate into feeds, with claims made that they will improve the preservation of nutrients, nutritive value, and/or palatability of the feed.

Primal cuts. Ham, loin, Boston butt, picnic, and bacon.

Probe. A device to measure backfat thickness in pigs.

Probiotics. Term meaning "in favor of life." They have an opposite effect to antibiotics on the microorganisms of the digestive tract. They increase the population of the desirable microorganisms rather than kill or inhibit undesirable organisms.

Produce. A female's offspring. The produce-of-dam commonly refers to the offspring of one dam (female).

Progeny testing. The evaluation of an animal on the basis of the performance of its offspring.

Prolapse. Abnormal protrusion of a part or organ.

Proprietorship (individual). Business that is owned and operated by one individual.

Prostaglandins. A large group of chemically related 20-carbon hydroxy fatty acids with variable physiological effects in the body.

Protein. From the Greek, meaning "of first rank, importance." Complex organic compounds made up chiefly of amino acids present in characteristic proportions for each specific protein. Twenty amino acids generally occur in combinations to form an almost limitless number of proteins. Protein always contains carbon, hydrogen, oxygen, and nitrogen; in addition, it usually contains sulfur and frequently phosphorus. Crude protein is determined by finding the nitrogen content and multiplying the result by 6.25. The nitrogen content of proteins averages about 16% ($100/16 = 6.25$). Proteins are essential in all plant and animal life as components of the active protoplasm of each living cell. Feed ingredients that contain more than 20% of their total weight in crude protein are generally classified as protein feeds.

Protein supplements. Products that contain more than 20% protein or protein equivalent.

Proud flesh. Excess flesh growing around a wound.

Provitamin. The material from which an animal may produce vitamins; for example, carotene (provitamin A) in plants is converted to vitamin A in animals.

Provitamin A. Carotene.

Proximate analysis. A chemical scheme for evaluating feedstuffs, in which a feedstuff is partitioned into the six fractions: (1) moisture (water) or dry matter (DM); (2) total (crude) protein (CP or TP − N × 6.25); (3) ether extract (EE) or fat; (4) ash (mineral salts); (5) crude fiber (CF)—the incompletely digested carbohydrates; and (6) nitrogen-free extract (NFE)—the more readily digested carbohydrates (calculated rather than measured chemically).

PSS. *See porcine stress syndrome.*

Puberty. The age at which the reproductive organs become functionally operative—sexual maturity.

Purebred. An animal of pure breeding, registered or eligible for registration in the herd book of the breed to which it belongs.

Purified diet. A mixture of the known essential dietary nutrients in a pure form that is fed to experimental (test) animals in nutrition studies.

Purulent. Consisting of or forming pus.

Pus. A liquid inflammatory product consisting of leukocytes (white blood cells), lymph, bacteria, dead tissue cells, and the fluid derived from their disintegration.

Q

Qualitative traits. Traits in which there is a sharp distinction between phenotypes, usually involving only one or two pairs of genes.

Quality. A term used to denote the desirability or acceptance of an animal or feed product.

Quality of protein. A term used to describe the amino acid balance of protein. A protein is said to be of good quality when it contains all the essential amino acids in the proper proportions and amounts needed by a specific animal, and it is said to be poor quality when it is deficient in either content or balance of essential amino acids.

Quantitative traits. Traits in which there is no sharp distinction between phenotypes, usually involving several genes and the environment. These include such economic traits as gestation length, birth weight, weaning weight, rate and efficiency of gain, and carcass quality.

Quarantine.
- Compulsory segregation of exposed susceptible animals for a period of time equal to the longest usual incubation period of the disease to which they have been exposed.
- An enforced regulation for the exclusion or isolation of an animal to prevent the spread of an infectious disease.

R

Radioactive. Giving off atomic energy in the form of alpha, beta, or gamma rays.

Rancid. Fats that have undergone partial decomposition.

Ranting. Characteristic behavior of an agitated boar—frothing, chomping, and nervousness.

Rate of passage. The time taken by undigested residues from a given meal to reach the feces. (A stained undigestible material is commonly used to estimate rate of passage.)

Ration(s). The amount of feed supplied to an animal for a definite period, usually for a 24-hour period.

Ratios. "Weight ratio," "gain ratio," and "conformation score ratio" are used to indicate the performance of an individual in relation to the average of all animals of the same group. Calculated as follows:

$$\frac{\text{Individual record}}{\text{Average of animals in group} \times 100}$$

It is a record or index of individual deviation from the group average expressed in terms of percentage. A ratio of 100 is average for a particular group. Thus, ratios above 100 indicate animals above average, whereas ratios below 100 indicate animals below average.

Recessive. Describes a gene that, when paired with its dominate allele, is not expressed in the phenotypic expression of that gene, unless both gene pairs are recessive.

Red meat. Meat that is red when raw because of the red coloration of myoglobin, the pigment of muscle. Red meats include beef, veal, pork, mutton, and lamb muscle tissue with attendant fat and bone.

Registered. Designating purebred animals whose pedigrees are recorded in the breed registry.

Regurgitation. The casting up (backward flow) of undigested food from the stomach to the mouth, as by ruminants.

Replacement. An animal selected to be kept for the breeding herd.

Ridgeling. Any male animal whose testicles fail to descend into the scrotum—a cryptorchid.

Rigor mortis. The stiffness of body muscles that is observed shortly after death.

Ring. To place a small ring in the snout of a swine for the purpose of discouraging rooting.

Roasting pig (roaster). Fat, plump, suckling pigs weighing 30 to 60 lb dressed with head on and not split at the chest or between the hams.

Roughage. Feed consisting of bulky and coarse plants or plant parts, containing a high-fiber content and low total digestible nutrients, arbitrarily defined as feed with more than 18% crude fiber. Roughage may be classed as either dry or green.

Runt. A piglet of small size in relation to its littermates.

S

Saccharides. Referring to sugars. The prefixes *mono-*, *di-*, *tri-*, and *poly-* denote the number of sugars contained in the saccharide.

Saliva. A clear, somewhat viscid solution secreted by glands within the mouth. It may contain the enzymes salivary amylase and salivary maltase.

Salmonella. A pathogenic, diarrhea-producing organism, of which there are more than 100 known strains, sometimes present in contaminated feeds.

Satiety. Full satisfaction of desire; may refer to satisfaction of appetite.

Saturated fat. A completely hydrogenated fat—each carbon atom is associated with the maximum number of hydrogens; there are no double bonds.

Scouring. One of the major problems facing livestock producers is scouring (diarrhea) in young animals. It may be caused by feeding practices, management practices, the environment, or disease.

Screened. A feedstuff that has been separated into various-sized particles by passing over or through screens.

Secondary infection. Infection following an infection already established by other pathogens.

Sectioned and formed ham. Ham made from pieces of meat trimmed from the hind leg and, after curing, formed into a loaf, placed in a casing, and cooked and smoked like normal ham.

Secular trends. Longtime trends that persist over a period of several cycles.

Seedstock suppliers. Suppliers who sell boars and gilts of specialized bloodlines. These lines of breeding, usually called *hybrids*, originate from crossing two or more breeds, then applying some specialized selection programs.

Segregated early weaning (SEW). An infectious disease control procedure, the primary objective of which is to improve productivity of the growing/finishing phase by preventing the transfer of disease. Generally, it involves removing the piglets from the sow and segregating (isolating) them from other pigs that were not farrowed and weaned within a similar time frame, usually about 1 week.

Selection. Determining which animals in a population will produce the next generation. Pork producers practice artificial selection, whereas nature practices natural selection.

Selenium. An element that functions with glutathione peroxidase, an enzyme that enables the tripeptide glutathione peroxidase to perform its role as a biological antioxidant in the body. This explains why deficiencies of selenium and vitamin E result in similar signs—loss of appetite and slow growth.

Self-fed. Provided with a part or all of the ration on a continuous basis, thereby permitting the animal to eat at will.

Self-feeder. A feed container by means of which animals can eat at will. (See also ad libitum.)

Semen. The fluid containing the sperm that is ejaculated by the male.

Separate-sex feeding. *See Split-sex feeding.*

Serum. The colorless fluid portion of blood remaining after clotting and removal of corpuscles. It differs from plasma in that the fibrinogen has been removed.

Service. Denotes the mating of a female by a male.

Settled. Indicates the animal has become pregnant.

Shoat (shote). A young pig, of either sex, after weaning—synonymous with *pig*.

Show box (tack box). A container in which to keep all show equipment and paraphernalia.

Shrinkage.
- A term indicating the amount of loss in body weight when animals are exposed to adverse conditions, such as being transported, severe weather, or shortage of feed.
- The loss in carcass weight during the aging process.

Sib. A brother or sister.

Sib testing. A method of selection in which an animal is selected on the basis of the performance of its brothers or sisters.

Sire.
- The male parent.
- To father or beget.

Slaughter hog pooling. Small hog producers joining together for the purpose of marketing hogs in truckload lots. Pooling is usually prompted because of lack of access to a nearby market.

Slotted floors. Floors with slots through which the feces and urine pass to a storage area below or nearby.

Sloughing. A mass of dead tissue separating from a surface.

Soap. A compound formed along with glycerol from the reaction of fat with alkali.

Soft pork. Feed fats are laid down in the body without undergoing much change. Thus, when finishing hogs are liberally fed on high-fat content feeds in which the fat is liquid at ordinary temperatures, soft pork results. This condition prevails when hogs are liberally fed such feeds as full-fat soybeans, peanuts, mast, or garbage.

Solubles. Liquids containing dissolved substances obtained from processing animal or plant materials. They may contain some fine suspended solids.

Solution. A uniform liquid mixture of two or more substances molecularly dispersed within one another.

Sow. A female swine that shows evidence of having produced pigs or that is in an evident state of pregnancy.

Sowbelly Salt pork; unsmoked fat bacon.

Soy protein concentrate. The protein produced by removing the water soluble sugars, ash, and other minor constituents from defatted soy flour. It contains 65 to 70% protein.

Soy protein isolate. The highest protein source; produced by removal of the insoluble fibrous material.

Specific dynamic action (SDA). The increased production of heat by the body as a result of a stimulus to metabolic activity caused by ingesting food.

Specific gravity. The ratio of the weight of a body to the weight of an equal volume of water.

$$\text{Specific gravity} = \frac{\text{Weight of body in air}}{\text{Weight of body in air} - \text{Weight in H}_2\text{O}}$$

Specific heat.
- The heat-absorbing capacity of a substance in relation to that of water.
- The heat expressed in calories required to raise the temperature of 1 g of a substance 1°C.

Specific pathogen-free (SPF) pigs. Pigs that are free of disease at birth.

Speculating. Risk-taking by anyone who hopes to make a profit in the advances or declines in the price of the futures contract.

Split-sex feeding. Sorting boars, gilts, and barrows and feeding each group separate diets.

Spray-dried blood meal. Product that contains both the plasma and red blood cell fractions of blood.

Spray-dried porcine plasma. Product made up of the albumin, globin, and globulin fractions of blood. It contains 68% protein and 6.1% lysine.

Stabilized. Made more resistant to chemical change by the addition of a particular substance.

Stag. A male that was castrated after the secondary sexual characteristics developed sufficiently to give the appearance of a mature male.

STAGES. See Swine Testing and Genetic Evaluation System.

Standing heat. Period in a sow or gilt's heat (estrus) during which she will stand still when being mounted or when pressure is applied to her back.

Sterile. Incapable of reproducing.

Stillborn. Born lifeless; dead at birth.

Straw. The plant residue remaining after separation of the seeds in threshing. It includes chaff.

Stress. Any physical or emotional factor to which an animal fails to make a satisfactory adaptation. Stress may be caused by excitement, temperament, fatigue, shipping, disease, heat or cold, nervous strain, number of animals together, previous nutrition, breed, age, or management. The greater the stress, the more

exacting the nutritive requirements.

Subcutaneous. Situated or occurring beneath the skin.

Suckle. To nurse at the breast or mammary glands.

Sugar. A sweet, crystallizable substance that consists essentially of sucrose and that occurs naturally in the most readily available amounts in sugarcane, sugar beet, sugar maple, sorghum, and sugar palm.

Supplement. A feed or feed mixture used to improve the nutritional value of basal feeds (e.g., protein supplement—soybean meal). Supplements are usually rich in protein, minerals, vitamins, antibiotics, or a combination of part or all of these, and they are usually combined with basal feeds to produce a complete feed.

Suppuration. Formation of pus.

Suspect animal. An animal that inspectors identify before slaughter because they suspect the animal or parts of the animal may be unfit for human consumption. These carcasses will receive additional scrutiny in the postmortem inspection.

Sustainable agriculture. Farming with reduced off-farm-purchased inputs of pesticides, herbicides, and fertilizers, along with reduced negative impact on natural resources and improved environmental quality and economic efficiency, while producing and distributing food and fiber.

Swine. Collective term for all age groups within the species.

Swine Testing and Genetic Evaluation System (STAGES). System of the breed registries for recording performance data provided by swine producers.

Symmetry. A balanced development of all parts.

Syndactylism. The union of two or more toes that is normal in many birds and some mammals.

Synthesis. The bringing together of two or more substances to form a new material.

Synthetics. Artificially produced products that may be similar to natural products.

T

Tail biting. An abnormal behavior, characterized by one pig biting the tail of another.

Tankage. A protein supplement consisting of ground meat by-products of animals that have been slaughtered.

Tattoo. Permanent identification of animals produced by placing indelible ink under the skin; generally put in the ears of young animals.

TDN. See total digestible nutrients.

Teart. Molybdenosis of farm animals caused by feeding on vegetation grown on soil that contains high levels of molybdenum.

Terminal market. Livestock trading centers that generally include several commission firms and an independent stockyards company for the sale of live slaughter animals. They are also referred to as terminals, central markets, public stockyards, and public markets.

Tetany. A condition in an animal in which there are localized, spasmodic muscular contractions.

Tether. To tie an animal with a rope or a chain to allow feeding but to prevent straying.

Thermal. Refers to heat.

Thermogenesis. The chemical production of heat in the body.

Thermoneutral. The state of thermal (heat) balance between an animal and its environment. The thermoneutral zone is referred to as the comfort zone.

Thrifty. Healthy and vigorous in appearance.

Thumps (thumping). Jerky breathing; a respiratory disturbance in which the pigs breathe with difficulty, rapidly, and spasmodically.

Tocopherol. Any of four different forms of an alcohol also known as vitamin E.

Tonic. A drug, medicine, or feed designed to stimulate the appetite.

Total digestible nutrients (TDN). The energy value of a feedstuff. It is computed using the following formula:

$$\% \text{ TDN} = \frac{\text{DCP} + \text{DCF} + \text{DNFE} + (\text{DEE} \times 2.25)}{\text{feed consumed}} \times 100$$

Where DCP = digestible crude protein; DCF = digestible crude fiber; DNFE = digestible nitrogen-free extract; and DEE = digestible ether extract. One lb of TDN = 2,000 kcal of digestible energy.

Toxic. Of a poisonous nature.

Toxoid. A killed vaccine produced from toxins.

Trace element. A chemical element used in minute amounts by organisms and held essential to their physiology. The essential trace elements are cobalt, copper, iodine, iron, manganese, selenium, and zinc.

Trace mineral. A mineral nutrient required by animals in microamounts only (measurable in mg/lb or parts per million).

Traceback. An animal identification system that makes it possible to trace animals back to their origin in order to locate the source of contamination.

Tuber. A short, thickened, fleshy stem or terminal portion of a stem or rhizome that is usually formed underground, bears minute scale leaves each with a bud capable under suitable conditions of developing into a new plant, and constitutes the resting stage of various plants, such as the potato and the Jerusalem artichoke.

Turn-around, comfort stalls. *See freedom stalls.*

Tusk. An elongated or greatly enlarged tooth.

Twenty-eight hour law in rail shipments. Law prohibiting transportation of livestock by rail for a longer period than 28 consecutive hours without unloading, feeding, watering, and resting 5 consecutive hours before resuming transportation. On request of the owner, the period can be extended to 36 hours.

Type.
- Physical conformation of an animal.
- All those physical attributes that contribute to the value of an animal for a specific purpose.

U

Udder. The encased group of mammary glands with each gland provided with a nipple or teat.

Ultrasonics. Electronic equipment that uses the "pulse echo" technique. Reflected sound waves are used to measure depth of backfat and muscle.

Underfeeding. Usually refers to not providing sufficient energy. The degree of lowered production therefrom is related to the extent of underfeeding and the length of time it exists.

Unidentified factors. *Unidentified* or *unknown* factors that have not yet been isolated or synthesized in the laboratory. There is evidence that the growth factors exist in dried whey, marine and packing house by-products, distillers' solubles, antibiotic fermentation residues, alfalfa meal, and certain green forages. Most of the unidentified factor sources are added to the diet at levels of 1 to 3%.

Unsaturated fat. A fat having one or more double bonds; not completely hydrogenated.

Unsaturated fatty acid. Any one of several fatty acids containing one or more double bonds, such as oleic, linoleic, linolenic, and arachidonic acids.

Unthriftiness. Lack of vigor, poor growth or development; the quality or state of being unthrifty in animals.

Upper critical temperature (UCT). The temperature above which the animal can no longer dissipate sufficient heat and body temperature starts to rise.

U.S. pharmacopoeia (USP) unit. A unit of measurement or potency of biologicals that usually coincides with an international unit. (Also see international unit.)

V

Vaccination (shot). An injection of vaccine, bacterin, antiserum, or antitoxin to produce immunity or tolerance to disease.

Vaccine. A suspension of attenuated or killed microorganisms (bacteria, viruses, or rickettsiae) administered for the prevention, improvement, or treatment of infectious diseases.

Vacuum-packed fresh pork. Often large boneless cuts of pork, referred to as subprimals, with a shelf life of about 21 days because the oxygen has been excluded. In the package the pork's color is more of a dull purple, but when the package is opened the pink returns.

Variety meats. Liver, brains, heart, kidney—all excellent sources of many essential nutrients.

Vectors. Living organisms that carry pathogens.

Vermifuge (vermicide). Any chemical substance given to animals to kill internal parasitic worms.

Vertical integration. In pork a structure in which, through alliances (complete ownership, joint ownership, or contract), an individual or company controls the product through two or more phases, for example, a meat packer owns hogs from birth through slaughter.

Veterinary feed directive (VFD). A written statement issued by a veterinarian that authorizes the client to obtain and use animal feed containing a VFD drug to treat their animals only in accordance with the Food and Drug Administration-approved directions for use.

Vietnamese pot-bellied pigs. Native Asian pigs, which are normally less than one-fifth the size of traditional U.S. hogs. They have a pot-belly, a swayed back, and a straight tail, which they wag when they are happy.

Virus. One of a group of minute infectious agents. They lack independent metabolism and can only multiply within living host cells.

Viscera. Internal organs of the body, particularly in the chest and abdominal cavities.

Vitamin product labels. When a product is marketed as a vitamin supplement per se, the quantitative guarantees (unit/lb) of vitamins A and D are expressed in USP units, vitamin E in IU, and of other vitamins in mg/lb.

Vitamin supplements. Rich synthetic or natural feed sources of one or more of the complex organic compounds, called vitamins, that are required in minute amounts by animals for normal growth, production, reproduction, and/or health.

Vitamins. Complex organic compounds that function as parts of enzyme systems essential for the transformation of energy and the regulation of metabolism of the body, and required in minute amounts by one or more animal species for normal growth, production, reproduction, and/or health. All vitamins must be present in the diet for normal functioning, except for B vitamins in the ruminants (cattle and sheep) and vitamin C.

Void. To evacuate feces and/or urine.

Vomiting. The forcible expulsion of the contents of the stomach through the mouth.

W

Wasty.
- A carcass with too much fat, requiring excessive trimming.
- Paunchy live animal.

Waxy corn. A variety of corn that is sometimes grown for industrial use because of its special type of starch.

Weaner. A pig that has reached the age that it can be weaned or that has been weaned.

Weaning. The stopping of young animals from suckling their mothers.

Wheat gluten. Spray-dried protein fraction of wheat remaining after the starch has been extracted for use in human food products.

White meat.
- The breast of broilers or turkeys.
- Most cooked pork—"the other white meat."

Wiltshire side. The entire half of a dressed pig, minus the head, shank, shoulder bone, and hip bone. All of the side, except the ham and shoulder, is sold as bacon.

Withdrawal period. The withholding of certain feed additives prior to slaughter in order to ensure that drug residues do not occur in carcasses.

Workers' compensation. Program designed to provide employees assured payment for medical expenses or lost income as a result of injury on the job.

Z

Zoning. Ordinances governing the keeping of animals or the type of businesses or residences.

INDEX